Methods in Cell Biology

The Zebrafish: Disease Models and Chemical Screens

Volume 138

Series Editors

Methods in Cell Biology

The Zebrafish: Disease Models and Chemical Screens

Volume 138

Edited by

H. William Detrich, III
Northeastern University
Marine Science Center
Nahant, MA

Monte Westerfield
University of Oregon
Institute of Neuroscience
Eugene, OR

Leonard I. Zon
Harvard University
Boston Children's Hospital
Boston, MA

Academic Press is an imprint of Elsevier
50 Hampshire Street, 5th Floor, Cambridge, MA 02139, United States
525 B Street, Suite 1800, San Diego, CA 92101-4495, United States
125 London Wall, London EC2Y 5AS, United Kingdom
The Boulevard, Langford Lane, Kidlington, Oxford OX5 1GB, United Kingdom

Fourth edition 2017

ISBN: 978-0-12-803473-6
ISSN: 0091-679X

For information on all Academic Press publications visit our website at https://www.elsevier.com

Publisher: Zoe Kruze
Acquisition Editor: Zoe Kruze
Editorial Project Manager: Hannah Colford
Production Project Manager: Surya Narayanan Jayachandran
Designer: Victoria Pearson

Typeset by TNQ Books and Journals

Len, Monte, and I dedicate the 4th Edition of Methods in Cell Biology: The Zebrafish *to the postdoctoral fellows and graduate students who conducted the genetic screens that established the zebrafish as a preeminent vertebrate model system for analysis of development.*

Contents

PART VII CANCER

PART VIII TRANSPLANTATION

PART IX CHEMICAL SCREENING

Contributors

S.L. Amacher
The Ohio State University, Columbus, OH, United States

M. Ammerman
Harvard Medical School, Boston, MA, United States

F. Argenton
Università degli Studi di Padova, Padova, Italy

K.R. Astell
University of Edinburgh, Edinburgh, United Kingdom

J.W. Astin
University of Auckland, Auckland, New Zealand

B. Blanco-Sánchez
University of Oregon, Eugene, OR, United States

T. Boehm
Max Planck Institute of Immunobiology and Epigenetics, Freiburg, Germany

J.W. Chen
Massachusetts General Hospital, Boston, MA, United States; Harvard Medical School, Boston, MA, United States

L. Chen
Leiden University, Leiden, The Netherlands

A. Clément
University of Oregon, Eugene, OR, United States

K.E. Crosier
University of Auckland, Auckland, New Zealand

P.S. Crosier
University of Auckland, Auckland, New Zealand

P.D. Currie
Monash University, Clayton, VIC, Australia

M. Dang
Harvard Medical School, Boston, MA, United States

M. D'Rozario
Washington University School of Medicine, St. Louis, MO, United States

L. Du
University of Auckland, Auckland, New Zealand

S.C. Eames Nalle
Genentech, Inc., South San Francisco, CA, United States

J.S. Eisen
University of Oregon, Eugene, OR, Unites States

B. Ewa Snaar-Jagalska
Leiden University, Leiden, The Netherlands

Y. Feng
University of Edinburgh, Edinburgh, United Kingdom

J.M. Frame
Beth Israel Deaconess Medical Center, Harvard Medical School, Boston, MA, United States

K.F. Franse
Appalachian State University, Boone, NC, United States

K.A. Gabor
National Institute of Environmental Health Sciences, Durham, NC, United States

J.L. Galloway
Massachusetts General Hospital, Boston, MA, United States; Harvard Medical School, Boston, MA, United States

J.M. Gansner
Harvard Medical School, Boston, MA, United States

S. Gomez De La Torre Canny
Duke University, Durham, NC, United States

D. Greenald
University of Sheffield, Sheffield, United Kingdom

A. Groenewoud
Leiden University, Leiden, The Netherlands

K. Guillemin
University of Oregon, Eugene, OR, Unites States; Canadian Institute for Advanced Research, Toronto, ON, Canada

C.J. Hall
University of Auckland, Auckland, New Zealand

M.P. Harris
Harvard Medical School, Boston, MA, United States

M.N. Hayes
Massachusetts General Hospital, Boston, MA, United States; Massachusetts General Hospital, Charlestown, MA, United States; Harvard Stem Cell Institute, Boston, MA, United States

S. He
Harvard Medical School, Boston, MA, United States

I. Hess
Max Planck Institute of Immunobiology and Epigenetics, Freiburg, Germany

K.J. Hromowyk
The Ohio State University, Columbus, OH, United States

A. Huysseune
Ghent University, Ghent, Belgium

N. Iwanami
Max Planck Institute of Immunobiology and Epigenetics, Freiburg, Germany

C.-B. Jing
Harvard Medical School, Boston, MA, United States

D. Jurczyszak
University of Maine, Orono, ME, United States

P. Keerthisinghe
University of Auckland, Auckland, New Zealand

L. Kelly
University of Edinburgh, Edinburgh, United Kingdom

M. Kelly
University of Oregon, Eugene, OR, Unites States

C.H. Kim
University of Maine, Orono, ME, United States

H.R. Kim
University of Sheffield, Sheffield, United Kingdom

M.D. Kinkel
Appalachian State University, Boone, NC, United States

M. Kruithof-de Julio
University of Bern, Bern, Switzerland; Leiden University Medical Centre, Leiden, The Netherlands

D.M. Langenau
Massachusetts General Hospital, Boston, MA, United States; Massachusetts General Hospital, Charlestown, MA, United States; Harvard Stem Cell Institute, Boston, MA, United States

D.W. Laux
University of Edinburgh, Edinburgh, United Kingdom

M. Li
Monash University, Clayton, VIC, Australia

S.-E. Lim
Beth Israel Deaconess Medical Center, Harvard Medical School, Boston, MA, United States

A.T. Look
Harvard Medical School, Boston, MA, United States

E. Markham
University of Sheffield, Sheffield, United Kingdom

M.A. Matty
Duke University School of Medicine, Durham, NC, United States

E. Melancon
University of Oregon, Eugene, OR, Unites States

P.J. Millard
University of Maine, Orono, ME, United States

J.E.N. Minchin
University of Edinburgh, Edinburgh, United Kingdom; Duke University, Durham, NC, United States

K.R. Monk
Washington University School of Medicine, St. Louis, MO, United States; Hope Center for Neurological Disorders, Washington University School of Medicine, St. Louis, MO, United States

T.E. North
Beth Israel Deaconess Medical Center, Harvard Medical School, Boston, MA, United States; Harvard Stem Cell Institute, Cambridge, MA, United States

M. Pack
University of Pennsylvania, Philadelphia, PA, United States

Y. Padberg
Institute for Cardiovascular Organogenesis and Regeneration, Faculty of Medicine, University of Münster, Münster, Germany; Cells-in-Motion Cluster of Excellence (EXC M 1003-CiM), University of Münster, Münster, Germany

Barry H. Paw
Brigham & Women's Hospital, Boston, MA, United States; Harvard Medical School, Boston, MA, United States; Dana-Farber Cancer Institute, Boston, MA, United States; Boston Children's Hospital, Boston, MA, United States

S.C. Petersen
Kenyon College, Gambier, OH, United States

J.B. Phillips
University of Oregon, Eugene, OR, United States

T. Ramezani
University of Edinburgh, Edinburgh, United Kingdom

J.F. Rawls
Duke University, Durham, NC, United States

S.E. Redfield
Stem Cell Program and Division of Hematology and Oncology, Childrens' Hospital Boston, Dana-Farber Cancer Institute, Howard Hughes Medical Institute and Harvard Medical School, Boston, MA, United States

I. Ribeiro Bravo
University of Edinburgh, Edinburgh, United Kingdom

L.E. Sanderson
Universite libre de Bruxelles, Gosselies, Belgium

K. Santhakumar
SRM University, Kattankulathur, India

M. Schorpp
Max Planck Institute of Immunobiology and Epigenetics, Freiburg, Germany

S. Schulte-Merker
Institute for Cardiovascular Organogenesis and Regeneration, Faculty of Medicine, University of Münster, Münster, Germany; Cells-in-Motion Cluster of Excellence (EXC M 1003-CiM), University of Münster, Münster, Germany

S. Sichel
University of Oregon, Eugene, OR, Unites States

D. Sieger
University of Edinburgh, Edinburgh, United Kingdom

C. Sullivan
University of Maine, Orono, ME, United States

D.M. Tobin
Duke University School of Medicine, Durham, NC, United States

C. Tulotta
Leiden University, Leiden, The Netherlands

G. van der Horst
Leiden University Medical Centre, Leiden, The Netherlands

G. van der Pluijm
Leiden University Medical Centre, Leiden, The Netherlands

Lisa N. van der Vorm
Brigham & Women's Hospital, Boston, MA, United States

F. van Eeden
University of Sheffield, Sheffield, United Kingdom

A. van Impel
Institute for Cardiovascular Organogenesis and Regeneration, Faculty of Medicine, University of Münster, Münster, Germany; Cells-in-Motion Cluster of Excellence (EXC M 1003-CiM), University of Münster, Münster, Germany

A. Vettori
Università degli Studi di Padova, Padova, Italy

M. Westerfield
University of Oregon, Eugene, OR, United States

T.J. Wiles
University of Oregon, Eugene, OR, Unites States

D.S. Wiley
Stem Cell Program and Division of Hematology and Oncology, Childrens' Hospital Boston, Dana-Farber Cancer Institute, Howard Hughes Medical Institute and Harvard Medical School, Boston, MA, United States

C. Winkler
National University of Singapore, Singapore, Singapore

P.E. Witten
Ghent University, Ghent, Belgium

X. Zhao
University of Pennsylvania, Philadelphia, PA, United States

L.I. Zon
Stem Cell Program and Division of Hematology and Oncology, Childrens' Hospital Boston, Dana-Farber Cancer Institute, Howard Hughes Medical Institute and Harvard Medical School, Boston, MA, United States

E. Zoni
University of Bern, Bern, Switzerland; Leiden University Medical Centre, Leiden, The Netherlands

Preface

Len, Monte, and I are pleased to introduce the Fourth Edition of *Methods in Cell Biology: The Zebrafish*. The advantages of the zebrafish, *Danio rerio*, are numerous, including its short generation time and high fecundity, external fertilization, and the optical transparency of the embryo. The ease of conducting forward genetic screens in the zebrafish, based on the pioneering work of George Streisinger, culminated in screens from the laboratories of Wolfgang Driever, Mark C. Fishman, and Christiane Nüsslein-Volhard, published in a seminal volume of *Development* (Volume 123, December 1, 1996) that described a "candy store" of mutants whose phenotypes spanned the gamut of developmental processes and mechanisms. Life for geneticists who study vertebrate development became *really* fine.

Statistics derived from ZFIN (The Zebrafish Model Organism Database; http://zfin.org/cgi-bin/webdriver?MIval=aa-ZDB_home.apg) illustrate the dramatic growth of research involving zebrafish. The zebrafish genome has been sequenced, and as of 2014, more than 25,000 genes have been placed on the assembly. Greater than 15,500 of these genes have been established as orthologs of human genes. The zebrafish community has grown from ~1400 researchers in 190 laboratories as of 1998 to ~7000 in 930 laboratories in 2014. The annual number of publications based on the zebrafish has risen from 1913 to 21,995 in the same timeframe. Clearly, the zebrafish has arrived as a vertebrate biomedical model system *par excellence*.

When we published the First Edition (Volumes 59 and 60) in 1998, our goal was to encourage biologists to adopt the zebrafish as a genetically tractable model organism for studying biological phenomena from the cellular through the organismal. Our goal today remains unchanged, but the range of subjects and the suite of methods have expanded rapidly and significantly in sophistication over the years. With the Second and Third Editions of *MCB: The Zebrafish* (Volumes 76 and 77 in 2004; Volumes 100, 101, 104, and 105 in 2010−11), we documented this extraordinary growth, again relying on the excellent chapters contributed by our generous colleagues in the zebrafish research community.

When Len, Monte, and I began planning the Fourth Edition, we found that the zebrafish community had once more developed and refined novel experimental systems and technologies to tackle challenging biological problems across the spectrum of the biosciences. We present these methods following the organizational structure of the Third Edition, with volumes devoted to *Cellular and Developmental Biology*, to *Genetics, Genomics, and Transcriptomics*, and to *Disease Models and Chemical Screens*. Here we introduce the fourth volume, *Disease Models and Chemical Screens*.

Disease Models and Chemical Screens covers major technical advances in development of the zebrafish as an important biomedical model organism. Nine sections are devoted to adipose tissue, the innate and adaptive immune systems, blood and lymph, visceral organs, the musculoskeletal system, central and sensory nervous systems, cancer, transplantation, and chemical screening. We anticipate that you, our readership, will apply these methods successfully in your own zebrafish research programs and will develop your own disease models that may be considered for a future edition of *Methods in Cell Biology: The Zebrafish.*

We thank the series editors, Leslie Wilson and Phong Tran, and the staff of Elsevier/Academic Press, especially Zoe Kruze and Hannah Colford, for their enthusiastic support of our Fourth Edition. Their help, patience, and encouragement are profoundly appreciated.

H. William Detrich, III
Monte Westerfield
Leonard I. Zon

PART

Adipose Tissue

I

CHAPTER

In vivo imaging and quantification of regional adiposity in zebrafish

1

J.E.N. Minchin[*,§,1], **J.F. Rawls**[§]

University of Edinburgh, Edinburgh, United Kingdom

[§]*Duke University, Durham, NC, United States*

[1]*Corresponding author: E-mail: james.minchin@ed.ac.uk*

CHAPTER OUTLINE

Abstract

Adipose tissues (ATs) are lipid-rich structures that supply and sequester energy-dense lipid in response to the energy status of an organism. As such, ATs provide an organism energetic insurance during periods of adverse physiological burden. ATs are deposited in diverse anatomical locations, and excessive accumulation of particular regional ATs modulates disease risk. Therefore, a model system that facilitates the visualization and quantification of regional adiposity holds significant biomedical promise. The zebrafish

Methods in Cell Biology, Volume 138, ISSN 0091-679X, http://dx.doi.org/10.1016/bs.mcb.2016.11.010

(*Danio rerio*) has emerged as a new model system for AT research in which the entire complement of regional ATs can be imaged and quantified in live individuals. Here we present detailed methods for labeling adipocytes in live zebrafish using fluorescent lipophilic dyes, and for identifying and quantifying regional zebrafish ATs.

INTRODUCTION

The prevalence of obesity and overweight has reached epidemic proportions worldwide (Yach, Stuckler, & Brownell, 2006). Obesity is recognized as a major risk factor for the development of insulin resistance, type II diabetes, nonalcoholic fatty liver disease, and cardiovascular disease (Morse, Gulati, & Reisin, 2010). Therefore, understanding how obesity increases disease risk is a primary research question and of central global public health concern.

Obesity develops when energy intake exceeds energy expenditure, resulting in increased storage of excess energy, in the form of triacylglycerides (TG), within adipose tissues (ATs) (Rosen & Spiegelman, 2006). The major cellular constituent of ATs is lipid-rich adipocytes (or "fat cells"). ATs have traditionally been classified into white AT—containing large, lipid-filled adipocytes and thermogenic brown AT—containing smaller, multilocular adipocytes (Berry, Jiang, & Graff, 2016; Peirce, Carobbio, & Vidal-Puig, 2014). Zebrafish and other ray-finned fishes are not considered to possess brown AT; therefore, we will not discuss brown AT further in this chapter. White adipocytes (hereafter referred to as adipocytes), within AT, store surplus energy in the form of TG and hydrolyze accumulated TG for use as a fuel source during times of nutrient deprivation (Redinger, 2009). Adipocytes are present in bony vertebrates (e.g., fish, amphibians, reptiles, birds, and mammals), and AT is considered the primary site of energy storage in these vertebrate taxa (Gesta et al., 2006). White adipocytes exhibit a unique morphology and routinely grow to sizes >100 μm in diameter due to stored TG within single, large lipid droplets (LDs) (Farese & Walther, 2009). This unique cellular morphology greatly aids the identification of adipocytes.

Mammalian in vivo and in vitro cell culture systems have contributed the majority of current knowledge on AT cell biology (Virtue & Vidal-Puig, 2010). However, the importance of AT to commercial aquaculture, and the increasing use of zebrafish as a biomedical model system, has provided a rapidly expanding knowledge base on AT in teleost fish (the largest subclass of Actinopterygii or ray-finned fish). Many teleost species analyzed to date accumulate lipid within AT (Bou et al., 2016; Flynn, Trent, & Rawls, 2009; Imrie & Sadler, 2010; Johansson, Morgenroth, Einarsdottir, Gong, & Bjornsson, 2016; Song & Cone, 2007). Further, deposition and mobilization of lipid within teleost AT are altered in response to nutritional manipulation, suggesting the energy storage functions of AT are conserved between teleosts and mammals (Albalat et al., 2007; Bou et al., 2014; Flynn et al., 2009; Imrie & Sadler, 2010; Salmeron et al., 2015). Furthermore, histological analysis reveals evolutionarily conserved morphological features of teleost adipocytes, including large cytoplasmic LDs, caveolae, and close association with capillaries

(Flynn et al., 2009; Imrie & Sadler, 2010; McMenamin, Minchin, Gordon, Rawls, & Parichy, 2013; Minchin et al., 2015). In addition, teleost adipocytes express genes associated with adipocyte differentiation (*fatty acid binding protein 11a*, *fabp11a*; *peroxisome proliferator–activated receptor gamma*, *pparg*; and *CCAAT/enhancer binding protein alpha*, *cebpa*) (Flynn et al., 2009; Ibabe, Bilbao, & Cajaraville, 2005; Imrie & Sadler, 2010; Oku & Umino, 2008; Vegusdal, Sundvold, Gjoen, & Ruyter, 2003), adipocyte lipolysis (*lipoprotein lipase*, *lpl*) (Oku, Tokuda, Okumura, & Umino, 2006), and adipocyte endocrine function (*leptin*, *lep*; *adiponectin*, *acrp30*; and *adipsin*, *cfd*) (Imrie & Sadler, 2010; Michel, Page-McCaw, Chen, & Cone, 2016; Vegusdal et al., 2003). Homologous to mammals, fish AT also possesses a stromal vascular fraction (defined as a heterogeneous population of stromal cells isolated by enzymatic digestion of AT), which contains adipocyte progenitors (Rodeheffer, Birsoy, & Friedman, 2008; Tang et al., 2008; Todorcevic, Skugor, Krasnov, & Ruyter, 2010; Vegusdal et al., 2003). Together, the considerable functional, morphological, and molecular homology between teleost and mammalian ATs suggests new insights into AT biology gained in the zebrafish system will be directly translatable to humans and other vertebrates.

AT is not a single homogenous tissue in mammals and fishes and is distributed at specific anatomical locations throughout the body (Fig. 1). In humans, the largest sites of AT deposition are either subcutaneous (defined as between muscle and skin) or intra-abdominal (within the abdominal cavity) (Gesta et al., 2006; Karpe & Pinnick, 2015; Shen et al., 2003). Anatomically distinct depots display different molecular and physiological characteristics (Gesta et al., 2006; Peinado et al., 2010; Vidal, 2001; Vohl et al., 2004) and have different risk associations for obesity-related disorders (Kissebah & Krakower, 1994). In particular, visceral AT (VAT, intra-abdominal AT surrounding internal organs) is associated with more adverse risk factors than subcutaneous AT (Despres, 1998; Fox et al., 2007). Teleost AT is also deposited in both subcutaneous and intra-abdominal positions, including VAT locations, raising the possibility that developmental programs responsible for AT anatomy have been maintained during vertebrate evolution (Flynn et al., 2009; Imrie & Sadler, 2010; McMenamin et al., 2013; Minchin et al., 2015; Weil, Sabin, Bugeon, Paboeuf, & Lefevre, 2009). Further, recent work from our lab suggested that the control of AT distribution may have a conserved molecular basis between humans and fish (Minchin et al., 2015). However, the molecular and cellular basis for AT regionality and disease risk associations is still largely unknown. Thus, the development of a novel model that facilitates the investigation of regional adiposity within whole animals is desperately needed.

1. RATIONALE

Current analysis of AT is predominantly conducted after fixation and histological sectioning of ATs. This often results in incomplete preservation of AT architecture and limited information on cellular interactions and dynamics (Xue, Lim,

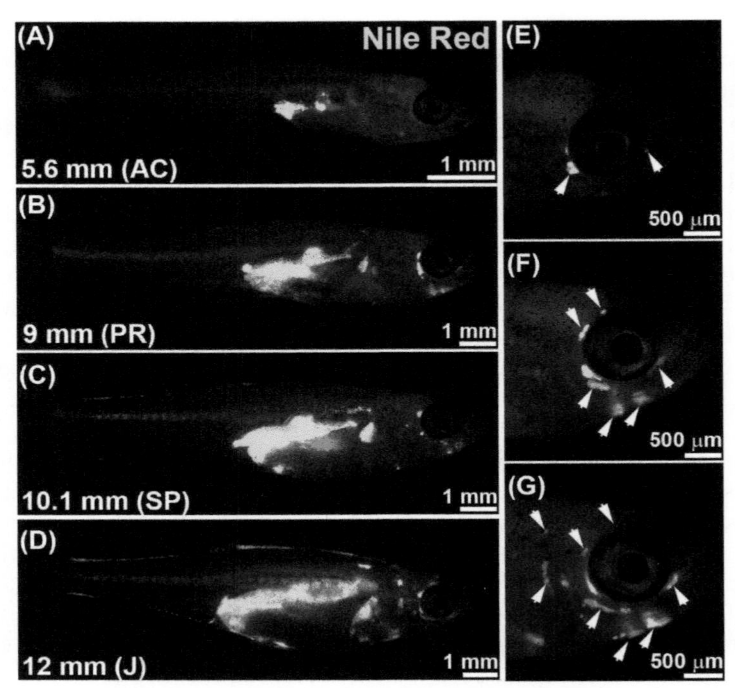

FIGURE 1 Fluorescent lipophilic dyes reveal lipid droplet accumulation in live zebrafish.

Live zebrafish stained with Nile Red (yellow) and imaged by fluorescence stereomicroscopy using the protocol described in Section 3.4. The standard length in millimeter of individual zebrafish is indicated in the bottom left corner of each image and the accompanying postembryonic stage is indicated in *parenthesis* as defined by Parichy et al. (2009). *Arrows* indicate the increasing complexity of regional lipid deposits in the zebrafish head.

Brakenhielm, & Cao, 2010). In addition, imaging of whole animal AT deposition in mammals is technically challenging, is typically restricted to low resolution views, and has only been undertaken on a limited scale (Shen & Chen, 2008). Moreover, most of our knowledge of mammalian ATs is derived from adult stages due in part to the difficulty of accessing ATs during the gestational stages when they initially develop (Ailhaud, Grimaldi, & Negrel, 1992). As a consequence, outstanding questions regarding the spatial and temporal dynamics of in vivo AT formation and growth remain understudied. Innovative approaches have been developed to address these gaps in our knowledge, such as high-resolution imaging of resected AT cultured in vitro (Nishimura et al., 2007) and in vivo imaging of adipocyte precursors introduced into mice fitted with an implanted cover slip (Nishimura et al., 2008). However, these approaches do not permit imaging of ATs within the intact physiological context of a living organism. Mathematical modeling has also been used to predict in vivo mechanisms of AT growth, but these models remain largely untested due to a paucity of suitable in vivo model systems

(Jo et al., 2009). There is therefore a pressing need for new experimental platforms for image analysis of AT formation and function in live animals.

The features of the zebrafish system are especially well suited to meet these needs. Zebrafish develop externally and are optically transparent from fertilization through the onset of adulthood, permitting in vivo imaging of dynamic cellular events during AT formation and growth (Fig. 1) (Flynn et al., 2009; Minchin et al., 2015). This provides new opportunities to investigate the earliest stages of AT morphogenesis, a process poorly understood in mammals with potentially high relevance for obesity and metabolic disease. The small size of the zebrafish also facilitates whole animal imaging of multiple adipose depots, unlike mammalian systems in which specific adipose depots are difficult to access (Fig. 1) (McMenamin et al., 2013; Minchin et al., 2015). Real-time imaging of living ATs is also possible in the zebrafish, enabling observation of molecular and cellular events over short time scales (Flynn et al., 2009; McMenamin et al., 2013). Furthermore, the amenability of the zebrafish to in vivo imaging permits longitudinal imaging of AT in individual animals, which can be used to mitigate complications from interindividual variation in adiposity (Flynn et al., 2009; McMenamin et al., 2013). As described earlier, the identification of extensive conserved homologies between teleost and mammalian AT suggests that insights gained in the zebrafish system could be applicable to humans and other vertebrates.

These diverse imaging strategies require robust methods for labeling the cellular constituents of AT in live animals. In this chapter, we present methods for labeling adipocytes in zebrafish using fluorescent lipophilic dyes (FLDs) that specifically incorporate into adipocyte LDs, for imaging ATs in live zebrafish using stereomicroscopy and guidelines on assessing the regional composition of zebrafish ATs.

2. MATERIALS

- Adult zebrafish. Any strain of adult zebrafish can be used for this protocol. Zebrafish lines may be obtained from the Zebrafish International Resource Center (ZIRC). All experiments should be performed in accordance with protocols approved by the user's Institutional Animal Care and Use Committee.
- Large nets (Aquatic Ecosystems, cat. no. AN8).
- Zebrafish aquarium (system) water.
- Breeding tanks (Laboratory Product Sales, cat. no. T233792).
- Plastic tea strainer, 7 cm (Comet Plastics, cat. no. strainer 0).
- Scienceware pipette pump (Fisher Scientific, cat. no. 13-683C).
- Wide-bore Pasteur pipettes (Kimble Chase, cat. no. 63A53WT).
- 100× 15 mm Petri dishes (Fisher Scientific, cat. no. 0875712).
- Methylene blue stock solution (0.01%) (Sigma, cat. no. M9140). Dissolve 50 mg methylene blue in 500 mL dH_2O. Dilute this stock solution 1:200 in fresh zebrafish aquarium system water to prevent growth of bacteria and mold during embryonic development.

- Distilled water (dH_2O).
- Fluorescence stereomicroscope (e.g., Leica MZ 16F or M205 FA) equipped with an eyepiece graticule and the following Leica emission filter sets: GFP2 (510LP) for the green FLDs (i.e., BODIPY 505/515, 500/510, NBT-Cholesterol, BODIPY FL C5 and the yellow-orange dye, Nile Red); YFP (535-630BP) for the yellow, orange, and orange-red dyes (i.e., BODIPY 530/550, 558/568, and Cholesteryl BODIPY 576/589); and Texas Red (610LP) for HCS LipidTOX Red/Deep Red. See Table 1 for a full description of FLDs. Equivalent fluorescence stereomicroscopes and filter sets can be used from alternative manufacturers.
- Air incubator set at 28.5°C (Powers Scientific Inc., cat. no. IS33SD).
- 2-L fish tanks (Marine Biotech, cat. no. 10198-00A).
- Assorted mesh drainage plugs for 2-L fish tanks (Marine Biotech, 425 μm, cat. no. 10222-01A; 600 μm, cat. no. 10222-02A; 1600 μm, cat. no. 10222-03A; 4000 μm, cat. no. 10222-04A).
- Brine shrimp (*Artemia franciscana*) cysts (Utah strain; Aquafauna Bio-Marine Inc., cat. no. ABMGSL-TIN90). Detailed brine shrimp hatchery methods are included in *The Zebrafish Book* (Westerfield, 1995). Briefly, 80 mL of brine shrimp cysts are momentarily immersed in bleach before rinsing with system water. After rinsing, the cysts are added to 12 L of system water supplemented with 10 g sodium bicarbonate and 155 g sodium chloride. The cysts are aerated vigorously for 24 h, under continuous light. The hatched brine shrimp are filtered through a 105-μm mesh sieve and diluted in 2 L of system water. Although brine shrimp hatching rates can vary, we typically find this procedure generates $\sim 4 \times 10^7$ brine shrimp per 24 h.
- Sodium bicarbonate (Aquatic Ecosystems, cat. no. SC12).
- Sodium chloride (Fisher Scientific, cat. no. S96860).
- Brine shrimp net (Aquatic Ecosystems, cat. no. BSN1).
- 15-mL conical tubes (polystyrene or polypropylene) (Becton Dickinson, cat. no. 35-2099).
- Plastic transfer pipettes (Samco Scientific, cat. no. 225).
- FLDs (see Table 1 for details). Chloroform:MeOH (2:1) is typically used as a solvent when making stock solutions of lipophilic dyes. However, chloroform: MeOH cannot be added directly to system water containing zebrafish. Therefore, the desired quantity of chloroform:MeOH stock solution containing lipophilic dye is air dried in a 1.6-mL microcentrifuge tube for ~10 min before being resuspended in 10 μL of 100% EtOH. The 10 μL of 100% EtOH can be added directly to system water containing zebrafish. Alternatively, dimethyl sulfoxide (DMSO) or acetone can be used as solvents when making stock solutions, and can be added directly to system water containing zebrafish. However, use of DMSO and acetone as solvents is not advised as the resulting stock solution is less stable over long periods of storage. Stock solutions of FLDs are kept in the dark at −20°C.
- Chloroform (Fisher Scientific, cat. no. BP1145-1).

Table 1 Lipophilic Fluorescent Dyes for Staining Lipid Droplets in Zebrafish

Color	Name	IUPAC Name	Cat. No.[a]	Solvent	Stock Conc.	Working Conc.[b]	Absorption/ Emission Maxima (nm)	Additional Notes[c]
Green	BODIPY 500/510	4,4-Difluoro-5-methyl-4-bora-3a, 4a-diaza-s-indacene-3-dodecanoic acid	D3823	Chloroform: MeOH	1 mg/mL	0.5 μg/mL (2000×)	500/510 (use filter sets appropriate for Alexa Fluor 488)	Very bright lipid droplet (LD) stain, weak gall bladder stain
Green	BODIPY 505/515	4,4-Difluoro-1,3,5, 7-tetramethyl-4-bora-3a,4a-diaza-s-Indacene	D3921	DMSO	1 mg/mL	1 μg/mL (1000×)	505/515 (use filter sets appropriate for Alexa Fluor 488)	Very bright LD stain, bright gall bladder stain, bright intestine stain, significant fluorescence emission in red channel after 488 nm laser excitation
Green	BODIPY FL C5	4,4-Difluoro-5,7-dimethyl-4-bora-3a,4a-diaza-s-indacene-3-pentanoic acid	D3834	Chloroform: MeOH	0.5 μg/mL	0.25 ng/mL (2000×)	503/512 (use filter sets appropriate for Alexa Fluor 488)	Weak LD stain, bright gall gladder stain, bright intestine stain
Green	NBT—cholesterol	22-(N-(7-nitrobenz-2-oxa-1,3-diazol-4-yl)amino)-23,24-bisnor-5-cholen-3β-ol	N1148	Chloroform: MeOH	5 μg/mL	0.5 ng/mL (10,000×)	Absorption/ emission maxima is dependent on solvent and environment. However, we use filter sets appropriate for Alexa Fluor 488)	Bright LD stain, weak gall bladder stain
Yellow	BODIPY 530/550	4,4-Difluoro-5,7-diphenyl-4-bora-3a,4a-diaza-s-indacene-3-dodecanoic acid	D3832	Chloroform: MeOH	1 mg/mL	10 μg/mL (100×)	530/550 (use filter sets appropriate for Alexa Fluor 532)	Very weak LD stain, high non-LD background

Continued

Table 1 Lipophilic Fluorescent Dyes for Staining Lipid Droplets in Zebrafish—cont'd

Color	Name	IUPAC Name	Cat. No.[a]	Solvent	Stock Conc.	Working Conc.[b]	Absorption/ Emission Maxima (nm)	Additional Notes[c]
Yellow-orange	Nile Red	9-Diethyiamino-5H-benzo[alpha] phenoxazine-5-one	N1142	Acetone	1.25 mg/mL	0.5 μg/mL	510/580 (excite with Argon 514 nm laser, collect emission with a long pass 530 filter)[d]	Very bright LD stain, red phospholipid background stain
Orange	BODIPY 558/568	4,4-Difluoro-5-(2-thienyl)-4-bora-3a,4a-diaza-s-indacene-3-dodecanoic acid	D3835	Chloroform: MeOH	1 mg/mL	2 μg/mL (500×)	558/568 (use filter sets appropriate for Alexa Fluor 546)	Very bright LD stain, very bright non-LD background
Red-orange	Cholesteryl BODIPY 576/589	Cholesteryl 4,4-difluoro-5-(2-pyrrolyl)-4-bora-3a,4a-diaza-s-indacene-3-undecanoate	C12681	Chloroform: MeOH	1 mg/mL	10 μg/mL (100×)	576/589 (use filter sets appropriate for Alexa Fluor 568)	Weak LD stain
Red	HCS LipidTOX red	–	H34476	DMSO	–	(5000×)	577/609 (use filter sets appropriate for Alexa Fluor 594 or Texas Red)	Very bright LD stain, bright gall bladder stain, zero non-LD background
Far-red	HCS LipidTOX deep red	–	H34477	DMSO	–	(5000×)	637/655 (use filter sets appropriate for Alexa Fluor 647 or Cy5 dye)	Very bright LD stain, bright gall bladder stain, zero non-LD background

[a] *All catalog numbers (cat. no.) are from Invitrogen.*
[b] *Dilution of stock solution to achieve working concentration is in parentheses.*
[c] *Staining patterns correspond to wild-type fish raised under the husbandry protocols described here. These staining patterns could potentially be altered as a function of fish genotype, dietary status, and other exposures.*
[d] *Nile Red bound to phospholipid bilayer has absorption/emission maxima of ~550/640. Alexa Fluor dyes are from Invitrogen.*

- Methanol (MeOH) (VWR, cat. no. BDH1135-4LP).
- Ethanol (EtOH) (Decon Labs, Inc., cat. no. 2716).
- DMSO (Fisher Scientific, cat. no. D128-1).
- Acetone (Mallinckrodt Chemicals, cat. no. 2440-02).
- Ethyl 3-aminobenzoate methanesulfonate salt (Tricaine or MS222) stock solution (24×) (Sigma, cat. no. A5040-110G). Combine 0.8 g of tricaine, 4.2 mL of 1 M Tris pH 9.0, and 195.8 mL of dH_2O. Adjust pH to between 7.0 and 7.5, and store at 4°C. Anesthetizing concentration is 1×, and euthanizing concentration is 5×.
- Methyl cellulose (4%) (Sigma, cat. no. MO387-100G). Dissolve 4 g methyl cellulose in 100 mL dH_2O. Make 1 mL aliquots in 1.6 mL microcentrifuge tubes and freeze at −20°C.
- Low melting point (LMP) agarose (1%) (Fisher Scientific, cat. no. BP165-25). Dissolve 1 g LMP agarose in 100 mL of 1× phosphate buffered saline (PBS). Make 1 mL aliquots in 1.6 mL microcentrifuge tubes and freeze at −20°C.
- 1.6-mL microcentrifuge tubes (Genesee Scientific, cat. no. 22-282A).
- Heat block set at 65 and 42°C (Denville Scientific, Inc., D1100).
- 24-well plastic culture plates (Greiner Bio-one, cat. no. 662160).
- Metal dissection probe (Fine Science Tools, cat. no. 10140-01).
- 30-mm Petri dish with glass cover slip as base (MatTek Corp., cat. no. P35G-1.5-10-C).
- Epinephrine stock solution (100 mg/mL) (Sigma, cat. no. E4375-5G). Dissolve 1 g epinephrine powder in 10 mL of dH_2O. Store at 4°C in the dark. Add 3 mL of stock solution to 30 mL system water (10 mg/mL final concentration) containing fish for 5 min to contract melanosomes (Rawls & Johnson, 2003).
- Paraformaldehyde (PFA) stock solution (4%) (Acros, cat. no. 30525-89-4). Dissolve 4 g PFA powder in 99 mL prewarmed 1× PBS. Once cooled, add 1 mL of DMSO.
- PBS stock solution (25×). Dissolve 200 g sodium chloride, 5 g potassium chloride, 36 g sodium dihydrogen phosphate, and 6 g monopotassium phosphate in 1 L dH_2O and autoclave.
- Tween-20 (Fisher Scientific, cat. no. BP337-500).
- Laser scanning confocal microscope (e.g., Zeiss LSM 510), equipped with Argon 488 nm, HeNe1 543 nm, and HeNe2 633 nm excitation lasers. Other platforms, such as light-sheet, spinning disk confocal, and multiphoton fluorescence microscopes, can also be used.

3. METHODS

3.1 OBTAINING ZEBRAFISH EMBRYOS

1. [Day 1, duration ~60 min] The day before embryos are required, use a large net to place suitable breeding pairs of adult zebrafish in specialized breeding tanks filled with fresh system water. Adults remain in breeding tanks overnight and

typically spawn once the aquarium lights turn on the following morning. Be sure to label each breeding tank with the genotype and stock number(s) of the respective breeding pair.

Optional: Specific fluorescent lipid probes can be used in conjunction with transgenic zebrafish lines expressing fluorescent proteins (FPs) to facilitate cell localization studies (see Table 1 for excitation/emission properties of lipid probes). If required, obtain transgenic zebrafish from ZIRC (see Section 2).

2. [Day 2, duration ~60 min] To collect fertilized embryos, remove adult zebrafish to a different tank. Maintain labeling system of adult fish to track parentage of embryos. Collect embryos by pouring through a tea strainer. Clean embryos by rinsing multiple times in fresh system water, and dispense groups of 20–40 embryos into each 100-mm Petri dish filled with 30 mL of fresh system water. View embryos on a light stereomicroscope and remove unfertilized embryos using a wide-bore Pasteur pipette and pipette pump. Place fertilized embryos within an air incubator at 28.5°C.

   Optional: If natural breeding is unsuccessful, embryos can alternatively be generated by in vitro fertilization (or "squeezing") using established protocols (Westerfield, 1995).

   Optional: Methylene blue stock solution can be added to system water (0.01% final concentration) to inhibit fungal and bacterial growth during zebrafish development.

   Optional: If using FP-expressing transgenic zebrafish, use a fluorescence stereomicroscope to screen for fluorescent embryos at a suitable developmental stage when FP expression is known to be observed. This procedure is preferentially done during embryonic stages to minimize rearing and feeding of unnecessary nontransgenic fish.

3.2 REARING ZEBRAFISH TO POSTEMBRYONIC STAGES IN PREPARATION FOR FLUORESCENT LIPID STAINING

1. [Days 2–6; duration 15–30 min/day] Continue to raise embryos/larvae at 28.5°C in 100 mm Petri dishes until 5 days postfertilization (dpf). During these first 5 days of development, check daily for, and remove, dead embryos using a wide-bore Pasteur pipette and pipette pump (dead embryos are typically white in appearance). In addition, replace ~50% of water with fresh system water once every 2–3 days using a plastic transfer pipette. Zebrafish do not need exogenous nutrition until 5 dpf, therefore do not feed during this period.
2. [Day 6; static tank stage; duration ~15 min] At 5 dpf, transfer ~40 larvae to 1 L of fresh system water contained within a clean 2-L tank and fitted with 425-μm mesh drainage plugs.
3. [Day 6 onwards; duration ~15 min/day] From 5 dpf, and once larvae are transferred to 2-L tanks, feeding can commence. It is routine procedure for

laboratory zebrafish facilities to feed young larvae (typically 5–10 dpf) either a *Paramecia multimicronucleatum* diet (http://zfin.org/zf_info/zfbook/chapt3/3.3.html) and/or commercial powdered food. We have also found feeding each 2-L tank containing 20–40 fish with 0.5 mL of ~1000 brine shrimp/mL concentration once per day can aid with larval survival. Dead brine shrimp and debris collecting at the bottom of the tank should be removed every few days with a plastic transfer pipette.

4. [Day 9; duration ~15 min] At 8 dpf, add 1 L of fresh system water to the existing 1 L containing larvae in each 2-L tank.
5. [Day 12; low flow tank stage; duration ~15 min/day] At 12 dpf, the 2-L tanks are placed on slow running system water with fine mesh (600 μm) drainage plugs. It is important to clean the drainage plugs every day to ensure unobstructed water flow. Continue feeding each tank with 0.5 mL of ~1000 brine shrimp/mL concentration once per day.
6. [Day 16; duration 15 min/day] Increase strength of water flow from 15 dpf. It is important to continue swapping mesh drainage plugs for ones with a larger mesh (typically 1600 and 4000 μm can be used) concomitant with growth of larvae/juveniles. Continue feeding each tank with 0.5 mL of ~1000 brine shrimp/mL concentration once per day.
7. Zebrafish that are fed using this protocol begin storing neutral lipid in adipocytes from ~10 to 15 dpf (Flynn et al., 2009). However, once independent feeding initiates (5 dpf), subsequent larval and juvenile growth rates vary considerably. Physical measurements such as standard length (SL; defined as distance from snout tip to caudal peduncle) provide more accurate metrics of postembryonic zebrafish development and growth (Parichy, Elizondo, Mills, Gordon, & Engeszer, 2009). AT development in wild-type zebrafish is robustly correlated with SL (Imrie & Sadler, 2010; McMenamin et al., 2013). SL can be measured directly on live larvae/juveniles using a stereomicroscope equipped with an eyepiece graticule. Alternatively, the specimen can be imaged on a stereomicroscope using known magnification and subsequently measured using suitable image analysis software (e.g., ImageJ or Adobe Photoshop).
8. [Day 16 onwards; duration 5 min] To undertake a lipid stain at a selected timepoint, larvae/juveniles must be transferred to a smaller vessel such as a 15-mL conical tube. Larval and juvenile zebrafish are very delicate; therefore, during this step, it is essential to handle them with care. To transfer larvae/juveniles from 2-L tanks to a suitable vessel, fill a clean 2-L tank with system water. Then place a brine shrimp net into the freshly prepared tank of system water, and gently pour your larvae/juvenile sample into the net partially submerged within the freshly prepared tank. Pour carefully so that the larvae are not buffeted by water, as vigorous pouring will damage sample. Individually remove each larvae/juvenile from the net while partially submerged in the tank of system water using a plastic transfer pipette, and place in a 15-mL conical tube filled with an appropriate volume of fresh system water.

3.3 STAINING LIVE ZEBRAFISH LARVAE/JUVENILES WITH FLUORESCENT LIPOPHILIC DYES

A distinguishing feature of white adipocytes in fish as well as mammals is the presence of large cytoplasmic neutral LDs that consist largely of TG (Hashemi & Goodman, 2015). As described below, these characteristic adipocyte organelles can be unambiguously labeled in live zebrafish using any of several commercially available fluorescent lipophilic probes (see Table 1).

1. [Duration 5 min] Wash unanesthetized larvae/juveniles in fresh system water three times at room temperature. This is accomplished by removing 80% of the system water from the 15-mL conical tube with a transfer pipette before adding fresh system water. This step removes debris included with larvae. After final wash, adjust volume of each 15-mL conical tube containing zebrafish up to 5 mL with fresh system water.
2. [Duration 5 min] Add lipid probe, at appropriate concentration (see Table 1), to 5 mL fresh system water containing larvae/juveniles. Lipid probe dissolved in DMSO can be added directly to system water. However, lipid probe dissolved in chloroform:MeOH must first be dried by chloroform:MeOH evaporation then resuspended in 100% EtOH (see Methods section). It is important to minimize the volume of 100% EtOH added to system water containing specimen, therefore we typically add 10 μL to 5 mL system water containing fish.
3. The experimental procedure for zebrafish LD staining varies dependent on lipid probe used. Based on FLDs that we have used for staining LDs in zebrafish (see Table 1), we have devised two main protocols (see Fig. 2 for details). Protocol 1 should be followed for BODIPY dyes (both nonpolar and fatty acid/cholesteryl conjugates) as these require extensive wash steps to reduce background staining. Protocol 2 should be followed for Nile Red and HCS LipidTOX stains as these dyes do not require a wash step. Subsequent to the lipid staining protocol, all specimens are imaged using standard techniques regardless of protocol followed (see Sections 3.4 and 3.5).

   Optional: It is common for ingested food and bile within the intestine to undergo autofluorescence. To reduce these unwanted fluorescence signals, fish can be starved overnight to "clear" the intestine of food before beginning the lipid stain.

3.4 IN VIVO IMAGING OF NEUTRAL LIPID ON A FLUORESCENCE STEREOMICROSCOPE

Imaging fluorescently labeled neutral lipid on a stereomicroscope allows for relatively high-throughput analysis of whole animal fat deposition (Fig. 2). Using the following imaging procedure, it is possible to acquire multiple images of distinct adipose depots in an individual fish within 5 min. Therefore, stereoscopic imaging of fluorescently stained zebrafish ATs is suitably quick for use as a viable phenotyping assay during chemical or genetic screens for factors that influence lipid storage in ATs.

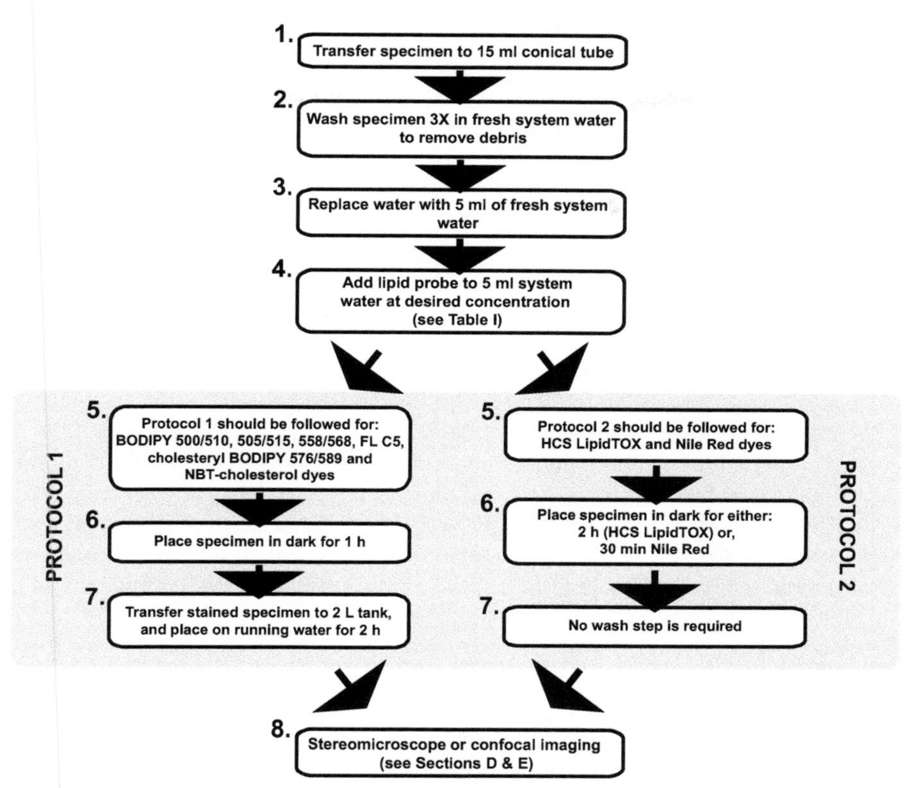

FIGURE 2 Flow diagram depicting protocols for staining zebrafish with fluorescent lipophilic dyes.

Protocol 1 includes wash steps and should be used for BODIPY dyes. Protocol 2 does not contain wash steps and should be used for HCS LipidTOX and Nile Red dyes. See Table 1 for information on lipophilic dyes.

To image and measure stained zebrafish, 1× tricaine is used as a standard anesthetic in zebrafish research. Embryonic and larval zebrafish are particularly amenable to dosing with 1× Tricaine, and recovery after 72 h of anesthesia is commonplace. However, older larval (larger than ~7 mm SL), juvenile and adult zebrafish are more difficult to recover after Tricaine anesthesia. Presumably this is due to increasing requirement for gill respiration as the zebrafish grows due to increased body size and epidermal thickness, combined with the absence of active gill respiration while under Tricaine anesthesia. The consequence of increased sensitivity to Tricaine anesthesia is higher rates of death. We try to minimize exposure to Tricaine by imaging as quickly as possible and increasing animal number to compensate for any death that might occur.

1. Fill a 100 mm Petri dish with 30 mL of fresh system water and add 1.25 mL of 24× Tricaine.
2. Place a 3 × 3 mm drop of 4% methyl cellulose at the center of a 100 mm Petri dish lid. Cover the 4% methyl cellulose drop with system water containing 1× Tricaine.
3. Transfer stained unanesthetized fish from 15 mL conical tube to fish water containing Tricaine. Allow fish to sit in Tricaine-containing system water for ~5 min, or until fish has symptoms of being under anesthesia (i.e., belly up swimming, reducing gill respiration). Do not keep fish in Tricaine for longer than required as this may impede recovery from anesthesia.
4. Immediately transfer anesthetized fish to Petri dish lid containing 4% methyl cellulose droplet. Gently position tail of larvae/juvenile in the methyl cellulose and orientate appropriately using a metal dissection probe. Do not completely embed specimen in 4% methyl cellulose as this increases likelihood of damaging fish when subsequently releasing it. In zebrafish, neutral lipid within adipocytes is first deposited in association with the pancreas at ~4.4 mm SL, which is asymmetrically located on the right-hand side of the visceral cavity larvae/juveniles (Fig. 1A) (Flynn et al., 2009). Therefore, it is usually important to orientate specimen so that right-hand side is observable to the microscope objective.
5. Once larvae/juveniles are positioned correctly, measure SL using an eyepiece graticule.

   Optional: The zebrafish pigment pattern can obscure imaging of adipose depots. Before imaging, confounding effects of the zebrafish pigment pattern can by reduced by treating animals with 10 mg/mL epinephrine for 5 min to contract melanosomes (Rawls & Johnson, 2003). This step is not optimal as epinephrine is known to stimulate lipolysis in adipocytes (Fain & Garcija-Sainz, 1983); however, our imaging procedure is sufficiently brief to prevent any salient effect on fat storage. Furthermore, for imaging visceral/pancreatic AT, melanin is usually not a problem during larval and early juvenile stages (i.e., through ~8 mm SL). Alternative solutions include using zebrafish mutants that fail to develop pigment cells; however, the potential effects of these mutations on adipose development and function have not been explored. Although inhibition of melanin synthesis during zebrafish development can be achieved by treatment with phenylthiourea, this is not practical for fish kept in tanks maintained on flowing recirculating water.

3.5 IN VIVO IMAGING OF NEUTRAL LIPID ON A CONFOCAL MICROSCOPE

Although considerably more time-consuming than stereomicroscope analysis, imaging fluorescently stained neutral lipid by confocal microscopy provides much greater resolution. Furthermore, if FP transgenic lines are used, exact colocalization of adipocyte LDs with fluorescent cell types of interest is achievable (Minchin et al., 2015). The following protocol takes ~35 min per zebrafish imaged (based on a 15 min Z-stack); however, exact timing will vary depending on length of confocal scan taken.

1. Anesthetize fish in Tricaine (see Section 3.4).
2. Measure SL of each fish to be imaged (see list 7 in Section 3.2).
3. Thaw 1% LMP agarose aliquots by placing 1 mL aliquot at 65°C until completely melted. Cool LMP agarose to 42°C using a heat block or water bath for 1 h until needed.
   Note: For confocal imaging there are multiple methods for stably mounting zebrafish in 1% LMP agarose during imaging. Depending on microscope design (inverted or upright) and the type of objective to be used (air, immersion, or dipping) we have found the following methods for mounting to be most successful (see Fig. 3).
4. Upright confocal microscopes. It is preferential to use a water dipping objective. First, deposit a 3 × 3 mm droplet of 4% methyl cellulose in the center of a 30 mm Petri dish. Anesthetize larvae/juveniles in 1× Tricaine (see Section 3.4), transfer to methyl cellulose droplet, and roughly orientate. Remove excess system water with plastic transfer pipette and replace with 1% LMP agarose cooled to 42°C. Using a stereoscope quickly orientate sample into correct position with a metal dissecting probe. Continuously observe specimen as LMP solidifies to ensure correct positioning. It is important for specimen to be only lightly covered by LMP agarose. If too much LMP is added, it will prevent objective focusing to deep regions within zebrafish AT. Once LMP agarose is solidified, add ~2 mL of 1× Tricaine diluted in system water to the Petri dish to keep fish anesthetized.
5. Inverted confocal microscopes. It is common to use either air or immersion objectives. We mount the specimen in 30-mm Petri dish with glass cover slip as base. When using this method it is important to orientate specimen as close to cover slip as possible. Anesthetize zebrafish in 1× Tricaine (see Section 3.4) and place on the glass cover slip within the 30 mm Petri dish. Remove excess system water carried over from transfer of specimen. Quickly add 1% LMP agarose cooled to 42°C to fish and orientate into correct position using a metal dissection probe. Continuously observe orientation of the fish as LMP solidifies to ensure correct positioning. Once LMP agarose is solidified, add ~2 mL of 1% Tricaine diluted in system water to the Petri dish.
6. Table 1 contains excitation and emission information for each fluorescent lipid stain used to visualize neutral lipid. Due to the size of late larvae/juveniles, best results are obtained using 10× or 20× objectives that have large working distances but a numerical aperture of ~1.

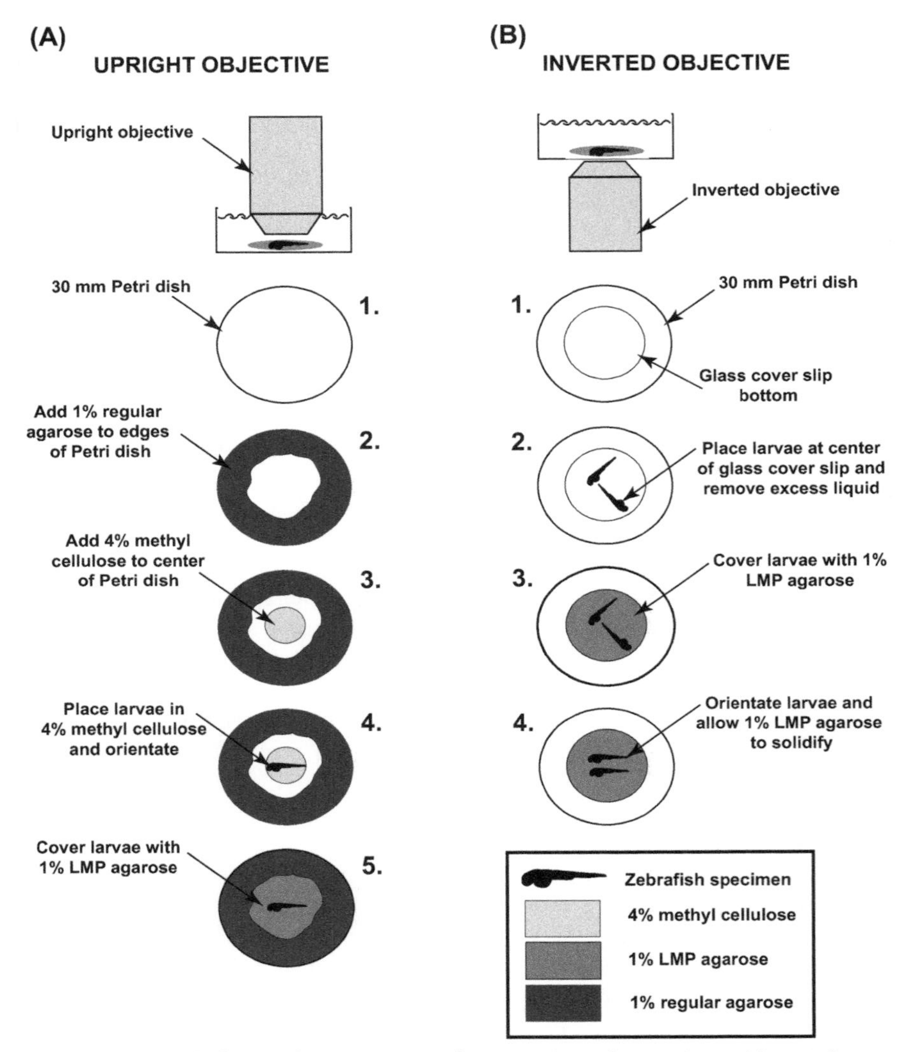

FIGURE 3 Schematic illustrating stable mounting procedures for regular and inverted microscopes.

(A) When imaging using an upright objective, a 30 mm Petri dish is used (step 1), and 1% regular agarose is used to fill in around the edges of the Petri dish (step 2). 4% methyl cellulose is then placed in the center of Petri dish (step 3), before the anesthetized animal is placed in 4% methyl cellulose and orientated (step 4). Once specimen is correctly orientated, it is covered with 1% low melting point (LMP) agarose and allowed to solidify (step 5). (B) For imaging using an inverted objective, a 30-mm Petri dish with a fitted glass cover slip as a base is used (step 1). Anesthetized zebrafish are place onto the glass cover slip base (step 2) and excess liquid is removed. The specimen is then covered with 1% LMP agarose (step 3) and quickly orientated while agarose is solidifying (step 4). For both methods, 2 mL of system water, containing 1× Tricaine, is added to specimen. When undertaking imaging using the upright objective method, be sure to use a water dipping objective.

3.6 RECOVERY OF SAMPLE AFTER IMAGING OF FLUORESCENT NEUTRAL LIPID

1. Zebrafish are amenable to longitudinal analyses of fat storage within individual fish (Flynn et al., 2009; McMenamin et al., 2013). Therefore, once imaging has been completed, it is often necessary to recover larvae and allow development to proceed. Under a dissecting microscope gently cut LMP agarose away from tail of larvae with a metal dissection probe. We find it is easier to recover fish from 1% LMP agarose if it is first immersed in fresh system water. Once tail is free, it may be possible to release larvae by gently squeezing clean system water over the specimen with a plastic transfer pipette. If necessary, carefully remove more of LMP agarose from anterior regions until larvae are released.
2. Once larvae are free and recovered from anesthesia, individually house each fish in a well of a 24-well plate filled with 1 mL system water to keep record of subsequent larvae/juvenile growth during longitudinal analysis. It is necessary to change 80% of system water within each well daily and to feed ~30 live brine shrimp per well per day.

3.7 GUIDELINES FOR ANALYZING REGIONAL ZEBRAFISH ADIPOSE TISSUES

As described above, the ability to image the regional deposition of AT in whole animals has significant potential for investigating genetic and environmental factors that regulate AT distribution and disease susceptibility. Zebrafish are particularly amenable to whole-animal imaging of AT regionality (Fig. 1); therefore, we recently established a preliminary classification system and nomenclature for zebrafish ATs to promote experimental reliability and precision. Our proposed classification system is based on an existing system used to classify human ATs (Table 2) (Shen et al., 2003), and first divides total AT into "internal" or "subcutaneous" compartments (Fig. 4). Internal AT is subsequently divided into visceral (VAT) and nonvisceral (NVAT) ATs (Fig. 4). Subcutaneous ATs are more numerous and deposited throughout the body (Fig. 4). We categorized SAT into cranial, trunk, and appendicular deposits (Fig. 4). Table 2 and Fig. 4 provide a detailed overview of the zebrafish adipose classification system. Using the new classification system it is possible to quantify the area of each AT within an FLD-labeled using the fluorescence stereomicroscopy techniques outlined in Section 3.4. Briefly, individual ATs can be segmented using fluorescence thresholding to create an ROI that accurately represents the 2D area of an AT (Fig. 4A). In turn it has been shown that AT area, as assessed by FLD staining and stereomicroscopy, is an accurate measure of lipid content (Tingaud-Sequeira, 2011 #3). The majority of zebrafish ATs do not touch or merge with another AT; therefore, pixel intensity-based thresholding is the simplest and quickest segmentation method. For ATs that do touch (e.g., PVAT and AVAT) an intersecting boundary was defined by a straight line connecting the two AT extremes. Using this methodology it is possible to obtain a picture of lipid deposition within zebrafish ATs across a large developmental timeframe (Fig. 5).

Table 2 Classification System for Zebrafish Adipose Tissues (ATs)

AT Classification				Acronym	AT Appearance (Standard Length (SL) mm)[a]
Internal				IAT	4.4
	Visceral			VAT	4.4
		Cardiac		CVAT	9.7
			Anterior	aCVAT	9.7
			Posterior	pCVAT	12.5
		Pancreatic		PVAT	4.4
		Abdominal		AVAT	5.5
		Renal		RVAT	7.9
	Nonvisceral			NVAT	8.7
		Paraosseal		POS	8.9
			Dorsal	dPOS	10
			Central	cPOS	9.1
			Ventral	vPOS	11.5
		Intermuscular		IM	10.2
			Caudal	cIM	10.1
			Dorsal	dIM	n.d.
			Ventral	vIM	n.d.
Subcutaneous				SAT	6.6
	Appendicular			APPSAT	9.3
		Caudal fin ray		CFRSAT	10.8
		Dorsal fin ray		DFRSAT	11.5
		Anal fin ray		AFRSAT	9.7
		Anal fin cluster		AFCSAT	10

		Pelvic fin		PELSAT	n.d.
		Pectoral fin		PECSAT	10.2
			Anterior	aPECSAT	12.5
			Posterior	pPECSAT	11.8
			Loose	lPECSAT	11.5
	Cranial			CSAT	6.6
		Ocular		OCU	7.2
		Opercular		OPC	9.5
			Dorsal	dOPC	10.3
			Ventral	vOPC	9.5
		Hyoid		HYD	6.9
			Basihyoid	BHD	7.3
			Ceratohyoid	CHD	7.7
			Urohyoid	UHD	n.d.
	Truncal			TSAT	8.2
		Lateral		LSAT	8.2
		Dorsal		DSAT	9.5
			Anterior	aDSAT	10
			Posterior	pDSAT	9.5
		Ventral		VSAT	9.8
		Abdominal		ASAT	10.5

[a] *The SL at which zebrafish ATs first appear ($P > 0.5$) was identified using logistic regression within a cohort of 362 Ekkwill wild-type fish.*

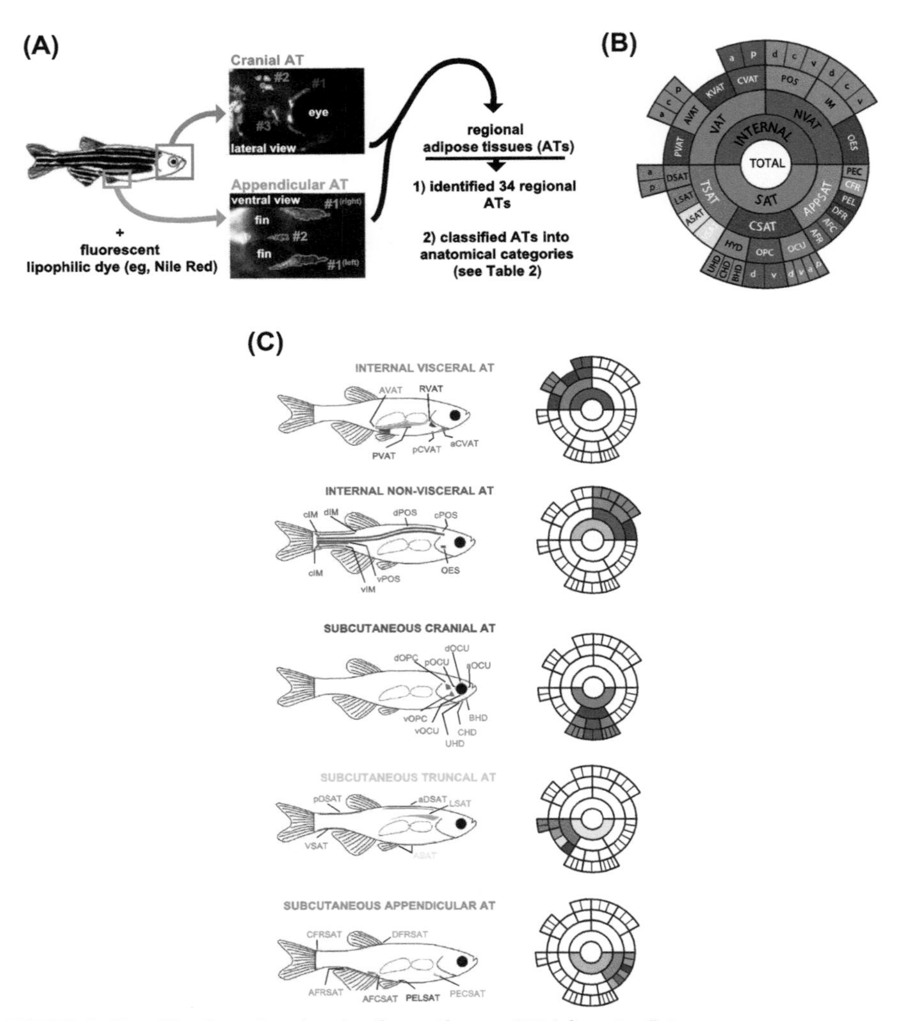

FIGURE 4 Classification of regional adipose tissues (ATs) in zebrafish.

(A) Schematic illustrating the experimental procedure used to identify 34 regionally distinct zebrafish ATs. (B) Circular graph illustrating the relationship between the 34 regionally distinct zebrafish ATs. Acronyms are defined in Table 2. (C). Schematic illustrating the anatomical location of zebrafish regional ATs. Circular graphs correspond to (B) and acronyms are defined in Table 2.

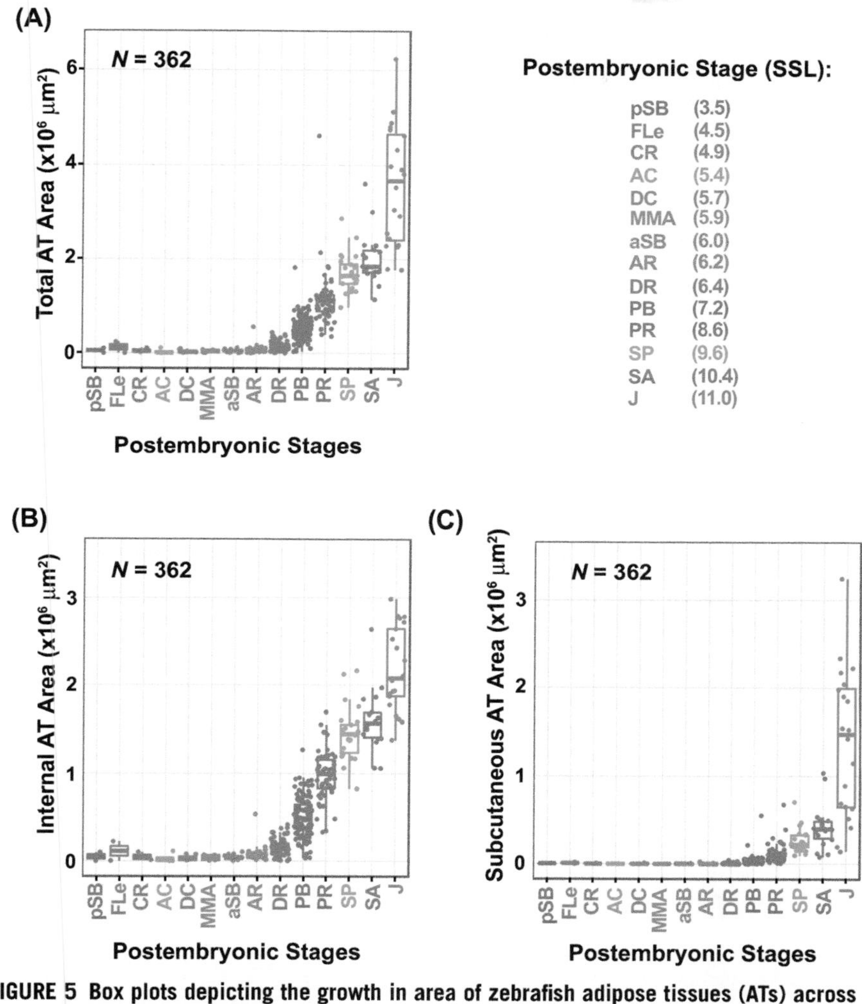

FIGURE 5 Box plots depicting the growth in area of zebrafish adipose tissues (ATs) across a range of postembryonic stages.

Box plots depicting the growth of total AT (A), internal AT (B), and subcutaneous AT (C) across postembryonic stages. The stages are defined in Parichy et al. (2009).

4. SUMMARY

Investigation of white ATs has only recently been initiated in the zebrafish, and consequently we have a relatively limited knowledge of AT development and physiology in this important vertebrate model system. However, the amenability of the zebrafish to high resolution in vivo imaging presents exciting opportunities to address longstanding questions about AT formation and function which have been difficult to resolve using available mammalian models. Moreover, systematic genetic and chemical tests in the zebrafish model could be used to identify novel factors regulating distinct aspects of AT biology.

Here we have provided detailed methods for in vivo imaging of zebrafish ATs to support the use of the zebrafish as a model for AT research. We anticipate that the fluorescence stereomicroscopy methods presented here will be useful for rapid phenotypic assessments required for genetic and chemical screens, while the confocal microscopy methods will facilitate high-resolution analysis of cellular and molecular events within ATs. Furthermore, we expect that these methods will also be generally applicable to in vivo imaging of ATs in other fish species. Deployment of these methods in the zebrafish system will be enriched by the use of existing and forthcoming zebrafish lines expressing FPs and other transgenes in adipocyte lineages and other cellular constituents of ATs. Use of zebrafish as a model for AT biology will also be enhanced by identification of genetic alterations, dietary and environmental manipulations, and chemicals that modify zebrafish AT formation and function.

ACKNOWLEDGMENTS

This work was supported by a British Heart Foundation Centre of Research Excellence/University of Edinburgh Fellowship to J.E.N. Minchin, NIH grants DK081426 and DK073695 to J.F. Rawls, NIH grant DK056350 to the University of North Carolina at Chapel Hill's Nutrition Obesity Research Center (NORC), and a Pew Scholars Program in the Biomedical Sciences Award to J.F. Rawls.

REFERENCES

Ailhaud, G., Grimaldi, P., & Negrel, R. (1992). Cellular and molecular aspects of adipose tissue development. *Annual Review of Nutrition, 12*, 207–233. http://dx.doi.org/10.1146/annurev.nu.12.070192.001231.

Albalat, A., Saera-Vila, A., Capilla, E., Gutierrez, J., Perez-Sanchez, J., & Navarro, I. (2007). Insulin regulation of lipoprotein lipase (LPL) activity and expression in gilthead sea bream (*Sparus aurata*). *Comparative Biochemistry and Physiology. Part B, Biochemistry & Molecular Biology, 148*(2), 151–159. http://dx.doi.org/10.1016/j.cbpb.2007.05.004.

Berry, D. C., Jiang, Y., & Graff, J. M. (2016). Emerging roles of adipose progenitor cells in tissue development, homeostasis, expansion and thermogenesis. *Trends in Endocrinology and Metabolism, 27*(8), 574–585. http://dx.doi.org/10.1016/j.tem.2016.05.001.

Bou, M., Todorcevic, M., Fontanillas, R., Capilla, E., Gutierrez, J., & Navarro, I. (2014). Adipose tissue and liver metabolic responses to different levels of dietary carbohydrates in gilthead sea bream (*Sparus aurata*). *Comparative Biochemistry and Physiology. Part A, Molecular & Integrative Physiology, 175*, 72–81. http://dx.doi.org/10.1016/j.cbpa.2014.05.014.

Bou, M., Todorcevic, M., Torgersen, J., Skugor, S., Navarro, I., & Ruyter, B. (2016). De novo lipogenesis in Atlantic salmon adipocytes. *Biochimica et Biophysica Acta, 1860*(1 Pt A), 86–96. http://dx.doi.org/10.1016/j.bbagen.2015.10.022.

Despres, J. P. (1998). The insulin resistance-dyslipidemic syndrome of visceral obesity: effect on patients' risk. *Obesity Research, 6*(Suppl. 1), 8S–17S.

Fain, J. N., & Garcija-Sainz, J. A. (1983). Adrenergic regulation of adipocyte metabolism. *Journal of Lipid Research, 24*(8), 945–966.

Farese, R. V., Jr., & Walther, T. C. (2009). Lipid droplets finally get a little R-E-S-P-E-C-T. *Cell, 139*(5), 855–860. http://dx.doi.org/10.1016/j.cell.2009.11.005.

Flynn, E. J., 3rd, Trent, C. M., & Rawls, J. F. (2009). Ontogeny and nutritional control of adipogenesis in zebrafish (*Danio rerio*). *Journal of Lipid Research, 50*(8), 1641–1652. http://dx.doi.org/10.1194/jlr.M800590-JLR200.

Fox, C. S., Massaro, J. M., Hoffmann, U., Pou, K. M., Maurovich-Horvat, P., Liu, C. Y., … O'Donnell, C. J. (2007). Abdominal visceral and subcutaneous adipose tissue compartments: association with metabolic risk factors in the Framingham Heart Study. *Circulation, 116*(1), 39–48. http://dx.doi.org/10.1161/CIRCULATIONAHA.106.675355.

Gesta, S., Bluher, M., Yamamoto, Y., Norris, A. W., Berndt, J., Kralisch, S., … Kahn, C. R. (2006). Evidence for a role of developmental genes in the origin of obesity and body fat distribution. *Proceedings of the National Academy of Sciences of the United States of America, 103*(17), 6676–6681. http://dx.doi.org/10.1073/pnas.0601752103.

Hashemi, H. F., & Goodman, J. M. (2015). The life cycle of lipid droplets. *Current Opinion in Cell Biology, 33*, 119–124. http://dx.doi.org/10.1016/j.ceb.2015.02.002.

Ibabe, A., Bilbao, E., & Cajaraville, M. P. (2005). Expression of peroxisome proliferator-activated receptors in zebrafish (*Danio rerio*) depending on gender and developmental stage. *Histochemistry and Cell Biology, 123*(1), 75–87. http://dx.doi.org/10.1007/s00418-004-0737-2.

Imrie, D., & Sadler, K. C. (2010). White adipose tissue development in zebrafish is regulated by both developmental time and fish size. *Developmental Dynamics, 239*(11), 3013–3023. http://dx.doi.org/10.1002/dvdy.22443.

Jo, J., Gavrilova, O., Pack, S., Jou, W., Mullen, S., Sumner, A. E., … Periwal, V. (2009). Hypertrophy and/or hyperplasia: dynamics of adipose tissue growth. *PLoS Computational Biology, 5*(3), e1000324. http://dx.doi.org/10.1371/journal.pcbi.1000324.

Johansson, M., Morgenroth, D., Einarsdottir, I. E., Gong, N., & Bjornsson, B. T. (2016). Energy stores, lipid mobilization and leptin endocrinology of rainbow trout. *Journal of Comparative Physiology B, 186*(6), 759–773. http://dx.doi.org/10.1007/s00360-016-0988-y.

Karpe, F., & Pinnick, K. E. (2015). Biology of upper-body and lower-body adipose tissue–link to whole-body phenotypes. *Nature Reviews. Endocrinology, 11*(2), 90–100. http://dx.doi.org/10.1038/nrendo.2014.185.

Kissebah, A. H., & Krakower, G. R. (1994). Regional adiposity and morbidity. *Physiological Reviews, 74*(4), 761–811.

McMenamin, S. K., Minchin, J. E., Gordon, T. N., Rawls, J. F., & Parichy, D. M. (2013). Dwarfism and increased adiposity in the gh1 mutant zebrafish vizzini. *Endocrinology, 154*(4), 1476–1487. http://dx.doi.org/10.1210/en.2012-1734.

Michel, M., Page-McCaw, P. S., Chen, W., & Cone, R. D. (2016). Leptin signaling regulates glucose homeostasis, but not adipostasis, in the zebrafish. *Proceedings of the National Academy of Sciences of the United States of America, 113*(11), 3084–3089. http://dx.doi.org/10.1073/pnas.1513212113.

Minchin, J. E., Dahlman, I., Harvey, C. J., Mejhert, N., Singh, M. K., Epstein, J. A., ... Rawls, J. F. (2015). Plexin D1 determines body fat distribution by regulating the type V collagen microenvironment in visceral adipose tissue. *Proceedings of the National Academy of Sciences of the United States of America, 112*(14), 4363–4368. http://dx.doi.org/10.1073/pnas.1416412112.

Morse, S. A., Gulati, R., & Reisin, E. (2010). The obesity paradox and cardiovascular disease. *Current Hypertension Reports, 12*(2), 120–126. http://dx.doi.org/10.1007/s11906-010-0099-1.

Nishimura, S., Manabe, I., Nagasaki, M., Hosoya, Y., Yamashita, H., Fujita, H., ... Sugiura, S. (2007). Adipogenesis in obesity requires close interplay between differentiating adipocytes, stromal cells, and blood vessels. *Diabetes, 56*(6), 1517–1526. http://dx.doi.org/10.2337/db06-1749.

Nishimura, S., Manabe, I., Nagasaki, M., Seo, K., Yamashita, H., Hosoya, Y., ... Sugiura, S. (2008). In vivo imaging in mice reveals local cell dynamics and inflammation in obese adipose tissue. *Journal of Clinical Investigation, 118*(2), 710–721. http://dx.doi.org/10.1172/JCI33328.

Oku, H., Tokuda, M., Okumura, T., & Umino, T. (2006). Effects of insulin, triiodothyronine and fat soluble vitamins on adipocyte differentiation and LPL gene expression in the stromal-vascular cells of red sea bream, Pagrus major. *Comparative Biochemistry and Physiology. Part B, Biochemistry & Molecular Biology, 144*(3), 326–333. http://dx.doi.org/10.1016/j.cbpb.2006.03.008.

Oku, H., & Umino, T. (2008). Molecular characterization of peroxisome proliferator-activated receptors (PPARs) and their gene expression in the differentiating adipocytes of red sea bream Pagrus major. *Comparative Biochemistry and Physiology. Part B, Biochemistry & Molecular Biology, 151*(3), 268–277. http://dx.doi.org/10.1016/j.cbpb.2008.07.007.

Parichy, D. M., Elizondo, M. R., Mills, M. G., Gordon, T. N., & Engeszer, R. E. (2009). Normal table of postembryonic zebrafish development: staging by externally visible anatomy of the living fish. *Developmental Dynamics, 238*(12), 2975–3015. http://dx.doi.org/10.1002/dvdy.22113.

Peinado, J. R., Jimenez-Gomez, Y., Pulido, M. R., Ortega-Bellido, M., Diaz-Lopez, C., Padillo, F. J., ... Malagon, M. M. (2010). The stromal-vascular fraction of adipose tissue contributes to major differences between subcutaneous and visceral fat depots. *Proteomics, 10*(18), 3356–3366. http://dx.doi.org/10.1002/pmic.201000350.

Peirce, V., Carobbio, S., & Vidal-Puig, A. (2014). The different shades of fat. *Nature, 510*(7503), 76–83. http://dx.doi.org/10.1038/nature13477.

Rawls, J. F., & Johnson, S. L. (2003). Temporal and molecular separation of the kit receptor tyrosine kinase's roles in zebrafish melanocyte migration and survival. *Developmental Biology, 262*(1), 152–161.

Redinger, R. N. (2009). Fat storage and the biology of energy expenditure. *Translational Research, 154*(2), 52–60. http://dx.doi.org/10.1016/j.trsl.2009.05.003.

Rodeheffer, M. S., Birsoy, K., & Friedman, J. M. (2008). Identification of white adipocyte progenitor cells in vivo. *Cell, 135*(2), 240–249. http://dx.doi.org/10.1016/j.cell.2008.09.036.

Rosen, E. D., & Spiegelman, B. M. (2006). Adipocytes as regulators of energy balance and glucose homeostasis. *Nature, 444*(7121), 847–853. http://dx.doi.org/10.1038/nature05483.

Salmeron, C., Johansson, M., Angotzi, A. R., Ronnestad, I., Jonsson, E., Bjornsson, B. T., ... Capilla, E. (2015). Effects of nutritional status on plasma leptin levels and in vitro regulation of adipocyte leptin expression and secretion in rainbow trout.

General and Comparative Endocrinology, 210, 114–123. http://dx.doi.org/10.1016/j.ygcen.2014.10.016.

Shen, W., & Chen, J. (2008). Application of imaging and other noninvasive techniques in determining adipose tissue mass. *Methods in Molecular Biology, 456*, 39–54. http://dx.doi.org/10.1007/978-1-59745-245-8_3.

Shen, W., Wang, Z., Punyanita, M., Lei, J., Sinav, A., Kral, J. G., … Heymsfield, S. B. (2003). Adipose tissue quantification by imaging methods: a proposed classification. *Obesity Research, 11*(1), 5–16. http://dx.doi.org/10.1038/oby.2003.3.

Song, Y., & Cone, R. D. (2007). Creation of a genetic model of obesity in a teleost. *FASEB Journal, 21*(9), 2042–2049. http://dx.doi.org/10.1096/fj.06-7503com.

Tang, W., Zeve, D., Suh, J. M., Bosnakovski, D., Kyba, M., Hammer, R. E., … Graff, J. M. (2008). White fat progenitor cells reside in the adipose vasculature. *Science, 322*(5901), 583–586. http://dx.doi.org/10.1126/science.1156232.

Tingaud-Sequeira, A., Ouadah, N., & Babin, P. J. (2011). Zebrafish obesogenic test: a tool for screening molecules that target adiposity. *Journal of Lipid Research, 52*(9), 1765–1772. http://dx.doi.org/10.1194/jlr.D017012.

Todorcevic, M., Skugor, S., Krasnov, A., & Ruyter, B. (2010). Gene expression profiles in Atlantic salmon adipose-derived stromo-vascular fraction during differentiation into adipocytes. *BMC Genomics, 11*, 39. http://dx.doi.org/10.1186/1471-2164-11-39.

Vegusdal, A., Sundvold, H., Gjoen, T., & Ruyter, B. (2003). An in vitro method for studying the proliferation and differentiation of Atlantic salmon preadipocytes. *Lipids, 38*(3), 289–296.

Vidal, H. (2001). Gene expression in visceral and subcutaneous adipose tissues. *Annals of Medicine, 33*(8), 547–555.

Virtue, S., & Vidal-Puig, A. (2010). Adipose tissue expandability, lipotoxicity and the Metabolic Syndrome—an allostatic perspective. *Biochimica et Biophysica Acta, 1801*(3), 338–349. http://dx.doi.org/10.1016/j.bbalip.2009.12.006.

Vohl, M. C., Sladek, R., Robitaille, J., Gurd, S., Marceau, P., Richard, D., … Tchernof, A. (2004). A survey of genes differentially expressed in subcutaneous and visceral adipose tissue in men. *Obesity Research, 12*(8), 1217–1222. http://dx.doi.org/10.1038/oby.2004.153.

Weil, C., Sabin, N., Bugeon, J., Paboeuf, G., & Lefevre, F. (2009). Differentially expressed proteins in rainbow trout adipocytes isolated from visceral and subcutaneous tissues. *Comparative Biochemistry and Physiology Part D Genomics Proteomics, 4*(3), 235–241. http://dx.doi.org/10.1016/j.cbd.2009.05.002.

Westerfield, M. (1995). *The Zebrafish Book – A guide for the laboratory use of zebrafish (*Danio rerio*)*. Eugene, OR: Univ. of Oregon Press.

Xue, Y., Lim, S., Brakenhielm, E., & Cao, Y. (2010). Adipose angiogenesis: quantitative methods to study microvessel growth, regression and remodeling in vivo. *Nature Protocols, 5*(5), 912–920. http://dx.doi.org/10.1038/nprot.2010.46.

Yach, D., Stuckler, D., & Brownell, K. D. (2006). Epidemiologic and economic consequences of the global epidemics of obesity and diabetes. *Nature Medicine, 12*(1), 62–66. http://dx.doi.org/10.1038/nm0106-62.

PART

II

Innate and Adaptive Immune Systems

CHAPTER

Innate immune cells and bacterial infection in zebrafish

2

J.W. Astin*, P. Keerthisinghe*, L. Du*, L.E. Sanderson[§], K.E. Crosier*, P.S. Crosier*, C.J. Hall*[,1]

**University of Auckland, Auckland, New Zealand*

[§]Universite libre de Bruxelles, Gosselies, Belgium

[1]Corresponding author: E-mail: c.hall@auckland.ac.nz

CHAPTER OUTLINE

Methods in Cell Biology, Volume 138, ISSN 0091-679X, http://dx.doi.org/10.1016/bs.mcb.2016.08.002

Abstract

The physical attributes of the zebrafish, including optical transparency during embryogenesis, large clutch sizes, external development, and rapid organogenesis were features that initially attracted developmental biologists to use this vertebrate as an experimental model system. With the progressive development of an extensive genetic "tool kit" and an ever-growing number of transgenic reporter lines, the zebrafish model has evolved into an informative system in which to mimic and study aspects of human disease, including those associated with bacterial infections. This chapter provides detailed protocols for microinjection of bacterial strains into zebrafish larvae and subsequent experiments to investigate single-larva bacterial burdens, live imaging of specific neutrophil and macrophage bactericidal functions, and how these protocols may be applied to drug discovery approaches to uncover novel immunomodulatory drugs.

INTRODUCTION

The utility of the zebrafish model to live image innate immune cells and investigate host–microbe interactions has been extensively reviewed elsewhere (Gratacap & Wheeler, 2014; Harvie & Huttenlocher, 2015; Meijer, van der Vaart, & Spaink, 2014; Tobin, May, & Wheeler, 2012; Torraca, Masud, Spaink, & Meijer, 2014). Here we briefly highlight some early studies that pioneered exploiting the optical transparency of embryonic zebrafish to live image innate immune cell behavior. We also provide some high-impact examples of more contemporary live imaging studies where the use of transgenic reporter lines has provided new insights into host–pathogen interactions that have extended our understanding of infectious disease in humans.

Similar to mammals, blood development in zebrafish occurs in two waves: a primitive wave that provides a transient supply of erythrocytes and myeloid lineages (macrophages and neutrophils) within the early embryo and a definitive wave that eventually populates the developing kidney (the site of definitive hematopoiesis in fish, analogous to the mammalian bone marrow) with hematopoietic stem cells (HSCs) that contribute to all blood lineages through the life of the host (Carroll & North, 2014; Davidson & Zon, 2004). The caudal hematopoietic tissue in embryonic zebrafish (a site analogous to the mammalian fetal liver) provides a transient site for newly emerged HSC to divide and contribute to erythroid and myeloid lineages to ensure a larval supply of these lineages before true definitive hematopoiesis is established in the kidney marrow (Bertrand et al., 2007). Adult zebrafish have both innate and adaptive (B and T lymphocytes) cellular arms of the immune system. However, during the first month of development (after which the adaptive immune system becomes functional) the zebrafish immune system is solely innate (Lam, Chua, Gong, Lam, & Sin, 2004). This temporal separation of the innate and adaptive immune responses makes the zebrafish embryo and larva an excellent model in which to exclusively examine the functional role of innate immune cells, without any confounding influence of adaptive immunity.

The potential of embryonic zebrafish to provide a window to live image innate immune cell behavior was first realized by Herbomel and colleagues using Differential Interference Contrast (DIC) microscopy (Herbomel, Thisse, & Thisse, 1999). Exploiting video-enhanced DIC microscopy the developmental ontogeny of a macrophage-like mononuclear phagocyte was established. These "early macrophages" were shown to arise from ventrolateral mesoderm in the head where they migrate onto the yolk sack by 24 hours post fertilization (hpf), from which a subset then colonize the head mesenchyme (Herbomel et al., 1999). Early macrophages were observed to phagocytize apoptotic cell corpses and could engulf and eliminate gram-negative *Escherichia coli* and gram-positive *Bacillus subtilis* injected into the circulation at 30 hpf. Subsequent studies using similar microscopy techniques characterized a "patrolling" behavior of early macrophages within the brain and epidermis and a phenotypic switch from an "early macrophage" to "early (amoeboid) microglia" phenotype following colonization of the developing brain (Herbomel, Thisse, & Thisse, 2001). These seminal studies identified early innate immune cells by assessing morphology and expression of lineage-specific genes (e.g., the macrophage-specific *colony stimulating factor 1 receptor a* (*csf1ra*) and the microglia-specific *apolipoprotein Eb* (*apoeb*) genes) by whole mount in situ hybridization (WMISH) of fixed specimens. Similar expression studies also helped with the developmental ontogeny and functional characterization of the zebrafish neutrophil lineage (Lieschke, Oates, Crowhurst, Ward, & Layton, 2001). Exploiting a histochemical assay and WMISH, Lieschke et al. (2001) demonstrated that *myeloperoxidase* (*mpx*) expression and activity was specific to the neutrophil lineage and marked cells that were recruited to sterile acute inflammation generated by tail fin transection.

With the application of transgenic tools researchers have leveraged off these early studies to target fluorescent proteins to specific immune cell lineages. Such transgenic reporter lines have provided researchers with a window to directly observe innate immune cell behavior within a completely intact vertebrate model

Table 1 Selected Reporter Lines Marking Myelomonocytic Cells

Reporter Line[a]	Gene Promoter	Myelomonocytic Cells Marked	References
Tg(gata1:DsRed)[sd2]	*gata1*	Erythromyeloid progenitors	Bertrand et al. (2007) and Traver et al. (2003)
Tg(lmo2:EGFP)[zf71]	*lmo2*	Erythromyeloid progenitors	Bertrand et al. (2007) and Zhu et al. (2005)
Tg(-5.3spi1b:EGFP)[gl21] *Tg(-9.0spi1b:EGFP)[zdf11]*	*spi1/pu.1*	Early myeloid progenitors	Hsu et al. (2004) and Ward et al. (2003)
Tg(mpx:GFP)[i113] *Tg(mpx:GFP)[uwm1]*	*mpx*	Neutrophils	Mathias et al. (2006) and Renshaw et al. (2006)
Tg(lyz:EGFP)[nz50]	*lyz*	Neutrophils	Hall et al. (2007)
Tg(myd88:EGFP)[zf163]	*myd88*	Neutrophils and macrophages	Hall et al. (2009)
Tg(fli1a:EGFP)[y1]	*fli1a*	Macrophages	Lawson and Weinstein (2002) and Redd, Kelly, Dunn, Way, and Martin (2006)
Tg(mpeg1:EGFP)[gl22]	*mpeg1*	Macrophage-lineage cells	Ellett, Pase, Hayman, Andrianopoulos, and Lieschke (2011)
Tg(csf1ra:GAL4-VP16)[i186]/Tg(UAS:EGFP-CAAX)[m1230]	*csf1ra*	Macrophages	Gray et al. (2011)
Tg(mfap4:mTurquoise)[xt27]	*mfap4*	Macrophages	Oehlers et al. (2015)
Tg(apoeb:LY-EGFP)[zf147]	*apoeb*	Microglia	Peri and Nusslein-Volhard (2008)
Tg(tnfa:EGFP-F)[ump5]	*tnfa*	Activated macrophages	Nguyen-Chi et al. (2015)
Tg(irg1:EGFP)[nz26]	*irg1*	Activated macrophages	Sanderson et al. (2015)

[a] *Examples of reporter lines relate to the first version of that line and references relate to the first version and/or their first use to live image myelomonocytic cells.*

(Table 1). These studies have helped validate the zebrafish system by revealing remarkable conservation of neutrophil and macrophage function between mammals and zebrafish and helped to uncover completely new disease-relevant behaviors and functions for these innate immune cells. In particular, using these reporter lines to examine host–pathogen interactions has provided fundamental insights into how innate immune cells help orchestrate the host response to pathogens and how this contributes to disease pathology. One particular area where zebrafish have made significant contributions in understanding infectious disease is in modeling tuberculosis (TB) infections in humans (Adams et al., 2011; Berg et al., 2016; Cambier et al., 2014; Davis & Ramakrishnan, 2009; Myllymaki, Bauerlein, & Ramet, 2016; Oehlers et al., 2015; Ramakrishnan, 2013; Roca & Ramakrishnan, 2013; Tobin et al., 2010; Tobin, Roca, et al., 2012; Volkman et al., 2010). *Mycobacterium tuberculosis* is the causative pathogen of TB that, following inhalation into the lungs, is phagocytosed by alveolar macrophages where both pro- and antiinflammatory mechanisms lead to the formation of granulomas (Frieden, Sterling, Munsiff, Watt, & Dye, 2003). One of the limitations of modeling TB using conventional animal models (including mice and nonhuman primates) has been the inability to directly observe the changing behaviors of innate immune cells (in particular macrophages) throughout the various stages of the disease. *Mycobacterium marinum* is closely related to *M. tuberculosis* and causes fish mycobacteriosis, a disease characterized by necrotic granulomatous lesions that are very similar in structure to those caused by *M. tuberculosis* (Prouty, Correa, Barker, Jagadeeswaran, & Klose, 2003). Infecting zebrafish larvae with *M. marinum* recapitulates many of the key features of TB (Davis & Ramakrishnan, 2009). Exploiting this model has resulted in fundamental mechanistic insights into the role of the macrophages (and neutrophils) during granuloma formation and dissemination and has uncovered new pathogen-driven strategies to evade the hosts immune response (Adams et al., 2011; Berg et al., 2016; Cambier et al., 2014; Davis & Ramakrishnan, 2009; Roca & Ramakrishnan, 2013; Tobin et al., 2010; Tobin, Roca, et al., 2012; Volkman et al., 2010). Of particular significance many of these insights have uncovered new therapeutic possibilities.

1. QUANTIFYING THE INNATE IMMUNE CELL RESPONSE TO BACTERIAL INFECTION

Here we describe experimental protocols we use in our group to infect zebrafish larvae and measure survival, bacterial burdens within individual infected larvae, and strategies to evaluate the relative contribution of specific innate immune cell subsets to clearing bacterial challenges.

1.1 TRANSGENIC REPORTER LINES FOR LIVE IMAGING INNATE IMMUNE CELLS

A number of transgenic lines have been generated that fluorescently mark cells of the myelomonocytic lineage at various stages of development and with various levels of specificity (see Table 1). Transgenic reporter lines have also been developed which label specific populations of innate immune cells including neutrophils (Hall, Flores, Storm, Crosier, & Crosier, 2007; Mathias et al., 2006; Renshaw et al., 2006) and macrophages (Ellett et al., 2011; Walton, Cronan, Beerman, & Tobin, 2015). By applying binary genetic expression systems (e.g., Gal4-UAS and Cre-Lox systems) and gateway-based modular cloning strategies [e.g., the

Tol2kit (Kwan et al., 2007)] new iterations of these transgenic lines have greatly expanded their utility. Traditionally, promoter regions used to generate reporter lines that mark innate immune cells typically drive constitutively expressed genes and therefore do not reflect the activation state of the cell. More recently, promoter regions of infection-responsive genes (e.g., *immunoresponsive gene 1* (*irg1*) and *tumor necrosis factor a* (*tnfa*)) have been successfully used to generate transgenic lines that report bacteria-driven activation of macrophages (Nguyen-Chi et al., 2015; Sanderson et al., 2015).

1.2 INFECTION OF ZEBRAFISH BY HINDBRAIN MICROINJECTION

Zebrafish support the growth and replication of a number of bacterial species and these can be delivered either naturally through static immersion or by microinjection (Table 2). The infection protocol outlined here involves injection of bacteria into the hindbrain ventricle as this typically leads to a localized infection. The hindbrain ventricle also provides a superficial location that is well suited to live imaging of the subsequent neutrophil and macrophage antibacterial response (Hall, Boyle, et al., 2014; Hall et al., 2012, 2013). Alternatively, bacteria can be injected into the circulation via the duct of Cuvier or the caudal vein, generating a more systemic infection (Milligan-Myhre et al., 2011). Of note, we typically inject into the hindbrain of 2 days post fertilization (dpf) larvae and perform our macrophage live imaging studies within the first few hours postinfection. This is to ensure we are imaging macrophages and not microglia (both of which are marked within the *Tg*(*mpeg1:EGFP*) and related lines) that differentiate from macrophages within the developing brain at approximately 60 hpf (Herbomel et al., 2001).

Table 2 Selected Examples of Bacterial Infection Models in Embryonic and Larval Zebrafish

Bacteria	Method of Delivery	References
Aeromonas hydrophila	Static immersion	Rawls, Samuel, and Gordon (2004)
Bacillus subtilis	Microinjection	Herbomel et al. (1999)
Edwardsiella tarda	Static immersion	Pressley, Phelan, Witten, Mellon, and Kim (2005)
Escherichia coli	Microinjection	Herbomel et al. (1999)
Listeria monocytogenes	Microinjection	Levraud et al. (2009)
Mycobacterium marinum	Microinjection	Davis et al. (2002)
Pseudomonas aeruginosa	Static immersion	Rawls et al. (2004)
Salmonella arizonae	Microinjection	Davis et al. (2002)
Salmonella typhimurium	Microinjection	van der Sar et al. (2003)
Staphylococcus aureus	Microinjection	Prajsnar, Cunliffe, Foster, and Renshaw (2008)
Vibio anguillarum	Static immersion	O'Toole, Von Hofsten, Rosqvist, Olsson, and Wolf-Watz (2004)

Considerations to make with respect to the suitability of embryonic/larval zebrafish for infection experiments include, the temperature sensitivity of the pathogen (zebrafish are typically kept at 28°C but this can be altered slightly to accommodate pathogen growth (Brudal et al., 2014)), its host range (will the zebrafish embryo/larvae support its growth/replication), and natural routes of infection.

1.2.1 Protocol for Salmonella *injection into the hindbrain ventricle*

Our group largely utilizes a GFP-expressing *Salmonella enterica* serovar Typhimurium (hereafter referred to as Sal-GFP) for our infection studies (Hall, Boyle, et al., 2014; Hall et al., 2012, 2013; Hall, Flores, Kamei, Crosier, & Crosier, 2010). Using fluorescently labeled bacteria enhances the utility of embryonic zebrafish as a live imaging tool when examining the antibacterial response of innate immune cells and also provides a simple live readout of bacterial burden. Our *Salmonella* strain also expresses an antibiotic resistance gene that is used to selectively quantify the bacterial burden within individual infected embryos and larvae.

1.2.1.1 Materials

- Luria—Bertani (LB) broth
- LB agar plates (with appropriate antibiotic for bacterial selection, e.g., kanamycin for Sal-GFP)
- Dulbecco's Modified Eagle Medium, DMEM (Gibco)
- E3 medium (5 mM NaCl, 0.17 mM KCl, 0.33 mM $CaCl_2$, 0.33 mM $MgCl_2$)
- 1-Phenyl 2-thiourea (PTU), 100X stock (0.3% in E3 medium)
- Phenol red
- Filter-sterilized phosphate-buffered saline (PBS)
- Tricaine (4 mg/mL stock solution)
- 3% Methyl cellulose (in E3 medium)
- Microinjection needles (borosilicate thin-walled needles, outer diameter 1.00 mm, inner diameter 0.78 mm, G100T-4, Warner Instruments)
- Eppendorf microloader tips
- Plastic cuvette for spectrophotometer
- Hemocytometer for microinjection needle calibration

1.2.1.2 Methods (see Fig. 1A)

Day 1:

1. Streak a new LB agar plate (supplemented with 25 μg/mL kanamycin for Sal-GFP selection) from a stored bacterial glycerol stock.
2. Incubate at 37°C overnight.

Day 2:

1. Pick a single fresh bacterial colony from a newly streaked LB agar plate and inoculate in 4 mL of LB broth.
2. Incubate at 28°C overnight.

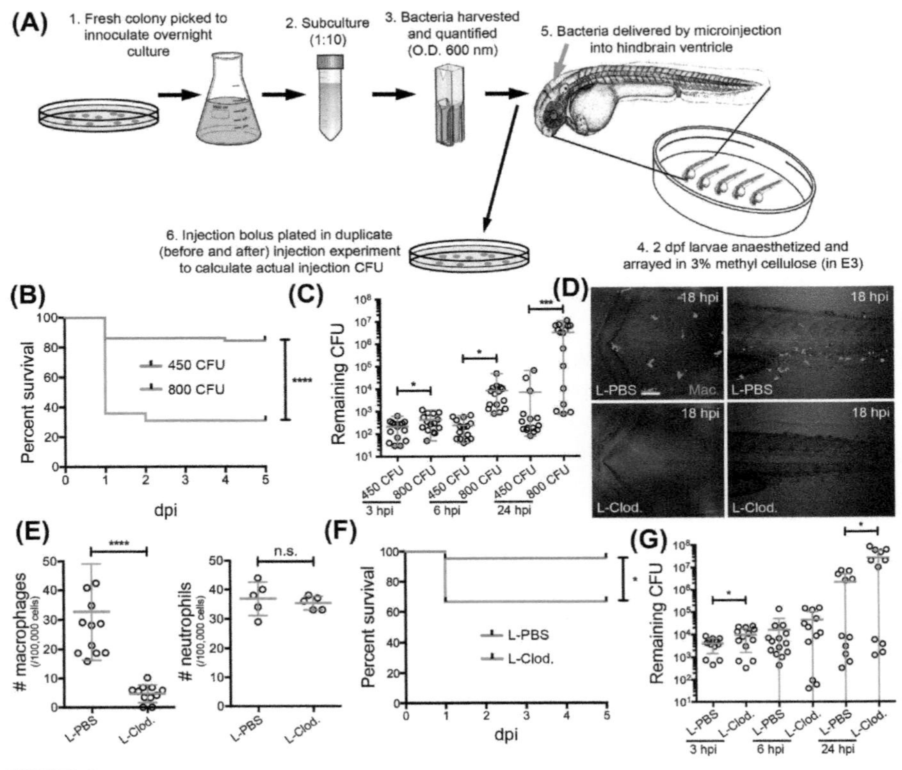

FIGURE 1

Hindbrain infection model and assessing the contribution of the macrophage lineage to clearing injected bacteria. (A) Schematic illustrating the protocol used to deliver bacteria (GFP-expressing *Salmonella enterica* serovar Typhimurium) into the hindbrain ventricle of 2 dpf zebrafish larvae. (1) A single colony of Sal-GFP (from a freshly streaked LB agar plate) is used to inoculate 4 mL of LB broth which is cultured overnight at 28°C. (2) The overnight culture is then diluted 10-fold in DMEM/LB and incubated for 45 min at 37°C. (3) The CFU of the culture is then estimated on a spectrophotometer (OD 600 nm) and an injection mixture made to give the desired injection dose in a 1 nL volume. (4) Anesthetized 2 dpf larvae are arrayed in 3% methyl cellulose. (5) Larvae are injected with a 1 nL dose of injection mixture into the hindrbain ventricle. (6) To calculate the actual injected CFU dose an injection bolus is injected into sterile PBS (in duplicate) before and after an injection experiment that is then plated at 1:10 and 1:100 dilution on LB agar. (B) Kaplan–Meier graph demonstrating percent survival of larvae injected with 450 and 800 CFU doses of Sal-GFP (assessed at 1, 2, 3, 4, and 5 dpi). (C) Quantification of remaining Sal-GFP (measured at 3, 6, and 24 hpi) from dissociated whole larvae injected with 450 and 800 CFU doses of Sal-GFP. (D) Live confocal images of the hindbrain and caudal hematopoietic tissue regions of *Tg(mpeg1:EGFP)* larvae 18 h following injection of liposomal-PBS (L-PBS) and liposomal-clodronate (L-Cod). (E) Flow cytometry quantification of macrophages and

Day 3:

1. Dilute the overnight culture 10-fold with 16 mL LB broth and 20 mL of DMEM.
2. Incubate at 37°C with shaking (225 rpm) for 45 min.
3. Place subculture on ice.
4. To calculate the bacterial CFU use 1 mL of the subculture to measure the optical density (OD) at 600 nm using a spectrophotometer.
5. Split the remaining subculture into two 50-mL Falcon tubes and harvest the bacterial cells by centrifugation at 4000 × g for 10 min at 4°C. Resuspend the bacterial pellet in 100 μL of filter-sterilized PBS and store on ice.
6. Generate an injection mixture by diluting the bacteria in filter-sterilized PBS supplemented with 0.25% phenol red to obtain the desired CFU injection dose (we typically inject a 1 nL volume).
7. Anesthetize larvae in E3 medium supplemented with tricaine (0.168 mg/mL) and 1X PTU.
8. Array anesthetized 2 dpf larvae in 3% methyl cellulose (in E3 medium) on the lid of a 35-mm petri dish.
9. Backload a microinjection needle with the injection mix using an Eppendorf microloader.
10. Cut the needle tip using a sharp pair of forceps.
11. To use a hemocytometer to calibrate the injection needle, place a drop of mineral oil on a hemocytometer and adjust the pressure of the microinjector such that an injection bolus has a diameter that correlates to a volume of 1 nL.
12. Carefully inject 1 nL of the injection mixture into the hindbrain ventricle of each larvae, taking care not to pierce the underlying neuroepithelium as this will result in the rapid spread of infection and death.
13. To quantify the actual CFU delivered, inject 1 nL of the injection mixture into a small quantity of filter-sterilized PBS (in duplicate) before and after each injection experiment. Plate at 1:10 and 1:100 dilutions onto LB agar (supplemented with 25 μg/mL kanamycin for Sal-GFP selection), and incubate overnight at 28°C. Colonies are counted and averaged the next day to determine the actual CFU dose delivered per larva.
14. Carefully wash infected larvae out of the 3% methyl cellulose by gentle agitation with E3 medium (supplemented with 1X PTU) using a plastic transfer pipette.

neutrophils (measured as EGFP positive cells/100,000 cells) from dissociated *Tg(mpeg1:EGFP)* and *Tg(lyz:EGFP)* larvae, respectively, 18 h following injection L-PBS and L-Clod. (F) Kaplan–Meier graph demonstrating percent survival of L-PBS- and L-Clod-injected larvae infected with a 450 CFU dose of Sal-GFP (assessed at 1, 2, 3, 4, and 5 dpi). (G) Quantification of remaining Sal-GFP (measured at 3, 6, and 24 hpi) from dissociated L-PBS- and L-Clod-injected larvae infected with a 450 CFU dose of Sal-GFP. *n.s.*, not significant; *, *p*-value < .05; ***, *p*-value < .001; ****, *p*-value < .0001. Scale bar, 50 μm in (D).

1.3 MEASURING SURVIVAL AND BACTERIAL BURDEN OF INFECTED LARVAE

We use these protocols to measure the survival of infected larvae and to enumerate the remaining bacterial burden within individual infected larvae at specific times postinfection (Hall et al., 2013; Hall, Boyle, et al., 2014; Oehlers et al., 2011) (Fig. 1B and C). It has been adapted from a previously described method (van der Sar et al., 2003).

1.3.1 Protocol for measuring survival of infected larvae

1.3.1.1 Methods

1. Maintain infected larvae at 28.5°C in E3 medium (supplemented with 1X PTU) at a density of no more than 50 larvae per 100 mm petri dish.
2. Monitor infected larvae several times a day to ensure that dead larval debris does not contaminate the E3 medium and impact the survival of any remaining infected larvae.
3. Transfer larvae displaying signs of excessive bacterial dissemination to another petri dish (when infected with Sal-GFP, bacterial burden can be live imaged by simple fluorescence microscopy).
4. Immediately remove and score any dead larvae (cardiac arrest).

1.3.2 Protocol for quantifying bacterial burdens within individual infected larvae

1.3.2.1 Materials

- Triton X-100
- Filter-sterilized PBS
- 200-μL pipette tips
- LB agar plates (with appropriate antibiotic for bacterial selection, e.g., kanamycin for Sal-GFP)

1.3.2.2 Methods

1. At selected time points postinfection, place individual larvae in 1 mL Eppendorf tubes containing 100 μL of filter-sterilized PBS supplemented with 1% Triton X-100.
2. Dissociate larvae by repeated passage through a 200-μL pipette tip and vortexing.
3. When completely dissociated, prepare a dilution series (1:10, $1{:}10^3$, $1{:}10^5$, and $1{:}10^7$) of the dissociated larval suspension and spot plate 10 μL (in triplicate) of each dilution onto LB agar plates (supplemented with 25 μg/mL kanamycin for Sal-GFP selection) and incubate overnight at 28°C.
4. Count and average the colonies from the dilution that gives the most easily quantifiable colonies to calculate the remaining bacterial burdens per larva.

1.4 ASSESSING THE CONTRIBUTION OF SPECIFIC INNATE IMMUNE SUBSETS TO CLEARING INJECTED BACTERIA

Liposomal clodronate has long been used to deplete professional phagocytic cells (such as macrophages) within mice (Weisser, van Rooijen, & Sly, 2012). More recently it has been employed to deplete zebrafish larvae of macrophage-lineage cells following microinjection (Carrillo et al., 2016; Pagan et al., 2015). Upon injection, macrophages phagocytize the clodronate-loaded liposomes and, following delivery of a sufficient quantity of the cytotoxic drug, undergo cell death. Here we present protocols for the delivery of liposomal clodronate into embryos and for quantifying the efficiency of liposomal clodronate-mediated ablation of macrophage-lineage cells, when injected into the macrophage-lineage marking *Tg(mpeg1:EGFP)* reporter line. The specificity of macrophage ablation can be determined by injection of liposomal clodronate into the neutrophil marking *Tg(lyz:EGFP)* reporter line (Fig. 1D–G).

1.4.1 Protocol for injecting liposomal clodronate to deplete macrophage-lineage cells

1.4.1.1 Materials

- Liposomal clodronate and liposomal PBS (supplied by ClodronateLiposomes.org)
- E3 medium (5 mM NaCl, 0.17 mM KCl, 0.33 mM $CaCl_2$, 0.33 mM $MgCl_2$)
- Tricaine (4 mg/mL stock solution)
- Microinjection needles (borosilicate thin-walled needles, outer diameter 1.00 mm, inner diameter 0.78 mm, G100T-4, Warner Instruments)

1.4.1.2 Methods

1. Dilute liposomal clodronate (as supplied by ClodronateLiposomes.org) in PBS (1:1) and vortex.
2. Anesthetize 30 hpf *Tg(mpeg1:EGFP)* embryos in E3 medium supplemented with tricaine (0.168 mg/mL) and 1X PTU.
3. Array anesthetized embryos in 3% methyl cellulose (in E3 medium) on the lid of a 35 mm petri dish.
4. Microinject ~3–5 nL liposomal clodronate (or liposomal PBS as a negative control) into either the hindbrain ventricle, the caudal hematopoietic tissue or directly into the circulation via the sinus venosa.
5. Carefully wash injected embryos out of the 3% methyl cellulose by gentle agitation with E3 medium (supplemented with 1X PTU) and a plastic transfer pipette.
6. Assess macrophage-lineage depletion by fluorescence (or confocal) microscopy at 18 hpi.
7. The above protocol can also be used with *Tg(lyz:EGFP)* larvae to ensure that the liposomal clodonate treatment does not affect neutrophil numbers.

1.4.2 Protocol for quantifying macrophage-lineage depletion by flow cytometry

1.4.2.1 Materials

- Calcium-free Ringer's solution (116 mM NaCl, 2.9 mM KCl, 5.0 mM HEPES, pH 7.2)
- 200 μL and 1000 μL pipette tips
- Filter-sterilized PBS
- 0.25% trypsin-EDTA [0.5% trypsin-EDTA (Gibco 15,400,054) diluted 1:1 in filter-sterilized PBS]
- 0.5 M $CaCl_2$
- Fetal bovine serum (FBS)
- 40 μm cell strainers (Falcon 352,340)

1.4.2.2 Methods

1. Rinse ~50 dechorionated liposomal clodronate- and liposomal PBS-injected larvae at ~18 hpi in calcium-free Ringer's solution for 15 min.
2. Remove yolks by repeated passage through a 200-μL pipette tip.
3. Transfer deyolked larvae into a 35-mm petri dish containing 4 mL of 0.25% trypsin-EDTA for 2 h at 28.5°C and manually dissociate every 10 min by passage through a 1000-μL pipette tip.
4. Inhibit the digestion with the addition of $CaCl_2$ to a concentration of 1 mM and 5% FBS.
5. Harvest dissociated cells by centrifugation at 3000 rpm for 5 min and resuspend in ice-cold 0.9X PBS supplemented with 5% FBS.
6. Pass cells through 40-μm cell strainers three times to remove noncellular debris.
7. Perform flow cytometry, gating for EGFP-expressing neutrophils (*Tg*(*lyz:EGFP*)) or macrophages (*Tg*(*mpeg1:EGFP*)).

Once the efficiency and specificity of macrophage-lineage depletion has been confirmed (Fig. 1D and E), liposomal clodronate- and liposomal PBS-injected larvae can be infected to assess the contribution of the macrophage-lineage to the clearance of a specific pathogen (Fig. 1F and G).

2. BIOASSAYS FOR ASSESSING NEUTROPHIL BACTERICIDAL FUNCTION

Injecting neutrophil-specific zebrafish reporter lines with spectrally nonoverlapping fluorescent bacteria provides an opportunity to directly investigate the fate of phagocytosed bacteria and measure antibacterial functions of neutrophils at the single cell level.

2.1 MEASURING THE BACTERIAL "KILLING RATE" OF INFECTED NEUTROPHILS

Using confocal time lapse microscopy to live image intracellular fluorescent bacteria within neutrophils and 4D image analysis software (such as Volcocity), a volume measurement can be applied to the fluorescent bacteria within individual neutrophils (Fig. 2A and B). By measuring the change in this volume over time, a "killing rate" can be calculated (Fig. 2C); a positive value indicates bacterial killing while a negative value shows intracellular bacterial growth. Below is the protocol we use to measure the bacterial "killing rates" of neutrophils following infection of *Tg(lyz:DsRED2)* larvae with Sal-GFP. It has been adapted from a previously described method (Yang et al., 2012).

2.1.1 Materials (in addition to those listed in Section 1.2)

- UltraPure low melting point agarose (Invitrogen 16520050)
- Scalpel blade
- Eyelash manipulator (superfine eyelash with handle, Ted Pella Inc.)

2.1.2 Methods

1. Infect 2 dpf *Tg(lyz:DsRED2)* larvae.
2. Generate mounting medium (0.75% low melting point agarose dissolved in E3 medium supplemented with 0.126 mg/mL tricaine and 1X PTU) and maintain in a water bath at 45–50°C.
3. Immediately following infection mount anesthetized larvae in mounting medium within a 35 mm petri dish.
4. To mount larvae first create an agarose bed approximately 5-mm deep and allow to set. Using a scalpel blade excavate a small trough sufficient to hold the ventral portion of the yolk. Place the infected larvae along with mounting medium (at 45–50°C) onto the agarose bed and immediately orientate the larva dorsal side up. Take care not to transfer too much mounting medium such that it fills the petri dish more than 1 mm above the agarose bed (this is to ensure sufficient space is left to accommodate the working distance of the microscope objective).
5. While the agarose is setting, maintain the orientation of the larva using an eyelash manipulator.
6. When the agarose has set, overlay with E3 medium supplemented with 0.126 mg/mL tricaine and 1X PTU.
7. Time lapse image the hindbrain ventricle using a confocal microscope equipped with a water immersion lens and incubation chamber to maintain the temperature at 28.5°C.
8. Locate *lyz:DsRED2*-expressing neutrophils that contain intracellular Sal-GFP.

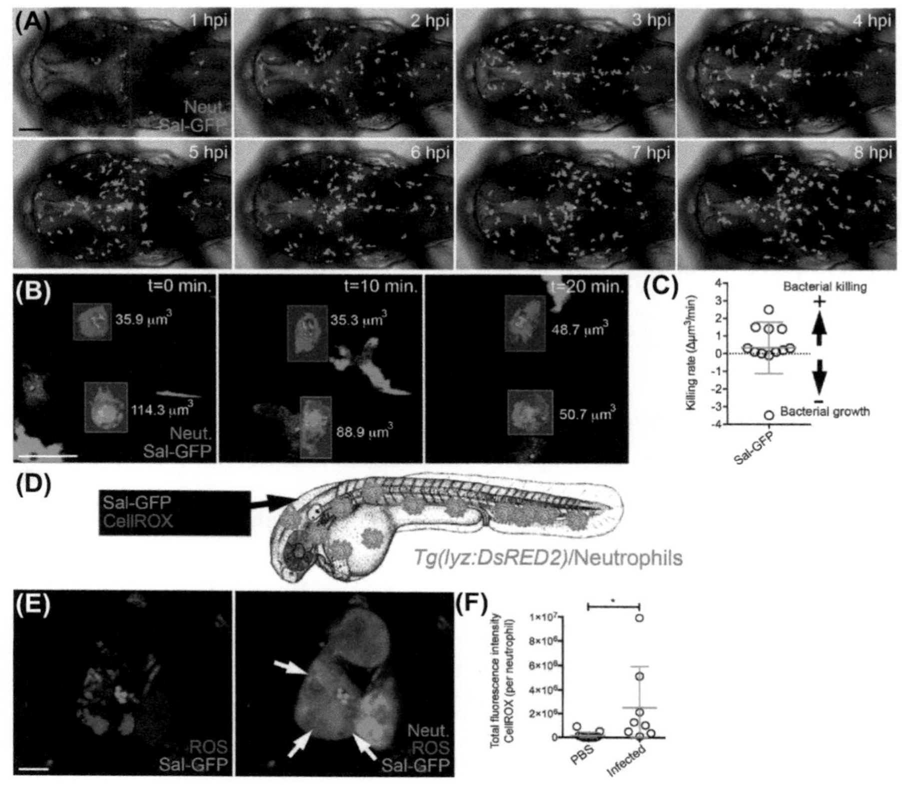

FIGURE 2

Live imaging the functional activity of neutrophils in response to *Salmonella* infection. (A) Live time lapse confocal imaging of the progressive recruitment of neutrophils to Sal-GFP injected into the hindbrain ventricle of *Tg*(*lyz:DsRED2*) larvae. (B) Live time lapse confocal imaging of individual infected neutrophils with volumetric measurements of intracellular Sal-GFP enables a "killing rate" (C) to be calculated per cell (the change in intracellular Sal-GFP volume per minute). A positive value is reflective of bacterial killing within the neutrophil while a negative value indicates bacterial growth. (D) Schematic illustrating the utility of CellROX fluorescent probe when coinjected with Sal-GFP into *Tg*(*lyz:DsRED2*) larvae to live image ROS production within individual neutrophils. (E) Live confocal imaging of ROS production (as detected using the CellROX deep red fluorescent probe) within individual neutrophils in response to PBS control and Sal-GFP injection. *Arrows* mark ROS within neutrophil containing phagocytozed Sal-GFP. (F) Quantification of ROS activity (measured as total fluorescence intensity of CellROX signal per neutrophil) within individual neutrophils containing intracellular Sal-GFP as shown in (E). *, p-value < .05. Scale bars, 100 μm in A, 25 μm in B, 5 μm in (E).

9. Capture a z-series (1 μm steps) through the entire neutrophil every 3 min ensuring sufficient space is allowed in the X, Y, and Z directions to accommodate any migration throughout the duration of the time lapse. For our imaging we use the following: 320 × 320 pixel dimension; 60X objective 1.4–1.6X zoom. Given the killing rate is measured as a change in the volume of bacterial fluorescence it is important to maintain identical imaging parameters (e.g., laser voltage, gain, offset, scanning speed, and resolution).
10. Ensure that only infected neutrophils that are not observed to phagocytize additional bacteria are included in the analysis and that the entire neutrophil is imaged through the time lapse (15–30 min).
11. Using 4D image analysis software (such as Volocity, PerkinElmer Inc.) measure the volume (μm^3) of bacterial (Sal-GFP) fluorescence within the infected neutrophils at each time point. Calculate the "killing rate" as (initial volume—final volume)/time (min) or $\Delta \mu m^3$/min, where a positive value is indicative of bacterial killing and a negative value reflects bacterial growth within the neutrophil.

2.2 QUANTIFICATION OF BACTERICIDAL REACTIVE OXYGEN SPECIES WITHIN INFECTED NEUTROPHILS

A fundamental mechanism used by neutrophils to kill intracellular bacteria is through the NADPH oxidase-driven production of bactericidal reactive oxygen species (ROS) within the phagosomal compartment (Dupre-Crochet, Erard, & Nubetae, 2013). Here we describe the use of the fluorogenic reagent CellROX to live image the generation of ROS within individual infected zebrafish neutrophils following coinjection of CellROX Deep Red Reagent with Sal-GFP into *Tg(lyz:DsRED2)* larvae (Fig. 2D–F). This cell-permeable probe, that is nonfluorescent when in a reduced state, demonstrates bright fluorescence following oxidation by ROS (Hall, Boyle, et al., 2014; Mugoni, Camporeale, & Santoro, 2014; Mugoni et al., 2013).

2.2.1 *Materials (in addition to those listed in Section 1.2)*

- CellROX Deep Red Reagent, for oxidative stress detection (Molecular Probes C10422)

2.2.2 *Methods*

1. CellROX Deep Red Reagent is supplied as a stabilized 2.5 mM solution (50 μL) in DMSO. Stock solutions of the CellROX Deep Red Reagent can be used to reliably detect ROS production within infected larvae for up to 6 months when stored at −20°C.
2. Infect 2 dpf *Tg(lyz:DsRED2)* larvae as previously described with the exception that the injection mixture is supplemented with 50 μM CellROX Deep Red

Reagent. Due to the light sensitivity of the CellROX reagent ensure infected larvae are protected form the light for the duration of the experiment.

3. Immediately mount larvae in mounting medium as previously described.
4. Live image neutrophils containing intracellular bacteria (Sal-GFP) by confocal microscopy within 3–4 h to detect ROS production (CellROX fluorescence signal at absorption/emission maxima of 640/665 nm). We use similar confocal settings to detect CellROX fluorescence within infected neutrophils as those used to measure the "killing rate" (as described above). Given neutrophil ROS production is measured as the total fluorescence intensity of CellROX fluorescence signal within infected neutrophils it is important to maintain identical imaging parameters (e.g., laser voltage, gain, offset, scanning speed, and resolution). This will ensure any measured change in CellROX fluorescence is not the result of altered image acquisition settings.
5. Use 4D image analysis software (such as Volocity, PerkinElmer Inc.) to measure the total fluorescence intensity of CellROX within individual infected neutrophils.

3. BIOASSAYS FOR ASSESSING MACROPHAGE BACTERICIDAL FUNCTION

Similar to above, injecting macrophage-specific zebrafish reporter lines with spectrally nonoverlapping fluorescent bacteria provides an opportunity to live image macrophage-mediated antibacterial activities.

3.1 QUANTIFICATION OF BACTERICIDAL MITOCHONDRIAL REACTIVE OXYGEN SPECIES PRODUCTION WITHIN MACROPHAGES

Mitochondrial reactive oxygen species (mROS) are important regulators of immune cell function (Hall, Sanderson, Crosier, & Crosier, 2014b; Nathan & Cunningham-Bussel, 2013; West, Shadel, & Ghosh, 2011). In addition to helping orchestrate a number of macrophage immune activities, mROS augments the bactericidal capacity of macrophages (West, Brodsky, et al., 2011). Delivered by the clustering of mitochondria around phagosomal compartments containing intracellular bacteria, this bactericidal mROS is believed to augment the antibacterial activity of the phagosome. Our group, and others, has demonstrated that zebrafish macrophages also produce bactericidal mROS that contributes to the clearance of intracellular bacteria (Hall et al., 2013; Roca & Ramakrishnan, 2013). The protocol explained later details how mROS can be quantified within individual macrophages following coinjection of MitoSOX and nonfluorescent *Salmonella* into *Tg(mpeg1:EGFP)* larvae (Fig. 3A–C). The MitoSOX Red Mitochondrial Superoxide Indicator is a cell-permeable fluorogenic dye, specifically targeted to the mitochondria, that emits bright red fluorescence following oxidation by superoxide (Hall et al., 2013). The protocol explained later can also be modified to live image the clustering of mitochondria

development, physiology, and evolution (McFall-Ngai et al., 2013). Advances in DNA sequencing technologies have enabled descriptions of animal-associated microbiota composition in exquisite detail. Indeed, large descriptive studies have identified numerous correlations between microbiota composition and host health (Human Microbiome Project Consortium, 2012). However, manipulative experiments are needed to test these relationships and to elucidate mechanisms by which microbiota affect their hosts. The term "gnotobiology" (*gnos*, known; *bios*, life) was coined to describe the study of animals raised in the absence of microorganisms or in the presence of known microbial strains or communities (Reyniers et al., 1949). In this way, much as a geneticist performs genetic loss- and gain-of-function experiments, gnotobiologists remove resident microbiota to test necessity and add back one or more microbial strains to test sufficiency.

Model organism choice for gnotobiotic studies is motivated by host processes under investigation, extent of experimental replication necessary to achieve statistical power, and feasibility to maintain sterility or defined microbial conditions. Gnotobiology has a rich history across many organisms, especially in laboratory mice. Our groups have established zebrafish as a gnotobiotic model (Bates et al., 2006; Rawls, Samuel, & Gordon, 2004) to take advantage of this animal's outstanding experimental attributes, including rapid external development, optical transparency, large brood size, and ease of housing (Grunwald & Eisen, 2002). These properties enable vertebrate gnotobiotic experimentation on a scale not possible with mammalian models.

We have used gnotobiotic zebrafish to study microbiota effects on the host as well as host effects on the microbiota. For example, we established that the resident microbiota are necessary for host intestinal epithelial maturation and turnover, lipid absorption, and immune homeostasis (Bates, Akerlund, Mittge, & Guillemin, 2007; Bates et al., 2006; Cheesman, Neal, Mittge, Seredick, & Guillemin, 2011; Kanther et al., 2011; Rawls et al., 2004; Rolig, Parthasarathy, Burns, Bohannan, & Guillemin, 2015; Semova et al., 2012). In some cases, generic bacterial products, such as lipopolysaccharide are sufficient to mediate microbiota effects, whereas in other cases microbial signals are more specific to particular bacterial lineages (Bates et al., 2006; Rawls, Mahowald, Ley, & Gordon, 2006). We have also demonstrated that the host modulates bacterial dynamics and competition within the intestine (Jemielita et al., 2014; Rawls, Mahowald, Goodman, Trent, & Gordon, 2007; Taormina et al., 2012; Wiles et al., 2016) and that gnotobiotic zebrafish can be used to identify determinants of microbial colonization, including diet (Semova et al., 2012; Stephens et al., 2015; Wong et al., 2015). Gnotobiotic zebrafish have also been used to study colonization resistance against pathogens (Rendueles et al., 2012). Future research directions are discussed in the Section 2 at the end of this chapter. We hope that this chapter inspires research endeavors that will enrich our collective understanding of animals as microbial ecosystems.

HISTORICAL PERSPECTIVE ON GNOTOBIOLOGY

The first concept of studying GF life is credited to correspondence in 1885 between Louis Pasteur (1885) and Émile Duclaux (1885). Stimulated by Duclaux's efforts to germinate beans and peas in the absence of microbes, Pasteur expressed interest in rearing animals from birth devoid of microbes, predicting that animal life without microbes would be "impossible,"hypothesizing that microbiota must be beneficial, or essential, to their animal hosts. Pasteur's famous prediction sparked theoretical debate and experimental tests of whether animal life was feasible in the absence of microbes. The first successful production of GF animals, in this case guinea pigs, was reported 10 years later by Nuttal and Thierfelder (Nuttall, 1896). This initial success was followed by reports of GF derivation of chickens, goats, and a range of other mammals, birds, and amphibians (Baker, Ferguson, & TenBroeck, 1942; Gordon, 1960). These early studies showed that deriving animals GF was relatively straightforward, but resulted in variable longevity and health status. The difficulties in maintaining healthy GF animals prolonged debate over the feasibility of GF animal life. In hindsight, this variability can be largely attributed to insufficient diets. Michel Cohendy (1912), working with chickens provided the first convincing demonstration that GF animals could be reared for prolonged periods with generally good health. These advances were supported by technological developments in GF gnotobiotic housing, first by Ersnt Küster (2013).

With these technical advances, the scientific inquiry of gnotobiology shifted from GF animal viability to the impact of microbes on animal life. These research questions motivated further developments in gnotobiotic isolator designs and materials (Gordon, 1960; Reyniers et al., 1949; Reyniers, 1959). Although current gnotobiotic isolators look similar to those of the mid-20th century, demands of the evolving gnotobiotic research field have motivated new innovations, such as methods for scalable gnotobiotics (Laukens, Brinkman, Raes, De Vos, & Vandenabeele, 2016).

The greatest challenge in gnotobiotic husbandry has been to develop sterile diets that support growth and survival of GF animals. A diet must meet nutritional requirements to effectively support animal growth and survival. Gnotobiotic research has revealed that the absence of microbiota and their metabolic capabilities significantly augments animal nutritional needs that must be met by diets even after undergoing sterilization, which may significantly reduce their nutritive content. This challenge became apparent early in gnotobiotic animal research, as variable growth and survival in early GF animal studies was attributed in part to inadequate diets. Advances in formulation and sterilization of GF diets, most notably discovery of vitamins (Semba, 2012) enabled gnotobiology to move forward. However, nutritional and biochemical contributions of microbiota to host biology continue to be discovered (Krishnan, Alden, & Lee, 2015), suggesting that even well-established standard diets for gnotobiotic mice and other animals may still have "insufficiencies" that could underlie observed differences between GF and colonized animals. This underscores the need to continue to study microbial contributions to host nutrition and to develop new diets that may more completely meet nutritional needs of GF animals.

Within this rich history of gnotobiotic research, fish models have played a small yet increasingly important role. The first fish species to be derived GF was platyfish (*Xiphophorus maculatus*) and survival on sterilized diets was achieved for several weeks after yolk resorption (Baker et al., 1942). Successful GF derivation was subsequently reported for many species including tilapia (Shaw, 1957), threespine stickleback (Milligan-Myhre et al., 2016), multiple salmonid (Lesel & Lesel, 1976; Trust, 1974) and Poecilidae species (Lesel, 1979), sheepshead minnow (Battalora, Ellender, & Martin, 1985), turbot (Munro, Barbour, & Birkbeck, 1995), Atlantic halibut (Verner-Jeffreys, Shields, & Birkbeck, 2003), sea bass (Dierckens et al., 2009), Atlantic cod (Forberg, Arukwe, & Vadstein, 2011), and threespine stickleback (Milligan-Myhre et al., 2016). These pioneering efforts confirmed that fish can survive under GF conditions for a limited time, but did not define nutritional requirements, explore GF phenotypes, or succeed in growing GF fish to reproductive maturity. As zebrafish emerged as a powerful model for biological and biomedical research, it was a natural choice to be developed as a gnotobiotic model. As described earlier, gnotobiotic zebrafish has become a powerful platform for mechanistic studies of associations between animals and their resident microbiota.

CONTROLS FOR GNOTOBIOTIC EXPERIMENTS

Model organisms with defined wild-type strains have enabled experimental reproducibility between laboratories across the world. However, all animals are microbial ecosystems, and individuals differ in the composition and function of their resident microbial consortia, making experimental reproducibility more elusive (Stappenbeck & Virgin, 2016). Indeed, mouse immunologists and infectious disease researchers have long known that mice purchased from different venders exhibit different immune phenotypes and susceptibilities to pathogens. More recently, many examples of mouse genetic models of chronic gastrointestinal disorders such as inflammation have been plagued with the problem that robust phenotypes observed in one laboratory are not reproducible when mice are shipped to another laboratory. Even within one experiment in a single laboratory we documented enormous inter-individual variability in microbiota composition in a group of identically reared sibling zebrafish (Stephens et al., 2016). How then, should an investigator approach the challenge of defining a colonized control group against which to compare GF or other gnotobiotic groups?

The gnotobiology field has struggled with this question for decades. One approach is to use a standardized, simplified microbiota as a surrogate for a natural community. In the mid-1960s Russell W. Schaedler developed a defined inoculum for GF mice, "Schaedler Flora," composed of specific mouse bacterial isolates (Dewhirst et al., 1999). This inoculum was further developed by Roger Orcutt in 1978 to an eight strain combination currently used by mouse researchers and referred to as the "Altered Schaedler Flora." Standardized reference microbiota have been further championed by Andrew Macpherson and Kathy McCoy (2015), who coined the term "isobiotic," analogous to "isogenic," as a standard for making experimental comparisons to infer microbiota function between individuals and across experiments. Macpherson and McCoy recognize the major logistical hurdles to performing experiments in the same microbiological "background," for example, requiring shipment of gnotobiotic animals around the world, but they assert that this level of replication is important for fields such as developmental immunology.

Another approach, articulated by Thad Stappenbeck and Skip Virgin (2016), is to accept but document the enormous variability in microbiomes within and between experiments. These authors stress the importance of characterizing the functional capacities of microbiota by performing microbiota transplantations as a strategy to determine whether a given phenotype in an experimental animal is explained more dominantly by its host genotype or its microbial genotype. Although this approach is labor intensive and costly, it highlights important considerations for the field.

We struggle with these questions as well. In practice, we favor focusing research on microbiota impacts that are robust across control groups and readily observed when comparing GF animals to a range of colonized animals. We typically use one of two control groups: conventional (CV) animals reared without any microbial manipulations, and conventionalized or ex-GF (CVZ or XGF) animals derived GF and then inoculated with a natural source of zebrafish-associated microbes. We still debate optimal timing of conventionalization [typically 3–4 days post fertilization (dpf) when zebrafish normally hatch from their protective chorions] and the source of inoculum (typically water from the parental crossing or "system water" that circulates in a conventional fish facility). We have experimented with zebrafish equivalent of a Schaedler Flora, referred to as a "Zebrafish-Associated Community" ("ZAC"), assembled based on knowledge of the major zebrafish microbiota constituents and composed of bacteria we isolated from healthy zebrafish and for which we have genome sequences (Stephens et al., 2016). However, we are reluctant to employ such a microbial community as our sole comparator, because we still have so much to learn about zebrafish microbiota. Indeed, our research has highlighted impacts of minor microbiota members (Rolig et al., 2015) as well as the importance of stochastic processes in zebrafish microbiota assembly (Burns et al., 2016) that call into question the feasibility of generating cohorts of zebrafish with stereotyped, defined microbial communities. We encourage other investigators to consider the best controls for their experiments, depending upon the questions they wish to address.

1. EXPERIMENTAL PROCEDURES

In this section, we describe the latest methodology for successful gnotobiotic zebrafish research.

1.1 IMPORTANT CONSIDERATIONS FOR GNOTOBIOTIC RESEARCH

The starting point for all gnotobiotic zebrafish research is derivation of GF embryos. The methodology for this procedure is straightforward, but successful derivation and maintenance of sterility require careful planning, repeated practice, and attention to detail. We discuss both requirements that should be considered before attempting to derive embryos GF and some challenges and common pitfalls to successful GF derivation and husbandry.

1.1.1 Physical space considerations for optimal sterility

Space considerations for GF derivation, bacterial associations, and long-term GF husbandry should include cleanliness, airflow, sink locations, human traffic patterns, and biosafety level status. All GF experimentation requires walls, floors, and work surfaces that can be easily cleaned and maintained. Surfaces should be bleach- and ethanol tolerant. Filtered air is preferable and friable ceilings and unfiltered ductwork should be avoided to prevent airborne contamination. For all gnotobiotic husbandry, we recommend working with GF and conventional zebrafish in separate, dedicated spaces with separate tools and supplies. Ideally, personnel complete their daily tasks with GF animals prior to performing any procedures with conventionally reared zebrafish or bacterial cultures to minimize potential for cross-contamination. Dedicated GF space should be uncluttered, storing only sterile materials and reagents critical for experiments, and sources of contamination such as human traffic, dust from room surfaces, and aerosolized materials, should be minimized. All items should be cleaned with 70% ethanol prior to room entry. We discourage sink use in the dedicated GF room due to splashing, aerosolization, and inadvertent spread of contaminants.

Because each of the experimental procedures we discuss—GF derivation, microbial association, and long-term husbandry—has unique requirements, if possible it is desirable to have separate rooms dedicated to each function, preferably within close proximity to the main fish facility and other necessary equipment such as incubators, autoclaves, and centrifuges.

Space for GF derivation of zebrafish embryos: GF derivation of embryos requires space devoted solely to that purpose. It should have a dedicated biosafety cabinet (BSC), a door that closes, and counter-space. Air pressure should be positive to the surrounding room or corridor to minimize external airborne contamination. Human traffic in the space should be minimal while the derivation procedure is in progress.

Space for microbial association of GF zebrafish: Microbial association of GF embryos and larvae should be performed in rooms that are easily decontaminated. Rooms should be Biosafety Level 2 (BSL-2) approved when using microorganisms that could be pathogenic to humans or fish or that carry antibiotic-resistance genes. Ideally, microbial association should be performed in a separate BSC from that used for GF derivations, or on a laboratory benchtop with careful sterile technique. Procedures for housing microbial-association experiments in temperature-controlled rooms and incubators should be clearly established, including proper signage for other users of the space.

Space for long-term husbandry of GF zebrafish: Long-term husbandry of GF and gnotobiotic zebrafish is an area of active exploration and our existing protocols for this procedure have not been as well tested as our short-term larval husbandry. When practical, long-term husbandry of GF zebrafish should occur in a dedicated room maintained at 28°C with a 14 h light/10 h dark photoperiod. Because cumulative risk of contamination increases with experiment duration, long-term GF husbandry requires a room designed to minimize contaminant introduction. In our experience, humans are the most common vectors for contamination, thus room access should be limited to trained investigators who always wear gloves and dedicated personal protective equipment (PPE) such as laboratory coats or smocks. Schedules should be arranged so that personnel only work in the dedicated GF room before interacting with conventionally reared fish or bacterial cultures.

1.1.2 Equipment considerations for preventing contamination

Gnotobiotic experimentation relies heavily on several types of equipment. It is vital that investigators receive standardized training on equipment functionality, use, care, testing, and certification requirements. In the following we discuss two pieces of equipment that are critical for GF zebrafish experimentation:

Biosafety cabinet: Zebrafish embryos are most reliably derived GF in a BSC. To use a BSC properly, investigators must understand how airflow protects both sample and user from contamination, how to work within a BSC environment to preserve sterility, the major routes of contamination, what can and cannot be achieved by UV illumination, how to clean the BSC, and how to maintain BSC certification. All supplies in the BSC should be cleaned with 70% ethanol and investigators should wear a laboratory coat, disposable smock, or sterile sleeves. The BSC should be supplied with dedicated pipettes, filter pipette tips, pipet aids, tube racks, and pens. It is a misconception that UV light is the best method for disinfection of the BSC and its contents. Surface disinfection of the BSC, its contents, and users' gloved hands and arms with 70% ethanol is the most reliable strategy to minimize contamination. BSCs work well for GF derivation of zebrafish embryos if used appropriately, but do not prevent contamination if used inappropriately.

Autoclave: Autoclaves play a crucial role in preparing reagents and supplies for gnotobiotic research. Therefore, it is imperative that users have a basic understanding of their instrument and know how to reliably test whether it is functioning properly. Sterilization requires the entire volume of liquid to reach temperatures effective at eliminating the full spectrum of microbes present. Temperature can be monitored with autoclave probes and manufactured spore tests can reveal autoclave sterilization effectiveness.

1.1.3 Importance of healthy zebrafish stocks for germ-free experiments

The complexity of GF experiments demands special attention to animal husbandry that maximizes outcomes of lengthy experiments. Zebrafish facilities benefit from regularly sampling sentinel fish and a variety of facility locations for known pathogens as well as adopting practices that limit pathogen transmission (Mason et al., 2016). To ensure starting GF experiments with large clutches of robust embryos, we recommend using young (<12 months old), healthy adult zebrafish with special attention to maintaining as much background diversity as possible, screening each generation for developmental anomalies, and selecting for fecundity. For experiments with mutant or transgenic lines, optimizing parental stock genotypes maximizes experimental success. For experiments requiring individuals carrying a homozygous recessive mutation, one strategy to minimize genetic variation between the mutant individuals and the wild-type control siblings is to generate them from a mating of heterozygous parents. In experiments requiring large numbers of individuals, for which genotyping offspring would be prohibitive, breeding separate lines of homozygous mutant carriers and wild-type stocks may be required.

1.1.4 Training is critical for gnotobiology success

Standardized training is crucial for successful GF derivation of zebrafish embryos. Because personnel turnover is common in academia, it is critical to develop consistent training protocols. Training should include principles of zebrafish husbandry, sterile tissue culture technique, and equipment care and use. We recommend three training sessions: in session one trainees shadow the mentor, in session two trainees work alongside the mentor, and in session three trainees perform procedures and the mentor critiques. The mentor then determines whether the trainee can perform procedures successfully without direct supervision. Because it is critical that all laboratory members understand the shared responsibility of maintaining sterility and preventing contamination, we recommend a "certification" so that everyone involved knows who has the appropriate skills and training to perform GF experiments.

1.2 BEST PRACTICES FOR GERM-FREE DERIVATION OF ZEBRAFISH EMBRYOS

Fig. 1 provides an overview of critical GF derivation procedures, Table 1 shows a sample time line, and Tables 2 and 3 list materials and specific procedures. Healthy adult zebrafish are placed in breeding cages with dividers separating males and females the afternoon or evening prior to spawning (Day −1). At the start of the light cycle the following morning (Day 0), the breeding insert is transferred to autoclaved water and adults allowed to spawn for no more than 60 min, to minimize exposure to microbes from the parents and to ensure a synchronized clutch of eggs for disinfection. Embryos are collected and chorions rinsed free of large particulates before transfer to a petri dish of filter-sterilized embryo medium containing antibiotics (ABEM) to reduce bacterial growth as embryos develop to the appropriate age for surface disinfection. Prior to surface sterilization, infertile and deformed embryos are removed, healthy embryos are sorted to a density of no more than 150–200 per dish, rinsed multiple times in fresh ABEM, and incubated for at least 5–6 h to maximize the antibiotic disinfection period. Between 50% epiboly and embryonic

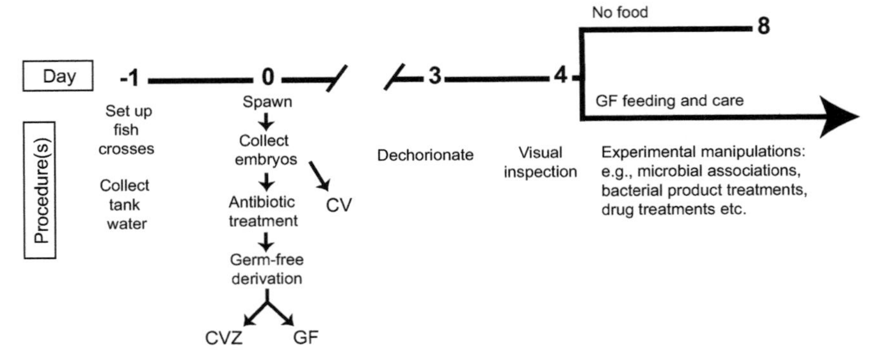

FIGURE 1

Overview of critical procedures for GF zebrafish derivation. Timeline depicts timing of procedures (see Tables 1–3 for specific details). Day refers to egg fertilization. Day −1: Adult fish are set up for breeding and tank water collected for conventionalization. Day 0: Adult zebrafish spawn and embryos are collected. If experiments demand conventional (CV) controls, embryos are segregated following collection and embryos destined for GF derivation are treated with antibiotics. Following GF derivation, embryos are placed in tissue culture flasks and conventionalized (CVZ) controls inoculated with a small amount of parental tank water collected on Day 0. Day 3: Flasks are inspected to ensure larvae have hatched; unhatched larvae are dechorionated by gently shaking the flask. Day 4: Flasks are visually inspected with phase optics to ensure sterility; Fig. 2 shows examples of contamination. Larval zebrafish can be kept in tissue culture flasks up to 8 dpf without feeding; for longer experiments axenic feeding and care should begin at 4–5 dpf.

shield stage (Kimmel, Ballard, Kimmel, Ullmann, & Schilling, 1995), embryos are surface disinfected with rinses in dilute polyvinylpyrrolidone-iodine (PVP-I) and bleach (sodium hypochlorite) and then incubated in sterile tissue culture flasks with vented lids. Flasks are checked daily for health and survival. If hatching has not occurred by 3 dpf, larvae are mechanically dechorionated by flicking the flask to knock embryos against the side of the container. Dechorionation prevents developmental deformities caused by physical constraints of the chorion.

Daily visual inspection of flask contents with phase contrast microscopy allows detection of adherent and motile bacteria and fungi, nonculturable microbes, and microbes present at very low density (Fig. 2). By 4 dpf, contamination can readily be detected by inspection of the flask using phase microscopy and culturing a sample on rich microbiological media such as tryptic soy (TSA), Luria broth (LB) agar, or Brain–Heart Infusion (BHI) at 28–30°C under aerobic and anaerobic conditions. We recommend practicing examining flasks containing fish tank water, conventionalized embryo medium (EM), and sterile EM, to learn the spectrum of organisms and debris within the experimental environment. Microorganisms within the water column are regularly shaped, dark, and may exhibit directional movement different from the Brownian motion or hydraulic flow of inert particles (Fig. 2A). This may include bacteria, mold, spores, and protozoans (Fig. 2A and B). Reflective,

Table 1 Example Zebrafish Embryo GF Derivation Time Table. Refer to Tables 2 and 3 for Materials and Specific Procedures. This Procedure Precedes Defined Microbial Associations and/or Long-term Husbandry Experiments

Day	Time	Procedure(s)
−1	4–5 p.m.	Adult zebrafish breeding preparation
0	8:30–9 a.m.	Adult zebrafish spawning
	9:30–10 a.m.	Embryo collection, rinse in ABEM
	11:00 a.m.	Sort embryos, rinse in ABEM
	1:00 p.m.	Sort embryos, rinse in ABEM
	2:30–3:30 p.m.	Disinfection solution, embryo, and biosafety cabinet preparation
	3:30–4:30 p.m.	Surface disinfection procedure in biosafety cabinet
	4:30 p.m.	Conventionalize control embryos
1	Any time	Health check
2	Any time	Health check
3	Any time	Dechorionate unhatched larvae, health check
4	Morning	Health check, visual inspection with phase optics, culture flask media

irregularly shaped objects are often debris such as pieces of chorions and microscopic plastic shavings present inside tissue culture flasks (Fig. 2C), and may occur even in GF flasks. Culturing samples from flasks aerobically and anaerobically on rich media such as TSA allows detection of culturable contaminants suspended in the water column. Plating also provides the ability to identify potential contaminants by colony morphology, colony polymerase chain reaction (PCR), and sequencing of small subunit ribosomal RNA genes. Tables 1–3 provide an overview of timing, supplies, and procedures for GF derivation of zebrafish embryos.

For GF experiments it is important to have a control group of zebrafish colonized by conventional microbiota. As described in the Introduction, these could be either CV or CVZ zebrafish, depending on the goals of the experiment.

1.3 MICROBIAL ASSOCIATION OF GERM-FREE ZEBRAFISH

Microbial colonization from the water column begins at about 3.5–4 dpf when the intestine is open at both the mouth and the anus. The density of fish influences bacterial colonization dynamics, so the density of fish per flask should be consistent, generally about 1 fish/mL. Prior to inoculation with microbes, sterile technique should be used to remove dead larvae, as they can be a source of nutrients for introduced microbes. We recommend inoculation at 4 dpf, to ensure all fish can be colonized, and because by this stage contaminated flasks can be easily detected and then

Table 2 GF Derivation Materials and Solutions. Refer to Table 1 for Example Time Schedule and Table 3 for Specific Procedures. Solution Recipes Follow Chapter. Prepare Shaded Solutions Immediately Prior to Use

Item(s)/Solutions	Special Considerations
Sterile screw top tissue culture flasks with filter caps	Short-term experiments: 25 cm^2 (15 mL with 10–15 embryos) or 75 cm^2 (50 mL with 25–50 embryos)
	Long-term experiments: 150 cm^2 (100 mL with 20–25 embryos)
50 mL glass beakers	Bleach, rinse, autoclave with foil covers
	2 beakers for every 200 eggs disinfected
Individually wrapped sterile transfer pipets	
25 or 50 mL wrapped, sterile serological pipets	
1–2 L beaker	
Pipet aid	
70% ethanol (EtOH) in spray bottle	
6% bleach, laboratory grade	Store at 4°C, replace every 3 months
10% polyvinylpyrrolidone (PVP-I)	Store in dark, replace every 3 months
100 × 20 mm polystyrene petri dishes	
Sterile gowns or sleeves	Wear when working in BSC
0.2 μM filtration units, 1 L, 250 mL	Clean aspiration tubing regularly to prevent backflow contamination
Sterile Embryo Medium (EM)	Make in advance
Antibiotic embryo medium (ABEM)	Make just prior to egg collection, filter sterilize
0.003% bleach solution	Make both just prior to surface sterilization, filter sterilize
0.1% PVP-I solution	
Digital timer	

Table 3 GF Derivation Procedure. Example Schedule and Materials Found in Tables 1 and 2, Respectively. Solution Recipes Follow Chapter. Perform Steps 11–24 Using Sterile Technique With Aattention to BSC Function. Shading Differentiates Days in the Procedure Sequence

Day	Procedure
−1	**1.** Set up adult zebrafish overnight in breeding cages with dividers to prevent natural spawning. **2.** Collect parental tank water for conventionalizing embryos into a sterile bottle. Incubate at 28°C overnight.
0	**3.** Transfer adult fish into clean breeding cages with fresh, autoclaved, or sterile filtered system water. Remove dividers to allow for breeding. Allow less than 1 h to ensure that embryos are of similar ages.

Table 3 GF Derivation Procedure. Example Schedule and Materials Found in Tables 1 and 2, Respectively. Solution Recipes Follow Chapter. Perform Steps 11–24 Using Sterile Technique With Aattention to BSC Function. Shading Differentiates Days in the Procedure Sequence—cont'd

Day	Procedure
	4. Collect embryos, thoroughly rinse to remove visible debris. **5.** Transfer embryos to petri dishes with fresh, sterile ABEM, ~1000 embryos per dish. **6.** Divide embryos further into 3–4 petri dishes with fresh ABEM, remove any debris. **7.** Incubate embryos at 28–30°C between 4 and 6 h depending on temperature. **8.** At ~2 h intervals, remove infertile embryos, transfer fertile embryos to clean dishes with fresh ABEM, ending with ~200 healthy embryos per dish. **9.** At 50% epiboly up to shield stage (see Kimmel et al., 1995), transfer up to 200 viable embryos into foil-covered, autoclaved 50 mL beakers with ~40 mL of ABEM. Replace foil cover. **10.** Prepare biosafety cabinet (BSC) for use. Put on gloves and gown or sterile sleeves, clean hands and arms thoroughly with 70% ethanol (EtOH). **11.** Clean the following with 70% EtOH. Do not block BSC vents. **a.** 50 mL sterile beakers with foil tops **b.** Individually wrapped sterile transfer pipets **c.** 1 L filter-sterilized EM **d.** 0.003% bleach solution **e.** 0.1% PVP-I solution **f.** Sterile flasks with filter screw caps of desired size **g.** 25 or 50 mL wrapped, sterile serological pipets **h.** Large beaker (1–2 L) for waste collection **i.** Pipet aid dedicated to BSC use **j.** Foil-covered beakers holding embryos. Do not allow ethanol seepage into beakers. **12.** Remove foil top from embryo beaker, slowly decant liquid into waste beaker. **13.** Rinse 3 times with sterile EM (*Note*: Do not use ABEM). **a.** Add ~40 mL sterile EM to beakers of embryos. Allow embryos to settle. **b.** Decant liquid into waste beaker, reserve embryos. **c.** Add ~40 mL sterile EM. **d.** After final rinse, decant all liquid into waste beaker. **14.** Immerse embryos in ~50 mL sterile 0.1% PVP-I for *exactly* 2 min. **15.** Decant 0.1% PVP-I solution into waste beaker. **16.** Rinse embryos 3 times with sterile EM as described in steps 13a–d. **17.** Transfer embryos into new sterile 50 mL beaker. **a.** Add ~20 mL sterile EM to beaker of embryos. **b.** Gently swirl embryos to resuspend, pour embryos into new sterile 50 mL beaker. **c.** Repeat a and b to collect any remaining embryos. Do not overfill beakers. **18.** Allow embryos to settle, decant liquid into waste beaker, reserve embryos.

Continued

Table 3 GF Derivation Procedure. Example Schedule and Materials Found in Tables 1 and 2, Respectively. Solution Recipes Follow Chapter. Perform Steps 11–24 Using Sterile Technique With Aattention to BSC Function. Shading Differentiates Days in the Procedure Sequence—cont'd

Day	Procedure
	19. Immerse embryos in ~50 mL 0.003% sterile bleach for 18–20 min. **20.** During bleach immersion, fill flasks with recommended volumes of sterile EM. **a.** Using pipet aid and sterile serological pipets, dispense sterile EM into flasks. **b.** For short-term experiments, dispense 13 mL sterile EM into 25 cm^2 flasks or 48 mL sterile EM into 75 cm^2 flasks. **c.** For long-term husbandry, dispense 98 mL sterile EM into 150 cm^2 flasks. **21.** Following bleach immersion, decant bleach into waste beaker. **22.** Rinse embryos 3 times with sterile EM as in steps 13a–c. **23.** Transfer embryos into flasks of sterile EM using sterile transfer pipets. **a.** 10–15 embryos per 25 cm^2 flasks. **b.** 25–50 embryos per 50 cm^2 flasks. **c.** 20–25 embryos per 150 cm^2 flasks. **24.** Tighten lids on all flasks and remove from BSC. **25.** Inoculate control flasks with reserved parental tank water. **26.** Incubate embryo flasks at 28°C with a 14 h light/10 h dark cycle.
1–2	**27.** Perform health check, noting sick or dead embryos.
3	**28.** Mechanically dechorionate unhatched larvae by flicking flask to knock embryos against the side of the container. **29.** Perform health check, noting sick or dead embryos.
4	**30.** Perform health check, noting sick or dead embryos. **31.** Begin feeding and/or microbial associations as experimental paradigm requires.

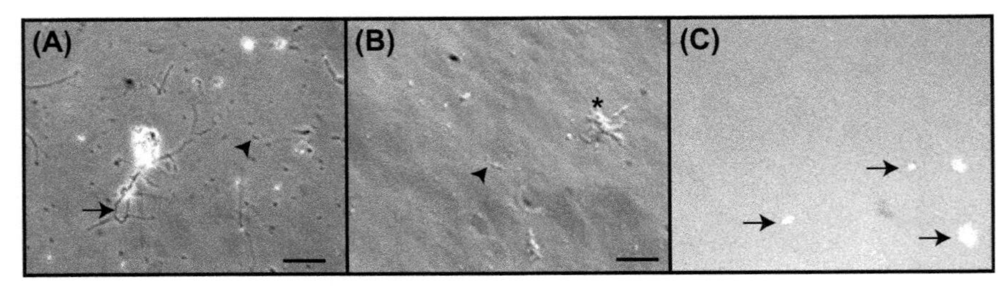

FIGURE 2

Phase microscopy visualization of potential flask contaminants and identification of sterile debris. Microorganisms present in tank water or CVZ flasks include bacteria (A, B *arrowheads*), filamentous microorganisms (A, *arrow*) and protozoans such as amoeba (B, *asterisk*). Reflective, irregularly shaped debris (*arrows*) observed in an otherwise sterile flask (C). Scale bar (A, C) 50 μM, (B) 20 μM.

removed from the experiment. Inoculum size should be determined empirically for each microbial species. For many bacteria, 10^4–10^6 colony-forming units (CFUs)/mL final concentration is sufficient to achieve intestinal colonization within 24 h at 1 fish/mL with minimal toxicity. These bacterial densities are selected to mimic the densities of culturable microbes observed in conventional aquaculture facilities. In the United States, experiments involving exposure of zebrafish to commensal microorganisms, especially those that could pose a health risk, may require Institutional Animal Care and Use Committee approval and should adhere to BSL-2 guidelines.

Fig. 3 describes methods for inoculating fish with commensal bacterial species that we have used for several bacterial isolates (e.g., Rolig et al., 2015). Table 4 describes the necessary materials and Table 5 provides the work flow and a sample calculation for a zebrafish isolate of an *Aeromonas* species. Optimal media and in vitro growth conditions vary for different bacterial species. Prior to mono-association with a single bacterial strain, or association with a consortium of bacterial strains, it is important to identify appropriate media and to optimize growth conditions to achieve an appropriate and reproducible bacterial density for colonization (Fig. 3A). Some bacterial strains will grow well in liquid culture directly from a frozen stock, whereas others may require streaking frozen stocks on appropriate agar plates, followed by 24 h incubation at 28–30°C before liquid culture inoculation with a single colony.

Once parameters for a particular inoculum are determined, prepare the culture for introduction into the GF flask as described in Table 5 and according to the specific experiment (Fig. 3B). First, flask sterility is verified by culturing on TSA (Fig. 3C). Inoculum should be introduced using sterile technique; actual bacterial density in the flask should be verified immediately following inoculation and again at the experimental end point.

1.4 LONG-TERM GERM-FREE ZEBRAFISH HUSBANDRY

Understanding the myriad ways resident microbes influence host development and physiology requires rearing zebrafish GF past early larval stages, with the ultimate goal of rearing them GF throughout their lifespan. The vast majority of gnotobiotic zebrafish research to date has focused on larval stages, in part due to the challenges of long-term GF zebrafish husbandry methods. These challenges mirror those of running a miniature, specialized zebrafish facility, thus animal care and use regulations should inform all practices. Long-term gnotobiotic zebrafish husbandry requires consideration of animal housing, water quality and exchange, waste removal, nutritional value of food, live food culturing, work space requirements, standardized work flow, record keeping, supplies, expenses, and labor (Fig. 4).

Procedures for long-term GF husbandry are still evolving; Tables 6 and 7 show current materials and work flow. We have successfully raised GF zebrafish for up to 1 month, but in all cases they were significantly smaller than their counterparts reared in the main zebrafish facility on conventional diets. Our current methodology is extremely labor intensive, and must be carried out every day of the experiment,

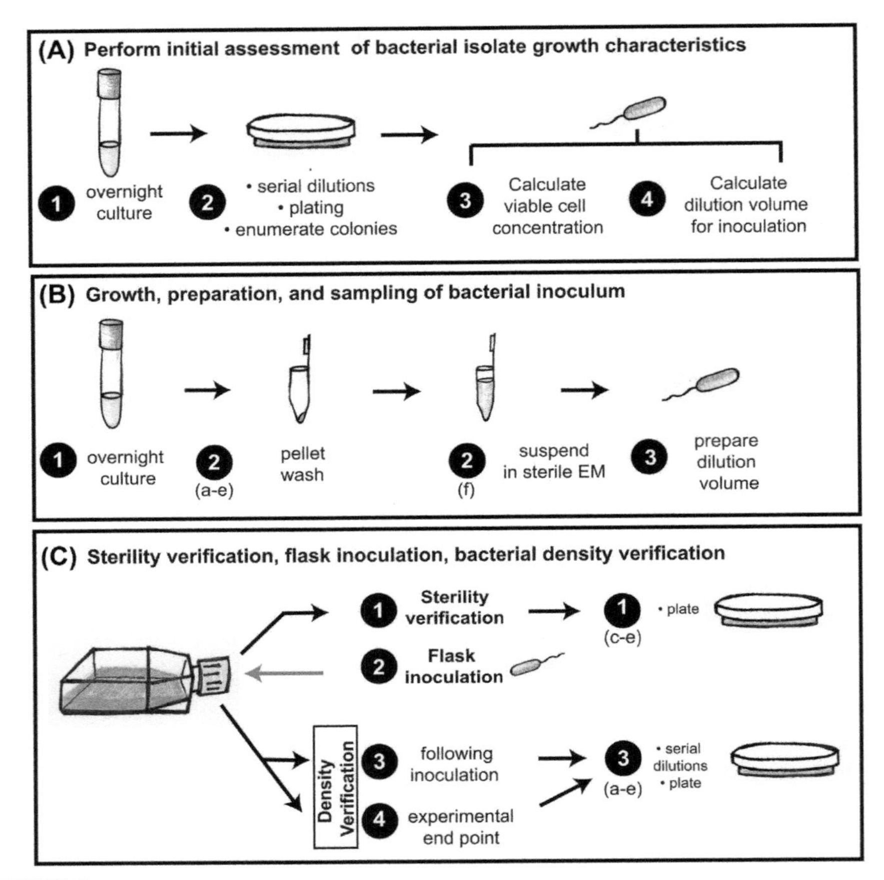

FIGURE 3

Microbial association of GF zebrafish procedure overview. Schematic diagram to be used with Tables 4 and 5 Number and letter sequences match procedures in Table 5. (A) Experiments with individual bacterial isolates to relate culture density to viable CFUs/mL to enable calculation of inoculation volume; see Table 5A for sample calculations. (B) For each experiment, grow and prepare bacterial inoculum according to specifications determined in (A). Pellet and wash bacteria with sterile EM, pellet again and suspend in sterile EM. Use washed bacteria for flask inoculation. (C) (1) Prior to inoculation, inspect flask for contamination under phase optics and remove media to plate any culturable bacterial contaminants. (2) Inoculate flask with washed bacteria and mix gently. Verify density by sampling flasks following inoculation (3) and at experimental end point (4).

thus we recommend practicing with conventional zebrafish and establishing a personnel schedule for the duration of the experiment before embarking on long-term GF husbandry. It is also important to calculate the number of embryos and flasks that is needed based on: (1) experimental design (number of time points,

Table 4 Bacterial Association of GF Zebrafish Materials

Material	Special Considerations
Bacterial strain of choice	Consider biosafety level, determine growth characteristics and optimal inoculum in advance
Liquid growth media	Appropriate for strain used
Sterile EM	Bacteria must be washed and suspended in EM prior to zebrafish association
Tryptic soy, LB, or BHI agar plates	Make in advance
Sterile test tubes with lids	Autoclave prior to use
Sterile 1.6 mL tubes	
Sterile plating beads	

Table 5 Microbial Association of GF Zebrafish Procedure. *Do Not* Perform Bacterial Work in BSC Dedicated to GF Derivation. Part A is Repeated Several Times for Each Bacterial Strain to Ensure Reproducibility Prior to an Experiment. Shading Indicates Separate Procedures. Numbered Sequences for Each Procedure Also Appear in Fig. 3

Part	Procedure	Instructions
A	Initial assessment of bacterial isolate growth characteristics	Determine the concentration of *viable* cells in overnight cultures for each new strain of bacteria used to calculate the experimental dilution volume for inoculation reproducibly. **1.** Measure the OD_{600} of an overnight culture. **2.** Perform serial dilutions, plating and enumerate colony growth. **3.** Calculate the culture density. Example: For *Escherichia coli*, an OD_{600} of 1 typically indicates that the concentration of viable cells is 10^9 colony-forming units (CFU/mL). *Note*: Sample calculation for *Aeromonas* overnight culture: *OD* × *dilution factor* × *conversion factor* = CFU/mL $0.765 \times 10 \times (1 \times 10^9) = 7.65 \times 10^9$ CFU/mL **4.** Calculate inoculation volume. Example: For *Aeromonas* inoculation in flask with 15 mL sterile EM at 1×10^6/mL: $(C_1)(V_1) = (C_2)(V_2)$ $(7.65 \times 10^9 \text{ CFU/mL})(X \text{ mL}) = (1 \times 10^6 \text{ CFU/mL})(15 \text{ mL})$ $x = 0.00196 \text{ mL} = 1.96\ \mu\text{L}$

Continued

Table 5 Microbial Association of GF Zebrafish Procedure. *Do Not* Perform Bacterial Work in BSC Dedicated to GF Derivation. Part A is Repeated Several Times for Each Bacterial Strain to Ensure Reproducibility Prior to an Experiment. Shading Indicates Separate Procedures. Numbered Sequences for Each Procedure Also Appear in Fig. 3—cont'd

Part	Procedure	Instructions
B	Growth and preparation of bacterial inoculum	**1.** Incubate bacteria overnight (~16 h) in 5 mL of growth medium in a 20 mL test tube with shaking at 28–30°C. **2.** Wash bacteria prior to association with zebrafish larvae. **a.** Pellet 1 mL of an overnight culture for 2 min at 7000xG. **b.** Discard supernatant by aspiration. **c.** Gently suspend pellet in 1 mL sterile EM by pipetting. **d.** Pellet for 2 min at 7000xG. **e.** Discard supernatant by aspiration. **f.** Gently suspend pellet in 1 mL sterile EM by pipetting.
C	Sterility verification, flask inoculation, bacterial density verification	In the following steps, minimize opening/closing of flask to reduce contamination risk. For consistency, we recommend dispensing small inoculation volumes by suspending them in 1 mL sterile EM and transferring this suspension to flasks using a 1 mL pipet. It is important to also maintain flask volume by first removing an equivalent volume of water from flasks (i.e., 1 mL). **1.** Sterility verification **a.** Inspect flasks using phase optics for bacterial contamination. **b.** Using sterile technique, remove 500 μL of media from each flask. **c.** Spread 100 μL of each tube in step b on TSA or LB plates with sterile beads. **d.** Incubate for 48 h. **e.** Inspect plates for growth of bacterial contaminants. **2.** Inoculate flask with calculated dilution of washed bacteria determined in step A4 to flask aseptically (e.g., 1.96 μL washed *Aeromonas* cells suspended in 1 mL sterile EM). **3.** Sample initial inoculation level by removing 500 μL from flask and adding to sterile EM in 1.6 mL tube: **a.** Perform serial dilutions of sample in sterile EM. **b.** Plate 100 μL from each dilution on TSA or LB plates. **c.** Incubate 24 h at 30°C. **d.** Count CFUs and calculate inoculation level. **4.** Sample experimental end point bacterial density by repeating step 3.

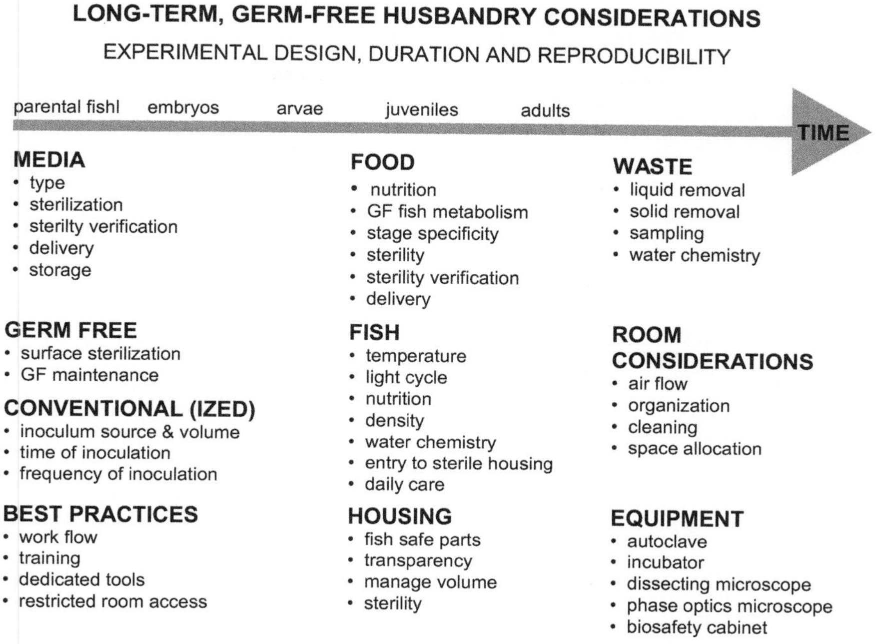

LONG-TERM, GERM-FREE HUSBANDRY CONSIDERATIONS

EXPERIMENTAL DESIGN, DURATION AND REPRODUCIBILITY

parental fishl embryos arvae juveniles adults TIME

MEDIA
- type
- sterilization
- sterilty verification
- delivery
- storage

GERM FREE
- surface sterilization
- GF maintenance

CONVENTIONAL (IZED)
- inoculum source & volume
- time of inoculation
- frequency of inoculation

BEST PRACTICES
- work flow
- training
- dedicated tools
- restricted room access

FOOD
- nutrition
- GF fish metabolism
- stage specificity
- sterility
- sterility verification
- delivery

FISH
- temperature
- light cycle
- nutrition
- density
- water chemistry
- entry to sterile housing
- daily care

HOUSING
- fish safe parts
- transparency
- manage volume
- sterility

WASTE
- liquid removal
- solid removal
- sampling
- water chemistry

ROOM CONSIDERATIONS
- air flow
- organization
- cleaning
- space allocation

EQUIPMENT
- autoclave
- incubator
- dissecting microscope
- phase optics microscope
- biosafety cabinet

FIGURE 4

Long-term GF husbandry considerations. Caring for GF zebrafish on a continuum from embryonic stages to adulthood resembles the process for assembling a small fish facility. Developing housing, appropriate, stage-specific nutrition, and efficient media delivery and waste removal under sterile conditions are paramount.

biological replicates, internal controls for developmental stage and maturation); (2) anticipated survival rates (from previous GF experiments); (3) expected contamination rates (from practice sessions); and (4) expected genotype ratios (from parental genetics). We also rear at least 25 zebrafish from the cross in facility main housing, as a control for general embryo health. In our experience, 12 flasks each of CVZ and GF animals require three people working about 4–6 h per day, 7 days a week, including general maintenance. We discuss key parameters for long-term GF husbandry in the following sections.

1.4.1 Germ-free live food

There is limited knowledge of zebrafish stage-specific nutritional requirements (Watts, Powell, & D'Abramo, 2012), thus the single greatest challenge for long-term GF zebrafish husbandry is supplying adequate nutrition. Our best success has been with GF cultures of *Tetrahymena thermophila*, ciliates that are highly motile, elicit normal hunting behavior of zebrafish larvae, and help maintain flask water quality by consuming fecal matter and dead fish. To improve their nutritional

Table 6 Additional Materials for Long-Term GF Husbandry. These Materials Supplement Those for GF Derivation. Refer to Tables 1–3 for GF Derivation Timeline, Materials and Procedures. Refer to Long-term Husbandry Procedures in Table 7. Solution Recipes Follow Chapter

Material	Special Considerations
Controlled temperature room or incubator (28°C) with 14 h light/10 h dark cycle	Control clutter, clean work spaces with 70% ethanol following use, empty trash daily
Filter-sterilized EM	Enough for 50–70% daily water exchange, verify sterility by plating on TSA prior to use
Sterile food cultures	Verify sterility by culturing on TSA prior to use
2 L autoclaved waste beakers	Separate beakers for GF and CVZ, cover with foil and autoclave prior to use
15 and 50 mL sterile conical tubes and rack	Autoclave racks regularly
Ethanol-resistant pens	Labeling tubes and plates
GF dedicated pipet aid, P-1000 and P-200	Keep in biosafety cabinet, clean with 70% ethanol following use
CVZ dedicated pipet aid, P-1000, P-200	Store away from GF supplies, clean with 70% ethanol following use
Sterile filter tips for P-1000 and P-200	Dedicated sets for GF and CVZ
Sterile 50 mL and 10 mL wrapped serological pipets	Used for media exchange, water sample collection, and removing dead larvae
Ammonia test kit (API, item# LR8600)	Follow manufacturer's instructions, 5 min incubation is critical
Trays	Use separate trays for storing GF and CVZ flasks
Daily data sheets	Record feeding volume, plating results, dead fish numbers, summary sheet
Dissecting scope	Clean stage and focus knobs with 70% ethanol following use
Phase contrast inverted microscope	
TSA plates	Have at least 2 days' supply ready in advance
Sterile plating beads	Autoclave before use

content, we developed methods to culture *Tetrahymena* on sterile, emulsified, chicken egg yolk (Fig. 5), resulting in engorgement with numerous internal vacuoles compared to media-reared *Tetrahymena* (Fig. 6). In controlled feeding trials, zebrafish reared on egg yolk–cultured *Tetrahymena* had better survival and were larger than siblings reared on media-reared *Tetrahymena* (Melancon, Sichel, and Kelly, unpublished). Tables 8–10 provide the reagents and protocols for GF *Tetrahymena* culture propagation and egg yolk nutritional enhancement.

Table 7 Current Long-Term GF Husbandry Procedures. Follow Procedure for GF Derivation of Embryos (Table 3). Live Food Culture Materials and Procedures Follow in Tables 8–10 Solution Recipes Follow Chapter. Beginning on Day 4, Continue With Long-term Husbandry Morning and Afternoon Procedures Outlined in the Following. This Procedure Includes Both GF and CVZ Conditions. All GF Procedures Should be Performed Using Sterile Technique in BSC. Record Data Recommended in Text and Table 6 Daily and Prepare Sterile Media, Food Cultures and TSA Plates in Advance. Table Shading Indicates Different Procedures.
Important: *Verify Sterility of all Solutions Prior to Use*

Procedure	a.m.	p.m.	Instructions
Check cultures	✓		**1.** Check bacterial plates to verify sterility of reagents and GF flask media. **2.** Get fresh TSA plates out of refrigerator and warm to 28°C.
Health check	✓	✓	**3.** Health check **a.** Inspect each flask, record fatalities. **b.** Remove inviable larvae from CVZ flasks only. Do not compromise sterility of GF flasks.
Visual inspection for contamination	✓		**4.** Visual inspection **a.** Inspect a CVZ flask with phase optics focusing on bacteria adhering to the bottom (see Fig. 2). Motile bacteria should be visible on the bottom and in the water column. **b.** Inspect GF flasks at the same focal plane as CVZ flask. Record results for each flask in data table and in notes. **c.** If bacteria are visible in GF flasks, remove flask from experiment. In ambiguous cases, segregate questionable flasks until sterility is determined.
Media exchange/ water sampling	✓	✓ **(CV/CVZ only)**	**5.** General guidelines for media exchange and water sampling. *Note*: Larvae are sensitive to turbulence and handling. **a.** Handle GF samples in BSC. Clean with 70% EtOH prior to entry, do not overcrowd workspace. Divide flasks into manageable groups for media exchange and feeding. **b.** Handle CV/CVZ samples in dedicated space. **c.** On Days 4–6 only remove small amounts of media for sampling, add ~50 mL sterile EM daily to dilute ammonia waste build up.

Continued

Table 7 Current Long-Term GF Husbandry Procedures. Follow Procedure for GF Derivation of Embryos (Table 3). Live Food Culture Materials and Procedures Follow in Tables 8–10 Solution Recipes Follow Chapter. Beginning on Day 4, Continue With Long-term Husbandry Morning and Afternoon Procedures Outlined in the Following. This Procedure Includes Both GF and CVZ Conditions. All GF Procedures Should be Performed Using Sterile Technique in BSC. Record Data Recommended in Text and Table 6 Daily and Prepare Sterile Media, Food Cultures and TSA Plates in Advance. Table Shading Indicates Different Procedures.—cont'd

Procedure	a.m.	p.m.	Instructions
			6. Waste media removal beginning 7 dpf. **a.** With a sterile pipet, remove 50–100 mL media from flasks. Replace lid, move flask out of the way. **b.** Pipet 5 mL of the removed volume into 50 mL conical tube for cumulative ammonia testing. **c.** Pipet 5 mL into a 15 mL conical tube for visual inspection or plating, replace lid. **d.** Pipet remaining liquid from media removal into large waste beaker. **e.** Repeat a–d for each flask.
Media addition	✓	✓ **(CV/CVZ only)**	**7.** Fresh media addition (begin day 4) **a.** Using a sterile 50 mL pipet, add sterile EM to match volume removed to each flask. **b.** Add fresh media to all flasks in BSC before feeding.
Feeding	✓	✓ **(CV/CVZ only)**	The volume of live food added depends on *Tetrahymena* culture density, age of fish, volume of media exchange, and overall volume of media in flask. Check *Tetrahymena* density by viewing flask on dissecting scope following feeding. **8.** Feeding **a.** Using a P-1000 with sterile filter tip, add 48 h, sterile 1% egg yolk + *Tetrahymena* culture to each flask. Replace lid and tighten. **b.** From Days 1–7, place flasks upright on trays for storage. Once volume reaches 250 mL, incubate flasks on side for better air exchange.
Ammonia testing	✓		**9.** Ammonia testing **a.** Perform ammonia testing on both cumulative GF and CVZ waste media according to manufacturer's instructions. Incubate colorimetric test for full 5 min. **b.** Ammonia levels must remain near 0 for optimal health.

Flask media plating	✓		**10.** Be aware of air flow in room that might blow contaminants onto plates. Use sterile benchtop technique. **11.** Pipette 200 μL of each sample below onto TSA plates and spread evenly with glass beads. **a.** Negative control sterile glass beads **b.** Positive control 1–3 CV/CVZ flasks **c.** Individual GF flasks **d.** Food cultures **e.** Sterile EM (test sterility prior to use)
Clean up	✓	✓	**12.** Maintain clean work environment to reduce widespread contamination. **a.** Empty trash. Remove dirty glassware for cleaning. **b.** Sweep or mop floor, if needed. **c.** Clean all work surfaces and microscope stages with 70% ethanol. **d.** Spray door and cabinet handles. **e.** Clean sink area by routinely treating with 10% bleach, rinsing and then spraying with 70% ethanol. **f.** Refill 70% ethanol spray bottles. **g.** Discard bacterial plates in biohazard waste.
Biosafety cabinet preparation	✓	✓	**13.** Morning set up for afternoon session (food culture preparation). Clean with 70% ethanol and place in BSC. **a.** 2 × 50 mL sterile conical tubes **b.** 2 × 50 mL wrapped, sterile serological pipets **c.** 250 mL bottle of validated sterile EM **d.** PPYE + *Tetrahymena* (4–7 days old) **e.** Aluminum foil to cover *Tetrahymena* cultures **14.** Afternoon set up for morning session (media exchange and feeding). Clean with 70% ethanol and place in BSC. **a.** Sterile EM for media exchange. **b.** 50 mL conical tube for ammonia testing of media removed. **c.** 15 mL conical tubes for bacterial surveillance of media removed. **d.** Wrapped sterile, serological pipets for media exchange.

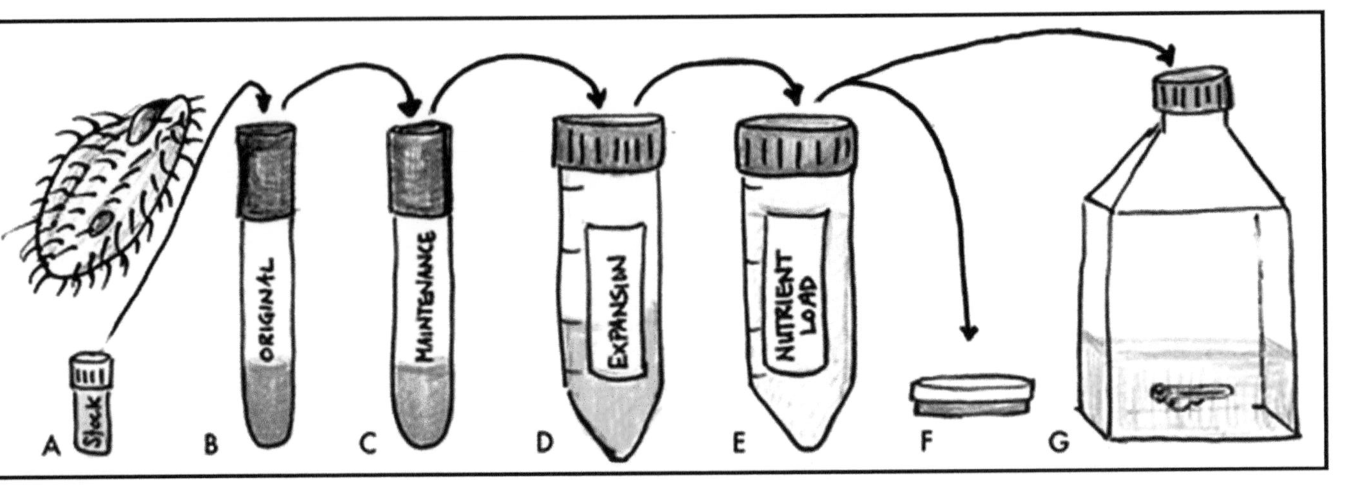

FIGURE 5

Tetrahymena culture maintenance overview. Schematic diagram to be used with Tables 8–10; letters in figure correspond specifically to steps in Table 8.

Table 8 Quick Guide for *Tetrahymena thermophila* Live Food Culture. Refer to Fig. 5 for Culture Maintenance Schematic. Refer to Tables 9 and 10 for Materials and Specific Procedures. Solution Recipes can be Found at the end of the Chapter. Shading for Visual Separation of Reagents and Incubation Details

Corresponding Element in Fig. 5	B	C	D	E	F	G
	Original culture	Maintenance culture	Expansion culture	Nutrient-loaded culture	Sterility verification	Zebrafish flask
Media	5 mL MNM	5 mL MNM	15 mL PPYE	37 mL Sterile EM	TSA	100–250 mL Sterile EM
Penicillin-streptomycin cocktail (Sigma P4458)	5 μL	5 μL	10 μL	–	–	–
Amphotericin B (8 mg/mL)	1 μL	1 μL	–	–	–	–

30% sterile egg yolk emulsion	–	–	–	1.4 mL	–	–
Tetrahymena Inoculation (see Fig. 5A, Table 9)	100 μL Stock center aliquot	100 μL Original culture	200 μL Maintenance Culture	2 mL Expansion culture	200 μL Nutrient-loaded culture	Nutrient-loaded culture volume added varies with zebrafish stage, media exchange and media volume
Incubation temperature	20°C	20°C	28°C	28°C	28–30°C	28°C
Incubation duration	1 week	2–4 weeks between new cultures	4–7 days	1 day—plate to verify sterility 2 days—feed if sterile	2 days	Add sterile nutrient-loaded culture daily

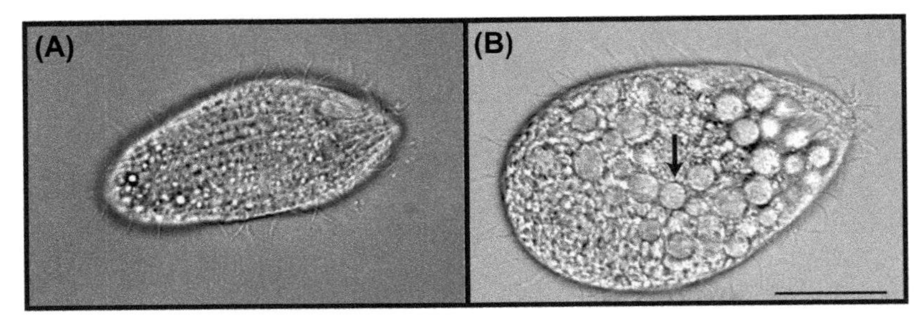

FIGURE 6

Tetrahymena nutrient enhancement. (A) *Tetrahymena* cultured in proteose peptone medium are small and have few vacuoles. (B) *Tetrahymena* cultured in chicken egg yolk–enriched medium are larger and packed with large vacuoles (*arrow*). Scale bar, 20 μm.

Table 9 *Tetrahymena thermophila* Culture Materials. Refer to Fig. 5 and Table 8 for Quick Guide and Table 10 for Specific Procedure. Solution Recipes Follow Chapter

Materials	Special Considerations
Tetrahymena thermophila stock	*Tetrahymena* Stock Center, Cornell University (stock ID SD00178, strain CU428), verify sterility
Penicillin–streptomycin cocktail (pen/strep)	Sigma–Aldrich, P4458
Amphotericin B	VWR, 97061-608 make 8 mg/mL stock solution in DMSO
Sterile glass culture test tubes with lids	Autoclave
Sterile 50 mL conical tubes	
Pipet aid	Dedicated to GF procedures
50 mL sterile wrapped serological pipets	
P-20, P-100, P-1000 pipettors and sterile filter tips	Dedicated to GF procedures
Sterile Modified Neff's Media (MNM)	
Sterile proteose peptone yeast extract (PPYE) media	
Sterile 30% organic egg emulsion in EM	Verify sterility of both solutions prior to use
Sterile EM	

The most common route of GF experiment contamination is daily feedings, thus sterile technique and surveillance of GF *Tetrahymena* stocks are imperative. GF *Tetrahymena* cultures should be started at least a month in advance of any long-term GF experiment and if maintenance stocks become contaminated, new GF stocks should be ordered from the *Tetrahymena* Stock Center (https://tetrahymena.vet.cornell.edu/).

Table 10 *Tetrahymena thermophila* Culture Procedures. Refer to Fig. 5 and Table 8 for Quick Guide and Table 9 for Materials. Solution Recipes Follow Chapter. *Use Sterile Technique When Working With* Tetrahymena *Stocks and Verify Sterility Regularly. Pipette* Tetrahymena *inoculum From Surface of Culture. Constant Shaking of Cultures is Unnecessary. Table Shading Indicates Different Procedures*

***Tetrahymena* culture type**	**Procedure**
Initiation	Inoculate four maintenance stocks upon receipt of the stock culture and four more 4 days later for a total of eight maintenance stocks using the procedure given as follows: **1.** To each glass test tube with lid, add 5 mL of sterile MNM, 5 μL of pen/strep cocktail and 1 μL of amphotericin B stock solutions using sterile technique at a laboratory bench. Shake tube gently from side to side to mix well. **2.** Inoculate media with 100 μL of the purchased *Tetrahymena* stock. **3.** Incubate stock culture tubes on benchtop at room temperature (~20°C) out of direct light. **4.** Incubate maintenance stock at room temperature on benchtop for 1 week. Inspect tube for *Tetrahymena* growth. Plate 200 μL of each stock tube on TSA plates for 48 h to test for bacterial growth and visually inspect tubes of *Tetrahymena* for cloudy appearance indicating bacterial growth.
Maintenance	This stock is kept in the laboratory at room temperature away from the fish facility and serially propagated for continued use to inoculate expansion cultures. **1.** Propagate maintenance stocks by serially transferring 100 μL of MNM culture to 5 mL of new, sterile media plus 5 μL pen/strep and 1 μL amphotericin B every 2–4 weeks. **2.** Incubate maintenance stocks at room temperature for at least 1 week prior to use in expansion stocks.
Expansion	Inoculate expansion culture on the benchtop and incubate for at least 4 days at 28°C prior to use for nutrient-enriched cultures. **1.** Aliquot 15 mL sterile PPYE into 50 mL conical tube. **2.** Add 10 μL pen/strep to 15 mL of sterile PPYE using benchtop sterile technique. **3.** Inoculate PPYE + pen/strep with 200 μL 1–2-week-old MNM culture. Leave lid slightly loose for air exchange. **4.** Sample for bacterial growth before and after inoculation of the nutrient-loaded culture.
Nutrient loaded	During an experiment, inoculate two cultures each day. **1.** Prepare nutrient-loaded culture in BSC. Wear gown or sterile sleeves, clean gloved hands and arms with 70% EtOH.

Continued

Table 10 *Tetrahymena thermophila* Culture Procedures. Refer to Fig. 5 and Table 8 for Quick Guide and Table 9 for Materials. Solution Recipes Follow Chapter.—cont'd

Tetrahymena culture type	Procedure
	2. Clean the following with 70% EtOH thoroughly, and place in BSC. **a.** 50 mL conical tube of 30% egg yolk emulsion **b.** 250 mL filter-sterilized EM **c.** 4–7-day-old PPYE *Tetrahymena* culture **d.** 25 mL wrapped sterile serological pipet **e.** 2 × 50 mL sterile conical tubes (one for GF, one for CVZ culture) **f.** 400 mL beaker (waste) **g.** Foil large enough to wrap around 50 mL conical tube (if using UV in BSC) **3.** Using serological pipet, fill each 50 mL sterile conical tube with 37 mL sterile EM. **4.** Add 1.4 mL 30% sterile egg yolk emulsion to each tube. **5.** Add 2.0 mL 4–7-day-old PPYE *Tetrahymena* culture. **6.** Swirl tubes gently to mix. **7.** Wrap the GF-dedicated *Tetrahymena* culture in aluminum foil to protect from UV irradiation. **8.** Incubate cultures at 28°C. Sample for sterility after 24 h incubation. Use sterile 48 h cultures for feeding.

1.4.2 Media exchange, water quality, and surveillance

Another major challenge for long-term GF zebrafish husbandry is maintaining water quality. We typically change about 75% of the media daily, and use removed media to monitor water quality and contamination. The water quality of GF zebrafish co-cultured with GF *Tetrahymena* is relatively stable with low ammonia levels. More challenging has been maintenance of CV controls because of the tendency of the conventionalized flasks to develop microbial overgrowth resulting in poor zebrafish health. We are developing continuous flow cultures (see Section 2), which we hope will keep microbial growth under control during long-term culture.

1.4.3 Project management and record keeping

Just as running a zebrafish facility requires planning, allocation of responsibilities, and careful record keeping, so too does undertaking long-term GF zebrafish experiments. We recommend allocating responsibilities among participating personnel, including media and plate preparation, *Tetrahymena* maintenance cultures, and supply logistics. All participating personnel must be cross-trained so they can rotate through daily tasks that occur not only throughout the week, but also on weekends and holidays.

It is invaluable to keep detailed records during long-term GF experiments, enabling continual improvement of procedures, pinpointing possible contamination sources, and improving reproducibility. Developing a standardized method of daily note taking, with established checklists, facilitates efficient record keeping and communication among rotating staff. We use a separate checklist for each husbandry task and a standardized platform for recording data and notes. For each flask, we record flask plating results, visual inspection results, and numbers of sick and dead animals. In addition, we maintain a daily summary sheet that records media exchange volumes, feeding volumes, and ammonia levels.

2. PROSPECTUS

Zebrafish are a powerful model for studying diverse aspects of animal development, physiology, immunity, toxicology, and evolution. Gnotobiotic methods described here have begun to elucidate the diverse contributions microorganisms make to zebrafish biology. By providing detailed methods, we hope to stimulate broader adoption of gnotobiotic investigations in the zebrafish model and to gain new insights into host–microbiota relationships. Looking forward, we anticipate several important challenges that this research community will need to address.

2.1 ZEBRAFISH NUTRITION

Zebrafish husbandry has historically used a combination of live and artificial diets (Lawrence, 2007). Although live diets are highly palatable and widely used, their macro- and micronutrient content is often poorly defined or unknown, and can vary considerably across facilities. By contrast, artificial diets are manufactured with defined chemical compositions in mind. Despite the prevalence of zebrafish as a model, specific dietary requirements remain undefined (Watts et al., 2012). As a result, there are no uniform feeding practices, thus zebrafish research cannot take full advantage of chemically defined diets. Because the impact of diet composition and specific nutrients on zebrafish development and physiology remains largely unexplored, we need to be aware of diet as a possible confounding factor across studies. The limitations of our current understanding of zebrafish nutrition will become more evident with the increased popularity of zebrafish as a model for physiological and pathophysiological processes where diet is a crucial environmental factor. Moreover, these limitations are amplified in GF husbandry, where in the absence of the nutritional contributions of microbes, missing nutrients need to be supplemented in the diet.

Development of GF diets has combined live and artificial diets. Initially, artificial powder diets that were partially defined were added to water to create a slurry before autoclaving (Pham, Kanther, Semova, & Rawls, 2008). Although these diets were consumed and evoked appropriate responses (Rawls et al., 2006), the impact of autoclaving and leaching on bioavailability and content, and potentially on water quality, was not studied in detail. To bypass these caveats, we next irradiated defined-composition, artificial powder diets (Semova et al., 2012). Despite presumed improvements in nutrient content and bioavailability, fish fed these diets display suboptimal, post-larval growth. While the zebrafish research community works to determine nutritional requirements, we focus our efforts on developing live diets for GF animals. Although it may prove to be more difficult to achieve reproducible nutritional content with live diets compared to artificial diets, preliminary studies indicate that live GF diets yield better growth. Several groups have succeeded in culturing GF zebrafish for up to a month, using diets based on *Tetrahymena* and *Artemia* (Milligan-Myhre et al., 2011; Rendueles et al., 2012). In the future, it will be important to refine protocols for reproducible culture of live GF diets and define their nutritional content, to more fully understand the impact of GF status on development and physiology. Broader efforts to define zebrafish nutritional requirements and development and implementation of a standardized artificial diet could be used as a platform for customized diets with targeted manipulations of specific nutrients. Only after these resources are established across zebrafish research will they be applicable to study zebrafish–microbiota interactions.

2.2 HOUSING

Use of tissue culture flasks or multiwell plates for rearing GF fish enables gnotobiotic zebrafish to be maintained outside of a gnotobiotic isolator, making GF husbandry accessible to laboratories that only have BSCs. However, scalability of growing gnotobiotic zebrafish in culture flasks and multiwell plates has limitations. The requirement for media changes increases daily husbandry labor and can become limiting as we strive to increase throughput and extend GF husbandry to adults.

We envision innovations that could help more easily and consistently rear gnotobiotic zebrafish through their life cycle. First, transitioning from flasks into bioreactor-like stand-alone units with various entry ports connected to reservoirs for dosing of fresh sterile media for water changes, feeding, and colonization. Units should also have outflow ports for used media, situated to prevent detritus accumulation. Similar vessels have already been developed for other fishes (Forberg et al., 2011; Lesel, 1979). Second, use of interconnected mini-bioreactors could permit large-scale screening of GF fish. This type of apparatus has already been developed for culturing microorganisms (Auchtung, Robinson, & Britton, 2015) and should be adaptable for zebrafish and other small aquatic animals. Together these proposed innovations could allow zebrafish to become the only gnotobiotic vertebrate model for which large-scale, long-term husbandry is possible.

2.3 POST-LARVAL GNOTOBIOTIC ZEBRAFISH

Our recent advances in gnotobiotic zebrafish husbandry allow us to begin studying later life cycle stages under GF conditions. This exciting frontier will greatly benefit from improved housing and diet. Although GF husbandry of post-embryonic stages remains challenging, we anticipate that advancing this part of the gnotobiotic zebrafish field will empower analysis of these and other topics.

2.3.1 Metamorphosis and post-embryonic growth and development

Post-embryonic zebrafish development and growth are accompanied by metamorphosis, during which irreversible morphological and physiological changes occur as the animal transitions from larval to juvenile and adult forms (McMenamin & Parichy, 2013). The diversity of traits that undergo metamorphic changes is vast, yet, only a limited number of them have been studied (McMenamin & Parichy, 2013). Concomitant changes of the microbiota and animal form have been shown during metamorphosis in invertebrates, amphibians, and jawless fishes (Johnston & Rolff, 2015; Kohl, Cary, Karasov, & Dearing, 2013; Tetlock, Yost, Stavrinides, & Manzon, 2012). Metamorphosis is tightly linked to the environment (Paris & Laudet, 2008). We previously reported that zebrafish intestinal microbiota changes during and after metamorphosis (Burns et al., 2016; Stephens et al., 2016; Wong et al., 2015), but it remains unclear whether metamorphic processes in zebrafish, or in other animals for that matter, are influenced by the microbiota.

2.3.2 Obesity and metabolic disease modeling

Zebrafish have been used to model genetic obesity (Song & Cone, 2007), body fat distribution (Minchin et al., 2015), and obesogenic molecules (Tingaud-Sequeira, Ouadah, & Babin, 2011). White adipose tissues develop in teleosts, but in zebrafish this tissue does not appear until around the onset of metamorphosis and then continues to increase in size and distribution as animals grow to adult stages (Flynn, Trent, & Rawls, 2009; Imrie & Sadler, 2010). Zebrafish white adipocytes have similar morphology and gene expression to mammalian adipocytes (Flynn et al., 2009; Imrie & Sadler, 2010), and gnotobiotic murine models have revealed microbial influences on adipose tissue accumulation and metabolic disease (Bäckhed et al., 2004; Cani et al., 2008; Ridaura et al., 2013). There is already evidence of microbiota contribution to intestinal fat absorption in zebrafish (Semova et al., 2012), and microbial modulation of host regulators of fat metabolism like Angptl4 is conserved in mice and zebrafish (Bäckhed et al., 2004; Camp, Jazwa, Trent, & Rawls, 2012). In the future, use of post-larval GF zebrafish can complement and extend existing efforts to understand microbial contributions to fat storage and energy balance.

2.3.3 Immunity

During the first few weeks post fertilization, zebrafish lack an adaptive immune system and instead rely exclusively on the innate immune system to respond to immune challenges (Kanther & Rawls, 2010). Development of tissues and organs involved in adaptive immunity and the humoral response is observed about 4–6 weeks post fertilization (Lam, Chua, Gong, Lam, & Sin, 2004; Langenau et al., 2004; Page et al., 2013). Zebrafish possess adaptive immune system cells, including B cells, T cells, and dendritic cells, however structures homologous to gut-associated lymphoid tissue and Peyer's patches have not yet been identified in teleosts (Renshaw & Trede, 2012). Both innate and adaptive arms of the immune system are required for normal host–microbe interactions, as demonstrated by the hypersusceptibility to infection of immunocompromised zebrafish lines lacking the innate immune signal transducer *myd88* or the adaptive immunity cell receptor recombinase *rag1* (Swaim et al., 2006; van der Vaart, van Soest, Spaink, & Meijer, 2013). Extending the GF state to later life cycle stages should enable testing of microbial influences on development and function of zebrafish adaptive immunity, as has been shown in mammals (Honda & Littman, 2016). Thus, GF post-larval zebrafish could greatly contribute to comparative and evolutionary studies of vertebrate immune responses.

2.3.4 Xenobiotic metabolism

The gut microbiota are known to modulate interactions of the host with foreign substances or xenobiotics (Haiser, Seim, Balskus, & Turnbaugh, 2014; Saha, Butler, Neu, & Lindenbaum, 1983; Zheng et al., 2013) and to regulate differential expression of genes involved in xenobiotic metabolism in livers of GF and specific pathogen-free mice (Bjorkholm et al., 2009). However, the role of microbiota in xenobiotic metabolism remains under studied. Post-embryonic GF zebrafish could be a platform to investigate microbial-dependent changes of host–xenobiotic interactions, as well as host–microbiota interactions in the context of the response, adaption, and long-term consequences of exposure to xenobiotics, including therapeutic drugs and environmental chemicals. GF zebrafish as a toxicology model provide the ability to regulate the microbial environment and the nominal concentration of chemicals, along with regulation of the host genetic background, to understand how the microbiota and genetics intersect to influence host responses to xenobiotics.

2.3.5 Behavior

Zebrafish has become a premier model for exploring cellular and molecular mechanisms underlying a variety of behaviors, including swimming, as well as social, sexual, cognitive, sleep, reward, and anxiety (Fetcho, Higashijima, & McLean, 2008; Kalueff et al., 2013). All of these behaviors are generated by neural circuitry that can be readily studied through the course of development, making zebrafish an ideal platform to learn the role of the microbiota in nervous system development and function. Coupling use of transgenic zebrafish in which morphology or activity of specific neuronal subpopulations can be visualized (Stewart, Braubach, Spitsbergen, Gerlai, & Kalueff, 2014) with use of genetically modified bacteria (Stephens et al., 2015) will enable unprecedented resolution of molecular mechanisms by which resident microbes affect the nervous system and in turn, by which the nervous system affects resident microbial communities.

In summary, there are many exciting areas in which the well-established genetic and imaging zebrafish toolkit can be combined with gnotobiotic husbandry to investigate mechanisms by which resident microbiota influence host development, physiology, metabolism, and behavior.

3. SOLUTIONS

Here we list solutions required for each procedure.

1. Embryo medium (EM)
   Stock solutions
   a. A: 250 mL RO H_2O, 20 g NaCl, 1 g KCl
   b. B: 100 mL RO H_2O, 0.358 g Na_2HPO_4 anhydrous, 0.6 g KH_2PO_4 monobasic
   c. C: 250 mL RO H_2O, 4.68 g $CaCl_2$ dihydrate
   d. D: 250 mL RO H_2O, 6.15 g $MgSO_4 \cdot 7H_2O$
   e. E: 250 mL RO H_2O, 8.75 g $NaHCO_3$
   Preparation (5205 mL)
   a. Add 5 L of RO water to a 5 L carboy
   b. Add 50 mL stock solution A
   c. Add 5 mL stock solution B
   d. Add 50 mL stock solution C
   e. Add 50 mL stock solution D
   f. Add 50 mL stock solution E
   g. Adjust pH to 7.2 with 1 M HCl
2. Antibiotic embryo medium (ABEM)
   a. 500 μL ampicillin (Fisher Scientific, BP1760-25) 100 mg/mL (100 μg/mL final)
   b. 50 μL kanamycin (Fisher Scientific, BP906-5) 50 mg/mL (5 μg/mL final)
   c. 15.6 μL amphotericin B (VWR, 97061-608) 8 mg/mL (250 ng/mL final)
   d. 500 mL EM
   e. Mix and filter sterilize with 0.2 μm filtration unit
3. 0.003% Bleach solution
   a. 125 μL 6.0% laboratory-grade bleach solution (Fisher Scientific, SS290-1)
   b. 250 mL EM
   c. Mix and filter sterilize with 0.2 μm filtration unit
4. 0.1% Polyvinylpyrrolidone-iodine (PVP-I) solution
   a. 2.5 mL 10% PVP-I stock (Ovadine, Western Chemical, Inc.) 5 g PVP-I in 50 mL nanopure water)
   b. 247.5 mL EM
   c. Mix and filter sterilize with 0.2 μm filtration unit
5. Modified Neff's media (MNM)
   a. Dissolve 3 mg $FeCl_3 \cdot 6H_2O$ in 500 mL RO water
   Add
   a. 10 g proteose peptone (VWR, CRITERION, Hardy Diagnostics, 89405-922)
   b. 2.5 g yeast extract (Fisher BioReagents, BP9727-500)
   c. 5 g glucose (Fisher Scientific, AC410950010)
   d. Add RO water to 1 L
   e. Autoclave
   Caution: $FeCl_3 \cdot 6H_2O$ is corrosive and hygroscopic.

6. Proteose peptone yeast media (PPYE)
   a. 0.75 g proteose peptone
   b. 0.75 g yeast extract
   c. Add RO water to 300 mL
   d. Autoclave
   e. Using sterile technique, dispense 15 mL media into 50 mL conical sterile tubes
7. Sterile 30% organic egg emulsion in embryo media

   Materials
   a. Organic chicken eggs
   b. Micro-90 concentrated cleaning solution (VWR, 89210-138)
   c. Double distilled water (dH_2O)
   d. 1 L polypropylene beaker for egg soaking
   e. Scotch-Brite abrasive (use once and dispose afterward)
   f. 250 mL autoclaved beakers
   g. 1 L autoclaved plastic waste beaker
   h. 70% ethanol
   i. Aluminum foil
   j. n × 50 mL conical tubes n = number of eggs
   k. P-1000 sterile pipet tips
   l. 250 mL sterile EM
   m. 1 × 50 mL sterile wrapped serological pipet

   Procedure
   1. Soak organic eggs in 1% Micro-90 concentrated cleaning solution in a 1 L plastic beaker for at least 3 h.
   2. Scrub shells of three eggs with an abrasive sponge.
   3. Rinse 3 times in dH_2O.
   4. Replace dH_2O with 70% EtOH, soak at least 1 h.
   5. Transfer each egg to a separate 250 mL beaker and cover with fresh 70% EtOH. Place foil over top of beaker.
   6. Clean the following with 70% EtOH and place in BSC:
      a. Large sheet of aluminum foil to protect surface of the BSC work surface. Do not block vents (*Note*: Egg is difficult to remove from stainless steel surfaces)
      b. Beakers containing eggs
      c. 1 × 1 L beaker (waste)
      d. n × 50 mL conical tubes n = number of eggs
      e. 250 mL sterile EM
   7. Label 50 mL conical tubes and loosen caps. Place upright in rack.
   8. Remove foil from one beaker. Carefully decant 70% EtOH into 1 L waste beaker.
   9. Place beaker on aluminum foil. Crack egg and separate egg white into beaker by passing the yolk from one half shell to the other. The goal is to have the

intact yolk with very little white in one of the half shells. Discard shell without yolk.

10. Holding the shell with intact yolk, use a sterile P-1000 pipet tip to pierce the yolk membrane. Gently tilt the shell to pour yolk out of membrane into a sterile 50 mL conical tube. Do not let membrane slide into conical tube, it does not emulsify. Put lid on conical tube.
11. Discard shell and P-1000 tip into waste beaker.
12. Spray hands and work with 70% EtOH. Change gloves as needed.
13. Repeat yolk isolation process for remaining eggs.
14. Using graduation on sides of 50 mL conical tubes, estimate volume of yolk collected. Calculate volume of sterile EM to add for a 30% solution.
15. Use pipet aid and 50 mL sterile wrapped serological pipet to dispense sterile EM to each conical tube containing egg yolk. Screw caps of tubes tightly and shake vigorously to emulsify.
16. Remove tubes of emulsion from BSC.
17. Incubate 30% egg yolk emulsion at 28°C for 24 h.
18. Plate 200 μL of emulsion on TSA plates. Spread evenly with sterile glass beads. Avoid bubble formation while spreading. Bubbles can look like bacterial colonies later.
19. Incubate plates at 30°C, check for bacterial growth at 24 and 48 h.

ACKNOWLEDGMENTS

Research reported in this publication was supported by the National Institutes of Health under award numbers P01HD22486, P50GM098911, R01GM095385, R21ES023369, and R24OD016761. The content is solely the responsibility of the authors and does not necessarily represent the official views of the NIH.

REFERENCES

Auchtung, J. M., Robinson, C. D., & Britton, R. A. (2015). Cultivation of stable, reproducible microbial communities from different fecal donors using minibioreactor arrays (MBRAs). *Microbiome, 3*, 42. PMCID: 4588258.

Bäckhed, F., Ding, H., Wang, T., Hooper, L. V., Koh, G. Y., Nagy, A., ... Gordon, J. I. (2004). The gut microbiota as an environmental factor that regulates fat storage. *Proceedings of the National Academy of Sciences of the United States of America, 101*, 15718–15723.

Baker, J. A., Ferguson, M. S., & TenBroeck, C. (1942). Growth of platyfish (Platypoecilus maculatus) free from bacteria and other micro'organisms. *Experimental Biology and Medicine, 51*, 116–119.

Bates, J. M., Akerlund, J., Mittge, E., & Guillemin, K. (2007). Intestinal alkaline phosphatase detoxifies lipopolysaccharide and prevents inflammation in zebrafish in response to the gut microbiota. *Cell Host and Microbe, 2*, 371–382. PMCID: 2730374.

Bates, J. M., Mittge, E., Kuhlman, J., Baden, K. N., Cheesman, S. E., & Guillemin, K. (2006). Distinct signals from the microbiota promote different aspects of zebrafish gut differentiation. *Developmental Biology, 297*, 374–386.

Battalora, M. S. J., Ellender, R. D., & Martin, B. J. (1985). Gnotobiotic maintenance of sheepshead minnow larvae. *The Progressive Fish-Culturist, 47*, 122–125.

Bjorkholm, B., Bok, C. M., Lundin, A., Rafter, J., Hibberd, M. L., & Pettersson, S. (2009). Intestinal microbiota regulate xenobiotic metabolism in the liver. *PLoS One, 4*, e6958. PMCID: PMC2734986.

Burns, A. R., Stephens, W. Z., Stagaman, K., Wong, S., Rawls, J. F., Guillemin, K., & Bohannan, B. J. M. (2016). Contribution of neutral processes to the assembly of gut microbial communities in the zebrafish over host development. *The ISME Journal, 10*, 655–664.

Camp, J. G., Jazwa, A. L., Trent, C. M., & Rawls, J. F. (2012). Intronic *cis*-regulatory modules mediate tissue-specific and microbial control of angptl4/fiaf transcription. *PLoS Genetics, 8*, e1002585.

Cani, P. D., Bibiloni, R., Knauf, C., Waget, A., Neyrinck, A. M., Delzenne, N. M., & Burcelin, R. (2008). Changes in gut microbiota control metabolic endotoxemia-induced inflammation in high-fat diet-induced obesity and diabetes in mice. *Diabetes, 57*, 1470–1481.

Cheesman, S. E., Neal, J. T., Mittge, E., Seredick, B. M., & Guillemin, K. (2011). Epithelial cell proliferation in the developing zebrafish intestine is regulated by the Wnt pathway and microbial signaling via Myd88. *Proceedings of the National Academy of Sciences of the United States of America, 108*(Suppl 1), 4570–4577. PMCID: 3063593.

Cohendy, M. (1912). Experiences sur la vie sans microbes. *Annales de l'Institut Pasteur, 26*, 106–137.

Dewhirst, F. E., Chien, C. C., Paster, B. J., Ericson, R. L., Orcutt, R. P., Schauer, D. B., & Fox, J. G. (1999). Phylogeny of the defined murine microbiota: altered Schaedler flora. *Applied and Environmental Microbiology, 65*, 3287–3292. PMCID: 91493.

Dierckens, K., Rekecki, A., Laureau, S., Sorgeloos, P., Boon, N., Van Den Broeck, W., & Bossier, P. (2009). Development of a bacterial challenge test for gnotobiotic sea bass (*Dicentrarchus labrax*) larvae. *Environmental Microbiology, 11*, 526–533.

Duclaux, M. E. (1885). Sur la germination dans un sol riche en matières organiques, mais exempt de microbes. *Comptes Rendus de Académie des Sciences, 100*, 66–68.

Fetcho, J. R., Higashijima, S., & McLean, D. L. (2008). Zebrafish and motor control over the last decade. *Brain Research Reviews, 57*, 86–93. PMCID: 2237884.

Flynn, E. J., Trent, C. M., & Rawls, J. F. (2009). Ontogeny and nutritional control of adipogenesis in zebrafish (*Danio rerio*). *The Journal of Lipid Research, 50*, 1641–1652.

Forberg, T., Arukwe, A., & Vadstein, O. (2011). A protocol and cultivation system for gnotobiotic Atlantic cod larvae (*Gadus morhua* L.) as a tool to study host microbe interactions. *Aquaculture, 315*, 222–227.

Gordon, H. A. (1960). The germ-free animal. Its use in the study of "physiologic" effects of the normal microbial flora on the animal host. *The American Journal of Digestive Diseases, 5*, 841–867.

Grunwald, D. J., & Eisen, J. S. (2002). Headwaters of the zebrafish – emergence of a new model vertebrate. *Nature Reviews. Genetics, 3*, 717–724.

Haiser, H. J., Seim, K. L., Balskus, E. P., & Turnbaugh, P. J. (2014). Mechanistic insight into digoxin inactivation by Eggerthella lenta augments our understanding of its pharmacokinetics. *Gut Microbes, 5*, 233–238. PMCID: PMC4063850.

Honda, K., & Littman, D. R. (2016). The microbiota in adaptive immune homeostasis and disease. *Nature, 535*, 75–84.

Human Microbiome Project Consortium. (2012). Structure, function and diversity of the healthy human microbiome. *Nature, 486*, 207–214. PMCID: 3564958.

Imrie, D., & Sadler, K. C. (2010). White adipose tissue development in zebrafish is regulated by both developmental time and fish size. *Developmental Dynamics, 239*, 3013–3023.

Jemielita, M., Taormina, M. J., Burns, A. R., Hampton, J. S., Rolig, A. S., Guillemin, K., & Parthasarathy, R. (2014). Spatial and temporal features of the growth of a bacterial species colonizing the zebrafish gut. *mBio, 5*.

Johnston, P. R., & Rolff, J. (2015). Host and symbiont jointly control gut microbiota during complete metamorphosis. *PLoS Pathogens, 11*, e1005246.

Kalueff, A. V., Gebhardt, M., Stewart, A. M., Cachat, J. M., Brimmer, M., Chawla, J. S., ... Zebrafish Neuroscience Research Consortium. (2013). Towards a comprehensive catalog of zebrafish behavior 1.0 and beyond. *Zebrafish, 10*, 70–86. PMCID: 3629777.

Kanther, M., & Rawls, J. F. (2010). Host–microbe interactions in the developing zebrafish. *Current Opinion in Immunology, 22*, 10–19.

Kanther, M., Sun, X., Muhlbauer, M., Mackey, L. C., Flynn, E. J., 3rd, Bagnat, M., Jobin, C., & Rawls, J. F. (2011). Microbial colonization induces dynamic temporal and spatial patterns of NF-kappaB activation in the zebrafish digestive tract. *Gastroenterology, 141*, 197–207. PMCID: 3164861.

Kimmel, C. B., Ballard, W. W., Kimmel, S. R., Ullmann, B., & Schilling, T. F. (1995). Stages of embryonic development of the zebrafish. *Developmental Dynamics, 203*, 253–310. PMID: 8589427.

Kohl, K. D., Cary, T. L., Karasov, W. H., & Dearing, M. D. (2013). Restructuring of the amphibian gut microbiota through metamorphosis. *Environmental Microbiology Reports, 5*, 899–903.

Krishnan, S., Alden, N., & Lee, K. (2015). Pathways and functions of gut microbiota metabolism impacting host physiology. *Current Opinion in Biotechnology, 36*, 137–145. PMCID: 4688195.

Küster, F. (2013). *Die Gewinnung, Haltung und Aufzucht keimfreier Tiere und ihre Bedeutung für die Erforschung natürlicher Lebensvorgänge*. Springer-Verlag.

Lam, S. H., Chua, H. L., Gong, Z., Lam, T. J., & Sin, Y. M. (2004). Development and maturation of the immune system in zebrafish, *Danio rerio*: a gene expression profiling, in situ hybridization and immunological study. *Developmental and Comparative Immunology, 28*, 9–28.

Langenau, D. M., Ferrando, A. A., Traver, D., Kutok, J. L., Hezel, J. P. D., Kanki, J. P., ... Trede, N. S. (2004). In vivo tracking of T cell development, ablation, and engraftment in transgenic zebrafish. *Proceedings of the National Academy of Sciences, 101*, 7369–7374.

Laukens, D., Brinkman, B. M., Raes, J., De Vos, M., & Vandenabeele, P. (2016). Heterogeneity of the gut microbiome in mice: guidelines for optimizing experimental design. *FEMS Microbiology Reviews, 40*, 117–132. PMCID: 4703068.

Lawrence, C. (2007). The husbandry of zebrafish (*Danio rerio*): a review. *Aquaculture, 269*, 1–20.

Lesel, R., & Lesel, M. (1976). Obtention d'alevins non vésiculés axéniques de salmonides. *Annales d'Hydrobiologie, 7*, 21–25.

Lesel, R. D. P. (1979). Obtention de poissons ovovivipares axénique; amélioration technique. *Annales de Zoologie Ecologie Aminale, 11*, 389–395.

Macpherson, A. J., & McCoy, K. D. (2015). Standardised animal models of host microbial mutualism. *Mucosal Immunology, 8*, 476–486. PMCID: 4424382.

Mason, T., Snell, K., Mittge, E., Melancon, E., Montgomery, R., McFadden, M., ... Peirce, J. (2016). Strategies to mitigate a *Mycobacterium marinum* outbreak in a zebrafish research facility. *Zebrafish, 13*(Suppl 1), S77–S87. PMCID: 4931754.

McFall-Ngai, M., Hadfield, M. G., Bosch, T. C., Carey, H. V., Domazet-Loso, T., Douglas, A. E., ... Wernegreen, J. J. (2013). Animals in a bacterial world, a new imperative for the life sciences. *Proceedings of the National Academy of Sciences of the United States of America, 110*, 3229–3236. PMCID: 3587249.

McMenamin, S. K., & Parichy, D. M. (2013). Metamorphosis in Teleosts. *Current Topics in Developmental Biology, 103*, 127–165.

Milligan-Myhre, K., Charette, J. R., Phennicie, R. T., Stephens, W. Z., Rawls, J. F., Guillemin, K., & Kim, C. H. (2011). Study of host-microbe interactions in zebrafish. *Methods in Cell Biology, 105*, 87–116.

Milligan-Myhre, K., Small, C. M., Mittge, E. K., Agarwal, M., Currey, M., Cresko, W. A., & Guillemin, K. (2016). Innate immune responses to gut microbiota differ between oceanic and freshwater threespine stickleback populations. *Disease Models and Mechanisms, 9*, 187–198. PMCID: PMC4770144.

Minchin, J. E., Dahlman, I., Harvey, C. J., Mejhert, N., Singh, M. K., Epstein, J. A., ... Rawls, J. F. (2015). Plexin D1 determines body fat distribution by regulating the type V collagen microenvironment in visceral adipose tissue. *Proceedings of the National Academy of Sciences of the United States of America, 112*, 4363–4368. PMCID: PMC4394244.

Munro, P. D., Barbour, A., & Birkbeck, T. H. (1995). Comparison of the growth and survival of larval turbot in the absence of culturable bacteria with those in the presence of *Vibrio anguillarum*, *Vibrio alginolyticus*, or a marine *Aeromonas* sp. *Applied and Environmental Microbiology, 61*, 4425–4428.

Nuttall, G. H. F. (1896). Thierisches Leben ohne Bakterien im Verdauungskanal. *Hoppe-Seyleŕ s Zeitschrift für physiologische Chemie, 21*, 109–121.

Page, D. M., Wittamer, V., Bertrand, J. Y., Lewis, K. L., Pratt, D. N., Delgado, N., ... Traver, D. (2013). An evolutionarily conserved program of B-cell development and activation in zebrafish. *Blood, 122*, e1–11.

Paris, M., & Laudet, V. (2008). The history of a developmental stage: metamorphosis in chordates. *Genesis, 46*, 657–672.

Pasteur, L. (1885). Observations relatives à la note précédente de M. *Comptes Rendus Des Seances De L'Academie Des Sciences, 100*, 68. http://visualiseur.bnf.fr/Visualiseur?Destination=Gallica&O=NUMM-3056.

Pham, L. N., Kanther, M., Semova, I., & Rawls, J. F. (2008). Methods for generating and colonizing gnotobiotic zebrafish. *Nature Protocols, 3*, 1862–1875.

Rawls, J. F., Mahowald, M. A., Goodman, A. L., Trent, C. M., & Gordon, J. I. (2007). In vivo imaging and genetic analysis link bacterial motility and symbiosis in the zebrafish gut. *Proceedings of the National Academy of Sciences of the United States of America, 104*, 7622–7627. PMCID: 1855277.

Rawls, J. F., Mahowald, M. A., Ley, R. E., & Gordon, J. I. (2006). Reciprocal gut microbiota transplants from zebrafish and mice to germ-free recipients reveal host habitat selection. *Cell, 127*, 423–433.

Rawls, J. F., Samuel, B. S., & Gordon, J. I. (2004). Gnotobiotic zebrafish reveal evolutionarily conserved responses to the gut microbiota. *Proceedings of the National Academy of Sciences of the United States of America, 101*, 4596–4601. PMCID: PMC384792.

Rendueles, O., Ferrieres, L., Fretaud, M., Begaud, E., Herbomel, P., Levraud, J. P., & Ghigo, J. M. (2012). A new zebrafish model of oro-intestinal pathogen colonization reveals a key role for adhesion in protection by probiotic bacteria. *PLoS Pathogens, 8*, e1002815. PMCID: 3406073.

Renshaw, S. A., & Trede, N. S. (2012). A model 450 million years in the making: zebrafish and vertebrate immunity. *Disease Models and Mechanisms, 5*, 38–47.

Reyniers, J., Trexler, P., Ervin, R., Wagner, M., Luckey, T., & Gordon, H. (1949). The need for a unified terminology in germfree life studies. *Lobund Reports, 2*, 151–162.

Reyniers, J. A. (1959). The pure-culture concept and gnotobiotics. *Annals of the New York Academy of Sciences, 78*, 3–16.

Ridaura, V. K., Faith, J. J., Rey, F. E., Cheng, J., Duncan, A. E., Kau, A. L., & Gordon, J. I. (2013). Gut microbiota from twins discordant for obesity modulate metabolism in mice. *Science (New York, N.Y.), 341*, 1241214.

Rolig, A. S., Parthasarathy, R., Burns, A. R., Bohannan, B. J., & Guillemin, K. (2015). Individual members of the microbiota disproportionately modulate host innate immune responses. *Cell Host and Microbe, 18*, 613–620.

Saha, J. R., Butler, V. P., Jr., Neu, H. C., & Lindenbaum, J. (1983). Digoxin-inactivating bacteria: identification in human gut flora. *Science, 220*, 325–327.

Semba, R. D. (2012). The discovery of the vitamins. *International Journal for Vitamin and Nutrition Research, 82*, 310–315.

Semova, I., Carten, J. D., Stombaugh, J., Mackey, L. C., Knight, R., Farber, S. A., & Rawls, J. F. (2012). Microbiota regulate intestinal absorption and metabolism of fatty acids in the zebrafish. *Cell Host and Microbe, 12*, 277–288. PMCID: 3517662.

Shaw, E. (1957). Potentially simple technique for rearing "germ-free" fish. *Science, 125*, 987–988.

Song, Y., & Cone, R. D. (2007). Creation of a genetic model of obesity in a teleost. *The FASEB Journal, 21*, 2042–2049.

Stappenbeck, T. S., & Virgin, H. W. (2016). Accounting for reciprocal host-microbiome interactions in experimental science. *Nature, 534*, 191–199.

Stephens, W. Z., Wiles, T. J., Martinez, E. S., Jemielita, M., Burns, A. R., Parthasarathy, R., … Guillemin, K. (2015). Identification of population bottlenecks and colonization factors during assembly of bacterial communities within the zebrafish intestine. *mBio, 6*.

Stephens, Z. W., Burns, A. R., Stagaman, K., Wong, S., Rawls, J. F., Guillemin, K., & Bohannan, B. J. M. (2016). The composition of the zebrafish intestinal microbial community varies across development. *The ISME Journal, 10*, 644–654.

Stewart, A. M., Braubach, O., Spitsbergen, J., Gerlai, R., & Kalueff, A. V. (2014). Zebrafish models for translational neuroscience research: from tank to bedside. *Trends in Neurosciences, 37*, 264–278. PMCID: 4039217.

Swaim, L. E., Connolly, L. E., Volkman, H. E., Humbert, O., Born, D. E., & Ramakrishnan, L. (2006). *Mycobacterium marinum* infection of adult zebrafish causes caseating granulomatous tuberculosis and is moderated by adaptive immunity. *Infection and Immunity, 74*, 6108–6117. PMCID: 1695491.

Taormina, M. J., Jemielita, M., Stephens, W. Z., Burns, A. R., Troll, J. V., Parthasarathy, R., & Guillemin, K. (2012). Investigating bacterial-animal symbioses with light sheet microscopy. *The Biological Bulletin, 223*, 7–20.

Tetlock, A., Yost, C. K., Stavrinides, J., & Manzon, R. G. (2012). Changes in the gut microbiome of the sea lamprey during metamorphosis. *Applied and Environmental Microbiology, 78*, 7638–7644.

Tingaud-Sequeira, A., Ouadah, N., & Babin, P. J. (2011). Zebrafish obesogenic test: a tool for screening molecules that target adiposity. *Journal of Lipid Research, 52*, 1765–1772.

Trust, T. J. (1974). Sterility of salmonid roe and practicality of hatching gnotobiotic salmonid fish. *Applied Microbiology, 28*, 340–341. PMCID: PMC186721.

van der Vaart, M., van Soest, J. J., Spaink, H. P., & Meijer, A. H. (2013). Functional analysis of a zebrafish myd88 mutant identifies key transcriptional components of the innate immune system. *Disease Models and Mechanisms, 6*, 841–854.

Verner-Jeffreys, D. W., Shields, R. J., & Birkbeck, T. H. (2003). Bacterial influences on Atlantic halibut *Hippoglossus hippoglossus* yolk sac larval survival and start feed response. *Diseases of Aquatic Organisms, 56*, 105–113.

Watts, S. A., Powell, M., & D'Abramo, L. R. (2012). Fundamental approaches to the study of zebrafish nutrition. *ILAR Journal / National Research Council, Institute of Laboratory Animal Resources, 53*, 144–160.

Wiles, T. J., Jemielita, M., Baker, R. P., Schlomann, B. H., Logan, S. L., Ganz, J., … Parthasarathy, R. (2016). Host gut motility promotes competitive exclusion within a model intestinal microbiota. *PLoS Biology, 14*, e1002517. PMCID: 4961409.

Wong, S., Stephens, W. Z., Burns, A. R., Stagaman, K., David, L. A., Bohannan, B. J. M., … Rawls, J. F. (2015). Ontogenetic differences in dietary fat influence microbiota assembly in the zebrafish gut. *mBio, 6*, e00687–e00715.

Zheng, X., Zhao, A., Xie, G., Chi, Y., Zhao, L., Li, H., … Luther, M. (2013). Melamine-induced renal toxicity is mediated by the gut microbiota. *Science Translational Medicine, 5*(172), 172ra22.

CHAPTER

Infectious disease models in zebrafish

C. Sullivan*, M.A. Matty§, D. Jurczyszak*, K.A. Gabor¶, P.J. Millard*, D.M. Tobin§, C.H. Kim*,[1]

**University of Maine, Orono, ME, United States*

§Duke University School of Medicine, Durham, NC, United States

¶National Institute of Environmental Health Sciences, Durham, NC, United States

[1]Corresponding author: E-mail: carolkim@maine.edu

CHAPTER OUTLINE

Methods in Cell Biology, Volume 138, ISSN 0091-679X, http://dx.doi.org/10.1016/bs.mcb.2016.10.005

Abstract

In recent years, the zebrafish (*Danio rerio*) has developed as an important alternative to mammalian models for the study of hostpathogen interactions. Because they lack a functional adaptive immune response during the first 4–6 weeks of development, zebrafish rely upon innate immune responses to protect against injuries and infections. During this early period of development, it is possible to isolate and study mechanisms of infection and inflammation arising from the innate immune response without the complications presented by the adaptive immune response. Zebrafish possess several inherent characteristics that make them an attractive option to study hostpathogen interactions, including extensive sequence and functional conservation with the human genome, optical clarity in larvae that facilitates the high-resolution visualization of host cell–microbe interactions, a fully sequenced and annotated genome, robust forward and reverse genetic tools and techniques (e.g., CRISPR-Cas9 and TALENs), and amenability to chemical studies and screens. Here, we describe methods for studying hostpathogen interactions both through systemic infections and through localized infections that allow analysis of host cell response, migration patterns, and behavior. Each of the methods described can be modified for use in downstream applications that include ecotoxicant studies and chemical screens.

INTRODUCTION

Infectious diseases have shaped the course of human history. They arise in human populations when disruptions occur in the balance between hosts, pathogens, and environment, leading to conditions that favor their emergence (or reemergence) and spread (Engering, Hogerwerf, & Slingenbergh, 2013). The constantly evolving host–pathogen relationship has necessitated the development of a number of animal infection models in order to gain insight into how human hosts mount effective immune responses to infections by pathogens, how pathogens, in turn, have evolved to subvert the human host response to infection, and how disruptions in environments can influence the human host–pathogen interaction. While a variety of animals have been adopted to study aspects of the host–pathogen interaction, it has become increasingly apparent that there is no perfect model organism. In choosing an animal model to study human infections, several considerations should be made, including the genetic and functional conservation of the immune response, the aspect of immunity to be tested (e.g., innate vs. adaptive), the inherent characteristics of the model system, including its advantages and limitations, and the availability and cost of reagents.

The zebrafish is a formidable model organism to study infectious diseases and serves an important complement to other model organisms (Sullivan & Kim, 2008). Zebrafish possess several inherent characteristics that make them a particularly useful model organism for biomedical research, including large clutches of embryos per mating pair, optical clarity throughout early development, rapid development to adulthood, and inexpensive costs for maintenance and husbandry. As a model organism for infection and immune function, the zebrafish has facilitated tremendous advancements in recent years due in large part to the conservation of gene sequence and function (Howe et al., 2013), the ability to perform forward and reverse genetics through high-throughput mutagenesis screens and the emergence of genome-editing techniques like ZFN, TALEN, and CRISPR-Cas9 (Gaj, Gersbach, & Barbas, 2013), and the generation of immune cell-specific transgenic lines that facilitate high-resolution visualization of host cell–microbe interactions in ways not currently possible in other models, including the mouse (Ellett, Pase, Hayman, Andrianopoulos, & Lieschke, 2011; Mathias et al., 2006; Renshaw et al., 2006). The zebrafish is a particularly powerful model for the study of innate immunity, as a functional and integrated adaptive immune response is not present until 4–6 weeks postfertilization (Novoa & Figueras, 2012). Elegant studies aimed at understanding host–pathogen interactions have been aided by the development of zebrafish models for a variety of human bacterial infections, including *Pseudomonas aeruginosa* (Brannon et al., 2009; Clatworthy et al., 2009; Phennicie, Sullivan, Singer, Yoder, & Kim, 2010), *Staphylococcus aureus* (Prajsnar, Cunliffe, Foster, & Renshaw, 2008), *Streptococcus* spp. (Neely, Pfeifer, & Caparon, 2002), and *Mycobacterium* spp. (Bernut et al., 2015; Davis et al., 2002; Tobin & Ramakrishnan, 2008), fungal infections, including *Candida albicans* (Brothers, Newman, & Wheeler, 2011; Brothers & Wheeler, 2012), *Mucor circinelloides* (Voelz, Gratacap, & Wheeler, 2015), and *Aspergillus fumigatus* (Knox et al., 2014); and viral infections, including influenza A (Gabor et al., 2014), chikungunya (Palha et al., 2013), herpes simplex (Burgos, Ripoll-Gomez, Alfaro, Sastre, & Valdivieso, 2008), and hepatitis C (Ding et al., 2011, 2015) (Table 1).

1. METHODS FOR SYSTEMIC BACTERIAL AND VIRAL INFECTIONS

Aspects of human host–pathogen interactions in the context of microbemia leading to septicemia can be effectively modeled in zebrafish by direct injection of bacteria, virus, or fungus into the circulatory system (Bojarczuk et al., 2016; Brannon et al., 2009; Brothers et al., 2011; Clatworthy et al., 2009; Gabor et al., 2014; McVicker et al., 2014; Phennicie et al., 2010; Prajsnar et al., 2008; van der Sar et al., 2003; Tenor, Oehlers, Yang, Tobin, & Perfect, 2015; Wiles, Bower, Redd, & Mulvey, 2009) (Fig. 1). The following protocols describe the procedures used to model bacterial, viral, and fungal infections (Fig. 2). Each of these protocols can be modified to accommodate different pathogens compatible with the zebrafish model.

1.1 QUANTIFYING DISSEMINATED *PSEUDOMONAS AERUGINOSA* INFECTIONS

1.1.1 Materials and reagents

1.1.1.1 Embryo collection

Autoclaved egg water with Instant Ocean (60-μg/mL final concentration) (Spectrum Brands).

1.1.1.2 Preparation of bacterial culture

P. aeruginosa (PA14) (p67T1) (constitutive dTomato expression)
Luria broth (Affymetrix 75854)
Bacteriological Agar (Affymetrix 10906)
Ampicillin (Sigma A9518)

Table 1 Recent Examples (Since 2011) of Viral, Bacterial, and Fungal Infection Models Developed in Adult and Larval Zebrafish

Pathogen	Infection Method	Adults/Larvae	Citation	Major Finding/Use/Aspect of Infection That is Unique/Importance
Virus				
Chikungunya virus	Caudal vein or aorta at 3 dpf	Larvae	Palha et al. (2013)	Neutrophil response is critical to containment of infection
Hepatitis C virus	Inject at 1–8 cell stage	Embryo	Ding et al. (2011, 2015)	Subreplicon system; anti-HCV drug treatment modeling
Herpes simplex virus -1	Hindbrain ventricle, caudal vein, i.p. injection	Adults; larvae	Ge et al. (2015)	Larval model to investigate cytosolic DNA sensing mechanism
Influenza A virus	Duct of Cuvier at 2 dpf; swimbladder at 5 dpf	Larvae	Gabor et al. (2014)	Zebrafish possess alpha-2,6-sialic acids; antiinfluenza drug treatment modeling
Bacteria				
Aeromonas hydrophila			Wang, Ren, Fu, and Su (2016)	Rarely infects humans, but causes gastric problems in young children. Modeled probiotic usage in zebrafish as a way to treat it
Anaerobic microbial community derived from a single human fecal sample.	Immersion and injection	Larvae	Toh, Goodyear, Daigneault, Allen-Vercoe, and Van Raay (2013)	Human microbial communities can grow in zebrafish
Cronobacter turicensis LMG 23827(T),	Yolk injection	Larvae	Fehr et al. (2015)	Dissemination and bacteremia of an opportunistic pathogen

Continued

Table 1 Recent Examples (Since 2011) of Viral, Bacterial, and Fungal Infection Models Developed in Adult and Larval Zebrafish—cont'd

Pathogen	Infection Method	Adults/Larvae	Citation	Major Finding/Use/ Aspect of Infection That is Unique/Importance
Escherichia coli (nonpathogenic)	Injection into notochord	Larvae	Nguyen-Chi et al. (2014)	Models cartilage and bone inflammation
E. coli CFT073	Pericardial cavity and the circulation valley	Larvae	Wiles et al. (2013)	Fitness and virulence of a *E. coli* strain CFT073
F. noatunensis subsp. *noatunensis*, *F. noatunensis* subsp. *orientalis*, and *F. tularensis* subsp. *novicida*	Duct of Cuvier, tail muscle, otic vesicle	Larvae	Brudal et al. (2014)	Model for human pathogen *F. tularensis*
Listeria monocytogenes	Oral immersion and injection via yolk sac, brain ventricle and blood island	Larvae	Shan et al. (2015, 2016)	Models for *L. monocytogenes* via different routes
Mycobacterium marinum	Parenchyma, caudal vein, hindbrain	Larvae	van Leeuwen et al. (2014), Berg et al. (2016) and Takaki, Davis, Winglee, and Ramakrishnan (2013)	TB meningitis modeling; high throughput imaging; modeling lysosomal storage disease
Mycobacterium marinum	IP	Adult	Oehlers et al. (2015), Oksanen et al. (2013, 2016), Takaki et al. (2013) and van Leeuwen et al. (2014)	Drug treatment; vaccine development, meningeal TB modeling
Mycobacterium abscessus	Caudal vein injection	Embryos and larvae	Bernut et al. (2015)	Modeling S and R infections
Myroides odoratimimus			Ravindran, Varatharajan, Raju, Vasudevan, and Anantha (2015)	Opportunistic pathogen in humans
Pseudomonas aeruginosa	Hindbrain	Larvae	Rocker, Weiss, Lam, Van Raay, and Khursigara (2015)	Visualized and quantified microcolony formation

Shigella flexneri	Injection	Larvae	Mostowy et al. (2013)	Modeling autophagy
Staphylococcus aureus	Pericardial cavity, eye, the fourth hindbrain ventricle, yolk circulation valley, caudal vein, yolk body, and duct of Cuvier	Larvae	Li and Hu (2012)	Multipoint infection models human disease
Staphylococcus epidermidis	Yolk injection	Larvae	Veneman et al. (2013, 2014)	Understanding biomaterial-associated infections
Stenotrophomonas maltophilia			Ferrer-Navarro et al. (2013)	Opportunistic pathogen in humans; models nonsocomial infection
Streptococcus iniae	Otic vesicle injection	Larvae	Harvie, Green, Neely, and Huttenlocher (2013)	A model for meningitis and sepsis
Streptococcus pneumoniae	IP and IM injection	Adult	Saralahti et al. (2014)	Models meningitis
Vibrio vulnificus			Jheng et al. (2015)	Can infect humans and cause septicemia. Found that zebrafish mutants can be more resistant
Fungi				
Candida albicans	Into the hindbrain	Larvae	Brothers et al. (2011)	Lethal disseminated disease that shares important traits with disseminated candidiasis in mammals
Cryptococcus neoformans	Caudal vein injection	Larvae	Bojarczuk et al. (2016) and Tenor et al. (2015)	*C. neoformans* can grow in macrophages, attenuated mutants in mammalian models are similarly attenuated in zebrafish, *Cryptococcus* can infect brain
Mucor circinelloides	Hindbrain and swim bladder	Larvae	Voelz et al. (2015)	Mucormycete infection

Earlier studies are presented in previous editions (Milligan-Myhre et al., 2011).

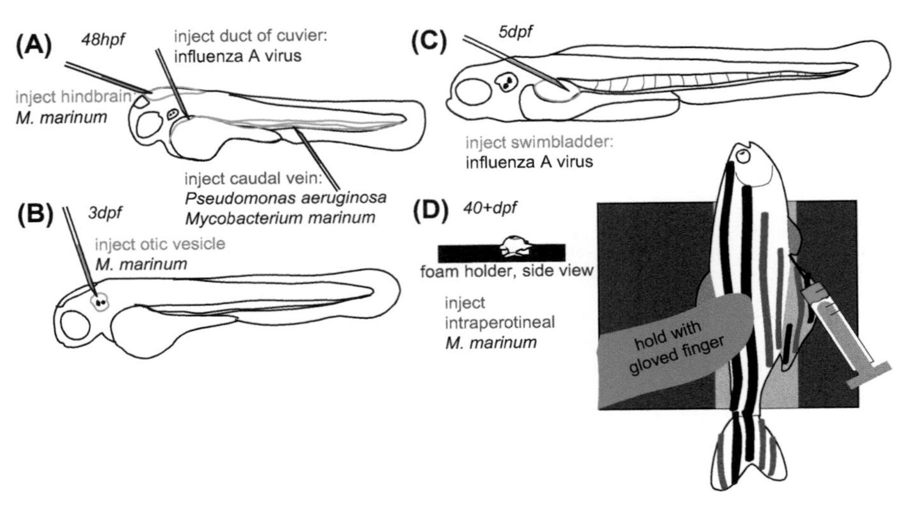

FIGURE 1

Injection sites for various microbial infections in the zebrafish. It is a good practice to perform mock injections in addition to microbial injections to ensure that the act of injection does not affect the phenotype of interest. (A) At 48 hpf, embryos are manually dechorionated and anesthetized with Tris-buffered Tricaine methane sulfonate (Tricaine 200 μg/mL). For hindbrain injections (A1), embryos are placed on glass depression slide and manipulated so that the sharp borosilicate needle is touching the widest part of the hindbrain. With the wire manipulator between the head and the heart, push the fish into the needle. For duct of Cuvier injections (A2), embryos are placed on 1.5% agarose plates so that the microinjection needle is aligned with the duct of Cuvier and the direction of circulation. Influenza A is injected into the duct of Cuvier at a position that is dorsal to the area where the circulation broadens. Care is taken not to nick the yolk sac. Caudal vein injections (A3) result in a systemic bacterial infection. The ventral side of the larva should be in contact with the needle. Push the fish from the dorsal side with the wire manipulator until the needle slips into the caudal vein, just posterior of the cloaca. (B) To inject the ear (B4), pick a side to consistently infect. Maneuver the larva so the anterior dorsal side ear is just below the needle. Using the wire manipulator, push from the area between the head and the heart while simultaneously lowering the needle with the micromanipulator onto and into the otic vesicle. (C) At 5 dpf, larvae are anesthetized with Tricaine (200 μg/mL) and placed on 1.5% agarose plates. Larvae are aligned on their sides in the same direction. Influenza A virus is injected into the lumen of the swimbladder, toward the posterior of the fish, so that the injection bolus collects at the back of the swimbladder and the air bubble is displaced to the anterior. (D) Adult zebrafish aged over 40 days must be anesthetized with Tricaine (100 μg/mL) in fish system water and manually infected with a small insulin needle. Using a foam (or agarose) holder, place fish as shown in D6 to hold the fish in place. Carefully restrain the fish with a gloved finger. Align the needle just anterior and between the pelvic fin. The needle should penetrate the intraperitoneal (IP) space and the cavity should puff out a bit as the syringe contents (15 μL) are injected into the fish.

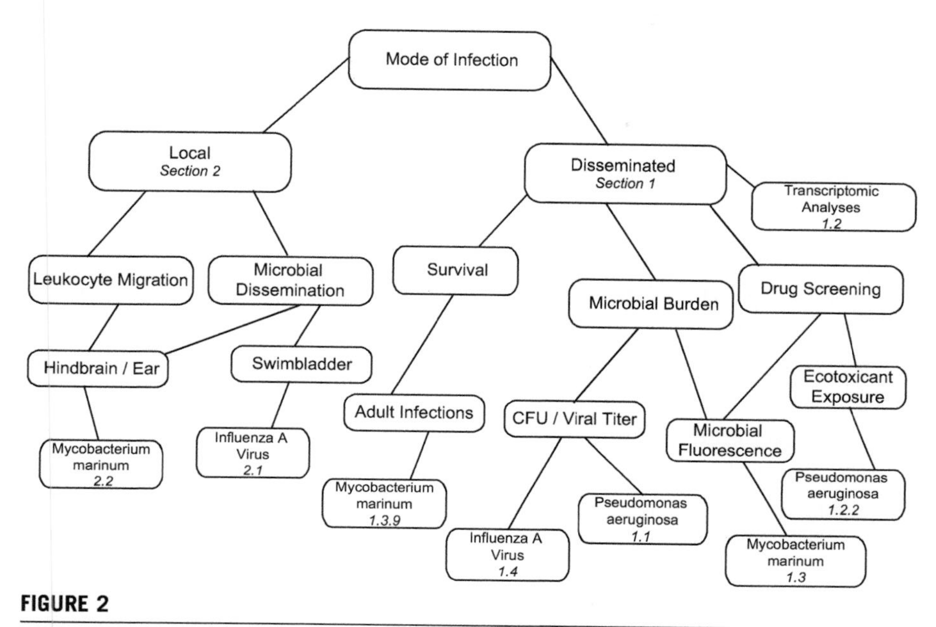

FIGURE 2

Overview of the choices of assays that can be made following the induction of microbial infection. Assays described herein are likely to be applicable, with varying degrees of modification, to many zebrafish pathogen models. The numbers shown denote the sections in which they are described.

Spectrophotometer (Beckman DU 640B)
Semimicro 1.5-mL cuvettes (Fisherbrand 14955127)
Allegra 25R centrifuge with TA 10.250 fixed angle (25 degrees) rotor (Beckman Coulter)
Phosphate buffered saline (PBS), pH 7.4
Phenol red (Sigma–Aldrich P-4758)

1.1.1.3 Preparation of embryos for injections

Dumont #5 forceps (Electron Microscopy Sciences 72700-D)
Transfer pipettes (Fisherbrand 13-711-7M)
Tricaine- S (MS-222) (Western Chemical)
Borosilicate glass with filament (Sutter Instrument BF120-69-10)

1.1.1.4 Injection of bacteria into zebrafish embryo to establish infection

Borosilicate glass with filament (Sutter Instrument BF120-69-10)
Flaming/Brown Micropipette Puller (Sutter Instrument P-97)
Microscope Stage Calibration Slide (AmScope MR095)
MPPI-3 Pressure Injector (Applied Scientific Instrumentation)
SZ61 Stereo Microscope (Olympus)

1.1.1.5 Bacterial burden assays

1.5-mL sterile microcentrifuge tubes (USA Scientific 1615-5510)
100 × 25-mm sterile disposable Petri dishes (VWR 89107-632)
50-mL conical tubes (VWR 21008-178)
Bullet Blender (Next Advance BBX-24)
3.2-mm diameter stainless steel beads (Next Advance SSB32)
Petri-spread disposable L-shaped spreader (USA Scientific 2977-5510)
Fluorescence Stereo Microscope (Zeiss Discovery V.12)
Cetrimide agar (Fluka Analytical 22470-500G)
Glycerol (Fisher Chemical G33-1)
Agarose (Lonza 50004)

1.1.2 Embryo collection

1. Zebrafish should be maintained in accordance with procedures, policies, and guidelines of an Institutional Animal Care and Use Committee.
2. Zebrafish should be maintained at 28°C in recirculating aquarium systems with a 14-h on/10-h off light cycle.
3. Place adult AB females into breeding tanks the night prior to embryo collections.
4. The next morning, add AB males to the breeding tanks just after the lights come on.
5. At 30-min intervals, collect age-matched embryos for use in infection experiments.
6. Maintain embryos at low density (fewer than 100 per Petri dish) in 50–60 mL autoclaved egg water at 28°C.
7. Thoroughly remove dead embryos and debris from dishes at 4–6 hpf (h post-fertilization) and each day thereafter. Remove egg water and replace each day.

1.1.3 Preparation of bacterial culture

1. Streak an LB plate containing 750-μg/mL ampicillin with *P. aeruginosa* (PA14) (p67T1) and incubate overnight at 37°C.
2. Inoculate 4 mL of LB broth containing 750-μg/mL ampicillin with an isolated colony and incubate overnight at 37°C with shaking at 250 rpm.
3. Determine the OD600 of the overnight culture using a spectrophotometer. Add one OD600 bacteria to 25-mL LB broth containing 750-μg/mL ampicillin and shake for 3–4 h to allow the bacteria to enter logarithmic growth phase.
4. Centrifuge bacteria at 5000 rpm for 15 min to pellet cells.
5. Wash and centrifuge bacterial pellets three times in 10-mL sterile, cold phosphate buffered saline at 5000 rpm for 5 min.
6. Thoroughly resuspend bacterial pellets in cold PBS and pass through a 30-G needle mounted on a 3-mL syringe one to two times to disaggregate cells.
7. Using a predetermined relationship between OD600 value and viable cell density, prepare bacteria for injection in a PBS solution containing 0.5% phenol red

(to assist in visualization of injection) at the preferred concentration (colony-forming units per mL).

1.1.4 Preparation of embryos for injections

1. Dechorionate embryos with Dumont #5 forceps, taking care not to injure the embryos.
2. Gently transfer live embryos to fresh egg water with a disposable transfer pipet.
3. Add tricaine to a final concentration of 4 mg/mL to anesthetize the fish.
4. Use a disposable transfer pipet to transfer 10–20 fish to a 1% agar plate.
5. Gently spread the fish across the plate with a flame-polished and sealed borosilicate glass and orient the fish on their sides for injection.
6. Gently wick away excess water from the plate with a Kimwipe to prevent movement of the embryo during injection.

1.1.5 Injection of bacteria into zebrafish embryo to establish infection

1. Load bacterial culture containing 0.5% phenol red into a microinjection needle created by pulling borosilicate glass with filament with a Sutter puller [settings: pressure setting = 100, heat = 550, pull = (no value), velocity = 130, time = 110]
2. Gently clip the needle with sharp Dumont #5 forceps and test the injection amount by measuring on a calibration micrometer. Adjust pressure and timing on the MPPI-3 microinjection apparatus and continue to clip the needle until the desired injection volume is achieved (typically 1–3 nL).
3. Inject embryos via the caudal vein (see Fig. 1). Multiple injections can be performed to achieve the desired inoculation amount.
4. Transfer injected embryos into a 100 × 25-mm Petri dish containing 50-mL fresh egg water and incubate at 28°C.
5. Remove at least five embryos from each injection group to determine the initial bacterial burden (i.e., inoculation dose).

1.1.6 Bacterial burden assays

1. Determine the number of fish to be included in the assay and set up the same number of microcentrifuge tubes.
2. Add two sterile stainless steel balls per microcentrifuge tube.
3. At a predetermined time postinjection, remove infected embryos from the egg water and examine by fluorescence stereomicroscopy to assess the development of the infection. Since bacteria are fluorescent red due to the expression of dTomato fluorescent protein, it is possible to remove infected embryos showing signs of infection at the site of injection and keep embryos exhibiting systemic infection.
4. Transfer systemically infected embryos to a clean Petri dish containing fresh egg water. Repeat two times to thoroughly wash the exteriors of the infected embryos.

5. Anesthetize the embryos in tricaine at a final concentration of 200-μg/mL egg water.
6. Transfer an embryo in 200 μL of egg water (with tricaine) to a microcentrifuge tube containing the stainless steel beads. Repeat for each embryo to be tested.
7. Place the tubes in a Bullet Blender and disrupt on a setting of 8 for 2 min.
8. For each tube, plate 40 μL of zebrafish lysate containing the *P. aeruginosa* bacteria on cetrimide agar. In cases where infection is obviously more robust, based on observations made under the fluorescence microscope, lysates should be diluted to the appropriate amount in sterile PBS.

1.1.7 Representative results

Zebrafish injected with a fluorescently labeled *P. aeruginosa* develop a systemic infection, as can be seen at 20-h postinjection (hpi) (Fig. 3). Bacterial burden assays performed on *Pseudomonas*-selective cetrimide agar clearly demonstrate the presence *P. aeruginosa* in the zebrafish homogenates. These bacterial burden assays can be performed at several times during an infection and can be coupled with other assays, including survival analyses, to measure the effects of drugs, ecotoxicants, and/or genetic manipulations.

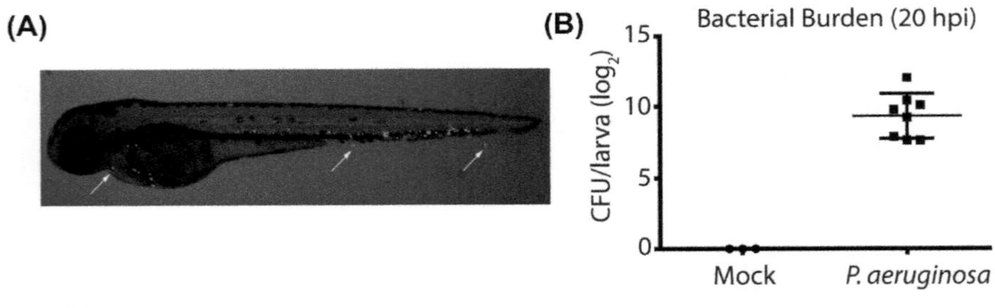

FIGURE 3

Representative data showing (A) a zebrafish exhibiting a systemic *Pseudomonas aeruginosa* infection at approximately 20 hpi and (B) the results of a bacterial burden assay. At 48 hpf, zebrafish are injected with 2–7 × 10^3 CFU per larva (depending on the assay). Evidence of a systemic infection can be observed by confocal laser microscopy when using fluorescent bacterial strains. (A) An overlapped and merged image of Z-stacks containing information from the red and brightfield channels showing a systemic, disseminated infection of a *P. aeruginosa* strain that expresses dTomato fluorescent protein (white areas). *White arrows* point to discrete areas of infection throughout the fish. (B) Representative bacterial burden data collected from infected zebrafish at 20 hpi. Mock infection represents fish that had been injected with PBS (n = 3). *P. aeruginosa* represents fish that had been injected with 6.4 × 10^3 CFU *P. aeruginosa* (n = 8) per larva.

1.2 MEASURING THE CHANGES IN TRANSCRIPT EXPRESSION UPON *PSEUDOMONAS AERUGINOSA* INFECTION IN THE PRESENCE OR ABSENCE OF THE ECOTOXICANT ARSENIC

Zebrafish have several inherent advantages that make them an attractive model for immunotoxicology studies and for drug screens, including high fecundity allowing for high-throughput screening; embryonic and larval transparency allowing for ease of visualization; amenability to genetic manipulations, including the creation of transgenic reporter lines that can provide an indication of fish stress or health; and the ability to add chemicals directly to the water (Kim & Tanguay, 2013; Lieschke & Currie, 2007; MacRae & Peterson, 2015). In ecotoxicant studies, several endocrine disrupting chemicals have been shown to perturb the immune response (Jin, Chen, Liu, & Fu, 2010; Keiter et al., 2012). Metals like titanium dioxide (Jovanovic, Ji, & Palic, 2011), silver (Myrzakhanova et al., 2013), gold (Truong et al., 2013), depleted uranium (Gagnaire, Bado-Nilles, & Sanchez, 2014), and arsenic (Hermann & Kim, 2005; Lage, Nayak, & Kim, 2006; Mattingly, Hampton, Brothers, Griffin, & Planchart, 2009; Nayak, Lage, & Kim, 2007) each can alter the immune response. In pharmacology studies, zebrafish have the potential to be used to identify effective drugs designed to counter infections (Meijer & Spaink, 2011). As a model for ecotoxicity and for drug efficacy, the zebrafish model system holds tremendous value, particularly in the context of the host–pathogen interaction.

1.2.1 Materials and reagents

1.2.1.1 Embryo collection

Same as described in Section 1.1.1
Sodium meta-arsenite (Fluka 71287)

1.2.1.2 Preparation of bacterial culture

Same as described in Section 1.1.1

1.2.1.3 Preparation of embryos for injections

Same as described in Section 1.1.1

1.2.1.4 Injection of bacteria into zebrafish embryo to establish infection

Same as described in Section 1.1.1

1.2.1.5 Total RNA collection for downstream uses (RNAseq and RT-qPCR validation)

TRIzol reagent (ThermoFisher Scientific 15596-026)
Chloroform (Sigma Aldrich 132950)
Linear acrylamide (ThermoFisher Scientific AM9520)
Isopropanol (JT Baker 9095-03)
Ethanol (Pharmco-AAPER 111ACS200)
Nuclease-free water (Integrated DNA Technologies 11-05-01-14)

Nanodrop 2000 Spectrophotometer (Thermo Scientific)
2100 Bioanalyzer (Agilent Genomics)

1.2.2 Embryo collection and arsenic exposure

1. Collect embryos as described in Section 1.1.2.
2. Distribute embryos into Petri dishes (~100/dish) containing 50 mL of 10-ppb sodium meta-arsenite in egg water (=10 μg/L). Establish egg water controls containing ~100 fish per dish.
3. Remove dead embryos at 4–6 hpf, and every 24 h for the length of experiment.
4. Change water every 24 h.

1.2.3 Preparation and injection of bacteria

1. Pull injection needles using borosilicate glass with filaments and a Sutter micropipette puller as described in list (1) in Section 1.1.5.
2. Prepare *P. aeruginosa* suspensions and inject into 48-hpf embryos through the caudal vein, as previously described in list (3) in Section 1.1.5.

1.2.4 Total RNA collection for downstream uses (RNAseq and RT-qPCR validation)

1. At 1, 3, 6, 12, and 24 hpi, anesthetize in Tricaine (200 μg/mL egg water) and transfer embryos in groups of at least 10 fish to microcentrifuge tubes.
2. Remove water from tubes with a pipet, taking care not to disrupt or destroy embryos.
3. Add 200 μL of TRIzol reagent to each tube and homogenize with a motorized mortar and pestle. Embryos tend to float in TRIzol, so care should be taken to ensure complete dissociation.
4. Add 600 μL of TRIzol to each tube and vortex to mix.
5. Add 160 μL of chloroform to each tube. Invert each tube vigorously for 1 min to mix thoroughly.
6. Incubate at room temperature for 5 min.
7. Centrifuge at 16,000 × g for 15 min.
8. Carefully transfer top, aqueous layer to another RNase-free microcentrifuge tube.
9. Add 2 μL of linear acrylamide as a coprecipitant and 1 volume of isopropanol. Rapidly invert 40 times to mix.
10. Incubate at room temperature for 5 min.
11. Centrifuge for 30 min at 16,000 × g.
12. Remove isopropanol, taking care not to disrupt pellet.
13. Wash pellet with 70% ethanol in RNase-free water.
14. Centrifuge for 5 min at 16,000 × g.
15. Remove as much 70% ethanol by pipet as possible.
16. Air dry pellet.

17. Add 15-μL nuclease-free water for every 10 embryos processed. Typical yield range is from 0.5 to 1.0 μg/fish between 48 and 72 hpf.
18. Quantify the amount of total RNA by Nanodrop and assess quality by Bioanalyzer, according to manufacturers' instructions.
19. Use total RNA in downstream applications, including RNAseq and RT-qPCR. RNA can be purified further, if necessary, using column-based procedures such as those offered by Zymo Research or Life Technologies.
20. Precipitate and store unused RNA under ethanol at −80°C.

1.3 SYSTEMIC *MYCOBACTERIUM MARINUM* INFECTION

Mycobacterium marinum, the causative agent of "fish tuberculosis" has been developed as a validated model for human tuberculosis (Davis et al., 2002; Tobin & Ramakrishnan, 2008). The zebrafish innate immune response to *M. marinum* is similar to that described in the human response to *Mycobacterium tuberculosis* (Cronan & Tobin, 2014). Further, *M. marinum* and *M. tuberculosis* share many pathogenicity genes (Cosma, Klein, Kim, Beery, & Ramakrishnan, 2006; Volkman et al., 2004) and many aspects of the host response are conserved (Cronan & Tobin, 2014; Torraca et al., 2015), making the zebrafish:*M. marinum* model suitable for drug screening for novel *Mtb* antibiotics in vivo.

1.3.1 Materials and reagents

1.3.1.1 Preparation of *Mycobacterium marinum* bacterial culture

7H9 liquid culture

Dissolve 4.7-g Middlebrook 7H9 Brother Base (Difco) in 500-mL distilled water. Add 4-mL 50% (v/v) glycerol. Bring to a final volume of 900 mL in distilled water. Autoclave. Store at room temperature.

1% Oleic Acid:

5 g oleic acid in 500-mL 0.2-N NaOH. Heat solutions to 55°C to dissolve completely. Store at −20°C in 50-mL aliquots. Thaw at 65°C prior to use with OADC.

10X oleic albumin dextrose catalase (OADC supplement)

Dissolve the following in 700-mL distilled water:

50-g BSA fraction V.
50-mL 1% oleic acid.
20-g dextrose.
8.5% NaCl.
Bring to 1 L final volume and filter sterilize using a 0.22-μm filter. Store at 4°C.

Polysorbate 80 (Tween 80)

7H9 Liquid Culture Medium

900-mL 7H9 Liquid Culture.
100-mL OADC stock.

2.5-mL 20% (v/v) Tween 80.
Store at 4°C.
Hygromycin B (Corning) use at final concentration of 50 μg/mL
M.marinum M strain containing pMSP12 promoter driving expression of a fluorophore
Freezing medium
180-mL 7H9 Liquid Culture.
20-mL 10X OADC.
27G needles with syringes (BD 1 mL TB Syringe 27G × ½)
5-μm filters (Pall Acrodisc 25-mm syringe filter with 5-μm versapor membrane)
Hemocytometer
Epifluorescent microscope
PCR tubes

1.3.1.2 IV infection of larvae with *Mycobacterium marinum* for systemic infection

TC Dish 100 Suspension (Sarstedt)
Forceps
5X Pronase
FemtoJet 4X (Eppendorf)
Micromanipulator (TriTech Research Inc)
Fire Polished Borosilicate Glass (Sutter Instruments #BF100-58-10)
Needle Puller (Sutter Instruments)
Microloader pipettes (Eppendorf)
Phenol Red Indicator (Sigma P0290)
7H9 medium (see above)
Glass Depression slide
Platinum wire manipulator
25X Tricaine-S (see Section 1.1.1)

1.3.1.3 Quantification of mycobacterial burden by fluorescence

Computer with image files
ImageJ (Schneider, Rasband, & Eliceiri, 2012)
Prism or other statistical analysis software

1.3.1.4 Testing the effects of drug exposure during mycobacterial infection

Chemicals (or chemical library)
Embryo medium (see Section 1.1.1)
Phenylthiourea (PTU) (Sigma–Aldrich): add to embryo medium at a final concentration of 45 μg/mL
Transfer pipet
Zebrafish larvae (see Sections 1.1.1 and 1.3.3)

1.3.2 Preparation of Mycobacterium marinum *bacterial culture (can be done in advance)*

1. Grow *M. marinum* in 7H9 Liquid Culture Medium +50-μg/mL Hygromycin B at 33°C until OD600 is near 1.
2. Pellet culture at 4000 × g for 20 min
3. Resuspend pellet in 1-mL Freezing Medium
4. Pass bacterial suspension through 27G needle 10 times into a 1.5-mL tube.
5. Spin tube(s) at 800 × g for 30 s to pellet large clumps.
6. Keep supernatant and pass through new 27G needle another 10 times.
7. Transfer suspension to syringe with a 5-μm filter for filtering.
8. Count fluorescent bacteria using a hemocytometer.
9. Adjust concentration to approximately 5×10^8 CFU/mL using freezing medium,
10. With this concentration, injecting 10 nL of thawed and diluted medium (see below Section 1.3.3) will deliver around 100 CFU per larva.
11. Aliquot volumes of 5 μL into PCR tubes and freeze at −80°C. These aliquots can be stored for at least 1 year without loss of viability.

1.3.3 Embryo collection

1. Same as Section 1.1.2.

1.3.4 Prepare larvae for infection with Mycobacterium marinum

1. Dechorionate 2 dpf larvae manually with forceps or with pronase (15-μL 5X pronase can be added to Petri dish of fish at 1 dpf and rinsed the next morning).
2. Anesthetize fish in 1.5X Tricaine-S.
3. Using a transfer pipette, move 15—30 larvae onto a glass depression slide.
4. Maneuver fish with platinum wire manipulator so that the ventral side of the larvae is facing the needle.

1.3.5 Prepare bacterial aliquots for infection

1. Dilute a 5-μL aliquot of bacteria in 15-μL 7H9 and 15-μL phenol red indicator.
2. Mix well.
3. Use microloader pipette to load about 3 μL of diluted bacteria into pulled borosilicate needle.
4. Load needle onto micromanipulator apparatus.
5. Break needle with forceps so that the needle will not clog but will still cleanly puncture through the larval skin.

1.3.6 Caudal vein infection of larvae with Mycobacterium marinum *for disseminated infection*

1. Use the micromanipulator to place the needle just on top of and next to the ventral fin of the larva, just posterior to the cloaca (Fig. 1).

2. Using platinum wire manipulator, push the fish onto the needle, using force on the dorsal side of the larvae. The needle should slip into the caudal vein; do not overpush! Overpushing results in trunk infection and muscle destruction.
3. Inject using a foot pedal or button on FemtoJet until a phenol red indicator is seen along the entire caudal vein.
4. Use the fluorescence microscope to ensure that the burden is at the desired level and consistency.
5. Change the needle every 50 fish or every hour, whichever is sooner.
6. Move infected fish to fresh E3 and incubate at 28.5°C for the desired times.
7. Change E3 daily, removing and recording the presence of any dead fish.

1.3.7 Quantification of mycobacterial burden by fluorescence

Most microscopy software programs have built-in analysis routines, but these are typically proprietary, and so we have outlined the simplest way to quantify infection burden by fluorescence using the free software package, ImageJ (Schneider et al., 2012).

1. Export images as TIFFs.
2. Open the fluorescence channel corresponding to bacteria.
3. Remove background signal using threshold tool (image → adjust → threshold). Do this enough to cut out yolk autofluorescence.
4. Outline the zebrafish using the selector tool to exclude fluorescent debris. If yolk autofluorescence is too high to cut out, pick a consistent designated area and size to quantify (see Fig. 4).
5. Quantify the area of the infected fish or region of interest (analyze → analyze particles).
6. ImageJ will record information for each selection in a pop-up table with an information including the area of fluorescence and number of fluorescent foci.
7. Insert this information into the graphing and statistical programs of your choice. See Fig. 4 for typical results.

1.3.8 Modeling the effects of drug exposure during mycobacterial infection

1. Place three to four systemically infected larvae (see Section 1.3.6) and one to two uninfected larvae in each well of a glass bottom 96-well plate with up to 200 μL of fish water.
2. Add chemical to each well to an appropriate concentration (e.g., 1–20 μM) for each chemical. Dose should be determined empirically, but an initial concentration of 5 μM is usually a good starting point. Maximum tolerated doses (MTDs) can be determined later, once a drug of interest has been identified.
3. Image the fish and infections prior to drug treatment for initial burden.
4. Image daily. Many microscopes have software to automate the imaging of 96-well plates.
5. If possible, carefully change water and chemical daily or every other day.
6. Remove any dead fish and record mortalities, taking note of morphological changes and toxicity.

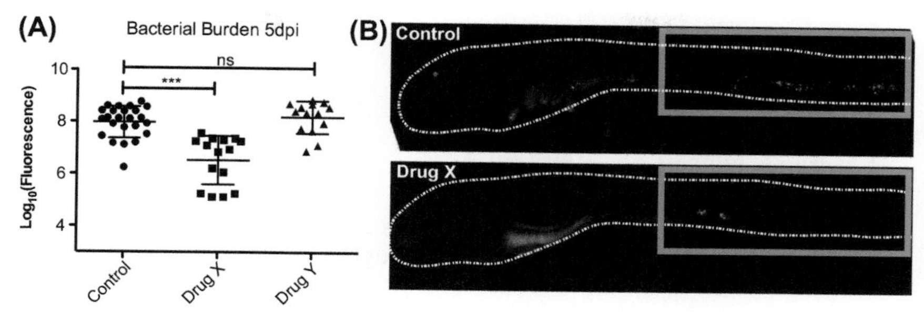

FIGURE 4

Representative data from drug treatments during mycobacterial infection. Fish have been soaking in E3 containing Drug X, Drug Y, or DMSO control for 5 days. (A) GraphPad Prism Column graph with each dot representing the fluorescent area and intensity of *M. marinum*: cerulean. The 1–1.5 $\log_{10}$ difference seen from treatment with Drug X for 5 days is a reasonable reduction in burden. Drug Y is ineffective at reducing bacterial burden at its current dose and treatment time. Statistical analysis performed in GraphPad Prism Version 5.0c: $^{***}p < .0001$, One-way analysis of variance with Tukey's multiple comparison test. *Error bars* are mean ± SD. (B) Representative images of 5 dpi larvae taken with Zeiss Axio Observer Z.1 microscope. *White dotted lines* outline the whole larvae and *red solid boxes* outline the region quantified with ImageJ software.

7. Use fluorescence intensity and area to determine effects of the drug on mycobacterial growth (see Section 1.3.7). You may also look for effects on granuloma size and formation (Davis & Ramakrishnan, 2009), macrophage recruitment and dissemination of bacteria (Clay et al., 2007) changes in vasculature (Oehlers et al., 2015), involvement of other immune cell types (Yang et al., 2012), or transcriptome changes in the zebrafish (van der Sar, Spaink, Zakrzewska, Bitter, & Meijer, 2009).

1.4 ASSESSING SURVIVAL DURING ADULT MYCOBACTERIAL INFECTION

Use of the adult zebrafish infection model enables analysis of infection in the context of both innate and adaptive immunities. Below is a protocol for mycobacterial infection in adult zebrafish modified from Swaim et al. (2006), which should be generalizable to other infections. Some common uses of adult infections include postinfection histology, CLARITY/PACT (Cronan et al., 2015), drug treatments (Oehlers et al., 2015), or testing of host/pathogen susceptibility or survival genes (Elks et al., 2014; Pagan et al., 2015; Tobin et al., 2010).

1.4.1 Materials and reagents

1.4.1.1 Prepare fish for injection

Zebrafish aged 40 days or older

25X Tricaine-S (see Section 1.1.1)
250-mL beakers

1.4.1.2 Prepare bacterial dilution

M. marinum aliquots (see Section 1.3.2)
Parafilm
Phosphate buffered saline (PBS), pH 7.4
A 28°C incubator with a 14:10-h light:dark cycle

1.4.1.3 Inject fish

Blunt-end tweezers
Foam holder:
Cut a 3-cm by 1-cm hole, about 0.25-cm deep into a piece of sponge or other foam. This foam piece should fit into a petri dish, for easy clean up.
31G 8-mm 3/10-mL insulin needle (BD)

1.4.1.4 Continued care and survival studies

500-mL glass beakers
Serological pipettes and Pipetman
Large carboy for system fish water
2–4-L plastic beakers
Bleach

1.4.2 Prepare fish for injection

Note that working with adult fish as well as larvae requires approval, in accordance with procedures, policies, and guidelines of an Institutional Animal Care and Use Committee.

1. Anesthetize five fish at a time
   a. Add 2 mL of 25X tricaine to 100-mL system fish water.
   b. Wait until fish are fully anesthetized (stop swimming), but work quickly with each fish to ensure survival.

1.4.3 Prepare bacterial dilution

1. Dilute *M. marinum* stock to desired concentration. We typically try to infect with between 150 and 1000 CFU per fish, depending on the experimental design.
   a. Ex: One 5-μL aliquot (typically 2×10^8 CFU/mL) should be diluted 1: 10,000 in 1X PBS.
2. Pipette 15-μL drops of this diluted bacterial medium onto parafilm.

1.4.4 Inject fish

1. Using blunt tweezers, pick up an anesthetized fish and place on the foam holder (Note: a foam holder can be any piece of material you can cut to hold a fish in place, see Fig. 1D).

2. Restrain the fish gently with one finger of nondominant hand.
3. Using small insulin needle (30G), inject 15 μL of bacterial medium (from parafilm drop) into the intraperitoneal (IP) space of the fish. You should see the IP space puff up a bit. See Fig. 1D for diagram.
4. Place fish in new 500-mL beaker with fresh system fish water.
5. Watch closely to ensure that the fish survive tricaine treatment and injection.

1.4.5 Continued care and survival studies

1. Keep fish in 500-mL beakers in 28.5°C incubator with no more than five fish per 200-mL fish water.
2. Change water completely every 2 days using a 50-mL serological pipette, cleaning out excrement every day. Discarded water should be disinfected using 10% bleach.
3. Feed once daily, which is typically less than fish fed in a flow-through system.
4. Monitor daily for survival, using proper judgment for morbidity and subsequent euthanasia.
5. Euthanized fish remains should be fully disinfected in 10% bleach unless being kept for genotyping, sectioning, or other purposes.

1.5 QUANTIFYING DISSEMINATED INFLUENZA A VIRUS INFECTIONS

1.5.1 Materials and reagents

1.5.1.1 Embryo collection

Same as described in Section 1.1.1

1.5.1.2 Preparation of influenza A virus

Influenza A/PR/8/34 (H1N1) virus (Charles River 490710) or Influenza A X-31, A/Aichi/68 (H3N2) (Charles River 490715)

1.5.1.3 Injection of influenza A into zebrafish embryo to establish infection

Borosilicate glass with filament (Sutter Instrument BF120-69-10)
Flaming/Brown Micropipette Puller (Sutter Instrument P-97)
Microscope Stage Calibration Slide (AmScope MR095)

1.5.1.4 Viral titer TCID50 assays

CELLSTAR 96-well tissue culture plates, sterile F-bottom with lid (Greiner Bio-One 655180)
MDCK cells (ATCC CCL-34)
MDCK cell culture medium
 Dulbecco's modified Eagle's medium (DMEM), high glucose (Gibco 11965-092)
 10% heat-inactivated fetal bovine serum (FBS) (Gibco 10082-147)

100-U/mL penicillin and 100 μg/mL streptomycin (Gibco 15140-122)
HEPES, 1 M (Gibco 15630-080)
Cordless pestle motor (VWR 47747-370)
Disposable pestles (VWR 47747-360)
Eagle's minimum essential medium (Gibco 11095-080)

1.5.2 Embryo collection

1. Collect age-matched embryos as described in Section 1.1.2.

1.5.3 Preparation of influenza A virus

1. Dilute virus to 3.3×10^6 EID50/mL in ice-cold Dulbecco's PBS supplemented with 0.25% phenol red to aid in visualization.

1.5.4 Injection of influenza A into zebrafish embryo to establish infection

1. Prepare embryos on agar plate for microinjections with virus as described in Section 1.1.4.
2. Pull injection needles using borosilicate glass with filaments and a Sutter micropipette puller.
3. Gently clip the needle with Dumont #5 forceps and test the injection volume by measuring on a calibration micrometer. Continue to clip needle until desired injection volume is achieved (typically 1.5 nL for 5×10^3 EID50).
4. Inject 1.5 nL of influenza A virus into the duct of Cuvier to initiate a systemic infection.
5. Transfer injected embryos into a Petri dish containing 50-mL fresh egg water and incubate at 33°C. Dead embryos should be removed every 24 h beginning at 24 hpi.

1.5.5 Viral titer TCID50 assays

1. At desired times post-infection (e.g., 0, 24, 48, 72, and 96 hpi), transfer the desired number of embryos to a clean Petri dish containing sterile egg water to wash. Typically, 20 fish per biological replicate should be used. Repeat two more times to thoroughly wash the infected embryos.
2. Collect 20 fish per biological replicate in 1.5-mL microcentrifuge tubes.
3. Thoroughly remove water and replace with 200-μL DMEM supplemented with sterile 25-mM HEPES, 0.2% bovine serum albumin, 100-U/mL penicillin and 100-μg/mL streptomycin.
4. Gently homogenize on ice with a motorized mortar and pestle, taking care not to overheat the samples or introduce bubbles.
5. Centrifuge homogenates at $15{,}000 \times g$ at 4°C for 10 min.
6. Transfer supernatants to prechilled microcentrifuge tubes. Cap and freeze samples at −80°C.
7. 24 hours prior to initiating the assay, trypsinize cells and resuspend to a concentration of 1.5×10^5 cells/mL in DMEM supplemented with 10%

heat-inactivated fetal bovine serum, 100-U/mL penicillin, 100-μg/mL streptomycin, and 2-mM glutamine.

8. Plate approximately 1.5×10^4 cells (100 μL) per well in each well of a 96-well plate.
9. On the day of infection, ensure that cells are ~70% confluent prior to initiating the assay.
10. At the time of assay, rapidly thaw virus and establish serial half-log dilutions in unsupplemented Eagle's minimum essential medium, making sure to avoid bubbling.
11. Gently add 100 μL of the virus dilution per well and incubate for 2 h at 37°C with 5% CO_2.
12. Add 100 μL of unsupplemented Eagle's minimum essential medium to the wells.
13. Seal plate with adhesive film and incubate for 3 days at 37°C with 5% CO_2.
14. Calculate the TCID50/mL according to the Reed—Muench method.

1.5.6 Representative results

As shown in Fig. 5, zebrafish infected with influenza A virus develop a profound disease phenotype characterized by edema in the pericardium and yolk sac that worsens with time. Zebrafish also demonstrate evidence of lordosis and alterations in pigmentation, as well as craniofacial and eye deformities.

2. METHODS FOR LOCALIZED BACTERIAL AND VIRAL INFECTIONS

One of the primary measures of immune response is infiltration of immune cells. Innate immune cells such as neutrophils (Lieschke, Oates, Crowhurst, Ward, & Layton, 2001; Rocha et al., 2016) and macrophages (Torraca et al., 2015; Wiegertjes, Wentzel, Spaink, Elks, & Fink, 2016) often respond to infectious and inflammatory cues within minutes after exposure to a pathogen. The number of immune cells present during an infection is often a sign of the strength of the immune response. Using zebrafish lines in which immune cells of interest are labeled, macrophages (Ellett et al., 2011; Walton, Cronan, Beerman, & Tobin, 2015), neutrophils (Hall, Flores, Storm, Crosier, & Crosier, 2007; Lieschke et al., 2001), and a mixture of all of these populations (Lieschke et al., 2002) can be enumerated at initial and final time points and even followed using time-lapse microscopy to determine characteristics like morphology, velocity, and directionality. Researchers can utilize zebrafish lines lacking specific cell populations, including WHIM (Walters, Green, Surfus, Yoo, & Huttenlocher, 2010) or nitroreductase (Curado et al., 2007) lines, ablating the activity of neutrophils or any cell of interest, respectively. It is important to note, however, that the chemical agent metronidazole, used to ablate nitroreductase-expressing cells, is also an antibiotic and may confound some infection studies. Using a cell-specific promoter driving

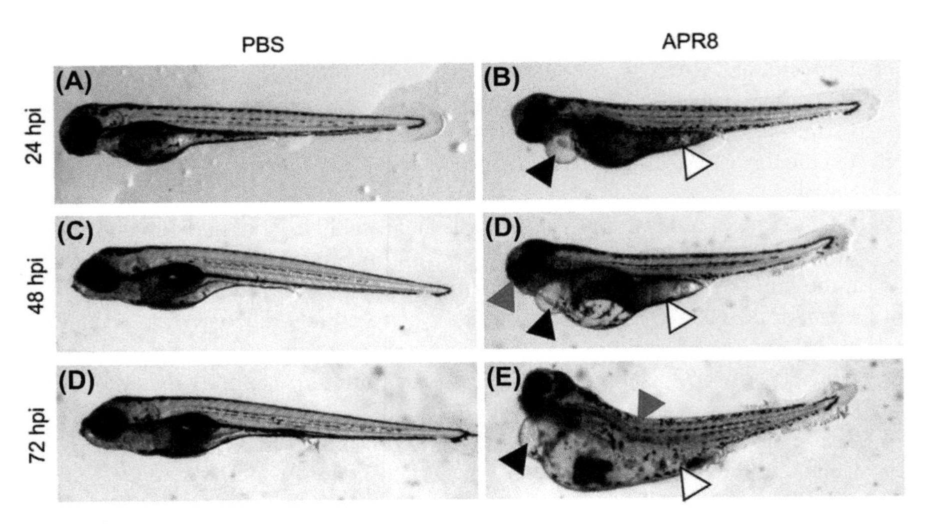

FIGURE 5

Zebrafish infected by influenza A exhibit a profound disease phenotype. Brightfield microscopy of live zebrafish, oriented on their sides, anterior to the left, dorsal to the top, magnification 25×. Representative PBS-injected controls at 24 (A), 48 (C), and 72 hpi (E) exhibits the hallmarks of normal development. In contrast, zebrafish infected with influenza A (B, D, E) exhibit evidence of a progressive disease that worsens with time. At 24 hpi (B), zebrafish infected with influenza A exhibit edema in the pericardium (*black arrowhead*) and yolk sac extension (*white arrowhead*). At 48 hpi (D), zebrafish infected with influenza A exhibit worsening edema and evidence of craniofacial deformities. By 72 hpi (F), zebrafish infected with influenza A demonstrate a generalized edema phenotype with lordosis, or spinal curvature.

This figure was provided as a courtesy by Michelle Goody (University of Maine) and originally appeared in Gabor K. A., Goody M. F., Mowel W. K., Breitbach M. E., Gratacap R. L., Witten P. E., & Kim C. H. (2014). Influenza A virus infection in zebrafish recapitulates mammalian infection and sensitivity to anti-influenza drug treatment. Disease Models and Mechanisms, *7(11), 1227–1237. http://dx.doi.org/10.1242/dmm.014746. It is reprinted under the Creative Commons Attribution 4.0 International Public License (https://creativecommons.org/licenses/by/4.0/legalcode).*

mammalian Trpv1 is a new method for ablating specific cells types using a chemical without antibiotic properties (Beerman et al., 2015; Chen, Chiu, McArthur, Fetcho, & Prober, 2016).

2.1 EPITHELIAL INFLUENZA A INFECTION LOCALIZED TO THE SWIMBLADDER

Zebrafish possess sialic acids with α-2, 6-linkages that enable their cells to bind influenza A virus hemagglutinin, which enables their infection. Recently, we demonstrated that influenza A virus can infect and replicate within zebrafish embryos and, in the case of systemic, disseminated infections, cause a pathological phenotype leading to death (Gabor et al., 2014). In addition, we demonstrated that we could model a localized, epithelial influenza A infection in the swimbladder, a structure that has been shown to bear notable similarities to the human lung (Perry, Wilson, Straus, Harris, & Remmers, 2001; Winata et al., 2009). The ability to model localized influenza A infections in the zebrafish enables investigations into host immune cell dynamics in the context of this important viral infection.

2.1.1 Materials and reagents

2.1.1.1 Embryo collection

Same as described in Section 1.1.1

2.1.1.2 Preparation of influenza A virus

Same as described in Section 1.4.1
Influenza A virus (NS1-GFP) (Manicassamy et al., 2010)

2.1.1.3 Injection of zebrafish larvae with influenza A to establish localized, epithelial infections

Same as described in Section 1.4.1

2.1.2 Embryo collection

1. Collected age-matched embryos as described in Section 1.1.2.
2. Maintain embryos at 28°C until 5-d postfertilization, when swimbladder is fully inflated. Remove dead embryos and larvae each day and change water.

2.1.3 Preparation of influenza A virus

1. Pull injection needles using borosilicate glass with filaments and a Sutter micropipette puller.
2. Dilute NS1-GFP influenza A virus to $\sim 1.5 \times 10^2$ plaque-forming units (PFUs) per nL in ice-cold Dulbecco's PBS supplemented with 0.25% phenol red to aid in visualization.
3. Load microinjection needle and prepare so that 4.0 nL is injected per pulse.

2.1.4 Injection of influenza A into zebrafish larvae to establish localized, epithelial infections

1. Puncture the swimbladder and inject toward the posterior with 4.0 nL of influenza A virus dilution ($\sim$600 PFU/larva). Also inject control fish with sterile PBS including 0.25% phenol red. The injected volume should collect at the posterior and the air bubble should be displaced anteriorly.
2. Maintain zebrafish at 33°C and track the course of the infection at regular intervals. Change the water daily.

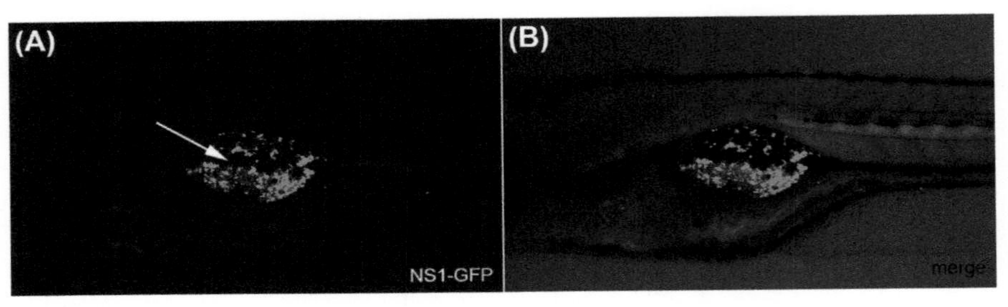

FIGURE 6

Representative data showing the infection of the zebrafish swimbladder with influenza A as a model for localized epithelial infection. At 5 dpf, zebrafish larvae were anesthetized and injected in the posterior swimbladder with the GFP-labeled influenza A virus NS1-GFP (Manicassamy et al., 2010). Larvae are side-mounted, with anterior to the left, dorsal to the top, magnification 10×. (A) Green fluorescence (NS1-GFP) and (B) green fluorescence, brightfield merge (merge) at 24 hpi. Punctate green fluorescence indicative of an active, replicating influenza A infection is observed in the swimbladder region (*white arrow*).

2.1.5 Representative results

Infections of the swimbladder are intended to model human lung infections. As shown in Fig. 6, zebrafish locally infected in the swimbladder with a fluorescent strain of influenza A virus develop an infection. This strategy is useful to study localized, epithelial infections that model those observed in the human lung.

2.2 LOCALIZED INFECTION IN THE HINDBRAIN AND OTIC VESICLE

Leukocyte migration is often used to understand the extent of inflammatory cues and wound response. It is therefore important to ensure that the site of injection is similarly injured in all infections and to include a mock-injected fish to normalize for the initial injury caused by the needle. To monitor inflammatory cues in vivo, in real time, specific transgenic reporter lines have been developed that allow live monitoring of tumor necrosis factor (TNF) transcription (Marjoram et al., 2015).

2.2.1 Materials

2.2.1.1 Embryo collection

Same as described in Section 1.1.1

To analyze leukocyte influx, there are many zebrafish lines available with labeled leukocytes (see Section 2). If these lines are unavailable, other alternatives include the use of Sudan Black staining (Le Guyader et al., 2008) for neutrophils or Neutral Red for macrophages (Herbomel, Thisse, & Thisse, 2001).

2.2.1.2 Preparation of mycobacteria

Same as described in Section 1.3.1

2.2.1.3 Infection of larvae

Same as described in Section 1.3.1

2.2.1.4 Optional

96-well glass bottom plates (Greiner Bio One)
0.75% low-melt agarose (BP165-25)
Prepared from stock 1%, diluted with tricaine E3
MatTek dishes (P35G-1.5-20-C)

2.2.2 Infecting hindbrain ventricle

1. Anesthetize 2 dpf zebrafish larvae. At 2 dpf, the zebrafish larvae have large and easily accessible hindbrains.
2. Place about five larvae on glass depression slide, with enough water to fill the depression.
3. Position fish accordingly (Fig. 1). There are three different orientations to try based on ease of injection, handedness of individual, and sharpness of the needle. In all cases, hold larvae still on the opposite side of the needle with a platinum wire manipulator, using the wire manipulator to push the fish onto the needle, using the area between the head and the heart as a point of contact. The phenol red indicator will verify that the bacterial suspension stays in the point of interest.
4. Inject using proper time and pressure. It helps to have constant pressure down as low as possible to ensure that no bacteria get into the area surrounding the injection site. For a fairly fine pointed needle, we typically use 0.10 s at 75 hPa on the FemtoJet 4X. These specifications must be determined empirically for needle sizes and the desired burden.
5. Pipette the injected fish into fresh PTU-containing embryo water. Wait for the appropriate amount of time to measure leukocyte migration. Typically, this is between 20 min and 4 h.

2.2.3 Infecting otic vesicle

1. Anesthetize 3-dpf zebrafish larvae. The otic vesicle (ear) is more easily accessed at 3 dpf. At this time point, the head has flattened a bit on the sides and the vesicle is larger and more easily penetrated.
2. Place about five larvae on glass depression slide with enough water to fill depression.
3. Position the fish accordingly (Fig. 1). It is important to consistently inject the same side of the head for imaging purposes. Align the larva so that the dorsal side is facing the needle and use the platinum wire manipulator to push the larva under and into the needle. The needle may come into contact with the otoliths, but we have not found that this influences recruitment.

2.2.4 Time course imaging the influx of leukocytes or the efflux of a pathogen

Given the transparency and size of zebrafish larvae, in-vivo imaging is easily performed in infected zebrafish larvae. If the pathogen of interest is fluorescently labeled, one can examine both initial recruitment of leukocytes and the dissemination of pathogen outside the infected compartment during the course of infection. After infecting either localized region, wait for a standardized period of time. Depending on the infection, this time may be 20 min, 4 h, 3 days or somewhere in between. Using a fluorescence microscope, analyze fluorescent foci found outside the localized compartment of initial infection.

1. Follow procedure for infection in Section 2.2.2 or Section 2.2.3.
2. Make 0.75% low-melt agarose
   a. 1% stock in fish water
      i. Weigh out 1 g of low-melt agarose (BP165-25) and add to 100-mL E3 in a round medium bottle. Microwave to dissolve completely. Can be stored at room temperature and reheated for 3 months, with lid screwed tightly.
   b. Dilute 1–0.75% in E3 containing PTU and tricaine to have a final concentration of 0.8X tricaine and 1.5X PTU.
   c. Keep at 42°C or above until use.
3. Pipette one to three infected larvae into 96-well plate (or 5–10 in MakTek dish).
4. Remove excess water.
5. Add 40-μL 0.75% agarose and quickly align the larvae for optimal imaging, keeping in mind the orientation of your microscope and injection site.
6. Wait 5 min for agarose to harden.
7. Add 80-μL PTU + tricaine water over the agarose embedded fish.
8. Set up imaging parameters
   a. For *M. marinum*, we typically image for 2–12 h, taking images in macrophage-fluorescent and *M. marinum*-fluorescent channels every 10 min, using a few Z-stacks to capture the area of interest. Turn down the laser intensity to limit the photobleaching of the larvae.
9. If using 96-well dish, set up multiple-point imaging using a proper software.
10. These images can be quantified for a variety of purposes that are beyond the scope of this methods chapter.

CONCLUSION

In recent years, an array of zebrafish models for host–pathogen interactions have been developed in an effort to address important questions related to systemic and localized responses to infections. These efforts will undoubtedly bear additional fruit as more bacterial, viral, and fungal models are generated. The power of these models has been amplified by the creation of myriad molecular and cellular tools that exploit many of the inherent advantages of the zebrafish model. By leveraging these advantages, it will be possible to gain insights into biological problems that are relatively inaccessible in other model systems.

REFERENCES

Beerman, R. W., Matty, M. A., Au, G. G., Looger, L. L., Choudhury, K. R., Keller, P. J., & Tobin, D. M. (2015). Direct in vivo manipulation and imaging of calcium transients in neutrophils identify a critical role for leading-edge calcium flux. *Cell Reports, 13*(10), 2107–2117. http://dx.doi.org/10.1016/j.celrep.2015.11.010. pii:S2211-1247(15)01310-8.

Berg, R. D., Levitte, S., O'Sullivan, M. P., O'Leary, S. M., Cambier, C. J., Cameron, J., … Ramakrishnan, L. (2016). Lysosomal disorders drive susceptibility to tuberculosis by compromising macrophage migration. *Cell, 165*(1), 139–152. http://dx.doi.org/10.1016/j.cell.2016.02.034. pii:S0092-8674(16)30136-2.

Bernut, A., Dupont, C., Sahuquet, A., Herrmann, J. L., Lutfalla, G., & Kremer, L. (2015). Deciphering and imaging pathogenesis and cording of *Mycobacterium abscessus* in zebrafish embryos. *Journal of Visualized Experiments*, (103)http://dx.doi.org/10.3791/53130.

Bojarczuk, A., Miller, K. A., Hotham, R., Lewis, A., Ogryzko, N. V., Kamuyango, A. A., … Johnston, S. A. (2016). *Cryptococcus neoformans* intracellular proliferation and capsule size determines early macrophage control of infection. *Scientific Reports, 6*, 21489. http://dx.doi.org/10.1038/srep21489.

Brannon, M. K., Davis, J. M., Mathias, J. R., Hall, C. J., Emerson, J. C., Crosier, P. S., … Moskowitz, S. M. (2009). *Pseudomonas aeruginosa* type III secretion system interacts with phagocytes to modulate systemic infection of zebrafish embryos. *Cellular Microbiology, 11*(5), 755–768. http://dx.doi.org/10.1111/j.1462-5822.2009.01288.x.

Brothers, K. M., Newman, Z. R., & Wheeler, R. T. (2011). Live imaging of disseminated candidiasis in zebrafish reveals role of phagocyte oxidase in limiting filamentous growth. *Eukaryotic Cell, 10*(7), 932–944. http://dx.doi.org/10.1128/EC.05005-11. pii: EC.05005-11.

Brothers, K. M., & Wheeler, R. T. (2012). Non-invasive imaging of disseminated candidiasis in zebrafish larvae. *Journal of Visualized Experiments*, (65)http://dx.doi.org/10.3791/4051.

Brudal, E., Ulanova, L. S., O Lampe, E., Rishovd, A. L., Griffiths, G., & Winther-Larsen, H. C. (2014). Establishment of three *Francisella* infections in zebrafish embryos at different temperatures. *Infection and Immunity, 82*(6), 2180–2194. http://dx.doi.org/10.1128/IAI.00077-14.

Burgos, J. S., Ripoll-Gomez, J., Alfaro, J. M., Sastre, I., & Valdivieso, F. (2008). Zebrafish as a new model for herpes simplex virus type 1 infection. *Zebrafish, 5*(4), 323–333. http://dx.doi.org/10.1089/zeb.2008.0552. pii:10.1089/zeb.2008.0552.

Chen, S., Chiu, C. N., McArthur, K. L., Fetcho, J. R., & Prober, D. A. (2016). TRP channel mediated neuronal activation and ablation in freely behaving zebrafish. *Nature Methods, 13*(2), 147–150. http://dx.doi.org/10.1038/nmeth.3691.

Clatworthy, A. E., Lee, J. S., Leibman, M., Kostun, Z., Davidson, A. J., & Hung, D. T. (2009). *Pseudomonas aeruginosa* infection of zebrafish involves both host and pathogen determinants. *Infection and Immunity, 77*(4), 1293–1303. http://dx.doi.org/10.1128/iai.01181-08.

Clay, H., Davis, J. M., Beery, D., Huttenlocher, A., Lyons, S. E., & Ramakrishnan, L. (2007). Dichotomous role of the macrophage in early *Mycobacterium marinum* infection of the zebrafish. *Cell Host and Microbe, 2*(1), 29–39. http://dx.doi.org/10.1016/j.chom.2007.06.004.

Cosma, C. L., Klein, K., Kim, R., Beery, D., & Ramakrishnan, L. (2006). *Mycobacterium marinum* Erp is a virulence determinant required for cell wall integrity and intracellular survival. *Infection and Immunity, 74*(6), 3125–3133. http://dx.doi.org/10.1128/IAI.02061-05.

Cronan, M. R., Rosenberg, A. F., Oehlers, S. H., Saelens, J. W., Sisk, D. M., Jurcic Smith, K. L., … Tobin, D. M. (2015). CLARITY and PACT-based imaging of adult zebrafish and mouse for whole-animal analysis of infections. *Disease Models and Mechanisms, 8*(12), 1643–1650. http://dx.doi.org/10.1242/dmm.021394.

Cronan, M. R., & Tobin, D. M. (2014). Fit for consumption: zebrafish as a model for tuberculosis. *Disease Models and Mechanisms, 7*(7), 777–784. http://dx.doi.org/10.1242/dmm.016089.

Curado, S., Anderson, R. M., Jungblut, B., Mumm, J., Schroeter, E., & Stainier, D. Y. (2007). Conditional targeted cell ablation in zebrafish: a new tool for regeneration studies. *Developmental Dynamics, 236*(4), 1025–1035. http://dx.doi.org/10.1002/dvdy.21100.

Davis, J. M., Clay, H., Lewis, J. L., Ghori, N., Herbomel, P., & Ramakrishnan, L. (2002). Real-time visualization of *Mycobacterium*-macrophage interactions leading to initiation of granuloma formation in zebrafish embryos. *Immunity, 17*(6), 693–702.

Davis, J. M., & Ramakrishnan, L. (2009). The role of the granuloma in expansion and dissemination of early tuberculous infection. *Cell, 136*(1), 37–49. http://dx.doi.org/10.1016/j.cell.2008.11.014.

Ding, C. B., Zhang, J. P., Zhao, Y., Peng, Z. G., Song, D. Q., & Jiang, J. D. (2011). Zebrafish as a potential model organism for drug test against hepatitis C virus. *PLoS One, 6*(8), e22921. http://dx.doi.org/10.1371/journal.pone.0022921. pii:PONE-D-11-02951.

Ding, C. B., Zhao, Y., Zhang, J. P., Peng, Z. G., Song, D. Q., & Jiang, J. D. (2015). A zebrafish model for subgenomic hepatitis C virus replication. *International Journal of Molecular Medicine, 35*(3), 791–797. http://dx.doi.org/10.3892/ijmm.2015.2063.

Elks, P. M., van der Vaart, M., van Hensbergen, V., Schutz, E., Redd, M. J., Murayama, E., … Meijer, A. H. (2014). Mycobacteria counteract a TLR-mediated nitrosative defense mechanism in a zebrafish infection model. *PLoS One, 9*(6), e100928. http://dx.doi.org/10.1371/journal.pone.0100928. pii:PONE-D-14-04677.

Ellett, F., Pase, L., Hayman, J. W., Andrianopoulos, A., & Lieschke, G. J. (2011). mpeg1 promoter transgenes direct macrophage-lineage expression in zebrafish. *Blood, 117*(4), e49–56. http://dx.doi.org/10.1182/blood-2010-10-314120.

Engering, A., Hogerwerf, L., & Slingenbergh, J. (2013). Pathogen-host-environment interplay and disease emergence. *Emerging Microbes and Infections, 2*, e5.

Fehr, A., Eshwar, A. K., Neuhauss, S. C., Ruetten, M., Lehner, A., & Vaughan, L. (2015). Evaluation of zebrafish as a model to study the pathogenesis of the opportunistic pathogen *Cronobacter turicensis*. *Emerging Microbes and Infections, 4*(5), e29. http://dx.doi.org/10.1038/emi.2015.29.

Ferrer-Navarro, M., Planell, R., Yero, D., Mongiardini, E., Torrent, G., Huedo, P., … Daura, X. (2013). Abundance of the quorum-sensing factor Ax21 in four strains of

Stenotrophomonas maltophilia correlates with mortality rate in a new zebrafish model of infection. *PLoS One, 8*(6), e67207. http://dx.doi.org/10.1371/journal.pone.0067207.

Gabor, K. A., Goody, M. F., Mowel, W. K., Breitbach, M. E., Gratacap, R. L., Witten, P. E., & Kim, C. H. (2014). Influenza A virus infection in zebrafish recapitulates mammalian infection and sensitivity to anti-influenza drug treatment. *Disease Models and Mechanisms, 7*(11), 1227–1237. http://dx.doi.org/10.1242/dmm.014746.

Gagnaire, B., Bado-Nilles, A., & Sanchez, W. (2014). Depleted uranium disturbs immune parameters in zebrafish, *Danio rerio*: an ex vivo/in vivo experiment. *Archives of environmental contamination and toxicology*. http://dx.doi.org/10.1007/s00244-014-0022-x.

Gaj, T., Gersbach, C. A., & Barbas, C. F., 3rd (2013). ZFN, TALEN, and CRISPR/Cas-based methods for genome engineering. *Trends in Biotechnology, 31*(7), 397–405. http://dx.doi.org/10.1016/j.tibtech.2013.04.004.

Ge, R., Zhou, Y., Peng, R., Wang, R., Li, M., Zhang, Y., … Wang, C. (2015). Conservation of the STING-mediated cytosolic DNA sensing pathway in zebrafish. *Journal of Virology, 89*(15), 7696–7706. http://dx.doi.org/10.1128/JVI.01049-15. pii:JVI.01049-15.

Hall, C., Flores, M. V., Storm, T., Crosier, K., & Crosier, P. (2007). The zebrafish lysozyme C promoter drives myeloid-specific expression in transgenic fish. *BMC Developmental Biology, 7*, 42. http://dx.doi.org/10.1186/1471-213X-7-42.

Harvie, E. A., Green, J. M., Neely, M. N., & Huttenlocher, A. (2013). Innate immune response to *Streptococcus iniae* infection in zebrafish larvae. *Infection and Immunity, 81*(1), 110–121. http://dx.doi.org/10.1128/IAI.00642-12.

Herbomel, P., Thisse, B., & Thisse, C. (2001). Zebrafish early macrophages colonize cephalic mesenchyme and developing brain, retina, and epidermis through a M-CSF receptor-dependent invasive process. *Developmental Biology, 238*(2), 274–288. http://dx.doi.org/10.1006/dbio.2001.0393.

Hermann, A. C., & Kim, C. H. (2005). Effects of arsenic on zebrafish innate immune system. *Marine Biotechnology, 7*(5), 494–505. http://dx.doi.org/10.1007/s10126-004-4109-7.

Howe, K., Clark, M. D., Torroja, C. F., Torrance, J., Berthelot, C., Muffato, M., … Stemple, D. L. (2013). The zebrafish reference genome sequence and its relationship to the human genome. *Nature, 496*(7446), 498–503. http://dx.doi.org/10.1038/nature12111.

Jheng, Y. H., Lee, L. H., Ting, C. H., Pan, C. Y., Hui, C. F., & Chen, J. Y. (2015). Zebrafish fed on recombinant *Artemia* expressing epinecidin-1 exhibit increased survival and altered expression of immunomodulatory genes upon *Vibrio vulnificus* infection. *Fish and Shellfish Immunology, 42*(1), 1–15. http://dx.doi.org/10.1016/j.fsi.2014.10.019. pii:S1050-4648(14)00393-3.

Jin, Y., Chen, R., Liu, W., & Fu, Z. (2010). Effect of endocrine disrupting chemicals on the transcription of genes related to the innate immune system in the early developmental stage of zebrafish (*Danio rerio*). *Fish and shellfish immunology, 28*(5–6), 854–861. http://dx.doi.org/10.1016/j.fsi.2010.02.009.

Jovanovic, B., Ji, T., & Palic, D. (2011). Gene expression of zebrafish embryos exposed to titanium dioxide nanoparticles and hydroxylated fullerenes. *Ecotoxicology and Environmental Safety, 74*(6), 1518–1525. http://dx.doi.org/10.1016/j.ecoenv.2011.04.012.

Keiter, S., Baumann, L., Farber, H., Holbech, H., Skutlarek, D., Engwall, M., & Braunbeck, T. (2012). Long-term effects of a binary mixture of perfluorooctane sulfonate (PFOS) and bisphenol A (BPA) in zebrafish (*Danio rerio*). *Aquatic Toxicology, 118–119*, 116–129. http://dx.doi.org/10.1016/j.aquatox.2012.04.003.

Kim, K. T., & Tanguay, R. L. (2013). Integrating zebrafish toxicology and nanoscience for safer product development. *Green Chemistry, 15*(4), 872–880. http://dx.doi.org/10.1039/C3GC36806H.

Knox, B. P., Deng, Q., Rood, M., Eickhoff, J. C., Keller, N. P., & Huttenlocher, A. (2014). Distinct innate immune phagocyte responses to *Aspergillus fumigatus* conidia and hyphae in zebrafish larvae. *Eukaryotic Cell, 13*(10), 1266–1277. http://dx.doi.org/10.1128/ec.00080-14.

Lage, C. R., Nayak, A., & Kim, C. H. (2006). Arsenic ecotoxicology and innate immunity. *Integrative and Comparative Biology, 46*(6), 1040–1054. http://dx.doi.org/10.1093/icb/icl048.

Le Guyader, D., Redd, M. J., Colucci-Guyon, E., Murayama, E., Kissa, K., Briolat, V., … Herbomel, P. (2008). Origins and unconventional behavior of neutrophils in developing zebrafish. *Blood, 111*(1), 132–141. http://dx.doi.org/10.1182/blood-2007-06-095398.

van Leeuwen, L. M., van der Kuip, M., Youssef, S. A., de Bruin, A., Bitter, W., van Furth, A. M., & van der Sar, A. M. (2014). Modeling tuberculous meningitis in zebrafish using *Mycobacterium marinum. Disease Models and Mechanisms, 7*(9), 1111–1122. http://dx.doi.org/10.1242/dmm.015453.

Li, Y. J., & Hu, B. (2012). Establishment of multi-site infection model in zebrafish larvae for studying *Staphylococcus aureus* infectious disease. *Journal of Genetics and Genomics, 39*(9), 521–534. http://dx.doi.org/10.1016/j.jgg.2012.07.006.

Lieschke, G. J., & Currie, P. D. (2007). Animal models of human disease: zebrafish swim into view. *Nature Reviews. Genetics, 8*(5), 353–367. http://dx.doi.org/10.1038/nrg2091.

Lieschke, G. J., Oates, A. C., Crowhurst, M. O., Ward, A. C., & Layton, J. E. (2001). Morphologic and functional characterization of granulocytes and macrophages in embryonic and adult zebrafish. *Blood, 98*(10), 3087–3096.

Lieschke, G. J., Oates, A. C., Paw, B. H., Thompson, M. A., Hall, N. E., Ward, A. C., … Layton, J. E. (2002). Zebrafish SPI-1 (PU.1) marks a site of myeloid development independent of primitive erythropoiesis: implications for axial patterning. *Developmental Biology, 246*(2), 274–295. http://dx.doi.org/10.1006/dbio.2002.0657.

MacRae, C. A., & Peterson, R. T. (2015). Zebrafish as tools for drug discovery. *Nature Reviews. Drug Discovery, 14*(10), 721–731. http://dx.doi.org/10.1038/nrd4627.

Manicassamy, B., Manicassamy, S., Belicha-Villanueva, A., Pisanelli, G., Pulendran, B., & García-Sastre, A. (2010). Analysis of in vivo dynamics of influenza virus infection in mice using a GFP reporter virus. *Proceedings of the National Academy of Sciences of the United States of America, 107*(25), 11531–11536.

Marjoram, L., Alvers, A., Deerhake, M. E., Bagwell, J., Mankiewicz, J., Cocchiaro, J. L., … Bagnat, M. (2015). Epigenetic control of intestinal barrier function and inflammation in zebrafish. *Proceedings of the National Academy of Sciences of the United States of America, 112*(9), 2770–2775. http://dx.doi.org/10.1073/pnas.1424089112. pii:1424089112.

Mathias, J. R., Perrin, B. J., Liu, T. X., Kanki, J., Look, A. T., & Huttenlocher, A. (2006). Resolution of inflammation by retrograde chemotaxis of neutrophils in transgenic zebrafish. *Journal of Leukocyte Biology, 80*(6), 1281–1288. http://dx.doi.org/10.1189/jlb.0506346.

Mattingly, C. J., Hampton, T. H., Brothers, K. M., Griffin, N. E., & Planchart, A. (2009). Perturbation of defense pathways by low-dose arsenic exposure in zebrafish embryos. *Environmental Health Perspectives, 117*(6), 981–987. http://dx.doi.org/10.1289/ehp.0900555.

McVicker, G., Prajsnar, T. K., Williams, A., Wagner, N. L., Boots, M., Renshaw, S. A., & Foster, S. J. (2014). Clonal expansion during *Staphylococcus aureus* infection dynamics reveals the effect of antibiotic intervention. *PLoS Pathogens, 10*(2), e1003959. http://dx.doi.org/10.1371/journal.ppat.1003959.

Meijer, A. H., & Spaink, H. P. (2011). Host-pathogen interactions made transparent with the zebrafish model. *Current Drug Targets, 12*(7), 1000–1017. http://dx.doi.org/10.2174/138945011795677809.

Milligan-Myhre, K., Charette, J. R., Phennicie, R. T., Stephens, W. Z., Rawls, J. F., Guillemin, K., & Kim, C. H. (2011). Study of host-microbe interactions in zebrafish. *Methods in Cell Biology, 105*, 87–116. http://dx.doi.org/10.1016/B978-0-12-381320-6.00004-7.

Mostowy, S., Boucontet, L., Maria, J., Moya, M., Sirianni, A., Boudinot, P., Hollinshead, M., … Colucci-Guyon, E. (2013). The zebrafish as a new model for the in vivo study of *Shigella flexneri* interaction with phagocytes and bacterial autophagy. *PLoS Pathogens, 9*(9), e1003588. http://dx.doi.org/10.1371/journal.ppat.1003588. pii: PPATHOGENS-D-13-00274.

Myrzakhanova, M., Gambardella, C., Falugi, C., Gatti, A. M., Tagliafierro, G., Ramoino, P., … Diaspro, A. (2013). Effects of nanosilver exposure on cholinesterase activities, CD41, and CDF/LIF-like expression in zebrafish (*Danio rerio*) larvae. *Biomed Research International, 2013*, 205183. http://dx.doi.org/10.1155/2013/205183.

Nayak, A. S., Lage, C. R., & Kim, C. H. (2007). Effects of low concentrations of arsenic on the innate immune system of the zebrafish (*Danio rerio*). *Toxicological Sciences, 98*(1), 118–124. http://dx.doi.org/10.1093/toxsci/kfm072.

Neely, M. N., Pfeifer, J. D., & Caparon, M. (2002). *Streptococcus*-zebrafish model of bacterial pathogenesis. *Infection and Immunity, 70*(7), 3904–3914.

Nguyen-Chi, M., Phan, Q. T., Gonzalez, C., Dubremetz, J. F., Levraud, J. P., & Lutfalla, G. (2014). Transient infection of the zebrafish notochord with *E. coli* induces chronic inflammation. *Disease Models and Mechanisms, 7*(7), 871–882. http://dx.doi.org/10.1242/dmm.014498. pii:7/7/871.

Novoa, B., & Figueras, A. (2012). Zebrafish: model for the study of inflammation and the innate immune response to infectious diseases. *Advances in Experimental Medicine and Biology, 946*, 253–275. http://dx.doi.org/10.1007/978-1-4614-0106-3_15.

Oehlers, S. H., Cronan, M. R., Scott, N. R., Thomas, M. I., Okuda, K. S., Walton, E. M., … Tobin, D. M. (2015). Interception of host angiogenic signalling limits mycobacterial growth. *Nature, 517*(7536), 612–615. http://dx.doi.org/10.1038/nature13967.

Oksanen, K. E., Halfpenny, N. J., Sherwood, E., Harjula, S. K., Hammaren, M. M., Ahava, M. J., … Ramet, M. (2013). An adult zebrafish model for preclinical tuberculosis vaccine development. *Vaccine, 31*(45), 5202–5209. http://dx.doi.org/10.1016/j.vaccine.2013.08.093. pii:S0264-410X(13)01205-X.

Oksanen, K. E., Myllymaki, H., Ahava, M. J., Makinen, L., Parikka, M., & Ramet, M. (2016). DNA vaccination boosts *Bacillus* Calmette-Guerin protection against mycobacterial infection in zebrafish. *Developmental and Comparative Immunology, 54*(1), 89–96. http://dx.doi.org/10.1016/j.dci.2015.09.001. pii:S0145-305X(15)30035-5.

Pagan, A. J., Yang, C. T., Cameron, J., Swaim, L. E., Ellett, F., Lieschke, G. J., & Ramakrishnan, L. (2015). Myeloid growth factors promote resistance to mycobacterial infection by curtailing granuloma necrosis through macrophage replenishment. *Cell Host and Microbe, 18*(1), 15–26. http://dx.doi.org/10.1016/j.chom.2015.06.008.

Palha, N., Guivel-Benhassine, F., Briolat, V., Lutfalla, G., Sourisseau, M., Ellett, F., ... Levraud, J. P. (2013). Real-time whole-body visualization of chikungunya virus infection and host interferon response in zebrafish. *PLoS Pathogens, 9*(9), e1003619. http://dx.doi.org/10.1371/journal.ppat.1003619. pii:PPATHOGENS-D-13-00510.

Perry, S. F., Wilson, R. J., Straus, C., Harris, M. B., & Remmers, J. E. (2001). Which came first, the lung or the breath? *Comparative Biochemistry and Physiology. Part A, Molecular and Integrative Physiology, 129*(1), 37–47.

Phennicie, R. T., Sullivan, M. J., Singer, J. T., Yoder, J. A., & Kim, C. H. (2010). Specific resistance to *Pseudomonas aeruginosa* infection in zebrafish is mediated by the cystic fibrosis transmembrane conductance regulator. *Infection and Immunity, 78*(11), 4542–4550. http://dx.doi.org/10.1128/iai.00302-10.

Prajsnar, T. K., Cunliffe, V. T., Foster, S. J., & Renshaw, S. A. (2008). A novel vertebrate model of *Staphylococcus aureus* infection reveals phagocyte-dependent resistance of zebrafish to non-host specialized pathogens. *Cellular Microbiology, 10*(11), 2312–2325. http://dx.doi.org/10.1111/j.1462-5822.2008.01213.x.

Ravindran, C., Varatharajan, G. R., Raju, R., Vasudevan, L., & Anantha, S. R. (2015). Infection and pathogenecity of *Myroides odoratimimus* (NIOCR-12) isolated from the gut of grey mullet (*Mugil cephalus* (Linnaeus, 1758)). *Microbial Pathogenesis, 88*, 22–28. http://dx.doi.org/10.1016/j.micpath.2015.08.001. pii:S0882-4010(15)00126-6.

Renshaw, S. A., Loynes, C. A., Trushell, D. M., Elworthy, S., Ingham, P. W., & Whyte, M. K. (2006). A transgenic zebrafish model of neutrophilic inflammation. *Blood, 108*(13), 3976–3978. http://dx.doi.org/10.1182/blood-2006-05-024075.

Rocha, O. P., Cesila, C. A., Christovam, E. M., Barros, S. B., Zanoni, M. V., & de Oliveira, D. P. (2016). Ecotoxicological risk assessment of the "Acid Black 210" dye. *Toxicology.* http://dx.doi.org/10.1016/j.tox.2016.04.002. pii:S0300-483X(16)30032-4.

Rocker, A. J., Weiss, A. R., Lam, J. S., Van Raay, T. J., & Khursigara, C. M. (2015). Visualizing and quantifying *Pseudomonas aeruginosa* infection in the hindbrain ventricle of zebrafish using confocal laser scanning microscopy. *Journal of Microbiological Methods, 117*, 85–94. http://dx.doi.org/10.1016/j.mimet.2015.07.013.

van der Sar, A. M., Musters, R. J., van Eeden, F. J., Appelmelk, B. J., Vandenbroucke-Grauls, C. M., & Bitter, W. (2003). Zebrafish embryos as a model host for the real time analysis of *Salmonella typhimurium* infections. *Cellular Microbiology, 5*(9), 601–611.

van der Sar, A. M., Spaink, H. P., Zakrzewska, A., Bitter, W., & Meijer, A. H. (2009). Specificity of the zebrafish host transcriptome response to acute and chronic mycobacterial infection and the role of innate and adaptive immune components. *Molecular Immunology, 46*(11–12), 2317–2332. http://dx.doi.org/10.1016/j.molimm.2009.03.024.

Saralahti, A., Piippo, H., Parikka, M., Henriques-Normark, B., Ramet, M., & Rounioja, S. (2014). Adult zebrafish model for pneumococcal pathogenesis. *Developmental and Comparative Immunology, 42*(2), 345–353. http://dx.doi.org/10.1016/j.dci.2013.09.009. pii:S0145-305X(13)00261-9.

Schneider, C. A., Rasband, W. S., & Eliceiri, K. W. (2012). NIH Image to ImageJ: 25 years of image analysis. *Nature Methods, 9*(7), 671–675.

Shan, Y., Fang, C., Cheng, C., Wang, Y., Peng, J., & Fang, W. (2015). Immersion infection of germ-free zebrafish with *Listeria monocytogenes* induces transient expression of innate immune response genes. *Frontiers in Microbiology, 6*, 373. http://dx.doi.org/10.3389/fmicb.2015.00373.

Shan, Y., Zhang, Y., Zhuo, X., Li, X., Peng, J., & Fang, W. (2016). Matrix metalloproteinase-9 plays a role in protecting zebrafish from lethal infection with *Listeria monocytogenes* by enhancing macrophage migration. *Fish and Shellfish Immunology, 54*, 179–187. http://dx.doi.org/10.1016/j.fsi.2016.04.003. pii:S1050-4648(16)30154-1.

Sullivan, C., & Kim, C. H. (2008). Zebrafish as a model for infectious disease and immune function. *Fish and Shellfish Immunology, 25*(4), 341–350. http://dx.doi.org/10.1016/j.fsi.2008.05.005. pii:S1050-4648(08)00131-9.

Swaim, L. E., Connolly, L. E., Volkman, H. E., Humbert, O., Born, D. E., & Ramakrishnan, L. (2006). *Mycobacterium marinum* infection of adult zebrafish causes caseating granulomatous tuberculosis and is moderated by adaptive immunity. *Infection and Immunity, 74*(11), 6108–6117. http://dx.doi.org/10.1128/IAI.00887-06.

Takaki, K., Davis, J. M., Winglee, K., & Ramakrishnan, L. (2013). Evaluation of the pathogenesis and treatment of *Mycobacterium marinum* infection in zebrafish. *Nature Protocols, 8*(6), 1114–1124. http://dx.doi.org/10.1038/nprot.2013.068.

Tenor, J. L., Oehlers, S. H., Yang, J. L., Tobin, D. M., & Perfect, J. R. (2015). Live imaging of host-parasite interactions in a zebrafish infection model reveals Cryptococcal determinants of virulence and central nervous system invasion. *mBio, 6*(5), e01425-15. http://dx.doi.org/10.1128/mBio.01425-15.

Tobin, D. M., & Ramakrishnan, L. (2008). Comparative pathogenesis of *Mycobacterium marinum* and *Mycobacterium tuberculosis*. *Cellular Microbiology, 10*(5), 1027–1039. http://dx.doi.org/10.1111/j.1462-5822.2008.01133.x. pii:CMI1133.

Tobin, D. M., Vary, J. C., Jr., Ray, J. P., Walsh, G. S., Dunstan, S. J., Bang, N. D., … Ramakrishnan, L. (2010). The lta4h locus modulates susceptibility to mycobacterial infection in zebrafish and humans. *Cell, 140*(5), 717–730. http://dx.doi.org/10.1016/j.cell.2010.02.013. pii:S0092-8674(10)00128-5.

Toh, M. C., Goodyear, M., Daigneault, M., Allen-Vercoe, E., & Van Raay, T. J. (2013). Colonizing the embryonic zebrafish gut with anaerobic bacteria derived from the human gastrointestinal tract. *Zebrafish, 10*(2), 194–198. http://dx.doi.org/10.1089/zeb.2012.0814.

Torraca, V., Cui, C., Boland, R., Bebelman, J. P., van der Sar, A. M., Smit, M. J., … Meijer, A. H. (2015). The CXCR3-CXCL11 signaling axis mediates macrophage recruitment and dissemination of mycobacterial infection. *Disease Models and Mechanisms, 8*(3), 253–269. http://dx.doi.org/10.1242/dmm.017756.

Truong, L., Tilton, S. C., Zaikova, T., Richman, E., Waters, K. M., Hutchison, J. E., & Tanguay, R. L. (2013). Surface functionalities of gold nanoparticles impact embryonic gene expression responses. *Nanotoxicology, 7*(2), 192–201. http://dx.doi.org/10.3109/17435390.2011.648225.

Veneman, W. J., Marin-Juez, R., de Sonneville, J., Ordas, A., Jong-Raadsen, S., Meijer, A. H., & Spaink, H. P. (2014). Establishment and optimization of a high throughput setup to study *Staphylococcus epidermidis* and *Mycobacterium marinum* infection as a model for drug discovery. *Journal of Visualized Experiments*, (88), e51649. http://dx.doi.org/10.3791/51649.

Veneman, W. J., Stockhammer, O. W., de Boer, L., Zaat, S. A., Meijer, A. H., & Spaink, H. P. (2013). A zebrafish high throughput screening system used for *Staphylococcus epidermidis* infection marker discovery. *BMC Genomics, 14*, 255. http://dx.doi.org/10.1186/1471-2164-14-255. pii:1471-2164-14-255.

Voelz, K., Gratacap, R. L., & Wheeler, R. T. (2015). A zebrafish larval model reveals early tissue-specific innate immune responses to *Mucor circinelloides*. *Disease Models and Mechanisms, 8*(11), 1375–1388. http://dx.doi.org/10.1242/dmm.019992.

Volkman, H. E., Clay, H., Beery, D., Chang, J. C., Sherman, D. R., & Ramakrishnan, L. (2004). Tuberculous granuloma formation is enhanced by a *Mycobacterium* virulence determinant. *PLoS Biology, 2*(11), e367. http://dx.doi.org/10.1371/journal.pbio.0020367.

Walters, K. B., Green, J. M., Surfus, J. C., Yoo, S. K., & Huttenlocher, A. (2010). Live imaging of neutrophil motility in a zebrafish model of WHIM syndrome. *Blood, 116*(15), 2803–2811. http://dx.doi.org/10.1182/blood-2010-03-276972.

Walton, E. M., Cronan, M. R., Beerman, R. W., & Tobin, D. M. (2015). The macrophage-specific promoter mfap4 allows live, long-term analysis of macrophage behavior during mycobacterial infection in zebrafish. *PLoS One, 10*(10), e0138949. http://dx.doi.org/10.1371/journal.pone.0138949.

Wang, Y., Ren, Z., Fu, L., & Su, X. (2016). Two highly adhesive lactic acid bacteria strains are protective in zebrafish infected with *Aeromonas hydrophila* by evocation of gut mucosal immunity. *Journal of Applied Microbiology, 120*(2), 441–451. http://dx.doi.org/10.1111/jam.13002.

Wiegertjes, G. F., Wentzel, A. S., Spaink, H. P., Elks, P. M., & Fink, I. R. (2016). Polarization of immune responses in fish: the "macrophages first" point of view. *Molecular Immunology, 69*, 146–156. http://dx.doi.org/10.1016/j.molimm.2015.09.026.

Wiles, T. J., Bower, J. M., Redd, M. J., & Mulvey, M. A. (2009). Use of zebrafish to probe the divergent virulence potentials and toxin requirements of extraintestinal pathogenic *Escherichia coli*. *PLoS Pathogens, 5*(12), e1000697. http://dx.doi.org/10.1371/journal.ppat.1000697.

Wiles, T. J., Norton, J. P., Smith, S. N., Lewis, A. J., Mobley, H. L., Casjens, S. R., & Mulvey, M. A. (2013). A phyletically rare gene promotes the niche-specific fitness of an *E. coli* pathogen during bacteremia. *PLoS Pathogens, 9*(2), e1003175. http://dx.doi.org/10.1371/journal.ppat.1003175. pii:PPATHOGENS-D-12-01433.

Winata, C. L., Korzh, S., Kondrychyn, I., Zheng, W., Korzh, V., & Gong, Z. (2009). Development of zebrafish swimbladder: the requirement of hedgehog signaling in specification and organization of the three tissue layers. *Developmental Biology, 331*(2), 222–236. http://dx.doi.org/10.1016/j.ydbio.2009.04.035.

Yang, C. T., Cambier, C. J., Davis, J. M., Hall, C. J., Crosier, P. S., & Ramakrishnan, L. (2012). Neutrophils exert protection in the early tuberculous granuloma by oxidative killing of mycobacteria phagocytosed from infected macrophages. *Cell Host and Microbe, 12*(3), 301–312. http://dx.doi.org/10.1016/j.chom.2012.07.009.

CHAPTER

Live imaging the earliest host innate immune response to preneoplastic cells using a zebrafish inducible KalTA4-ER^{T2}/UAS system

D.W. Laux[a,b], L. Kelly[a], I. Ribeiro Bravo, T. Ramezani, Y. Feng[1]

University of Edinburgh, Edinburgh, United Kingdom

[1]Corresponding author: E-mail: yi.feng@ed.ac.uk

CHAPTER OUTLINE

[a]These authors contributed equally to this work.

[b]Present address: Washington College, Chestertown, MD, Unites States.

Methods in Cell Biology, Volume 138, ISSN 0091-679X, http://dx.doi.org/10.1016/bs.mcb.2016.10.002

Abstract

As cancers develop, transformed cells hijack various host mechanisms and manipulate them to create a dynamic tumor microenvironment, which supports tumor growth. This protumorigenic microenvironment is made up of many different cell types, including transformed cells, fibroblasts, inflammatory cells, and endothelial cells, the interactions of which have been shown to play a role in sustaining tumor growth. Multiple reports implicate the inflammatory cells of the tumor microenvironment as having both pro- and antitumorigenic roles, the balance of which is vital for the progression of the tumor, and while our understanding of established cancers has vastly increased since the turn of the 21st Century, our knowledge of these cellular interactions at the earliest stages of cancer initiation and development remains relatively limited. This is largely due to difficulties in monitoring these processes in vivo and in real time. Since the late nineties, the zebrafish (*Danio rerio*) has emerged as a vital model organism, allowing studies of previously unattainable stages of tumor initiation in a vertebrate model system. Using genetic and live-imaging approaches, this model system can be used both independently to monitor stages of tumor progression from the earliest initiation stages and incorporated into previously established systems to investigate the interactions between cancer cells and the various cell types of the tumor microenvironment, including inflammatory cells. Here, we describe the use of an inducible KalTA4-ER^{T2}/UAS expression system in zebrafish, which allows spatial and temporal control of preneoplastic cell (PNC) growth and monitoring of innate immune cells in response to the developing PNC microenvironment.

INTRODUCTION

The development of cancer involves complex interactions between preneoplastic cells (PNCs) and their host environment. Only a small number of initial PNCs, which are altered cells that have the potential to become cancerous, may eventually develop into a tumor and progress toward aggressive cancer. To design effective cancer preventive strategies, it is important to understand how PNCs interact with their host, how host factors might have both positive and negative influences on the development of a cancerous lesion, and how such host: PNC interactions may determine the outcome of cancer development. The interaction between cancer and host is an active area of research in the field of cancer biology. Many advanced new imaging techniques employed in murine cancer models have been instrumental for advancing our understanding of the interactions between cancer cells and host stromal cells in the tumor microenvironment, both at steady state and under various therapeutic conditions (Prescher & Contag, 2012). Such techniques can also be used in other model systems. Here we describe how we use an inducible zebrafish system to live-image the host: PNC interactions during the earliest stages of cancer development and the impact of this system on our understanding of cancer biology.

Despite its importance in development, the initial phase of cancer progression is inaccessible in most commonly used model systems in vivo. The zebrafish (*Danio rerio*) is now a well-established model organism for live-imaging studies due to its transparency throughout development, the possibility for genetic manipulation, and accessibility for water-soluble drugs (Feng & Martin, 2015). Many oncogenic pathways that are frequently expressed in human cancers are also conserved in zebrafish, and based on this, many types of cancers have been successfully modeled in fish through transgenic overexpression of known human oncogenes (Lam et al., 2006; Langenau et al., 2007; Langenau et al., 2003; Liu & Leach, 2011; Mione & Trede, 2010; Park et al., 2008; Patton et al., 2005; Zhu & Look, 2016). When the oncogenic protein is fluorescently tagged, the development of cancer can be observed from the initiation stage through to a full blown tumor (Ignatius & Langenau, 2011; Kaufman et al., 2016). The field of live imaging of the inflammatory response has been explored among the zebrafish community in recent years (Davis et al., 2002; Gray et al., 2011; Henry, Loynes, Whyte, & Renshaw, 2013; Lam, Fischer, Shin, Waterman, & Huttenlocher, 2014; Novoa & Figueras, 2012), with an increasing list of transgenic reporter fish for visualizing immune cell lineages being generated (Table 1). This has created an opportunity to establish

Table 1 Commonly Used Neutrophil Reporter Zebrafish Transgenic Lines

Function	Transgenic Zebrafish Line
Early myeloid precursors	Tg(-9.0spi1:eGFP)zdf11 (Hsu et al., 2004)
	Tg(-5.3spi1:eGFP)gl21 (Hsu et al., 2004)
Neutrophil reporter	Tg(LysC:DsRed2)nz50 (Hall, Flores, Storm, Crosier, & Crosier, 2007)
	Tg(LysC:eGFP)nz117 (Hall et al., 2007)
	Tg(BACmpo:eGFP)^{i113}/Tg(BACmpo: eGFP)^{i114} (Renshaw et al., 2006)
	Tg(zMPO:GFP)^{uw} (Mathias et al., 2006)
Macrophage reporter	Tg(fli1a:eGFP)y1 (Lawson & Weinstein, 2002)
	Tg(mpeg1:eGFP) (Ellett, Pase, Hayman, Andrianopoulos, & Lieschke, 2011)
	Tg(mpeg1:Gal4-VP16/UAS:Kaede) (Ellett et al., 2011)
	Tg(mfap4:tdTomato-CAAX) (Walton, Cronan, Beerman, & Tobin, 2015)
	Tg(mfap4:Turquoise2) (Walton et al., 2015)
	Tg(mfap4:dLanYFP-CAAX) (Walton et al., 2015)
	Tg(fms:Gal4.VP16)il86:Tg(UAS: nfsB.mCherry)il49 (Gray et al., 2011)
Neutrophil and macrophage reporter	Tg(coro1a:eGFP) (Li, Yan, Shi, Zhang, & Wen, 2012)
Activated macrophage reporter	Tg(irg1:eGFP) (Sanderson et al., 2015)
Eosinophil reporter	Tg(gata2:eGFP) (Balla et al., 2010)
T cell reporter	Tg(lck:eGFP) (Langenau et al., 2004)
B cell reporter	Tg(IgM1:eGFP) (Page et al., 2013)

a novel model in which the earliest interaction between PNCs and host innate immune cells can be observed and manipulated in real time at the cellular/subcellular level. Recent studies, by our group and others, using a larval zebrafish cancer model overexpressing the human oncogene $HRAS^{G12V}$ in skin tissue, revealed that PNC derived H_2O_2 and also IL-8 led to the recruitment of host innate immune cells (Feng, Santoriello, Mione, Hurlstone, & Martin, 2010; Freisinger & Huttenlocher, 2014). Interestingly, the recruited innate immune cells provide trophic support to PNC progression (Feng, Renshaw, & Martin, 2012), as well as leading to upregulation of epithelial−mesenchymal transition genes in PNCs (Freisinger & Huttenlocher, 2014), which demonstrates the active involvement of host components during the earliest stages of cancer development and provides potential new targets for cancer prevention.

Zebrafish possess both innate and adaptive immune systems, similar to their mammalian counterparts (Traver et al., 2003), and the availability of transgenic zebrafish strains allows for live-imaging studies of PNCs interacting with all immune cells lineages (Table 1). The zebrafish larval model is ideal for investigating the interaction of host innate immune cells with PNCs without complication from adaptive immunity interaction, although the involvement of the adaptive immune system can be studied at a later stage, using adult zebrafish cancer models (>3 weeks) (Traver et al., 2003). The development of transparent adult zebrafish strains has made it possible to image the initiation of melanoma from a single cell in an adult fish (Kaufman et al., 2016), but transgenic reporter fish that label various lymphocytic lineages are needed for imaging adaptive immune involvement in the tumor microenvironment.

The use of the Gal4/UAS binary expression system in zebrafish allows fast and effective testing of multiple gene products in a tissue-specific manner. This system has been successfully used to generate cancer models that arise from different tissues (Alghisi et al., 2013; Burger et al., 2014; Santoriello et al., 2010) and to study the effect of diverse KRAS mutations on the zebrafish pancreatic tumorigenesis system (Park et al., 2015). However, using this system, the expression of the oncogene starts as soon as the Gal4 is expressed, which can occur very early in development, depending on the promoter used. For example, in our case keratin 4 is expressed at high level in EVL from as early as 6 hpf (Thisse et al., 2001). Therefore, it may not be a feasible stage for the study of downstream consequences of oncogene expression. To achieve a precise temporal and spatial control of oncogene expression, we have developed an inducible system, in which a tamoxifen inducible KalTA-ER^{T2} activator is used instead of Gal4 (Ramezani, Laux, Bravo, Tada, & Feng, 2015). This allows us to observe downstream signaling events of oncogene activation, as well as host: PNC interaction immediately following the oncogene expression.

In this chapter we will be focusing on using this inducible system to generate $HRAS^{G12V}$ over-expressing PNCs in the skin tissue of zebrafish larvae and subsequent live imaging of the interaction of the host innate immune system with PNCs.

1. GENERATION OF PRENEOPLASTIC CELL CLONES USING THE KalTA4-ER^{T2}/UAS SYSTEM IN ZEBRAFISH LARVAE

Here we describe the tissue-specific generation of PNCs in the superficial skin layer of zebrafish larvae by expression of the fluorescently tagged oncogene under the control of the *krt4* promoter. However, this system can be easily modified to specifically target any tissue of interest, simply by the use of an appropriate promoter. Microinjection of the genetic construct into the one-cell stage of fertilized embryos allows for mosaic expression of the oncogene in the tissue of interest. This permits investigations of not only the PNCs themselves, but also of interactions between PNCs and the various cells of the developing microenvironment.

1.1 GENERATION OF UPSTREAM ACTIVATING SEQUENCE–REGULATED eGFP-HRAS^{G12V} TRANSGENE CONSTRUCT

1. Using the Gateway cloning system (Invitrogen), insert the eGFP-HRAS^{G12V} transgene downstream of a 6× UAS element in a pDestTol2CG backbone, according to the manufacturers guidelines, to generate the UAS:eGFP-HRAS^{G12V} expression plasmid DNA clone.
2. Purify the plasmid DNA using a commercial DNA purification kit, according to the manufacturer's instructions and dilute the expression clone to a final stock solution concentration of 100 ng/μL.

1.2 MICROINJECTION OF UAS:eGFP-HRAS^{G12V} PLASMID DNA INTO ONE-CELL STAGE Tg(cmlc2:eGFP; Krt4:KalTA4-ER^{T2}; LysC: DsRed) EMBRYOS

1. Prepare an aliquot of injection solution: 1–2 μL of plasmid DNA (final concentration of 10–20 ng/μL), 2 μL of transposase mRNA (final concentration 20 ng/μL), 2 μL of 10 mg/μL Rhodamine B isothiocyanate-Dextran, 2 μL of 5× Danieau's solution (290 mM NaCl, 3.5 mM KCl, 2 mM $MgSO_4$, 3 mM $Ca(NO_3)_2$, 25 mM HEPES, pH 7.6) to a final volume of 10 μL in nuclease-free water.
   Note: We synthetize tol2 transposase mRNA from the transposase-containing plasmid, pT3TS/Tol2, linearized with BamHI, using a commercial in vitro mRNA transcription kit according to the manufacturer's protocol. mRNA is purified using a standard phenol: chloroform extraction followed by ethanol precipitation.
2. Prepare injection needles using Micropipette puller (Sutter Instrument Co.) and Kwik-Fill Borosilicate Glass Capillaries, 1-mm diameter (World Precision Instruments).

Note: We use the following settings for injection needle preparation (Heat: 530; Pull200; Velocity: 80; Time: 150). Settings will depend on both machine and filament and must therefore be optimized each time the inner filament is replaced.

3. Load 2–5 μL of injection solution into needle, taking care not to break the tip of the needle. Take care to avoid bubbles in the injection needle by slowly withdrawing the microloader tip from the capillary tube while gently expelling liquid. Once loaded, carefully break the tip of the needle using a pair of forceps.
4. Regulate the injection drop size with a pneumatic pico pump (Warner Instruments) and ensure uniform drop size/volume by measuring drop size with an ocular micrometer while injecting under a stereoscope into dimethylpolysiloxane.
5. Place eggs in wells of a premade agarose injection mold or along a glass slide fitted in a lid of 90 mm petri dish.
6. Inject a premeasured drop of 100–150 μm in diameter into the blastodisc of embryos prior to the first cell cleavage to ensure uniform distribution of DNA and to enhance probability of germline incorporation.

1.3 CONDITIONAL SKIN TRANSFORMATION BY 4-HYDROXYTAMOXIFEN

KalTA4-ER^{T2}, the modified version of the transcriptional activator Gal4, allows for conditional induction of the upstream activating sequence (UAS) transgene. KalTA4-ER^{T2} has been modified with the ligand-binding domain of the human estrogen receptor α (ER^{T2}), which retains KalTA4 within the cytoplasm through interactions with heat shock proteins. Introduction of 4-hydroxytamoxifen (4-OHT) leads to dissociation of the heat shock protein-KalTA4-ER^{T2} complex allowing KalTA4 to enter the nucleus and activate UAS–controlled genes. It should be noted that this reaction is reversible, allowing researchers to turn off KalTA4-ER^{T2}–controlled genes with the removal of tamoxifen.

1. Transfer injected embryos into E3 medium and grow at 28.5°C.
2. Remove unfertilized embryos at the shield stage.
3. At 24 hpf, *cmlc2:eGFP* expression may be used to select embryos with plasmid incorporation. Continue growing *cmlc2:eGFP* positive embryos until desired developmental time point. Monitor water quality and change when necessary.
4. For induction of KalTA4-ER^{T2} transgene, add 5 μL of 10 mM 4-hydroxytamoxifen to 20 mL of E3 medium (final concentration: 2.5 μM). Embryos can be treated in standard 90 mm petri dishes with roughly 40–60 embryos per dish.
5. After addition of induction solution, dishes should be kept in dark to limit degradation of tamoxifen.
   Note: (Z)-4-hydroxytamoxifen undergoes a cis-trans (E-Z) interconversion when exposed to light. (E)-4-hydrotamoxifen is found to be 100× less antiestrogenic compared to its (Z) counterpart. Due to this fact, exposure to light must be

limited. We recommend storing embryo dishes in boxes during treatment and changing induction solution at least once daily.

6. In our experiments, the first signs of expression can be detected 10–12 hours post induction (Fig. 1).

2. LIVE IMAGING PRENEOPLASTIC CELL: NEUTROPHIL INTERACTION USING CONFOCAL MICROSCOPY

2.1 MOUNTING EMBRYOS FOR IN VIVO IMAGING

1. Prepare a 60 mm petri dish for imaging by introducing a circular window (18–20 mm in diameter) in the bottom of the dish.
2. Seal the hole of the dish with high-vacuum silicone grease and a ∅25 mm coverslip.
3. Raise embryos to desired developmental stage at 28.5°C in E3 medium, supplemented with tamoxifen for the desired time frame.
4. Before imaging, anesthetize embryos using 0.02% buffered 3- aminobenzoic acid ethyl ester (Tricane), then dechorinate with a fine forceps (Dupont no. 55) or pronase solution (1 mg/mL in Embryo medium), taking care not to damage the embryo.
5. Embed embryos in 1% low melting temperature agarose (prepared in E3 medium) directly on the coverslip of imaging plate.
6. Position embryos, in desired orientation for imaging, with forceps, eyelash tool, or pipette tip.
7. Once agarose has set, submerge embedded embryos in induction solution supplemented with tricaine to keep embryos anesthetized during imaging.

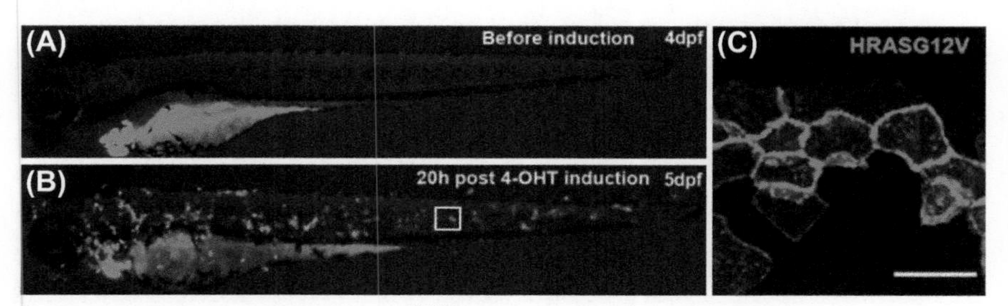

FIGURE 1 Mosaic expression of HRAS^{G12V}-eGFP in larval zebrafish skin.

Confocal images of (A) a Tg(Krt4:KalTA4-ER^{T2};UAS:eGFP-HRAS) embryo at 4 dpf, before the addition of 4-hydroxytamoxifen (4-OHT). (B) Same embryo as in A at 24 h after addition of 4-OHT, showing mosaic expression of HRAS^{G12V}-eGFP in the cells of superficial skin layer. An area of the trunk near the cloaca (*white box*) is typically selected for imaging. (C) Zoom in image from B showing membrane bound HRAS^{G12V}-eGFP expression of preneoplastic cells can be observed by approximately 10–12 hours post induction. Scale bar 20 μm.

Note: Most time-lapse experiments are taken for 4–5 h. For experiments longer than these, we recommend changing the embryo medium due to (Z)-4-hydroxytamoxifen decay and potential toxicity caused by tricaine.

2.2 LIVE CONFOCAL IMAGING OF NEUTROPHIL: PRENEOPLASTIC CELL INTERACTIONS

1. Visualize host: PNC interactions with an inverted fluorescent, laser scanning, or spinning disc confocal microscope (or equivalent). Typically, we chose the skin area at the trunk region near the cloaca (Fig. 1) We find 40× or 63× objectives with long working distances provide the best resolution for imaging host: PNC interactions (Fig. 2). Using "Zoom in" function on confocal we can achieve very high subcellular resolution of detailed interaction between host neutrophils and PNCs using 63× objective.
2. Typical scanning settings using the Leica SP5 laser scanning confocal microscope are as follows:
   - objective: APO L 40×/1.10 W,
   - sequential scan,
   - multiline argon laser (488 nm) set at 30%, helium neon (543 nm),
   - scan speed: 400 Hz point scanner,
   - image format: 512 × 512,
   - z-series, step size: 0.5–3 μm,
   - if background: noise ratio affects image quality, kalman averaging (line) is set at three, and
   - time interval set to minimal for duration of 4–5 h.

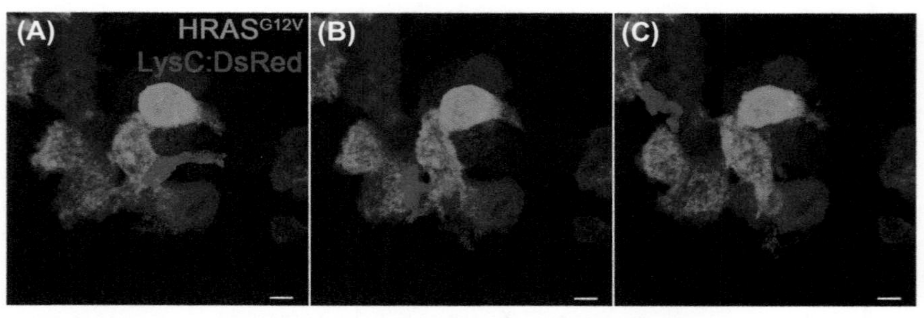

FIGURE 2 Time-lapse imaging of host: preneoplastic cell (PNC) interactions.

Still images taken from a confocal time-lapse movie of a PNC clone in a Tg(Krt4:KalTA4-ER^{T2}; UAS:eGFP-HRAS; lysC:DsRed) embryo at 2 dpf, 24 hours post induction. (A–C) Consecutive frames from a time-lapse movie taken using Leica SP5 laser scanning confocal microscope fitted with 40×/1.10 W objective lens that allows subcellular resolution of cellular interactions. Neutrophils [magenta] migrate to developing PNCs [green], where they exhibit crawling behavior as they interact with the PNC clones. Scale bar 10 μm.

Note:

a. Images for this study were collected on a TCS SP5 Leica Confocal Microscope outfitted with an APO L 40×/1.10 W objective (Leica Microsystems). Optimal settings for these systems were determined empirically and were dependent on the fluorescent signal from the embryo. In our experience, fluorescent intensity can vary widely between embryos, and therefore care must be taken to ensure settings are optimized for each embryo to ensure the highest quality image.
b. Due to migration speed of neutrophils, time between z-series projections should be minimized and should not exceed 2–3 min to best monitor neutrophil: PNC interactions.

3. IMAGE ANALYSIS AND 4D RECONSTRUCTION

3.1 IMAGE ANALYSIS AND LIVE CELL TRACKING OF NEUTROPHILS AROUND PRENEOPLASTIC CELLS

Cell Tracks of neutrophils visiting PNCs are prepared from time-lapse confocal images using Imaris 8.0.0 3-D and 4-D Image Analysis Suite. There are other free softwares available that could also be used for data analysis, such as ImagJ, Fiji, or Icy.

1. Open 3D View (Fig. 3A) of selected time-lapse image and select "Add new surfaces" from the Properties menu.
2. In the Surface Creation window select "Track Surfaces (over Time)" and click the "Next" arrow.

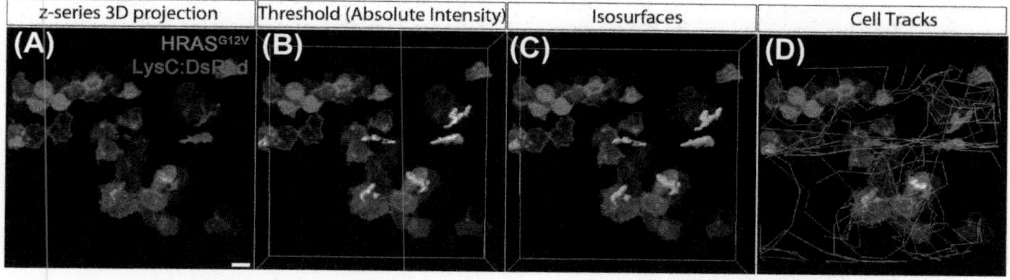

FIGURE 3 Cell tracking of neutrophil behavior in response to developing preneoplastic cells.

Snapshot images from Imaris 8.0.0 3-D and 4-D Image Analysis Suite, showing step by step guide of basic cell tracking analysis. (A) 3D view of a still image from a time-lapse sequence showing preneoplastic cell [green] and neutrophils [magenta]. (B) Surfaces are generated for neutrophils [magenta] and absolute Intensity is adjusted to account for varying cell size and fluorescent intensity. (C) Rendered isosurfaces can be edited or deleted before final cell tracking. (D) Example of tracks generated by setting up parameters appropriate for neutrophil migration tracking.

3. A region of interest can be selected in x,y,z, and t. To highlight a region of interest in x,y, and z choose "Select" under "Pointer" options. Clicking on the solid arrows on the 3D image allows the user to highlight a region of interest. A specific timeframe can be selected in the "Create" menu by inputting "Min" and "Max" time points. Alternatively, the entire time-lapse can be analyzed by inputting the total time point value for the "Max" value. When a region of interest has been highlighted, click the "Next" arrow.
4. In the next window, select the proper Source Channel from the drop down menu for tracking neutrophils. The "Surfaces Area Detail Level" should be set for a value approximately 1/10th the size of a single neutrophil. To determine neutrophil size, go into "Slice" View. Find an appropriate time point and z position to view a neutrophil. On the image, left click once on either side of the neutrophil. A solid white line should appear between two (+) signs. The distance is displayed on the right side of the screen under the "Measure" tab. Return to "3D View" to continue tracking analysis. Ensure under Thresholding, whether "Absolute Intensity" is selected. Click the "Next" Arrow
5. The "Threshold" should be set such that solid gray surfaces nearly encapsulate imaged neutrophils (Fig. 3B). The threshold can be adjusted by clicking on the solid yellow line and sliding left or right to include weaker or stronger fluorescent signal respectively. Check that these surfaces adequately cover migrating neutrophils at both the beginning and end of the time-lapse sequence. Click the "Next" arrow.
6. Rendered isosurfaces (Fig. 3C) can be deleted in the next creation menu by adding filters under the "Classify Surfaces" menu and selecting different "Filter Types". After erroneous points have been deleted select the "Next" arrow. Click the "Next" arrow in the next "Edit Surfaces" window.
7. The next window allows one to choose a Tracking Algorithm. "Autoregressive Motion" is recommended for neutrophil migration as it permits a certain degree of predictive migration. Under "Parameters" a Max Distance can be set to allow for larger distance jumps for neutrophils. We find that for a time-lapse experiment in which z-series projections are taken every 2.5–3 min, a Max Distance of 30.0 μm is ideal. Adjustment of "Max Gap Size" permits Imaris to recognize the same neutrophil that may travel outside the z-stack between timepoints. Although the user will need to test different "Max Gap Size" parameters, we find that a size of 5–10 will typically yield successful tracks. Click the "Next" button.
8. The next screen will yield cell tracks for neutrophils (Fig. 3D). Scroll through time series to view whether neutrophil migrations are properly followed. The "Back" arrow may always be used to change parameters to produce more accurate tracks. When satisfied with tracks, click the "Finish" arrow.
9. The "Settings" window allows for modification of track presentation features.

CONCLUSION

The use of an inducible KalTA4-ER^{T2}/UAS–mediated expression system allows for temporal and spatial control of the expression of fluorescently labeled oncogenes in an in vivo model organism, the zebrafish larvae. This has allowed in vivo live imaging of cell behavior to investigate the interactions between various cell types, including preneoplastic cells and host innate immune cells, in the tumor initiation niche. We saw very rapid immune cell recruitment toward PNCs during tumor initiation which is driven by reactive oxygen species generated from PNCs and their normal neighbors. We have live imaged at high resolution the intimate interaction between recruited innate immune cells and PNCs, which has revealed cytoplasmic "nano-tube" like structures bridging the two cell types as they interact. Our live imaging has also caught neutrophils engulfing fragments of cytoplasmic material from PNCs over long periods of time without their killing of the PNCs (Feng et al., 2010). Finally, combining zebrafish genetic tools and a live-imaging approach, we have showed that recruited innate immune cells promote PNC proliferation, which is beneficial for the development of cancer (Feng et al., 2012). Using a Vertebrate Automated Screening Technology (VAST) BioImager platform (Pardo-Martin et al., 2010), the live-imaging model we described here would provide a possible drug screening platform for small molecules that target the tumor initiation and host response.

The model system outlined above is a useful tool for in vivo studies of the earliest stages of cancer development, which is often not accessible in other in vivo systems and provides an opportunity for investigations of possible cancer preventative strategies.

ACKNOWLEDGMENTS

Y. F. is funded by a Wellcome Trust Sir Henry Dale Fellowship WT100104AIA, a Wellcome Trust ISSF funding and a Royal Society research grant; Y. F. is also funded by a Chancellor's Fellow Start up fund from the University of Edinburgh; L. K. receives a Principal's Career Development Scholarship from the University of Edinburgh.

REFERENCES

Alghisi, E., Distel, M., Malagola, M., Anelli, V., Santoriello, C., Herwig, L., … Mione, M. C. (2013). Targeting oncogene expression to endothelial cells induces proliferation of the myelo-erythroid lineage by repressing the Notch pathway. *Leukemia, 27*(11), 2229–2241. http://dx.doi.org/10.1038/leu.2013.132.

Balla, K. M., Lugo-Villarino, G., Spitsbergen, J. M., Stachura, D. L., Hu, Y., Bañuelos, K., … Traver, D. (2010). Eosinophils in the zebrafish: prospective isolation, characterization, and eosinophilia induction by helminth determinants. *Blood, 116*(19), 3944–3954. http://dx.doi.org/10.1182/blood-2010-03-267419.

Burger, A., Vasilyev, A., Tomar, R., Selig, M. K., Nielsen, G. P., Peterson, R. T., … Haber, D. A. (2014). A zebrafish model of chordoma initiated by notochord-driven expression of HRASV12. *Disease Models and Mechanisms, 7*(7), 907–913. http://dx.doi.org/10.1242/dmm.013128.

Davis, J. M., Clay, H., Lewis, J. L., Ghori, N., Herbomel, P., & Ramakrishnan, L. (2002). Real-time visualization of mycobacterium-macrophage interactions leading to initiation of granuloma formation in zebrafish embryos. *Immunity, 17*(6), 693–702. http://dx.doi.org/10.1016/S1074-7613(02)00475-2.

Ellett, F., Pase, L., Hayman, J. W., Andrianopoulos, A., & Lieschke, G. J. (2011). mpeg1 promoter transgenes direct macrophage-lineage expression in zebrafish. *Blood, 117*(4), e49–56. http://dx.doi.org/10.1182/blood-2010-10-314120.

Feng, Y., & Martin, P. (2015). Imaging innate immune responses at tumour initiation: new insights from fish and flies. *Nature Reviews Cancer, 15*(9), 556–562. http://dx.doi.org/10.1038/nrc3979.

Feng, Y., Renshaw, S., & Martin, P. (2012). Live imaging of tumor initiation in zebrafish larvae reveals a trophic role for leukocyte-derived PGE_2. *Current Biology, 22*(13), 1253–1259. http://dx.doi.org/10.1016/j.cub.2012.05.010.

Feng, Y., Santoriello, C., Mione, M., Hurlstone, A., & Martin, P. (2010). Live imaging of innate immune cell sensing of transformed cells in zebrafish larvae: parallels between tumor initiation and wound inflammation. *PLoS Biology, 8*(12), e1000562. http://dx.doi.org/10.1371/journal.pbio.1000562.

Freisinger, C. M., & Huttenlocher, A. (2014). Live imaging and gene expression analysis in zebrafish identifies a link between neutrophils and epithelial to mesenchymal transition. *PLoS One, 9*(11), e112183. http://dx.doi.org/10.1371/journal.pone.0112183.

Gray, C., Loynes, C. A., Whyte, M. K. B., Crossman, D. C., Renshaw, S. A., & Chico, T. J. A. (2011). Simultaneous intravital imaging of macrophage and neutrophil behaviour during inflammation using a novel transgenic zebrafish. *Thrombosis and Haemostasis, 105*(5), 811–819. http://dx.doi.org/10.1160/TH10-08-0525.

Hall, C., Flores, M. V., Storm, T., Crosier, K., & Crosier, P. (2007). The zebrafish lysozyme C promoter drives myeloid-specific expression in transgenic fish. *BMC Developmental Biology, 7*, 42. http://dx.doi.org/10.1186/1471-213X-7-42.

Henry, K. M., Loynes, C. A., Whyte, M. K. B., & Renshaw, S. A. (2013). Zebrafish as a model for the study of neutrophil biology. *Journal of Leukocyte Biology, 94*(4), 633–642. http://dx.doi.org/10.1189/jlb.1112594.

Hsu, K., Traver, D., Kutok, J. L., Hagen, A., Liu, T.-X., Paw, B. H., … Look, A. T. (2004). The pu.1 promoter drives myeloid gene expression in zebrafish. *Blood, 104*(5), 1291–1297. http://dx.doi.org/10.1182/blood-2003-09-3105.

Ignatius, M. S., & Langenau, D. M. (2011). Fluorescent imaging of cancer in zebrafish. In *Methods in cell biology* (Vol. 105, pp. 437–459). http://dx.doi.org/10.1016/B978-0-12-381320-6.00019-9.

Kaufman, C. K., Mosimann, C., Fan, Z. P., Yang, S., Thomas, A. J., Ablain, J., … Zon, L. I. (2016). A zebrafish melanoma model reveals emergence of neural crest identity during melanoma initiation. *Science, 351*(6272), aad2197. http://dx.doi.org/10.1126/science.aad2197.

Lam, P.-Y., Fischer, R. S., Shin, W. D., Waterman, C. M., & Huttenlocher, A. (2014). Spinning disk confocal imaging of neutrophil migration in zebrafish. *Methods in Molecular Biology, 1124*, 219–233. http://dx.doi.org/10.1007/978-1-62703-845-4_14.

Lam, S. H., Wu, Y. L., Vega, V. B., Miller, L. D., Spitsbergen, J., Tong, Y., … Gong, Z. (2006). Conservation of gene expression signatures between zebrafish and human liver tumors and tumor progression. *Nature Biotechnology, 24*(1), 73–75. http://dx.doi.org/10.1038/nbt1169.

Langenau, D. M., Ferrando, A. A., Traver, D., Kutok, J. L., Hezel, J.-P. D., Kanki, J. P., … Trede, N. S. (2004). In vivo tracking of T cell development, ablation, and engraftment in transgenic zebrafish. *Proceedings of the National Academy of Sciences of the United States of America, 101*(19), 7369–7374. http://dx.doi.org/10.1073/pnas.0402248101.

Langenau, D. M., Keefe, M. D., Storer, N. Y., Guyon, J. R., Kutok, J. L., Le, X., … Zon, L. I. (2007). Effects of RAS on the genesis of embryonal rhabdomyosarcoma. *Genes and Development, 21*(11), 1382–1395. http://dx.doi.org/10.1101/gad.1545007.

Langenau, D. M., Traver, D., Ferrando, A. A., Kutok, J. L., Aster, J. C., Kanki, J. P., … Look, A. T. (2003). Myc-induced T cell leukemia in transgenic zebrafish. *Science, 299*(5608), 887–890. http://dx.doi.org/10.1126/science.1080280.

Lawson, N. D., & Weinstein, B. M. (2002). In vivo imaging of embryonic vascular development using transgenic zebrafish. *Developmental Biology, 248*(2), 307–318. Retrieved from: http://www.ncbi.nlm.nih.gov/pubmed/12167406.

Li, L., Yan, B., Shi, Y.-Q., Zhang, W.-Q., & Wen, Z.-L. (2012). Live imaging reveals differing roles of macrophages and neutrophils during zebrafish tail fin regeneration. *Journal of Biological Chemistry, 287*(30), 25353–25360. http://dx.doi.org/10.1074/jbc.M112.349126.

Liu, S., & Leach, S. D. (2011). Zebrafish models for cancer. *Annual Review of Pathology: Mechanisms of Disease, 6*, 71–93. http://dx.doi.org/10.1146/annurev-pathol-011110-130330.

Mathias, J. R., Perrin, B. J., Liu, T.-X., Kanki, J., Look, A. T., & Huttenlocher, A. (2006). Resolution of inflammation by retrograde chemotaxis of neutrophils in transgenic zebrafish. *Journal of Leukocyte Biology, 80*(6), 1281–1288. http://dx.doi.org/10.1189/jlb.0506346.

Mione, M. C., & Trede, N. S. (2010). The zebrafish as a model for cancer. *Disease Models and Mechanisms, 3*(9–10), 517–523. http://dx.doi.org/10.1242/dmm.004747.

Novoa, B., & Figueras, A. (2012). Zebrafish: model for the study of inflammation and the innate immune response to infectious diseases. In *Advances in experimental medicine and biology* (Vol. 946, pp. 253–275). http://dx.doi.org/10.1007/978-1-4614-0106-3_15.

Page, D. M., Wittamer, V., Bertrand, J. Y., Lewis, K. L., Pratt, D. N., Delgado, N., … Traver, D. (2013). An evolutionarily conserved program of B-cell development and activation in zebrafish. *Blood, 122*(8), e1–11. http://dx.doi.org/10.1182/blood-2012-12-471029.

Pardo-Martin, C., Chang, T.-Y., Koo, B. K., Gilleland, C. L., Wasserman, S. C., & Yanik, M. F. (2010). High-throughput in vivo vertebrate screening. *Nature Methods, 7*(8), 634–636. http://dx.doi.org/10.1038/nmeth.1481.

Park, J. T., Johnson, N., Liu, S., Levesque, M., Wang, Y. J., Ho, H., … Leach, S. D. (2015). Differential in vivo tumorigenicity of diverse KRAS mutations in vertebrate pancreas: a comprehensive survey. *Oncogene, 34*(21), 2801–2806. http://dx.doi.org/10.1038/onc.2014.223.

Park, S. W., Davison, J. M., Rhee, J., Hruban, R. H., Maitra, A., & Leach, S. D. (2008). Oncogenic KRAS induces progenitor cell expansion and malignant transformation in zebrafish exocrine pancreas. *Gastroenterology, 134*(7), 2080–2090. http://dx.doi.org/10.1053/j.gastro.2008.02.084.

Patton, E. E., Widlund, H. R., Kutok, J. L., Kopani, K. R., Amatruda, J. F., Murphey, R. D., ... Zon, L. I. (2005). BRAF mutations are sufficient to promote nevi formation and cooperate with p53 in the genesis of melanoma. *Current Biology, 15*(3), 249–254. http://dx.doi.org/10.1016/j.cub.2005.01.031.

Prescher, J. A., & Contag, C. H. (2012). Imaging mouse models of human cancer. In *Genetically engineered mice for cancer research* (pp. 235–260). New York, NY: Springer New York. http://dx.doi.org/10.1007/978-0-387-69805-2_11.

Ramezani, T., Laux, D. W., Bravo, I. R., Tada, M., & Feng, Y. (2015). Live imaging of innate immune and preneoplastic cell interactions using an inducible Gal4/UAS expression system in larval zebrafish skin. *Journal of Visualized Experiments, 96*, 1–7. http://dx.doi.org/10.3791/52107.

Renshaw, S. A., Loynes, C. A., Trushell, D. M. I., Elworthy, S., Ingham, P. W., & Whyte, M. K. B. (2006). A transgenic zebrafish model of neutrophilic inflammation. *Blood, 108*(13), 3976–3978. http://dx.doi.org/10.1182/blood-2006-05-024075.

Sanderson, L. E., Chien, A.-T., Astin, J. W., Crosier, K. E., Crosier, P. S., & Hall, C. J. (2015). An inducible transgene reports activation of macrophages in live zebrafish larvae. *Developmental and Comparative Immunology, 53*(1), 63–69. http://dx.doi.org/10.1016/j.dci.2015.06.013.

Santoriello, C., Gennaro, E., Anelli, V., Distel, M., Kelly, A., Köster, R. W., ... Mione, M. (2010). Kita driven expression of oncogenic HRAS leads to early onset and highly penetrant melanoma in zebrafish. *PLoS One, 5*(12), e15170. http://dx.doi.org/10.1371/journal.pone.0015170.

Thisse, B., Pfumio, S., Fürthauer, M. B. L., Heyer, V., Degrave, A., ... Thisse, C. (2001). *Expression of the zebrafish genome during embryogenesis*. ZFIN Online Publication. Retrieved from: https://zfin.org/ZDB-PUB-010810-1.

Traver, D., Herbomel, P., Patton, E. E., Murphey, R. D., Yoder, J. A., Litman, G. W., ... Trede, N. S. (2003). The zebrafish as a model organism to study development of the immune system. *Advances in Immunology, 81*, 254–330. http://dx.doi.org/10.1016/S0065-2776(03)81007-6.

Walton, E. M., Cronan, M. R., Beerman, R. W., & Tobin, D. M. (2015). The macrophage-specific promoter MFAP4 allows live, long-term analysis of macrophage behavior during mycobacterial infection in zebrafish. *PLoS One, 10*(10), e0138949. http://dx.doi.org/10.1371/journal.pone.0138949.

Zhu, S., & Look, A. T. (2016). Neuroblastoma and its zebrafish model. *Advances in Experimental Medicine and Biology, 916*, 451–478. http://dx.doi.org/10.1007/978-3-319-30654-4_20.

CHAPTER

Studying the adaptive immune system in zebrafish by transplantation of hematopoietic precursor cells

N. Iwanami, I. Hess, M. Schorpp, T. Boehm[1]

Max Planck Institute of Immunobiology and Epigenetics, Freiburg, Germany

[1]*Corresponding author: E-mail: boehm@immunbio.mpg.de*

CHAPTER OUTLINE

Abstract

Traditionally, transplantation has been a major experimental procedure to study the development and function of hematopoietic and immune systems.

Here, we describe the use of a zebrafish strain lacking definitive hematopoiesis (cmyb^{I181N}) for interspecific analysis of hematopoietic and immune cell development. Without conditioning prior to transplantation, allogeneic and xenogeneic hematopoietic progenitor cells stably engraft in adult zebrafish homozygous for the *cmyb* mutation. This unique animal model can be used to genetically and functionally disentangle universal and species-specific contributions of the microenvironment to hematopoietic progenitor cell maintenance and development.

Methods in Cell Biology, Volume 138, ISSN 0091-679X, http://dx.doi.org/10.1016/bs.mcb.2016.08.003

INTRODUCTION

Recent work has indicated that the general design of the vertebrate hematopoietic system has been surprisingly well conserved over the course of more than 500 million years (Boehm, 2011). More specifically, evidence has been accumulating that, similar to the situation in mammals, the hematopoietic system of lower vertebrates originates from specialized progenitor cells (Hartenstein, 2006); in addition, the evolutionarily conserved dichotomy of lymphocyte lineages is a general feature of the immune systems of all vertebrates (Hirano, Das, Guo, & Cooper, 2011). In contrast to the extraordinarily stable cellular framework of hematopoiesis, its molecular underpinnings are much more plastic. For instance, while the two sister branches of vertebrates—jawless fishes and jawed vertebrates—share various T and B cell lineages, the molecular structures of the antigen receptors and the mode of their somatic diversification differ substantially (Boehm, 2011; Hirano et al., 2011) (Table 1). While the structural building block of antigen receptors in jawed vertebrates is the immunoglobulin fold (Flajnik, 2002), jawless vertebrates instead rely on the structural versatility of the leucine-rich repeat (Boehm et al., 2012). With respect to the mechanisms of somatic diversification of antigen repertoires, there are distinct differences between the two branches. Whereas jawed vertebrates employ the RAG (recombination activation gene) proteins to achieve the so-called VDJ (variable; diversity; joining element) recombination (Agrawal, Eastman, & Schatz, 1998) (with AID [activation induced cytidine deaminase]) playing a role in secondary somatic diversification of immunoglobulins), jawless fishes appear to rely on the activity of cytidine deaminases for the generation of combinatorial diversity of variable lymphocyte receptors (Rogozin et al., 2007) (Table 1).

Collectively, these findings have broadened the spectrum of animal models available to researchers aiming at understanding the function of vertebrate hematopoietic and immune systems. In fact, it would be highly desirable to be able to examine many more of the tens of thousands of vertebrate species inhabiting the planet, to develop an even more detailed view on the evolutionarily selected variations on

Table 1 Deep Evolutionary Roots of Hematopoietic and Immune Systems

Vertebrate Group	Lymphocyte Lineages	Types of Antigen Receptors	Somatic Diversification
Agnatha	T; B	VLRA (variable lymphocyte receptor type A); VLRC (variable lymphocyte receptor type C); VLRB (variable lymphocyte receptor type B)	Yes CDA [cytidine deaminase]
Gnathostomata	T; B	TCR (T cell receptor); BCR (B cell receptor)	Yes (RAG [recombination activating genes]; AID [activation induced cytidine deaminase])

the theme of immune facilities. In a dramatic example of such noncanonical designs, it was recently shown that the cod lacks the entire MHC class II pathway (Star et al., 2011). Indeed, when considered only from the point of view of mammalian immunity, this constellation would be considered to be indicative of severe immunodeficiency (Steimle, Otten, Zufferey, & Mach, 1993). Yet, the cod must have developed a functionally sufficient compensatory mechanism(s) supporting an average life span of many decades (Malmstrøm, Jentoft, Gregers & Jakobsen, 2013). It is to be expected that many more such noncanonical solutions to the same problem of immune defense will be found among vertebrate species living in the diverse ecosystems known to exist on earth. Collectively, these considerations underscore the urgent need to develop methods and suitable animal models for the functional interrogation of such alternative hematopoietic/immune systems.

Judged from a historical perspective, cell transplantation is likely to play an important role in this endeavor. Ideally, such a system would consist of a genetically modifiable recipient accepting allogeneic and xenogeneic grafts not requiring prior conditioning (for example, see Waskow et al., 2009). Humanized mouse strains (Doulatov, Notta, Laurenti, & Dick, 2012) are used to great effect in the study of the human hematopoietic/immune system in an experimentally amenable physiological context (Brehm, Shultz, Luban, & Greiner, 2013).

In recent years, the zebrafish model has risen to prominence as a suitable animal model for biomedical research (see other chapters in this volume). Indeed, the zebrafish is notable both for its ease of handling and the availability of a highly diversified genetic toolbox; moreover, zebrafish belongs to the teleosts, the most species-rich group of vertebrates inhabiting our planet, comprising almost half of all vertebrates. Teleosts are morphologically diverse and populate extraordinarily different in ecological niches and, as mentioned earlier, can therefore be expected to exhibit diversity in many functional traits, including immunity.

In addition to the exploration of natural diversity, recent advances in zebrafish genetics also call for a facile method of supporting functional analyses of altered hematopoietic/immune systems (Avagyan & Zon, 2016). Indeed, a vast number of mutant strains have emerged from several forward genetic screens after chemically induced mutagenesis (Boehm, Bleul, & Schorpp, 2003; Ransom et al., 1996; Trede et al., 2008); moreover, rapid advances in site-directed mutagenesis, notably using the CRISPR/Cas9 system, have added to the growing list of interesting phenotypes (Fujii et al., 2016).

Several features of zebrafish biology impede progress toward developing versatile transplantation models across histocompatibility barriers. The phenomenon of inbreeding depression (McCune, Houle, McMillan, Annable, & Kondrashov, 2004) and the fact that the genes of the MHC are scattered throughout the genome (Sambrook, Figueroa, & Beck, 2005), essentially prevent the development of genetically homogenous zebrafish stocks. Hence, unless early embryos and larvae are used as recipients (with the associated practical problems of small graft size) (Traver et al., 2003), conditioning is required for the successful engraftment of cells and organs in the host (De Jong & Zon, 2012). Inevitably, such preconditioning (most commonly performed by irradiation) damages the host organism. This might result in a general reduction in fitness and/or damage to the recipient microenvironmental niches that are important for the homing and maintenance of hematopoietic progenitor cells in the kidney and/or thymus.

Several strategies can be used to overcome the problems associated with histoincompatibility. Traver et al. (2003) achieved long-term reconstitution of hematopoietic function by transplantation of whole kidney marrow cells into *gata1*$^{-/-}$ bloodless mutant zebrafish. These authors used 48-hpf larvae as recipients; at this point of development, fish have not yet developed a functional immune system. However, while effective, this procedure is hampered by the low number of cells that can be transplanted into the recipients. Recently, it was demonstrated that the failing immune system of *rag2* mutants, which lack B and T cells, could be reconstituted by transplantation of whole kidney marrow cells (Tang et al., 2014). However, as with transplantations using wild-type adult fish, the *rag2*$^{-/-}$ fish had to be irradiated to achieve successful engraftment. In the case of *rag2*$^{-/-}$ fish, irradiation likely eliminated residual innate immune activity and/or also vacated occupied hematopoietic progenitor niches. In our own efforts, we have examined the possibility of using a zebrafish mutant lacking definitive hematopoiesis as a universal recipient (Hess, Iwanami, Schorpp, & Boehm, 2013). Zebrafish homozygous for a missense mutation in the *cmyb* gene (cmyb^{I181N}) exhibit failure of definitive hematopoiesis (Soza-Ried, Hess, Netuschil, Schorpp, & Boehm, 2010). We have exploited this property to develop these *cmyb* mutant fish as a transplantation model not requiring conditioning prior to transplantation of allogeneic and xenogeneic hematopoietic cells (Hess & Boehm, 2016; Hess et al., 2013) (Table 2).

Table 2 Zebrafish Models for Cell/Tissue Transplantation

Recipient Genotype	Requirement for Preconditioning	Age of Host Organism	Source of Transplant	Comment
Wild type	Yes	Adult	Allogeneic; xenogeneic	High frequency of transplantation failure
gata1$^{-/-}$	No	Larva	Allogeneic	Only low numbers of cells can be transplanted
rag2$^{-/-}$	Yes	Adult	Allogeneic; xenogeneic	Preconditioning possibly associated with tissue damage; risk of graft versus host disease
cmyb$^{-/-}$	No	Adult	Allogeneic; xenogeneic	Delayed reconstitution may lead to premature death of host

In this chapter, we describe the basic procedure of transplantation into *cmyb* mutants. We then discuss avenues of future research aimed at exploring the range of foreign species that could serve as donors for the *cmyb* mutant zebrafish recipient.

1. METHODOLOGY FOR THE TRANSPLANTATION OF HEMATOPOIETIC CELLS

1.1 MATERIALS

Mesab (tricaine, MS-222, ethyl-*m*-aminobenzoate methanesulphonate). Safety note: Prepare solutions under a fume hood.

Beaker
Dissecting microscope
Tweezers
Scalpel
Petrie dish
40-μm cell strainer (BD Falcon)
0.9× PBS supplemented with 5% (vol/vol) FCS
Piston of 1-ml syringe
Hemacytometer
Sponge
Syringe (Hamilton X034.1, microliter syringe model 702N, 25 μL, needle inside diameter 0.15 mm, outside diameter 0.72 mm)
Fish water

The equipment for the harvest of kidney marrow and the injection procedure is sterilized in 70% EtOH and rinsed with sterile PBS before use.

1.2 CHOICE OF RECIPIENT STRAIN

Homozygous *cmyb* mutant zebrafish lack definitive hematopoiesis owing to a cell-intrinsic block. The median survival time of *cmyb* mutants is in the order of 12–14 weeks, particularly when raised together with sentinel wild-type fish such as albino fish (for facile distinction, use slc45a2 mutant (White et al., 2008)). In our experience, fish are best used for transplantation at 6 weeks of age, when they can be easily distinguished from wild-type and heterozygous littermates by their smaller size and pale complexion. At that age, fish are usually large enough to allow the transplantation to be safely carried out, yet still robust enough to survive the inevitable stress associated with the procedure. If live imaging of transplanted cells is desired, it may be advantageous to transfer the *cmyb* mutation onto the Casper mutant background (White et al., 2008), as this compound genotype renders *cmyb* mutants transparent.

1.3 HARVESTING OF KIDNEY MARROW CELLS

Adult zebrafish are anesthetized in fish water supplemented with 0.02% (v/v) Mesab.

The ventral peritoneum is opened using a scalpel, the fish is placed in a Petrie dish under the dissecting microscope with the ventral side facing upward. All internal organs that are attached to the intestine are removed using tweezers. The kidney remains attached to the dorsal peritoneum. Both head and posterior kidney are harvested. The reader is referred to a video illustrating this procedure (Gerlach, Schrader, & Wingert, 2011).

The kidney tissue is placed on a 40-μm mesh that is situated on a Petrie dish together with 0.5 mL of 0.9× PBS; 5% (v/v) FCS and minced. The tissue is disintegrated on the mesh using the piston of a 1-ml syringe resulting in a single-cell suspension. This cell suspension is transferred into a sterile 1.5-ml test tube. The mesh and piston are rinsed with 0.9× PBS; 5% (v/v) FCS, and the cell suspensions are combined. In our hands, the average cell count in such preparations is in the order of 500,000 cells, of which about 400,000 cells are used per transplant. The remainder of the cells can be used for analytical purposes, for example, flow cytometry. We prefer to transplant cells from one donor only to prevent any adverse effects that could arise from undesirable graft versus graft reactions in situations of histoincompatible donors. Cell suspensions are centrifuged (400× g, 8 min, 4°C), the supernatant is removed, and the cells are resuspended in 0.9× PBS; 5% (v/v) FCS at 20,000–40,000 cells/μL.

1.4 INJECTION OF DONOR CELLS INTO THE BLOOD CIRCULATION

Recipient fish are briefly anesthetized in fish water supplemented with 0.01% Mesab and placed on a wet sponge, immobilized by mild pressure, and injected retroorbitally with up to 400,000 cells in a maximum volume of 10 μL using a Hamilton syringe. The reader is referred to a video illustrating this procedure (LeBlanc, Bowman, & Zon, 2007). After the transplantation, the fish are kept in tanks without water circulation for 1 day; during this period, they are monitored regularly for signs of ill health (slow movements, bleeding, etc.). Most procedure-related complications become obvious during the first 24 h after the transplantation. After that, transplanted fish are returned to the system. In the hands of a skilled experimenter, the procedure-associated fatality rate should be less than 10–15%; this loss should be considered when planning the number of fish required for an experiment.

As a matter of principle, fish that are transplanted with cells harvested from fish not raised in our own fish facility are kept in a quarantine unit to avoid possible complications arising from the introduction of foreign microflora.

In the case of xenogeneic transplantations, it is important to consider the water temperature to which transplanted fish are exposed after the procedure. In our experiments using goldfish donor cells, we keep transplanted zebrafish at 22°C, instead of the usual 28°C. To this end, we acclimatize the recipients to this lower temperature in a step-wise fashion, by lowering the temperature from 28 to 22°C by 2°C every day. Likewise, donor fish are acclimatized to 22°C well before the removal of cells.

1.5 ASSESSMENT OF SHORT- AND LONG-TERM RECONSTITUTION

When mutant fish are reconstituted with wild-type zebrafish kidney marrow, within 7–10 days the transplanted fish assume a reddish complexion (Fig. 1A). Reconstitution of the lymphoid and myeloid compartments soon follows, resulting in full multilineage reconstitution after 3 weeks. The extent and quality of the reconstitution are best analyzed by flow cytometry (Fig. 1B), which distinguishes cell types based on their forward and side light scatter characteristics (Traver et al., 2003). Histological analysis additionally provides information on the location and morphology of hematopoietic cells (Fig. 1C); this assay is complemented by the analysis of Giemsa-stained blood smears. As a rule of thumb, we have found that a reddish/whitish appearance of the kidney at autopsy is indicative of successful reconstitution of *cmyb* mutant fish.

If hematopoietic cells originate from an appropriate reporter line, the kinetics and extent of reconstitution can be monitored by fluorescence microscopy, best carried out in a light-transparent recipient. In the example illustrated in Fig. 1D, an accumulation of EGFP+ cells is detectable in the thymus; this fluorescence emanates from lymphoid cells marked with an *ikaros:eGFP* transgene (Hess & Boehm, 2012).

The aforementioned analyses provide an indication of successful reconstitution. To prove that renewing hematopoietic stem cells (HSCs) were transplanted and that they had subsequently engrafted in the host organism, secondary transplantations should be carried out (Hess & Boehm, 2016; Hess et al., 2013).

Successful hematopoietic reconstitution is usually accompanied by rapid growth of *cmyb* mutants and an increase in agility and swimming speed. Moreover, they reach sexual maturation and become fertile, which allows the generation of large clutches of pure *cmyb* homozygous mutant fish. Hence, establishing a colony of mutant fish is advantageous when planning large-scale transplantation experiments.

2. DISCUSSION

The reconstitution of blood lineage cells without the necessity for preconditioning such as irradiation is a powerful method for gene function analysis in adult hematopoiesis. This is of particular concern in quantitative studies, because conditioning inevitably causes harm to the recipient, most likely also affecting the microenvironmental niches for incoming hematopoietic precursors. The consideration of this fact makes the *cmyb* model arguably the most physiological transplant model in zebrafish described to date. Using this model, we have been able to estimate that the kidney marrow of an adult zebrafish contains about 10 transplantable HSCs (Hess et al., 2013).

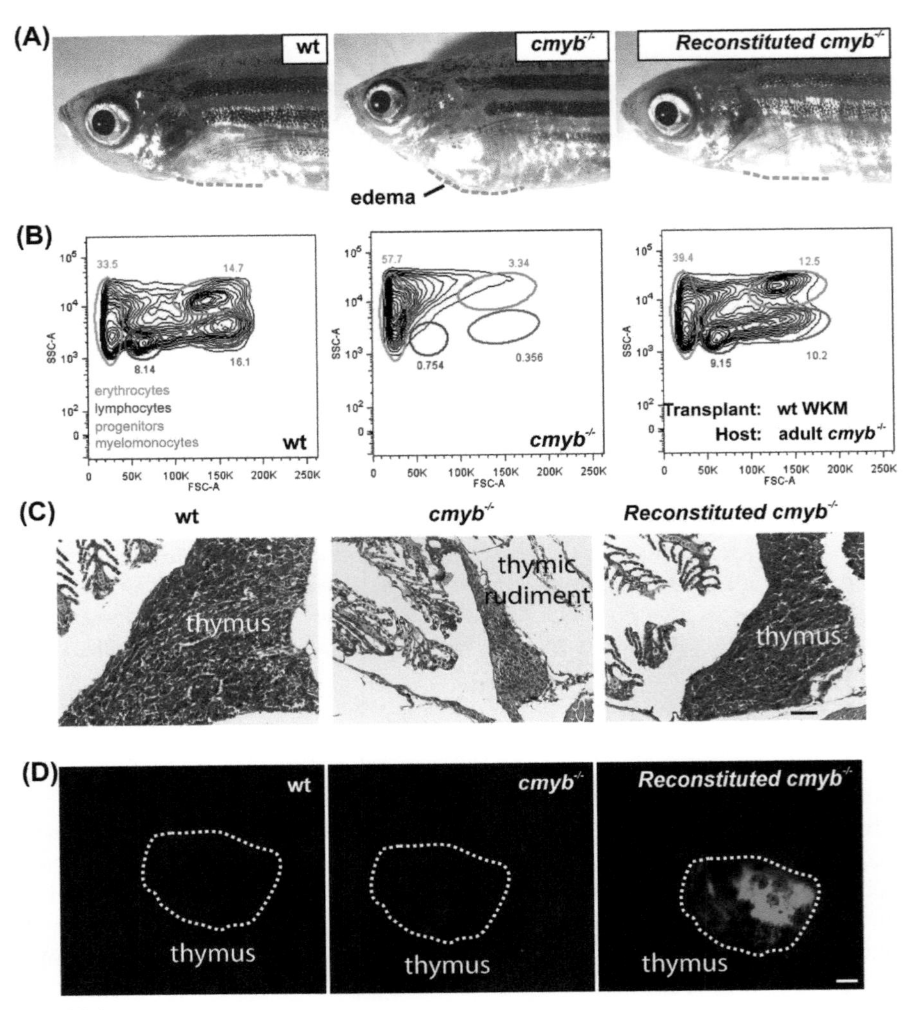

FIGURE 1

Multilineage hematopoietic reconstitution of *cmyb* mutant zebrafish (Hess & Boehm, 2016; Hess et al., 2013). (A) Macroscopic appearance of wild-type fish (wt), *cmyb* mutant (*cymb*$^{-/-}$), and transplanted *cmyb* mutant (reconstituted *cymb*$^{-/-}$; 5 weeks after transplantation) fish. Note the pale appearance of the mutant fish and the edema owing to the absence of red blood cells. (B) Flow cytometric analysis of cells in whole kidney marrow. The light scatter characteristics (*SSC*, side scatter; *FSC*, forward scatter) of individual cell populations are identified by their color code (Traver et al., 2003). The source of cells is indicated and corresponds to panel (A). (C) Colonization of the thymus after reconstitution. Sections are stained with hematoxylin/eosin. Note the alymphoid thymic rudiment in *cmyb* mutants. Scale bar, 50 μm. (D) Colonization of the thymus after reconstitution. Colonization is detectable by fluorescence emanating from an *ikaros:eGFP* reporter allele (Hess & Boehm, 2012). Scale bar, 50 μm.

Our recent demonstration that *cmyb* mutants also sustain hematopoietic progenitor cells originating from goldfish (Hess & Boehm, 2016) suggests that the range of xenogeneic donors might be expanded to evolutionarily more distant species of Cyprinidae, which comprise more than 2000 freshwater species. Given this success, we consider it worthwhile to examine whether cells from more distantly related fishes, or even representatives of tetrapod species, could be used in this model.

If the latter possibility could be realized, the *cmyb* model would provide a fresh impetus to examine the properties of vertebrate HSCs via large-scale chemical screens (Trompouki & Zon, 2010). An exciting prospect in this context is the possibility of live imaging of hematopoietic development and maintenance with the potential to establish an in vivo readout by live imaging of fluorescently marked cells (Bertrand et al., 2010; Kissa & Herbomel, 2010).

One further obvious use of the *cmyb* model is the reconstitution of mutant hematopoietic systems, as previously demonstrated for *il7r* and *jak3* mutant cells (Hess et al., 2013). In an extension of this approach, we envisage that the *cmyb* model could be used to study the properties of hematopoiesis in mutants in which pleiotropic effects of the affected gene cause a reduction in life span or even impede survival past early larval stages. In such a way, the model described here could be used to establish tissue chimeras with a mutated hematopoietic system.

3. FUTURE DIRECTIONS

In addition to the use of the *cmyb* model to establish allografts of wild-type and mutant zebrafish, and xenografts with hematopoietic cells of other species, we envisage further refinements of the *cmyb* model as follows. We found that zebrafish cells eventually out-competed the goldfish contribution; moreover, in the case of hematopoiesis reconstituted by pure goldfish donors, it was observed that the expected ratio of lymphoid and myeloid cells was skewed towards lymphoid cells (Hess & Boehm, 2016). These findings were interpreted to indicate only partial compatibility of the zebrafish microenvironment. Recent advances in genome engineering open up the possibility to specifically add candidate environmental factors or even replace them with their xenogeneic counterparts. Prime targets for such an approach are cytokines and chemokines, a group of communication components whose exquisite species specificity has been described in other systems (Chen, Khoury, & Chen, 2009).

ACKNOWLEDGMENTS

We gratefully acknowledge the contributions of Cristian Soza, who identified the *cmyb* mutant in our forward screen, and the entire zebrafish group for their valuable advice and encouragement. Financial support for these studies was provided by the Max Planck Society.

REFERENCES

Agrawal, A., Eastman, Q. M., & Schatz, D. G. (1998). Transposition mediated by RAG1 and RAG2 and its implications for the evolution of the immune system. *Nature, 394*, 744–751.

Avagyan, S., & Zon, L. I. (2016). Fish to learn: insights into blood development and blood disorders from zebrafish hematopoiesis. *Human Gene Therapy, 27*, 287–294.

Bertrand, J. Y., Chi, N. C., Santoso, B., Teng, S., Stainier, D. Y., & Traver, D. (2010). Haematopoietic stem cells derive directly from aortic endothelium during development. *Nature, 464*, 108–111.

Boehm, T. (2011). Design principles of adaptive immune systems. *Nature Reviews Immunology, 11*, 307–317.

Boehm, T., Bleul, C. C., & Schorpp, M. (2003). Genetic dissection of thymus development in mouse and zebrafish. *Immunological Reviews, 195*, 15–27.

Boehm, T., McCurley, N., Sutoh, Y., Schorpp, M., Kasahara, M., & Cooper, M. D. (2012). VLR-based adaptive immunity. *Annual Review of Immunology, 30*, 203–220.

Brehm, M. A., Shultz, L. D., Luban, J., & Greiner, D. L. (2013). Overcoming current limitations in humanized mouse research. *The Journal of Infectious Diseases, 208*(Suppl. 2), S125–S130.

Chen, Q., Khoury, M., & Chen, J. (2009). Expression of human cytokines dramatically improves reconstitution of specific human-blood lineage cells in humanized mice. *Proceedings of the National Academy of Sciences of the United States of America, 106*, 21783–21788.

De Jong, J. L., & Zon, L. I. (2012). Histocompatibility and hematopoietic transplantation in the zebrafish. *Advances in Hematology, 2012*, 282318.

Doulatov, S., Notta, F., Laurenti, E., & Dick, J. E. (2012). Hematopoiesis: a human perspective. *Cell Stem Cell, 10*, 120–136.

Flajnik, M. F. (2002). Comparative analyses of immunoglobulin genes: surprises and portents. *Nature Reviews Immunology, 2*, 688–698.

Fujii, T., Tsunesumi, S., Sagara, H., Munakata, M., Hisaki, Y., Sekiya, T., … Watanabe, S. (2016). Smyd5 plays pivotal roles in both primitive and definitive hematopoiesis during zebrafish embryogenesis. *Scientific Reports, 6*, 29157.

Gerlach, G. F., Schrader, L. N., & Wingert, R. A. (2011). Dissection of the adult zebrafish kidney. *Journal of Visualized Experiments*, e2839.

Hartenstein, V. (2006). Blood cells and blood cell development in the animal kingdom. *Annual Review of Cell and Developmental Biology, 22*, 677–712.

Hess, I., & Boehm, T. (2012). Intravital imaging of thymopoiesis reveals dynamic lympho-epithelial interactions. *Immunity, 36*, 298–309.

Hess, I., & Boehm, T. (2016). Stable multilineage xenogeneic replacement of definitive hematopoiesis in adult zebrafish. *Scientific Reports, 6*, 19634.

Hess, I., Iwanami, N., Schorpp, M., & Boehm, T. (2013). Zebrafish model for allogeneic hematopoietic cell transplantation not requiring preconditioning. *Proceedings of the National Academy of Sciences of the United States of America, 110*, 4327–4332.

Hirano, M., Das, S., Guo, P., & Cooper, M. D. (2011). The evolution of adaptive immunity in vertebrates. *Advances in Immunology, 109*, 125–157.

Kissa, K., & Herbomel, P. (2010). Blood stem cells emerge from aortic endothelium by a novel type of cell transition. *Nature, 464*, 112–115.

LeBlanc, J., Bowman, T. V., & Zon, L. (2007). Transplantation of whole kidney marrow in adult zebrafish. *Journal of Visualized Experiments*, e159.

Malmstrøm, M., Jentoft, S., Gregers, T. F., & Jakobsen, K. S. (2013). Unraveling the evolution of the Atlantic cod's (*Gadus morhua* L.) alternative immune strategy. *PLoS One, 8*, e74004.

McCune, A. R., Houle, D., McMillan, K., Annable, R., & Kondrashov, A. S. (2004). Two classes of deleterious recessive alleles in a natural population of zebrafish, Danio rerio. *Proceedings. Biological Sciences, 271*, 2025–2033.

Ransom, D. G., Haffter, P., Odenthal, J., Brownlie, A., Vogelsang, E., Kelsh, R. N., … Nüsslein-Volhard, C. (1996). Characterization of zebrafish mutants with defects in embryonic hematopoiesis. *Development, 123*, 311–319.

Rogozin, I. B., Iyer, L. M., Liang, L., Glazko, G. V., Liston, V. G., Pavlov, Y. I., … Pancer, Z. (2007). Evolution and diversification of lamprey antigen receptors: evidence for involvement of an AID-APOBEC family cytosine deaminase. *Nature Immunology, 8*, 647–656.

Sambrook, J. G., Figueroa, F., & Beck, S. (2005). A genome-wide survey of Major Histocompatibility Complex (MHC) genes and their paralogues in zebrafish. *BMC Genomics, 6*, 152.

Soza-Ried, C., Hess, I., Netuschil, N., Schorpp, M., & Boehm, T. (2010). Essential role of *c-myb* in definitive hematopoiesis is evolutionarily conserved. *Proceedings of the National Academy of Sciences of the United States of America, 107*, 17304–17308.

Star, B., Nederbragt, A. J., Jentoft, S., Grimholt, U., Malmstrøm, M., Gregers, T. F., … Jakobsen, K. S. (2011). The genome sequence of Atlantic cod reveals a unique immune system. *Nature, 477*, 207–210.

Steimle, V., Otten, L. A., Zufferey, M., & Mach, B. (1993). Complementation cloning of an MHC class II transactivator mutated in hereditary MHC class II deficiency (or bare lymphocyte syndrome). *Cell, 75*, 135–146.

Tang, Q., Abdelfattah, N. S., Blackburn, J. S., Moore, J. C., Martinez, S. A., Moore, F. E., … Langenau, D. M. (2014). Optimized cell transplantation using adult *rag2* mutant zebrafish. *Nature Methods, 11*, 821–824.

Traver, D., Paw, B. H., Poss, K. D., Penberthy, W. T., Lin, S., & Zon, L. I. (2003). Transplantation and in vivo imaging of multilineage engraftment in zebrafish bloodless mutants. *Nature Immunology, 4*, 1238–1246.

Trede, N. S., Ota, T., Kawasaki, H., Paw, B. H., Katz, T., Demarest, B., … Zon, L. I. (2008). Zebrafish mutants with disrupted early T-cell and thymus development identified in early pressure screen. *Developmental Dynamics, 237*, 2575–2584.

Trompouki, E., & Zon, L. I. (2010). Small molecule screen in zebrafish and HSC expansion. *Methods in Molecular Biology, 636*, 301–316.

Waskow, C., Madan, V., Bartels, S., Costa, C., Blasig, R., & Rodewald, H.-R. (2009). Hematopoietic stem cell transplantation without irradiation. *Nature Methods, 6*, 267–269.

White, R. M., Sessa, A., Burke, C., Bowman, T., Leblanc, J., Ceol, C., … Zon, L. I. (2008). Transparent adult zebrafish as a tool for in vivo transplantation analysis. *Cell Stem Cell, 2*, 183–189.

PART

Blood and Lymph

CHAPTER

7

Hematopoietic stem cell development: using the zebrafish to identify extrinsic and intrinsic mechanisms regulating hematopoiesis

J.M. Frame*[,a], S.-E. Lim*[,a], T.E. North*[,§,1]

Beth Israel Deaconess Medical Center, Harvard Medical School, Boston, MA, United States

[§]*Harvard Stem Cell Institute, Cambridge, MA, United States*

[1]*Corresponding author: E-mail: tnorth@bidmc.harvard.edu*

CHAPTER OUTLINE

[a]Equal contribution.

Methods in Cell Biology, Volume 138, ISSN 0091-679X, http://dx.doi.org/10.1016/bs.mcb.2016.08.004

Abstract

Hematopoietic stem cells (HSCs) reside at the apex of the hematopoietic hierarchy, giving rise to each of the blood lineages found throughout the lifetime of the organism. Since the genetic programs regulating HSC development are highly conserved between vertebrate species, experimental studies in zebrafish have not only complemented observations reported in mammals but have also yielded important discoveries that continue to influence our understanding of HSC biology and homeostasis. Here, we summarize findings that have established zebrafish as an important conserved model for the study of hematopoiesis, and describe methods that can be utilized for future investigations of zebrafish HSC biology.

INTRODUCTION TO HEMATOPOIETIC DEVELOPMENT

There are several phases of blood cell formation, referred to as waves, which occur during embryonic development (Fig. 1). The first wave of hematopoiesis, termed primitive, occurs in the zebrafish in the intermediate cell mass (ICM); it is the functional equivalent of the blood islands in the mammalian yolk sac (at murine embryonic days E7.5–9.5) (Detrich et al., 1995) and gives rise primarily to erythrocytes. In the second, or definitive, wave, multipotent progenitors are produced; intriguingly, in fish, chick, and murine embryos, this ability to contribute to multiple lineages appears to be acquired in reverse. In the mouse, erythromyeloid progenitors (EMPs) are formed in the blood islands of the yolk sac from E8.25 (Palis, Robertson, Kennedy, Wall, & Keller, 1999). Subsequently, adult-repopulating hematopoietic stem cells (HSCs) emerge from the ventral wall of the dorsal aorta (VDA) in the aorta-gonad mesonephros (AGM) region, as well as from the umbilical

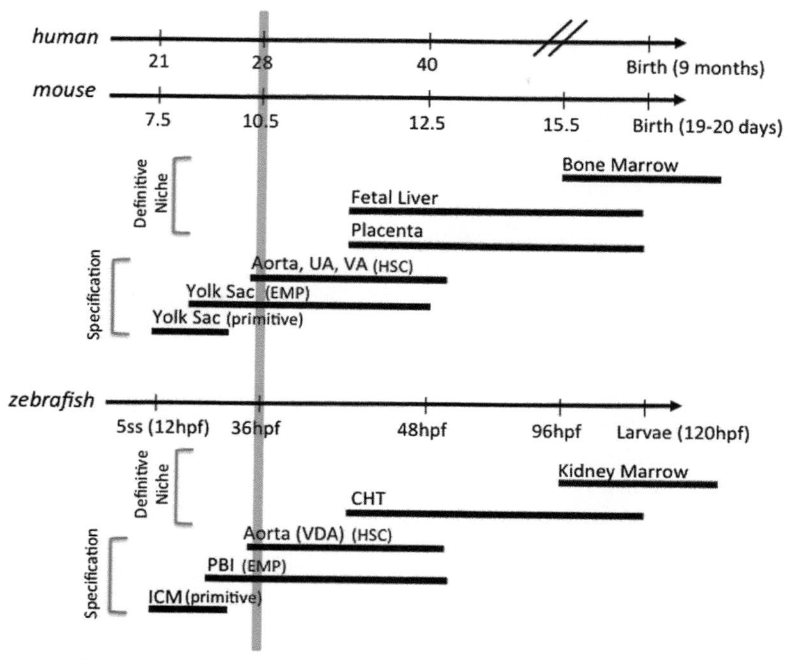

FIGURE 1

Comparison of conserved spatiotemporal regulation of hematopoiesis across vertebrate species. Red line denotes the emergence of the first adult-repopulating definitive hematopoietic stem cells. Developmental time is in days unless otherwise indicated. *UA*, umbilical artery; *VA*, vitelline artery.

and vitelline arteries, from E10.5–E12.5 (Fig. 1) (Medvinsky & Dzierzak, 1996; de Bruijn, Speck, Peeters, & Dzierzak, 2000). As in other vertebrates, this sequence is maintained in the zebrafish: EMPs, non–self-renewing progenitors with limited multilineage differentiation potential, are formed prior to the generation of HSCs (Bertrand et al., 2007). It is unclear if multiple sites can produce zebrafish HSCs; however, they do migrate to seed different tissues for the purpose of proliferation and differentiation as in mammals. HSCs initially emerge in the ventral aspect of the dorsal aorta (VDA) between 30 and 36 hours postfertilization (hpf). Subsequently, they localize in the distal region of the tail, termed the caudal hematopoietic tissue (CHT) by 48 hpf, before seeding the kidney marrow and thymus to continue lifelong hematopoiesis (Fig. 1). The conservation of these waves of hematopoietic specification has proven zebrafish to be an excellent and tractable model for discovering and characterizing the signaling networks and biophysical forces regulating vertebrate hematopoietic development.

HEMANGIOBLAST FORMATION

During early somitogenesis, the ventral lateral mesoderm is specified into hematopoietic, endothelial, and kidney progenitors. Within this region, the anterior portion of the lateral plate mesoderm (ALM) gives rise to primitive myeloid cells, and the posterior portion of the lateral plate mesoderm (PLM) produces primarily erythrocytes, in addition to some cells of the myeloid lineage. Endothelial and hematopoietic precursors are initially colocalized in the lateral plate mesoderm and coexpress transcription factors such as *tal1* (previously *scl*), *lmo2*, *fli1*, *gata2*, and *etv2* (previously *etsrp*), indicative of either blood or endothelial fate. These findings support the existence of a common precursor for blood and vessels during the primitive wave of hematopoiesis, which has been termed the hemangioblast. The hemangioblast concept was initially postulated in the early 1920s based on studies in the chick embryo (Sabin, 1920). Hemangioblasts are believed to be transient and are regulated by signals that are antagonistic to cardiac development (Gering, Yamada, Rabbitts, & Patient, 2003; Schoenebeck, Keegan, & Yelon, 2007). Further evidence for the existence of a common hematopoietic and endothelial precursor comes from the *cloche* mutant, which lacks both blood and blood vessels but demonstrates no other defect in mesodermal organ formation (Stainier, Weinstein, Detrich, Zon, & Fishman, 1995). Notably, Vogeli, Jin, Martin, and Stainier (2006) used a fate-mapping approach to demonstrate that a cell population, identified during gastrulation, can develop into both endothelial precursors and HSCs, but no other mesodermal structures, providing in vivo evidence for the existence of a hemangioblast. Analyses of *Kdr* (formerly *Flk1*; Mouse Genome Informatics) deficient mice have further argued for the existence of the hemangioblast because these mice do not develop blood islands or yolk sac endothelial cells and produce very few hematopoietic progenitors (Shalaby et al., 1995, 1997). In contrast, murine blastocyst transplantation studies suggest that the development of primitive hematopoietic cells and blood vessels from common hemangioblast precursors is at best a very rare event in the early yolk sac, indicating that not all endothelium is hemangioblast-derived (Ueno & Weissman, 2006). However, analyses of hematopoietic differentiation in murine embryonic stem cells have demonstrated that rounded blood cells form from a cell with endothelial characteristics (Eilken, Nishikawa, & Schroeder, 2009; Lancrin et al., 2009). The observation of these specialized hemogenic endothelial cells has revised the original model of the hemangioblast (Lancrin et al., 2010). Recent studies indicate that mammalian hemogenic endothelial cells have distinct antigenic and gene expression profiles from non–blood-forming endothelium (Ditadi et al., 2015; Swiers et al., 2013). Intriguingly, cells giving rise to both blood and endothelium in zebrafish exclusively contributed to endothelium in the dorsal aorta and posterior cardinal vein, an area associated with hematopoietic specification (Vogeli et al., 2006). Thus, the importance and relative contribution of hemangioblasts to both vasculogenesis and hematopoiesis is an area of active and intense study.

PRIMITIVE HEMATOPOIESIS

During midsomitogenesis, cells in both the ALM and PLM begin to express genes of myeloid differentiation, such as *spi1* and *lcp1* (also known as *pu.1* and *l-plastin*, respectively) (Bennett et al., 2001), whereas *gata1*-expressing erythroid progenitors are mainly formed in the PLM. The hematopoietic cells in the PLM, together with their endothelial counterparts, converge to the midline and subsequently form the ICM region, the main site of primitive hematopoiesis. Several genes are involved in these early steps of blood formation. The zinc finger transcription factor *gata1* is expressed in the PLM and is considered the principal regulator of erythropoiesis, as evident in the *gata1* mutant *vlad tepes* (vlt^{m651}), which has few circulating red blood cells, but normal to enhanced production of myeloid cell types (Lyons et al., 2002). Knockdown studies have shown that *pbx2/pbx4* and *meis1* act upstream of *gata1* to affect primitive erythropoiesis (Pillay, Forrester, Erickson, Berman, & Waskiewicz, 2010). Spi1 is a central regulator of myeloid fate during the primitive wave (Hsu et al., 2004; Lieschke et al., 2002; Ward et al., 2003). Together, Gata1 and Spi1 function in a mutually antagonistic manner, such that lack of *gata1* expression leads to a conversion to myeloid fate (Galloway, Wingert, Thisse, Thisse, & Zon, 2005), while *spi1* knockdown causes ectopic *gata1* expression (Rhodes et al., 2005).

DEFINITIVE HEMATOPOIESIS

The definitive wave of hematopoiesis initially produces multipotent progenitors, and is followed by the production of HSCs that can give rise to all hematopoietic lineages for the lifetime of the organism. The anatomical locations and genes involved in HSC development during embryogenesis are highly conserved among vertebrate species (Fig. 1). Definitive hematopoiesis begins at ~24 hpf with the production of EMPs in the posterior blood island (Bertrand et al., 2007; Jin, Xu, & Wen, 2007). The first HSCs arise shortly thereafter from endothelium located in the VDA, also termed the AGM for its mammalian equivalent. Aortic hemogenic endothelium is characterized both by expression of endothelial genes (such as *kdrl*, formerly known as *flk1*) and transcription factors proven essential for the genesis and/or function of HSCs: *runx1*, *myb*, and subsequently, *itga2b* (previously called *CD41*). As in the mouse and chick embryo, there are successive anatomical sites that support definitive hematopoietic development following the initial specification of HSCs in the AGM. At ~44 hpf, myb^+ cells localize further distally in the tail, in the CHT, where they are also marked by expression of the surface marker *itga2b*; it has been argued that the CHT represents the equivalent of the mammalian fetal liver (Murayama et al., 2006), as both niches primarily support expansion of HSCs rather than de novo production. Eventually, from 96 hpf, HSCs seed the pronephros, either directly from the AGM or via the CHT (Bertrand, Kim, Teng, & Traver, 2008) and remain localized primarily in the kidney marrow (mammalian bone marrow equivalent) for the lifetime of the adult zebrafish.

As indicated above, several genes considered essential for definitive hematopoiesis are well-characterized markers of definitive HSC maturation. Runx1 (also known as acute myeloid leukemia protein 1 and core-binding factor a2) is a member of the Runt-domain containing family of transcription factors, which are highly evolutionarily conserved. Runt family members are involved in multiple developmental processes and often play a role in carcinogenesis. *RUNX1* is involved as a fusion gene partner in a significant fraction of cases of acute myelogenous and lymphoblastic leukemia (Gilliland, Jordan, & Felix, 2004). *Runx1* is expressed in hematopoietic clusters within the ventral endothelial layer of the mouse dorsal aorta (North et al., 1999), and knockout studies revealed its requirement for definitive hematopoiesis (Okuda, van Deursen, Hiebert, Grosveld, & Downing, 1996; Wang et al., 1996). In the zebrafish, *runx1* is similarly detected in HSCs emerging from the ventral endothelium of the dorsal aorta (Burns et al., 2002; Kalev-Zylinska et al., 2002). As in mice (Chen, Yokomizo, Zeigler, Dzierzak, & Speck, 2009; North et al., 2002), *runx1* knockdown in zebrafish does not significantly alter primitive erythropoiesis, indicating that it is not required for the first wave of hematopoiesis, but completely impairs HSC budding from the endothelial layer (Burns et al., 2002; Kalev-Zylinska et al., 2002; Kissa & Herbomel, 2010). Interestingly, analysis of a *runx1* truncation mutant ($runx1^{w84x}$) revealed that some homozygous fish can overcome the initial bloodless phase, with *itga2b:GFP*^{lo} cells emerging in the AGM and migrating to the kidney at a later time point (Sood et al., 2010). These data not only highlight the unique ability of the zebrafish to survive hematopoietic deficiency due to passive diffusion of oxygen from the aqueous environment but also suggest the presence of a "salvage" mechanism for hematopoietic stem and progenitor cells (HSPC) formation that does not require *runx1* function. In the zebrafish, *myb* is coexpressed with *runx1* in the AGM at 36 hpf and labels the HSPC population; its expression is dependent on Runx1, suggesting a transcriptional hierarchy. Myb belongs to a proto-oncogene family of transcription factors, and in the mouse, its absence leads to embryonic death due to lack of hepatic erythropoiesis (Mucenski et al., 1991), indicating its primary role in definitive (formerly called fetal liver) hematopoiesis. Myb is also essential for myeloid differentiation (Sakamoto et al., 2006). *myb:GFP* reporter fish (in combination with *lmo2*, and most recently, *kdrl*) have been used to identify and quantify effects on HSPCs in vivo (North et al., 2007; Bertrand, Chi, et al., 2010). The characterization of a *myb* mutant zebrafish (myb^{t25127}) demonstrated the high degree of evolutionary conservation in definitive HSC production; *myb* mutant embryos have no discernible defects in the primitive wave of hematopoiesis, but by 20 dpf (days post fertilization), they exhibit defects in all hematopoietic lineages, with stunted growth and premature death at 2–3 months (Soza-Ried, Hess, Netuschil, Schorpp, & Boehm, 2010). HSC maturation, including the ability to migrate and home to the niche, is indicated by the expression of *itga2b* (Bertrand et al., 2008; Jin et al., 2007), first shown to be associated with HSPC function in the mouse (Mikkola, Fujiwara, Schlaeger, Traver, & Orkin, 2003). Similarly, the pan-leukocyte marker Ptprc (previously called CD45) can be accurately utilized in the zebrafish, as in all vertebrates, to denote cells of all hematopoietic lineages (with the exception of erythrocytes) as they differentiate from the HSC population (Bertrand et al., 2008).

1. USE OF ZEBRAFISH TO INVESTIGATE HEMATOPOIETIC STEM CELL DEVELOPMENT

Zebrafish embryos have become extraordinarily useful for high-throughput screening analyses, due to the robust numbers of synchronous embryos obtained, rapid external development, and their ability to better survive "bloodless" phases that would cause rapid lethality in mammalian embryos (Rombough & Drader, 2009). Here, we highlight experimental approaches and findings from zebrafish studies that have contributed to our overall knowledge of HSC formation and differentiation.

1.1 FORWARD AND REVERSE GENETICS

An array of hematopoietic mutants have arisen as the result of spontaneous mutations, N-ethyl-N-nitrosourea (ENU) mutagenesis screens, targeted deletions, and induced transgenesis. With the improvement of gene targeting using zinc finger nucleases (Foley, Maeder, et al., 2009; Foley, Yeh, et al., 2009), TALENs (transcription activator-like effector nucleases) (Cade et al., 2012) and the CRISPR (clustered regularly interspaced short palindromic repeat) system for gene targeting with Cas9 nuclease, an increasing number of investigators are employing reverse genetics. Of note, CRISPR-Cas9 technology has been successfully used to recapitulate established hematopoietic phenotypes in a tissue-specific fashion (Ablain, Durand, Yang, Zhou, & Zon, 2015). Importantly, comparison of mutants generated from both targeted and unbiased screening approaches with their murine counterparts continues to illustrate the high level of conservation across vertebrates. Representative examples are highlighted in the following section; more exhaustive lists can be found elsewhere (de Jong & Zon, 2005; Martin, Moriyama, & Zon, 2011).

1.1.1 tal1

A mutant in *tal1* was found through a genetic screen to identify embryos with abnormal vascular development (Habeck, Odenthal, Walderich, Maischein, & Schulte-Merker, 2002). In addition to a severe reduction in endothelial alkaline phosphatase activity at 4 dpf, *tal1* mutants lack all blood lineages (Bussmann, Bakkers, & Schulte-Merker, 2007), mirroring the phenotypes reported in knockout mice (Robb et al., 1995; Shivdasani, Mayer, & Orkin, 1995). These fish also display defects in endocardial precursor migration and subsequent failure of endocardial development. These findings provide further support for the existence of a common vascular-hematopoietic precursor cell.

1.1.2 cloche

The mutant *cloche* (*clo*), also discovered via large-scale ENU screening, is characterized by gross cardiac edema and an enlarged atrium due to endocardial loss, giving it its name, which is French for "bell". *clo* mutants exhibit disrupted differentiation of both endothelial and hematopoietic structures, with almost complete loss of expression of *tal1*, *lmo2*, *gata1*, *mpx*, *lcp1*, and *kdrl* (Liao et al., 1997; Stainier et al., 1995). The effect of *clo* was shown to have both cell-autonomous and non—cell-autonomous features in blastula transplantation studies (Parker & Stainier, 1999). In contrast, other mesodermal organs do not show any defects, further indicating that *clo* might play a role in hemangioblast formation. Due to the telomeric location of the mutation on chromosome 13, the gene responsible for the *cloche* phenotype had not been elucidated for over 20 years. Using TALENs and CRISPR technologies, Reischauer et al. (2016) were able to assess the possible contribution of several candidate genes to the *clo* phenotype and conclusively revealed a novel basic helix-loop-helix-PAS transcription factor, *npas4l*, as the *cloche* gene; excitingly, the *clo* phenotype is partially rescued with a mammalian *npas4l* ortholog, indicating that *npas4l* function is well conserved.

1.1.3 bloodless

The bloodless (*bls*) mutant was identified as a spontaneous mutation. *bls* embryos exhibit a failure of the initial wave of primitive erythropoiesis, demonstrating no erythrocytes until 5 dpf. However, they can survive via passive diffusion of oxygen into all tissues due to ex vivo aqueous development and subsequently produce (definitive) erythrocytes (Liao et al., 2002). In *bls* mutants, *tal1* expression initiated in the PLM disappears by 23 hpf, and *gata1* expression is absent; in contrast, ALM myelopoiesis is not affected, indicating that this mutation specifically affects the primitive wave. The gene affected in the bloodless mutant has not yet been cloned.

1.1.4 vlad tepes

The *vlad tepes* (*vlt*) mutant carries a mutation in the *gata1a* transcription factor (Lyons et al., 2002; Weinstein et al., 1996), which is essential for both murine and human erythropoiesis (Fujiwara, Browne, Cunniff, Goff, & Orkin, 1996; Nichols et al., 2000). Although *vlt* mutants initiate expression of erythroid genes such as α-globin during somitogenesis, they are no longer expressed after 26 hpf; the loss of binding and function of Gata1a at the regulatory regions of erythroid target genes presumably causes mesodermal cells to lose the ability to maintain erythropoiesis, resulting in a switch to a myeloid fate (Galloway et al., 2005). The mutants eventually die between 8 and 15 dpf, although rare adults of small size have been reported when maintained under high tank aeration. *Vlt* continues to be a useful tool for the study of hematopoietic specification; mutants carrying one *vlt* allele and a second hypomorphic allele ($gata1^{T301K}$) completely lack primitive hematopoiesis but do exhibit definitive erythropoiesis, suggesting that primitive hematopoiesis is more reliant on Gata1 function (Belele et al., 2009). Most recently, in combination with *clo*, *vlt* zebrafish were used to characterize a novel factor involved in both erythroid and endothelial differentiation (Kawahara, Endo, & Dawid, 2012).

1.1.5 runx1 and cbfb

Runx1, along with its requisite binding partner Cbfb, has thus far been shown to be universally required for HSC formation from hemogenic endothelium during the definitive wave of hematopoiesis. As described above, Sood et al. identified a *runx1* truncation mutant (*runx1*^{w48x}) through TILLING (Targeting-Induced Local Lesions in Genomes). Interestingly, a subset of *runx1*^{w48x} zebrafish regain circulating blood cells at 15–21 dpf after a bloodless phase, allowing them to develop into adulthood; this finding has led the authors to postulate the existence of a secondary or "rescue" mechanism by which HSCs can be generated if Runx1 is not present; however, the factor(s) controlling this process have not yet been identified (Sood et al., 2010). Notably, targeted mutation of *cbfb* with the use of zinc finger nucleases has also revealed a lack of definitive hematopoiesis, with embryonic death reported to be at 14 dpf (Bresciani et al., 2014). At this time, it is unclear if definitive hematopoiesis is recovered in *cbfb* mutants as reported in *runx1*^{w48x} zebrafish.

1.1.6 mindbomb

The characterization of the hematopoietic defects in the mindbomb (*mib*) mutant confirmed the role of Notch signaling in HSC formation (Burns, Traver, Mayhall, Shepard, & Zon, 2005), which was initially described in the mouse (Kumano et al., 2003). As the name implies, *mib* mutants were originally described for their neural defects. The *mib* mutation causes a defect in the *mindbomb E3 ubiquitin protein ligase 1* gene, which is required for the processing of the Notch ligands Delta and Jagged. *mib* mutants have normal primitive erythropoiesis, but exhibit defects in definitive hematopoiesis with significantly reduced *runx1* and *myb* expression in the VDA. Recent data investigating the formation of EMPs as a hematopoietic wave preceding HSC development found no Notch receptors on Gata3-expressing EMPs, and no defect in this cell population in *mib* mutants (Bertrand, Cisson, Stachura, & Traver, 2010). This phenotype is in agreement with analyses of *Notch1*$^{-/-}$ murine embryos, and chimeric embryos containing *Notch1*$^{-/-}$ *cells*, which maintain functional EMPs but fail to produce long-term HSCs (Hadland et al., 2004; Kumano et al., 2003).

1.1.7 myb

As described previously, *myb*^{t25127} zebrafish exhibit no discernible defects in primitive hematopoiesis, but by 20 dpf, defects in all hematopoietic lineages are apparent, resulting in premature death at 2–3 months (Soza-Ried et al., 2010). The differential requirement for Myb in definitive but not primitive hematopoiesis reiterates the conservation of key genes required for normal definitive hematopoiesis in vertebrates. Significantly, the prolonged survival of *myb*^{t25127} zebrafish makes them an ideal recipient model for engraftment studies that do not require preconditioning (Hess, Iwanami, Schorpp, & Boehm, 2013).

1.1.8 rumba, samba, and tango

rumba, *samba*, and *tango* are ENU-induced mutants that were identified by the loss of *rag1* expression in the developing thymus. Subsequently, these mutants were discovered to exhibit specific defects in the proliferation of HSCs in the CHT (Du et al., 2011; Li, Lan, Xu, Zhang, & Wen, 2012). *rumba* and *samba* encode for a novel C2H2 zinc finger factor with unknown function and a protein homologous to the human augmin complex subunit 3, respectively. *tango*^{hkz5} zebrafish harbor a point mutation in *small ubiquitin-related modifier-activating enzyme subunit 1* (*sae1*), causing reduced affinity for its binding partner, Sae2. All of these mutants are characterized by normal *myb* expression at 30–36 hpf but exhibit progressive loss of HSC markers and myeloid cell types at 3–5 dpf, with *rumba* affecting the early phase of CHT population at 3 days and *samba* and *tango* modifying the subsequent expansion phase. These mutants all display a loss of cell cycle progression, while *tango* mutants also exhibit increased HSC apoptosis, resulting in death by 10 dpf. All three mutants are cell-autonomously required to maintain the HSC pool in the CHT.

1.1.9 *cebpa*

TALEN-mediated targeting of the zebrafish transcriptional regulator *CCAAT/enhancer binding protein alpha (c/EBPa)* confirmed that sumolyation of c/EBPa regulates *runx1* expression and HSPC proliferation during embryogenesis. Interestingly, *cebpa* mutants rescue defects in HSPC emergence caused by the knockdown of all genes in the sumolyation pathway. Thus, it appears that zebrafish *cebpa* negatively regulates HSC proliferation, as was reported in mice (Heath et al., 2004; Yuan et al., 2015).

1.1.10 *vhl*

Zebrafish lacking *vhl*, an endogenous inhibitor of HIF1α activity in normoxic conditions, were identified from a library of ENU-induced mutants through the use of TILLING. HIF1α activity is tightly controlled and is known to be required for initiation and regulation of hematopoiesis in mammals (Adelman, Maltepe, & Simon, 1999; Ramírez-Bergeron et al., 2004, 2006). Complete loss of *Vhl* during murine development also leads to early lethality, however, conditional alleles do not exhibit all pathologies associated with human *VHL* mutations. In contrast, *vhl* mutant zebrafish appear to recapitulate clinical features of the human VHL-associated disease Chuvash polycythemia (van Rooijen et al., 2009). *vhl*-deficient zebrafish are viable until 11–14 dpf, providing a new vertebrate model for the study of hypoxia signaling. Recently, this model was used to confirm the role of HIF1α signaling in definitive hematopoiesis (Harris et al., 2013), which was subsequently demonstrated in conditional mutant mice (Imanirad et al., 2014).

1.2 CHEMICAL SCREENS

Over the past several years, chemical screening approaches have also yielded great insight into the biology of hematopoiesis. Pioneered in the 2000s, chemical screening provides the ability to simultaneously interrogate the influence of a broad array of compounds on a biological process (Peterson, Link, Dowling, & Schreiber, 2000; for review, see Rennekamp & Peterson, 2015). Furthermore, unlike mutation, knockdown, or over expression analysis, the timing and duration of exposure can be

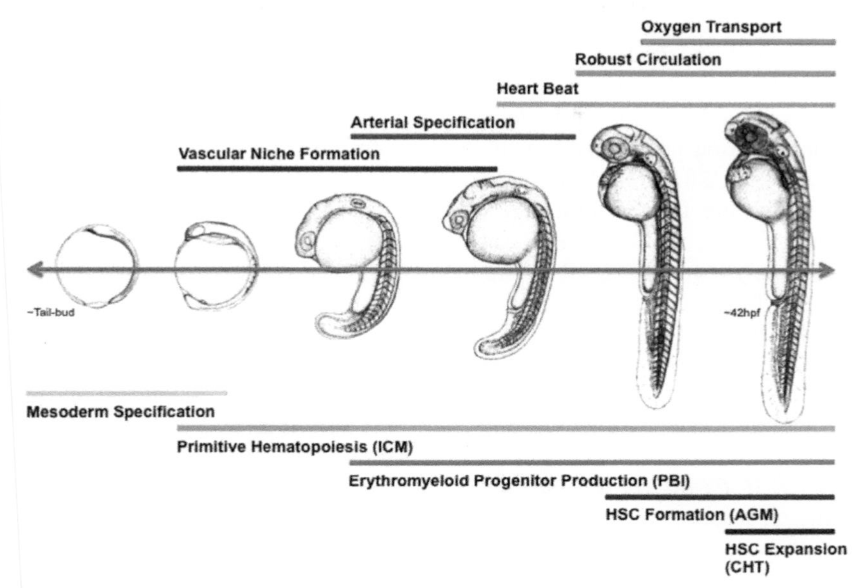

FIGURE 2

Depiction of developmental milestones for determination of exposure windows during chemical screens to impact hematopoiesis.

altered to more precisely impact the population of interest during hematopoietic development (Fig. 2). Screening can be conducted on wild-type fish scored for alterations in developmental expression patterns or performed with lineage-specific hematopoietic reporters; it can likewise be used to modify or rescue alterations in the hematopoietic program caused by mutation or transgenesis. Although we will focus on developmental hematopoiesis, chemical screens have also successfully identified compounds that resolve inflammation (Liu et al., 2013) and exhibit anti-leukemia properties (Gutierrez et al., 2014; Ridges et al., 2012).

1.2.1 PGE2 and wnt

The initial hematopoietic regulatory pathway discovered by this approach is that of prostaglandin E2 (PGE2). In a chemical genetic screen in zebrafish embryos for modulators of the expression of HSC markers *runx1* and *myb*, exposure to exogenous PGE2, the most ubiquitous prostanoid produced during development (Grosser, Yusuff, Cheskis, Pack, & FitzGerald, 2002), significantly enhanced HSC formation (North et al., 2007). PGE2 also positively influences adult zebrafish hematopoiesis (North et al., 2007) and plays a role in lymphoid development (Villablanca et al., 2007). Through the use of multiple inducible transgenic fish to modify Wnt signaling, PGE2 was found to affect HSC formation by promoting β-catenin activation via cyclic adenosine monophosphate (cAMP)-mediated activation of protein kinase A (PKA) downstream of EP2/EP4 stimulation (Goessling et al., 2009). Eicosanoid stimulation of the related cannabinoid receptor, CNR2, also regulates HSPC production and function in part by upregulating endogenous PGE2 synthesis (Esain et al., 2015). In mammals, macrophage-derived PGE2 was found to promote retention and expansion of HSCs in the marrow through EP4 signaling (Hoggatt et al., 2013; Ikushima et al., 2013; Ludin et al., 2012). These findings, together with preclinical studies using human umbilical cord blood (UCB) (Goessling et al., 2011), established PGE2 as a regulator of HSCs across vertebrate species and ultimately led to the first Food and Drug Administration-approved clinical trial originating from a chemical screen in zebrafish. In phase I trials, ex vivo stimulation of human UCB stem cells with dmPGE2 (dimethyl PGE2, a long-acting derivative of PGE2) enhanced and accelerated engraftment in human UCB transplantation (Cutler et al., 2013); phase 2 trials are ongoing, highlighting the potential utility of the zebrafish model for therapeutic compound screening.

1.2.2 Nitric oxide and blood flow

A chemical screening approach also resulted in the identification of several groups of compounds that had the common ability to regulate blood flow by affecting heart rate, contractility, or vascular diameter (North et al., 2009). The chemicals included agents acting on α- and β-adrenergic receptors and those affecting calcium flux and angiotensin converting enzyme signaling; however, the action of each ultimately impacted blood flow in the VDA. The only compounds able to affect HSC formation before the onset of a heart beat at 24 hpf were those modulating nitric oxide (NO) signaling. NO donors, such as S-nitroso-N-acetylpenicillamine, enhanced the population of *runx1*- and *myb*-positive HSCs in the AGM at 36 hpf, whereas the NO synthesis inhibitor N-nitro-L-arginine methyl ester diminished HSC formation. The mutant *silent heart* (*sih*), which lacks a heart beat and blood flow due to a mutation in *troponin T2a* (*tnnt2a*), exhibits diminished HSC formation, indicating the importance of blood flow for HSC formation. Interestingly, NO donors rescued the HSC defect in *sih* mutant embryos, indicating that NO signaling is the downstream signal linking blood flow to HSC development in the AGM. Similarly, mimicking blood flow in vitro induced hematopoietic commitment in differentiating murine embryonic stem cells (Adamo et al., 2009). More recently, blood flow was found to stimulate mechanosensors that trigger calcium flux, inducing upregulation of *Ptgs2* (also known as *Cox2*), which is responsible for PGE2 production (Diaz et al., 2015). Subsequent increases in PGE2 biosynthesis and cAMP-PKA-CREB activity induced by blood flow were necessary for HSC emergence (Kim et al., 2015). Blood flow also induces adenosine release in zebrafish, which was found to be important for HSC emergence (Jing et al., 2015).

1.2.3 Epoxyeicosatrienoic acids

A screen for bioactive compounds that improve adult HSC engraftment uncovered a role for epoxyeicosatrienoic acids (EETs). Transient treatment of fluorescent zebrafish HSCs with EETs increased hematopoietic repopulation in transparent *casper* fish. Like PGE2, EETs also increased *runx1* expression during HSC specification. Although EETs and PGE2 are both derived from arachadonic acid, they function through distinct signaling cascades, with EETs affecting activator protein 1 activity through PI3Kγ (Li et al., 2015), suggesting endogenous synthesis of both mediators could synergize to influence hematopoietic activity in vivo.

1.2.4 Rescue screens

Recent chemical screens in the zebrafish have focused on the rescue of a mutant phenotype or a transgenically induced alteration. Paik, de Jong, Pugach, Opara, and Zon (2010) used a bioactive library to identify chemicals that rescue *gata1* expression in *kgg* embryos, which exhibit defects in both primitive and definitive hematopoiesis. Two psoralens, bergapten and 8-methoxypsoralen, that rescued the *cdx4* mutant defect were identified. In separate studies, Yeh et al. (2008) sought to isolate modulators of the hematopoietic differentiation defect induced by the leukemogenic transgenic fusion construct *runx1−mtg8* (also known as *aml1-eto*). The transgenic fish typically have an arrest in erythropoietic differentiation; the screen revealed the Cyclooxygenase 2 inhibitor nimesulide as capable of preventing this block, further documenting the importance of prostaglandin signaling in hematopoietic regulation (Yeh et al., 2009).

Taken together, chemical screens have proven to be a powerful tool for elucidating the molecular mechanisms required for HSC formation and for maintenance in adult vertebrates. Screens can be adapted to not only examine normal developmental hematopoiesis but can also be performed in the presence of a mutant phenotype or model of leukemia. With the improvement of gene-targeting methods, increased utility of rescue or modifier screens will surely continue to impact our knowledge of HSC biology.

1.3 ENVIRONMENTAL REGULATORS OF HEMATOPOIETIC STEM CELL DEVELOPMENT

Because of their external development, zebrafish embryos serve as an ideal model to investigate the effects of environmental factors that are induced or rescued by maternal contribution in mammals. Recent chemical screens have begun to uncover modifiers of common hormonal and metabolic cascades, and analyses of these targets suggest that heightened exposure or activation of these pathways during embryogenesis impart lasting effects on the hematopoietic system.

1.3.1 Nuclear hormone receptor signaling

One of the first nuclear hormone receptors recognized to influence hematopoiesis in mammals is that induced by retinoic acid (RA). RA signaling regulates HSC homeostasis and differentiation in the adult (reviewed by Purton, 2007), and its synthesis is required for the onset of definitive hematopoiesis in mice (Chanda, Ditadi, Iscove, & Keller, 2013; Goldie, Lucitti, Dickinson, & Hirschi, 2008). Interestingly, RA signaling appears to dampen primitive hematopoiesis in mice and zebrafish (de Jong et al., 2010). Recently, 17β-estradiol and vitamin D were each identified as environmental regulators of HSPC formation in the aforementioned chemical screen (Carroll et al., 2014; Cortes et al., 2015). As HSCs arise from endothelial cells, correct establishment of the artery is required for HSC formation. The Hedgehog-vascular endothelial growth factor (VEGF)-Notch signaling axis is essential for the formation of this arterial hemogenic niche and subsequent HSC formation (Lawson, Vogel, & Weinstein, 2002). 17β-estradiol antagonizes VEGF signaling to alter arterial endothelial identity and thereby preclude HSPC formation; endogenous estrogen in the yolk sac was found to serve as a ventral-limiting factor to confine the hemogenic vascular niche to the artery (Carroll et al., 2014). Interestingly, while the active form of vitamin D positively influences hematopoietic specification through stimulation of the vitamin D nuclear hormone receptor (Cortes et al., 2016), the precursor cholecalciferol (D3) was found to antagonize the Hh pathway, acting on the extracellular sterol-binding domain of smoothened, thus impairing arterial identity and HSPC production in a similar fashion as estrogen (Cortes et al., 2015).

1.3.2 Inflammatory signals in hematopoietic stem cell emergence

Recent studies have identified several proinflammatory cytokines that are necessary and sufficient for vertebrate HSC emergence (Espín-Palazón et al., 2014; He et al., 2015; Li et al., 2014; Sawamiphak, Kontarakis, & Stainier, 2014) and continue to influence HSC homeostasis in the adult (Baldridge, King, Boles, Weksberg, & Goodell, 2010; Pietras et al., 2016; Takizawa, Regoes, Boddupalli, Bonhoeffer, & Manz, 2011). Induction of tumor necrosis factor α is required for the specification and emergence of HSCs through activation of Notch signaling and nuclear factor kappa-light-chain-enhancer of activated B cells (NFκB) (Espín-Palazón et al., 2014). The Tlr4bb-MyD88-NFkB signaling axis was subsequently found to promote HSC emergence by acting upstream of Notch signaling (He et al., 2015), while the proinflammatory cytokine Interferon-γ and its receptor Crfb17 were shown to be downstream effectors of Notch signaling that positively influence endothelial-to-hematopoietic transition and HSC specification in fish (Li et al., 2014; Sawamiphak et al., 2014). Importantly, murine embryos lacking Interferon-γ produced significantly fewer AGM HSPCs, suggesting conservation of function between species (Kanz, Konantz, Alghisi, North, & Lengerke, 2016; Li et al., 2014). Together, these data highlight the molecular complexity of Notch activation in ECs and/or HSCs during HSC generation. Future studies will be required to explore whether inflammatory signaling can be utilized to enhance in vitro HSC specification from pluripotent stem cells or to modulate Notch activity for HSC expansion. Furthermore, as the implicated source of cytokines is the primitive myeloid population, these studies have also introduced the novel concept that earlier waves of hematopoiesis may regulate de novo HSC production (Espín-Palazón et al., 2014; Li et al., 2014).

1.3.3 Metabolic regulation of hematopoietic stem cell development

Interestingly, the emergence of HSCs temporally coincides with fluctuations in nutrient availability and energy. In a targeted screen to identify novel environmental factors impacting embryonic HSC development, exposure of zebrafish embryos to 1% D-glucose was found to promote the production of Runx1+ HSCs during embryogenesis without systemic alterations to growth or vascular development (Harris et al., 2013). Increased glycolysis and oxidative phosphorylation led to an increase in reactive oxygen species, which in turn stabilize Hif1α (Harris et al., 2013; Pan et al., 2007). The role of Hif1α as the mediator of the glucose response was confirmed by the loss of *runx1* expression in the VDA of *hif1α* morphants. HIF1α signaling has long been known to play a functional role in early hematopoietic development, as loss of this pathway leads to a lack of primitive hematopoiesis (Adelman et al., 1999; Ramírez-Bergeron et al., 2004, 2006). Conditional loss of *Hif1α* confirmed that it also regulates HSC development in the mammalian embryo (Imanirad et al., 2014) and maintains quiescence in adult HSCs (Simsek et al., 2010; Takubo et al., 2010). In more recent studies, adenosine, a purinergic signaling molecule involved in biochemical processes such as energy transfer, was found to regulate HSPC emergence by regulating Scl-mediated hematopoietic commitment from endothelium (Jing et al., 2015), further highlighting a role for metabolic signaling in HSC development.

2. ZEBRAFISH TOOLS AND PROTOCOLS

In recent years, great effort has been put forth to enhance the ability to quantify the effects of genetic (or chemical) hematopoietic regulation in the zebrafish. The use of transgenic reporters is now widespread and offers unprecedented ability for in vivo characterization; these lines can be further used to quantify effects of interest by fluorescence-activated cell sorting (FACS) analysis, as is standard in mammalian systems, or to enrich populations for genomic or functional analysis. Both flow cytometric analysis and functional approaches, such as transplantation (Traver et al., 2003) and in vitro culture (Stachura et al., 2009), were modified from the murine system and are now available in the zebrafish; while much of the work to date has focused on the adult (de Jong et al., 2011), current protocols are being adapted to study embryonic hematopoiesis (Bertrand, Chi, et al., 2010; Lam, Hall, Crosier, Crosier, & Flores, 2010; North et al., 2007).

2.1 TOOLS—TRANSGENIC REPORTER LINES

The generation of transgenic reporter lines (Table 1) to facilitate study of organs and cell types of interest has substantially increased due to the development of enhanced transgenesis methodologies, as well as the use of recombination-based strategies for fate mapping. These reporter lines have been used to highlight developmental cell populations and processes in vivo, to isolate cell populations for gene expression studies, and for cell transplantation experiments.

Table 1 Compilation of Currently Published Reporter Zebrafish Lines to Highlight Blood and Vessel Development

Transgenic Line	Target Cell Population
Tg(kdrl:mCherry)is5,Tg(kdrl:RFP)la4, Tg(kdrl:GRCFP) zn10, Tg(kdrl:GFP)la116	Vasculature
Tg(fli1a:EGFP)y5,Tg(fli1a.ep:DsRedEx)um13, Tg(fli1a: nEGFP)y7	Vasculature
Tg(lmo2:DsRed)zf73, Tg(lmo2:EGFP)zf71	Vasculature, HSC
Tg(gata1a:GFP)la781,Tg(-8.1gata1:gata1-EGFP)zf100, Tg(gata1:DsRed)sd2	Erythrocytes
Tg(ptprc:DsRed)sd3	Myeloid cells
Tg(mpx:GFP)i114, Tg(-8mpx:Dendra2)uwm4, Tg(-8mpx: mCherry)uwm5	Neutrophils, monocytes
Tg(-6.0itga2b:EGFP)la2	Thrombocytes, HSC, EMP
Tg(rag2:GFP)la6; Tg(rag2:dsRed)	Lymphoid precursors
Tg(lck:lck-EGFP)cz2	T cells
Tg(cmyb:EGFP)zf169	HSC
Tg(runx1P2:EGFP)zf188	HSC
Tg(-5.0tal1:EGFP)sq1	HSC
TgBAC(gata2b:KalTA4)sd32 × *Tg(4xUAS:GFP)hzm3*	Hemogenic endothelium
Tg(Mmu.Runx1:NLS-mCherry); Tg(Mmu.Runx1:EGFP)	HSPC
Tg(mpeg:eGFP)gl22	Macrophages

2.2 PROTOCOL—CHEMICAL SCREEN FOR HEMATOPOIETIC STEM CELL FORMATION

In addition to its advantages for forward genetics in mutagenesis screens, zebrafish are uniquely amenable to chemical interrogation of developmental processes. In our experience, zebrafish are best used in this context for a global assessment of in vivo processes, using chemical libraries of well-characterized compounds that serve as indicators of signaling pathways of interest. Although recent advances have enabled enhanced throughput, such as in situ robots or automated fluorescence imaging, traditional screening approaches may provide the biggest advantage due to the ability to assess phenotypic changes in the context of whole organism modifications.

1. Harvest and pool stage–matched embryos; about 250–500 embryos are required for each 48-well plate.
2. Slow embryo development in the incubator at 22°C until the five somite stage.
3. Prepare multiwell plates: fill each well with 1 mL E3 embryo buffer. Note: we have found that 48-well plates allow better embryo survival (and the ability to remove dead embryos) than 96-well plates when filled with 5–10 embryos.
4. Prepare chemical compounds from library. Most premade libraries will have stocks of 10–20 mM compounds, dissolved in DMSO. Transfer desired

amount of each compound into each well. Important: in our experience, combinatorial exposure to two or more bioactive compounds at the same time may result in excessive toxicity.

5. Using a small spatula, aliquot 5–10 embryos/well, minimizing transfer of additional water. Note: this can be achieved most easily by removing excess embryo buffer from the embryo plate before aliquoting the embryos via the "Caviar" method (pick up relatively dry embryos in chorions with spatula and flick against the top side wall of each well).
6. Incubate embryos in drug at 28.5°C until 36 hpf for HSC screening. Note: use the developmental stage of control wells to assess the stopping point; do not wait for delayed embryos to catch up and fix separately.
7. Process for in situ hybridization. We found the combination of *runx1* and *myb* to be more robust as HSC markers than either one alone for screening purposes. Alteration: if using HSC reporters, skip to #8; we recommend fixing embryos to confirm phenotype if scoring fluorescent reporters for analysis.
8. Score embryos for HSC alterations, using untreated and/or DMSO-treated embryos from the same clutch as reference. Note: in early stages of the screen, it is helpful to seed each plate with a positive and/or negative control in addition to the untreated well.
9. Identify compound wells that affect organ formation and correlate with chemical library; group-related hits according to the mechanism of action and/or biological function.
10. Retest "hits" to confirm phenotype. Note: for known bioactives, you do not need to use the exact compound from the screen in all cases if a prototypical or classically used related compound from the same pathway/target class is more readily available.

2.3 PROTOCOL—CONFOCAL MICROSCOPY FOR ANALYSIS OF HEMATOPOIETIC STEM CELL FORMATION AND BLOOD FLOW

Using the optical clarity of the developing zebrafish and innovative transgenic constructs, several groups have recently visualized the emergence and migration of HSCs and hematopoietic progenitors in vivo (Fig. 3). Bertrand et al. (2008) visualized colonization of the pronephros with *itga2b:GFP*$^+$ HSCs. To demonstrate the direct formation of HSCs from vascular precursors, the same group performed confocal time lapse imaging of *myb:EGFP*; *kdrl:mCherry* double transgenic zebrafish embryos (Bertrand, Chi, et al., 2010). Lineage tracing confirmed that *myb:EGFP*$^+$ cells developed directly from *kdrl:mCherry*$^+$ cells in the ventral wall of the dorsal aorta, forming buds extending into the vascular lumen. Similarly, Kissa and Herbomel (2010) used a *kdrl:GFP* transgenic line to observe the interaction and movements of endothelial cells in the dorsal aorta, including the emergence and release of HSCs from the vessel wall, which was defective in *runx1* mutants. Using the recently established *runx1P2:GFP* transgenic line, Lam et al. (2010) also visualized the direct maturation and emergence of HSCs from

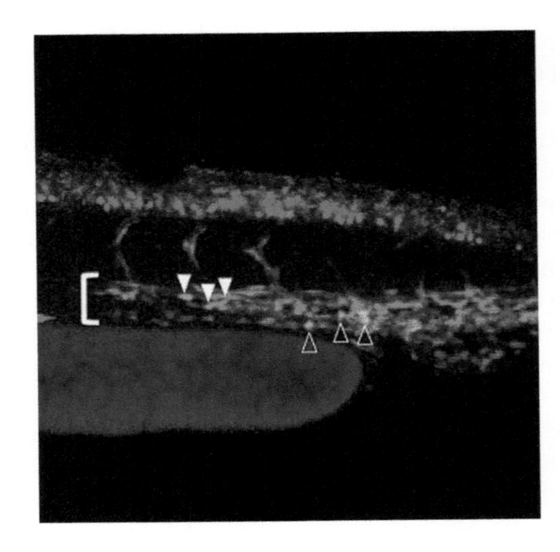

FIGURE 3

Confocal image of the aorta (bracket) in a *kdrl:DsRed*; *myb:GFP* embryo at 48 hpf. Blue staining denotes Edu + nuclei. Closed arrowheads indicate examples of $Kdrl^{+}$ Myb^{lo} hemogenic endothelium. Open arrowheads indicate examples of $Kdrl^{+}$ Myb^{+} HSCs.

$Kdrl^{+}$ endothelial cells and confirmed the conserved use of the distal *runx1* promoter for HSC regulation. Together, these seminal studies confirmed long-standing hypotheses that vertebrate HSCs emerge from endothelium, demonstrating the importance of the zebrafish model for analyzing complex and essential developmental processes in vivo.

We have previously demonstrated that *myb:GFP*$^{+}$ *kdrl:DsRed*$^{+}$ HSCs can be enumerated (within a selected field) for the purpose of quantifying an effect of a pathway of interest. Other transgenic lines may be utilized as desired (Table 1).

1. Harvest appropriate transgenic embryos.
2. Expose embryos to chemicals from five somite stage until 36 hpf.
3. Using fluorescence stereomicroscopy, identify transgenic embryos.
4. Anesthetize embryos in tricaine solution.
5. Embed embryos in 1% low-melting agarose (in E3 embryo water) using glass-bottom optical dishes.
6. Cover agarose with cover glass.
7. For HSC imaging, confocal microscopy using lasers with appropriate wavelength for the most commonly used fluorophores is performed focused on the tail region; we use a 40× long-distance lens for high-resolution imaging.
   Alteration: to assess blood flow, a spinning disc confocal microscope, such as the Perkin Elmer UltraView VoX, can be utilized.

8. Following image acquisition and processing with confocal software, HSCs can be enumerated in a fixed 3-D field; several embryos should be imaged and counted to produce statistically meaningful data. Note: we use the end of the yolk sac extension in each embryo as a fixed point of interest to center the image for counting.

2.4 PROTOCOL—USE OF FLOW CYTOMETRY TO QUANTIFY ALTERATIONS IN HSPC NUMBER OR FUNCTION

The availability of a vast array of antibodies has greatly facilitated the characterization of hematopoietic cell types in humans and mouse. Although this resource is largely absent in the zebrafish, flow cytometry can be used to analyze the zebrafish hematopoietic system by taking advantage of the transgenic reporter lines. The FACS-based approach to identify hematopoietic cell types in the adult kidney marrow was pioneered by Traver et al. (2003). Here, we describe the procedure for characterizing and quantifying changes in embryonic hematopoietic cell types in mutants, morphants, and/or compound-treated embryos.

1. Use appropriate transgenic reporter fish as highlighted in Table 1; it is necessary to cross line to highlight your cell type of interest if each labels multiple populations (e.g., *myb:GFP* × *kdrl:DsRed* to mark HSCs). Alteration: crossing reporters can also help you identify modifications of multiple lineages or stages of differentiation in single embryos (e.g., itga2b × ptprc); care must be taken to ensure each embryo expresses all reporter constructs prior to analysis for accurate comparison.
2. Perform appropriate manipulations and treatments to analyze embryos at desired time point.
3. Separate nonfluorescent and fluorescent embryos under the microscope; singly fluorescent embryos (of a given color) and wild-type embryos will serve as gating controls.
4. Pool equal numbers of control (as above) or fluorescent embryos into separate 1.5 mL microfuge tubes. Note: At least five embryos are recommended per sample.
5. Remove all excess fish water by pipet.
6. Rinse with 0.9× phosphate buffered saline (PBS).
7. Remove PBS and add 500 μL of liberase solution (50 μg/mL in 0.9× PBS).
8. Incubate for 1 h at 33°C in a thermomixer (850 rpm). Note: Instead of a thermomixer, a water bath can be used to dissociate embryos. In this case, we recommend dissociating the embryos more often (every 20–30 min).
9. Gently dissociate the embryos by pipetting up and down with a P1000 pipette (approximately 15–20 times). If embryos are not fully dissociated, place them back at 33°C and repeat the dissociation process every 30 min. For embryos younger than 48 hpf, up to 2 h may be required; older embryos typically digest in about 1 h.

10. Strain homogenate through a 30 μm mesh filter (Celltrics) into a 5 mL FACS tube to remove any tissue aggregates.
11. Immediately add 3 mL of 0.9× PBS through the filter to rinse the filter and dilute the liberase.
12. Centrifuge tubes at 300 g for 5 min at 4°C.
13. Carefully remove the supernatant.
14. Resuspend in a small amount (approximately 100 μL) of 0.9× PBS containing 5 nM Sytox Red (or other live/dead indicator of choice) to exclude dead cells from analysis.
15. Collect data on a flow cytometer, using negative and single-color embryos to set up proper voltages, gating, and compensation.
16. Examine data using available software (such as FlowJo) (Esain, Cortes, & North, 2016) and calculate alterations in cell number across biological replicates.

REFERENCES

Ablain, J., Durand, E. M., Yang, S., Zhou, Y., & Zon, L. I. (2015). A CRISPR/Cas9 vector system for tissue-specific gene disruption in zebrafish. *Developmental Cell, 32*, 756–764.

Adamo, L., Naveiras, O., Wenzel, P. L., McKinney-Freeman, S., Mack, P. J., Gracia-Sancho, J., … Daley, G. Q. (2009). Biomechanical forces promote embryonic haematopoiesis. *Nature, 459*, 1131–1135.

Adelman, D. M., Maltepe, E., & Simon, M. C. (1999). Multilineage embryonic hematopoiesis requires hypoxic ARNT activity. *Genes and Development, 13*, 2478–2483.

Baldridge, M. T., King, K. Y., Boles, N. C., Weksberg, D. C., & Goodell, M. A. (2010). Quiescent haematopoietic stem cells are activated by IFN-gamma in response to chronic infection. *Nature, 465*, 793–797.

Belele, C. L., English, M. A., Chahal, J., Burnetti, A., Finckbeiner, S. M., Gibney, G., … Liu, P. P. (2009). Differential requirement for Gata1 DNA binding and transactivation between primitive and definitive stages of hematopoiesis in zebrafish. *Blood, 114*, 5162–5172.

Bennett, C. M., Kanki, J. P., Rhodes, J., Liu, T. X., Paw, B. H., Kieran, M. W., … Look, A. T. (2001). Myelopoiesis in the zebrafish, *Danio rerio*. *Blood, 98*, 643–651.

Bertrand, J. Y., Chi, N. C., Santoso, B., Teng, S., Stainier, D. Y. R., & Traver, D. (2010). Haematopoietic stem cells derive directly from aortic endothelium during development. *Nature, 464*, 108–111.

Bertrand, J. Y., Cisson, J. L., Stachura, D. L., & Traver, D. (2010). Notch signaling distinguishes 2 waves of definitive hematopoiesis in the zebrafish embryo. *Blood, 115*, 2777–2783.

Bertrand, J. Y., Kim, A. D., Teng, S., & Traver, D. (2008). CD41+ cmyb+ precursors colonize the zebrafish pronephros by a novel migration route to initiate adult hematopoiesis. *Development, 135*, 1853–1862.

Bertrand, J. Y., Kim, A. D., Violette, E. P., Stachura, D. L., Cisson, J. L., & Traver, D. (2007). Definitive hematopoiesis initiates through a committed erythromyeloid progenitor in the zebrafish embryo. *Development, 134*, 4147–4156.

Bresciani, E., Carrington, B., Wincovitch, S., Jones, M., Gore, A. V., Weinstein, B. M., ... Liu, P. P. (2014). CBF and RUNX1 are required at 2 different steps during the development of hematopoietic stem cells in zebrafish. *Blood, 124*, 70–78.

de Bruijn, M. F., Speck, N. A., Peeters, M. C., & Dzierzak, E. (2000). Definitive hematopoietic stem cells first develop within the major arterial regions of the mouse embryo. *The EMBO Journal, 19*, 2465–2474.

Burns, C. E., DeBlasio, T., Zhou, Y., Zhang, J., Zon, L., & Nimer, S. D. (2002). Isolation and characterization of runxa and runxb, zebrafish members of the runt family of transcriptional regulators. *Experimental Hematology, 30*, 1381–1389.

Burns, C. E., Traver, D., Mayhall, E., Shepard, J. L., & Zon, L. I. (2005). Hematopoietic stem cell fate is established by the Notch-Runx pathway. *Genes and Development, 19*, 2331–2342.

Bussmann, J., Bakkers, J., & Schulte-Merker, S. (2007). Early endocardial morphogenesis requires Scl/Tal1. *PLoS Genetics, 3*, e140.

Cade, L., Reyon, D., Hwang, W. Y., Tsai, S. Q., Patel, S., Khayter, C., ... Yeh, J.-R. J. (2012). Highly efficient generation of heritable zebrafish gene mutations using homo- and heterodimeric TALENs. *Nucleic Acids Research, 40*, 8001–8010.

Carroll, K. J., Esain, V., Garnaas, M. K., Cortes, M., Dovey, M. C., Nissim, S., ... North, T. E. (2014). Estrogen defines the dorsal-ventral limit of VEGF regulation to specify the location of the hemogenic endothelial niche. *Developmental Cell, 29*, 437–453.

Chanda, B., Ditadi, A., Iscove, N. N., & Keller, G. (2013). Retinoic acid signaling is essential for embryonic hematopoietic stem cell development. *Cell, 155*, 215–227.

Chen, M. J., Yokomizo, T., Zeigler, B. M., Dzierzak, E., & Speck, N. A. (2009). Runx1 is required for the endothelial to haematopoietic cell transition but not thereafter. *Nature, 457*, 887–891.

Cortes, M., Chen, M. J., Stachura, D. L., Liu, S. Y., Kwan, W., Wright, F., ... North, T. E. (2016). Developmental vitamin D availability impacts hematopoietic stem cell production. *Cell Reports, 17*, 458–468.

Cortes, M., Liu, S. Y., Kwan, W., Alexa, K., Goessling, W., & North, T. E. (2015). Accumulation of the vitamin D precursor cholecalciferol antagonizes hedgehog signaling to impair hemogenic endothelium formation. *Stem Cell Reports, 5*, 471–479.

Cutler, C., Multani, P., Robbins, D., Kim, H. T., Le, T., Hoggatt, J., ... Shoemaker, D. D. (2013). Prostaglandin-modulated umbilical cord blood hematopoietic stem cell transplantation. *Blood, 122*, 3074–3081.

Detrich, H. W., Kieran, M. W., Chan, F. Y., Barone, L. M., Yee, K., Rundstadler, J. A., ... Zon, L. I. (1995). Intraembryonic hematopoietic cell migration during vertebrate development. *Proceedings of the National Academy of Sciences of the United States of America, 92*, 10713–10717.

Diaz, M. F., Li, N., Lee, H. J., Adamo, L., Evans, S. M., Willey, H. E., ... Wenzel, P. L. (2015). Biomechanical forces promote blood development through prostaglandin E2 and the cAMP-PKA signaling axis. *The Journal of Experimental Medicine, 212*, 665–680.

Ditadi, A., Sturgeon, C. M., Tober, J., Awong, G., Kennedy, M., Yzaguirre, A. D., ... Keller, G. (2015). Human definitive haemogenic endothelium and arterial vascular endothelium represent distinct lineages. *Nature Cell Biology, 17*, 580–591.

Du, L., Xu, J., Li, X., Ma, N., Liu, Y., Peng, J., ... Wen, Z. (2011). Rumba and Haus3 are essential factors for the maintenance of hematopoietic stem/progenitor cells during zebrafish hematopoiesis. *Development, 138*, 619–629.

Eilken, H. M., Nishikawa, S.-I., & Schroeder, T. (2009). Continuous single-cell imaging of blood generation from haemogenic endothelium. *Nature, 457*, 896–900.

Esain, V., Cortes, M., & North, T. E. (2016). Enumerating hematopoietic stem and progenitor cells in zebrafish embryos. *Methods in Molecular Biology, 1451*, 191–206.

Esain, V., Kwan, W., Carroll, K. J., Cortes, M., Liu, S. Y., Frechette, G. M., ... North, T. E. (2015). Cannabinoid receptor-2 regulates embryonic hematopoietic stem cell development via prostaglandin E2 and P-Selectin activity: CNR2 impacts embryonic HSC development. *Stem Cells, 33*, 2596–2612.

Espín-Palazón, R., Stachura, D. L., Campbell, C. A., García-Moreno, D., Del Cid, N., Kim, A. D., ... Traver, D. (2014). Proinflammatory signaling regulates hematopoietic stem cell emergence. *Cell, 159*, 1070–1085.

Foley, J. E., Maeder, M. L., Pearlberg, J., Joung, J. K., Peterson, R. T., & Yeh, J.-R. J. (2009). Targeted mutagenesis in zebrafish using customized zinc-finger nucleases. *Nature Protocols, 4*, 1855–1867.

Foley, J. E., Yeh, J.-R. J., Maeder, M. L., Reyon, D., Sander, J. D., Peterson, R. T., & Joung, J. K. (2009). Rapid mutation of endogenous zebrafish genes using zinc finger nucleases made by Oligomerized Pool ENgineering (OPEN). *PLoS One, 4*, e4348.

Fujiwara, Y., Browne, C. P., Cunniff, K., Goff, S. C., & Orkin, S. H. (1996). Arrested development of embryonic red cell precursors in mouse embryos lacking transcription factor GATA-1. *Proceedings of the National Academy of Sciences of the United States of America, 93*, 12355–12358.

Galloway, J. L., Wingert, R. A., Thisse, C., Thisse, B., & Zon, L. I. (2005). Loss of Gata1 but not Gata2 converts erythropoiesis to myelopoiesis in zebrafish embryos. *Developmental Cell, 8*, 109–116.

Gering, M., Yamada, Y., Rabbitts, T. H., & Patient, R. K. (2003). Lmo2 and Scl/Tal1 convert non-axial mesoderm into haemangioblasts which differentiate into endothelial cells in the absence of Gata1. *Development, 130*, 6187–6199.

Gilliland, D. G., Jordan, C. T., & Felix, C. A. (2004). The molecular basis of leukemia. *Hematology Education Program of the American Society of Hematology*, 80–97.

Goessling, W., Allen, R. S., Guan, X., Jin, P., Uchida, N., Dovey, M., ... North, T. E. (2011). Prostaglandin E2 enhances human cord blood stem cell xenotransplants and shows long-term safety in preclinical nonhuman primate transplant models. *Cell Stem Cell, 8*, 445–458.

Goessling, W., North, T. E., Loewer, S., Lord, A. M., Lee, S., Stoick-Cooper, C. L., ... Zon, L. I. (2009). Genetic interaction of PGE2 and Wnt signaling regulates developmental specification of stem cells and regeneration. *Cell, 136*, 1136–1147.

Goldie, L. C., Lucitti, J. L., Dickinson, M. E., & Hirschi, K. K. (2008). Cell signaling directing the formation and function of hemogenic endothelium during murine embryogenesis. *Blood, 112*, 3194–3204.

Grosser, T., Yusuff, S., Cheskis, E., Pack, M. A., & FitzGerald, G. A. (2002). Developmental expression of functional cyclooxygenases in zebrafish. *Proceedings of the National Academy of Sciences of the United States of America, 99*, 8418–8423.

Gutierrez, A., Pan, L., Groen, R. W. J., Baleydier, F., Kentsis, A., Marineau, J., ... Aster, J. C. (2014). Phenothiazines induce PP2A-mediated apoptosis in T cell acute lymphoblastic leukemia. *The Journal of Clinical Investigation, 124*, 644–655.

Habeck, H., Odenthal, J., Walderich, B., Maischein, H., & Schulte-Merker, S. (2002). Analysis of a zebrafish VEGF receptor mutant reveals specific disruption of angiogenesis. *Current Biology, 12*, 1405–1412.

Hadland, B. K., Huppert, S. S., Kanungo, J., Xue, Y., Jiang, R., Gridley, T., … Longmore, G. D. (2004). A requirement for Notch1 distinguishes 2 phases of definitive hematopoiesis during development. *Blood, 104*, 3097–3105.

Harris, J. M., Esain, V., Frechette, G. M., Harris, L. J., Cox, A. G., Cortes, M., … North, T. E. (2013). Glucose metabolism impacts the spatiotemporal onset and magnitude of HSC induction in vivo. *Blood, 121*, 2483–2493.

He, Q., Zhang, C., Wang, L., Zhang, P., Ma, D., Lv, J., & Liu, F. (2015). Inflammatory signaling regulates hematopoietic stem and progenitor cell emergence in vertebrates. *Blood, 125*, 1098–1106.

Heath, V., Suh, H. C., Holman, M., Renn, K., Gooya, J. M., Parkin, S., … Keller, J. (2004). C/EBPalpha deficiency results in hyperproliferation of hematopoietic progenitor cells and disrupts macrophage development in vitro and in vivo. *Blood, 104*, 1639–1647.

Hess, I., Iwanami, N., Schorpp, M., & Boehm, T. (2013). Zebrafish model for allogeneic hematopoietic cell transplantation not requiring preconditioning. *Proceedings of the National Academy of Sciences of the United States of America, 110*, 4327–4332.

Hoggatt, J., Mohammad, K. S., Singh, P., Hoggatt, A. F., Chitteti, B. R., Speth, J. M., … Pelus, L. M. (2013). Differential stem- and progenitor-cell trafficking by prostaglandin E2. *Nature, 495*, 365–369.

Hsu, K., Traver, D., Kutok, J. L., Hagen, A., Liu, T.-X., Paw, B. H., … Look, A. T. (2004). The pu.1 promoter drives myeloid gene expression in zebrafish. *Blood, 104*, 1291–1297.

Ikushima, Y. M., Arai, F., Hosokawa, K., Toyama, H., Takubo, K., Furuyashiki, T., … Suda, T. (2013). Prostaglandin E(2) regulates murine hematopoietic stem/progenitor cells directly via EP4 receptor and indirectly through mesenchymal progenitor cells. *Blood, 121*, 1995–2007.

Imanirad, P., Solaimani Kartalaei, P., Crisan, M., Vink, C., Yamada-Inagawa, T., de Pater, E., … Dzierzak, E. (2014). HIF1α is a regulator of hematopoietic progenitor and stem cell development in hypoxic sites of the mouse embryo. *Stem Cell Research, 12*, 24–35.

Jin, H., Xu, J., & Wen, Z. (2007). Migratory path of definitive hematopoietic stem/progenitor cells during zebrafish development. *Blood, 109*, 5208–5214.

Jing, L., Tamplin, O. J., Chen, M. J., Deng, Q., Patterson, S., Kim, P. G., … Zon, L. I. (2015). Adenosine signaling promotes hematopoietic stem and progenitor cell emergence. *The Journal of Experimental Medicine, 212*, 649–663.

de Jong, J. L. O., Burns, C. E., Chen, A. T., Pugach, E., Mayhall, E. A., Smith, A. C. H., … Zon, L. I. (2011). Characterization of immune-matched hematopoietic transplantation in zebrafish. *Blood, 117*, 4234–4242.

de Jong, J. L. O., Davidson, A. J., Wang, Y., Palis, J., Opara, P., Pugach, E., … Zon, L. I. (2010). Interaction of retinoic acid and scl controls primitive blood development. *Blood, 116*, 201–209.

de Jong, J. L. O., & Zon, L. I. (2005). Use of the zebrafish system to study primitive and definitive hematopoiesis. *Annual Review of Genetics, 39*, 481–501.

Kalev-Zylinska, M. L., Horsfield, J. A., Flores, M. V. C., Postlethwait, J. H., Vitas, M. R., Baas, A. M., … Crosier, K. E. (2002). Runx1 is required for zebrafish blood and vessel development and expression of a human RUNX1-CBF2T1 transgene advances a model for studies of leukemogenesis. *Development, 129*, 2015–2030.

Kanz, D., Konantz, M., Alghisi, E., North, T. E., & Lengerke, C. (2016). Endothelial-to-hematopoietic transition: Notch-ing vessels into blood. *Annals of the New York Academy of Sciences, 1370*(1), 97–108.

Kawahara, A., Endo, S., & Dawid, I. B. (2012). Vap (Vascular Associated Protein): a novel factor involved in erythropoiesis and angiogenesis. *Biochemical and Biophysical Research Communications, 421*, 367–374.

Kim, P. G., Nakano, H., Das, P. P., Chen, M. J., Rowe, R. G., Chou, S. S., … Daley, G. Q. (2015). Flow-induced protein kinase A-CREB pathway acts via BMP signaling to promote HSC emergence. *The Journal of Experimental Medicine, 212*, 633–648.

Kissa, K., & Herbomel, P. (2010). Blood stem cells emerge from aortic endothelium by a novel type of cell transition. *Nature, 464*, 112–115.

Kumano, K., Chiba, S., Kunisato, A., Sata, M., Saito, T., Nakagami-Yamaguchi, E., … Hirai, H. (2003). Notch1 but not Notch2 is essential for generating hematopoietic stem cells from endothelial cells. *Immunity, 18*, 699–711.

Lam, E. Y. N., Hall, C. J., Crosier, P. S., Crosier, K. E., & Flores, M. V. (2010). Live imaging of Runx1 expression in the dorsal aorta tracks the emergence of blood progenitors from endothelial cells. *Blood, 116*, 909–914.

Lancrin, C., Sroczynska, P., Serrano, A. G., Gandillet, A., Ferreras, C., Kouskoff, V., … Lacaud, G. (2010). Blood cell generation from the hemangioblast. *Journal of Molecular Medicine, 88*, 167–172.

Lancrin, C., Sroczynska, P., Stephenson, C., Allen, T., Kouskoff, V., & Lacaud, G. (2009). The haemangioblast generates haematopoietic cells through a haemogenic endothelium stage. *Nature, 457*, 892–895.

Lawson, N. D., Vogel, A. M., & Weinstein, B. M. (2002). Sonic hedgehog and vascular endothelial growth factor act upstream of the notch pathway during arterial endothelial differentiation. *Developmental Cell, 3*, 127–136.

Li, P., Lahvic, J. L., Binder, V., Pugach, E. K., Riley, E. B., Tamplin, O. J., … Zon, L. I. (2015). Epoxyeicosatrienoic acids enhance embryonic haematopoiesis and adult marrow engraftment. *Nature, 523*, 468–471.

Li, X., Lan, Y., Xu, J., Zhang, W., & Wen, Z. (2012). SUMO1-activating enzyme subunit 1 is essential for the survival of hematopoietic stem/progenitor cells in zebrafish. *Development, 139*, 4321–4329.

Li, Y., Esain, V., Teng, L., Xu, J., Kwan, W., Frost, I. M., … Speck, N. A. (2014). Inflammatory signaling regulates embryonic hematopoietic stem and progenitor cell production. *Genes and Development, 28*, 2597–2612.

Liao, E. C., Trede, N. S., Ransom, D., Zapata, A., Kieran, M., & Zon, L. I. (2002). Non-cell autonomous requirement for the bloodless gene in primitive hematopoiesis of zebrafish. *Development, 129*, 649–659.

Liao, W., Bisgrove, B. W., Sawyer, H., Hug, B., Bell, B., Peters, K., … Stainier, D. Y. (1997). The zebrafish gene cloche acts upstream of a flk-1 homologue to regulate endothelial cell differentiation. *Development, 124*, 381–389.

Lieschke, G. J., Oates, A. C., Paw, B. H., Thompson, M. A., Hall, N. E., Ward, A. C., … Layton, J. E. (2002). Zebrafish SPI-1 (PU.1) marks a site of myeloid development independent of primitive erythropoiesis: implications for axial patterning. *Developmental Biology, 246*, 274–295.

Liu, Y.-J., Fan, H.-B., Jin, Y., Ren, C.-G., Jia, X.-E., Wang, L., … Ren, R. (2013). Cannabinoid receptor 2 suppresses leukocyte inflammatory migration by modulating the JNK/c-Jun/Alox5 pathway. *The Journal of Biological Chemistry, 288*, 13551–13562.

Ludin, A., Itkin, T., Gur-Cohen, S., Mildner, A., Shezen, E., Golan, K., … Zon, L. I. (2012). Monocytes-macrophages that express α-smooth muscle actin preserve primitive hematopoietic cells in the bone marrow. *Nature Immunology, 13*, 1072–1082.

Lyons, S. E., Lawson, N. D., Lei, L., Bennett, P. E., Weinstein, B. M., & Liu, P. P. (2002). A nonsense mutation in zebrafish gata1 causes the bloodless phenotype in vlad tepes. *Proceedings of the National Academy of Sciences of the United States of America, 99*, 5454–5459.

Martin, C. S., Moriyama, A., & Zon, L. I. (2011). Hematopoietic stem cells, hematopoiesis and disease: lessons from the zebrafish model. *Genome Medicine, 3*, 83.

Medvinsky, A., & Dzierzak, E. (1996). Definitive hematopoiesis is autonomously initiated by the AGM region. *Cell, 86*, 897–906.

Mikkola, H. K., Fujiwara, Y., Schlaeger, T. M., Traver, D., & Orkin, S. H. (2003). Expression of CD41 marks the initiation of definitive hematopoiesis in the mouse embryo. *Blood, 101*, 508–516.

Mucenski, M. L., McLain, K., Kier, A. B., Swerdlow, S. H., Schreiner, C. M., Miller, T. A., … Potter, S. S. (1991). A functional c-myb gene is required for normal murine fetal hepatic hematopoiesis. *Cell, 65*, 677–689.

Murayama, E., Kissa, K., Zapata, A., Mordelet, E., Briolat, V., Lin, H.-F., … Herbomel, P. (2006). Tracing hematopoietic precursor migration to successive hematopoietic organs during zebrafish development. *Immunity, 25*, 963–975.

Nichols, K. E., Crispino, J. D., Poncz, M., White, J. G., Orkin, S. H., Maris, J. M., & Weiss, M. J. (2000). Familial dyserythropoietic anaemia and thrombocytopenia due to an inherited mutation in GATA1. *Nature Genetics, 24*, 266–270.

North, T., Gu, T. L., Stacy, T., Wang, Q., Howard, L., Binder, M., … Speck, N. A. (1999). Cbfa2 is required for the formation of intra-aortic hematopoietic clusters. *Development, 126*, 2563–2575.

North, T. E., de Bruijn, M. F. T., Stacy, T., Talebian, L., Lind, E., Robin, C., … Speck, N. A. (2002). Runx1 expression marks long-term repopulating hematopoietic stem cells in the midgestation mouse embryo. *Immunity, 16*, 661–672.

North, T. E., Goessling, W., Peeters, M., Li, P., Ceol, C., Lord, A. M., … Zon, L. I. (2009). Hematopoietic stem cell development is dependent on blood flow. *Cell, 137*, 736–748.

North, T. E., Goessling, W., Walkley, C. R., Lengerke, C., Kopani, K. R., Lord, A. M., … Zon, L. I. (2007). Prostaglandin E2 regulates vertebrate haematopoietic stem cell homeostasis. *Nature, 447*, 1007–1011.

Okuda, T., van Deursen, J., Hiebert, S. W., Grosveld, G., & Downing, J. R. (1996). AML1, the target of multiple chromosomal translocations in human leukemia, is essential for normal fetal liver hematopoiesis. *Cell, 84*, 321–330.

Paik, E. J., de Jong, J. L. O., Pugach, E., Opara, P., & Zon, L. I. (2010). A chemical genetic screen in zebrafish for pathways interacting with cdx4 in primitive hematopoiesis. *Zebrafish, 7*, 61–68.

Palis, J., Robertson, S., Kennedy, M., Wall, C., & Keller, G. (1999). Development of erythroid and myeloid progenitors in the yolk sac and embryo proper of the mouse. *Development, 126*, 5073–5084.

Pan, Y., Mansfield, K. D., Bertozzi, C. C., Rudenko, V., Chan, D. A., Giaccia, A. J., & Simon, M. C. (2007). Multiple factors affecting cellular redox status and energy metabolism modulate hypoxia-inducible factor prolyl hydroxylase activity in vivo and in vitro. *Molecular and Cellular Biology, 27*, 912–925.

Parker, L., & Stainier, D. Y. (1999). Cell-autonomous and non-autonomous requirements for the zebrafish gene cloche in hematopoiesis. *Development, 126*, 2643–2651.

Peterson, R. T., Link, B. A., Dowling, J. E., & Schreiber, S. L. (2000). Small molecule developmental screens reveal the logic and timing of vertebrate development. *Proceedings of the National Academy of Sciences of the United States of America, 97*, 12965–12969.

Pietras, E. M., Mirantes-Barbeito, C., Fong, S., Loeffler, D., Kovtonyuk, L. V., Zhang, S., ... Passegué, E. (2016). Chronic interleukin-1 exposure drives haematopoietic stem cells towards precocious myeloid differentiation at the expense of self-renewal. *Nature Cell Biology, 18*, 607–618.

Pillay, L. M., Forrester, A. M., Erickson, T., Berman, J. N., & Waskiewicz, A. J. (2010). The Hox cofactors Meis1 and Pbx act upstream of gata1 to regulate primitive hematopoiesis. *Developmental Biology, 340*, 306–317.

Purton, L. E. (2007). Roles of retinoids and retinoic acid receptors in the regulation of hematopoietic stem cell self-renewal and differentiation. *PPAR Research, 2007*, 1–7.

Ramírez-Bergeron, D. L., Runge, A., Adelman, D. M., Gohil, M., & Simon, M. C. (2006). HIF-dependent hematopoietic factors regulate the development of the embryonic vasculature. *Developmental Cell, 11*, 81–92.

Ramírez-Bergeron, D. L., Runge, A., Dahl, K. D. C., Fehling, H. J., Keller, G., & Simon, M. C. (2004). Hypoxia affects mesoderm and enhances hemangioblast specification during early development. *Development, 131*, 4623–4634.

Reischauer, S., Stone, O. A., Villasenor, A., Chi, N., Jin, S.-W., Martin, M., ... Stainier, D. Y. (2016). Cloche is a bHLH-PAS transcription factor that drives haemato-vascular specification. *Nature, 535*, 294–298.

Rennekamp, A. J., & Peterson, R. T. (2015). 15 years of zebrafish chemical screening. *Current Opinion in Chemical Biology, 24*, 58–70.

Rhodes, J., Hagen, A., Hsu, K., Deng, M., Liu, T. X., Look, A. T., & Kanki, J. P. (2005). Interplay of pu.1 and gata1 determines myelo-erythroid progenitor cell fate in zebrafish. *Developmental Cell, 8*, 97–108.

Ridges, S., Heaton, W. L., Joshi, D., Choi, H., Eiring, A., Batchelor, L., ... Trede, N. S. (2012). Zebrafish screen identifies novel compound with selective toxicity against leukemia. *Blood, 119*, 5621–5631.

Robb, L., Lyons, I., Li, R., Hartley, L., Köntgen, F., Harvey, R. P., ... Begley, C. G. (1995). Absence of yolk sac hematopoiesis from mice with a targeted disruption of the scl gene. *Proceedings of the National Academy of Sciences of the United States of America, 92*, 7075–7079.

Rombough, P., & Drader, H. (2009). Hemoglobin enhances oxygen uptake in larval zebrafish (*Danio rerio*) but only under conditions of extreme hypoxia. *The Journal of Experimental Biology, 212*, 778–784.

van Rooijen, E., Voest, E. E., Logister, I., Korving, J., Schwerte, T., Schulte-Merker, S., ... van Eeden, F. J. (2009). Zebrafish mutants in the von Hippel-Lindau tumor suppressor display a hypoxic response and recapitulate key aspects of Chuvash polycythemia. *Blood, 113*, 6449–6460.

Sabin, F. (1920) [Carnegie Inst Wash Pub Num 272 Contrib Embryol]. *Studies on the origin of blood vessels and of red blood corpuscles as seen in the living blastoderm of chicks during the second day of incubation* (Vol. 9, p. 214).

Sakamoto, H., Dai, G., Tsujino, K., Hashimoto, K., Huang, X., Fujimoto, T., ... Ogawa, M. (2006). Proper levels of c-Myb are discretely defined at distinct steps of hematopoietic cell development. *Blood, 108*, 896–903.

Sawamiphak, S., Kontarakis, Z., & Stainier, D. Y. R. (2014). Interferon gamma signaling positively regulates hematopoietic stem cell emergence. *Developmental Cell, 31*, 640–653.

Schoenebeck, J. J., Keegan, B. R., & Yelon, D. (2007). Vessel and blood specification override cardiac potential in anterior mesoderm. *Developmental Cell, 13*, 254–267.

Shalaby, F., Ho, J., Stanford, W. L., Fischer, K.-D., Schuh, A. C., Schwartz, L., ... Rossant, J. (1997). A requirement for Flk1 in primitive and definitive hematopoiesis and vasculogenesis. *Cell, 89*, 981–990.

Shalaby, F., Rossant, J., Yamaguchi, T. P., Gertsenstein, M., Wu, X.-F., Breitman, M. L., & Schuh, A. C. (1995). Failure of blood-island formation and vasculogenesis in Flk-1-deficient mice. *Nature, 376*, 62–66.

Shivdasani, R. A., Mayer, E. L., & Orkin, S. H. (1995). Absence of blood formation in mice lacking the T-cell leukaemia oncoprotein tal-1/SCL. *Nature, 373*, 432–434.

Simsek, T., Kocabas, F., Zheng, J., DeBerardinis, R. J., Mahmoud, A. I., Olson, E. N., ... Sadek, H. A. (2010). The distinct metabolic profile of hematopoietic stem cells reflects their location in a hypoxic niche. *Cell Stem Cell, 7*, 380–390.

Sood, R., English, M. A., Belele, C. L., Jin, H., Bishop, K., Haskins, R., ... Liu, P. P. (2010). Development of multilineage adult hematopoiesis in the zebrafish with a runx1 truncation mutation. *Blood, 115*, 2806–2809.

Soza-Ried, C., Hess, I., Netuschil, N., Schorpp, M., & Boehm, T. (2010). Essential role of c-myb in definitive hematopoiesis is evolutionarily conserved. *Proceedings of the National Academy of Sciences of the United States of America, 107*, 17304–17308.

Stachura, D. L., Reyes, J. R., Bartunek, P., Paw, B. H., Zon, L. I., & Traver, D. (2009). Zebrafish kidney stromal cell lines support multilineage hematopoiesis. *Blood, 114*, 279–289.

Stainier, D. Y., Weinstein, B. M., Detrich, H. W., Zon, L. I., & Fishman, M. C. (1995). Cloche, an early acting zebrafish gene, is required by both the endothelial and hematopoietic lineages. *Development, 121*, 3141–3150.

Swiers, G., Baumann, C., O'Rourke, J., Giannoulatou, E., Taylor, S., Joshi, A., ... de Bruijn, M. F. (2013). Early dynamic fate changes in haemogenic endothelium characterized at the single-cell level. *Nature Communications, 4*, 2924.

Takizawa, H., Regoes, R. R., Boddupalli, C. S., Bonhoeffer, S., & Manz, M. G. (2011). Dynamic variation in cycling of hematopoietic stem cells in steady state and inflammation. *The Journal of Experimental Medicine, 208*, 273–284.

Takubo, K., Goda, N., Yamada, W., Iriuchishima, H., Ikeda, E., Kubota, Y., ... Suda, T. (2010). Regulation of the HIF-1α level is essential for hematopoietic stem cells. *Cell Stem Cell, 7*, 391–402.

Traver, D., Paw, B. H., Poss, K. D., Penberthy, W. T., Lin, S., & Zon, L. I. (2003). Transplantation and in vivo imaging of multilineage engraftment in zebrafish bloodless mutants. *Nature Immunology, 4*, 1238–1246.

Ueno, H., & Weissman, I. L. (2006). Clonal analysis of mouse development reveals a polyclonal origin for yolk sac blood islands. *Developmental Cell, 11*, 519–533.

Villablanca, E. J., Pistocchi, A., Court, F. A., Cotelli, F., Bordignon, C., Allende, M. L., ... Russo, V. (2007). Abrogation of prostaglandin E2/EP4 signaling impairs the development of rag1+ lymphoid precursors in the thymus of zebrafish embryos. *The Journal of Immunology, 179*, 357–364.

Vogeli, K. M., Jin, S.-W., Martin, G. R., & Stainier, D. Y. R. (2006). A common progenitor for haematopoietic and endothelial lineages in the zebrafish gastrula. *Nature, 443*, 337–339.

Wang, Q., Stacy, T., Binder, M., Marin-Padilla, M., Sharpe, A. H., & Speck, N. A. (1996). Disruption of the Cbfa2 gene causes necrosis and hemorrhaging in the central nervous system and blocks definitive hematopoiesis. *Proceedings of the National Academy of Sciences of the United States of America, 93*, 3444–3449.

Ward, A. C., McPhee, D. O., Condron, M. M., Varma, S., Cody, S. H., Onnebo, S. M. N., ... Lieschke, G. J. (2003). The zebrafish spi1 promoter drives myeloid-specific expression in stable transgenic fish. *Blood, 102*, 3238–3240.

Weinstein, B. M., Schier, A. F., Abdelilah, S., Malicki, J., Solnica-Krezel, L., Stemple, D. L., ... Fishman, M. C. (1996). Hematopoietic mutations in the zebrafish. *Development, 123*, 303–309.

Yeh, J.-R. J., Munson, K. M., Chao, Y. L., Peterson, Q. P., MacRae, C. A., & Peterson, R. T. (2008). AML1-ETO reprograms hematopoietic cell fate by downregulating scl expression. *Development, 135*, 401–410.

Yeh, J.-R. J., Munson, K. M., Elagib, K. E., Goldfarb, A. N., Sweetser, D. A., & Peterson, R. T. (2009). Discovering chemical modifiers of oncogene-regulated hematopoietic differentiation. *Nature Chemical Biology, 5*, 236–243.

Yuan, H., Zhang, T., Liu, X., Deng, M., Zhang, W., Wen, Z., ... Zhu, J. (2015). Sumoylation of CCAAT/enhancer-binding protein α is implicated in hematopoietic stem/progenitor cell development through regulating runx1 in zebrafish. *Scientific Reports, 5*, 9011.

CHAPTER

Studying disorders of vertebrate iron and heme metabolism using zebrafish

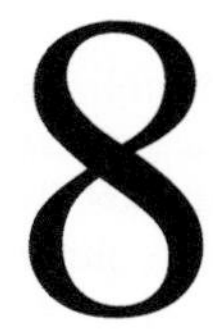

Lisa N. van der Vorm[*,a], **Barry H. Paw**[*,¶,||,#,1]

*Brigham & Women's Hospital, Boston, MA, United States

¶Harvard Medical School, Boston, MA, United States

||Dana-Farber Cancer Institute, Boston, MA, United States

#Boston Children's Hospital, Boston, MA, United States

[1]Corresponding author: E-mail: bpaw@rics.bwh.harvard.edu

CHAPTER OUTLINE

Abstract

Iron is a crucial component of heme- and iron-sulfur clusters, involved in vital cellular functions such as oxygen transport, DNA synthesis, and respiration. Both excess and insufficient levels of iron and heme-precursors cause human disease, such as iron-deficiency anemia, hemochromatosis, and porphyrias. Hence, their levels must be tightly regulated, requiring a complex network of transporters and feedback mechanisms. The use of zebrafish to study these pathways and the underlying genetics offers

[a]Present address: Maastricht Universitair Medisch Centrum, Maastricht, The Netherlands.

Methods in Cell Biology, Volume 138, ISSN 0091-679X, http://dx.doi.org/10.1016/bs.mcb.2016.10.008

many advantages, among others their optical transparency, ex-vivo development and high genetic and physiological conservations. This chapter first reviews well-established methods, such as large-scale mutagenesis screens that have led to the initial identification of a series of iron and heme transporters and the generation of a variety of mutant lines. Other widely used techniques are based on injection of RNA, including complementary morpholino knockdown and gene overexpression. In addition, we highlight several recently developed approaches, most notably endonuclease-based gene knockouts such as TALENs or the CRISPR/Cas9 system that have been used to study how loss of function can induce human disease phenocopies in zebrafish. Rescue by chemical complementation with iron-based compounds or small molecules can subsequently be used to confirm causality of the genetic defect for the observed phenotype. All together, zebrafish have proven to be – and will continue to serve as an ideal model to advance our understanding of the pathogenesis of human iron and heme-related diseases and to develop novel therapies to treat these conditions.

1. OVERVIEW OF VERTEBRATE CELLULAR IRON AND HEME METABOLISM

Iron is a trace metal essential for survival of almost all organisms. As a crucial component of heme- and iron sulfur (Fe-S) cluster-containing proteins, iron is involved in a number of vital cellular functions, such as oxygen transport, DNA synthesis, and respiration (Loreal et al., 2014; Muckenthaler & Lill, 2012). However, it can also have deleterious effects when present in excess by participation in the Fenton reaction, leading to the formation of reactive oxygen species (ROS) that result in lipid peroxidation and damage to DNA and proteins (Fenton, 1894). Fortunately, under physiological conditions, cellular iron levels are well regulated by a number of mechanisms that govern its uptake, mobilization, storage, and recycling (Andrews & Schmidt, 2007). The vast majority of iron is required for synthesis of the oxygen-carrier hemoglobin in red blood cells. This need is fulfilled mostly (90%) by recycled iron from degradation of senescent RBCs by macrophages in the spleen. Iron absorbed from the diet in the duodenum accounts for the remaining 10% (Zhang, Ghosh, & Rouault, 2014). Iron absorption is facilitated by a ferrireductase, whose exact identity remains controversial (Gunshin et al., 2005; McKie et al., 2001). After reduction of its ferric (Fe^{3+}) form to ferrous (Fe^{2+}), the iron is taken up into the enterocyte by the divalent metal transporter-1 (DMT-1/SLC11A2) (Fleming et al., 1997; Gunshin et al., 1997). Once inside the cell, iron-regulatory proteins (IRPs) and iron-responsive elements (IREs) control iron levels by regulating the expression of a series of proteins involved in uptake, storage, and utilization of iron (Muckenthaler, Galy, & Hentze, 2008). In the case of low intracellular iron, IRPs bind with high affinity to IREs to inhibit mRNA translation when in the 5′UTR, e.g., for ferritins, and simultaneously stabilize mRNA when in the 3′UTR, e.g., for the transferrin receptor 1 (*TfR1*), thereby facilitating uptake and utilization. Conversely, when iron excess occurs, ferritin translation will take place uninhibited and TfR1 expression is repressed (Torti & Torti, 2002; Zhang et al., 2014). Whereas this elegant IRE/IRP system is present in nonerythroid cells, different regulatory mechanisms may be in place in erythroid cells (Anderson, Shen, Eisenstein, & Leibold, 2012; Ponka, 1997). The iron demand of these cells is many orders of magnitude higher than for other mammalian tissues because iron is required not only for mitochondrial cytochromes but also as prosthetic groups for hemoglobin, which contains 70% of the total iron content of a normal adult (Ponka, 1997).

The only known cellular iron exporter is Ferroportin-1 (FPN1/SLC40A1) (Abboud & Haile, 2000; Donovan et al., 2000; McKie et al., 2000). Release of iron by FPN1 is controlled by the peptide hormone hepcidin (Ganz & Nemeth, 2012). Hepcidin regulates iron availability by binding to FPN1, promoting its internalization and degradation, thereby preventing duodenal iron uptake and release from stores into the circulation (Nemeth et al., 2004). Upon release by FPN1, Fe^{2+} is oxidized by multicopper ferroxidases [ceruloplasmin for macrophages (Mukhopadhyay, Attieh, & Fox, 1998) and hephaestin for enterocytes (Vulpe et al., 1999)] so that it can be bound by transferrin. These Fe^{3+}transferrin (Tf) complexes then bind to Tf receptors on the cell surface to be endocytosed (Dautry-Varsat, Ciechanover, & Lodish, 1983). After reduction by STEAP3 ferrireductase (Ohgami et al., 2005) the Fe^{2+} enters the cytosol via the endosomal transporter DMT1 or ZIP14, which can take up both endosomal transferrin- and extracellular nontransferrin-bound iron (Liuzzi, Aydemir, Nam, Knutson, & Cousins, 2006; Zhao, Gao, Enns, & Knutson, 2010). The iron may then be transferred directly from endosomes to mitochondria via so-called kiss-and-run (Sheftel, Zhang, Brown, Shirihai, & Ponka, 2007; Zhang, Sheftel, & Ponka, 2005) or enters the cytosolic pool of metabolically active "labile" iron that is loosely coordinated by water, small molecules, and proteins (Shvartsman & Ioav Cabantchik, 2012). Most of the labile iron pool (LIP) is directed to mitochondria (Shvartsman & Ioav Cabantchik, 2012), where it can be used in the assembly of Fe–S clusters (Rouault, 2015) or inserted into protoporphyrin IX to form heme (Ajioka, Phillips, & Kushner, 2006).

In vertebrates, erythroid cells have by far the greatest need for iron, which is required for the synthesis of heme in the oxygen transport protein hemoglobin. Heme synthesis is a highly complex process involving eight nuclear-encoded enzymes, four of which are cytoplasmic and four of which localize to mitochondria (Ajioka et al., 2006). The first step, taking place in the mitochondrial matrix, comprises condensation of succinyl-CoA and glycine by aminolevulinic acid synthase (ALAS) to generate D-aminolevulinic acid (ALA). Of note, erythroid and nonerythroid cells express distinct isoforms of ALAS: ALAS1 is expressed ubiquitously, whereas ALAS2 is specific to differentiating erythroid cells. ALA is then exported across the mitochondrial outer membrane into the cytoplasm by an unknown mechanism, where it is converted in a series of four biosynthetic reactions to coproporphyrin III (CPIII). This intermediate is transported back into the mitochondrial intermembrane space, where it is converted to protoporphyrinogen III and then to protoporphyrin IX. Finally, heme is formed by metalation of protoporphyrin IX with Fe^{2+} by the enzyme ferrochelatase (FECH) (Ajioka et al., 2006; Chung, Chen, & Paw, 2012; Dailey & Meissner, 2013).

Cytosolic iron that is not directed to mitochondria may be used to metalate nonheme iron enzymes in the cytosol and nucleus, to assemble cytosolic Fe–S clusters, or to be stored in ferritin. The fate of cytosolic iron is likely determined by the activity of iron transporters, such as mitochondrial iron importers [mitoferrin 1 and 2 (Shaw et al., 2006)], and iron efflux pumps (e.g., ferroportin) (Abboud & Haile, 2000; Donovan et al., 2000; McKie et al., 2000), whose discovery by positional cloning will be discussed in this chapter. For more details on mechanisms of intracellular iron trafficking the reader is referred to this excellent review (Philpott & Ryu, 2014).

Although the regulation and mechanisms of iron metabolism and heme synthetic enzymes are well characterized, significant knowledge gaps remain regarding the pathways that govern transport and trafficking of iron and heme intermediates (Nilsson et al., 2009). The rest of this chapter presents an overview of past and present approaches and methodologies that employ zebrafish as a model to unravel new genes involved in cellular iron and heme synthesis and transport.

2. ADVANTAGEOUS PROPERTIES OF ZEBRAFISH TO STUDY GENETICS

The potential of the zebrafish as a vertebrate model system for genetic screening first became apparent over 2 decades ago by the work of George Streisinger (Streisinger, Walker, Dower, Knauber, & Singer, 1981). An important advantageous characteristic is that zebrafish development takes place externally. In addition, the transparency of the embryos allows for visualization of developmental processes from gastrulation to organogenesis and monitoring of gene expression in vivo without having to sacrifice the subjects. Furthermore, many genetic pathways underlying the development and the regulation thereof are fairly well conserved throughout evolution, with approximately 70% of human genes having at least one obvious zebrafish orthologue (Howe et al., 2013). Hence, similar phenotypes are often observed between zebrafish mutants and human disease (Shin & Fishman, 2002). From a practical perspective, compared to rodents and other mammalian systems, large numbers of zebrafish can be relatively easily maintained in the laboratory setting. Spawning, which can be controlled by light, can take place frequently and does not depend on spawning season. Spawns are generally large (average 100 embryos) and embryos can be raised to (sexual) maturity within 2–3 months, which is less labor-intensive and cheaper than for other commonly used laboratory animals. Also, routine techniques used in genetic research, including knock-in, knockdown, and knockout, are well-established in zebrafish (Lin, Chiang, & Tsai, 2016). Hence, zebrafish are particularly suitable vertebrates for high-frequency mutagenesis and large-scale genetic screening studies.

Roughly, two approaches to screening can be distinguished. Genetic screens start with a phenotype and then attempts to identify the gene(s) responsible, whereas in candidate screens a known gene is disrupted, and subsequently, the phenotype is analyzed to determine its function. Both strategies will be discussed below in the light of their contribution to the study of iron and heme metabolism in zebrafish.

3. TOOLS TO STUDY IRON AND HEME METABOLISM USING ZEBRAFISH

3.1 GENETIC SCREENING

The approach of large-scale genetic screens starts with treating male fish with a mutagen, e.g., ethylnitrosurea (ENU) or gamma irradiation, which cause a large number of point mutations in the male germ cells (spermatogonia). Subsequently, these male F0 fish are crossed to wild-type females to produce the F1 generation. F1 fish are then in-crossed with each other to create the F2 generation, in which 50% of the fish will be heterozygous for a specific mutation and 50% will be wild-type. Finally, by in-crossing the heterozygous F2 siblings the F3 generation is created, in which 25% are wild-type, 50% heterozygous and 25% homozygous for recessive mutations (Patton & Zon, 2001). New developmental mutants are then characterized phenotypically, after which the responsible gene and its function can be identified.

Although other useful model systems (yeast, mouse, human) exist, zebrafish has contributed in especially meaningful ways to our understanding of iron and heme metabolism. The advantageous characteristics of zebrafish to study genetics led two labs in Tübingen (Germany) and Boston (USA) (Driever et al., 1996; Haffter et al., 1996) to initiate large-scale ENU mutagenesis screens. This resulted in the identification of ~2000 mutants affecting a large variety of processes such as early development, organ formation, behavior, and axonal pathfinding (Haffter et al., 1996). A number of the ~2000 mutated developmental genes that were identified in these two screens have been cloned, and by selecting those mutants with an anemic phenotype (identified as having low blood cell counts) several zebrafish lines with defects in iron or heme metabolism were established (see Table 1 for an overview).

For instance, positional cloning of the *weissherbst* (*weh*) mutant resulted in the identification of the, so far only known, cellular iron exporter FPN1 (Slc40A1) (Donovan et al., 2000). *Weissherbst* fish were found to suffer from hypochromic

Table 1 Overview of Zebrafish Mutants That Gave Insight Into Iron Metabolism and Heme Biosynthesis

	Name Mutant	Gene or Protein	Human Disease	Reference Zebrafish Mutant
Heme biosynthesis	*montalcino (mno)*	*PPOX*	Variegate porphyria	Dooley et al. (2008)
	sauternes (sau)	*ALAS2*	X-linked dominant erythropoietic protoporphyria	Brownlie et al. (1998)
	yquem (yqe)	*UROD*	Porphyria cutanea tarda/hepato-erythropoietic porphyria	Wang et al. (1998)
	freixenet (frx)/dracula (drc)	*FECH*	Erythropoietic protoporphyria	Ransom et al. (1996) and Childs et al. (2000)
	"alad"	*ALAD*	ALA dehydratase deficient porphyria	Sakamoto et al. (2004) (in medaka fish)
	"cpox"	*CPOX*	Hereditary coproporphyria *readers are referred to* Puy, Gouya, and Deybach (2010) *for details on these porphyrias*	
Iron metabolism	*gavi*	*TF*	Iron-deficiency anemia	Fraenkel et al. (2009)
	shiraz (sir)	*GRX5*	Nonketotic hyperglycinemia and severe neurodegeneration (Baker et al., 2014); Sideroblastic anemia (Camaschella et al., 2007)	Wingert et al. (2005)
	frascati (frs)	*MFRN1*	Erythropoietic protoporphyria associated with abnormal mitoferrin expression (Wang et al., 2011)	Shaw et al. (2006)
	chardonnay (cdy)	*DMT-1*	Hypochromic microcytic anemia, hepatic iron overload (Mims et al., 2005)	Donovan et al. (2002)
	chianti (cia)	*TFR1a*	Iron-deficiency anemia (experimentally induced by TfR antibody), (Larrick & Hyman, 1984). Combined immunodeficiency, intermittent neutropenia, thrombocytopenia, and mild anemia (rare phenotype) (Jabara et al., 2016)	Wingert et al. (2004)

	weissherbst (wei)	*FPN1*	Ferroportin disease (Zoller et al., 2005)	Donovan et al. (2000) and Fraenkel, Traver, Donovan, Zahrieh, and Zon (2005)
Erythropoiesis general	*vlad tepes (vlt)* *cabernet (cab)* *grenache (gre)* *thunderbird (tbr)*	*GATA-1*	Familial dyserythropoietic anemia and thrombocytopenia (Nichols et al., 2000)	Lyons et al. (2002) and Ransom et al. (1996)
	retsina (ret)	band-3 (Slc4A1)	*Readers are referred to this review* (Mohandas & Gallagher, 2008)	Paw et al. (2003)
	chablis/merlot (mot)	*protein 4.1*		Shafizadeh et al. (2002)
	riesling (ris)	*erythroid β spectrin*		Liao et al. (2000)
	malbec	*CDH5*		Anderson et al. (2015)
	pinotage (pnt)	*ATPIF1*		Shah et al. (2012)
	moonshine (mon)	*TIF1 gamma*		Ransom et al. (2004)
	zinfandel	*globin*	Hemoglobinopathies (Cao & Galanello, 2010; Piel & Weatherall, 2014)	Brownlie et al. (2003)
	cloche (clo)	*bHLH-PAS transcription factor*		Stainier, Weinstein, Detrich, Zon, and Fishman (1995), Wang et al. (2007), Reischauer et al. (2016),
	spadetail (spt) and TBX16			Griffin, Amacher, Kimmel, and Kimelman (1998) and Thompson et al. (1998)
	bloodless (bls)	unknown		Liao et al. (2002)

anemia, as evident from decreased hemoglobin levels, blocked erythroid maturation, and reduced numbers of erythrocytes (Donovan et al., 2000; Ransom et al., 1996). Erythroid cells of mutant embryo zebrafish had significantly lower iron concentrations compared to wild-type embryos. This suggests a circulatory iron deficiency as the cause of the anemic phenotype, which was later confirmed by positional cloning, as discussed later in this chapter. Furthermore, the *FPN1* gene is highly conserved throughout evolution. In mammals, FPN1 is expressed mainly at sites of iron transport, such as the placenta, duodenum and liver. Murine FPN1 protein is predominantly present in the basolateral membrane of enterocytes, indicative of a role as an intestinal iron transporter (Donovan et al., 2000). This was subsequently confirmed by mouse knockout studies (Donovan et al., 2005). Taken together, these data led to the identification of FPN1 as the, to date only, iron exporter present in all eukaryotic organisms interrogated for this gene. Hence, these studies exemplify how genetic screens in zebrafish can lead to the identification of novel genes.

Positional cloning of *frascati* (*frs*) zebrafish mutants identified missense mutations in the mitoferrin *MFRN* (*Slc25A37*) gene. Previous studies in yeast (Foury & Roganti, 2002; Li & Kaplan, 2004; Mühlenhoff et al., 2003) had uncovered the role of the MRS3 and MRS4 yeast homologs in mitochondrial iron uptake, as they were found to facilitate both heme formation and Fe−S cluster biosynthesis, the two major iron-consuming mitochondrial processes (Mühlenhoff et al., 2003). Later, *frs* mutants were observed to develop hypochromic anemia and erythroid maturation arrest (Shaw et al., 2006). Positional cloning in zebrafish further supported that MFRN1 is the principal mitochondrial iron importer in vertebrate erythroblasts (Shaw et al., 2006), which was later validated by mouse knockouts (Chung et al., 2014; Troadec et al., 2011). The MFRN2 (Slc25A28) paralogue functions in mitochondrial iron import in nonerythroid tissues and erythroid progenitors (Shaw et al., 2006).

The protein DMT1 was originally isolated in the rat duodenum as a divalent ion transporter that is upregulated by dietary iron deficiency (Fleming et al., 1997; Gunshin et al., 1997). Several years later, the *chardonnay* (*cdy*) zebrafish mutation, which presents with reduced hemoglobin levels and delayed erythrocyte maturation, was found to have a mutation in *DMT1* (*Slc11A2*) (Donovan et al., 2002), leading to a severely truncated, nonfunctional protein. The DMT1 protein localizes to erythroid cells and the intestine, suggestive of its role in iron absorption. More direct evidence was obtained with overexpression experiments discussed under "Candidate gene approaches." Of note, whereas the *weh* mutants die during early development (Donovan et al., 2000), anemic *cdy* homozygotes survive, which might be explained by yet-to be unraveled redundant modes of iron absorption (Donovan et al., 2002). In humans, mutations in DMT1 cause a phenotype of hypochromic microcytic anemia combined with iron overload (Mims et al., 2005). This supports the possible existence of alternative iron absorption mechanisms in the duodenum that can bypass DMT1 (Mims et al., 2005).

The zebrafish strain *gavi* (*gav*) was shown to have mutations in transferrin-a (*Tf-a*), which encodes the principal iron carrier in all vertebrates (Fraenkel et al., 2009).

Mutants exhibit reduced Tf-a expression and impaired hemoglobin production, which causes hypochromic anemia and embryonic lethality by 14 days of development (Fraenkel et al., 2009). Interestingly, in humans, the phenotype of congenital hypotransferrinemia as a result of Tf mutations is highly similar to that of the *gav* mutants, including hypochromic anemia and embryonic death (Goldwurm et al., 2000; Hayashi, Wada, Suzuki, & Shimizu, 1993). The *gav* mutant is thus an example of a zebrafish mutant that serves as an ideal model to study corresponding human pathologies.

Transferrin-bound iron is taken up into the cell by binding to the transferrin receptor 1 (TFR1). In four different zebrafish *chianti* (*cia*) mutants with varying degrees of hypochromic anemia and defective erythroid differentiation, positional cloning revealed missense and splicing mutations in the *TFR1a* gene (Ransom et al., 1996; Wingert et al., 2004). Remarkably, the cloning of *TFR1a* revealed a second *TFR1* gene, *TFR1b* (Wingert et al., 2004), whose presence probably resulted from the teleost whole genome duplication (Amores et al., 1998; Postlethwait et al., 1998). Whereas *TFR1a* is expressed in erythrocytes during early development and *cia* mutants are anemic, the *TFR1b* gene is expressed ubiquitously throughout embryogenesis and mutation leads to growth delay and brain necrosis. Together, TFR1a (*cia*) and TFR1b deficient zebrafish embryos recapitulate the phenotype of *TFR1−/−* mice (Levy, Jin, Fujiwara, Kuo, & Andrews, 1999). Hence, the *cia* mutant allows study of the function of TFR1 in erythropoiesis unbiased by developmental abnormalities in other tissues.

Phenotypic analysis of *shiraz* (*sir*) mutant zebrafish revealed a novel connection between heme biosynthesis and Fe−S cluster formation, the two main iron-demanding processes in mitochondria that were previously believed to be independent processes in vertebrates. In *sir* mutants, hypochromic anemia, in the context of normal mitochondrial iron and oxidative stress levels, was shown to be caused by a deletion in the *glutaredoxin 5* (*GRX5*) gene (Wingert et al., 2005). *GRX5* is evolutionarily conserved, and studies in yeast (Rodriguez-Manzaneque, Tamarit, Belli, Ros, & Herrero, 2002) revealed that GRX5 is required for the synthesis of Fe−S clusters in mitochondria, where GRX5 also localizes in zebrafish. Fe−S clusters are known to negatively regulate binding of IRP to IREs. Hence, the observed anemia in the absence of reduced mitochondrial iron levels was proposed to be caused by decreased Fe−S cluster assembly, leading to increased IRP1 activity, which in turn inhibits the expression of IRE-regulated target genes involved in heme biosynthesis. In these studies a role for GRX5 in heme synthesis regulation was confirmed in human patients carrying this mutation (Camaschella et al., 2007). These findings highlight the advantages of using zebrafish as a vertebrate model system for discovering novel biological pathways that might not be present, or may be different, in lower organisms.

In addition to iron deficiency or reduced Fe−S cluster assembly, defects in the availability of heme substrates or in the catalytic activity of enzymes involved in heme biosynthesis can cause human congenital anemias and/or porphyrias. Two well-characterized zebrafish mutants with porphyria are the *yquem* (*yqe*) and *freixenet* (*frx*) [also called *dracula* (*drc*)] mutants, which have mutations later in the heme synthesis pathway and therefore present with porphyria instead of congenital anemia. Both mutants have autofluorescent red blood cells that photoablate when exposed directly to light (Ransom et al., 1996). The gene mutated in *yqe* fish encodes the enzyme uroporphyrinogen decarboxylase (UROD), which converts uroporphyrinogen to coproporphyrinogen. Homozygous mutation of the *UROD* gene leads to two forms of porphyrias, porphyria cutanea tarda (PCT) and hepatoerythropoietic porphyria (HEP) in humans, similar to the phenotype in fish (Wang, Long, Marty, Sassa, & Lin, 1998). The most prevalent form of porphyria is the autosomal dominant disease PCT. These patients have a slightly less severe phenotype, characterized by photosensitive skin and excessive excretion of protoheme biosynthesis intermediates, uroporphyrin and 7-decarboxylate porphyrin, in their urine. The homozygous *yqe* mutant represents the first true animal model for HEP (Wang et al., 1998), which has been proposed for studying the pathogenesis of HEP and for the development of new therapeutics.

The zebrafish mutant *frx*/*drc* was found to bear mutations in the gene encoding for FECH, the final enzyme that catalyzes metalation of protoporphyrin IX (PPIX) with Fe^{2+}. Mutation of FECH in humans presents as erythropoietic protoporphyria due to accumulation of light-sensitive, toxic porphyrins (Childs et al., 2000). Mutation of the mitochondrial ATPase inhibitory factor 1 (*ATPIF1*) gene, which regulates the catalytic efficiency of FECH, causes heme-deficient phenotype. As is the case in *frx* mutants with direct mutation of FECH, loss of ATPIF1 in the anemic *pinotage* (*pnt*) mutants impairs heme synthesis as a consequence of diminished FECH activity combined with elevated mitochondrial pH. Thus, using cloning strategies and data from zebrafish genetic screens, the *pnt* mutant has provided insight into a novel role of ATPIF1 as a regulator of heme synthesis, thereby advancing the understanding of mitochondrial heme homeostasis and erythropoiesis (Shah et al., 2012). Of note, in a mouse knockout of ATPIF1, no obvious phenotype was found (Nakamura, Fujikawa, & Yoshida, 2013); the explanation for this difference in phenotype between these two models is not clear.

The mutants discussed above are just a few examples of the many identified in the Boston and Tübingen screens that have significantly contributed to the general understanding of iron and heme metabolism. In addition, several candidate-screening techniques were used in many of the studies described above to obtain confirmation that the identified gene is indeed causative of the observed phenotype. These approaches will be discussed below and illustrated with exemplary studies performed on the zebrafish mutants introduced above.

3.2 CANDIDATE GENE APPROACHES

One of the major limitations of performing large-scale mutagenesis screens is that they are time-consuming and laborious. Recently, methods for targeted editing of specific candidate genomic loci in been vertebrates have developed to analyze gene and protein function in vivo. The workflow toward prioritization of candidate genes derived from expression studies, microarrays, or RNA sequencing starts with performing whole mount in-situ hybridization (ISH) studies to verify tissue-specific expression. Products of genes involved in hematopoiesis are expected to localize to the intermediate cell mass (ICM) of the zebrafish, the functional equivalent of mammalian yolk-sac blood islands (Detrich et al., 1995). See Fig. 1A–H for an example of screening by ISH for localization of gene products

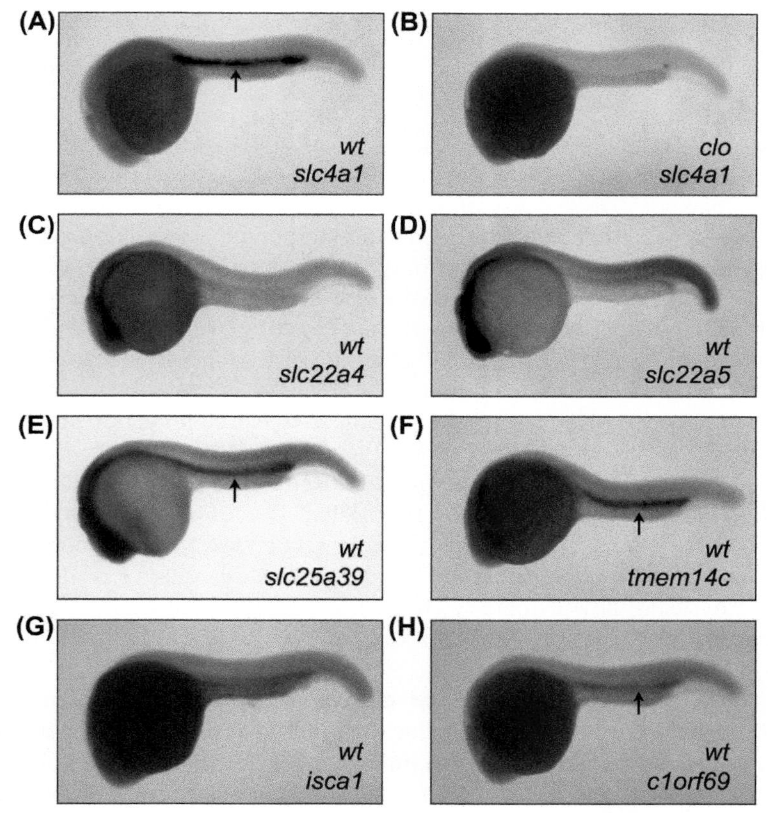

FIGURE 1

(A–H) Expression of candidate genes identified in a large-scale computational screen to identify mitochondrial proteins whose transcripts coexpress with the heme biosynthesis machinery. Whole embryo in-situ hybridization was performed on embryos at 24 hpf. (A) *Slc4a1* was used as control to delineate the intermediate cell mass (indicated by *arrows*). (B) *Cloche* (*clo*) embryos were used to show specificity of the hybridizations. (C–G) Candidates identified in this study.

Adapted with permission from Nilsson, R., Schultz, I. J., Pierce, E. L., Soltis, K. A., Naranuntarat, A., Ward, D. M., ..., Mootha, V. K. (2009). Discovery of genes essential for heme biosynthesis through large-scale gene expression analysis. Cell Metabolism, *10(2), 119–130, Cell Press.*

to the ICM (Nilsson et al., 2009). The candidate genes that pass this first "triage" are then screened for possible causality of the observed erythroid phenotype by a variety of techniques, of which the ones most commonly used to study heme and iron metabolism in zebrafish are discussed below.

3.2.1 Microinjection techniques

The ex-utero development of zebrafish enables researchers to manipulate gene expression by injection of antisense oligomers [morpholino oligonucleotides (MOs) or cRNA]. This strategy can be employed to either block or drive overexpression of the candidate gene of interest.

3.2.1.1 Gene knockdown

The first viable sequence-specific gene inactivation method in zebrafish was knockdown using MOs (Ekker & Larson, 2001). MOs are synthetic oligonucleotides analogous to DNA, but with a morpholine ring instead of the ribose sugar moiety and a neutral charge backbone with phosphorodiamidate linkage resistant to phosphodiesterases. Their high affinity for RNA and no known cellular binding proteins or nucleases makes them very well suitable as antisense reagents (Summerton, 1999). The mechanism of gene targeting by MOs can be either alteration of transcriptional processing or inhibition of translation (Draper, Morcos, & Kimmel, 2001). Compared to conventional ENU mutagenesis, which targets only zygotic transcripts, MOs can potentially knock down both zygotic and maternal transcripts (Nasevicius & Ekker, 2000). Fig. 2A–G (Nilsson et al., 2009) shows clear examples in which MO disruption of mRNA translation of candidate genes identified through in situ (Fig. 1A–H) resulted in profoundly anemic zebrafish embryos. Here the read-out is qualitative, namely a lack of hemoglobinized cells after staining with *o*-dianisidine for hemoglobin. A quantitative read-out can be obtained by using flow cytometry (Kardon et al., 2015) on transgenic zebrafish lines that produce fluorescent erythrocytes, such as *Tg(globin-LCR:eGFP)* (Ganis et al., 2012).

Although MOs have been and are still commonly used to knock down gene expression during early development due to the relative ease of injecting developing embryos, they have some caveats that limit their use. Because MOs are short-lived, they only temporarily block translation or alter alternative splicing and therefore are unsuitable for analysis of late onset phenotypes. Moreover, gene knockdown by this approach has potential off-target effects, although they can be avoided with rigorous controls (Bill, Petzold, Clark, Schimmenti, & Ekker, 2009; Eisen & Smith, 2008). Nevertheless, controversy exists on the specificity of MO knockdowns compared to engineered knockouts using emerging methods, such as CRISPR/cas9 genome editing (Kok et al., 2015; Rossi et al., 2015).

3.2.1.2 Gene overexpression

As an example of microinjection of cRNA encoding the protein of interest can be employed to drive its overexpression. The correct timing and target of injection

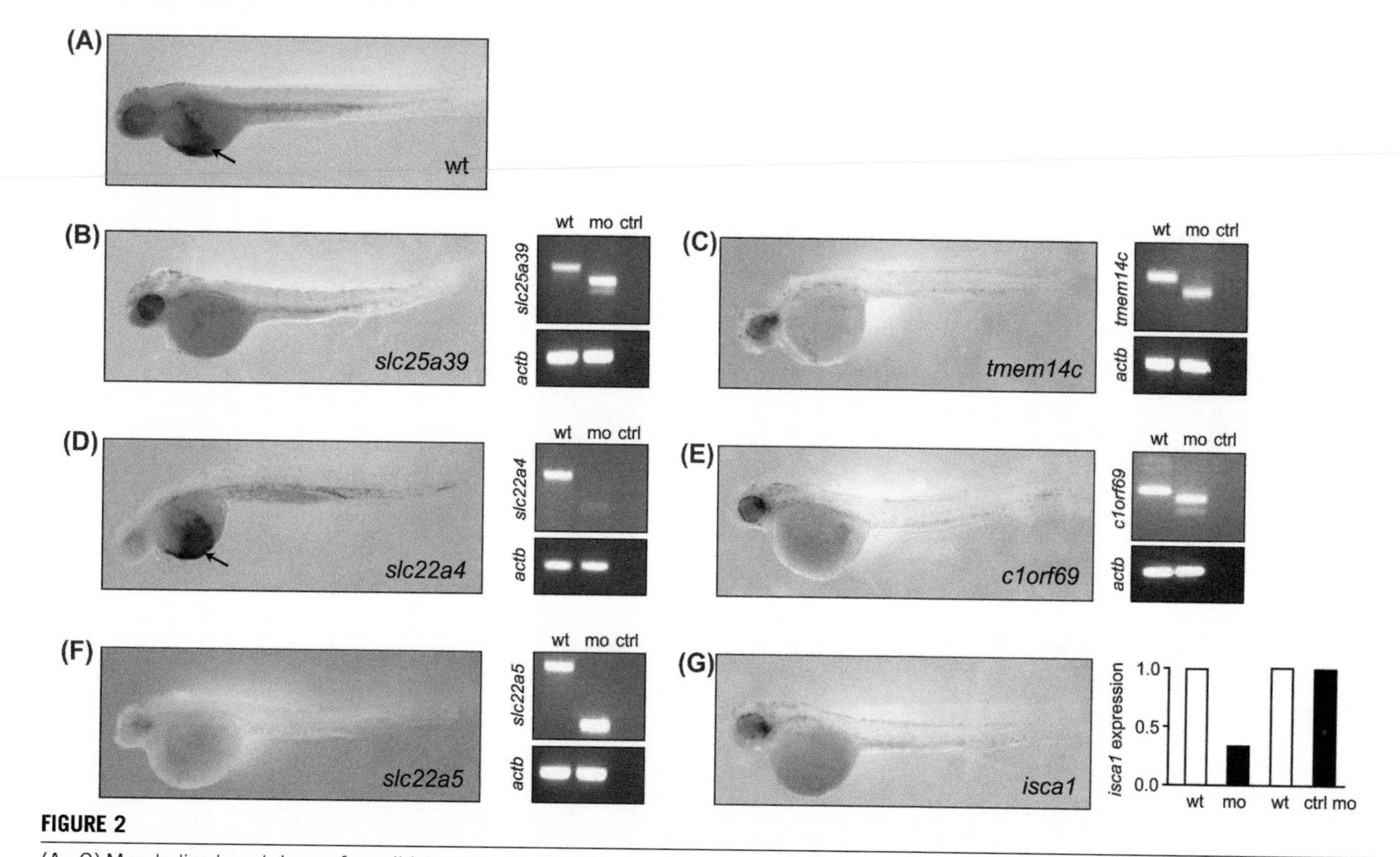

FIGURE 2

(A–G) Morpholino knockdown of candidate genes in zebrafish results in profound anemia. Wild-type (WT) zebrafish embryos were injected at the 1-cell stage with the respective morpholino oligonucleotides (MOs) and stained at 48 hpf with *o*-dianisidine to detect hemoglobinized cells. (A) Uninjected (WT) embryos show normal hemoglobinization as indicated by dark-brown staining on the yolk sac (*arrow*). (B–G) MO-injected embryos. Accurate MO gene targeting was verified by RT-PCR (*slc25a39*, *slc22a4*, *slc22a5*, *tmem14c*, and *c1orf69*) or real-time quantitative RT-PCR (*isca1*) on cDNA from uninjected (WT) or morpholino-injected (MO) embryos. β-actin (*actb*) was used as a control for off-target effects in the RT-PCR. For RT-PCR, (ctrl) indicates no cDNA template control. For real-time quantitative RT-PCR, (ctrl MO) indicates embryos injected with a standard control morpholino.

Adapted with permission from Nilsson, R., Schultz, I. J., Pierce, E. L., Soltis, K. A., Naranuntarat, A., Ward, D. M., ..., Mootha, V. K. (2009). Discovery of genes essential for heme biosynthesis through large-scale gene expression analysis. Cell Metabolism, *10(2), 119–130, Cell Press.*

depends on the desired expression pattern: injection of the mRNA into the yolk of the one-cell stage embryo will drive protein expression in every cell of the developing embryo, whereas injection into a single blastomere of a multicell embryo generates mosaic expression in a subpopulation of cells (Rosen, Sweeney, & Mably, 2009). The injected cRNA transcript is commonly stabilized by regulatory mRNA regions, such as 3′ untranslated regions (UTRs) (e.g., the poly-A-tail present in the beta-globin gene from *Xenopus laevis* (Brown, Zipkin, & Harland, 1993; Jiang, Xu, & Russell, 2006).

As a study using cRNA to verify hypothesized gene function, Shaw et al. (2006) introduced *MFRN* cRNA by microinjection into *frs* embryos at the one-cell stage and assessed whether hemoglobinization was restored. The injected embryos were allowed to develop to 72 hpf, stained with *o*-dianisidine to detect hemoglobinized cells, and genotyped. Half of the genotyped *frs* mutant embryos were rescued. Microinjection with a mutant *MFRN* cRNA did not rescue any *frs* embryos. Together, these findings indicate that *MFRN* is the specific gene whose disruption causes the severe chronic anemia in *frs* embryos (Shaw et al., 2006).

Usage of knockdown and overexpression strategies provides compelling evidence that a candidate gene is indeed causative of an observed phenotype. For instance, MO knockdown of *TfR1a* in wild-type zebrafish embryos prevented erythrocyte hemoglobin synthesis, resulting in embryos that displayed hypochromic erythrocytes that were *o*-dianisidine negative. The ability of TfR1a to rescue hemoglobin production in *cia* mutants was then confirmed by overexpression. Wild-type *TfR1a* mRNA was injected into *cia* homozygous embryos, resulting in a partial rescue of *o*-dianisidine positive cells at the 36–40 hpf stage. These experiments demonstrated that mutations in *TfR1a* were causal for the *cia* phenotype (Wingert et al., 2004).

Whereas transient overexpression is sufficient to study protein localization and function during early zebrafish development (up to 3 days), it is less suitable to study later developmental stages or when the protein must be expressed in a specific tissue or cell type (Finckbeiner et al., 2011).

3.2.2 Emerging techniques

3.2.2.1 Gene knockout

The recent development of sequence-specific endonucleases, namely zinc-finger nucleases (ZFNs), transcription activator-like effector nucleases (TALENs), and RNA-guided nucleases (RGNs), has provided revolutionary approaches to editing the genome in situ. As ZFNs have not been used to study iron and heme metabolism in zebrafish, they are not further discussed here; readers are referred to (Amacher, 2008) and the ZFIN website ("Zinc Finger Consortium. Available from: zincfingers.org/default2.htm,") for protocols and additional background. TALENs and RGNs are also widely used (Dahlem et al., 2012; Huang et al., 2011; Hwang et al., 2013; Jao, Wente, & Chen, 2013; Sung et al., 2014) in zebrafish for targeted mutagenesis. Readers are referred to these reviews (Hoban & Bauer, 2016; Hsu, Lander, & Zhang, 2014; Peng et al., 2014) for details on these techniques.

Genome-editing using TALENs has been demonstrated to be successful in a variety of model systems, including zebrafish (Joung & Sander, 2013). However, precise and efficient target-specific gene knock-in and knockout in vivo can be problematic. Both precision and efficiency are especially important for investigating endogenous gene expression or generating conditional gene knockouts in model organisms such as zebrafish, as gene targeting in embryonic stem cells is not available (Bedell et al., 2012). Recently, the clustered regularly interspaced palindromic repeats/CRISPR-associated (CRISPR/Cas) system has emerged as a promising alternative genome-editing approach. Site-specific cuts introduced by CRISPR/Cas9 can be used to engineer nonhomologous end-joining (NHEJ)-mediated deletions, discussed below, and introduction of small foreign DNA fragments via homology-directed repair (HDR) (Chang et al., 2013), which will be discussed in the next section on gene knock-in.

An exemplary study on the use of the CRISPR/Cas9 system to generate a zebrafish knockout line to study heme metabolism is the recently published study by Ablain, Durand, Yang, Zhou, and Zon (2015). A CRISPR-based vector system containing the *GATA-1* promoter driving Cas9 expression was used to silence the gene encoding the UROD enzyme. UROD-deficient erythrocytes, like those present in the *yquem* mutant zebrafish, have a clear phenotype of red autofluorescent erythrocytes due to accumulation of porphyrin precursors. Indeed, UROD targeting by the sgRNA yielded red fluorescent erythrocytes in zebrafish embryos (Fig. 3A–D). Co-injection of mRNA encoding UROD partially rescued this phenotype, while injection of cRNA bearing the *yquem* mutation did not, demonstrating the specificity of the targeting. Importantly, whereas F0 embryos displayed mosaic gene disruption, the phenotype was highly penetrant in F1 fish. Hence, this seminal study demonstrates how the CRISPR/Cas9 vector system can be employed to spatially control tissue-restricted gene knockout and is thereby an example for other loss-of-function studies in zebrafish.

3.2.2.2 Gene knock-in

TALENS and the CRISPR/Cas9 system can also be used to insert small sequence stretches using single-stranded oligonucleotides (ssDNA) (Bedell et al., 2012; Chang et al., 2013; Hruscha et al., 2013; Hwang et al., 2013), bigger fragments like open reading frames (Zu et al., 2013), or even whole plasmid vectors (Auer, Duroure, De Cian, Concordet, & Del Bene, 2014). To generate a bioassay for potassium channel function in the heart, for example, Hoshijima, Jurynec, and Grunwald (2016) used TALENs to replace the coding sequences of the *KCNH6a* gene with GFP-coding sequences. The *KCNH6a*GFP/+ heterozygous embryos, which were fully viable, expressed the GFP reporter protein exclusively in the heart. Sequence analysis of the resulting *KCNH6a*GFP alleles from the genomes of F1 embryos confirmed that donor sequences had been perfectly integrated by HR (Hoshijima et al., 2016).

Nonhomologous end-joining (NHEJ) DNA repair mechanism, which is at least ten-fold more active than HR during early zebrafish development (Dai, Cui, Zhu,

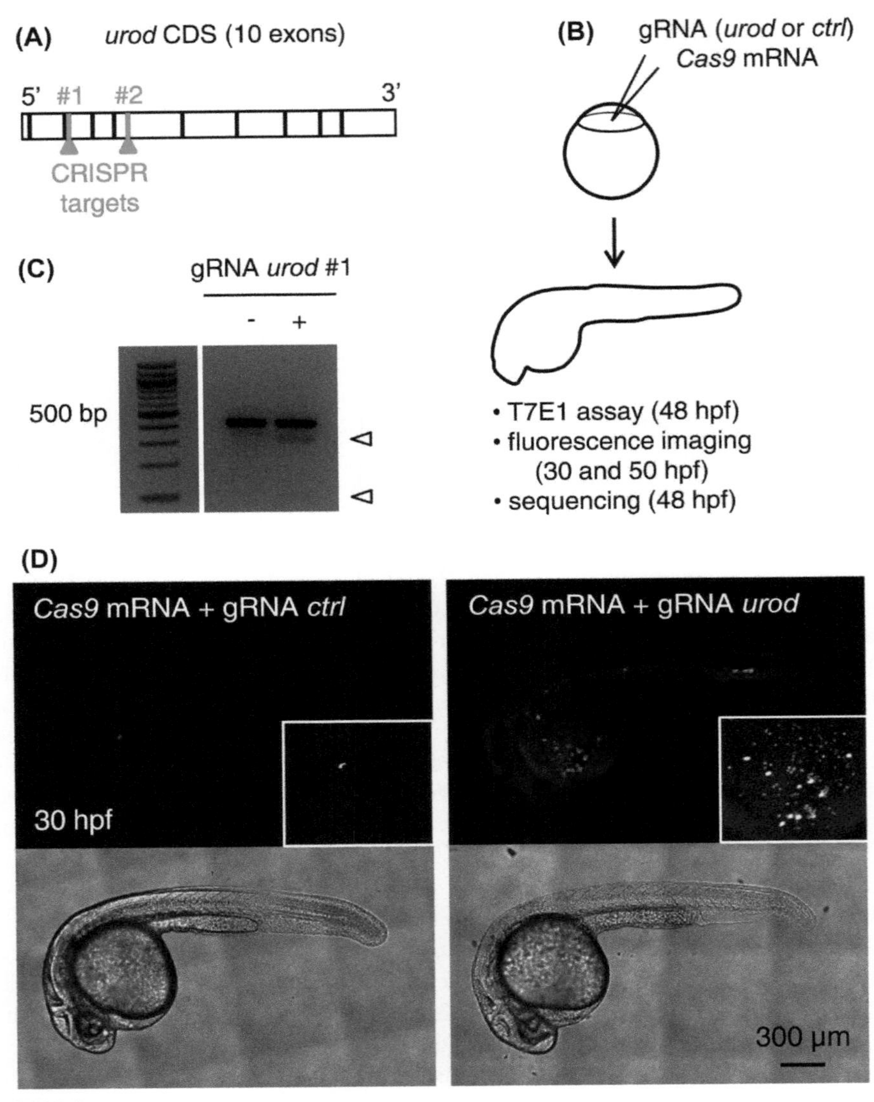

FIGURE 3

(A–D) Fluorescent phenotype resulting from CRISPR targeting of *urod*. (A) Schematic representation of *urod* coding sequence. The position of the CRISPR target sequences at the beginning of exons three and five are indicated. (B) Outline of the experiment, *Cas9* mRNA is injected into one-cell stage embryos along with a gRNA targeting either *urod* or *p53* as a negative control (ctrl). Targeted mutation efficiency was assessed by sequencing, and the presence of red fluorescent erythrocytes by confocal microscopy. (C) T7E1 mutagenesis assay at the CRISPR target site in the *urod* gene. The assay was performed on

& Hu, 2010; Liu et al., 2012), can also be exploited to integrate DNA at specific target sites. Auer et al. (2014) used TALENs and RGNs to introduce the *KalTA4* sequence into the *Tg(neurod:eGFP)* locus via NHEJ. This approach relies on a bait sequence mimicking the endogenous cutting site, located upstream of a functional cassette. Both the endogenous locus and the donor plasmid are linearized by the (TALE or RG) nuclease. The double strand breaks in the genome and the donor plasmid are religated by DNA repair, resulting in targeted integration of the donor vector DNA (Auer & Del Bene, 2014).

3.2.2.3 Transplantation

Transplantation, which generates chimeric embryos (Lin, Long, Chen, & Hopkins, 1992), is a powerful technique that can be exploited for studies of blood formation, including heme and iron disorders. For a detailed description of this technique the reader is referred to Chapter: Transplantation in Zebrafish by Gansner, Dang, Ammerman, and Zon (2016) of this volume. An important issue addressed using transplantation is whether a gene function is cell-autonomous. To this end, mutant cells lacking the functional gene of interest are transplanted into a wild-type embryo and, conversely, wild-type cells are introduced into mutant embryos, after which the phenotypes of the transplanted cells and the host embryo cells are compared. Using transplantation, Parker and Stainier (1999) investigated the cell autonomy of the blood deficiency of *cloche* mutants, and Liao et al. (2002) assessed the cell autonomy of the *bloodless* gene in hematopoiesis.

3.2.2.4 Rescue/chemical complementation

In-vivo complementation involves the selection of an enzyme or molecule based on its ability to complement an essential activity that has been deleted from the wild-type cell or organism (Baker et al., 2002). This strategy can be used in zebrafish to provide compelling evidence for the involvement of a specific protein in a biological pathway. For instance, Chen et al. (2013) reported that sorting nexin 3 (SNX3) regulates recycling of the TfR to the cell membrane in zebrafish embryos and is thus required for delivery of iron to erythroid progenitors. Silencing of *SNX3* with MO resulted in anemia and hemoglobin defects in zebrafish, attributable to impaired TfR-mediated iron uptake and its accumulation in early endosomes. The role of SNX3 in iron uptake was confirmed by treatment of *SNX3*-knockdown embryos with non-Tf iron chelate supplements followed by an assay using flow

genomic DNA from 2 dpf embryos injected at the one-cell stage with *Cas9* mRNA and either a gRNA against *p53* (negative control) or a gRNA against *urod*. Cleavage bands (*arrowheads*) indicate the presence of mutations at the target site. (D) Confocal images reveal the presence of fluorescent blood cells in *urod* mosaic knockout embryos at 30 hpf. The black and white insets show a 2× magnification of the fluorescent signal in the yolk region.
Adapted with permission from Ablain, J., Durand, E. M., Yang, S., Zhou, Y., & Zon, L. I. (2015). A CRISPR/Cas9 vector system for tissue-specific gene disruption in zebrafish. Developmental Cell, 32*(6), 756–764, Cell Press.*

cytometry. Supplementation with the non-Tf iron chelate FeSIH was found to increase the number of erythrocytes in *SNX3*-knockdown *Tg(globin-LCR:eGFP)* embryos (Fig. 4B). These results strongly reinforced the proposed model that SNX3 specifically regulates TfR-dependent iron assimilation in red cell development (Chen et al., 2013).

Kardon et al. (2015) reported (patho)physiological insight regarding the role of the mitochondrial chaperone CLPX in heme metabolism by treating *CLPX* morphant *Tg(globin-LCR:eGFP)* embryos with ALA, the product of the ALAS enzyme (Fig. 5A–C). These morphants suffered from severe anemia, as shown by flow cytometry to quantify GFP + erythrocytes (Fig. 5A,B). Interestingly, supplementation with 2 mM ALA from 24 to 72 hpf rescued the anemia in ALAS2 mutant embryos (*sauternes*; Brownlie et al., 1998), but not in mitochondrial iron transporter (*MFRN1*) mutants (*frascati*; Shaw et al., 2006) (Fig. 5C), which indicates that ALA specifically rescues defects in ALA synthesis only; defects in later steps of the heme biosynthesis pathway are not rescued. Clearly, CLP stimulation of ALA synthesis is important for efficient heme synthesis during erythropoiesis (Kardon et al., 2015).

Because heme or iron metabolism pathways are closely linked in by Fe–S clusters, assignment of mutated genes to these two pathways can be problematic. Supplementation with FeSIH, a lipid soluble iron chelate, has been the "gold standard" for identifying genes exclusively involved in iron metabolism. However, high doses of FeSIH are also able to rescue embryos with heme-based TMEM14c deficiency (Y.Y. Yien, unpublished data), which indicates that FeSIH is not absolutely specific to iron metabolism. Recently, the small molecule-iron transporter, *hinokitiol* (*CAS #499-44-5*), was shown by Grillo et al. (2016) to selectively

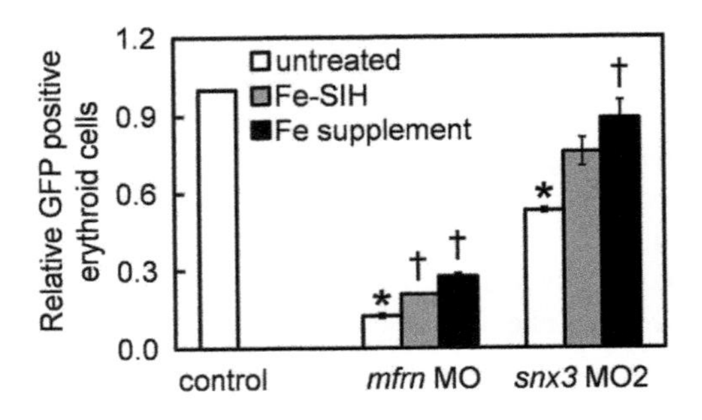

FIGURE 4

Complementation of hemoglobinization defects with non-Tf-bound iron supplements. Fe-SIH, and a chemically defined iron supplement increase hemoglobin production in *snx3*-and *mfrn*-silenced zebrafish morphants. *Error bars* represent SEM from three independent experiments. $^*p < .001$ compared with the control clone; $^\dagger p < .05$ compared with the control, untreated embryos within the group.

Adapted with permission from Chen, C., Garcia-Santos, D., Ishikawa, Y., Seguin, A., Li, L., Fegan, K. H., ..., Paw, B.H. (2013). Snx3 regulates recycling of the transferrin receptor and iron assimilation. Cell Metabolism, *17(3), 343–352, Cell Press.*

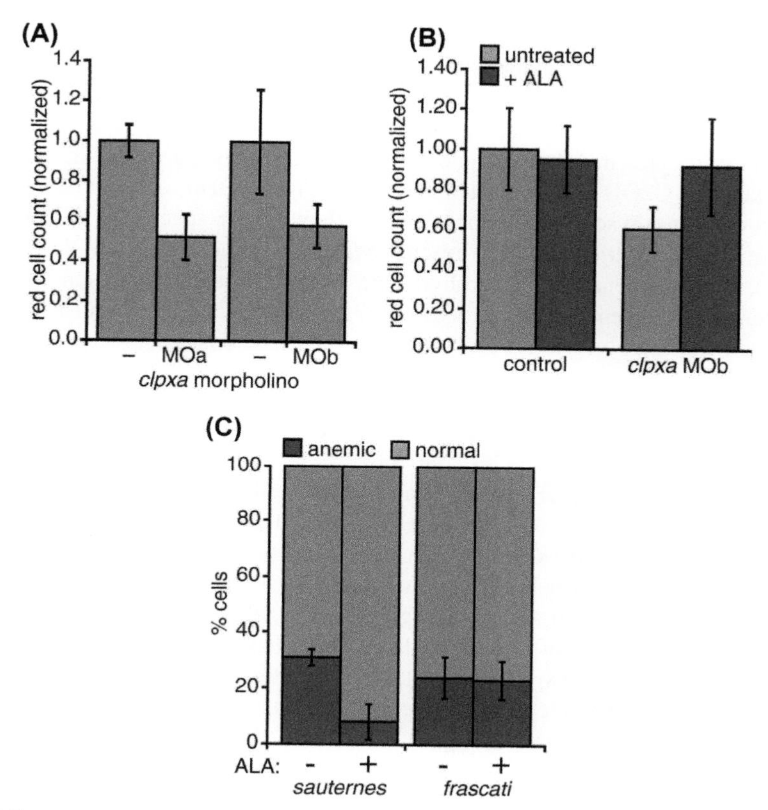

FIGURE 5

Morpholino knockdown of mtClpX can be rescued with aminolevulinic acid (ALA) supplementation. Embyros were grown from zygotes injected at the one- to two-cell stage with *clpxa*-targeting morpholinos (MOa and MOb) or uninjected zygotes (control). (A) Erythrocyte development at 72 hpf was quantified by flow cytometry, using dissociated cells from *Tg(globin-LCR:eGFP)* zebrafish. $p \leq .01$ for erythrocyte reduction by clpxa knockdown with either morpholino. (B) Rescue of clpxa MOb-induced anemia by ALA supplementation. *Tg(globin-LCR:eGFP)* zebrafish embryos were supplemented with 2-mM ALA from 24 to 72 hpf, upon which GFP + erythrocytes were quantified by flow cytometry. $p = .025$ for rescue of anemia in clpxa knockdown embryos by ALA supplementation. (C) Heterozygous *sauternes* (ALAS2 mutant) or *frascati* (MFRN mutant) zebrafish were crossed, and progeny were grown for 72 hpf, with or without ALA supplementation as in (B). Anemia was assayed by *o*-dianisidine staining. $p = .04$ for rescue of anemia in *sauternes* $\pm$ progeny by ALA. n = 52 for *sauternes* – ALA; n = 43 for *sauternes* + ALA; n = 98 for *frascati* – ALA; n = 122 for *frascati* + ALA. *Error bars* represent mean $\pm$ SD.

Adapted with permission from Kardon, J. R., Yien, Y. Y., Huston, N. C., Branco, D. S., Hildick-Smith, G. J., Rhee, K. Y., ..., Baker, T. A. (2015). Mitochondrial ClpX activates a key enzyme for heme biosynthesis and erythropoiesis. Cell, 161*(4), 858–867, Cell Press.*

rescue zebrafish mutant embryos and cell lines with defects in transporters involved in iron, but not heme, metabolism. In MFRN1-deficient *frs* zebrafish, addition of hinokitiol rescued the anemic phenotype, as observed qualitatively by restored *o*-dianisidine staining for hemoglobin and quantitatively by the number of GFP + erythrocytes in Mfrn1-deficient *Tg(globinLCR:eGFP)* zebrafish. A derivative lacking the C2OH group, C2-deoxy hinokitiol (C2deOHino), required to chelate the iron could not rescue hemoglobinization, showing the specificity of hinokitiol. Furthermore, hinokitiol did not rescue *sauternes* (*sau*) zebrafish deficient in ALAS2, the initial enzyme in porphyrin biosynthesis. Hence, this small molecule iron transporter opens up new possibilities to efficiently distinguish between causative genes involved in the iron versus heme metabolism pathways.

CONCLUSIONS AND FUTURE DIRECTIONS

As illustrated in this chapter, zebrafish provide an ideal molecular-genetic model for investigating diseases of iron and heme metabolism. New, revolutionary technologies, such as the CRISPR/Cas9 system, have greatly facilitated the generation of targeted loss-of-function fish mutants that affect these pathways, thereby augmenting our repertoire of experimental strategies. Hence, zebrafish allow one to study the functional consequences of genetic mutations that have been identified in humans and other mammals at the systemic level. Future studies in this field will likely focus on identifying and characterizing transporter proteins of iron and heme intermediates, which are to date largely unknown. Ultimately, the zebrafish provides an ideal research platform for developing new treatments for patients suffering from defects in iron and heme metabolism.

ACKNOWLEDGMENTS

L.N. van der Vorm was supported by fellowships from the Nora Baart Foundation (the Netherlands Society for Biochemistry and Molecular Biology) and the Dutch Stomach Liver Bowel Foundation (Maag Lever Darm Stichting). B.H. Paw was supported by National Institutes of Health grants, R01 DK070838 and P01 HL032262.

REFERENCES

Abboud, S., & Haile, D. J. (2000). A novel mammalian iron-regulated protein involved in intracellular iron metabolism. *The Journal of Biological Chemistry, 275*(26), 19906–19912.

Ablain, J., Durand, E. M., Yang, S., Zhou, Y., & Zon, L. I. (2015). A CRISPR/Cas9 vector system for tissue-specific gene disruption in zebrafish. *Developmental Cell, 32*(6), 756–764.

Ajioka, R. S., Phillips, J. D., & Kushner, J. P. (2006). Biosynthesis of heme in mammals. *Biochimica et Biophysica Acta, 1763*(7), 723–736.

Amacher, S. L. (2008). Emerging gene knockout technology in zebrafish: zinc-finger nucleases. *Briefings in Functional Genomics and Proteomics, 7*(6), 460–464.

Amores, A., Force, A., Yan, Y. L., Joly, L., Amemiya, C., Fritz, A., ... Postlethwait, J. H. (1998). Zebrafish hox clusters and vertebrate genome evolution. *Science, 282*(5394), 1711–1714.

Anderson, C. P., Shen, M., Eisenstein, R. S., & Leibold, E. A. (2012). Mammalian iron metabolism and its control by iron regulatory proteins. *Biochimica et Biophysica Acta, 1823*(9), 1468–1483.

Anderson, H., Patch, T. C., Reddy, P. N., Hagedorn, E. J., Kim, P. G., Soltis, K. A., ... Shah, D. I. (2015). Hematopoietic stem cells develop in the absence of endothelial cadherin 5 expression. *Blood, 126*(26), 2811–2820.

Andrews, N. C., & Schmidt, P. J. (2007). Iron homeostasis. *Annual Review of Physiology, 69*, 69–85.

Auer, T. O., & Del Bene, F. (2014). CRISPR/Cas9 and TALEN-mediated knock-in approaches in zebrafish. *Methods, 69*(2), 142–150.

Auer, T. O., Duroure, K., De Cian, A., Concordet, J. P., & Del Bene, F. (2014). Highly efficient CRISPR/Cas9-mediated knock-in in zebrafish by homology-independent DNA repair. *Genome Research, 24*(1), 142–153.

Baker, K., Bleczinski, C., Lin, H., Salazar-Jimenez, G., Sengupta, D., Krane, S., & Cornish, V. W. (2002). Chemical complementation: a reaction-independent genetic assay for enzyme catalysis. *Proceedings of the National Academy of Sciences of the United States of America, 99*(26), 16537–16542.

Baker, P. R., 2nd, Friederich, M. W., Swanson, M. A., Shaikh, T., Bhattacharya, K., Scharer, G. H., ... Van Hove, J. L. (2014). Variant non ketotic hyperglycinemia is caused by mutations in LIAS, BOLA3 and the novel gene GLRX5. *Brain, 137*(Pt 2), 366–379.

Bedell, V. M., Wang, Y., Campbell, J. M., Poshusta, T. L., Starker, C. G., Krug, R. G., 2nd, ... Ekker, S. C. (2012). In vivo genome editing using a high-efficiency TALEN system. *Nature, 491*(7422), 114–118.

Bill, B. R., Petzold, A. M., Clark, K. J., Schimmenti, L. A., & Ekker, S. C. (2009). A primer for morpholino use in zebrafish. *Zebrafish, 6*(1), 69–77.

Brown, B. D., Zipkin, I. D., & Harland, R. M. (1993). Sequence-specific endonucleolytic cleavage and protection of mRNA in *Xenopus* and *Drosophila*. *Genes and Development, 7*(8), 1620–1631.

Brownlie, A., Donovan, A., Pratt, S. J., Paw, B. H., Oates, A. C., Brugnara, C., ... Zon, L. I. (1998). Positional cloning of the zebrafish sauternes gene: a model for congenital sideroblastic anaemia. *Nature Genetics, 20*(3), 244–250.

Brownlie, A., Hersey, C., Oates, A. C., Paw, B. H., Falick, A. M., Witkowska, H. E., ... Zon, L. (2003). Characterization of embryonic globin genes of the zebrafish. *Developmental Biology, 255*(1), 48–61.

Camaschella, C., Campanella, A., De Falco, L., Boschetto, L., Merlini, R., Silvestri, L., ... Iolascon, A. (2007). The human counterpart of zebrafish shiraz shows sideroblastic-like microcytic anemia and iron overload. *Blood, 110*(4), 1353–1358.

Cao, A., & Galanello, R. (2010). Beta-thalassemia. *Genetics in Medicine, 12*(2), 61–76.

Chang, N., Sun, C., Gao, L., Zhu, D., Xu, X., Zhu, X., ... Xi, J. J. (2013). Genome editing with RNA-guided Cas9 nuclease in zebrafish embryos. *Cell Research, 23*(4), 465–472.

Chen, C., Garcia-Santos, D., Ishikawa, Y., Seguin, A., Li, L., Fegan, K. H., ... Paw, B. H. (2013). Snx3 regulates recycling of the transferrin receptor and iron assimilation. *Cell Metabolism, 17*(3), 343–352.

Childs, S., Weinstein, B. M., Mohideen, M. A., Donohue, S., Bonkovsky, H., & Fishman, M. C. (2000). Zebrafish dracula encodes ferrochelatase and its mutation provides a model for erythropoietic protoporphyria. *Current Biology, 10*(16), 1001–1004.

Chung, J., Anderson, S. A., Gwynn, B., Deck, K. M., Chen, M. J., Langer, N. B., ... Paw, B. H. (2014). Iron regulatory protein-1 protects against mitoferrin-1-deficient porphyria. *The Journal of Biological Chemistry, 289*(11), 7835–7843.

Chung, J., Chen, C., & Paw, B. H. (2012). Heme metabolism and erythropoiesis. *Current Opinion in Hematology, 19*(3), 156–162.

Dahlem, T. J., Hoshijima, K., Jurynec, M. J., Gunther, D., Starker, C. G., Locke, A. S., ... Grunwald, D. J. (2012). Simple methods for generating and detecting locus-specific mutations induced with TALENs in the zebrafish genome. *PLoS Genetics, 8*(8), e1002861.

Dai, J., Cui, X., Zhu, Z., & Hu, W. (2010). Non-homologous end joining plays a key role in transgene concatemer formation in transgenic zebrafish embryos. *International Journal of Biological Sciences, 6*(7), 756–768.

Dailey, H. A., & Meissner, P. N. (2013). Erythroid heme biosynthesis and its disorders. *Cold Spring Harbor Perspectives in Medicine, 3*(4), a011676.

Dautry-Varsat, A., Ciechanover, A., & Lodish, H. F. (1983). pH and the recycling of transferrin during receptor-mediated endocytosis. *Proceedings of the National Academy of Sciences of the United States of America, 80*(8), 2258–2262.

Detrich, H. W., 3rd, Kieran, M. W., Chan, F. Y., Barone, L. M., Yee, K., Rundstadler, J. A., ... Zon, L. I. (1995). Intraembryonic hematopoietic cell migration during vertebrate development. *Proceedings of the National Academy of Sciences of the United States of America, 92*(23), 10713–10717.

Donovan, A., Brownlie, A., Dorschner, M. O., Zhou, Y., Pratt, S. J., Paw, B. H., ... Zon, L. I. (2002). The zebrafish mutant gene chardonnay (cdy) encodes divalent metal transporter 1 (DMT1). *Blood, 100*(13), 4655–4659.

Donovan, A., Brownlie, A., Zhou, Y., Shepard, J., Pratt, S. J., Moynihan, J., ... Zon, L. I. (2000). Positional cloning of zebrafish ferroportin1 identifies a conserved vertebrate iron exporter. *Nature, 403*(6771), 776–781.

Donovan, A., Lima, C. A., Pinkus, J. L., Pinkus, G. S., Zon, L. I., Robine, S., & Andrews, N. C. (2005). The iron exporter ferroportin/Slc40a1 is essential for iron homeostasis. *Cell Metabolism, 1*(3), 191–200.

Dooley, K. A., Fraenkel, P. G., Langer, N. B., Schmid, B., Davidson, A. J., Weber, G., ... Tubingen Screen, C. (2008). montalcino, A zebrafish model for variegate porphyria. *Experimental Hematology, 36*(9), 1132–1142.

Draper, B. W., Morcos, P. A., & Kimmel, C. B. (2001). Inhibition of zebrafish fgf8 pre-mRNA splicing with morpholino oligos: a quantifiable method for gene knockdown. *Genesis, 30*(3), 154–156.

Driever, W., Solnica-Krezel, L., Schier, A. F., Neuhauss, S. C., Malicki, J., Stemple, D. L., ... Boggs, C. (1996). A genetic screen for mutations affecting embryogenesis in zebrafish. *Development, 123*, 37–46.

Eisen, J. S., & Smith, J. C. (2008). Controlling morpholino experiments: don't stop making antisense. *Development, 135*(10), 1735–1743.

Ekker, S. C., & Larson, J. D. (2001). Morphant technology in model developmental systems. *Genesis, 30*(3), 89–93.

Fenton, H. J. H. (1894). Oxidation of tartaric acid in presence of iron. *Journal of the Chemical Society, Transactions, 65*, 899–910.

Finckbeiner, S., Ko, P. J., Carrington, B., Sood, R., Gross, K., Dolnick, B., ... Liu, P. (2011). Transient knockdown and overexpression reveal a developmental role for the zebrafish enosf1b gene. *Cell and Bioscience, 1*, 32.

Fleming, M. D., Trenor, C. C., 3rd, Su, M. A., Foernzler, D., Beier, D. R., Dietrich, W. F., & Andrews, N. C. (1997). Microcytic anaemia mice have a mutation in Nramp2, a candidate iron transporter gene. *Nature Genetics, 16*(4), 383–386.

Foury, F., & Roganti, T. (2002). Deletion of the mitochondrial carrier genes MRS3 and MRS4 suppresses mitochondrial iron accumulation in a yeast frataxin-deficient strain. *The Journal of Biological Chemistry, 277*(27), 24475–24483.

Fraenkel, P. G., Gibert, Y., Holzheimer, J. L., Lattanzi, V. J., Burnett, S. F., Dooley, K. A., ... Zon, L. I. (2009). Transferrin-a modulates hepcidin expression in zebrafish embryos. *Blood, 113*(12), 2843–2850.

Fraenkel, P. G., Traver, D., Donovan, A., Zahrieh, D., & Zon, L. I. (2005). Ferroportin1 is required for normal iron cycling in zebrafish. *The Journal of Clinical Investigation, 115*(6), 1532–1541.

Ganis, J. J., Hsia, N., Trompouki, E., de Jong, J. L., DiBiase, A., Lambert, J. S., ... Zon, L. I. (2012). Zebrafish globin switching occurs in two developmental stages and is controlled by the LCR. *Developmental Biology, 366*(2), 185–194.

Gansner, J. M., Dang, M., Ammerman, M., & Zon, L. I (2016). Transplantation in zebrafish. In T. Lecuit (Ed.), *Cell polarity and morphogenesis* (Vol. 138, pp. 629–648).

Ganz, T., & Nemeth, E. (2012). Hepcidin and iron homeostasis. *Biochimica et Biophysica Acta, 1823*(9), 1434–1443.

Goldwurm, S., Casati, C., Venturi, N., Strada, S., Santambrogio, P., Indraccolo, S., ... Biondi, A. (2000). Biochemical and genetic defects underlying human congenital hypotransferrinemia. *The Hematology Journal, 1*(6), 390–398.

Griffin, K. J., Amacher, S. L., Kimmel, C. B., & Kimelman, D. (1998). Molecular identification of spadetail: regulation of zebrafish trunk and tail mesoderm formation by T-box genes. *Development, 125*(17), 3379–3388.

Grillo, A. S., SantaMaria, A. M., Kafina, M. D., Huston, N. C., Cioffi, A. G., Han, M., ... Burke, M. D. (2016). Restored iron transport by a small molecule promotes gut absorption and hemoglobinization. *Science* (in press).

Gunshin, H., Mackenzie, B., Berger, U. V., Gunshin, Y., Romero, M. F., Boron, W. F., ... Hediger, M. A. (1997). Cloning and characterization of a mammalian proton-coupled metal-ion transporter. *Nature, 388*(6641), 482–488.

Gunshin, H., Starr, C. N., Direnzo, C., Fleming, M. D., Jin, J., Greer, E. L., ... Andrews, N. C. (2005). Cybrd1 (duodenal cytochrome b) is not necessary for dietary iron absorption in mice. *Blood, 106*(8), 2879–2883.

Haffter, P., Granato, M., Brand, M., Mullins, M. C., Hammerschmidt, M., Kane, D. A., ... Nusslein-Volhard, C. (1996). The identification of genes with unique and essential functions in the development of the zebrafish, *Danio rerio*. *Development, 123*, 1–36.

Hayashi, A., Wada, Y., Suzuki, T., & Shimizu, A. (1993). Studies on familial hypotransferrinemia: unique clinical course and molecular pathology. *American Journal of Human Genetics, 53*(1), 201–213.

Hoban, M. D., & Bauer, D. E. (2016). A genome editing primer for the hematologist. *Blood, 127*(21), 2525–2535.

Hoshijima, K., Jurynec, M. J., & Grunwald, D. J. (2016). Precise editing of the zebrafish genome made simple and efficient. *Developmental Cell, 36*(6), 654–667.

Howe, K., Clark, M. D., Torroja, C. F., Torrance, J., Berthelot, C., Muffato, M., … Stemple, D. L. (2013). The zebrafish reference genome sequence and its relationship to the human genome. *Nature, 496*(7446), 498–503.

Hruscha, A., Krawitz, P., Rechenberg, A., Heinrich, V., Hecht, J., Haass, C., & Schmid, B. (2013). Efficient CRISPR/Cas9 genome editing with low off-target effects in zebrafish. *Development, 140*(24), 4982–4987.

Hsu, P. D., Lander, E. S., & Zhang, F. (2014). Development and applications of CRISPR-Cas9 for genome engineering. *Cell, 157*(6), 1262–1278.

Huang, P., Xiao, A., Zhou, M., Zhu, Z., Lin, S., & Zhang, B. (2011). Heritable gene targeting in zebrafish using customized TALENs. *Nature Biotechnology, 29*(8), 699–700.

Hwang, W. Y., Fu, Y., Reyon, D., Maeder, M. L., Tsai, S. Q., Sander, J. D., … Joung, J. K. (2013). Efficient genome editing in zebrafish using a CRISPR-Cas system. *Nature Biotechnology, 31*(3), 227–229.

Jabara, H. H., Boyden, S. E., Chou, J., Ramesh, N., Massaad, M. J., Benson, H., … Geha, R. S. (2016). A missense mutation in TFRC, encoding transferrin receptor 1, causes combined immunodeficiency. *Nature Genetics, 48*(1), 74–78.

Jao, L. E., Wente, S. R., & Chen, W. (2013). Efficient multiplex biallelic zebrafish genome editing using a CRISPR nuclease system. *Proceedings of the National Academy of Sciences of the United States of America, 110*(34), 13904–13909.

Jiang, Y., Xu, X. S., & Russell, J. E. (2006). A nucleolin-binding 3′ untranslated region element stabilizes beta-globin mRNA in vivo. *Molecular and Cellular Biology, 26*(6), 2419–2429.

Joung, J. K., & Sander, J. D. (2013). TALENs: a widely applicable technology for targeted genome editing. *Nature Reviews. Molecular Cell Biology, 14*(1), 49–55.

Kardon, J. R., Yien, Y. Y., Huston, N. C., Branco, D. S., Hildick-Smith, G. J., Rhee, K. Y., … Baker, T. A. (2015). Mitochondrial ClpX activates a key enzyme for heme biosynthesis and erythropoiesis. *Cell, 161*(4), 858–867.

Kok, F. O., Shin, M., Ni, C. W., Gupta, A., Grosse, A. S., van Impel, A., … Lawson, N. D. (2015). Reverse genetic screening reveals poor correlation between morpholino-induced and mutant phenotypes in zebrafish. *Developmental Cell, 32*(1), 97–108.

Larrick, J. W., & Hyman, E. S. (1984). Acquired iron-deficiency anemia caused by an antibody against the transferrin receptor. *The New England Journal of Medicine, 311*(4), 214–218.

Levy, J. E., Jin, O., Fujiwara, Y., Kuo, F., & Andrews, N. C. (1999). Transferrin receptor is necessary for development of erythrocytes and the nervous system. *Nature Genetics, 21*(4), 396–399.

Li, L., & Kaplan, J. (2004). A mitochondrial-vacuolar signaling pathway in yeast that affects iron and copper metabolism. *The Journal of Biological Chemistry, 279*(32), 33653–33661.

Liao, E. C., Paw, B. H., Peters, L. L., Zapata, A., Pratt, S. J., Do, C. P., … Zon, L. I. (2000). Hereditary spherocytosis in zebrafish riesling illustrates evolution of erythroid beta-spectrin structure, and function in red cell morphogenesis and membrane stability. *Development, 127*(23), 5123–5132.

Liao, E. C., Trede, N. S., Ransom, D., Zapata, A., Kieran, M., & Zon, L. I. (2002). Non-cell autonomous requirement for the bloodless gene in primitive hematopoiesis of zebrafish. *Development, 129*(3), 649–659.

Lin, C. Y., Chiang, C. Y., & Tsai, H. J. (2016). Zebrafish and Medaka: new model organisms for modern biomedical research. *Journal of Biomedical Science, 23*(1), 19.

Lin, S., Long, W., Chen, J., & Hopkins, N. (1992). Production of germ-line chimeras in zebrafish by cell transplants from genetically pigmented to albino embryos. *Proceedings of the National Academy of Sciences of the United States of America, 89*(10), 4519–4523.

Liu, J., Gong, L., Chang, C., Liu, C., Peng, J., & Chen, J. (2012). Development of novel visual-plus quantitative analysis systems for studying DNA double-strand break repairs in zebrafish. *Journal of Genetics and Genomics, 39*(9), 489–502.

Liuzzi, J. P., Aydemir, F., Nam, H., Knutson, M. D., & Cousins, R. J. (2006). Zip14 (Slc39a14) mediates non-transferrin-bound iron uptake into cells. *Proceedings of the National Academy of Sciences of the United States of America, 103*(37), 13612–13617.

Loreal, O., Cavey, T., Bardou-Jacquet, E., Guggenbuhl, P., Ropert, M., & Brissot, P. (2014). Iron, hepcidin, and the metal connection. *Frontiers in Pharmacology, 5*, 128.

Lyons, S. E., Lawson, N. D., Lei, L., Bennett, P. E., Weinstein, B. M., & Liu, P. P. (2002). A nonsense mutation in zebrafish gata1 causes the bloodless phenotype in vlad tepes. *Proceedings of the National Academy of Sciences of the United States of America, 99*(8), 5454–5459.

McKie, A. T., Barrow, D., Latunde-Dada, G. O., Rolfs, A., Sager, G., Mudaly, E., … Simpson, R. J. (2001). An iron-regulated ferric reductase associated with the absorption of dietary iron. *Science, 291*(5509), 1755–1759.

McKie, A. T., Marciani, P., Rolfs, A., Brennan, K., Wehr, K., Barrow, D., … Simpson, R. J. (2000). A novel duodenal iron-regulated transporter, IREG1, implicated in the basolateral transfer of iron to the circulation. *Molecular Cell, 5*(2), 299–309.

Mims, M. P., Guan, Y., Pospisilova, D., Priwitzerova, M., Indrak, K., Ponka, P., … Prchal, J. T. (2005). Identification of a human mutation of DMT1 in a patient with microcytic anemia and iron overload. *Blood, 105*(3), 1337–1342.

Mohandas, N., & Gallagher, P. G. (2008). Red cell membrane: past, present, and future. *Blood, 112*, 3939–3948.

Muckenthaler, M. U., Galy, B., & Hentze, M. W. (2008). Systemic iron homeostasis and the iron-responsive element/iron-regulatory protein (IRE/IRP) regulatory network. *Annual Review of Nutrition, 28*, 197–213.

Muckenthaler, M. U., & Lill, R. (2012). Cellular iron physiology. In G. D. McLaren, & G. J. Anderson (Eds.), *Iron physiology and pathophysiology in humans* (pp. 27–50). New York, NY: Humana Press.

Mühlenhoff, U., Stadler, J. A., Richhardt, N., Seubert, A., Eickhorst, T., Schweyen, R. J., … Wiesenberger, G. (2003). A specific role of the yeast mitochondrial carriers MRS3/4p in mitochondrial iron acquisition under iron-limiting conditions. *The Journal of Biological Chemistry, 278*(42), 40612–40620.

Mukhopadhyay, C. K., Attieh, Z. K., & Fox, P. L. (1998). Role of ceruloplasmin in cellular iron uptake. *Science, 279*(5351), 714–717.

Nakamura, J., Fujikawa, M., & Yoshida, M. (2013). IF1, a natural inhibitor of mitochondrial ATP synthase, is not essential for the normal growth and breeding of mice. *Bioscience Reports, 33*(5).

Nasevicius, A., & Ekker, S. C. (2000). Effective targeted gene 'knockdown' in zebrafish. *Nature Genetics, 26*(2), 216–220.

Nemeth, E., Tuttle, M. S., Powelson, J., Vaughn, M. B., Donovan, A., Ward, D. M., … Kaplan, J. (2004). Hepcidin regulates cellular iron efflux by binding to ferroportin and inducing its internalization. *Science, 306*(5704), 2090–2093.

Nichols, K. E., Crispino, J. D., Poncz, M., White, J. G., Orkin, S. H., Maris, J. M., & Weiss, M. J. (2000). Familial dyserythropoietic anaemia and thrombocytopenia due to an inherited mutation in GATA1. *Nature Genetics, 24*(3), 266–270.

Nilsson, R., Schultz, I. J., Pierce, E. L., Soltis, K. A., Naranuntarat, A., Ward, D. M., ... Mootha, V. K. (2009). Discovery of genes essential for heme biosynthesis through large-scale gene expression analysis. *Cell Metabolism, 10*(2), 119–130.

Ohgami, R. S., Campagna, D. R., Greer, E. L., Antiochos, B., McDonald, A., Chen, J., ... Fleming, M. D. (2005). Identification of a ferrireductase required for efficient transferrin-dependent iron uptake in erythroid cells. *Nature Genetics, 37*(11), 1264–1269.

Parker, L., & Stainier, D. Y. (1999). Cell-autonomous and non-autonomous requirements for the zebrafish gene cloche in hematopoiesis. *Development, 126*(12), 2643–2651.

Patton, E. E., & Zon, L. I. (2001). The art and design of genetic screens: zebrafish. *Nature Reviews. Genetics, 2*(12), 956–966.

Paw, B. H., Davidson, A. J., Zhou, Y., Li, R., Pratt, S. J., Lee, C., ... Zon, L. I. (2003). Cell-specific mitotic defect and dyserythropoiesis associated with erythroid band 3 deficiency. *Nature Genetics, 34*(1), 59–64.

Peng, Y., Clark, K. J., Campbell, J. M., Panetta, M. R., Guo, Y., & Ekker, S. C. (2014). Making designer mutants in model organisms. *Development, 141*(21), 4042–4054.

Philpott, C. C., & Ryu, M. S. (2014). Special delivery: distributing iron in the cytosol of mammalian cells. *Frontiers in Pharmacology, 5*, 173.

Piel, F. B., & Weatherall, D. J. (2014). The alpha-thalassemias. *The New England Journal of Medicine, 371*(20), 1908–1916.

Ponka, P. (1997). Tissue-specific regulation of iron metabolism and heme synthesis: distinct control mechanisms in erythroid cells. *Blood, 89*(1), 1–25.

Postlethwait, J. H., Yan, Y. L., Gates, M. A., Horne, S., Amores, A., Brownlie, A., ... Talbot, W. S. (1998). Vertebrate genome evolution and the zebrafish gene map. *Nature Genetics, 18*(4), 345–349.

Puy, H., Gouya, L., & Deybach, J. C. (2010). Porphyrias. *Lancet, 375*(9718), 924–937.

Ransom, D. G., Bahary, N., Niss, K., Traver, D., Burns, C., Trede, N. S., ... Zon, L. I. (2004). The zebrafish moonshine gene encodes transcriptional intermediary factor 1gamma, an essential regulator of hematopoiesis. *PLoS Biology, 2*(8), E237.

Ransom, D. G., Haffter, P., Odenthal, J., Brownlie, A., Vogelsang, E., Kelsh, R. N., ... Nusslein-Volhard, C. (1996). Characterization of zebrafish mutants with defects in embryonic hematopoiesis. *Development, 123*, 311–319.

Reischauer, S., Stone, O. A., Villasenor, A., Chi, N., Jin, S.-W., Martin, M., ... Stainer, D. Y. R. (2016). Cloche is a bHLH-PAS transcription factor that drives haemato-vascular specification. *Nature, 535*(7611), 294–298.

Rodriguez-Manzaneque, M. T., Tamarit, J., Belli, G., Ros, J., & Herrero, E. (2002). Grx5 is a mitochondrial glutaredoxin required for the activity of iron/sulfur enzymes. *Molecular Biology of the Cell, 13*(4), 1109–1121.

Rosen, J. N., Sweeney, M. F., & Mably, J. D. (2009). Microinjection of zebrafish embryos to analyze gene function. *Journal of Visualized Experiments, 25*.

Rossi, A., Kontarakis, Z., Gerri, C., Nolte, H., Holper, S., Kruger, M., & Stainier, D. Y. (2015). Genetic compensation induced by deleterious mutations but not gene knockdowns. *Nature, 524*(7564), 230–233.

Rouault, T. A. (2015). Mammalian iron-sulphur proteins: novel insights into biogenesis and function. *Nature Reviews. Molecular Cell Biology, 16*(1), 45–55.

Sakamoto, D., Kudo, H., Inohaya, K., Yokoi, H., Narita, T., Naruse, K., ... Kudo, A. (2004). A mutation in the gene for delta-aminolevulinic acid dehydratase (ALAD) causes hypochromic anemia in the medaka, *Oryzias latipes*. *Mechanisms of Development, 121*(7–8), 747–752.

Shafizadeh, E., Paw, B. H., Foott, H., Liao, E. C., Barut, B. A., Cope, J. J., ... Lin, S. (2002). Characterization of zebrafish merlot/chablis as non-mammalian vertebrate models for severe congenital anemia due to protein 4.1 deficiency. *Development, 129*(18), 4359–4370.

Shah, D. I., Takahashi-Makise, N., Cooney, J. D., Li, L., Schultz, I. J., Pierce, E. L., ... Paw, B. H. (2012). Mitochondrial Atpif1 regulates haem synthesis in developing erythroblasts. *Nature, 491*(7425), 608–612.

Shaw, G. C., Cope, J. J., Li, L., Corson, K., Hersey, C., Ackermann, G. E., ... Paw, B. H. (2006). Mitoferrin is essential for erythroid iron assimilation. *Nature, 440*(7080), 96–100.

Sheftel, A. D., Zhang, A. S., Brown, C., Shirihai, O. S., & Ponka, P. (2007). Direct interorganellar transfer of iron from endosome to mitochondrion. *Blood, 110*(1), 125–132.

Shin, J. T., & Fishman, M. C. (2002). From zebrafish to human: modular medical models. *Annual Review of Genomics and Human Genetics, 3*, 311–340.

Shvartsman, M., & Ioav Cabantchik, Z. (2012). Intracellular iron trafficking: role of cytosolic ligands. *Biometals, 25*(4), 711–723.

Stainier, D. Y., Weinstein, B. M., Detrich, H. W., 3rd, Zon, L. I., & Fishman, M. C. (1995). Cloche, an early acting zebrafish gene, is required by both the endothelial and hematopoietic lineages. *Development, 121*(10), 3141–3150.

Streisinger, G., Walker, C., Dower, N., Knauber, D., & Singer, F. (1981). Production of clones of homozygous diploid zebra fish (*Brachydanio rerio*). *Nature, 291*(5813), 293–296.

Summerton, J. (1999). Morpholino antisense oligomers: the case for an RNase H-independent structural type. *Biochimica et Biophysica Acta, 1489*(1), 141–158.

Sung, Y. H., Kim, J. M., Kim, H. T., Lee, J., Jeon, J., Jin, Y., ... Kim, J. S. (2014). Highly efficient gene knockout in mice and zebrafish with RNA-guided endonucleases. *Genome Research, 24*(1), 125–131.

Thompson, M. A., Ransom, D. G., Pratt, S. J., MacLennan, H., Kieran, M. W., Detrich, H. W., 3rd, ... Zon, L. I. (1998). The cloche and spadetail genes differentially affect hematopoiesis and vasculogenesis. *Developmental Biology, 197*(2), 248–269.

Torti, F. M., & Torti, S. V. (2002). Regulation of ferritin genes and protein. *Blood, 99*(10), 3505–3516.

Troadec, M. B., Warner, D., Wallace, J., Thomas, K., Spangrude, G. J., Phillips, J., ... Kaplan, J. (2011). Targeted deletion of the mouse mitoferrin1 gene: from anemia to protoporphyria. *Blood, 117*(20), 5494–5502.

Vulpe, C. D., Kuo, Y. M., Murphy, T. L., Cowley, L., Askwith, C., Libina, N., ... Anderson, G. J. (1999). Hephaestin, a ceruloplasmin homologue implicated in intestinal iron transport, is defective in the sla mouse. *Nature Genetics, 21*(2), 195–199.

Wang, C., Faloon, P. W., Tan, Z., Lv, Y., Zhang, P., Ge, Y., ... Xiong, J. W. (2007). Mouse lysocardiolipin acyltransferase controls the development of hematopoietic and endothelial lineages during in vitro embryonic stem-cell differentiation. *Blood, 110*(10), 3601–3609.

Wang, H., Long, Q., Marty, S. D., Sassa, S., & Lin, S. (1998). A zebrafish model for hepatoerythropoietic porphyria. *Nature Genetics, 20*(3), 239–243.

Wang, Y., Langer, N. B., Shaw, G. C., Yang, G., Li, L., Kaplan, J., ... Bloomer, J. R. (2011). Abnormal mitoferrin-1 expression in patients with erythropoietic protoporphyria. *Experimental Hematology, 39*(7), 784–794.

Wingert, R. A., Brownlie, A., Galloway, J. L., Dooley, K., Fraenkel, P., Axe, J. L., ... Zon, L. I. (2004). The chianti zebrafish mutant provides a model for erythroid-specific disruption of transferrin receptor 1. *Development, 131*(24), 6225–6235.

Wingert, R. A., Galloway, J. L., Barut, B., Foott, H., Fraenkel, P., Axe, J. L., ... Tubingen Screen, C. (2005). Deficiency of glutaredoxin 5 reveals Fe-S clusters are required for vertebrate haem synthesis. *Nature, 436*(7053), 1035–1039.

Zhang, A. S., Sheftel, A. D., & Ponka, P. (2005). Intracellular kinetics of iron in reticulocytes: evidence for endosome involvement in iron targeting to mitochondria. *Blood, 105*(1), 368–375.

Zhang, D. L., Ghosh, M. C., & Rouault, T. A. (2014). The physiological functions of iron regulatory proteins in iron homeostasis – an update. *Frontiers in Pharmacology, 5*, 124.

Zhao, N., Gao, J., Enns, C. A., & Knutson, M. D. (2010). ZRT/IRT-like protein 14 (ZIP14) promotes the cellular assimilation of iron from transferrin. *The Journal of Biological Chemistry, 285*(42), 32141–32150.

Zinc Finger Consortium. Available from: zincfingers.org/default2.htm.).

Zoller, H., McFarlane, I., Theurl, I., Stadlmann, S., Nemeth, E., Oxley, D., ... Vogel, W. (2005). Primary iron overload with inappropriate hepcidin expression in V162del ferroportin disease. *Hepatology, 42*(2), 466–472.

Zu, Y., Tong, X., Wang, Z., Liu, D., Pan, R., Li, Z., ... Lin, S. (2013). TALEN-mediated precise genome modification by homologous recombination in zebrafish. *Nature Methods, 10*(4), 329–331.

CHAPTER

9 The lymphatic vasculature revisited—new developments in the zebrafish

Y. Padberg*[§], S. Schulte-Merker*[§], A. van Impel*[§,1]

*Institute for Cardiovascular Organogenesis and Regeneration, Faculty of Medicine, University of Münster, Münster, Germany

[§]Cells-in-Motion Cluster of Excellence (EXC M 1003-CiM), University of Münster, Münster, Germany

[1]Corresponding author: E-mail: vanimpel@uni-muenster.de

CHAPTER OUTLINE

Abstract

The lymphatic system is lined by endothelial cells and part of the vasculature. It is essential for tissue fluid homeostasis, absorption of dietary fats, and immune surveillance in vertebrates. Misregulation of lymphatic vessel formation and dysfunction of the lymphatic system have been indicated in a number of pathological conditions including lymphedema formation, obesity or chronic inflammatory diseases such as rheumatoid arthritis. In zebrafish, lymphatics were discovered about 10 years ago, and the underlying molecular pathways involved in its development have since been studied in detail. Due to its superior live cell imaging possibilities and the broad tool kit for forward and reverse genetics, the zebrafish has become an important model organism to study the development of the lymphatic system during early embryonic development. In the current review, we will focus on the key players during zebrafish lymphangiogenesis and compare the roles of these genes to their mammalian counterparts. In particular, we will focus on novel findings that shed new light on the molecular mechanisms of lymphatic cell fate specification, as well as sprouting and migration of lymphatic precursor cells.

Methods in Cell Biology, Volume 138, ISSN 0091-679X, http://dx.doi.org/10.1016/bs.mcb.2016.11.001

The lymphatic vasculature is a highly branched, endothelial lined network which functions as a blind-ended, unidirectional system. Its primary function is the maintenance of tissue fluid homeostasis of the organism, but it also plays crucial roles in immune cell trafficking and fat absorption in the intestine (Alitalo, 2011; Tammela & Alitalo, 2010). Malformation or dysfunction of the lymphatics results in severely compromised immune function and lymphedema formation (Schulte-Merker, Sabine, & Petrova, 2011). During the past 15 years, we have gained considerable knowledge about the molecular pathways involved in the development of the lymphatic network in mice and zebrafish, and the work accomplished in zebrafish has had a significant impact on the lymphatic field. The use of zebrafish embryos has emerged as a sophisticated and efficient approach to study the lymphatic vasculature because transgenic lines and knockout fish can be easily generated. Furthermore, cellular migration can be studied in the zebrafish by live cell imaging, which enables researchers to unravel key molecular players in a dynamic model. In this review we will summarize the main mechanisms and key regulators involved in lymphangiogenesis in fish.

1. DEVELOPMENT OF THE LYMPHATIC SYSTEM IN THE ZEBRAFISH TRUNK

The trunk lymphatic system is formed via a dynamic process called lymphangiogenesis. Analogous to the angiogenic formation of new blood vessels, the initial lymphatic structures are formed by sprouting and migration of endothelial cells from existing endothelial structures. As is the case in amphibians and amniotes, the first pool of lymphatic endothelial cells (LECs) in fish is derived from the venous vasculature. In contrast to other systems, however, lymphangiogenesis in zebrafish occurs simultaneously with the formation of venous structures in the trunk, and it is therefore important to consider the events and mechanisms governing arterial and venous development when studying lymphatic development. In the zebrafish trunk there are two angiogenic waves that trigger the formation of arteries and veins. At about 22 hours postfertilization (hpf), when the two main blood vessels—the dorsal aorta (DA) and the posterior cardinal vein (PCV)—have already been formed through vasculogenesis (Ellertsdottir et al., 2010), approximately 30 arterial sprouts emerge from the DA on each side of the embryo (Isogai, Lawson, Torrealday, Horiguchi, & Weinstein, 2003). These primary sprouts will eventually give rise to the stereotypic patterns of intersegmental vessels (ISV) and the dorsal longitudinal anastomotic vessel, all connected to the arterial system at this stage (Fig. 1A-1). Slightly later, at about 32–34 hpf, a second round of angiogenic sprouting occurs, exclusively from the PCV (Bussmann et al., 2010; Isogai et al., 2003). During this (as it turns out to be) lympho-venous sprouting phase, ECs emerge from the PCV (Fig. 1A-2), migrate dorsally, and display one or the other of two distinct cellular behaviors. Approximately half of these venous sprouts connect to and

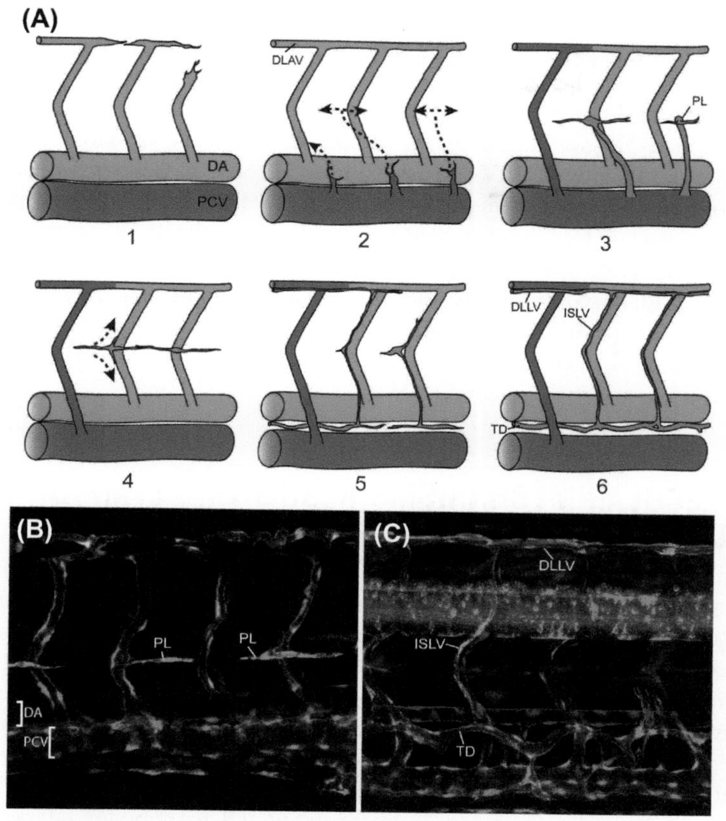

FIGURE 1 The formation of the embryonic vasculature.

(A) Schematic representation of the main angiogenic steps during the formation of the embryonic vasculature: 1: arterial sprouting and formation of intersegmental vessels; 2: lympho-venous sprouting; 3: arterio-venous patterning and PL accumulation at the horizontal myoseptum; 4: PL migration along arteries; 5: formation of lymphatic vessels; 6: mature embryonic vasculature; (B) confocal image of a *flt4*:mCitrine [marks venous and lymphatic endothelial cells (ECs) in green]; *flt1*:tdTomato (stains arterial ECs in red) double transgenic embryo at 48 hpf; (C) triple transgenic embryo at 5 dpf in which the lymphatic vasculature is marked by coexpression of a *prox1a* (red) and *flt4* (green) reporter, while all blood ECs are highlighted by *kdr-l*:mTurquoise expression (blue). *DA*, dorsal aorta; *DLAV*, dorsal longitudinal anastomotic vessel; *DLLV*, dorsal longitudinal lymphatic vessel; *ISLV*, intersegmental lymphatic vessel; *PCV*, posterior cardinal vein; *PL*, parachordal lymphangioblast; *TD*, thoracic duct.

form stable connections with nearby arterial ISVs, which results eventually in detachment of these ISVs from the DA and their remodeling into intersegmental veins. The remaining sprouts, however, do not stably connect to ISVs but rather migrate more dorsally, turn, and enter the horizontal myoseptum (HM) region

(Fig. 1A-3 and B), a process dependent on Netrin1a provided by muscle pioneer cells (Epting et al., 2010; Lim et al., 2011). Once these sprout cells populate the HM and detach from the PCV, they are referred to as parachordal lymphangioblasts (PLs), which constitute a pool of lymphatic precursors in the trunk of the zebrafish embryo (Hogan, Bos, et al., 2009). At 72 hpf, the PLs leave the HM by migrating exclusively along arterial ISVs (Bussmann et al., 2010; Geudens et al., 2010) (Fig. 1A-4). The PLs either migrate ventrally to form the major lymphatic vessel, the thoracic duct (TD), situated between the DA and the PCV, or they migrate dorsally to form the dorsal longitudinal lymphatic vessel (DLLV) (Fig. 1A-5). The DLLV will be connected to the TD via a set of intersegmental lymphatic vessels that are positioned in close proximity to the intersegmental arteries (Bussmann et al., 2010) (Fig. 1A-6 and C).

2. MOLECULAR MECHANISMS REGULATING LYMPHATIC CELL FATE SPECIFICATION

In the mouse, lymphatic cell fate specification occurs in the cardinal vein when definitive veins have already formed. At around E8.5, the transcription factors CoupTF-II (Srinivasan et al., 2010) and Sox18 (Francois et al., 2008) induce Prox-1 expression in a subpopulation of venous endothelial cells, which will differentiate into lymphatic precursor cells (Wigle & Oliver, 1999). Only these cells are responsive to the morphogen Vegfc and will migrate out of the vein to eventually give rise to the lymphatic vasculature (Srinivasan et al., 2014; Wigle et al., 2002).

In zebrafish, a number of studies have investigated the origin of the PLs but how sprouting venous cells of the PCV are programmed to remodel intersegmental arteries into veins, or alternatively to become PL cells, has not been settled. While knockdown studies reported lymphatic defects for all main players in murine lymphatic specification (Aranguren et al., 2011; Cermenati et al., 2013; Del Giacco, Pistocchi, & Ghilardi, 2010; Yaniv et al., 2006), generation of knockout zebrafish models did not consistently support a strong evolutionary conservation of the events which specify lymphatic cell fate (van Impel et al., 2014; Koltowska, Lagendijk, et al., 2015; Tao et al., 2011).

The transcription factors CoupTFII and Sox18 bind to the regulatory region of *Prospero-related homeobox gene 1* (*Prox1*) in mouse. *Prox1* was one of the first genes that was found to be essential for lymphatic development in mouse, and it has been shown to be the key regulatory gene that defines the lymphatic identity of murine venous endothelial cells. Knockout of *Prox1* in the mouse results in the complete absence of embryonic lymphatic structures and early embryonic death (Wigle & Oliver, 1999). In mice, CoupTFII was shown to promote venous cell identity during specification of the blood endothelial cells (You et al., 2005). Endothelial-specific knockout of *CoupTFII* results in abnormal venous specification (You et al., 2005) and, therefore, a lack of LECs in mouse (Srinivasan et al., 2007).

Mechanistically, CoupTFII directly induces *Prox1* expression in the cardinal vein, and it also participates in a positive feedback loop that maintains *Prox1* expression in LECs (Srinivasan et al., 2010). Whereas MO knockdown of *coupTFII* in zebrafish embryos has been reported to lead to a reduction of Prox1a in endothelial cells (48 hpf) and an impairment of PL formation (Aranguren et al., 2011), *coupTFII* knockout fish show normal blood vasculature, have no lymphatic defects, and are viable beyond 6 days postfertilization (dpf). Arteriovenous identity also appears to be unaffected in *coupTFII* mutants, which led to the conclusion that zebrafish *coupTFII* is not essential during venous specification and lymphatic development (van Impel et al., 2014).

The transcription factor Sox18 is expressed in a subset of cardinal vein cells in mice, and it has been shown to be involved in the induction of *Prox1* expression in future lymphatic ECs within the cardinal vein (Francois et al., 2008). Initial morpholino knockdown studies of *sox18* in zebrafish suggested that it might also be essential during early lymphatic differentiation, despite the fact that it is expressed in both the DA and the PCV during lympho-venous sprouting stages (Cermenati et al., 2013; Herpers, van de Kamp, Duckers, & Schulte-Merker, 2008). Although *sox7;sox18* double mutants in zebrafish fully recapitulate previously reported arteriovenous defects in double morphants, *sox18* single mutants do not show any lymphatic or vascular phenotype (Hermkens et al., 2015; Herpers et al., 2008; van Impel et al., 2014). Some of these discrepancies might be attributable to the use of morpholino knockdowns versus mutant lines—there has been an animated debate regarding the use of morpholinos and their reliability for studying gene functions in zebrafish (Kok et al., 2015; Schulte-Merker & Stainier, 2014). In this context, it is important to note that the phenotypes of stable mutations can be camouflaged by genetic compensation effects (Rossi et al., 2015). It is conceivable that similar masking occurs in *coupTFII* or *sox18* mutants, a possibility that should be addressed in the future.

In contrast to mice, zebrafish possess two *Prox1*-like genes: *prox1a* and *prox1b*. *prox1b* has been reported not to be expressed in the lymphatic vasculature and to be dispensable for lymphatic development (Tao et al., 2011). Zebrafish *prox1a*, however, is expressed in LECs and specifically marks the lymphatic vasculature, but its role in lymphatic specification has been contentious (van Impel et al., 2014; Koltowska, Lagendijk, et al., 2015; Yaniv et al., 2006). Considering the severe lymphatic phenotype of the knockout in mice, one would have expected *prox1a* mutant fish to have a similar defect. However, *prox1a* mutant fish show that zygotic *prox1a* function is not essential for lymphatic specification of endothelial cells and that there is no overt lymphatic phenotype [even though there is a mild reduction of LEC numbers (Koltowska, Lagendijk, et al., 2015)]. In *prox1a; prox1b* double mutants, PL cells continue to sprout from the vein and migrate to the HM. Although a moderate but significant reduction in the size of the TD can be seen at 5 dpf, lymphatic specification is not blocked (van Impel et al., 2014). Furthermore, mosaic overexpression of *prox1a* in venous and arterial ECs has no effect on the fate or behavior of the respective cells. The number of secondary sprouts is not altered, and the positioning of the overexpressing cells in the vasculature does not indicate a change in cell behavior (van Impel et al., 2014). However, Koltowska, Langendijk, et al. (2015) show that maternally contributed *prox1a* transcripts exist. Maternal-zygotic *prox1a* mutants show a reduction in the number of lymphatic nuclei in the trunk when compared to *prox1a* zygotic mutants at 4 dpf. An additional loss of zygotic *prox1b* does not cause a more severe phenotype in *prox1a* (maternal zygotic); *prox1b* (zygotic) double mutants, again suggesting that *prox1b* is not involved in lymphatic specification. The authors argue that maternally provided *prox1a* might be required for the initiation of zygotic *prox1a* expression in zebrafish (Koltowska, Lagendijk, et al., 2015). However, the finding that significant numbers of endothelial cells still undergo lymphatic specification and migrate to the HM in maternal-zygotic *prox1a* mutants suggests that Prox1a might be involved in lymphatic cell specification but cannot constitute the sole regulator of lymphatic fate in the fish.

Koltowska, Langendijk, et al. (2015) also followed Prox1 protein expression during development using antibody staining. They showed that larger numbers of Prox1-positive cells are detectable within the PCV as early as 32 hpf, with a strong enrichment in the dorsal part of the vessel. Quantification of embryonic secondary sprouts for Prox1 protein expression revealed that 65% of cells are positive, which indicates that Prox1 expression at this time is not restricted solely to the 50% of venous sprouts that will ultimately migrate to the HM and adopt a lymphatic fate (Bussmann et al., 2010). At 2 dpf, however, Prox1 staining is restricted to PL cells at the HM, whereas sprouts that form venous ISVs and cells of the PCV do not express the transcription factor indicating that Prox1 protein distribution reliably marks the lymphatic linage at this point in time.

Prior to lympho-venous sprouting, *prox1a* expressing cells appear to divide asymmetrically in a majority of cases. One daughter cell remains in the PCV and keeps its venous identity while downregulating Prox1a expression, whereas the other cell upregulates Prox1a, sprouts out of the PCV, and eventually gives rise to a lymphatic precursor cell. In rare cases, both daughter cells of a *prox1a* expressing cell end up at the HM (Koltowska, Lagendijk, et al., 2015). This suggests that the amount of Prox1 protein determines, or is at least indicative for, the cell fate of these two daughter cells within the PCV. Interestingly, the expression of *prox1a* within the cells of the PCV depends on Vegfc signaling (see later), since impairment of Vegfc/Flt4 signaling leads to a reduction in the number of Prox1a-positive cells within the PCV. *vegfc* overexpression consistently results in a dramatic increase in proliferation of venous ECs and induction of Prox1a expression (Koltowska, Lagendijk, et al., 2015). According to these findings, Vegfc/Flt4 occurs upstream of *prox1a* gene expression in the zebrafish embryo.

In another study, Nicenboim et al. (2015) performed cell tracing experiments within the lymphatic lineage and show that the majority of PL cells at the HM actually originates from the ventral side of the PCV. There, a specialized subpopulation of angioblasts (negative for *lyve-1* but positive for *flt1* expression) undergoes asymmetric cell divisions between 24 and 34 hpf, thereby generating progeny cells that translocate to the dorsal site of the PCV and subsequently bud off the PCV during secondary sprouting. They further report that this specialized population in the ventral side of the PCV depends on Wnt signaling, as *wnt5b* seems to be required and sufficient for the specification of these cardinal vein cells toward a lymphatic fate prior to the onset of lympho-venous sprouting. Wnt5b is secreted by the endoderm, which is positioned ventral to the PCV at 18 hpf. Overexpression of *wnt5b* in vivo results in an increase in *prox1a* reporter activity and an increase in PL cells, while morpholino knockdown of the gene has opposing effects, which suggests that *prox1a* acts downstream of Wnt5b during the differentiation of LECs already present in the floor of the PCV (Nicenboim et al., 2015).

The bone morphogenetic protein pathway has also been implicated as influencing *prox1* gene expression. Inhibition of Bmp signaling results in an increase of *prox1a* expression, whereas overexpression of *bmp2a* leads to lymphatic defects and a reduction in *prox1a* expression. Mechanistically, Bmp signaling can negatively regulate expression of *prox1a* via a micro RNA–mediated mechanism (Dunworth et al., 2014).

In summary, Prox1a expression in PL cells at 48 hpf appears to reliably mark future LECs in the zebrafish. Questions about earlier steps of lymphatic specification remain, including how maternal Prox1a protein exerts its function and to what extent the *flt1*$^+$ *lyve1*$^-$ cell population in the ventral aspect of the PCV (Nicenboim et al., 2015) is congruent with Prox1a$^+$ cells at the dorsal roof of the PCV a few hours later (Koltowska, Lagendijk, et al., 2015).

3. VEGFC SIGNALING AND SPROUTING FROM THE POSTERIOR CARDINAL VEIN

Exogenous signals also influence lymphatic development. The vascular endothelial growth factor-c (Vegfc) is a chemoattractant that is indispensable for secondary sprouting in zebrafish and mouse. Its main receptor is vascular endothelial growth factor receptor-3 (Vegfr3 or Flt4), which is expressed on both lymphatic and blood endothelial cells (Hogan, Bos, et al., 2009; van Impel et al., 2014; Tammela et al., 2008), but it can also bind to Vegfr2. Deletion of either *Vegfr3* or *Vegfc* causes a dramatic lymphatic phenotype in both zebrafish and mice due to the inability of endothelial cells to migrate out of the cardinal vein (Hagerling et al., 2013; Hogan, Herpers, et al., 2009; Karkkainen et al., 2004; Küchler et al., 2006; Le Guen et al., 2014; Villefranc et al., 2013; Yaniv et al., 2006). In fish, lympho-venous sprouting is blocked in *vegfr3* and *vegfc* mutants, resulting in a strong reduction in intersegmental veins and a complete loss of PL cells and lymphatic vessels in the trunk (Hogan, Herpers, et al., 2009; Le Guen et al., 2014; Villefranc et al., 2013) (Fig. 2A and C). It has been shown that Vegfc/Vegfr3 signaling is highly conserved because humans also suffer from reduced lymphatic drainage and lymphedema due to mutations in VEGFC and VEGFR3 (Milroy and Milroy-like disease) (Gordon, Schulte, et al., 2013; Gordon, Spiden, et al., 2013;

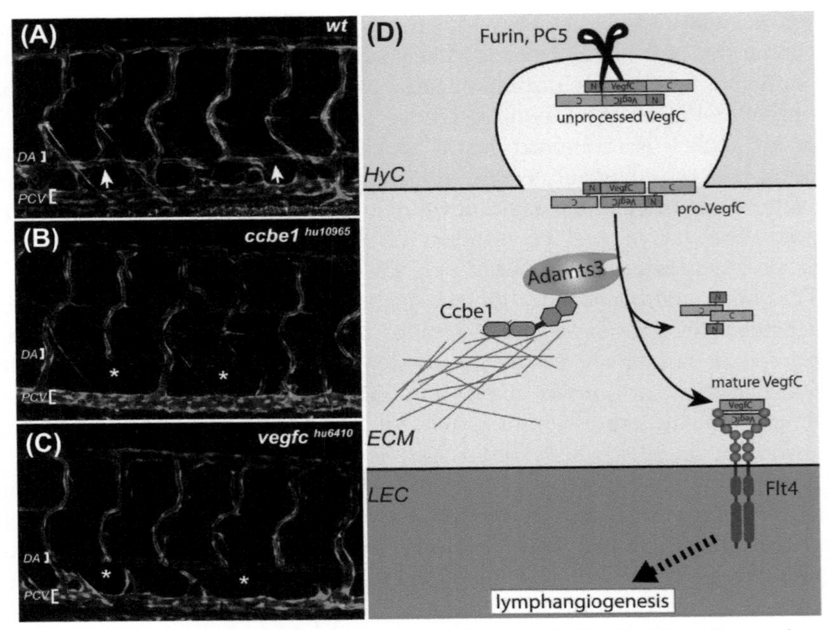

FIGURE 2 The Ccbe1/Vegfc/Flt4 signaling pathway is essential for lymphangiogenesis.

In contrast to the wild-type situation (A, thoracic duct marked with *arrows*), embryos homozygous for a *ccbe1* mutation (B) or for a *vegfc* mutation (C) completely lack lymphatic structures (see regions marked by *asterisks* in B and C) in the trunk at 5 dpf. Embryos are positive for *flt4*:mCitrine (veins and lymphatics in green) and *flt1*:tdTomato (arteries in red); (D) Cartoon summarizing the main steps and molecular players involved in Vegfc processing, which is essential for activation of the Flt4 signaling pathway in lymphatic precursor cells. The Vegfc protein produced in hypochord cells is cleaved by Furin or PC5 protein convertases on secretion. The second cleavage, which releases the fully active form of Vegfc, occurs in the extracellular space and is mediated by Ccbe1 and the Adamts3 protease. Fully mature Vegfc protein can bind and activate the Flt4 receptor on lymphatic precursor cells, inducing lymphangiogenic cell behavior. *DA*, dorsal aorta; *ECM*, extracellular matrix; *HyC*, hypochord cell; *LEC*, lymphatic endothelial cell; *PCV*, posterior cardinal vein.

Karkkainen et al., 2000). In vitro studies show that before Vegfc can bind to its receptor and initiate the sprouting process, it first needs to undergo two proteolytic processing steps. First, secreted pre-pro-Vegfc is cleaved at its C-terminus by Furin/PC5 (or PC7) producing pro-Vegfc (Joukov et al., 1997; Siegfried et al., 2003). Second, the fully active, mature form of Vegfc is generated by a second, N-terminal cleavage event carried out by the matrix-metalloproteinase ADAMTS3. Mature Vegfc protein initiates signaling via VEGFR3 resulting in venous sprouting (Jeltsch et al., 2014) (Fig. 2D). Consistent with these in vitro findings, the *Adamts3* gene knockout in mice causes the same sprouting defect as described for *Vegfc* mutants (Janssen et al., 2016) supporting the importance of Vegfc cleavage for Flt4 signaling in vitro and in vivo.

The *collagen- and calcium-binding EGF domains 1* (*ccbe1*) gene has been found to be another key regulator of embryonic lymphangiogenesis in both fish and mice (Bos et al., 2011; Hogan, Bos, et al., 2009). Ccbe1 is a secreted protein composed of N-terminal calcium-binding EGF (epidermal growth factor)-like and EGF domains and two C-terminal collagen repeat domains. Originally identified in a forward genetic screen in zebrafish, *ccbe1* mutants do not form secondary sprouts from the PCV, which consequently prevents PLs from populating the HM region and forming the TD (Hogan, Bos, et al., 2009) (Fig. 2A and B). Analysis of *Ccbe1* mutant mice confirms an evolutionarily conserved role for Ccbe1 as mutant embryos also show defects in the initial sprouting from the PCV (Hagerling et al., 2013). Significantly, mutation of *CCBE1* in humans cause primary generalized lymph vessel dysplasia (Hennekam syndrome), further highlighting the importance of this gene in vertebrate lymphatic development (Alders et al., 2009).

Recent work has shown that Ccbe1 has a direct influence on the Vegfc/Vegfr3 pathway (Jeltsch et al., 2014; Le Guen et al., 2014; Roukens et al., 2015) by binding to the aforementioned Adamts3, thereby increasing the amount of bioactive Vegfc (Jeltsch et al., 2014; Le Guen et al., 2014). The functions of the different domains of Ccbe1 have been analyzed in vitro and in vivo. The second collagen domain is of key importance for CCBE1 function during VEGFC processing in vitro, since a protein consisting of only the signal peptide, the second collagen domain, and the C-terminus of CCBE1 remains capable of VEGFC processing (Roukens et al., 2015). Deletion of the collagen domains in both the mouse and fish genes recapitulates the phenotype of the corresponding knockouts causing sprouting defects at the level of the cardinal vein. Since the CCBE1 EGF domains can bind to different types of collagens (Bos et al., 2011), they may function to localize CCBE1 to extracellular matrix components (Fig. 2D). Although all domains of the Ccbe1 protein are required for its biological function, the critical domain for enhancing Vegfc processing by Adamts3 seems to be the second collagen repeat domain of the protein (Roukens et al., 2015).

4. PARACHORDAL LYMPHANGIOBLAST MIGRATION AT THE LEVEL OF THE HORIZONTAL MYOSEPTUM

While the molecular mechanisms of early sprouting are appreciated in some detail, the molecular pathways regulating PL behavior in the HM region are less well understood. In particular, the factors that govern the complex migratory behavior of PL cells, which reside in the HM region for a certain time before they move dorsally and ventrally along arterial ISVs, are largely unknown.

Vegfc is not only involved in the process of secondary sprouting and proliferation of venous endothelial cells, but it also regulates transcription of the *mafba* gene, which encodes the Mafba transcription factor in fish (Koltowska, Paterson, et al., 2015) and mice (Dieterich et al., 2015). Zebrafish *mafba* is initially expressed in the PCV, and its expression is prominent in secondary sprouts at 48 hpf. Analysis of *mafba* mutants revealed that the transcription factor is required for the PL migration away from the HM region in a cell autonomous fashion. In the absence of Mafba, PLs no longer migrate ventrally or dorsally to give rise to the main lymphatic vessels, although the initial sprouting from the PCV and the migration to the HM region are not affected. *mafba* expression is also dependent on the activity of Sox18 and Sox7 (Koltowska, Paterson, et al., 2015). How these transcription factors interact is not yet clear, and their role during lymphatic morphogenesis still needs to be investigated in more detail.

The *pkd1a* gene also appears to play a role at the level of the HM. *pkd1a* is initially expressed ubiquitously but becomes largely restricted to the trunk of zebrafish embryos by 32 hpf. Using fluorescence-activated cell sorting analysis, venous and lymphatic expression of *pdk1a* during PL migration was demonstrated. Mutant embryos show a defect in PL cell migration at the HM, resulting in a block of lymphatic vessel formation. Precisely how Pkd1a functions at the level of the HM and how it can influence PL cell migration remains enigmatic. Knockout of *Pkd1* in mice also leads to a lymphatic defect, but the exact stage at which Pkd1 is required has yet to be unraveled (Coxam et al., 2014). The earliest defect that can be observed in mice is a change in lymphatic cell morphology. How this phenotype contributes to the lymphatic defect seen in zebrafish requires further investigation.

Cha et al. (2012) showed that chemokine signaling plays a central role in migration of lymphatic cells in the zebrafish trunk by acting as guidance cues for endothelial cells. This is a significant finding, since chemokines hitherto have only been associated with blood cell migration. LECs express the chemokine receptors Cxcr4a and Cxcr4b, and their ligand Cxcl12a is initially expressed at the HM. Knockdown of Cxcl12a or Cxcr4a leads to a defect in migration of PLs to the HM region. Sprouts still grow dorsally in the direction of the HM, but they fail to reach the midline of the zebrafish embryo, resulting in the absence of PL cells. Cxcl12b knockdown, on the other hand, does not cause any defects during the initial migration events but still leads to loss of the TD. Cxcl12b is exclusively expressed on arterial ISVs and the DA at the time when PL cells migrate away from the HM, providing the *cxcr4*-expressing PLs with a migration path to move to the ventral and dorsal part of the trunk. At the same stage, *cxcl12a* is expressed in the dorsal part of the PCV, which seems to guide LECs to assemble the TD between the two main blood vessels (Cha et al., 2012). For this later process, Vegfc provided by the DA and VEGFR3-expressing secondary motoneurons might also be important (Kwon et al., 2013).

Taken together, these studies serve as an entry point to better understand lymphatic cell migration. We have a good appreciation of the actual cellular events, but only a few genes have been shown to contribute to PL cell migration at the HM. We still do not understand the molecular events occurring at the midline of the zebrafish embryo. Further investigations are needed to fully appreciate the interplay of signaling pathways and different tissues during PL migration.

5. DEVELOPMENT OF LYMPHATIC STRUCTURES IN THE HEAD AND THE GUT

While the formation of the trunk lymphatics has been studied in some detail and has become a well-established model of lymphangiogenesis in fish, there are other lymphatic structures, such as the facial lymphatic network in the head and the gut lymphatics, which form by distinct cellular mechanisms.

The facial lymphatics of zebrafish are formed by three different pools of lymphangioblasts located at the common cardinal vein (CCV), at the primary head sinus (PHS), and at the ventral aorta (VA) (Fig. 3A and B). At 36 hpf, the stage

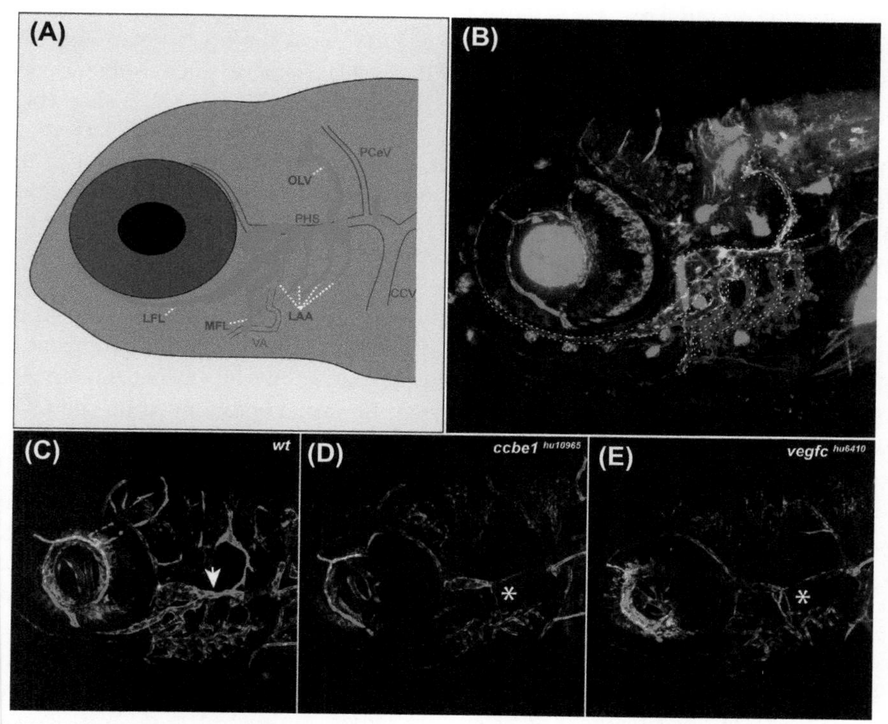

FIGURE 3 The facial lymphatic system.

(A) Schematic overview of the main lymphatic structures in the head of a 5 dpf embryo. (B) Confocal image of the facial lymphatic network (outlined by *white lines*) in a *prox1a* (red), *flt4* (green), and *kdrl-l* (blue) triple transgenic reporter line. (C–E) The facial lymphatic network (*arrow* in C) fails to form in mutants for *ccbe1* and *vegfc* (*asterisks* in D and E, respectively), which demonstrates the importance of this pathway for establishing the cranial lymphatic vasculature. The 5 dpf embryos are positive for *flt4*:mCitrine (green, veins and lymphatics) and *flt1*:tdTomato (red, arteries). *CCV*, common cardinal vein; *LAA*, lymphatic branchial arches; *LFL*, lateral facial lymphatics; *MFL*, medial facial lymphatic; *OLV*, otholithic lymphatic vessel; *PCeV*, posterior cerebral vein; *PHS*, primary head sinus; *VA*, ventral aorta.

when the trunk lymphatics start to form, the first indication of the facial lymphatic network can be appreciated at the level of the CCV. A large vascular sprout, referred to as the facial lymphatic sprout (FLS), emerges from the CCV and extends in the direction of the PHS, where a second pool of lymphangioblasts is located. The FLS is a critical structure—in its absence the facial lymphatic network does not become functional (Okuda et al., 2012). Yet another (and third) pool of lymphangioblasts is situated close to the VA, and these lymphangioblasts (VA-L) migrate dorsally and fuse with the FLS, which eventually gives rise to the lateral facial lymphatics (LFL). The LFL in turn branch off toward the mandibular arch (AA1) and migrate further toward the eye. At 3 dpf, the otolithic lymphatic vessels (OLVs) develop by sprouting from the LFL, parallel to the posterior cerebral vein over the otolith. From the LFL the medial facial lymphatics (MFL) and the four lymphatic branchial arches emerge at 4 dpf. Twenty-four hours later, the LFL migrates along the lateral DA and fuses with the TD. Because three different pools of lymphangioblasts give rise to the facial lymphatic network, whereas a single pool produces the trunk lymphatics, it is likely that the necessary growth factors act differently in the head compared to the trunk. Although lymphangiogenesis in fish is clearly complex, recent studies in mice indicate that the mammalian lymphatic system is equally complex (Klotz et al., 2015; Martinez-Corral et al., 2015). It will be extremely interesting to compare, between fish and mice, the different modes of establishing functional lymphatic networks in different organ systems.

The Ccbe1/Vegfc/Flt4 pathway not only triggers lymphatic development in the trunk of the zebrafish embryo but also is indispensable for the development of the facial lymphatic system (Fig. 3C–E). Ccbe1 knockdown blocks facial lymphatic development and seems to be crucial during the initial sprouting from the CCV, which results in a complete loss of the LFL at 4 dpf (Okuda et al., 2012). Injection of *flt4* morpholinos also leads to a defect in the initial sprouting from the CCV, resulting in the absence of the FLS and the LAA, but not of the MLV and the OLV at 5 dpf (Okuda et al., 2012). *Vegfc*, however, was reported not to be essential for initial development of the facial lymphatic system, as normal LFL and also FLS are formed in both *vegfc* morphants and *vegfc*^{hu5055} mutant embryos [a less penetrant *vegfc* allele (Le Guen et al., 2014)] (Astin et al., 2014).

VEGFC and VEGFD can both bind to the Flt4 receptor (Achen et al., 1998). Since *vegfc* was found not to be essential but *flt4* was shown to be of great importance during the development of facial lymphatics, the role of *vegfd* was investigated. *vegfd* expression appears to be absent in the trunk but can initiate lymphatic cell migration in the trunk when it is ectopically expressed (Astin et al., 2014). *vegfc*, *vegfd*, *ccbe1*, and *flt4* are expressed in the head region between 1.5 and 5 dpf, which would suggest that *vegfd* and *vegfc* might play an important role in the context of head lymphangiogenesis. Knockdown of both *vegfc* and *vegfd* leads to a severe defect of the facial lymphatics, including a reduction of LFL length in the head and the absence of OLV, MFL, and LAA at 5 dpf (Astin et al., 2014), which indicates that *vegfd* can compensate for loss of *vegfc* to some extent in this region of the embryo.

The gut lymphatics of zebrafish, which were discovered recently, are fully developed at 15 dpf and are dependent on *ccbe1*, *vegfc*, and *flt4*. This complex lymphatic network consists of the right and the left supraintestinal lymphatic vessels (R-SIL and L-SIL), which are positioned at both sides of the PCV, and the right and left intestinal lymphatics (R-IL and L-IL). The origin of this network is not entirely clear but likely stems from the CCV (Coffindaffer-Wilson, Craig, & Hove, 2011; Okuda et al., 2012).

The lymphatic vasculature becomes functional around 4–5 dpf in the zebrafish embryo, as defined by uptake of extracellular particles (Küchler et al., 2006; Yaniv et al., 2006) from the tissue surrounding the TD. At this point in time, or slightly later (6 dpf), edema become apparent in mutants that lack a lymphatic system, but not before. As a consequence, edema formation around the eyes or the heart earlier than 5 dpf is not diagnostic of a failing lymphatic system but rather due to other specific or unspecific defects.

CONCLUDING REMARKS

Considering that the molecular and cellular characterization of the lymphatics in zebrafish was initiated only a decade ago (Küchler et al., 2006; Yaniv et al., 2006), we have gained considerable insight into the mechanisms regulating lymphatic development. Forward genetic screens have uncovered essential genes, and a large number of mutant phenotypes have (not surprisingly) confirmed that many genetic pathways are conserved between fish and mammals—of note, mutations in Ccbe1/Vegfc/Vegfr3 all result in phenotypes in zebrafish, mice and humans. The emergence of the TALEN technology and the CRISPR-Cas9 system puts us in a position to mutate any gene of choice, which will be a strong driver for generating more mutants and analyzing them with single cell resolution in one of the many transgenic lines that empower us to distinguish between arteries, veins, and lymphatics within the same embryo.

There are areas where our understanding and our toolset needs to be improved. Research has mostly focused on the trunk lymphatic system in late stage embryos, and more work needs to be carried out to understand the lymphatic endothelium in other organs, such as the gut and the brain. It is unclear, for example, whether fish have a "glymphatic system" as can be found in mammals (Xie et al., 2013), and there are other adult structures that warrant our attention. One of the advantages of the zebrafish is, of course, its superior optical properties at early developmental stages, but it should also be possible to establish functional adult assays and to investigate the reaction of the lymphatic system to different environmental and physiological stimuli.

REFERENCES

Achen, M. G., Jeltsch, M., Kukk, E., Makinen, T., Vitali, A., Wilks, A. F., ... Stacker, S. A. (1998). Vascular endothelial growth factor D (VEGF-D) is a ligand for the tyrosine kinases VEGF receptor 2 (Flk1) and VEGF receptor 3 (Flt4). *Proceedings of the National Academy of Sciences of the United States of America, 95*(2), 548–553.

Alders, M., Hogan, B. M., Gjini, E., Salehi, F., Al-Gazali, L., Hennekam, E. A., … Hennekam, R. C. (2009). Mutations in CCBE1 cause generalized lymph vessel dysplasia in humans. *Nature Genetics, 41*(12), 1272–1274. http://dx.doi.org/10.1038/ng.484.

Alitalo, K. (2011). The lymphatic vasculature in disease. *Nature Medicine, 17*(11), 1371–1380. http://dx.doi.org/10.1038/nm.2545.

Aranguren, X. L., Beerens, M., Vandevelde, W., Dewerchin, M., Carmeliet, P., & Luttun, A. (2011). Transcription factor COUP-TFII is indispensable for venous and lymphatic development in zebrafish and *Xenopus laevis*. *Biochemical and Biophysical Research Communications, 410*(1), 121–126. http://dx.doi.org/10.1016/j.bbrc.2011.05.117.

Astin, J. W., Haggerty, M. J., Okuda, K. S., Le Guen, L., Misa, J. P., Tromp, A., … Crosier, P. S. (2014). Vegfd can compensate for loss of Vegfc in zebrafish facial lymphatic sprouting. *Development, 141*(13), 2680–2690. http://dx.doi.org/10.1242/dev.106591.

Bos, F. L., Caunt, M., Peterson-Maduro, J., Planas-Paz, L., Kowalski, J., Karpanen, T., … Schulte-Merker, S. (2011). CCBE1 is essential for mammalian lymphatic vascular development and enhances the lymphangiogenic effect of vascular endothelial growth factor-C in vivo. *Circulation Research, 109*(5), 486–491. http://dx.doi.org/10.1161/CIRCRESAHA.111.250738.

Bussmann, J., Bos, F. L., Urasaki, A., Kawakami, K., Duckers, H. J., & Schulte-Merker, S. (2010). Arteries provide essential guidance cues for lymphatic endothelial cells in the zebrafish trunk. *Development, 137*(16), 2653–2657. http://dx.doi.org/10.1242/dev.048207.

Cermenati, S., Moleri, S., Neyt, C., Bresciani, E., Carra, S., Grassini, D. R., … Beltrame, M. (2013). Sox18 genetically interacts with VegfC to regulate lymphangiogenesis in zebrafish. *Arteriosclerosis, Thrombosis, and Vascular Biology, 33*(6), 1238–1247. http://dx.doi.org/10.1161/ATVBAHA.112.300254.

Cha, Y. R., Fujita, M., Butler, M., Isogai, S., Kochhan, E., Siekmann, A. F., & Weinstein, B. M. (2012). Chemokine signaling directs trunk lymphatic network formation along the preexisting blood vasculature. *Developmental Cell, 22*(4), 824–836. http://dx.doi.org/10.1016/j.devcel.2012.01.011.

Coffindaffer-Wilson, M., Craig, M. P., & Hove, J. R. (2011). Determination of lymphatic vascular identity and developmental timecourse in zebrafish (*Danio rerio*). *Lymphology, 44*(1), 1–12.

Coxam, B., Sabine, A., Bower, N. I., Smith, K. A., Pichol-Thievend, C., Skoczylas, R., … Hogan, B. M. (2014). Pkd1 regulates lymphatic vascular morphogenesis during development. *Cell Reports, 7*(3), 623–633. http://dx.doi.org/10.1016/j.celrep.2014.03.063.

Del Giacco, L., Pistocchi, A., & Ghilardi, A. (2010). prox1b Activity is essential in zebrafish lymphangiogenesis. *PLoS One, 5*(10), e13170. http://dx.doi.org/10.1371/journal.pone.0013170.

Dieterich, L. C., Klein, S., Mathelier, A., Sliwa-Primorac, A., Ma, Q., Hong, Y. K., … Detmar, M. (2015). DeepCAGE transcriptomics reveal an important role of the transcription factor MAFB in the lymphatic endothelium. *Cell Reports, 13*(7), 1493–1504. http://dx.doi.org/10.1016/j.celrep.2015.10.002.

Dunworth, W. P., Cardona-Costa, J., Bozkulak, E. C., Kim, J. D., Meadows, S., Fischer, J. C., … Jin, S. W. (2014). Bone morphogenetic protein 2 signaling negatively modulates lymphatic development in vertebrate embryos. *Circulation Research, 114*(1), 56–66. http://dx.doi.org/10.1161/circresaha.114.302452.

Ellertsdottir, E., Lenard, A., Blum, Y., Krudewig, A., Herwig, L., Affolter, M., & Belting, H. G. (2010). Vascular morphogenesis in the zebrafish embryo. *Developmental Biology, 341*(1), 56–65. http://dx.doi.org/10.1016/j.ydbio.2009.10.035.

Epting, D., Wendik, B., Bennewitz, K., Dietz, C. T., Driever, W., & Kroll, J. (2010). The Rac1 regulator ELMO1 controls vascular morphogenesis in zebrafish. *Circulation Research, 107*(1), 45–55. http://dx.doi.org/10.1161/CIRCRESAHA.109.213983.

Francois, M., Caprini, A., Hosking, B., Orsenigo, F., Wilhelm, D., Browne, C., … Koopman, P. (2008). Sox18 induces development of the lymphatic vasculature in mice. *Nature, 456*(7222), 643–647. http://dx.doi.org/10.1038/nature07391. pii:nature07391.

Geudens, I., Herpers, R., Hermans, K., Segura, I., Ruiz de Almodovar, C., Bussmann, J., … Dewerchin, M. (2010). Role of delta-like-4/Notch in the formation and wiring of the lymphatic network in zebrafish. *Arteriosclerosis, Thrombosis, and Vascular Biology, 30*(9), 1695–1702. http://dx.doi.org/10.1161/ATVBAHA.110.203034.

Gordon, K., Schulte, D., Brice, G., Simpson, M. A., Roukens, M. G., van Impel, A., … Ostergaard, P. (2013). Mutation in vascular endothelial growth factor-C, a ligand for vascular endothelial growth factor receptor-3, is associated with autosomal dominant milroy-like primary lymphedema. *Circulation Research, 112*(6), 956–960. http://dx.doi.org/10.1161/CIRCRESAHA.113.300350.

Gordon, K., Spiden, S. L., Connell, F. C., Brice, G., Cottrell, S., Short, J., … Ostergaard, P. (2013). FLT4/VEGFR3 and Milroy disease: novel mutations, a review of published variants and database update. *Human Mutation, 34*(1), 23–31. http://dx.doi.org/10.1002/humu.22223.

Hagerling, R., Pollmann, C., Andreas, M., Schmidt, C., Nurmi, H., Adams, R. H., … Kiefer, F. (2013). A novel multistep mechanism for initial lymphangiogenesis in mouse embryos based on ultramicroscopy. *The EMBO Journal, 32*(5), 629–644. http://dx.doi.org/10.1038/emboj.2012.340.

Hermkens, D. M., van Impel, A., Urasaki, A., Bussmann, J., Duckers, H. J., & Schulte-Merker, S. (2015). Sox7 controls arterial specification in conjunction with hey2 and efnb2 function. *Development, 142*(9), 1695–1704. http://dx.doi.org/10.1242/dev.117275.

Herpers, R., van de Kamp, E., Duckers, H. J., & Schulte-Merker, S. (2008). Redundant roles for sox7 and sox18 in arteriovenous specification in zebrafish. *Circulation Research, 102*(1), 12–15. http://dx.doi.org/10.1161/CIRCRESAHA.107.166066.

Hogan, B. M., Bos, F. L., Bussmann, J., Witte, M., Chi, N. C., Duckers, H. J., & Schulte-Merker, S. (2009). ccbe1 is required for embryonic lymphangiogenesis and venous sprouting. *Nature Genetics, 41*(4), 396–398. http://dx.doi.org/10.1038/ng.321.

Hogan, B. M., Herpers, R., Witte, M., Helotera, H., Alitalo, K., Duckers, H. J., & Schulte-Merker, S. (2009). Vegfc/Flt4 signalling is suppressed by Dll4 in developing zebrafish intersegmental arteries. *Development, 136*(23), 4001–4009. http://dx.doi.org/10.1242/dev.039990.

van Impel, A., Zhao, Z., Hermkens, D. M., Roukens, M. G., Fischer, J. C., Peterson-Maduro, J., … Schulte-Merker, S. (2014). Divergence of zebrafish and mouse lymphatic cell fate specification pathways. *Development, 141*(6), 1228–1238. http://dx.doi.org/10.1242/dev.105031.

Isogai, S., Lawson, N. D., Torrealday, S., Horiguchi, M., & Weinstein, B. M. (2003). Angiogenic network formation in the developing vertebrate trunk. *Development, 130*(21), 5281–5290. http://dx.doi.org/10.1242/dev.00733.

Janssen, L., Dupont, L., Bekhouche, M., Noel, A., Leduc, C., Voz, M., … Colige, A. (2016). ADAMTS3 activity is mandatory for embryonic lymphangiogenesis and regulates placental angiogenesis. *Angiogenesis, 19*(1), 53–65. http://dx.doi.org/10.1007/s10456-015-9488-z.

Jeltsch, M., Jha, S. K., Tvorogov, D., Anisimov, A., Leppanen, V. M., Holopainen, T., … Alitalo, K. (2014). CCBE1 enhances lymphangiogenesis via A disintegrin and metalloprotease with thrombospondin motifs-3-mediated vascular endothelial growth factor-C activation. *Circulation, 129*(19), 1962–1971. http://dx.doi.org/10.1161/CIRCULATIONAHA.113.002779.

Joukov, V., Sorsa, T., Kumar, V., Jeltsch, M., Claesson-Welsh, L., Cao, Y., … Alitalo, K. (1997). Proteolytic processing regulates receptor specificity and activity of VEGF-C. *The EMBO Journal, 16*(13), 3898–3911. http://dx.doi.org/10.1093/emboj/16.13.3898.

Karkkainen, M. J., Ferrell, R. E., Lawrence, E. C., Kimak, M. A., Levinson, K. L., McTigue, M. A., … Finegold, D. N. (2000). Missense mutations interfere with VEGFR-3 signalling in primary lymphoedema. *Nature Genetics, 25*(2), 153–159. http://dx.doi.org/10.1038/75997.

Karkkainen, M. J., Haiko, P., Sainio, K., Partanen, J., Taipale, J., Petrova, T. V., … Alitalo, K. (2004). Vascular endothelial growth factor C is required for sprouting of the first lymphatic vessels from embryonic veins. *Nature Immunology, 5*(1), 74–80. http://dx.doi.org/10.1038/ni1013.

Klotz, L., Norman, S., Vieira, J. M., Masters, M., Rohling, M., Dube, K. N., … Riley, P. R. (2015). Cardiac lymphatics are heterogeneous in origin and respond to injury. *Nature, 522*(7554), 62–67. http://dx.doi.org/10.1038/nature14483.

Kok, F. O., Shin, M., Ni, C. W., Gupta, A., Grosse, A. S., van Impel, A., … Lawson, N. D. (2015). Reverse genetic screening reveals poor correlation between morpholino-induced and mutant phenotypes in zebrafish. *Developmental Cell, 32*(1), 97–108. http://dx.doi.org/10.1016/j.devcel.2014.11.018.

Koltowska, K., Lagendijk, A. K., Pichol-Thievend, C., Fischer, J. C., Francois, M., Ober, E. A., … Hogan, B. M. (2015). Vegfc regulates bipotential precursor division and Prox1 expression to promote lymphatic identity in zebrafish. *Cell Reports, 13*(9), 1828–1841. http://dx.doi.org/10.1016/j.celrep.2015.10.055.

Koltowska, K., Paterson, S., Bower, N. I., Baillie, G. J., Lagendijk, A. K., Astin, J. W., … Hogan, B. M. (2015). mafba is a downstream transcriptional effector of Vegfc signaling essential for embryonic lymphangiogenesis in zebrafish. *Genes Dev, 29*(15), 1618–1630. http://dx.doi.org/10.1101/gad.263210.115.

Küchler, A. M., Gjini, E., Peterson-Maduro, J., Cancilla, B., Wolburg, H., & Schulte-Merker, S. (2006). Development of the zebrafish lymphatic system requires VEGFC signaling. *Current Biology, 16*(12), 1244–1248. http://dx.doi.org/10.1016/j.cub.2006.05.026.

Kwon, H. B., Fukuhara, S., Asakawa, K., Ando, K., Kashiwada, T., Kawakami, K., … Mochizuki, N. (2013). The parallel growth of motoneuron axons with the dorsal aorta depends on Vegfc/Vegfr3 signaling in zebrafish. *Development, 140*(19), 4081–4090. http://dx.doi.org/10.1242/dev.091702.

Le Guen, L., Karpanen, T., Schulte, D., Harris, N. C., Koltowska, K., Roukens, G., … Hogan, B. M. (2014). Ccbe1 regulates Vegfc-mediated induction of Vegfr3 signaling during embryonic lymphangiogenesis. *Development, 141*(6), 1239–1249. http://dx.doi.org/10.1242/dev.100495.

Lim, A. H., Suli, A., Yaniv, K., Weinstein, B., Li, D. Y., & Chien, C. B. (2011). Motoneurons are essential for vascular pathfinding. *Development, 138*(17), 3847–3857. http://dx.doi.org/10.1242/dev.068403.

Martinez-Corral, I., Ulvmar, M. H., Stanczuk, L., Tatin, F., Kizhatil, K., John, S. W., … Makinen, T. (2015). Nonvenous origin of dermal lymphatic vasculature. *Circulation Research, 116*(10), 1649–1654. http://dx.doi.org/10.1161/CIRCRESAHA.116.306170.

Nicenboim, J., Malkinson, G., Lupo, T., Asaf, L., Sela, Y., Mayseless, O., … Yaniv, K. (2015). Lymphatic vessels arise from specialized angioblasts within a venous niche. *Nature, 522*(7554), 56–61. http://dx.doi.org/10.1038/nature14425.

Okuda, K. S., Astin, J. W., Misa, J. P., Flores, M. V., Crosier, K. E., & Crosier, P. S. (2012). lyve1 expression reveals novel lymphatic vessels and new mechanisms for lymphatic vessel development in zebrafish. *Development, 139*(13), 2381–2391. http://dx.doi.org/10.1242/dev.077701.

Rossi, A., Kontarakis, Z., Gerri, C., Nolte, H., Holper, S., Kruger, M., & Stainier, D. Y. (2015). Genetic compensation induced by deleterious mutations but not gene knockdowns. *Nature, 524*(7564), 230–233. http://dx.doi.org/10.1038/nature14580.

Roukens, M. G., Peterson-Maduro, J., Padberg, Y., Jeltsch, M., Leppanen, V. M., Bos, F. L., … Schulte, D. (2015). Functional Dissection of the CCBE1 protein: a crucial requirement for the collagen repeat domain. *Circulation Research, 116*(10), 1660–1669. http://dx.doi.org/10.1161/CIRCRESAHA.116.304949.

Schulte-Merker, S., Sabine, A., & Petrova, T. V. (2011). Lymphatic vascular morphogenesis in development, physiology, and disease. *The Journal of Cell Biology, 193*(4), 607–618. http://dx.doi.org/10.1083/jcb.201012094.

Schulte-Merker, S., & Stainier, D. Y. (2014). Out with the old, in with the new: reassessing morpholino knockdowns in light of genome editing technology. *Development, 141*(16), 3103–3104. http://dx.doi.org/10.1242/dev.112003.

Siegfried, G., Basak, A., Cromlish, J. A., Benjannet, S., Marcinkiewicz, J., Chretien, M., … Khatib, A. M. (2003). The secretory proprotein convertases furin, PC5, and PC7 activate VEGF-C to induce tumorigenesis. *The Journal of Clinical Investigation, 111*(11), 1723–1732. http://dx.doi.org/10.1172/jci17220.

Srinivasan, R. S., Dillard, M. E., Lagutin, O. V., Lin, F. J., Tsai, S., Tsai, M. J., … Oliver, G. (2007). Lineage tracing demonstrates the venous origin of the mammalian lymphatic vasculature. *Genes and Development, 21*(19), 2422–2432. http://dx.doi.org/10.1101/gad.1588407.

Srinivasan, R. S., Escobedo, N., Yang, Y., Interiano, A., Dillard, M. E., Finkelstein, D., … Oliver, G. (2014). The Prox1-Vegfr3 feedback loop maintains the identity and the number of lymphatic endothelial cell progenitors. *Genes and Development, 28*(19), 2175–2187. http://dx.doi.org/10.1101/gad.216226.113.

Srinivasan, R. S., Geng, X., Yang, Y., Wang, Y., Mukatira, S., Studer, M., … Oliver, G. (2010). The nuclear hormone receptor Coup-TFII is required for the initiation and early maintenance of Prox1 expression in lymphatic endothelial cells. *Genes and Development, 24*(7), 696–707. http://dx.doi.org/10.1101/gad.1859310.

Tammela, T., & Alitalo, K. (2010). Lymphangiogenesis: molecular mechanisms and future promise. *Cell, 140*(4), 460–476. http://dx.doi.org/10.1016/j.cell.2010.01.045.

Tammela, T., Zarkada, G., Wallgard, E., Murtomaki, A., Suchting, S., Wirzenius, M., … Alitalo, K. (2008). Blocking VEGFR-3 suppresses angiogenic sprouting and vascular network formation. *Nature, 454*(7204), 656–660. http://dx.doi.org/10.1038/nature07083.

Tao, S., Witte, M., Bryson-Richardson, R. J., Currie, P. D., Hogan, B. M., & Schulte-Merker, S. (2011). Zebrafish prox1b mutants develop a lymphatic vasculature, and prox1b does not specifically mark lymphatic endothelial cells. *PLoS One, 6*(12), e28934. http://dx.doi.org/10.1371/journal.pone.0028934.

Villefranc, J. A., Nicoli, S., Bentley, K., Jeltsch, M., Zarkada, G., Moore, J. C., … Lawson, N. D. (2013). A truncation allele in vascular endothelial growth factor c reveals distinct modes of signaling during lymphatic and vascular development. *Development, 140*(7), 1497–1506. http://dx.doi.org/10.1242/dev.084152.

Wigle, J. T., Harvey, N., Detmar, M., Lagutina, I., Grosveld, G., Gunn, M. D., … Oliver, G. (2002). An essential role for Prox1 in the induction of the lymphatic endothelial cell phenotype. *The EMBO Journal, 21*(7), 1505–1513. http://dx.doi.org/10.1093/emboj/21.7.1505.

Wigle, J. T., & Oliver, G. (1999). Prox1 function is required for the development of the murine lymphatic system. *Cell, 98*(6), 769–778. http://dx.doi.org/10.1016/S0092-8674(00)81511-1.

Xie, L., Kang, H., Xu, Q., Chen, M. J., Liao, Y., Thiyagarajan, M., … Nedergaard, M. (2013). Sleep drives metabolite clearance from the adult brain. *Science, 342*(6156), 373–377. http://dx.doi.org/10.1126/science.1241224.

Yaniv, K., Isogai, S., Castranova, D., Dye, L., Hitomi, J., & Weinstein, B. M. (2006). Live imaging of lymphatic development in the zebrafish. *Nature Medicine, 12*(6), 711–716. http://dx.doi.org/10.1038/nm1427. pii:nm1427.

You, L.-R., Lin, F.-J., Lee, C. T., DeMayo, F. J., Tsai, M.-J., & Tsai, S. Y. (2005). Suppression of Notch signalling by the COUP-TFII transcription factor regulates vein identity. *Nature, 435*(7038), 98–104. http://www.nature.com/nature/journal/v435/n7038/suppinfo/nature03511_S1.html.

PART

Visceral Organs IV

CHAPTER

Modeling intestinal disorders using zebrafish

10

X. Zhao, M. Pack[1]

University of Pennsylvania, Philadelphia, PA, United States

[1]*Corresponding author: E-mail: mpack@mail.med.upenn.edu*

CHAPTER OUTLINE

Abstract

Although the zebrafish was initially developed as a model system to study embryonic development, it has gained increasing attention as an advantageous system to investigate human diseases, including intestinal disorders. Zebrafish embryos develop rapidly, and their digestive system is fully functional and visible by 5 days post fertilization. There is a large degree of homology between the intestine of zebrafish and higher vertebrate organisms in terms of its cellular composition and function as both a digestive and immune organ. Furthermore, molecular pathways regulating injury and immune responses are highly conserved. In this chapter, we provide an overview of studies addressing developmental and physiological processes relevant to human intestinal disease. These studies include those related to congenital disorders, host–microbiota interactions, inflammatory diseases, motility disorders, and intestinal cancer. We also highlight the utility of zebrafish to functionally validate candidate genes identified through mutational analyses and genome-wide association studies, and discuss methodologies to investigate the intestinal biology that are unique to zebrafish.

Methods in Cell Biology, Volume 138, ISSN 0091-679X, http://dx.doi.org/10.1016/bs.mcb.2016.11.006

1. INTESTINAL DEVELOPMENT, MORPHOLOGY, AND PHYSIOLOGY

Intestinal development and function have been studied extensively in zebrafish, beginning with the original forward genetic screens and continuing through to present day work focused on physiology and disease. The works discussed in this chapter highlight the many ways in which the zebrafish have enhanced or refined discoveries made in mammalian models, and in other instances, generated novel insights, largely owing to its experimental tractability. Collectively, these examples demonstrate the many ways in which the zebrafish can contribute to translational biomedical research. Nonetheless, it remains important to recognize how intestinal development and morphology differ in fish and humans.

From a developmental perspective, perhaps the most striking difference between zebrafish and mammals is that in fish the anlagen of the intestine, esophagus, and pharynx are independent structures comprised of endodermal cells arranged as either sheet or cord. Following lumenization, the anlagen join to form a patent digestive tract. In contrast, mammalian digestive anlagen are contiguous within the primitive gut tube that forms during ventral folding of the embryo in early development. In zebrafish, intestinal development proceeds rapidly beginning with initial polarization of endodermal cells at about18 hours post fertilization (hpf). This is followed by segmental lumen formation and progressive morphogenesis and differentiation of the epithelium and stromal elements that are largely completed by 5 days post fertilization (dpf), when all the major intestinal functions are supported (Alvers, Ryan, Scherz, Huisken, & Bagnat, 2014; Wallace & Pack, 2003). Although the relative timing of formation of the intestinal anlagen occurs at different developmental stages in zebrafish and mammals (18 somites vs. 1–2 somites), the temporal sequence remains the same, with the rostral portion of the gut developing first, followed by hindgut and midgut (Wallace & Pack, 2003).

From a morphological perspective, the most obvious difference between the zebrafish and mammalian digestive tract is that zebrafish are stomach-less, like other *Cyprinid* family fish, and do not express genes that encode specific gastric functions (Wallace, Akhter, Smith, Lorent, & Pack, 2005). Instead, the intestine is joined directly to the esophagus, forming a tapered tube that is folded into three segments, the large caliber anterior intestinal bulb, the mid-intestine, and the small diameter posterior intestine (Fig. 1A). The anterior and mid-intestine, characterized by long epithelial intestinal folds that shorten in length posteriorly, are primarily involved in nutrient absorption (Figs. 1B–D). Histologically, abundant columnar-shaped absorptive enterocytes can be found (Wallace, Akhter, et al., 2005). Microarray analyses of these segments revealed high expression levels of genes involved in carbohydrate metabolism and lipid transport including *fatty acid binding protein 2* (*fabp2*), *apolipoprotein A1*, and *A4* (*apoa1*, *apoa4*) (Wang et al., 2010). Collectively, these data suggest that the anterior and mid gut segments in zebrafish are functional analogs of the mammalian small intestine. The posterior segment shares features with the mammalian large intestine. It exhibits short epithelial folds and can be distinguished in gene profiling analysis by the expression of *aquaporin 3* (*aqp3m*), which is a well-known large intestinal marker involved in water absorption (Wallace, Akhter, et al., 2005; Wang et al., 2010).

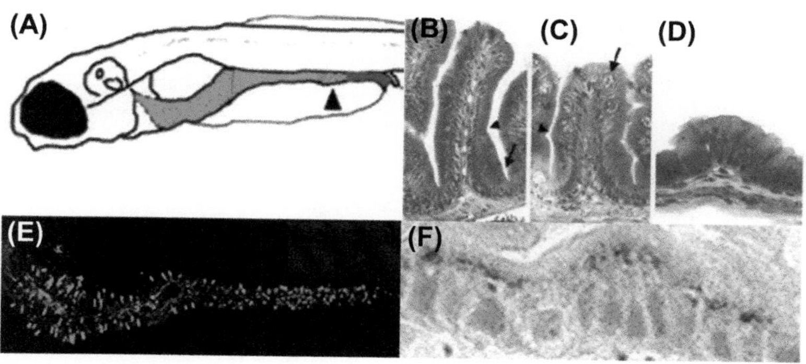

FIGURE 1

Zebrafish intestine: (A) Cartoon depicting anterior (red), mid (green), and posterior (blue) segments of the 5 dpf larval intestine. Distal region of the mid-intestine that contains specialized enterocytes is stippled (arrowhead). (B–D) Histological cross-section through the folds of the anterior (B), mid- (C), and posterior (D) intestine. Note progressive diminution of the height of mid- and posterior folds. A connective tissue core containing blood vessels and nerves underlies the epithelium of each fold. Note absence of crypts at base of folds (arrow in B). (E) Lateral view of the 5 dpf intestine showing regionalized distribution of differentiated epithelial cells. Pancreatic polypeptide containing enteroendocrine cells (green) are restricted to the anterior intestine whereas goblet cells identified with wheat germ agglutinin (red) are restricted to the mid-intestine. (F) Sagittal histological section through the mid-intestine of 4 dpf larva reveals horseradish peroxidase within the apical enterocyte cytoplasm.

In addition to enterocytes, most of the other differentiated epithelial cell types found in mammals are also identified in zebrafish. Goblet cells, which secrete mucins that serve as a protective barrier to pathogens and chemical injury, are the second most populous cell type and are present primarily in the mid- to distal intestine at larval stages (Fig. 1E) (Wallace, Akhter, et al., 2005). Enteroendocrine cells can be visualized using immunostaining against cytoplasmic pancreatic polypeptide hormone (Langer, Van Noorden, Polak, & Pearse, 1979), and their location in larvae is restricted to the anterior segment (Fig. 1E) (Wallace, Akhter, et al., 2005). While there is no Paneth cell within the zebrafish intestinal epithelium, defensin gene expression can be detected in the mid-intestine (Oehlers, Flores, Chen, et al., 2011). Similarly, while microfold (M) cells have not been identified in zebrafish, a related cell type characterized by its ability to pinocytose intact proteins is located in the posterior portion of the midgut (Fig. 1F) (Rombout, Lamers, Helfrich, Dekker, & Taverne-Thiele, 1985). These M-like cells appear to be involved in antigen delivery, and thus may be analogous to epithelial cells overlying Peyer's patches and other gastrointestinal-associated lymphoid tissues in the mammalian ileum and colon (Mabbott, Donaldson, Ohno, Williams, & Mahajan, 2013).

The architectural organization of the intestinal tract is also conserved between zebrafish and mammals. Similar to its mammalian counterparts, zebrafish intestinal mucosa is comprised of a single layer of epithelium that rests upon a connective tissue core containing blood vessels and enteric nerves. However, unlike the mammalian intestinal tract, there is no submucosa layer in zebrafish. Instead, only a thin layer of connective tissue (the lamina propria) separates the mucosa from

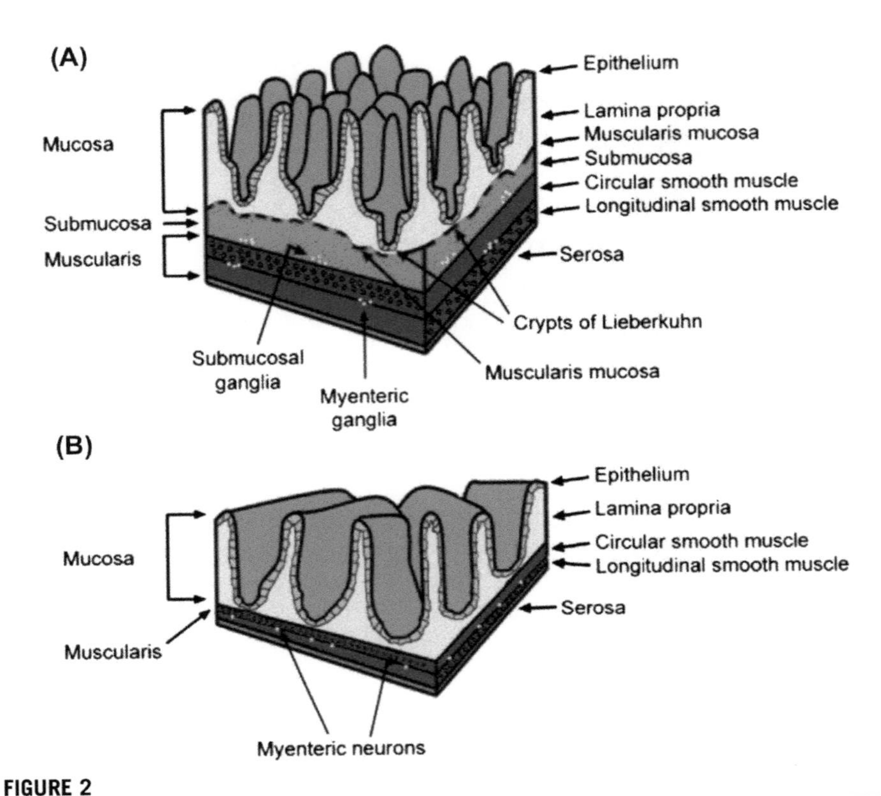

FIGURE 2

Mammalian and teleost intestinal architecture: this schematic compares intestinal layers within the mammalian (A) and teleost (B) intestines.

the underlying muscularis propria, which is comprised of an inner circular and outer longitudinal smooth muscle layer. Enteric nerve cell bodies reside between these two muscle layers but are not detected within the overlying connective tissue as in mammals (Fig. 2) (Wallace, Akhter, et al., 2005).

The intestinal epithelium in zebrafish undergoes continuous cell renewal. In adult fish, epithelial cells migrate from the base to the tip of the intestinal folds where they undergo apoptosis and are shed following their detachment from the underlying stroma and neighboring epithelial cells. As in mammals, it is not known whether shedding is triggered by apoptosis, or if cells undergo apoptosis after they are shed (anoikis). A role for epidermal growth factor in epithelial shedding has been reported in both species (Miguel et al., 2016). While the process of epithelial cell renewal and shedding appears similar in zebrafish and mammals, the organization of the progenitor pool differs. In zebrafish, these cells are diffusely distributed at the base of the intestinal folds, rather than in the discrete gland-like crypts of Lieberkühn that are present in mammals. The proliferating cells comprise about 20% of epithelial cells in both larval and adult zebrafish, and bromodeoxyuridine (BrdU) pulse-labeling experiments showed that these proliferative cells transverse the folds within 5–10 days (Wallace, Akhter, et al., 2005). The mechanisms controlling renewal and differentiation of gut epithelium appear to be highly conserved. For example, Delta-Notch signaling is essential for maintaining a balanced mixture of absorptive and secretory cells within the intestinal epithelium of both zebrafish and mammals (Crosnier et al., 2005). Differentiation into the secretory lineage (i.e., goblet cells) occurs by default in the absence of Notch activation. And like its mammalian counterpart, epithelial renewal and differentiation are dependent on microbial colonization (Bates et al., 2006; Rawls, Samuel, & Gordon, 2004).

1.1 CONGENITAL INTESTINAL DISEASES

Congenital diseases affecting the intestine most often arise from structural abnormalities such as atresia, webs, malrotation, hypoplasia (short gut syndrome), or epithelial defects that cause either malabsorptive or secretory diarrheal syndromes. Zebrafish models for several of these disorders have been reported. Microvillous inclusion disease is a severe malabsorptive syndrome arising from recessive mutation of *MYOVB*. Mutation of zebrafish *myovb* recapitulates many of the features of this disorder including epithelial atrophy, enterocytes microvillous inclusions, and altered function of the secretory pathway, thus providing a novel model to study disease pathogenesis and therapeutics (Sidhaye et al., 2016). Congenital short bowel syndrome (CSBS) is a rare recessive disorder that has been linked to recessive mutations in the gene encoding the Coxsackie- and adenovirus receptor-like membrane protein (CLMP), a tight junction–associated protein. Morpholino knockdown of zebrafish *clmp* recapitulates several features of human CSBS and is anticipated to aid in understanding the pathogenesis of this enigmatic condition (Van Der Werf et al., 2012). Recently, a surgical model of short bowel syndrome was developed using adult zebrafish (Schall et al., 2015).

Abetalipoproteinemia, hypobetalipoproteinemia, and chylomicron retention disease (CMRD; also known as Anderson disease) are congenital recessive disorders of intestinal lipid transport affecting either chylomicron assembly or maturation (CMRD). Knockdown of zebrafish *mtp*, the ortholog of the gene encoding Microsomal triglyceride transfer protein, which is mutated in abetalipoproteinemia, recapitulates many features of this condition (Schlegel, 2015). Similarly, depletion of zebrafish *sarb1*, the ortholog of the gene responsible for CMRD in humans, also phenocopies the human disease (Levic et al., 2015). Interestingly, the zebrafish knockdown phenotypes cause growth retardation at an earlier developmental stage, possibly reflecting differences in energy metabolism during teleost and human embryogenesis (Schlegel, 2015). These disease models offer unique opportunities to study intestinal lipid transport and complement related studies of lipid absorption in larval zebrafish (Clifton et al., 2010; Ho, Lorent, Pack, & Farber, 2006; Otis et al., 2015).

2. INTESTINAL MICROBIOTA AND HOST–MICROBE INTERACTIONS

The vertebrate digestive tract is host to a diverse population of microbes, which have co-evolved alongside their host under selective environmental pressures. This dynamic microbial ecosystem provides multiple benefits to its host, including, but not limited to, nutrient metabolism, detoxification, and maintenance of epithelial integrity. It also serves essential immunomodulatory functions and can have significant impact on the development of the gut immune system. It is now generally accepted that aberrations of the human intestinal microbiota can either directly cause or serve as risk factors for infectious, inflammatory, and metabolic diseases (Boulange, Neves, Chilloux, Nicholson, & Dumas, 2016). The development of gnotobiotic techniques in zebrafish (Bates et al., 2006; Rawls et al., 2004) has fostered studies that have uniquely examined the role of microbial–host interactions using methods that are either impractical or not feasible in mammalian models.

Like other vertebrate organisms, zebrafish harbor a complex population of commensal bacteria within their digestive tract. Acquisition of the "gut microbiome" occurs after hatching and the indigenous microbial community arising from maternal and environmental microbes diverges during development (Yan, van der Gast, & Yu, 2012). Regional differences of the microbiota have been detected, with the intestinal bulb harboring a higher bacterial diversity than the caudal intestine. Genomic profiling showed that the zebrafish gut microbiota share six bacterial divisions with mice and five with humans, however marked differences exist in the relative abundance of these phyla. The microbiota of laboratory-reared zebrafish intestine is dominated by *Proteobacteria*, whereas *Firmicutes* and *Bacteroidetes* dominate in mice and humans (Rawls et al., 2004).

Despite differences in the composition of their microbiota, the responses of zebrafish and mammals to microbial colonization are similar. As an example, microarray analysis comparing the digestive tracts of germ-free (GF) versus conventionally reared (CR) zebrafish revealed differential expression of over 200 genes, of which nearly one-third were conserved in mice (Rawls et al., 2004). The majority of these genes can be linked to epithelial proliferation, nutrient metabolism, and innate immune responses. Interestingly, their expression appears to be selectively regulated by specific bacterial species, as observed in mice (Hooper et al., 2001; Rawls et al., 2004). The microbiota are also required for normal intestinal development in zebrafish (Bates et al., 2006) as evidenced by the lack of brush border intestinal alkaline phosphatase (IAP) activity, immature glycan expression, and a paucity of goblet and enteroendocrine cells in GF larvae. These maturation defects can be entirely reversed by re-introducing the microbiota from CR fish, or selectively reversed via exposure to lipopolysaccharides (LPS) or heat-killed bacteria, suggesting the existence of specific signaling modules. The microbiota also promote epithelial proliferation in the larval intestine (Cheesman, Neal, Mittge, Seredick, & Guillemin, 2011). Mechanistically, this involves activation of canonical Wnt signaling by enhancing the stability of β-catenin in intestinal epithelial cells (IEC). The resident bacteria also transduce proliferative signals to the epithelium via Myd88, a key mediator of the innate immune toll-like receptor (TLR) signaling pathway (Cheesman et al., 2011).

While the composition of the gut microbiota is largely dependent on environmental exposure, transplantation studies have shown that host factors also help sculpt microbial community structure (Rawls, Mahowald, Ley, & Gordon, 2006). These experiments showed that the gut microbiota of GF zebrafish that had been transplanted with mouse intestinal microbes resemble the gut microbes of the CR zebrafish over time, rather than those of mice. The converse was also found to be true. Collectively, these findings suggest that selective pressures imposed within the gut habitat of each host determine the core microbial composition (Roeselers et al., 2011).

Studies in both human and mice have highlighted the importance of the gut microbiota in modulating host metabolism. Work on zebrafish has provided further mechanistic insights, through experiments that have utilized fluorescently tagged lipids that were initially used to monitor lipid metabolism in living larvae (Farber et al., 2001; Ho et al., 2006). By comparing the rate and characteristics of lipid droplet formation in enterocytes of GF and CR zebrafish, this work identified bacterial species that can uniquely influence dietary lipid metabolism (Semova et al., 2012). It also showed that modification of diet influences microbial composition, with caloric intake promoting the relative abundance of bacteria from the phylum *Firmicutes*, a finding consistent with prior mice studies (Turnbaugh, Backhed, Fulton, & Gordon, 2008).

The microbiota play an important role in immunity and the host response to pathogens. In zebrafish, the intestinal commensals have been shown to prime neutrophils for recruitment and induce the expression of proinflammatory cytokines and antiviral mediators via TLR- and NF-kB signaling pathways (Galindo-Villegas, Garcia-Moreno, de Oliveira, Meseguer, & Mulero, 2012; Kanther et al., 2011). Studies in zebrafish were also the first to identify the importance of IAP in promoting mucosal tolerance to commensal bacteria by dephosphorylating and detoxifying the endotoxin component of LPS. Zebrafish lacking IAP were hypersensitive to LPS toxicity and exhibited elevated levels of intestinal neutrophils (Bates, Akerlund, Mittge, & Guillemin, 2007). Subsequent studies in mice further confirm the importance of IAP in enhancing host–microbe symbiosis (Yang, Millan, Mecsas, & Guillemin, 2015).

3. INTESTINAL INFLAMMATORY CONDITIONS

The optical transparency of the zebrafish and its suitability for chemical screens provide unique opportunities to study the fundamental mechanisms underlying intestinal inflammation and injury. Like other vertebrate organisms, zebrafish possess both innate and adaptive immunity. The major mammalian blood cell lineages, including macrophages, neutrophils, eosinophils, and natural killer-like cells, B- and T-cell lymphocytes have been identified in zebrafish. A population of antigen presenting cells that share morphological and functional features of mammalian dendritic cells has also been found (Lugo-Villarino et al., 2010; Wittamer, Bertrand, Gutschow, & Traver, 2011). Pathways regulating microbial recognition and activation of the innate immune response are also greatly conserved. Orthologs of 10 mammalian TLR genes have been identified in the zebrafish genome, along with several fish-specific TLRs. Key mediator proteins involved in innate immune signaling such as MyD88 (van der Vaart, van Soest, Spaink, & Meijer, 2013) and Nod1/Nod2 (Oehlers, Flores, Hall, et al., 2011) also appear to be functionally conserved. Cells of the adaptive immune system can be detected in zebrafish larvae, however adaptive humoral immunity does not fully mature until about 4 weeks post fertilization (Lam, Chua, Gong, Lam, & Sin, 2004). This distinct feature allows the role of innate immunity in intestinal inflammation to be studied in zebrafish larvae without interference from the adaptive immune system.

One of the biggest advantages of using the zebrafish to study immune processes is the ability to visualize immune cells in real time. To date, a large number of transgenic lines using cell-specific enhancers to drive the expression of fluorescent proteins have been generated to label specific blood cells, including but not limited to neutrophils, macrophages, antigen-presenting cells, and T cells (Renshaw et al., 2006; Traver et al., 2003; Wittamer et al., 2011; Zimmerman, Romanowski, & Maddox, 2011). In vivo imaging of these reporter fish larvae allows for dynamic spatiotemporal monitoring of immune cells in the intestine and other tissues.

Inflammatory bowel disease (IBD) refers to a group of chronic diseases that can affect all organs of the digestive tract. While the exact etiology of IBD remains elusive, accumulating evidences suggest that dysregulation of the mucosal immune responses toward commensal microbiota in a genetically susceptible host may play a key role in driving these diseases. The mainstay medical therapy centers around immunosuppression; surgical treatments are often employed but not always curative (Vatn, 2009). Chemical and genetic models are widely used to study IBD in mammalian models and have been adapted to zebrafish. These approaches mimic multiple aspects of the disease, including disrupted epithelial integrity and impaired innate and adaptive immune responses. Here, we discuss key findings from zebrafish IBD models that can be broadly categorized into two groups: the chemically induced and the genetically engineered models.

3.1 CHEMICALLY INDUCED MODELS OF INTESTINAL INFLAMMATION

Four chemically induced models of intestinal injury and inflammation have been described in zebrafish (Table 1). The earliest agent used was the hapten oxazolone, which was administered as a single intra-rectal dose to adult fish. Oxazolone-mediated injury was manifested as disruption of the intestinal folds, depletion of goblet cells, and a pronounced inflammatory cell infiltrate consisting of neutrophils and eosinophils resulting in pronounced thickening of the intestinal wall (Fig. 3) (Brugman et al., 2009). These findings were present 5 h after the treatment and persisted for a week, with the most severe phenotype observed in the posterior mid-intestine. Molecular analyses showed that oxazolone up-regulated intestinal expression of both proinflammatory cytokines (*il-1* and *tnf-α*) and anti-inflammatory cytokines (*il-10*). Interestingly, pretreatment with an antibiotic, vancomycin, that has a selective effect on gram-positive flora, prevented the development of overt inflammation by significantly reducing neutrophil infiltration. In contrast, this effect

Table 1 Four Chemically Induced Models of Intestinal Injury and Inflammation

Treatment	Age	Dose	Phenotype	Ref.
Oxazolone	Adult	Single intra-rectal injection, 0.2% in 50% ethanol (w/v)	• Disruption of the intestinal folds • Depletion of goblet cells • Increased intestinal neutrophils and eosinophils • Up-regulation of proinflammatory cytokines	Brugman et al. (2009)
TNBS	3–8 dpf	Immersion, 75 μg/ml	• Expanded gut lumen • Disappearance of villi • Goblet cell proliferation • Loss of intestinal peristalsis	Fleming et al. (2010)
TNBS	3–6 dpf	Immersion, 50 μg/ml	• Shortening of the mid-intestine • Disruption of the intestinal vasculature • Up-regulation of pro-inflammatory cytokines • Increased intestinal neutrophils • Increased NO production • Altered lipid metabolism	Oehlers, Okuda, et al. (2011)
TNBS	Adult	Single intra-rectal injection, 40–320 mM in 30% ethanol	• Epithelial disruption with increased intestinal neutrophils	Geiger et al. (2013)

Continued

Table 1 Four Chemically Induced Models of Intestinal Injury and Inflammation—cont'd

Treatment	Age	Dose	Phenotype	Ref.
			• No change in goblet cells • Up-regulation of both proinflammatory and anti-inflammatory cytokines • Increased intestinal expression of MCH	
DSS	3–6 dpf	Immersion, 0.5% (w/v)	• Bacterial overgrowth • Increased intestinal neutrophils • Up-regulation of pro-inflammatory cytokines • Decrease in overall cellular proliferation with no change in goblet cell number • Mucus accumulation	Oehlers et al. (2012)
Glafenine	5.5–6 dpf	Immersion, 25 μM	• IEC apoptosis • ER stress in IEC	Goldsmith et al. (2013)

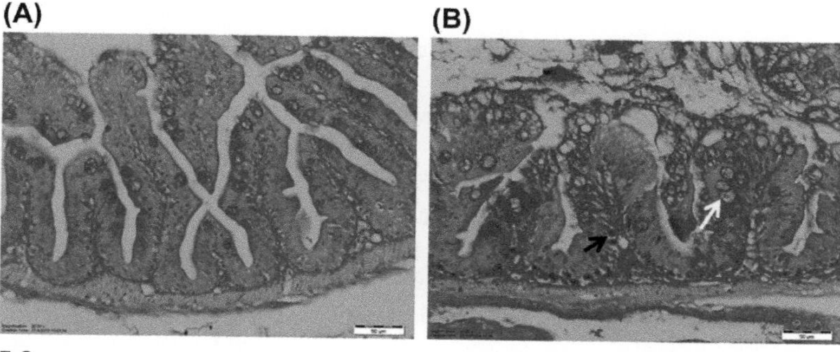

FIGURE 3

Healthy (A) and inflamed (B) proximal midgut in adult zebrafish intestines. Alcian Blue-Pas staining on paraffin embedded sections. Note the thickening of the villi and infiltrating eosinophils (*pink cells, black arrow*), as well as increased mucus production (*blue staining*) and enlarged goblet cells (*white arrow*) (B).

Reprinted from Brugman, S., Liu, K. Y., Lindenbergh-Kortleve, D., Samsom, J. N., Furuta, G. T., Renshaw, S. A., ... Nieuwenhuis, E. E. (2009). Oxazolone-induced enterocolitis in zebrafish depends on the composition of the intestinal microbiota. Gastroenterology, *137(5), 1757–1767 e1751. doi:10.1053/j.gastro.2009.07.069.*

was not seen when fish were pretreated with colistin, which targets gram-negative bacteria. These findings strongly suggest that components of the gut microbiota have differential capacity in mediating intestinal inflammation, a phenomenon also observed in mice (Hoentjen et al., 2003). Although the zebrafish oxazolone-induced enterocolitis model largely phenocopies the corresponding mouse model (Boirivant, Fuss, Chu, & Strober, 1998), its use has been restricted to adult, nontransparent zebrafish.

The second chemical agent used to induce intestinal inflammation in zebrafish was 2,4,6-trinitrobenzenesulfonic acid (TNBS). Fluorescent imaging of live zebrafish larvae immersed in TNBS from 3 dpf to 8 dpf revealed changes in gut architecture that are characteristic of TNBS-induced colitis in mice (Fleming, Jankowski, & Goldsmith, 2010). Exposure of larvae to a lower dose of TNBS for shorter duration led to shortening of the mid-intestine and disruption of the intestinal vasculature without overt morphological damage. These changes were accompanied by upregulation of pro-inflammatory cytokines and the degradative enzyme *metalloproteinase 9* (*mmp9*). There was significant neutrophil infiltration of the intestine, as visualized by the fluorescent neutrophil reporter strain (Tg(*mpx:EGFP*)) (Oehlers, Flores, Okuda, et al., 2011). Impaired gut function was also observed in both dosing regimens, as evidenced by the loss of peristalsis (Fleming et al., 2010) and altered lipid metabolism (Oehlers, Okuda, et al., 2011).

TNBS has also been used to induce enterocolitis in adult zebrafish (Geiger et al., 2013). Six hours after intra-rectal administration of this agent, there was significant influx of neutrophils in the intestine along with upregulation of both proinflammatory and anti-inflammatory cytokines. TNBS exposure also led to an increase in the intestinal expression of melanin-concentrating hormone (MCH), an evolutionarily conserved appetite-regulating neuropeptide that has been implicated in the pathogenesis of human IBD (Kokkotou et al., 2009). In an experimental mouse colitis model, MCH deficient mice showed an attenuated response to TNBS (Kokkotou et al., 2009).

Similar to the oxazolone model, TNBS-induced enterocolitis is dependent on the microbiota (Geiger et al., 2013; He, Wang, Wang, & Li, 2014; He et al., 2013; Oehlers, Okuda, et al., 2011). GF larvae treated with TNBS develop less intestinal inflammation than CR larvae, as evidenced by the lack of stimulation of the TLR signaling pathway and induction of pro-inflammatory cytokines in treated CR fish (He et al., 2014). TNBS enterocolitis can be ameliorated by antibiotic and anti-inflammatory drug treatments used to treat human IBD (Prednisolone and 5-ASA) (Oehlers, Okuda, et al., 2011). Interestingly, knockdown of *myd88* resulted in increased mortality upon TNBS exposure, suggesting a protective function of this gene that is also a key mediator of the innate immune TLR signaling pathway (Oehlers, Okuda, et al., 2011).

Dextran sodium sulfate (DSS) is perhaps the most widely used chemical agent employed in rodent colitis models (Mizoguchi, 2012). Immersion of zebrafish larvae in DSS from 3 dpf to 6 dpf resulted in bacterial overgrowth, neutrophil infiltration into the gut, up-regulation of pro-inflammatory genes, as well as global reduction in cellular proliferation (Oehlers, Flores, Hall, Crosier, & Crosier, 2012). DSS exposure also led to enhanced retention of acidic mucin within the intestinal bulb goblet cell, which was dependent on the microbiota as co-administration of broad-spectrum antibiotics prevented this phenotype. This is considered a protective mechanism that reduced DSS-induced tissue damage. Indeed, DSS pretreatment protected zebrafish larvae from TNBS-induced intestinal inflammation and improved overall survival rates. In contrast, suppression of intestinal mucin production by retinoic acid rendered the fish more susceptible to DDS-induced enterocolitis (Oehlers et al., 2012). Mucin depletion rather than enrichment typically was detected in the vicinity of inflammatory infiltrates in murine DSS-induced colitis (Faure et al., 2003). Because of this discrepancy, the zebrafish DSS colitis model provides an ideal system to study mucin regulations by injurious agents.

Nonsteroidal anti-inflammatory drugs (NSAIDs) impair mucosal barrier function and lead to intestinal epithelial injury in humans and rodent models (Morteau, 2000). Administration of glafenine induced acute intestinal injury in 5 dpf zebrafish larvae, characterized by epithelial cell apoptosis and sloughing without disruption of epithelial barrier function (Goldsmith, Cocchiaro, Rawls, & Jobin, 2013). Ultrastructural analysis revealed endoplasmic reticulum (ER) stress within the epithelium. Strikingly, resolution of ER stress by treatment with DALDA, an opioid receptor agonist that activates the unfolded protein response (UPR) and inhibits DSS-induced colitis in mice (Goldsmith, Uronis, & Jobin, 2011), protected larvae from glafenine-induced intestinal injury. These studies illustrate the usefulness of the zebrafish model in investigating the role of ER stress in mediating intestinal injury and inflammation.

3.2 GENETIC MODELS OF INTESTINAL INFLAMMATION

To date, more than 160 IBD-associated genomic loci have been identified and this list continues to expand (Anderson et al., 2011; Franke et al., 2008, 2010). The zebrafish has proven to be a useful and novel system to validate potential genetic determinants underlying intestinal immune disorders and address roles of specific genes in controlling host–microbe interactions. In humans, mutations in *NOD2* have the highest disease-specific risk association for Crohn's disease (Hugot et al., 2001; Ogura et al., 2001). A related gene, *NOD1*, is associated with ulcerative colitis. NOD genes encode for intracellular bacterial sensor proteins that act as innate immune triggers. Although ligands of zebrafish Nod proteins have not been identified in zebrafish, morpholino-mediated depletion of these proteins enhanced susceptibility of zebrafish larvae to systemic *Salmonella* infection (Oehlers, Hall, et al., 2011). The impaired ability of Nod-depleted larvae to neutralized intracellular bacteria most likely resulted from reduced epithelial expression of *dual oxidase* (*duox2*), which also plays a role in antibacterial immunity in the mammalian intestine (Corcionivoschi et al., 2012) and has recently been causatively linked to early onset IBD (Hayes et al., 2015). Altogether, these data highlight the feasibility of using zebrafish to functionally validate IBD susceptibility genes.

Genome-wide association studies (GWAS) have implicated genes encoding macrophage-stimulating protein (MSP) and its receptor RON (Recepteur d'Origine Nantais) as IBD susceptibility factors (Barrett et al., 2008). While mouse models have not clearly demonstrated a role for MSP–RON signaling in intestinal inflammation, juvenile zebrafish lacking a functional Msp developed spontaneous intestinal inflammation as evidenced by intestinal eosinophilia, increased expression of inflammatory marker *mmp9*, and alteration of goblet cells (Witte, Huitema, Nieuwenhuis, & Brugman, 2014). Moreover, Msp mutants exhibited heightened susceptibility toward ethanol-induced epithelial damage, which led to a prolonged intestinal proinflammatory cytokine response in some mutant fish.

GWAS have also identified genomic variants in IBD patients linked to DNA methyltransferases and their interacting proteins (Jostins et al., 2012), suggesting that IBD can be triggered by epigenetic changes in gene expression (Low, Mizoguchi, & Mizoguchi, 2013). One of these interacting proteins is E3 ubiquitin-protein ligase UHRF1, which recruits DNA (cytosine-5)-methyltransferase 1 to newly synthesized DNA in replicating cells, thus enabling the daughter cell to re-establish the methylation pattern of the maternal cell. Zebrafish homozygous for an inactivating mutation in *uhrf1* exhibited reduced methylation at the *tnf-α* promoter site, leading to an increase in TNF-α expression in IEC that can be detected via live imaging of a fluorescent *tnf-α* reporter (Marjoram et al., 2015). This elevation in TNF levels was microbe dependent and led to widespread epithelial degeneration and intestinal permeability defects that were reversed by *tnf-α* knockdown.

In addition to studies involving candidate IBD genes, forward genetic screens have identified mutations that result in intestinal epithelial disruption, a hallmark pathological feature of IBD (Abraham & Cho, 2009). One mutant, *cdipt*^{hi559}, which is deficient in de novo phosphatidylinositol (PI) synthesis (Thakur, Davison, Stuckenholz, Lu, & Bahary, 2014; Thakur et al., 2011), exhibited a complex intestinal phenotype including disorganized proliferation of epithelial cells, apoptosis of goblet cells, inflammation, and bacterial overgrowth, all pathologies reminiscent of IBD. Furthermore, these mutants displayed signs of ER stress that was ameliorated by chemical chaperones, but not by antibiotics and anti-inflammatory drugs, indicating that inflammation arises in response to ER stress. This study demonstrated a novel link between IP signaling and ER stress-mediated mucosal inflammation, which is significant given that dysregulated PI synthesis has been implicated in a number of gastrointestinal malignancies and inflammatory conditions (van Dieren et al., 2011).

4. ENTERIC NERVOUS SYSTEM AND MOTILITY DISORDERS

4.1 DEVELOPMENT OF THE ENTERIC NERVOUS SYSTEM

The enteric nervous system (ENS) provides intrinsic innervation to the digestive tract and hence plays a crucial role in regulating intestinal motility and secretion. The ease of genetic manipulation and in vivo imaging has allowed ENS development and pathology to be modeled in zebrafish. Like all vertebrate organisms, the zebrafish ENS is derived from the neural crest (Kelsh, Schmid, & Eisen, 2000). Vagal neural crest cells (NCCs) enter the rostral gut tube at about 36 hpf, migrate caudally, and fully colonize the intestine by 66 hpf. Unlike amniotes, where both vagal and sacral NCCs contribute to the ENS, there is no evidence for any sacral crest contribution in zebrafish. Differentiation of NCCs into enteric neurons occurs along the rostrocaudal axis, as in mammals. Neuronal differentiation markers are first detected in the rostral intestine at around 2.25 dpf, and by 7 dpf, several hundred neurons reside with the ENS (Holmberg, Olsson, & Hennig, 2007; Olden, Akhtar, Beckman, & Wallace, 2008). The overall organization of ENS is less complex in zebrafish larvae than in mammals. Instead of having two distinct layers of enteric ganglia, the zebrafish ENS lacks the submucosal plexus and is comprised of myenteric neurons arranged as single neurons or small groups of neurons (Fig. 2) (Wallace, Dolan, et al., 2005; Wallace & Pack, 2003). Glial cells have also not been definitively identified in the larval zebrafish ENS.

The mammalian ENS is comprised of multiple neuronal subtypes whose function can be distinguished by the combinatorial expression pattern of neurotransmitters. Like mammals, zebrafish enteric neurons express both excitatory and inhibitory neurotransmitters, including adenylate cyclase-activating polypeptide (PACAP), vasoactive intestinal polypeptide (VIP), calcitonin gene-related polypeptide (CGRP), neuronal nitric oxide synthase (nNOS) neurokinin-A (NKA), substance P, acetylcholine, and serotonin. Extensive immunohistochemical analyses of neuronal markers revealed at least five neuronal subtypes in the larval and adult zebrafish intestine (Uyttebroek et al., 2010; Uyttebroek, Shepherd, Hubens, Timmermans, & Van Nassauw, 2013). Furthermore, the overall distribution pattern and timing of appearance of these different neurochemical markers is comparable to that seen in amniotes.

4.2 MOLECULAR MECHANISMS OF ZEBRAFISH ENS DEVELOPMENT

Despite differences in ENS structure and complexity between zebrafish and mammals, molecular mechanisms regulating ENS development remain highly conserved. These include the role of sonic hedgehog (shh) in regulating the proliferation of vagal NCC and ENS precursors. Both *shh* and *ihh* mutants have ENS defects similar to those seen in their corresponding mouse mutants (Korzh et al., 2011; Reichenbach et al., 2008). Additionally, the function of the RET tyrosine kinase signaling pathway and its regulators (i.e., Sox10), which are well known for their critical role in ENS development, also appears to be conserved (Dutton et al., 2001; Kelsh, Dutton, Medlin, & Eisen, 2000).

To identify novel genes that direct ENS development, two chemical mutagenesis screens have been performed using an immunofluorescence-based quantification of enteric neuron census (Kuhlman & Eisen, 2007; Pietsch et al., 2006). A secondary motility assay was also used in one of the screens, enabling functional analyses of the new mutations (Kuhlman & Eisen, 2007). Although the genetic basis of most of these mutants has not yet been determined, one ENS mutant, *lessen*, was found to encode for a null mutation in the *med24* gene (Pietsch et al., 2006). Transplantation experiments indicate that endodermal Med24 functions in a cell nonautonomous manner to promote enteric neuron proliferation. As the MED24 null mutation is embryonic lethal in mice (Ito, Okano, Darnell, & Roeder, 2002), zebrafish $med24^{lessen}$ offers a novel opportunity to understand Med24 function in vivo. Recently, the *lessen* mutant has been used as an experimental model to investigate congenital enteric neuropathies (CEN) of the distal intestine (Uyttebroek et al., 2016).

4.3 INTESTINAL MOTILITY DISORDERS

Intestinal motility involves the coordinated function of three different cell populations, enteric neurons, interstitial cells of Cajal (ICC), and smooth muscle cells. Predictably, intestinal motility is altered in zebrafish larvae by disrupting their development and/or function (Abrams, Davuluri, Seiler, & Pack, 2012; Abrams et al., 2016; Kuhlman & Eisen, 2007). An advantage of studying intestinal motility in zebrafish larvae is that it can be quantified using simple in vivo visual assays (Fig. 4) (Abrams et al., 2016; Davuluri, Seiler, Abrams, Soriano, & Pack, 2010; Field, Kelley, Martell, Goldstein, & Serluca, 2009). Using this approach, early spontaneous intestinal contractions were shown to be myogenic in nature, in that they occur in the absence of enteric nerves (Fig. 4) (Abrams et al., 2012; Davuluri et al., 2010), and ICC (Rich et al., 2013), or with pharmacologic inhibition of enteric nerve signaling (Holmberg et al., 2007). These findings indicate that intrinsic

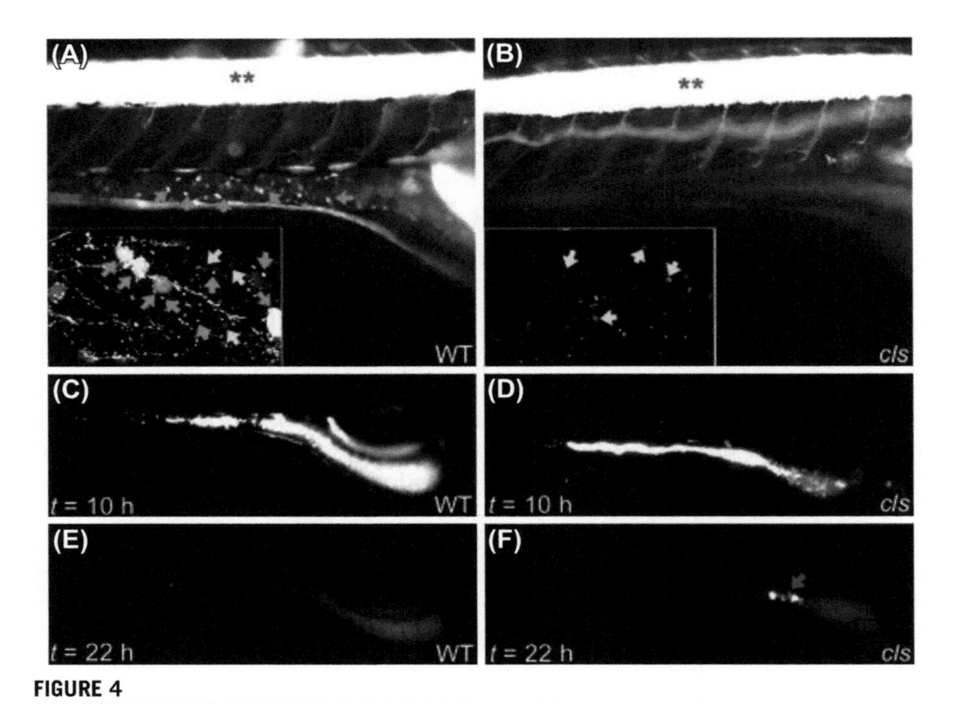

FIGURE 4

Enteric neurons in zebrafish intestine. (A, B) Lateral confocal projections of live 6 dpf wild-type [WT: *sox10* +/+; *Tg*(*NBT:dsRed*)] and *colorless* (*cls*) [*sox10*$^{m148/m148}$; *Tg*(*NBT:dsRed*)] larvae. *Red arrows*: cell bodies of enteric neurons in wild-type larvae that are absent in *cls*. Inset shows high magnification image of intestine with enteric neuron cell bodies (*red arrows*), axon vesicles containing dsRed protein (*green arrows*), and overlying skin neurons (*yellow arrows*). (C–F) Lateral fluorescent images of live 6 dpf wild-type and *cls* larvae 10 and 22 h after ingestion of fluorescent beads.

smooth muscle acto-myosin interactions are sufficient to stimulate early peristaltic contractions. This physiology appears to be evolutionarily conserved as initial intestinal contractions in fetal mice are not mediated by neurons or ICCs.

Recent studies indicate that the microbiota also influence intestinal motility. GF zebrafish larvae exhibited significantly faster and more regular intestinal contractions relative to CR larvae, a trait reversed by microbiota exposure (Bates et al., 2006). Interestingly, the converse is also true. Using a fluorescently labeled two-species microbiota, a recent study demonstrated that intestinal motility shapes the composition of the microbiota (Wiles et al., 2016).

Gastrointestinal motility disorders are common in both pediatric and adult patient populations and significantly impact the quality of life of affected individuals. Zebrafish studies have been used to functionally validate genes linked to heritable and acquired motility disorders that have been identified using genetic analyses or

GWAS. The best studied condition is Hirschsprung's disease (HSCR), a heritable form of intestinal agangliosis of the distal colon that is the most common cause of neonatal intestinal obstruction. HSCR is a complex disorder that exhibits incomplete penetrance with variable expressivity due to interactions among multiple susceptibility genes. To date, nine HSCR susceptibility genes have been identified with nearly half of HSCR patients carrying mutations in the *RET* proto-oncogene (Heanue & Pachnis, 2007). Functionally, RET functions in a receptor complex that mediates signals from glial cell line–derived neurotrophic factor (GDNF)-family ligands (Airaksinen & Saarma, 2002). Like mammals, zebrafish have two *ret* isoforms (*ret9* and *ret51*), of which *ret9* is sufficient to support complete colonization of the gut by ENS precursors (as in mice) (Heanue & Pachnis, 2008). Morpholino knock-down experiments demonstrated that Ret has a dose-dependent effect on enteric precursor proliferation, survival and migration (Heanue & Pachnis, 2008; Shepard et al., 2007). These findings are consistent with human studies implicating reduced *RET* levels as a key HSCR risk factor.

In addition to *RET*, HSCR can arise from mutations in genes encoding proteins that regulate RET-mediated signaling or contribute to the RET receptor complex (*sox10*, *phox2b* and *gfar*, *gndf*). Perturbation of these genes in zebrafish resulted in the loss of enteric neurons (Fig. 4), as observed in mice (Dutton et al., 2001; Elworthy, Pinto, Pettifer, Cancela, & Kelsh, 2005; Kelsh & Eisen, 2000; Minchin & Rawls, 2011; Montero-Balaguer et al., 2006; Stewart et al., 2006). Recently, a noncoding variant within the class 3 semaphorin gene family has been identified as an HSCR susceptibility factor (Jiang et al., 2015; Luzon-Toro et al., 2013). Knockdown of zebrafish *semaphorin 3c* (*sema3c*) or *semaphorin 3d* (*sema3d*) led to an HSCR-like phenotype with fewer enteric neurons in the posterior intestine. Furthermore, co-injection of *ret* and *sema3d* morpholinos resulted in a more severe phenotype, suggesting an epistatic interaction between *RET* and the *SEMA3* genes in humans (Jiang et al., 2015). Collectively, these studies confirm the importance of Ret signaling in vertebrate ENS development and pathology, and highlight the power of the zebrafish model in validating candidate genes uncovered from GWAS studies.

Going forward, we anticipate that the zebrafish will also be valuable in evaluating the significance of patient-derived noncoding variants that reside within suspected regulatory domains using transient and germline transgenic expression studies. These analyses have already led to the identification of key functional domains within the regulatory regions of *ret*, *sox10*, and *phox2b* (Antonellis et al., 2008; Fisher, Grice, Vinton, Bessling, & McCallion, 2006; McGaughey et al., 2008). Conversely, we also anticipate that studying missense variants in zebrafish genes that modulate larval enteric neuromuscular function may lead to the identification of novel clinically relevant variants in humans (Abrams et al., 2016).

Chronic intestinal pseudo-obstruction (CIPO) is another motility disorder that is an important cause of delayed intestinal transit in pediatric and adult patient populations. Whole exome sequencing of patients with familial CIPO identified a missense mutation in *RAD21*, a DNA double-strand break-repair gene that also appears to act as a transcriptional regulator (Bonora et al., 2015). Morpholino knockdown of *rad21* in zebrafish larvae recapitulated the CIPO phenotype, as evidenced by a significant reduction of enteric neurons in the distal intestine as well as delayed intestinal motility. Co-injection studies also revealed a previously unappreciated epistatic interaction between *rad21* and *ret* during enteric neuron development. Interestingly, overexpression of human *RET* mRNA in *rad21* morphants did not rescue the phenotype, suggesting that these two genes act in parallel pathways to regulate ENS development (Bonora et al., 2015). These studies demonstrate the power of using zebrafish as a tool to investigate diseases with complex genetic and phenotypic characteristics.

5. INTESTINAL TUMORIGENESIS AND CANCER

Sporadic intestinal tumors (both benign and malignant) are infrequently detected in laboratory zebrafish strains (Paquette et al., 2013). Nonetheless, a significant number of zebrafish studies devoted to analyses of cancer signaling pathways and aspects of epithelial biology that are relevant to cancer biology have been published (Lobert, 2016 #263). These studies fall into several broad categories. First, mutagenesis screens have identified cancer-related phenotypes arising from mutations in genes not previously linked to tumor formation. As an example, our group identified a novel role for the *elys* gene in the cellular response to DNA replication stress that is independent of its evolutionarily conserved role in nuclear pore formation (Davuluri et al., 2010; Galy, Askjaer, Franz, Lopez-Iglesias, & Mattaj, 2006; de Jong-Curtain et al., 2009; Rasala, Orjalo, Shen, Briggs, & Forbes, 2006). This study suggested that Elys influences the number of DNA replication origins that can be activated when replication forks stall. Our group also identified mutations in *myh11* that trigger invadopodia formation and cell invasion of epithelial cells of the posterior intestine through activation of a novel redox-sensitive physical signaling mechanism (Wallace, Dolan, et al., 2005) (Abrams et al., 2016; Seiler et al., 2012). Studies from our group and others indicate that these phenotypes are evolutionarily conserved, as DNA damage (but not pore defects) was observed with *Elys* disruption in mouse IEC (Gao et al., 2011), and somatic *MYH11* mutations were discovered in a large percentage of hereditary polyps that had advanced to invasive cancer (Alhopuro et al., 2008). In related works, our group and others have identified mutations in genes critical for differentiation, growth, proliferation, and survival of IEC (Boglev et al., 2013; Markmiller et al., 2014; Mayer & Fishman, 2003; Yee et al., 2007), thus identifying RNA Polymerase-3, the small unit processome, the U12 spliceosome, and RNA binding proteins as potential targets for cancer therapeutics.

A second category of cancer-relevant studies has identified novel features of established cancer genes and signaling pathways. This work includes a large number of studies related to the *apc* tumor suppressor, including its effects on retinol metabolism and signaling (Eisinger et al., 2006, 2007; Nadauld et al., 2006), its role in cell differentiation and cell transformation (Phelps et al., 2009; Rai et al., 2010), and a novel in vivo interaction with the mTOR pathway mediated by glycogen synthase kinase-3, which was also observed in mice (Valvezan, Huang, Lengner, Pack, & Klein, 2014). A novel link between mTOR and Lkb1, a tumor suppressor mutated in Peutz–Jeghers syndrome (hamartomatous polyposis) has also been identified using zebrafish (Marshall, Tomasini, Makky, Kumar, & Mayer, 2010; van der Velden et al., 2011).

A third category of studies involves development of zebrafish intestinal cancer models. These studies include a zebrafish model of adenomatous polyposis coli (Apc) in which adult fish heterozygous for a truncating mutation develop intestinal adenomatous polyps (Haramis et al., 2006), mirroring the effects of Apc heterozygosity in mammals. Zebrafish *apc* heterozygotes form fewer polyps than their human or rodent counterparts, however treatment of heterozygous larvae with the carcinogen 7,12-dimethylbenz[a]anthracene (DMBA) increased the frequency of adenoma formation and the presence of dysplasia within these polyps. A second study examined the effect of inducible expression of mutant human *KRAS* ($KRAS^{G12D}$) in zebrafish larvae (Le et al., 2007). Over one-third of surviving fish developed intestinal hyperplasia, suggesting that zebrafish can be used to model the effects of mutant KRAS in colorectal cancer. Studies from our group also showed that co-activation of KRAS and Wnt signaling (*axin* mutants) markedly enhanced cell invasion in *myh11* mutants (Abrams et al., 2016; Seiler et al., 2012).

A fourth category of studies has analyzed the metastatic potential of human cancer cells transplanted into zebrafish larvae. Zebrafish larvae are a highly tractable host for these experiments, owing to their optical transparency and lack of a mature adaptive immune system, which obviates the need for concomitant immunosuppression. The single study that analyzed colorectal cancer cells showed that their degree of metastasis in fish is proportional to their invasion potential in vitro (Teng et al., 2013). A second study examined the behavior of patient-derived gastrointestinal tumors, however, correlation with clinical outcomes was not provided (Marques et al., 2009).

CONCLUDING REMARKS

The large degree of functional conservation of intestinal anatomy and physiology between zebrafish and mammals makes zebrafish an attractive animal system to model human intestinal physiology and disease. The availability of forward and reverse genetic approaches, and the development of gnotobiotic techniques, in conjunction with outstanding imaging modalities to track cells in real time, position the zebrafish system at the forefront of experimental models to study host–microbe interactions, intestinal inflammation, as well as gut motility. With the advent of high efficiency genome editing techniques, we expect that zebrafish will be used more frequently as a platform to functionally validate disease susceptibility genes uncovered in GWAS or other genetic screens. The translational potential of zebrafish is also clear as it is an animal system uniquely suitable for high-throughput drug screens that can potentially lead to the discovery and development of new therapeutic agents.

ACKNOWLEDGMENTS

MP is supported by the NIH NIH/NIDDK grant R01 DK092111 and the Fred and Suzanne Biesecker Pediatric Liver Center. XZ is supported by the CTSA KL2 Mentored Career Development Award (1KL2TR001879-01). The authors are grateful to Kristin Lorent for comments on the manuscript.

REFERENCES

Abraham, C., & Cho, J. H. (2009). Inflammatory bowel disease. *The New England Journal of Medicine, 361*(21), 2066–2078. http://dx.doi.org/10.1056/NEJMra0804647.

Abrams, J., Davuluri, G., Seiler, C., & Pack, M. (2012). Smooth muscle caldesmon modulates peristalsis in the wild type and non-innervated zebrafish intestine. *Neurogastroenterology and Motility: the Official Journal of the European Gastrointestinal Motility Society, 24*(3), 288–299. http://dx.doi.org/10.1111/j.1365-2982.2011.01844.x.

Abrams, J., Einhorn, Z., Seiler, C., Zong, A. B., Sweeney, H. L., & Pack, M. (2016). Graded effects of unregulated smooth muscle myosin on intestinal architecture, intestinal motility and vascular function in zebrafish. *Disease Models & Mechanisms, 9*(5), 529–540. http://dx.doi.org/10.1242/dmm.023309.

Airaksinen, M. S., & Saarma, M. (2002). The GDNF family: signalling, biological functions and therapeutic value. *Nature Reviews. Neuroscience, 3*(5), 383–394. http://dx.doi.org/10.1038/nrn812.

Alhopuro, P., Phichith, D., Tuupanen, S., Sammalkorpi, H., Nybondas, M., Saharinen, J., … Aaltonen, L. A. (2008). Unregulated smooth-muscle myosin in human intestinal neoplasia. *Proceedings of the National Academy of Sciences of the United States of America, 105*(14), 5513–5518. http://dx.doi.org/10.1073/pnas.0801213105.

Alvers, A. L., Ryan, S., Scherz, P. J., Huisken, J., & Bagnat, M. (2014). Single continuous lumen formation in the zebrafish gut is mediated by smoothened-dependent tissue remodeling. *Development, 141*(5), 1110–1119. http://dx.doi.org/10.1242/dev.100313.

Anderson, C. A., Boucher, G., Lees, C. W., Franke, A., D'Amato, M., Taylor, K. D., … Rioux, J. D. (2011). Meta-analysis identifies 29 additional ulcerative colitis risk loci, increasing the number of confirmed associations to 47. *Nature Genetics, 43*(3), 246–252. http://dx.doi.org/10.1038/ng.764.

Antonellis, A., Huynh, J. L., Lee-Lin, S. Q., Vinton, R. M., Renaud, G., Loftus, S. K., … Pavan, W. J. (2008). Identification of neural crest and glial enhancers at the mouse Sox10 locus through transgenesis in zebrafish. *PLoS Genetics, 4*(9), e1000174. http://dx.doi.org/10.1371/journal.pgen.1000174.

Barrett, J. C., Hansoul, S., Nicolae, D. L., Cho, J. H., Duerr, R. H., Rioux, J. D., … Daly, M. J. (2008). Genome-wide association defines more than 30 distinct susceptibility loci for Crohn's disease. *Nature Genetics, 40*(8), 955–962. http://dx.doi.org/10.1038/ng.175.

Bates, J. M., Akerlund, J., Mittge, E., & Guillemin, K. (2007). Intestinal alkaline phosphatase detoxifies lipopolysaccharide and prevents inflammation in zebrafish in response to the gut microbiota. *Cell Host & Microbe, 2*(6), 371–382. http://dx.doi.org/10.1016/j.chom.2007.10.010.

Bates, J. M., Mittge, E., Kuhlman, J., Baden, K. N., Cheesman, S. E., & Guillemin, K. (2006). Distinct signals from the microbiota promote different aspects of zebrafish gut differentiation. *Developmental Biology, 297*(2), 374–386. http://dx.doi.org/10.1016/j.ydbio.2006.05.006.

Boglev, Y., Badrock, A. P., Trotter, A. J., Du, Q., Richardson, E. J., Parslow, A. C., … Heath, J. K. (2013). Autophagy induction is a Tor- and Tp53-independent cell survival response in a zebrafish model of disrupted ribosome biogenesis. *PLoS Genetics, 9*(2), e1003279. http://dx.doi.org/10.1371/journal.pgen.1003279.

Boirivant, M., Fuss, I. J., Chu, A., & Strober, W. (1998). Oxazolone colitis: a murine model of T helper cell type 2 colitis treatable with antibodies to interleukin 4. *The Journal of Experimental Medicine, 188*(10), 1929–1939.

Bonora, E., Bianco, F., Cordeddu, L., Bamshad, M., Francescatto, L., Dowless, D., … De Giorgio, R. (2015). Mutations in RAD21 disrupt regulation of APOB in patients with chronic intestinal pseudo-obstruction. *Gastroenterology, 148*(4), 771–782 e711. http://dx.doi.org/10.1053/j.gastro.2014.12.034.

Boulange, C. L., Neves, A. L., Chilloux, J., Nicholson, J. K., & Dumas, M. E. (2016). Impact of the gut microbiota on inflammation, obesity, and metabolic disease. *Genome Medicine, 8*(1), 42. http://dx.doi.org/10.1186/s13073-016-0303-2.

Brugman, S., Liu, K. Y., Lindenbergh-Kortleve, D., Samsom, J. N., Furuta, G. T., Renshaw, S. A., … Nieuwenhuis, E. E. (2009). Oxazolone-induced enterocolitis in zebrafish depends on the composition of the intestinal microbiota. *Gastroenterology, 137*(5), 1757–1767 e1751. http://dx.doi.org/10.1053/j.gastro.2009.07.069.

Cheesman, S. E., Neal, J. T., Mittge, E., Seredick, B. M., & Guillemin, K. (2011). Epithelial cell proliferation in the developing zebrafish intestine is regulated by the Wnt pathway and microbial signaling via Myd88. *Proceedings of the National Academy of Sciences of the United States of America, 108*(Suppl. 1), 4570–4577. http://dx.doi.org/10.1073/pnas.1000072107.

Clifton, J. D., Lucumi, E., Myers, M. C., Napper, A., Hama, K., Farber, S. A., … Pack, M. (2010). Identification of novel inhibitors of dietary lipid absorption using zebrafish. *PLoS One, 5*(8), e12386. http://dx.doi.org/10.1371/journal.pone.0012386.

Corcionivoschi, N., Alvarez, L. A., Sharp, T. H., Strengert, M., Alemka, A., Mantell, J., … Bourke, B. (2012). Mucosal reactive oxygen species decrease virulence by disrupting Campylobacter jejuni phosphotyrosine signaling. *Cell Host & Microbe, 12*(1), 47–59. http://dx.doi.org/10.1016/j.chom.2012.05.018.

Crosnier, C., Vargesson, N., Gschmeissner, S., Ariza-McNaughton, L., Morrison, A., & Lewis, J. (2005). Delta-Notch signalling controls commitment to a secretory fate in the zebrafish intestine. *Development, 132*(5), 1093–1104. http://dx.doi.org/10.1242/dev.01644.

Davuluri, G., Seiler, C., Abrams, J., Soriano, A. J., & Pack, M. (2010). Differential effects of thin and thick filament disruption on zebrafish smooth muscle regulatory proteins. *Neurogastroenterology and Motility: the Official Journal of the European Gastrointestinal Motility Society, 22*(10), 1100–e1285. http://dx.doi.org/10.1111/j.1365-2982.2010.01545.x.

van Dieren, J. M., Simons-Oosterhuis, Y., Raatgeep, H. C., Lindenbergh-Kortleve, D. J., Lambers, M. E., van der Woude, C. J., … Nieuwenhuis, E. E. (2011). Anti-inflammatory actions of phosphatidylinositol. *European Journal of Immunology, 41*(4), 1047–1057. http://dx.doi.org/10.1002/eji.201040899.

Dutton, K. A., Pauliny, A., Lopes, S. S., Elworthy, S., Carney, T. J., Rauch, J., … Kelsh, R. N. (2001). Zebrafish colourless encodes sox10 and specifies non-ectomesenchymal neural crest fates. *Development, 128*(21), 4113–4125.

Eisinger, A. L., Nadauld, L. D., Shelton, D. N., Peterson, P. W., Phelps, R. A., Chidester, S., … Jones, D. A. (2006). The adenomatous polyposis coli tumor suppressor gene regulates expression of cyclooxygenase-2 by a mechanism that involves retinoic acid. *Journal of Biological Chemistry, 281*(29), 20474–20482. http://dx.doi.org/10.1074/jbc.M602859200.

Eisinger, A. L., Nadauld, L. D., Shelton, D. N., Prescott, S. M., Stafforini, D. M., & Jones, D. A. (2007). Retinoic acid inhibits beta-catenin through suppression of Cox-2: a role for truncated adenomatous polyposis coli. *Journal of Biological Chemistry, 282*(40), 29394–29400. http://dx.doi.org/10.1074/jbc.M609768200.

Elworthy, S., Pinto, J. P., Pettifer, A., Cancela, M. L., & Kelsh, R. N. (2005). Phox2b function in the enteric nervous system is conserved in zebrafish and is sox10-dependent. *Mechanisms of Development, 122*(5), 659–669. http://dx.doi.org/10.1016/j.mod.2004.12.008.

Farber, S. A., Pack, M., Ho, S. Y., Johnson, I. D., Wagner, D. S., Dosch, R., … Halpern, M. E. (2001). Genetic analysis of digestive physiology using fluorescent phospholipid reporters. *Science, 292*(5520), 1385–1388. http://dx.doi.org/10.1126/science.1060418.

Faure, M., Moennoz, D., Montigon, F., Mettraux, C., Mercier, S., Schiffrin, E. J., … Boza, J. (2003). Mucin production and composition is altered in dextran sulfate sodium-induced colitis in rats. *Digestive Diseases and Sciences, 48*(7), 1366–1373.

Field, H. A., Kelley, K. A., Martell, L., Goldstein, A. M., & Serluca, F. C. (2009). Analysis of gastrointestinal physiology using a novel intestinal transit assay in zebrafish. *Neurogastroenterology and Motility: the Official Journal of the European Gastrointestinal Motility Society, 21*(3), 304–312. http://dx.doi.org/10.1111/j.1365-2982.2008.01234.x.

Fisher, S., Grice, E. A., Vinton, R. M., Bessling, S. L., & McCallion, A. S. (2006). Conservation of RET regulatory function from human to zebrafish without sequence similarity. *Science, 312*(5771), 276–279. http://dx.doi.org/10.1126/science.1124070.

Fleming, A., Jankowski, J., & Goldsmith, P. (2010). In vivo analysis of gut function and disease changes in a zebrafish larvae model of inflammatory bowel disease: a feasibility study. *Inflammatory Bowel Disease, 16*(7), 1162–1172. http://dx.doi.org/10.1002/ibd.21200.

Franke, A., Balschun, T., Karlsen, T. H., Sventoraityte, J., Nikolaus, S., Mayr, G., … Schreiber, S. (2008). Sequence variants in IL10, ARPC2 and multiple other loci contribute to ulcerative colitis susceptibility. *Nature Genetics, 40*(11), 1319–1323. http://dx.doi.org/10.1038/ng.221.

Franke, A., McGovern, D. P., Barrett, J. C., Wang, K., Radford-Smith, G. L., Ahmad, T., … Parkes, M. (2010). Genome-wide meta-analysis increases to 71 the number of confirmed Crohn's disease susceptibility loci. *Nature Genetics, 42*(12), 1118–1125. http://dx.doi.org/10.1038/ng.717.

Galindo-Villegas, J., Garcia-Moreno, D., de Oliveira, S., Meseguer, J., & Mulero, V. (2012). Regulation of immunity and disease resistance by commensal microbes and chromatin modifications during zebrafish development. *Proceedings of the National Academy of Sciences of the United States of America, 109*(39), E2605–E2614. http://dx.doi.org/10.1073/pnas.1209920109.

Galy, V., Askjaer, P., Franz, C., Lopez-Iglesias, C., & Mattaj, I. W. (2006). MEL-28, a novel nuclear-envelope and kinetochore protein essential for zygotic nuclear-envelope assembly in *C. elegans*. *Current Biology: CB, 16*(17), 1748–1756. http://dx.doi.org/10.1016/j.cub.2006.06.067.

Gao, N., Davuluri, G., Gong, W., Seiler, C., Lorent, K., Furth, E. E., … Pack, M. (2011). The nuclear pore complex protein Elys is required for genome stability in mouse intestinal epithelial progenitor cells. *Gastroenterology, 140*(5), 1547–1555 e1510. http://dx.doi.org/10.1053/j.gastro.2011.01.048.

Geiger, B. M., Gras-Miralles, B., Ziogas, D. C., Karagiannis, A. K., Zhen, A., Fraenkel, P., & Kokkotou, E. (2013). Intestinal upregulation of melanin-concentrating hormone in TNBS-induced enterocolitis in adult zebrafish. *PLoS One, 8*(12), e83194. http://dx.doi.org/10.1371/journal.pone.0083194.

Goldsmith, J. R., Cocchiaro, J. L., Rawls, J. F., & Jobin, C. (2013). Glafenine-induced intestinal injury in zebrafish is ameliorated by mu-opioid signaling via enhancement of Atf6-dependent cellular stress responses. *Disease Models & Mechanisms, 6*(1), 146–159. http://dx.doi.org/10.1242/dmm.009852.

Goldsmith, J. R., Uronis, J. M., & Jobin, C. (2011). Mu opioid signaling protects against acute murine intestinal injury in a manner involving Stat3 signaling. *The American Journal of Pathology, 179*(2), 673–683. http://dx.doi.org/10.1016/j.ajpath.2011.04.032.

Haramis, A. P., Hurlstone, A., van der Velden, Y., Begthel, H., van den Born, M., Offerhaus, G. J., & Clevers, H. C. (2006). Adenomatous polyposis coli-deficient zebrafish are susceptible to digestive tract neoplasia. *EMBO Reports, 7*(4), 444–449. http://dx.doi.org/10.1038/sj.embor.7400638.

Hayes, P., Dhillon, S., O'Neill, K., Thoeni, C., Hui, K. Y., Elkadri, A., … Knaus, U. G. (2015). Defects in NADPH oxidase genes NOX1 and DUOX2 in very early onset inflammatory bowel disease. *Cellular and Molecular Gastroenterology and Hepatology, 1*(5), 489–502. http://dx.doi.org/10.1016/j.jcmgh.2015.06.005.

Heanue, T. A., & Pachnis, V. (2007). Enteric nervous system development and Hirschsprung's disease: advances in genetic and stem cell studies. *Nature Reviews. Neuroscience, 8*(6), 466–479. http://dx.doi.org/10.1038/nrn2137.

Heanue, T. A., & Pachnis, V. (2008). Ret isoform function and marker gene expression in the enteric nervous system is conserved across diverse vertebrate species. *Mechanisms of Development, 125*(8), 687–699. http://dx.doi.org/10.1016/j.mod.2008.04.006.

He, Q., Wang, L., Wang, F., & Li, Q. (2014). Role of gut microbiota in a zebrafish model with chemically induced enterocolitis involving toll-like receptor signaling pathways. *Zebrafish, 11*(3), 255–264. http://dx.doi.org/10.1089/zeb.2013.0917.

He, Q., Wang, L., Wang, F., Wang, C., Tang, C., Li, Q., … Zhao, Q. (2013). Microbial fingerprinting detects intestinal microbiota dysbiosis in Zebrafish models with chemically-induced enterocolitis. *BMC Microbiology, 13*, 289. http://dx.doi.org/10.1186/1471-2180-13-289.

Ho, S. Y., Lorent, K., Pack, M., & Farber, S. A. (2006). Zebrafish fat-free is required for intestinal lipid absorption and Golgi apparatus structure. *Cell Metabolism, 3*(4), 289–300. http://dx.doi.org/10.1016/j.cmet.2006.03.001.

Hoentjen, F., Harmsen, H. J., Braat, H., Torrice, C. D., Mann, B. A., Sartor, R. B., & Dieleman, L. A. (2003). Antibiotics with a selective aerobic or anaerobic spectrum have different therapeutic activities in various regions of the colon in interleukin 10 gene deficient mice. *Gut, 52*(12), 1721–1727.

Holmberg, A., Olsson, C., & Hennig, G. W. (2007). TTX-sensitive and TTX-insensitive control of spontaneous gut motility in the developing zebrafish (Danio rerio) larvae. *The Journal of Experimental Biology, 210*(Pt 6), 1084–1091. http://dx.doi.org/10.1242/jeb.000935.

Hooper, L. V., Wong, M. H., Thelin, A., Hansson, L., Falk, P. G., & Gordon, J. I. (2001). Molecular analysis of commensal host-microbial relationships in the intestine. *Science, 291*(5505), 881–884. http://dx.doi.org/10.1126/science.291.5505.881.

Hugot, J. P., Chamaillard, M., Zouali, H., Lesage, S., Cezard, J. P., Belaiche, J., … Thomas, G. (2001). Association of NOD2 leucine-rich repeat variants with susceptibility to Crohn's disease. *Nature, 411*(6837), 599–603. http://dx.doi.org/10.1038/35079107.

Ito, M., Okano, H. J., Darnell, R. B., & Roeder, R. G. (2002). The TRAP100 component of the TRAP/Mediator complex is essential in broad transcriptional events and development. *EMBO Journal, 21*(13), 3464–3475. http://dx.doi.org/10.1093/emboj/cdf348.

Jiang, Q., Arnold, S., Heanue, T., Kilambi, K. P., Doan, B., Kapoor, A., … Chakravarti, A. (2015). Functional loss of semaphorin 3C and/or semaphorin 3D and their epistatic interaction with ret are critical to Hirschsprung disease liability. *American Journal of Human Genetics, 96*(4), 581–596. http://dx.doi.org/10.1016/j.ajhg.2015.02.014.

de Jong-Curtain, T. A., Parslow, A. C., Trotter, A. J., Hall, N. E., Verkade, H., Tabone, T., … Heath, J. K. (2009). Abnormal nuclear pore formation triggers apoptosis in the intestinal epithelium of elys-deficient zebrafish. *Gastroenterology, 136*(3), 902–911. http://dx.doi.org/10.1053/j.gastro.2008.11.012.

Jostins, L., Ripke, S., Weersma, R. K., Duerr, R. H., McGovern, D. P., Hui, K. Y., … Cho, J. H. (2012). Host-microbe interactions have shaped the genetic architecture of inflammatory bowel disease. *Nature, 491*(7422), 119–124. http://dx.doi.org/10.1038/nature11582.

Kanther, M., Sun, X., Muhlbauer, M., Mackey, L. C., Flynn, E. J., 3rd, Bagnat, M., … Rawls, J. F. (2011). Microbial colonization induces dynamic temporal and spatial patterns of NF-kappaB activation in the zebrafish digestive tract. *Gastroenterology, 141*(1), 197–207. http://dx.doi.org/10.1053/j.gastro.2011.03.042.

Kelsh, R. N., Dutton, K., Medlin, J., & Eisen, J. S. (2000). Expression of zebrafish fkd6 in neural crest-derived glia. *Mechanisms of Development, 93*(1–2), 161–164.

Kelsh, R. N., & Eisen, J. S. (2000). The zebrafish colourless gene regulates development of non-ectomesenchymal neural crest derivatives. *Development, 127*(3), 515–525.

Kelsh, R. N., Schmid, B., & Eisen, J. S. (2000). Genetic analysis of melanophore development in zebrafish embryos. *Developmental Biology, 225*(2), 277–293. http://dx.doi.org/10.1006/dbio.2000.9840.

Kokkotou, E., Espinoza, D. O., Torres, D., Karagiannides, I., Kosteletos, S., Savidge, T., … Pothoulakis, C. (2009). Melanin-concentrating hormone (MCH) modulates C difficile toxin A-mediated enteritis in mice. *Gut, 58*(1), 34–40. http://dx.doi.org/10.1136/gut.2008.155341.

Korzh, S., Winata, C. L., Zheng, W., Yang, S., Yin, A., Ingham, P., … Gong, Z. (2011). The interaction of epithelial Ihha and mesenchymal Fgf10 in zebrafish esophageal and swimbladder development. *Developmental Biology, 359*(2), 262–276. http://dx.doi.org/10.1016/j.ydbio.2011.08.024.

Kuhlman, J., & Eisen, J. S. (2007). Genetic screen for mutations affecting development and function of the enteric nervous system. *Developmental Dynamics: an Official Publication of the American Association of Anatomists, 236*(1), 118–127. http://dx.doi.org/10.1002/dvdy.21033.

Lam, S. H., Chua, H. L., Gong, Z., Lam, T. J., & Sin, Y. M. (2004). Development and maturation of the immune system in zebrafish, Danio rerio: a gene expression profiling, in situ hybridization and immunological study. *Developmental and Comparative Immunology, 28*(1), 9–28.

Langer, M., Van Noorden, S., Polak, J. M., & Pearse, A. G. (1979). Peptide hormone-like immunoreactivity in the gastrointestinal tract and endocrine pancreas of eleven teleost species. *Cell and Tissue Research, 199*(3), 493–508.

Le, X., Langenau, D. M., Keefe, M. D., Kutok, J. L., Neuberg, D. S., & Zon, L. I. (2007). Heat shock-inducible Cre/Lox approaches to induce diverse types of tumors and hyperplasia in transgenic zebrafish. *Proceedings of the National Academy of Sciences of the United States of America, 104*(22), 9410–9415. http://dx.doi.org/10.1073/pnas.0611302104.

Levic, D. S., Minkel, J. R., Wang, W. D., Rybski, W. M., Melville, D. B., & Knapik, E. W. (2015). Animal model of Sar1b deficiency presents lipid absorption deficits similar to Anderson disease. *Journal of Molecular Medicine (Berlin), 93*(2), 165–176. http://dx.doi.org/10.1007/s00109-014-1247-x.

Low, D., Mizoguchi, A., & Mizoguchi, E. (2013). DNA methylation in inflammatory bowel disease and beyond. *World Journal of Gastroenterology: WJG, 19*(32), 5238–5249. http://dx.doi.org/10.3748/wjg.v19.i32.5238.

Lugo-Villarino, G., Balla, K. M., Stachura, D. L., Banuelos, K., Werneck, M. B., & Traver, D. (2010). Identification of dendritic antigen-presenting cells in the zebrafish. *Proceedings of the National Academy of Sciences of the United States of America, 107*(36), 15850–15855. http://dx.doi.org/10.1073/pnas.1000494107.

Luzon-Toro, B., Fernandez, R. M., Torroglosa, A., de Agustin, J. C., Mendez-Vidal, C., Segura, D. I., … Borrego, S. (2013). Mutational spectrum of semaphorin 3A and semaphorin 3D genes in Spanish Hirschsprung patients. *PLoS One, 8*(1), e54800. http://dx.doi.org/10.1371/journal.pone.0054800.

Mabbott, N. A., Donaldson, D. S., Ohno, H., Williams, I. R., & Mahajan, A. (2013). Microfold (M) cells: important immunosurveillance posts in the intestinal epithelium. *Mucosal Immunology, 6*(4), 666–677. http://dx.doi.org/10.1038/mi.2013.30.

Marjoram, L., Alvers, A., Deerhake, M. E., Bagwell, J., Mankiewicz, J., Cocchiaro, J. L., … Bagnat, M. (2015). Epigenetic control of intestinal barrier function and inflammation in zebrafish. *Proceedings of the National Academy of Sciences of the United States of America, 112*(9), 2770–2775. http://dx.doi.org/10.1073/pnas.1424089112.

Markmiller, S., Cloonan, N., Lardelli, R. M., Doggett, K., Keightley, M. C., Boglev, Y., … Heath, J. K. (2014). Minor class splicing shapes the zebrafish transcriptome during development. *Proceedings of the National Academy of Sciences of the United States of America, 111*(8), 3062–3067. http://dx.doi.org/10.1073/pnas.1305536111.

Marques, I. J., Weiss, F. U., Vlecken, D. H., Nitsche, C., Bakkers, J., Lagendijk, A. K., … Bagowski, C. P. (2009). Metastatic behaviour of primary human tumours in a zebrafish xenotransplantation model. *BMC Cancer, 9*, 128. http://dx.doi.org/10.1186/1471-2407-9-128.

Marshall, K. E., Tomasini, A. J., Makky, K., Kumar, S. N., & Mayer, A. N. (2010). Dynamic Lkb1-TORC1 signaling as a possible mechanism for regulating the endoderm-intestine transition. *Developmental Dynamics: an Official Publication of the American Association of Anatomists, 239*(11), 3000–3012. http://dx.doi.org/10.1002/dvdy.22437.

Mayer, A. N., & Fishman, M. C. (2003). Nil per os encodes a conserved RNA recognition motif protein required for morphogenesis and cytodifferentiation of digestive organs in zebrafish. *Development, 130*(17), 3917–3928.

McGaughey, D. M., Vinton, R. M., Huynh, J., Al-Saif, A., Beer, M. A., & McCallion, A. S. (2008). Metrics of sequence constraint overlook regulatory sequences in an exhaustive analysis at phox2b. *Genome Research, 18*(2), 252–260. http://dx.doi.org/10.1101/gr.6929408.

Miguel, J. C., Maxwell, A. A., Hsieh, J. J., Harnisch, L. C., Al Alam, D., Polk, D. B., … Frey, M. R. (2016). Epidermal growth factor suppresses intestinal epithelial cell shedding through a MAPK-dependent pathway. *Journal of Cell Science*. http://dx.doi.org/10.1242/jcs.182584.

Minchin, J. E., & Rawls, J. F. (2011). In vivo analysis of white adipose tissue in zebrafish. *Methods in Cell Biology, 105*, 63–86. http://dx.doi.org/10.1016/B978-0-12-381320-6.00003-5.

Mizoguchi, A. (2012). Animal models of inflammatory bowel disease. *Progress in Molecular Biology and Translational Science, 105*, 263–320. http://dx.doi.org/10.1016/B978-0-12-394596-9.00009-3.

Montero-Balaguer, M., Lang, M. R., Sachdev, S. W., Knappmeyer, C., Stewart, R. A., De La Guardia, A., … Knapik, E. W. (2006). The mother superior mutation ablates foxd3 activity in neural crest progenitor cells and depletes neural crest derivatives in zebrafish. *Developmental Dynamics: an Official Publication of the American Association of Anatomists, 235*(12), 3199–3212. http://dx.doi.org/10.1002/dvdy.20959.

Morteau, O. (2000). Prostaglandins and inflammation: the cyclooxygenase controversy. *Archivum Immunologiae Et Therapiae Experimentalis, 48*(6), 473–480.

Nadauld, L. D., Phelps, R., Moore, B. C., Eisinger, A., Sandoval, I. T., Chidester, S., … Jones, D. A. (2006). Adenomatous polyposis coli control of C-terminal binding protein-1 stability regulates expression of intestinal retinol dehydrogenases. *Journal of Biological Chemistry, 281*(49), 37828–37835. http://dx.doi.org/10.1074/jbc.M602119200.

Oehlers, S. H., Flores, M. V., Chen, T., Hall, C. J., Crosier, K. E., & Crosier, P. S. (2011). Topographical distribution of antimicrobial genes in the zebrafish intestine. *Developmental and Comparative Immunology, 35*(3), 385–391. http://dx.doi.org/10.1016/j.dci.2010.11.008.

Oehlers, S. H., Flores, M. V., Hall, C. J., Crosier, K. E., & Crosier, P. S. (2012). Retinoic acid suppresses intestinal mucus production and exacerbates experimental enterocolitis. *Disease Models & Mechanisms, 5*(4), 457–467. http://dx.doi.org/10.1242/dmm.009365.

Oehlers, S. H., Flores, M. V., Hall, C. J., Swift, S., Crosier, K. E., & Crosier, P. S. (2011). The inflammatory bowel disease (IBD) susceptibility genes NOD1 and NOD2 have conserved anti-bacterial roles in zebrafish. *Disease Models & Mechanisms, 4*(6), 832–841. http://dx.doi.org/10.1242/dmm.006122.

Oehlers, S. H., Flores, M. V., Okuda, K. S., Hall, C. J., Crosier, K. E., & Crosier, P. S. (2011). A chemical enterocolitis model in zebrafish larvae that is dependent on microbiota and responsive to pharmacological agents. *Developmental Dynamics: an Official Publication of the American Association of Anatomists, 240*(1), 288–298. http://dx.doi.org/10.1002/dvdy.22519.

Ogura, Y., Bonen, D. K., Inohara, N., Nicolae, D. L., Chen, F. F., Ramos, R., … Cho, J. H. (2001). A frameshift mutation in NOD2 associated with susceptibility to Crohn's disease. *Nature, 411*(6837), 603–606. http://dx.doi.org/10.1038/35079114.

Olden, T., Akhtar, T., Beckman, S. A., & Wallace, K. N. (2008). Differentiation of the zebrafish enteric nervous system and intestinal smooth muscle. *Genesis, 46*(9), 484–498. http://dx.doi.org/10.1002/dvg.20429.

Otis, J. P., Zeituni, E. M., Thierer, J. H., Anderson, J. L., Brown, A. C., Boehm, E. D., … Farber, S. A. (2015). Zebrafish as a model for apolipoprotein biology: comprehensive expression analysis and a role for ApoA-IV in regulating food intake. *Disease Models & Mechanisms, 8*(3), 295–309. http://dx.doi.org/10.1242/dmm.018754.

Paquette, C. E., Kent, M. L., Buchner, C., Tanguay, R. L., Guillemin, K., Mason, T. J., & Peterson, T. S. (2013). A retrospective study of the prevalence and classification of intestinal neoplasia in zebrafish (Danio rerio). *Zebrafish, 10*(2), 228–236. http://dx.doi.org/10.1089/zeb.2012.0828.

Phelps, R. A., Chidester, S., Dehghanizadeh, S., Phelps, J., Sandoval, I. T., Rai, K., … Jones, D. A. (2009). A two-step model for colon adenoma initiation and progression caused by APC loss. *Cell, 137*(4), 623–634. http://dx.doi.org/10.1016/j.cell.2009.02.037.

Pietsch, J., Delalande, J. M., Jakaitis, B., Stensby, J. D., Dohle, S., Talbot, W. S., … Shepherd, I. T. (2006). Lessen encodes a zebrafish trap100 required for enteric nervous system development. *Development, 133*(3), 395–406. http://dx.doi.org/10.1242/dev.02215.

Rai, K., Sarkar, S., Broadbent, T. J., Voas, M., Grossmann, K. F., Nadauld, L. D., … Jones, D. A. (2010). DNA demethylase activity maintains intestinal cells in an undifferentiated state following loss of APC. *Cell, 142*(6), 930–942. http://dx.doi.org/10.1016/j.cell.2010.08.030.

Rasala, B. A., Orjalo, A. V., Shen, Z., Briggs, S., & Forbes, D. J. (2006). ELYS is a dual nucleoporin/kinetochore protein required for nuclear pore assembly and proper cell division. *Proceedings of the National Academy of Sciences of the United States of America, 103*(47), 17801–17806. http://dx.doi.org/10.1073/pnas.0608484103.

Rawls, J. F., Mahowald, M. A., Ley, R. E., & Gordon, J. I. (2006). Reciprocal gut microbiota transplants from zebrafish and mice to germ-free recipients reveal host habitat selection. *Cell, 127*(2), 423–433. http://dx.doi.org/10.1016/j.cell.2006.08.043.

Rawls, J. F., Samuel, B. S., & Gordon, J. I. (2004). Gnotobiotic zebrafish reveal evolutionarily conserved responses to the gut microbiota. *Proceedings of the National Academy of Sciences of the United States of America, 101*(13), 4596–4601. http://dx.doi.org/10.1073/pnas.0400706101.

Reichenbach, B., Delalande, J. M., Kolmogorova, E., Prier, A., Nguyen, T., Smith, C. M., … Shepherd, I. T. (2008). Endoderm-derived Sonic hedgehog and mesoderm Hand2 expression are required for enteric nervous system development in zebrafish. *Developmental Biology, 318*(1), 52–64. http://dx.doi.org/10.1016/j.ydbio.2008.02.061.

Renshaw, S. A., Loynes, C. A., Trushell, D. M., Elworthy, S., Ingham, P. W., & Whyte, M. K. (2006). A transgenic zebrafish model of neutrophilic inflammation. *Blood, 108*(13), 3976–3978. http://dx.doi.org/10.1182/blood-2006-05-024075.

Rich, A., Gordon, S., Brown, C., Gibbons, S. J., Schaefer, K., Hennig, G., & Farrugia, G. (2013). Kit signaling is required for development of coordinated motility patterns in zebrafish gastrointestinal tract. *Zebrafish, 10*(2), 154–160. http://dx.doi.org/10.1089/zeb.2012.0766.

Roeselers, G., Mittge, E. K., Stephens, W. Z., Parichy, D. M., Cavanaugh, C. M., Guillemin, K., & Rawls, J. F. (2011). Evidence for a core gut microbiota in the zebrafish. *ISME Journal, 5*(10), 1595–1608. http://dx.doi.org/10.1038/ismej.2011.38.

Rombout, J. H., Lamers, C. H., Helfrich, M. H., Dekker, A., & Taverne-Thiele, J. J. (1985). Uptake and transport of intact macromolecules in the intestinal epithelium of carp (Cyprinus carpio L.) and the possible immunological implications. *Cell and Tissue Research, 239*(3), 519–530.

Schall, K. A., Holoyda, K. A., Grant, C. N., Levin, D. E., Torres, E. R., Maxwell, A., … Grikscheit, T. C. (2015). Adult zebrafish intestine resection: a novel model of short bowel syndrome, adaptation, and intestinal stem cell regeneration. *American Journal of Physiology. Gastrointestinal and Liver Physiology, 309*(3), G135–G145. http://dx.doi.org/10.1152/ajpgi.00311.2014.

Schlegel, A. (2015). Studying lipoprotein trafficking in zebrafish, the case of chylomicron retention disease. *Journal of Molecular Medicine (Berlin), 93*(2), 115–118. http://dx.doi.org/10.1007/s00109-014-1248-9.

Seiler, C., Davuluri, G., Abrams, J., Byfield, F. J., Janmey, P. A., & Pack, M. (2012). Smooth muscle tension induces invasive remodeling of the zebrafish intestine. *PLoS Biology, 10*(9), e1001386. http://dx.doi.org/10.1371/journal.pbio.1001386.

Semova, I., Carten, J. D., Stombaugh, J., Mackey, L. C., Knight, R., Farber, S. A., & Rawls, J. F. (2012). Microbiota regulate intestinal absorption and metabolism of fatty acids in the zebrafish. *Cell Host & Microbe, 12*(3), 277–288. http://dx.doi.org/10.1016/j.chom.2012.08.003.

Shepard, J. L., Amatruda, J. F., Finkelstein, D., Ziai, J., Finley, K. R., Stern, H. M., … Zon, L. I. (2007). A mutation in separase causes genome instability and increased susceptibility to epithelial cancer. *Genes & Development, 21*(1), 55–59. http://dx.doi.org/10.1101/gad.1470407.

Sidhaye, J., Pinto, C. S., Dharap, S., Jacob, T., Bhargava, S., & Sonawane, M. (2016). The zebrafish goosepimples/myosin Vb mutant exhibits cellular attributes of human microvillus inclusion disease. *Mechanisms of Development*. http://dx.doi.org/10.1016/j.mod.2016.08.001.

Stewart, R. A., Arduini, B. L., Berghmans, S., George, R. E., Kanki, J. P., Henion, P. D., & Look, A. T. (2006). Zebrafish foxd3 is selectively required for neural crest specification, migration and survival. *Developmental Biology, 292*(1), 174–188. http://dx.doi.org/10.1016/j.ydbio.2005.12.035.

Teng, Y., Xie, X., Walker, S., White, D. T., Mumm, J. S., & Cowell, J. K. (2013). Evaluating human cancer cell metastasis in zebrafish. *BMC Cancer, 13*, 453. http://dx.doi.org/10.1186/1471-2407-13-453.

Thakur, P. C., Davison, J. M., Stuckenholz, C., Lu, L., & Bahary, N. (2014). Dysregulated phosphatidylinositol signaling promotes endoplasmic-reticulum-stress-mediated intestinal mucosal injury and inflammation in zebrafish. *Disease Models & Mechanisms, 7*(1), 93–106. http://dx.doi.org/10.1242/dmm.012864.

Thakur, P. C., Stuckenholz, C., Rivera, M. R., Davison, J. M., Yao, J. K., Amsterdam, A., … Bahary, N. (2011). Lack of de novo phosphatidylinositol synthesis leads to endoplasmic reticulum stress and hepatic steatosis in cdipt-deficient zebrafish. *Hepatology, 54*(2), 452–462. http://dx.doi.org/10.1002/hep.24349.

Traver, D., Paw, B. H., Poss, K. D., Penberthy, W. T., Lin, S., & Zon, L. I. (2003). Transplantation and in vivo imaging of multilineage engraftment in zebrafish bloodless mutants. *Nature Immunology, 4*(12), 1238–1246. http://dx.doi.org/10.1038/ni1007.

Turnbaugh, P. J., Backhed, F., Fulton, L., & Gordon, J. I. (2008). Diet-induced obesity is linked to marked but reversible alterations in the mouse distal gut microbiome. *Cell Host & Microbe, 3*(4), 213–223. http://dx.doi.org/10.1016/j.chom.2008.02.015.

Uyttebroek, L., Shepherd, I. T., Harrisson, F., Hubens, G., Blust, R., Timmermans, J. P., & Van Nassauw, L. (2010). Neurochemical coding of enteric neurons in adult and embryonic zebrafish (Danio rerio). *Journal of Comparative Neurology, 518*(21), 4419–4438. http://dx.doi.org/10.1002/cne.22464.

Uyttebroek, L., Shepherd, I. T., Hubens, G., Timmermans, J. P., & Van Nassauw, L. (2013). Expression of neuropeptides and anoctamin 1 in the embryonic and adult zebrafish intestine, revealing neuronal subpopulations and ICC-like cells. *Cell and Tissue Research, 354*(2), 355–370. http://dx.doi.org/10.1007/s00441-013-1685-8.

Uyttebroek, L., Shepherd, I. T., Vanden Berghe, P., Hubens, G., Timmermans, J. P., & Van Nassauw, L. (2016). The zebrafish mutant lessen: an experimental model for congenital enteric neuropathies. *Neurogastroenterology and Motility: the Official Journal of the European Gastrointestinal Motility Society, 28*(3), 345–357. http://dx.doi.org/10.1111/nmo.12732.

van der Vaart, M., van Soest, J. J., Spaink, H. P., & Meijer, A. H. (2013). Functional analysis of a zebrafish myd88 mutant identifies key transcriptional components of the innate immune system. *Disease Models & Mechanisms, 6*(3), 841–854. http://dx.doi.org/10.1242/dmm.010843.

Valvezan, A. J., Huang, J., Lengner, C. J., Pack, M., & Klein, P. S. (2014). Oncogenic mutations in adenomatous polyposis coli (Apc) activate mechanistic target of rapamycin complex 1 (mTORC1) in mice and zebrafish. *Disease Models & Mechanisms, 7*(1), 63–71. http://dx.doi.org/10.1242/dmm.012625.

Van Der Werf, C. S., Wabbersen, T. D., Hsiao, N. H., Paredes, J., Etchevers, H. C., Kroisel, P. M., … Hofstra, R. M. (2012). CLMP is required for intestinal development, and loss-of-function mutations cause congenital short-bowel syndrome. *Gastroenterology, 142*(3), 453–462 e453. http://dx.doi.org/10.1053/j.gastro.2011.11.038.

Vatn, M. H. (2009). Natural history and complications of IBD. *Current Gastroenterology Reports, 11*(6), 481–487.

van der Velden, Y. U., Wang, L., Zevenhoven, J., van Rooijen, E., van Lohuizen, M., Giles, R. H., … Haramis, A. P. (2011). The serine-threonine kinase LKB1 is essential for survival under energetic stress in zebrafish. *Proceedings of the National Academy of Sciences of the United States of America, 108*(11), 4358–4363. http://dx.doi.org/10.1073/pnas.1010210108.

Wallace, K. N., Akhter, S., Smith, E. M., Lorent, K., & Pack, M. (2005). Intestinal growth and differentiation in zebrafish. *Mechanisms of Development, 122*(2), 157–173. http://dx.doi.org/10.1016/j.mod.2004.10.009.

Wallace, K. N., Dolan, A. C., Seiler, C., Smith, E. M., Yusuff, S., Chaille-Arnold, L., … Pack, M. (2005). Mutation of smooth muscle myosin causes epithelial invasion and cystic expansion of the zebrafish intestine. *Developmental Cell, 8*(5), 717–726. http://dx.doi.org/10.1016/j.devcel.2005.02.015.

Wallace, K. N., & Pack, M. (2003). Unique and conserved aspects of gut development in zebrafish. *Developmental Biology, 255*(1), 12–29.

Wang, Z., Du, J., Lam, S. H., Mathavan, S., Matsudaira, P., & Gong, Z. (2010). Morphological and molecular evidence for functional organization along the rostrocaudal axis of the adult zebrafish intestine. *BMC Genomics, 11*, 392. http://dx.doi.org/10.1186/1471-2164-11-392.

Wiles, T. J., Jemielita, M., Baker, R. P., Schlomann, B. H., Logan, S. L., Ganz, J., … Parthasarathy, R. (2016). Host gut motility promotes competitive exclusion within a model intestinal microbiota. *PLoS Biology, 14*(7), e1002517. http://dx.doi.org/10.1371/journal.pbio.1002517.

Wittamer, V., Bertrand, J. Y., Gutschow, P. W., & Traver, D. (2011). Characterization of the mononuclear phagocyte system in zebrafish. *Blood, 117*(26), 7126–7135. http://dx.doi.org/10.1182/blood-2010-11-321448.

Witte, M., Huitema, L. F., Nieuwenhuis, E. E., & Brugman, S. (2014). Deficiency in macrophage-stimulating protein results in spontaneous intestinal inflammation and increased susceptibility toward epithelial damage in zebrafish. *Zebrafish, 11*(6), 542–550. http://dx.doi.org/10.1089/zeb.2014.1023.

Yan, Q., van der Gast, C. J., & Yu, Y. (2012). Bacterial community assembly and turnover within the intestines of developing zebrafish. *PLoS One, 7*(1), e30603. http://dx.doi.org/10.1371/journal.pone.0030603.

Yang, Y., Millan, J. L., Mecsas, J., & Guillemin, K. (2015). Intestinal alkaline phosphatase deficiency leads to lipopolysaccharide desensitization and faster weight gain. *Infection and Immunity, 83*(1), 247–258. http://dx.doi.org/10.1128/IAI.02520-14.

Yee, N. S., Gong, W., Huang, Y., Lorent, K., Dolan, A. C., Maraia, R. J., & Pack, M. (2007). Mutation of RNA Pol III subunit rpc2/polr3b leads to deficiency of subunit Rpc11 and disrupts zebrafish digestive development. *PLoS Biology, 5*(11), e312. http://dx.doi.org/10.1371/journal.pbio.0050312.

Zimmerman, A. M., Romanowski, K. E., & Maddox, B. J. (2011). Targeted annotation of immunoglobulin light chain (IgL) genes in zebrafish from BAC clones reveals kappa-like recombining/deleting elements within IgL constant regions. *Fish and Shellfish Immunology, 31*(5), 697–703. http://dx.doi.org/10.1016/j.fsi.2010.09.015.

CHAPTER

Analysis of pancreatic disease in zebrafish

11

S.C. Eames Nalle*, K.F. Franse§, M.D. Kinkel§,[1]

*Genentech, Inc., South San Francisco, CA, United States

§Appalachian State University, Boone, NC, United States

[1]Corresponding author: E-mail: kinkelmd@appstate.edu

CHAPTER OUTLINE

Methods in Cell Biology, Volume 138, ISSN 0091-679X, http://dx.doi.org/10.1016/bs.mcb.2016.08.005

Abstract

One of the appeals of the zebrafish model is the relative ease of studying disease progression from embryonic or larval stages through to adulthood. Because of this, the zebrafish has become an important model for postembryonic pancreatic disease, particularly diabetes and pancreatic cancer. Here we present methods for using the adult zebrafish to analyze pancreas function and structure, with an emphasis on the endocrine pancreas and the beta cells. The methods include fasting, weighing adults, and anesthetizing adults, and intraperitoneal injection of glucose based on body weight. We also present dissection methods for removing the pancreas intact for histological studies and for sterile dissection of the principal islet followed by dissociation for cell culture-based studies of beta-cell function.

INTRODUCTION

It is well established that the zebrafish is an excellent model for diabetes and for pancreatic cancer. An advantage of the model is that the zebrafish pancreas is structurally and functionally similar to the human pancreas. In both organisms, the pancreas consists of numerous hormone-producing islets that are distributed throughout the acinar tissue, and in both organisms the islets are responsible for maintaining blood glucose homeostasis. Another advantage is the relative ease of maintaining zebrafish in the laboratory for long-term studies of disease progression. For example, pancreatic adenocarcinoma was recently studied from the first appearance of tumors at 1 week postfertilization to 2 years (Schiavone et al., 2014), Type 2 diabetes progression was studied using an insulin resistance model in which glucose tolerance was assessed at 3 months and at 1 year (Maddison, Joest, Kammeyer, & Chen, 2015), and neonatal diabetes progression was studied using a model for a human insulin gene mutation in which endocrine pancreas structure and function were assessed from larval stages to 1 year (Eames, Kinkel, Rajan, Prince, & Philipson, 2013). The utility of the zebrafish as a model for diabetes and beta-cell biology studies has been reviewed extensively (Kimmel & Meyer, 2016; Prince, Anderson, & Dalgin, 2016), and its utility for pancreatic cancer studies has been recently reviewed in Hwang and Goessling (2016).

Here we present several methods that are useful for studying the endocrine pancreas in adult zebrafish. Methods include procedures for fasting, which are useful prior to measuring blood glucose and are also useful prior to dissection, as fasting empties the intestine. We also describe methods for weighing zebrafish without anesthetizing, delivering glucose via intraperitoneal injection, dissecting the pancreas for histological analyses, and isolating the principal islet for cell culture–based studies. These methods were developed in the context of studying beta-cell function in models of diabetes (Eames et al., 2013; Eames, Philipson, Prince, & Kinkel, 2010), but we believe they are broadly applicable for studies in which adult zebrafish are the model.

1. METHOD 1. FASTING ADULT ZEBRAFISH

Studies of pancreas function typically require fasting for a defined period to establish a baseline blood glucose level. The protocol described here is suitable for relatively short fasts of up to several hours, such as following a meal or glucose challenge, as well as for longer fasts that span several days. For zebrafish, fasting for 3 days is sufficient to lower blood glucose to a baseline concentration (Eames et al., 2010). Our method includes steps for fasting adult zebrafish offline, i.e., the tanks are disconnected from the recirculating aquarium system.

It is difficult to fast zebrafish when they are kept on a recirculating facility system with other nonfasting fish, as the fasting tanks may receive food particles that circulate through the system. On many systems the water flow to individual fasting tanks can be turned off, but it is possible that fish care staff may inadvertently feed the fish. Fasting offline, with tanks on a countertop, avoids these problems and offers several advantages. First, it increases the confidence that no food has been delivered to the tanks. Second, it allows easier access for daily cleaning of the tank bottoms and for daily tank water changes. Third, it allows easier monitoring of the fish and easier observation of the tank bottoms.

Pancreatic beta-cell function can be affected by stress, which can alter blood glucose concentration. To avoid stressing the fish, they should be maintained in mixed-sex groups in marbled tanks or crossing tanks. Either tank style will sequester eggs, which are a rich nutrient source. We previously reported a method for using marbled tanks for fasting zebrafish (Kinkel, Eames, Philipson, & Prince, 2010). Here, we describe using custom crossing tanks, for a simpler, less time-consuming method. The crossing tanks allow easier detection of eggs and much faster siphoning of the tank bottoms.

1.1 MATERIALS

Fish tank with lid and baffle, 9.5 L (Aquaneering #ZT950)
Custom tank insert, made from 1/8″ polycarbonate and wire screen (10 mesh stainless steel), secured with fishing line (e.g., 2.7 kg/6 lb. test)
Plastic mesh supports, two per tank, cut to fit (Pentair #TN524F)
Siphon, custom-made from flexible aquarium tubing and rigid airline tubing (both 3/16″ diameter)
Scissors or tubing cutter
Squeeze bottle of facility water
Waste container, 1–2 L
Fish net

1.2 PROCEDURE

1. Make a five-sided tank insert: The insert consists of four polycarbonate walls and a wire mesh bottom. The dimensions of the insert shown in Fig. 1A are 25.4 cm × 14.6 cm × 16.5 cm. The polycarbonate sides can be cut and glued/welded together by a machine shop or by the supplier (e.g., TAP Plastics). The

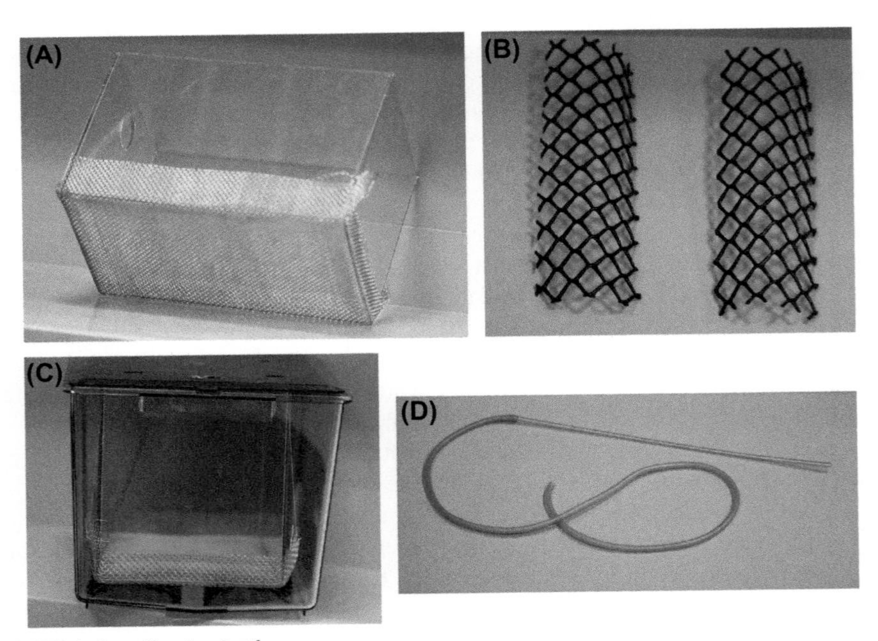

FIGURE 1 Supplies for fasting.

(A) Mesh-bottomed tank insert. The mesh is folded over the insert walls and secured with fishing line. (B) Plastic supports for elevating the insert above the tank bottom. (C) Fasting tank assembled with two supports and an insert. (D) Siphon for cleaning the tank bottom.

wire mesh bottom can be permanently welded to the bottom edges of the walls, or it can simply be held in place by fishing line. For this simpler approach, cut the wire mesh larger than the insert, (approximately 28.4 cm × 17.6 cm) so that there is an overhang of roughly 3 cm along each side. Center the mesh under the polycarbonate sides, fold up each overhanging edge, and make sure the mesh bottom fits snugly against the insert. Secure the mesh at each corner by looping fishing line around the insert walls and through the mesh, then tying tightly.

2. Make two supports for the insert by cutting the plastic mesh to a suitable size. The supports shown in Fig. 1B are 24 cm × 8.25 cm. After cutting the mesh, shape it into an arch by squeezing the long sides together.
3. Assemble the fasting tank: Set two plastic mesh supports into the 9.5 L tank, and rest the polycarbonate insert on the supports, as in Fig. 1C. Install the tank baffle and fill the tank with facility water. The water volume of the insert is ~4.75 L.
4. Transfer a mixed-sex population of approximately eight adults to the tank insert.
5. Make a cleaning siphon: Cut the rigid tubing to ~34 cm and cut the flexible tubing to ~74 cm. Insert one end of the rigid tubing into the flexible tubing. The end result is shown in Fig. 1D.
6. Daily, remove any waste or eggs off of the tank bottom using the siphon. The siphon can be started by filling it from the squeeze bottle of facility water. Use the 1–2 L container to collect the waste. After siphoning, restore the tank volume with fresh facility water.
7. Daily, monitor fish for signs of stress.
8. Fast the fish for 3 days, to lower blood glucose to a baseline level.

2. METHOD 2. WEIGHING LIVE, SWIMMING ZEBRAFISH WITHOUT ANESTHETIC

Determining the weight of individual fish allows you to adjust the concentration of a reagent based on body weight and thereby delivers the same relative dose to each individual. This approach can reduce the variability in results, and thus can reduce the number of fish required for an experiment. The method described here avoids using chemical anesthetics, as beta-cell function (and therefore blood glucose concentration) is sensitive to their use. It also avoids the use of cold-water anesthesia, to prevent anesthetizing the same fish twice in 1 day when downstream procedures require anesthetic. An important consideration for weighing a fish is the mesh size of the net used for transfer to the weighing beaker. Fine-meshed nets can trap excess water that is then transferred to the weighing beaker along with the fish. The weight of the transferred water can cause a significant increase in the measured weight. Unlike fine-meshed nets, a net with a relatively large mesh size allows the water to be wicked away efficiently. This results in minimal transfer of excess water to the weighing beaker. With this method, we have found that the measured weights are not significantly different from anesthetizing the fish, blotting it dry, then weighing it on a dry weigh boat.

2.1 MATERIALS

Balance, Mettler Toledo, New Classic MF model ML303E
Fish tanks, 1–2 L, one per fish
3″ fish net with 1/16″ mesh (Pentair #AN4)
Paper towels
Plastic beaker, 500 mL
Fish facility water
Lab tape and marker, for tank labels

2.2 PROCEDURE

1. Follow the manufacturer's instructions to set up the "dynamic weighing" application of the balance. Set the time interval to 120 s. Set the units to milligrams.
2. Transfer each fish to a separate tank, and label each tank, as appropriate.
3. Fill the 500 mL plastic beaker to approximately 1/3 with warm facility water. This is the weighing beaker.
4. Transfer the weighing beaker to the balance, tare it, and leave it on the balance.
5. Prepare a short stack of paper towels, and set it near the balance and fish tanks.
6. Net the first fish. Close the net to prevent the fish from escaping, and promptly wick away excess water by gently and quickly blotting the bottom of the net on the paper towel stack 2–3 times.
7. Transfer the fish to the tared weighing beaker.
8. Weigh the fish and record the value.
9. Transfer the fish back to its tank by gently pouring it from the beaker.
10. Repeat Steps 3–9, as necessary.
11. For each fish, calculate the reagent volume per milligram of body weight. Record this number on the tank label.
12. If the downstream step will be performed in another room, promptly transfer the tanks to that room to allow the fish time to acclimate.

3. METHOD 3. GLUCOSE DELIVERY TO ADULT ZEBRAFISH USING INTRAPERITONEAL INJECTION

This method uses a 10 μL syringe to inject volumes in the range of several hundred nanoliters. Since it is difficult to accurately inject small volumes manually, a microprocessor is used to control the syringe plunger. The syringe is mounted on a pump (UltraMicroPump III), and the syringe needle extends from the barrel via flexible tubing. This allows manual control of the needle, which is more efficient than using a micromanipulator. Manual control of the needle, combined with digital control of injection volume, allows precise delivery of small volumes.

To perform intraperitoneal injection of glucose by body weight, the fish should be fasted as described in Method 1 and weighed as described in Method 2. These procedures can potentially cause stress, and stress is known to alter blood glucose concentration, often by raising it. Therefore the potential effects of stress should be controlled for by using a group of fish that undergo the same handling procedures as experimental fish. Control fish should be fasted, weighed, and anesthetized in parallel with the experimental fish, and they should be vehicle injected, as described below.

The glucose solution is viscous and sticky, and the instruments require careful cleaning to prevent clogging. Cleaning is a lengthier process than injecting, as detailed in Section 3.6, and therefore it is best to inject with vehicle before glucose. This allows you to fill the syringe and tubing with glucose solution without cleaning the instruments in between.

The general set up and operation of the syringe pump and controller are detailed in the manufacturer's manual for the UltraMicroPump III. Here we consider the additional supplies and specific methods required for injecting adult zebrafish with glucose.

3.1 PREPARE THE INJECTION SOLUTIONS

Prior to injection, glucose is dissolved in a modified Cortland salt solution (CS) (Perry, Davie, Daxboeck, Ellis, & Smith, 1984; Wolf, 1963). The Cortland solution is also used minus glucose for injecting control fish. Both solutions must be filtered to remove any particles that might clog the Silflex tubing or injection needle. Likewise, the distilled water that will be used to prewet the tubing and needle must be prefiltered.

3.1.1 Materials

Glucose solution, 1 mg/μL
Cortland salt solution
Distilled H_2O
Disposable syringes (1 or 5 cc) and syringe filters
Microtubes, 1.5 or 2 mL

3.1.2 Procedure

1. Prepare the CS using 124.1 mM NaCl, 5.1 mM KCl, 2.9 mM Na_2HPO_4, 1.9 mM $MgSO_4$ $7H_2O$, 1.4 mM $CaCl_2$ $2H_2O$, 11.9 mM $NaHCO_3$, 4% polyvinylpyrrolidone, and heparin at 10,000 USP units/L. Adjust the pH to 7.4 if necessary, filter sterilize, and store at 4°C. A suggested stock volume is 100 mL.
2. Prepare the glucose solution: Dissolve 1 mg/μL D-glucose (Sigma) in CS, filter sterilize, and store at 4°C. A suggested stock volume is 10 mL.

3. Use a syringe filter to filter small volumes (1–2 mL) of the glucose solution, vehicle, and dH_2O. Store them in microtubes at 4°C.

3.2 ASSEMBLE SUPPLIES FOR ANESTHETIZING

The zebrafish must be anesthetized to allow handling for injection. Because chemical anesthetics affect blood glucose concentration in vertebrates, including zebrafish, we use cold facility water instead as a safe alternative (Collymore, Tolwani, Lieggi, & Rasmussen, 2014; Eames et al., 2010). The water temperature is gradually lowered to bring the fish to a surgical plane of anesthesia. For the injection itself, the fish is on the microscope stage and is held in a foam plug saturated with cold facility water, as demonstrated in Kinkel et al. (2010). The fish should be anesthetized in a location immediately adjacent to the injection equipment. This will allow swift transfer between the cold water container and microscope stage and back to the original tank of warm facility water for postinjection recovery.

3.2.1 Materials

Ice cube trays, two or more
Ice bucket with lid
Ice cube crusher, e.g., Waring Pro model IC70
Foam plug for *Drosophila* bottle (Jaece #L800-D)
Razor blade or scissors
Petri dish bottom, wider than foam plug (e.g., 50 mm)
Shallow pipet tip box lid, e.g., 1.5 cm deep
Food-safe storage container, e.g., 2.4 L
Thermometer
Fish facility H_2O, (~2 L warm water)
Metal plate or surface, optional

3.2.2 Procedure

1. Thoroughly clean all items that will come into direct contact with facility water or ice. Do not use detergent.
2. The evening before injections or earlier, fill ice cube trays with fish facility water and transfer to −20°C.
3. The morning of injections, run the ice cubes through the ice cube crusher. Keep the crushed ice in the ice bucket, with lid, on the bench top.
4. Trim the foam plug: Cut it so that it is just slightly higher than the Petri dish bottom. On the flat surface, cut a slit down the center from side to side, ~15 mm deep. The slit should be deep enough to support the fish body and accommodate the gills.
5. Nest the following items: Plug (slit facing up), into Petri dish bottom, into tip box lid, into food container. The first three items constitute the surgical table that will be transferred back and forth between the food container, for anesthesia, and the microscope stage, for injection.

6. Fill the Petri dish and tip box lid with facility water and partially fill the food container. Insert the thermometer into the food container. If desired, hold the set of containers on a metal surface to help maintain water temperature.
7. Use crushed ice to bring the temperature to ~17°C.
8. Press the sponge to thoroughly saturate it with cold water.
9. When you are ready to anesthetize a fish, proceed to Section 3.5.

3.3 SET UP THE INJECTION EQUIPMENT

Note: These steps assume that the person injecting is right-handed. For left-handed operation, the syringe pump should be located to the left of the microscope, and the controller should be placed to the right.

3.3.1 Materials

Small V-base (World Precision Instruments #503084)
250 mm stainless steel post (World Precision Instruments #503073)
Clamp (World Precision Instruments #14073)
Syringe pump with controller:
UltraMicroPump III (UMP3) and Micro4 controller (World Precision Instruments #UMPS-1)
Controller foot switch (World Precision Instruments #15867)
Dissecting microscope with external light source
Silicone mat (World Precision Instruments #504203)
Plastic wrap, optional
Lab tape, optional

3.3.2 Procedure

1. Assemble the V-base, steel post, and clamp, as shown in Fig. 2A. Attach the syringe pump to the clamp and place it to the right of the microscope base. Note: Once the pump has been attached to the post, it can be permanently left there. The pump should be covered when not in use, as dust will damage it.
2. Place the Micro4 controller to the left of the microscope, within easy reach.
3. Run the pump cable behind the microscope and connect it to channel 1 of the controller. Connect the foot pedal to the controller.
4. If the microscope has a light-transmitting base, drape it with plastic wrap. Secure the plastic wrap with lab tape.
5. Put a paper towel on top of the plastic wrap.
6. Set the silicone mat to the immediate right of the microscope, in front of the syringe pump stand.

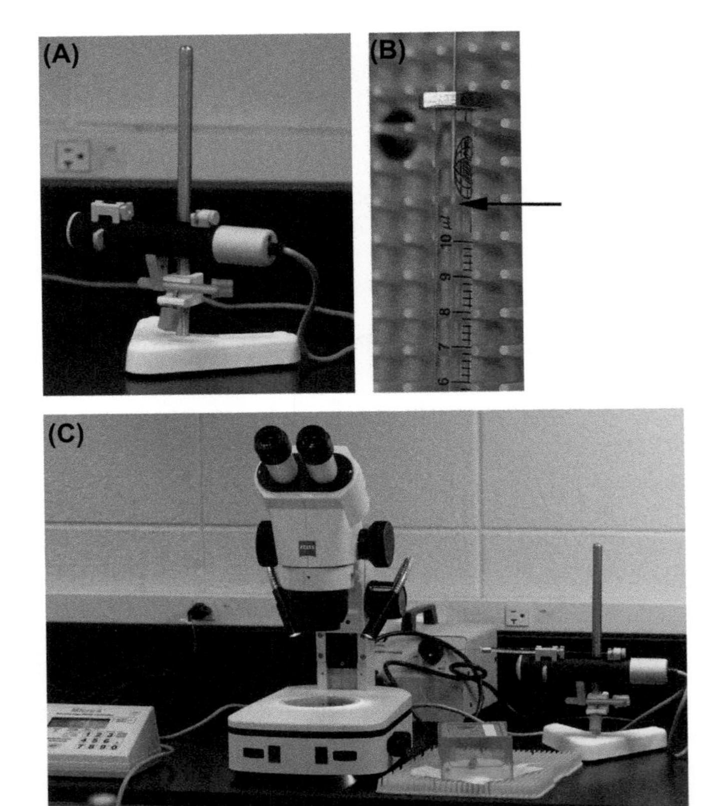

FIGURE 2 Injection equipment.

(A) Syringe pump mounted on steel post. (B) NanoFil syringe filled to ~11 μL. The *arrow* indicates the tip of the plunger. (C) Arrangement of the injection equipment. The needle is shown covered and resting on moist Kimwipes to prevent drying prior to the first injection.

3.4 PREPARE THE NANOFIL SYRINGE, SILFLEX TUBING, AND INJECTION NEEDLE

Before using the NanoFil syringe for the first time, the Teflon-coated plunger must be "formed" following the manufacturer's instructions found in the pump manual (not the syringe manual). This initial preparation requires completely removing the plunger from the barrel of the syringe. After this procedure, the plunger should not be removed again. The plunger of the NanoFil syringe is especially delicate and should not be touched without clean gloves. Avoid touching the Teflon-coated tip of the plunger, even with gloves.

The Silflex tubing and 35 G injection needle require careful handling. The tubing is semiflexible quartz and can be broken if bent too far. The needle is especially delicate and it is best to have a spare on hand.

The manufacturer's manual describes using the 35 G injection needle and the microprocessor (run in reverse) to fill the syringe directly. However, we have found that this does not work well for filling the syringe with glucose solution or vehicle. Instead, we manually fill the syringe using the 26 G filling needle, then change to the injection needle, as described below.

3.4.1 Materials

10 μL NanoFil syringe with 26 G filling needle (World Precision Instruments #NANOFIL)
35 G injection needle with beveled tip (World Precision Instruments #NF35BV-2)
Silflex tubing (World Precision Instruments #SILFLEX-2)
NanoFil injection holder (World Precision Instruments #NFINHLD)
Ruler, in millimeters
Instrument cover made from a plastic box, described below
Petri dish (e.g., 50 mm) of distilled H_2O, to soak filling needle
Kimwipes or paper towels

3.4.2 Procedure

1. Prewet the syringe: Attach the 26 G filling needle to the NanoFil syringe. Submerge the tip of the needle in filtered dH_2O and very slowly fill the syringe. Add an extra 1 μL by filling past the 10 μL mark, as shown in Fig. 2B.
2. Use the microscope to check the syringe for air bubbles. If there are any air bubbles, they must be removed. To do so, pump the syringe plunger while keeping the needle tip immersed in the dH_2O. When all bubbles have been removed, set the syringe aside on the silicone mat. (Do not mount the syringe onto the pump yet.)
3. Attach Silflex tubing and 35 G injection needle to the holder: Under magnification, first insert the tubing, then insert the injection needle, following the manufacturer's instructions. Briefly, inside the holder, the end of the tubing should be very close to the needle, but not touching. The gap should be smaller than 1 mm (check with a ruler).
4. Set aside the assembled holder, tubing, and needle. These should be kept on the silicone mat when not in use.
5. Under the microscope, very slowly remove the 26 G filling needle from the NanoFil syringe. Removing the needle slowly helps to prevent bubbles from forming at the needle-end of the syringe.
6. Under the microscope, insert the free end of the Silflex tubing into the NanoFil syringe. Be careful to keep the 35 G injection needle on the silicone mat during this procedure.
7. Prewet the Silflex tubing by manually pushing most of the water out of the syringe. It is useful to monitor the injection needle during this step, to confirm

that water is visible at the needle tip. This procedure also tests for clogs in the tubing and needle. If there is a clog, skip to Section 3.6 for cleaning methods.

8. After prewetting the tubing, slowly remove it from the syringe and set it aside on the silicone mat. Reinsert the 26 G filling needle.
9. Draw up the injection solution into the syringe. As in Step 1, overfill the syringe by 1 μL. Remove any air bubbles, as described in Step 2.
10. Slowly remove the 26 G filling needle and reinsert the Silflex tubing. Put the filling needle into a small dish of distilled water, submerged, until it can be cleaned. This will prevent clogging.
11. Manually push the plunger to the 9 μL mark, so that ~2 μL of solution is pushed through the tubing. Note: The Silflex tubing has a dead space of ~3 μL. Just prior to the first injection, an additional 1 μL will be pushed out.
12. Mount the syringe on the pump, following the manufacturer's instructions.
13. Program the syringe pump controller, in channel 1, as follows:
    a. Set rate units to "S."
    b. Set device type to "L."
    c. Set grouping to "N."
    d. Set rate to 170, for maximum speed.
    e. Program the injection volume by entering four digits, with a decimal point.

    Note: The manufacturer's UMP3 manual has detailed instructions for programming the controller for multiple scenarios. The protocol here assumes that only one 10 μL NanoFil syringe is being controlled, and that a foot pedal is being used.
14. If there is a delay before injections begin, prevent the injection needle from drying, as this will cause clogging. Make a needle cover from a small plastic box, such as the bottom part of a pipet tip box, by removing one of the short sides to make a four-sided box. Rest the needle on a moist paper towel or Kimwipe, on top of the silicone mat. Cover the needle with the plastic box, using the open short side to accommodate the Silflex tubing. By this point, the set up should be similar to that shown in Fig. 2C.
15. When you are ready to inject, set channel 1 of the Micro4 controller to I (for Infuse/Inject).
16. Use the controller to push out 1 μL of solution. (Program the volume as "1000." including the decimal point after the four digits.) This will push out the final microliter to fill the ~3 μL dead space of the Silflex tubing.
17. Put the surgical table on the microscope and adjust the focus so that a finger resting on the sponge is in sharp focus. Prefocusing will allow swifter injection, so that fish can be on the surgical table as briefly as possible.
18. Program the injection volume for the fish that will be anesthetized next.

3.5 ANESTHETIZE FISH AND INJECT

These procedures are described together because injection is performed as soon as the fish has been anesthetized. The anesthetized fish is transferred to the sponge, the surgical table is transferred to the microscope stage, and the needle is inserted as shown in Fig. 3. Prior to injection, the fish tanks should have been transferred to a location adjacent to the injection microscope, as described in Method 2.

3.5.1 Materials

As listed in Section 3.2 materials.

3.5.2 Procedure

1. Check that the water temperature in the plastic food container is ~17°C. Adjust if necessary, by adding crushed ice or warm facility water.
2. Transfer a fish to the food container of cool facility water.
3. Anesthetize the fish by gradually adding crushed ice over the course of several minutes to lower the water temperature to ~12°C. Gently swirl the water during this time.
4. While lowering the temperature in the food container, make sure the cooler water is also transferred to the tip box lid, and make sure to depress the foam sponge to saturate it with the cooler water.
5. Monitor fish behavior: As the temperature is lowered, swimming will stop and gill movements will noticeably slow. When the fish no longer responds to handling, transfer it to the trough of the saturated foam plug, belly up. Make sure the gills are within the trough. Notes: Keep your fingers cold for handling the fish—warm fingers can bring the fish out of anesthesia. We prefer not using gloves for this procedure, as gloveless fingers allow gentler handling.
6. Immediately transfer the fish and surgical table to the microscope stage.
7. Immediately insert the needle into the ventral midline, posterior to the pelvic fin girdle, and anterior to the anus, as indicated in Fig. 3. The tip of the needle should point toward the head and should be inserted shallowly.
8. Immediately inject by depressing the foot pedal.
9. Immediately withdraw the needle and set it aside on the silicone mat.

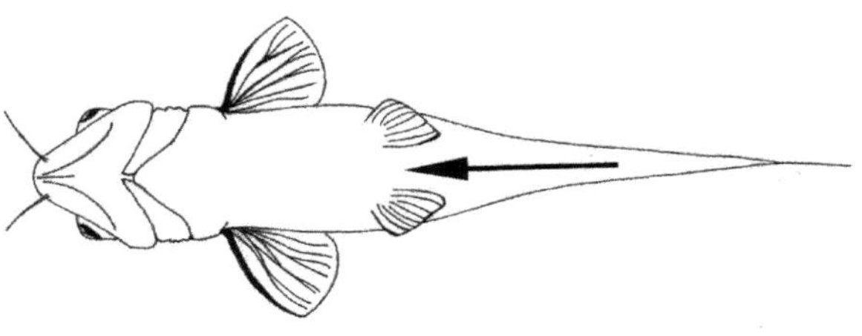

FIGURE 3 Injection site.

The *arrow* indicates a point on the ventral body wall posterior to the pelvic girdle and anterior to the anus, roughly midway along the length of a pelvic fin. The tip of the needle should point rostrally and should be inserted shallowly.

10. Immediately return the fish to its tank for recovery: Pick up the foam plug by the two edges that are parallel with the trough, invert the plug and pull the trough open to release the fish into the tank.
11. Gill movements should start immediately. If they do not, gently swirl water toward the gills to speed recovery.
12. Under the microscope, inspect the end of the injection needle. Sometimes, it will pick up a fish scale. If this is the case, carefully remove it by brushing it off with a Kimwipe.
13. Program the controller with the injection volume for the next fish.
14. Bring the water temperature of the food container back to ~17°C with warm facility water, as in Step 1, and repeat Steps 2–13 as needed.

3.6 DISASSEMBLE AND CLEAN THE SYRINGE, TUBING, AND INJECTION NEEDLE

Clean all items as soon as possible, on the same day as injection, to prevent clogging. The cleaning process is lengthy, in most cases lengthier than that of the injections. To effectively clean the tubing and injection needle, both items must be attached to a syringe. The syringe is then used to push filtered water through the tubing and needle. The NanoFil syringe could be used for this purpose, but its small volume requires multiple rounds of emptying and refilling, which shortens the life of the plunger's Teflon tip. Additionally, the NanoFil plunger is delicate and cannot travel the length of the barrel very quickly without bending. To limit wear and tear on the plunger, we use a 250 μL Hamilton syringe to clean the tubing and injection needle. This saves time, as the Hamilton syringe needs to be filled only once, and its plunger is far easier to manually operate than the NanoFil plunger.

3.6.1 Materials

Silicone mat, as used in previous steps
25 G NanoFil filling needle (supplied with the NanoFil syringe)
Acetone
Filtered dH_2O (prepared in Section 3.1)
250 μL Hamilton syringe number 825, #7646-01
Hamilton large hub removable needle, #7780-03
Silicone gaskets, one whole, one cut in half transversely (World Precision Instruments #NFGSK)
Small container such as a tip box lid, for soaking
Kimwipes

3.6.2 Procedure

1. Disassemble the injection components:
   a. Remove the NanoFil syringe from the pump and set it on the silicone mat.
   b. Loosen the screw cap (nut) and remove the Silflex tubing from the syringe.

c. Carefully transfer the Silflex tubing, with the holder and injection needle still attached, to a dish of distilled water. Keep them submerged until they are cleaned.

2. Clean the NanoFil syringe:
   a. Attach the 25 G filling needle to the NanoFil syringe.
   b. Slowly draw up water through the filling needle to maximally fill the syringe. Flush 5–10 times to dissolve the glucose.
   c. Flush once with acetone.
   d. Flush at least once with water.
3. Clean the Silflex tubing and injection needle:
   a. Attach the Hamilton large hub needle to the Hamilton syringe.
   b. Fill the Hamilton syringe with water, then remove its needle.
   c. Attach the Silflex tubing to the Hamilton syringe in the following manner: In order, thread the Silflex tubing through the metal nut of the Hamilton syringe, the ½ silicone gasket, and then the whole silicone gasket, as shown in Fig. 4. Finally, tighten the nut.
   d. Slowly push the syringe plunger to flush the tubing and needle. Flushing one time should be sufficient.
   e. Remove the injection needle from the holder and return it to the foam support in its original box.
   f. Separate the Silflex tubing from the holder. Carefully insert each end in the foam holders of the original box.
   g. Dry the holder with a Kimwipe and return it to its box.

4. METHOD 4. DISSECTION OF THE PANCREAS, EN BLOC, FOR HISTOLOGY

The adult zebrafish pancreas has multiple lobes that are closely associated with the intestinal loops (Chen, Li, Yuan, & Xie, 2007). The lobes are thin and translucent and appear to partially wrap around the intestine. It is therefore quite

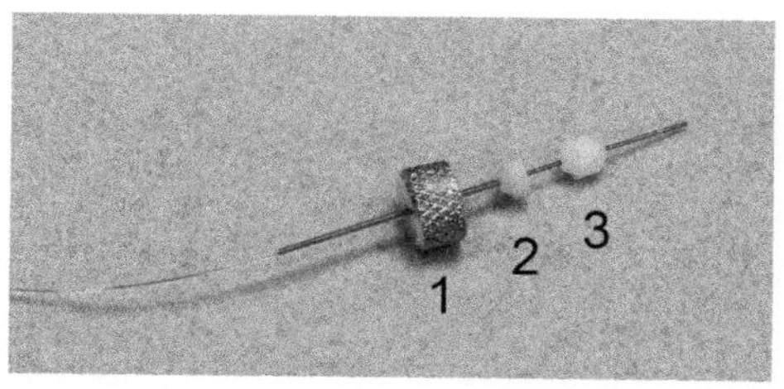

FIGURE 4 Method to adapt the Hamilton syringe for holding the Silflex tubing.

The tubing is threaded through the Hamilton syringe nut 1, a ½ gasket 2, and a whole gasket 3.

challenging to free the lobes from the intestine. We take the approach of dissecting the pancreas together with the intestine to preserve their three-dimensional relationship and to prevent damage to the pancreas. This is achieved by first removing all of the organs of the body cavity together as a unit (i.e., en bloc), then dissecting away all but the pancreas and intestine.

We have found that the tissue is best preserved when the organs are dissected immediately after euthanizing the fish. Further, when working with fresh tissue, we have found that dissection should be performed under liquid to maintain the integrity of the anatomical relationships of the internal organs, which otherwise begin to appear to "slump" shortly after opening the body cavity. To achieve excellent tissue preservation, the organs should be dissected with the entire specimen submerged in fixative. To dissect under fixative, we use a dissection dish whose bottom is covered in silicone elastomer. The elastomer can be pierced by pins multiple times without damaging it, and thus the dish can be used repeatedly.

4.1 MATERIALS

Dissecting microscope and external light source
Box cutter
800 mL plastic beaker, e.g., Fisher Scientific #S01974F
Disposable beaker, e.g., Fisher Scientific #FB012915
Disposable stir rod
Silicone elastomer Sylgard 184 (Dow Corning)
10% neutral buffered formalin (NBF) or 2–4% paraformaldehyde (PFA)
Mayo scissors, 15 cm, serrated (Fine Science Tools #14110-15)
Curved forceps: Dumont #7b (Fine Science Tools #11270-20)
Straight forceps: Dumont #5
Vannas spring scissors, serrated (Fine Science Tools #15007-08)
Iris spatula, curved (#10092-12, Fine Science Tools)
Stainless Steel Insect Pins, Size 3 (#26001-50 Fine Science Tools)
Materials for euthanizing
Vacuum jar and vacuum, optional

4.2 PROCEDURE

4.2.1 Make the dissection plate

1. Make a dish that is about 25 mm deep, as shown in Fig. 5, by cutting across the 800 mL beaker. The inner diameter will be about 88 mm at the rim of the dish.
2. Mix the Sylgard in the disposable beaker, following the manufacturer's instructions.
3. Pour about 59 mL of the Sylgard into the shallow dish.
4. If there are bubbles, put the dish under a vacuum until they are removed.

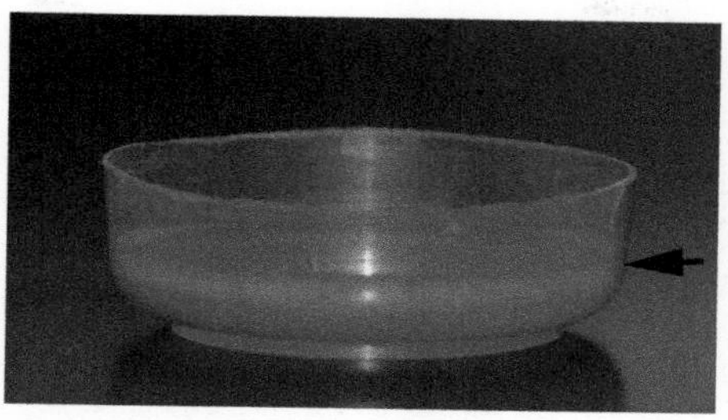

FIGURE 5 Reuseable dissection plate.

The *arrow* indicates the fill level for Sylgard.

4.2.2 Dissect the pancreas

1. Fast the zebrafish for 24 h to empty the intestinal bulb, as described in Method 1.
2. Fill the dissection dish with 10% NBF or 4% PFA.
3. Euthanize the fish using any method approved by your institution.
4. After euthanizing, remove the head: Use the Mayo scissors to make a swift ventral-to-dorsal cut through the gill region. Hold the fish gently for this procedure. Afterward, transfer the scissors to a small beaker of water so that the blades are submerged to prevent body fluids from drying on the blades.
5. Place the specimen on the dissection dish with the rostral end nearest your dominant hand. Secure its position with an insect pin through the fleshy part of the body dorsal to the anal fin. For easier dissection, angle the ventral side of the fish slightly upward, as shown in Fig. 6A. Alternatively, use forceps to orient and stabilize the body during dissection, in lieu of pinning.
6. If necessary, add more fixative to the dissection dish to completely submerge the specimen.
7. Use spring scissors to make three cuts as follows:
   a. Cut 1: Insert a blade of the scissors in the open end of the body and cut along the ventral midline just past the pelvic girdle, as indicated in Fig. 6A. Cut carefully to avoid damaging the internal organs.
   b. Cut 2: At the caudal end of the incision, cut dorsally on one side of the body to widen the first cut. Extend cut 2 into the axial muscles.
   c. Cut 3: Make a parallel cut on the opposite side of the body.
8. Hold the specimen steady with forceps and pull the body flaps apart to widen the opening.
9. Use the spatula to free the organs from the body wall. Use the spring scissors to cut the posterior-most end of the gut tube. The organs should now be isolated from the body cavity, as shown in Fig. 6B.
10. Remove the swim bladder: Use forceps to pull the swim bladder away from the intestine to reveal the duct. Cut the duct using the spring scissors.

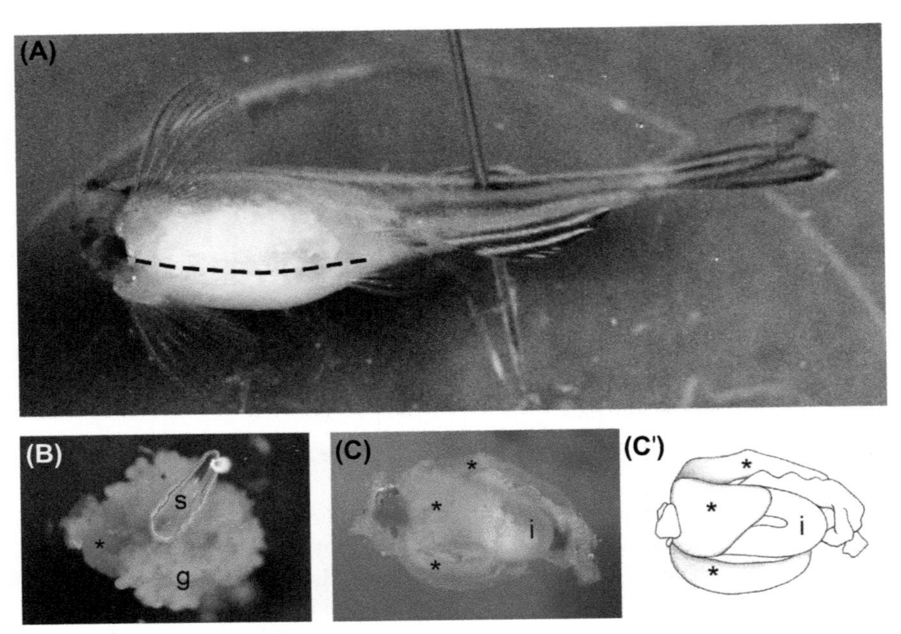

FIGURE 6 Dissection of the pancreas.

(A) Specimen is pinned semilaterally for easier access to the ventral body wall. *Dashed line* indicates the first cut line for opening the body cavity. (B) Organs of the body cavity are removed. Gonads (ovaries) indicated by "*g*," liver indicated by *asterisk*, and one chamber of the swim bladder is intact, indicated by "*s*." When the swim bladder is inflated, the organs rotate into dorsal view, as shown here, anterior to the left. (C) Gonads and swim bladder have been removed. The first loop of the intestine is indicated by "*i*." The intestinal bulb is covered by the liver, indicated by *asterisks*. The tissue is shown in lateral view, anterior to the left. (C′) Drawing of C, outlining three lobes of the liver, indicated by *asterisks*. The pancreas is deep to the liver, closely associate with the intestine and obscured by fat.

11. Remove the gonads: Carefully pull them away or trim them using blunt dissection. The specimen will look similar to Fig. 6C.
12. If desired, remove the heart, spleen, gall bladder, and liver.
13. Transfer the intestine with pancreas to an Eppendorf tube with fixative.

5. METHOD 5. STERILE DISSECTION AND CULTURE OF THE PRINCIPAL ISLET

The principal islet of adult zebrafish is normally visible under a dissection microscope at low magnification, using white light. Transgenic lines, such as *Tg(ins:GFP)*, in which the pancreatic beta cells are labeled with green fluorescent protein, allow easy identification of the islet, which is deep within the body cavity. However, with some experience, the islet can be easily located without the aid of fluorescence.

The major cell type within the principal islet is the beta cell, which constitutes the majority of the cell mass (Chen et al., 2007; Li,Wen, Peng, Korzh, & Gong, 2009). Therefore, culturing the dissociated islet cells is potentially useful for a wide variety of beta-cell studies. The following method describes how to locate the islet by dissection and how to isolate it for downstream procedures including cell culture. As zebrafish islets are relatively small, we recommend isolating five to six principal islets and pooling them. This can be done in ~1 h. After collecting the islets, they should be immediately dissociated and plated for cell culture, with no stopping point until the plates are in an incubator. The Liebovitz L-15 culture medium is diluted to 67%, following Peppelenbosch, Tertoolen, de Laat, and Zivkovic (1995).

5.1 MATERIALS

Dissection microscope with external light source, fluorescence optional
Inverted microscope
Ethanol spray bottle
Kimwipes
Plastic beaker, 800 mL
Ice bucket, with ice
Thermometer
Dissection dish, made as described in Method 4
Stainless steel insect pins, size 3 (Fine Science Tools #26001-50)
Mayo scissors, 15 cm, serrated (Fine Science Tools #14110-15)
Vannas spring scissors, serrated (Fine Science Tools #15007-08)
Curved forceps, Dumont #7b (Fine Science Tools #11270-20)
Straight forceps, Dumont #5
Silicone mats, cut as described in Step 1
Autoclavable box lid, such as a pipet tip box lid
Plastic beaker, 250 mL, autoclavable
Kimwipes or instrument wipes
Hot bead sterilizer (Fine Science Tools #18000-45)
Pasteur pipettes, fire polished, autoclaved
Pipet bulb for Pasteur pipettes
P1000 pipette
Wide-orifice P1000 pipette tips, sterile
37°C water bath or heat block
Leibovitz's L-15 culture medium (Gibco/ThermoFisher #11415), see preparation below
CS, prepared as in Method 3
Fetal bovine serum

D-glucose
L-glutamine
Dulbecco's Phosphate Buffered Saline (PBS), modified, without calcium and magnesium, sterile
Penicillin-streptomycin (Sigma #P0781)
Trypsin-Ethylenediaminetetraacetic acid EDTA (Gibco/ThermoFisher #25300)
Versene (Gibco/ThermoFisher #15040)
Collagenase P (Sigma #11213857001)
Water, sterile
Aluminum weigh boat
Two Eppendorf tubes, each with 1 mL prepared culture medium
35 mm MatTek culture plates, with 10 mm glass well
10 cm culture plates, optional
28.5°C sterile humidified incubator, without CO_2 supply
Cell culture hood (laminar flow biosafety cabinet)

5.2 PROCEDURE

5.2.1 Preparations for dissection

1. Prepare autoclaved supplies:
   a. Cut a piece of silicone mat to fit inside a pipette tip box lid or other autoclavable shallow box.
   b. Cut a piece of silicone mat to a convenient size for holding dissection instruments adjacent to the microscope, e.g., 16 cm × 18 cm.
   c. Flame polish several Pasteur pipettes. A narrow opening, e.g., 800 μm, is desirable.
   d. Autoclave the items listed in a-c (both silicone mats, the shallow box, and the flame-polished pipettes) as well as the 250 mL beaker, the PBS, and the water.
2. Dilute the L-15 medium to 67% using CS. For 100 mL of 67% L-15, add 33.4 mL CS to 66.6 mL L-15.
3. Make 50 mL of supplemented medium: To 42.4 mL of 67% L-15, add 7.5 mL fetal bovine serum, 1× pen-strep, 2 mM D-glucose, and 2 mM L-glutamine. Then, filter sterilize. Prewarm to 28.5°C before using.
4. Make up collagenase on the day it is needed, to avoid loss of activity. Weigh out using an antistatic weigh boat, e.g., aluminum. Dissolve 0.375 mg/mL collagenase in PBS.
5. Fast five to six fish for 24 h to empty the intestinal bulb, as described in Method 1.
6. At least 10 min before euthanizing the fish, add antibiotics to the tank water.
7. Prepare for dissection by adding sterile water to the plastic box with silicone mat and setting it within reach of the microscope. The water level should be just enough to cover the spring scissor blades when the scissors are on the mat. Add sterile water to the 250 mL beaker. There should be enough water to cover the blades of the Mayo scissors.

8. Add antibiotics to the sterile PBS, and just prior to euthanizing, flood the dissection dish with the PBS. Stick one insect pin into the Sylgard of the dish.

5.2.2 Islet dissection, dissociation, and culture

1. Transfer one fish to a beaker of 4–7°C facility water. Keep the fish in cold water for at least 1 min.
2. Meanwhile, sterilize both pairs of scissors and both pairs of forceps using the bead sterilizer.
3. Use the larger (Mayo) scissors to cut completely and swiftly through the gill region, from ventral to dorsal. This cut is anterior to the pectoral girdle. The fish should be held gently during this procedure (avoid squeezing). Note: If the fish is quite small, then finer scissors may be necessary, e.g., Fine Science Tools #14958-11. Transfer the scissors to the beaker of water to prevent body fluids from drying on the blades.
4. Pin the specimen to the plate, left side down: Push the pin through the fleshy body dorsal to the anal fin. The tail should be off-center on the plate, for easier dissection of the body cavity. Do not squeeze the fish during pinning.
5. Add more sterile PBS if necessary: The specimen should remain submerged during dissection.
6. With one hand, hold the specimen steady using curved forceps. With the other hand, use the spring scissors to make a series of four cuts through the body wall (Fig. 7A):
   a. Cut 1: Cut through the body wall at the ventral midline, from posterior to anterior. Begin the cut by inserting the scissor blade approximately midway between the anus and the pelvic girdle. Extend the cut to just behind the pectoral girdle.
   b. Cut 2: At the rostral-most end of the first cut, make a perpendicular cut (from ventral to dorsal) that extends into the region of the axial muscles.
   c. Cut 3: At the caudal-most end of the first cut, make a ventral-to-dorsal cut, parallel to the second cut. You now have a section of body wall that is free on three sides.
   d. Cut 4: Cut across the axial muscles, parallel with the first cut. The scissors should go deeper than the previous cuts, to cut through the muscle.
7. Remove the flap: Grasp and lift the ventral border of the flap with curved forceps. With straight forceps, use blunt dissection to free the flap from the underlying connective tissues and vessels. Work from ventral to dorsal, lifting the flap as you proceed. The flap may still be tightly adherent to the underlying axial muscles. If so, use straight forceps to strip the flap off of the muscles, or use spring scissors to cut it away.
8. Use forceps to grasp the right pectoral fin and pull cranially to remove. The entire pectoral girdle may be removed by this action.

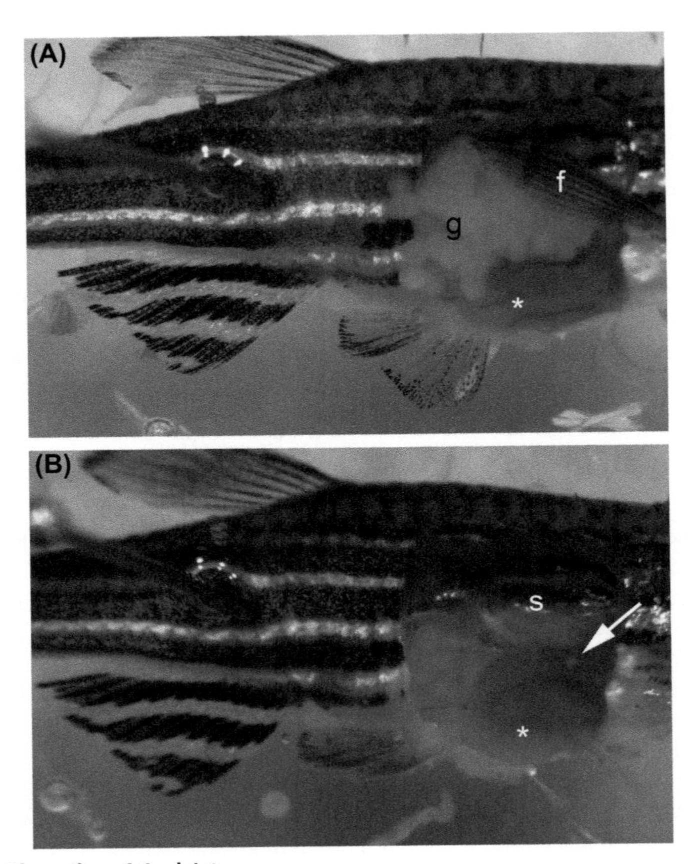

FIGURE 7 Dissection of the islet.

Specimen in lateral view, head to the right. (A) A window is opened in the right body wall. Gonads and liver are encountered immediately after removing the flap. (B) The principal islet, indicated by *arrow*, is visible after removing the right ovary and part of the liver. The right pectoral fin has been removed for easier access to the internal organs. *f*, pectoral fin; *g*, gonads; *s*, swim bladder; *, liver.

9. If the body cavity contains eggs, carefully remove the right ovary. If the specimen is male, the right testis may sometimes be in the way. Remove the testis, if necessary.
10. Carefully remove the right lobe of the liver: Use blunt dissection from underneath the lobe to free it from the underlying organs (gall bladder, pancreas, and intestine). Pull or cut the lobe to remove it completely.
11. Locate the principal islet (Fig. 7B): Typically, the principal islet is found in close association with the gall bladder. Often, it is located ventral to the gall bladder and dorsal to the intestinal bulb. Find the gall bladder first, then find

the intestinal bulb, which is the first intestinal loop encountered when dissecting from the right side. The intestinal bulb may have to be pushed ventrally to reveal the pancreas. Note: Occasionally, the principal islet is on the left side of the body. If you cannot find the islet on the right, flip the specimen over and repeat the dissection. Rarely, a visible, compact principal islet cannot be found because diffuse endocrine tissue is in its place.

12. Remove the principal islet: Use straight forceps to grasp the translucent exocrine pancreas tissue immediately adjacent to the islet. Use spring scissors to trim away the surrounding tissue and free the islet. If desired, search through the lobes of the pancreas to find additional large islets for isolation. Transfer the spring scissors to the submerged silicone mat. This will prevent tissues or fluid from drying on the scissors.
13. Use a sterile, fire-polished Pasteur pipette to transfer the islet to an Eppendorf containing culture medium.
14. Remove the PBS and all dissection debris from the dissection dish and repeat Steps 1–13 for each fish. Before resterilizing dissection instruments in the bead sterilizer, clean off any debris and wipe dry using a Kimwipe.
15. Put no more than 2–3 principal islets into each Eppendorf tube.
16. When the last islet has been added, allow the islets to settle to the bottom of the tube. From this point, maintain sterility by working in a laminar-flow hood using sterile techniques.
17. Remove the medium.
18. To each tube, add 0.375 mg/mL collagenase P in PBS.
19. Incubate at 37°C, for at least 10 min.
20. During the incubation, prewarm 0.005% trypsin-EDTA at 37°C. This is made up by diluting 100 μL of trypsin-EDTA (0.05%) in 900 μl versene.
21. Put tubes on ice briefly, ~1 min, to inhibit the collagenase.
22. Remove medium.
23. To each tube, add 1 mL of prewarmed 0.005% trypsin-EDTA.
24. Set the P1000 pipette to 1000 μL and use a wide-orifice tip. Pipet up and down vigorously, at least 25 times. The islets should start to dissociate during this trituration.
25. If the trituration is not effective, allow the islets to settle to the bottom (or spin down briefly) and repeat the process starting with Step 20.
26. When the islets have been dissociated, spin at 1600 rcf for 1 min.
27. Remove supernatant.
28. Examine the cells: If the cells have been well dissociated, they will adhere along the side of the tube. If this is the case, proceed to the next step. If the cells are not dissociated, they will form a tight pellet, and the process should be repeated from Step 22.
29. Add a maximum of 1.5 mL fresh supplemented culture medium and resuspend the cells. Use less medium when working with fewer/small islets.
30. Distribute ~200 μL to the central glass portion of each 35 mm MatTek plate. The medium should form a bubble.

31. Incubate at 28.5°C in a humidified sterile incubator, with no CO_2 supply, overnight.
32. The next morning, use an inverted microscope to check whether the cells have attached. For ease of handling on the microscope stage, the MatTek dish can be held in a larger culture dish (e.g., 10 cm).
33. If the cells are not attached, incubate overnight again. If the cells are attached, slowly add 2 mL culture medium to the dish, dropwise, along the dish periphery rather than directly on the central bubble so as not to detach the cells. Avoid touching the pipette tip to the plate and avoid jostling the plate.
34. Return the dish to the incubator or proceed immediately with downstream analysis.

ACKNOWLEDGMENTS

We thank Rachel Krizek for the drawings and Dr. Ted Zerucha for generously donating zebrafish. We thank our colleagues Dr. Victoria Prince, Dr. Ted Zerucha, and Dr. Cortney Bouldin for reading the manuscript and providing constructive criticism. We thank Dr. David Jacobson for providing beta-cell expertise that was critical for successful cell culture. The Kinkel lab is supported by the Office of Student Research at Appalachian State University and by startup funds from Appalachian State University.

REFERENCES

Chen, S., Li, C., Yuan, G., & Xie, F. (2007). Anatomical and histological observation on the pancreas in adult zebrafish. *Pancreas, 34*(1), 120–125. http://dx.doi.org/10.1097/01.mpa.0000246661.23128.8c.

Collymore, C., Tolwani, A., Lieggi, C., & Rasmussen, S. (2014). Efficacy and safety of 5 anesthetics in adult zebrafish (*Danio rerio*). *Journal of the American Association for Laboratory Animal Science, 53*(2), 198–203.

Eames, S. C., Kinkel, M. D., Rajan, S., Prince, V. E., & Philipson, L. H. (2013). Transgenic zebrafish model of the C43G human insulin gene mutation. *Journal of Diabetes Investigation, 4*(2), 157–167. http://dx.doi.org/10.1111/jdi.12015.

Eames, S. C., Philipson, L. H., Prince, V. E., & Kinkel, M. D. (2010). Blood sugar measurement in zebrafish reveals dynamics of glucose homeostasis. *Zebrafish, 7*(2), 205–213. http://dx.doi.org/10.1089/zeb.2009.0640.

Hwang, K. L., & Goessling, W. (2016). Baiting for cancer: using the zebrafish as a model in liver and pancreatic cancer. *Advances in Experimental Medicine and Biology, 916*, 391–410. http://dx.doi.org/10.1007/978-3-319-30654-4_17.

Kimmel, R. A., & Meyer, D. (2016). Zebrafish pancreas as a model for development and disease. *Methods in Cell Biology, 134*, 431–461. http://dx.doi.org/10.1016/bs.mcb.2016.02.009.

Kinkel, M. D., Eames, S. C., Philipson, L. H., & Prince, V. E. (2010). Intraperitoneal injection into adult zebrafish. *Journal of Visualized Experiments, 42*. http://dx.doi.org/10.3791/2126.

Li, Z., Wen, C., Peng, J., Korzh, V., & Gong, Z. (2009). Generation of living color transgenic zebrafish to trace somatostatin-expressing cells and endocrine pancreas organization. *Differentiation, 77*, 128–134.

Maddison, L. A., Joest, K. E., Kammeyer, R. M., & Chen, W. (2015). Skeletal muscle insulin resistance in zebrafish induces alterations in beta-cell number and glucose tolerance in an age- and diet-dependent manner. *American Journal of Physiology. Endocrinology and Metabolism, 308*, E662–E669. http://dx.doi.org/10.1152/ajpendo.00441.2014.

Peppelenbosch, M. P., Tertoolen, L. G., de Laat, S. W., & Zivkovic, D. (1995). Ionic responses to epidermal growth factor in zebrafish cells. *Experimental Cell Research, 218*(1), 183–188. http://dx.doi.org/10.1006/excr.1995.1146.

Perry, S., Davie, P., Daxboeck, C., Ellis, A., & Smith, D. (1984). Perfusion methods for the study of gill physiology. In W. Hoar, & D. Randall (Eds.), *Fish physiology volume X: Gills, part B: Ion and water transfer* (pp. 325–388). Orlando, FL: Academic Press.

Prince, V. E., Anderson, R., & Dalgin, G. (2016). Zebrafish pancreas development and regeneration: fishing for diabetes therapies. *Current Topics in Developmental Biology* (in press).

Schiavone, M., Rampazzo, E., Casari, A., Battilana, G., Persano, L., Moro, E., … Argenton, F. (2014). Zebrafish reporter lines reveal in vivo signaling pathway activities involved in pancreatic cancer. *Disease Models & Mechanisms, 7*, 883–894. http://dx.doi.org/10.1242/dmm.014969.

Wolf, K. (1963). Physiological salines for fresh-water teleosts. *The Progressive Fish-Culturist, 25*, 135–140.

PART

Musculoskeletal System

CHAPTER

Using the zebrafish to understand tendon development and repair

12

J.W. Chen[*,§], **J.L. Galloway**[*,§,1]
Massachusetts General Hospital, Boston, MA, United States
[§]*Harvard Medical School, Boston, MA, United States*
[1]*Corresponding author: E-mail: jgalloway@mgh.harvard.edu*

CHAPTER OUTLINE

Methods in Cell Biology, Volume 138, ISSN 0091-679X, http://dx.doi.org/10.1016/bs.mcb.2016.10.003

Abstract

Tendons are important components of our musculoskeletal system. Injuries to these tissues are very common, resulting from occupational-related injuries, sports-related trauma, and age-related degeneration. Unfortunately, there are few treatment options, and current therapies rarely restore injured tendons to their original function. An improved understanding of the pathways regulating their development and repair would have significant impact in stimulating the formulation of regenerative-based approaches for tendon injury. The zebrafish provides an ideal system in which to perform genetic and chemical screens to identify new pathways involved in tendon biology. Until recently, there had been few descriptions of tendons and ligaments in the zebrafish and their similarity to mammalian tendon tissues. In this chapter, we describe the development of the zebrafish tendon and ligament tissues in the context of their gene expression, structure, and interactions with neighboring musculoskeletal tissues. We highlight the similarities with tendon development in higher vertebrates, showing that the craniofacial tendons and ligaments in zebrafish morphologically, molecularly, and structurally resemble mammalian tendons and ligaments from embryonic to adult stages. We detail methods for fluorescent in situ hybridization and immunohistochemistry as an assay to examine morphological changes in the zebrafish musculoskeleton. Staining assays such as these could provide the foundation for screen-based approaches to identify new regulators of tendon development, morphogenesis, and repair. These discoveries would provide new targets and pathways to study in the context of regenerative medicine–based approaches to improve tendon healing.

INTRODUCTION

Tendons transmit force generated by the contracting muscle to the bone, enabling movement; ligaments connect bone to bone, maintaining stability. Each year, millions of individuals in the United States suffer from tendon and ligament injuries (Gomoll, Katz, Warner, & Millett, 2004). Injuries to these tissues have a slow and limited healing potential, and the current treatment options, which include anti-inflammatory treatments, surgery, and/or physical therapy, are often plagued by complications such as rerupture of the repair site, pain, and limited mobility (Sharma & Maffulli, 2005). In recent years, many groups have examined tissue engineering and stem cell–based approaches for treating these diseases. Although some approaches appear promising, there are currently no widespread accepted treatments for many tendon and ligament injuries (Guerquin et al., 2013; Kiapour, Fleming, & Murray, 2015; Nourissat et al., 2010; Pelled et al., 2012). Limitations likely stem from an incomplete understanding of the fundamental pathways that regulate tendon formation, growth, maturation, and healing.

The molecular signals involved in tendon development have been largely unexplored due to the absence of molecular markers. Accordingly, the discovery of *Scleraxis* as the earliest known marker of tendon and ligament progenitors in the mouse and chick paved the way for investigating the early formation of tendons and ligaments (Schweitzer et al., 2001). Subsequent studies in the mouse and chick have identified several factors involved in tendon development, and henceforth, provided fruitful insights into the molecular mechanisms governing tendon formation in the developing embryo (Schweitzer, Zelzer, & Volk, 2010). However, many questions remain regarding the molecular networks and cellular behaviors that regulate multiple stages of tendon formation and maturation. With the advent of high-throughput chemical genetic screens in zebrafish, this model offers a vertebrate genetic system amendable to discovery of novel genetic pathways relevant to tendon biology (Kaufman, White, & Zon, 2009). Such discoveries will greatly advance our understanding of the cellular and molecular mechanisms involved in tendon development and have the potential to impact regenerative medicine–based approaches to tendon injury and repair.

1. TENDON STRUCTURE

A tendon's capacity to tolerate tensile stress is unrivaled to that of other tissues: human tendons are capable of withstanding up to 9 kN of force, the equivalence of 12.5 times the body weight (Komi, 1990). Their ability to function under such forces is due to their molecular structure. Tendons and ligaments are hierarchically organized tissues with the fundamental structural unit being the collagen fibril (Prockop & Kivirikko, 1995). Macroscopically, the collagen fibrils organize to form collagen fibers (primary bundle), fascicles (secondary bundles), tertiary bundles, and finally, the tendon itself (Silver, Freeman, & Seehra, 2003). The biomechanical properties of tendons and ligaments are dependent on the intra- and intermolecular bonds of the collagen network (Butler, Grood, Noyes, & Zernicke, 1978). In adult mice with mutations in collagen or the tendon-enriched proteoglycans, the tendons have abnormal collagen fibrillogenesis (Connizzo, Yannascoli, & Soslowsky, 2013; Gaut & Duprez, 2016), further underscoring the importance of the matrix in tendon development and maturation of the adult tissue.

2. TENDON FORMATION AND DIFFERENTIATION

Tendon and ligament development is intimately connected with the development of the adjacent muscle, cartilage, and bone of the musculoskeletal system (Wortham, 1948). Cells are specified toward a tendon fate and then organize, initially as loose cellular aggregates and then as structurally distinct tendons, at the anatomical interface of the differentiating muscle and skeletal tissues. The earliest marker of tendon and ligament progenitors in the embryo is *Scleraxis*, a basic helix-loop-helix transcription factor (Fig. 1). *Scleraxis* is robustly expressed in the tendon and ligament lineages from the progenitor to the differentiated state (Brent, Schweitzer, & Tabin, 2003; Grenier, Teillet, Grifone, Kelly, & Duprez, 2009; Schweitzer et al., 2001). Scleraxis can activate the *Collagen type I alpha 1* proximal promoter in vitro (Lejard et al., 2007), and upregulate the expression levels of proteoglycans enriched in the tendon matrix (Alberton et al., 2012).

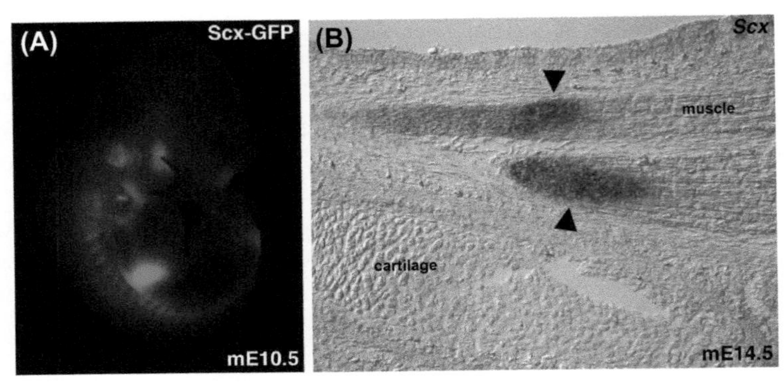

FIGURE 1

Expression of *Scleraxis* (Scx) in wild-type mouse embryos. (A) Scx-green fluorescent protein (GFP) embryo at mouse embryonic day 10.5 (mE10.5). (B) Section in situ hybridization of *Scx* expression (*arrowheads*) between cartilage and muscle in the forelimb at E14.5.

Moreover, Scleraxis positively regulates the expression of Tenomodulin, a marker of differentiated tenocytes, in vitro and in vivo (Shukunami, Takimoto, Oro, & Hiraki, 2006). Scleraxis null ($Scx^{-/-}$) mice are viable as adults but have impaired mobility. In $Scx^{-/-}$ embryos, the force-transmitting and intermuscular tendons are severely disrupted, whereas the ligaments and muscle-anchoring tendons are present (Murchison et al., 2007). The presence of some populations of tendons and ligaments in $Scx^{-/-}$ embryos indicates that while Scleraxis is a faithful marker of the lineage, it is not required for the development of all tendons and ligaments.

After specification of the tendon lineage, the progenitor cells organize and aggregate into primordia that differentiate in coordination with the neighboring muscle and skeletal tissues. Tenomodulin is a type II transmembrane protein that is expressed in the axial and limb tendons and ligaments, at a more differentiated stage of the lineage compared to *Scleraxis* (Shukunami et al., 2006). Loss of Tenomodulin results in abnormal collagen fibril structure (Docheva, Hunziker, Fassler, & Brandau, 2005), indicating a later role for Tenomodulin in collagen matrix maturation. As the tendons and ligaments mature, the cells secrete Collagen type I fibers that initially grow in number and subsequently, in length and diameter, eventually creating the predominantly extracellular matrix (ECM) tissue that is characteristic of adult tendons and ligaments (Birk, Zycband, Woodruff, Winkelmann, & Trelstad, 1997; Kalson et al., 2015; Zhang et al., 2005).

The craniofacial tendons and ligaments of the zebrafish (Fig. 2A and B) share similar morphological, molecular, and structural characteristics from embryonic to adult stages with that of amniotes (Chen & Galloway, 2014). The tendon and ligament progenitors initiate at the junctional interface between the developing muscle and cartilage or between cartilage segments, respectively, and express *scleraxisa* (*scxa*) (Fig. 2C and E). Due to genome duplication events, there exists a second *scleraxis* gene in the zebrafish, *scleraxisb* (*scxb*); however, its expression

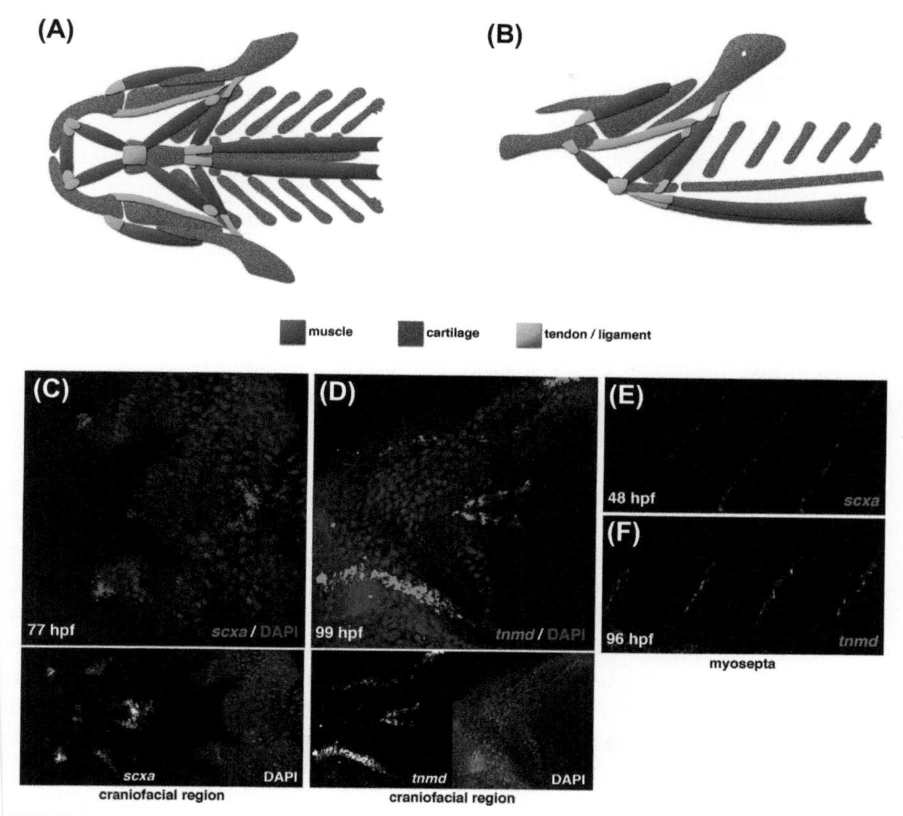

FIGURE 2

Expression of tendon genes in the zebrafish musculoskeleton. Schematic (A) ventral and (B) lateral views of the craniofacial region at 72 hpf with muscle, cartilage, and tendon/ligament depicted. Expression of *scleraxisa (scxa)* (C, E) and *tenomodulin (tnmd)* (D, F) in the craniofacial region (77 and 99 hpf, respectively) and myosepta (48 and 96 hpf, respectively). Ventral (C, D) and lateral (E, F) views of flat-mounted embryos were processed for the respective transcripts by fluorescent in situ hybridization.

is only detected by reverse transcription—polymerase chain reaction (RT-PCR) after 54 hours post-fertilization (hpf). At slightly later stages, following *scxa* expression, the expression of ECM components *tenomodulin* (Fig. 2D and F), Thrombospondin-4b, and *collagen type I alpha 2* can be observed (Chen & Galloway, 2014; Subramanian & Schilling, 2014). Moreover, *xirp2a,* an actin-binding protein, is expressed in the myosepta and all muscle attachment sites in the head and fin (Chen & Galloway, 2014; Otten et al., 2012). In the adult zebrafish, the tendon and ligaments are characterized by a *D*-periodicity, which results from a highly organized network of collagen fibrils and resembles the mammalian tendon structure (Chen & Galloway, 2014; Docheva et al., 2005).

3. TISSUE INTERACTIONS WITHIN THE DEVELOPING MUSCULOSKELETAL SYSTEM

Tendons and ligaments develop in multiple anatomic locations: the craniofacial, limb/fin, and axial regions. The induction and maintenance of tendons and ligaments are similar across vertebrate species, with some differences in the developmental program between the craniofacial, limb/fin, and axial tendons, which arise from distinct embryonic origins. Nevertheless, tendons and ligaments in all locations originate from common progenitor populations that also will give rise to cartilage and bone (Soeda et al., 2010) but are distinct from muscle-forming regions.

The jaw develops into a functioning musculoskeletal system early in the zebrafish, with cartilage and muscle progenitors present in the pharyngeal arches by 48 hpf and feeding evident by 96 hpf. The pharyngeal cartilage and muscle progenitor populations develop in synchrony and in close proximity to each other, as is observed in musculoskeletal development of amniotes (Schilling, 1997; Schilling & Kimmel, 1997). Fate mapping experiments have determined that the cartilage and connective tissue of the jaw in the zebrafish and higher vertebrates originate from the cranial neural crest, a vertebrate-specific multipotent cell population (Chen & Galloway, 2014; Couly, Coltey, & Le Douarin, 1993; Kontges & Lumsden, 1996; Le Lievre, 1978; Noden, 1978; Schilling & Kimmel, 1994). Similarly, the limb tendons/ligaments and skeletal structures of amniotes originate from the lateral plate mesoderm-derived limb mesenchyme (Pearse, Scherz, Campbell, & Tabin, 2007; Wachtler, Christ, & Jacob, 1981). Although both skeletal and tendon/ligament lineages arise from common progenitor populations, our understanding of the dynamics of their interactions has been limited until recently. In zebrafish, the specification of tendons/ligaments occurs independently of the cartilage program, though a properly formed cartilage template is critical for proper tendon/ligament organization (Chen & Galloway, 2014). This finding gains support from studies in the mouse, wherein the tendons of the middle limb segment, the zeugopod, form in *Sox9*-deficient limbs lacking cartilage, although the digit tendons in the distal autopod segment are dependent upon cartilage (Huang et al., 2015). In the context of bone eminences, the initiation of the Sox9-positive cartilage and Scleraxis-positive tendon populations is independent of each other as well as the enthesis domain (Blitz, Sharir, Akiyama, & Zelzer, 2013; Sugimoto et al., 2013).

The molecular cross talk between the muscle and the associated tendon has been well characterized in vertebrates. Studies in mice revealed that, in the head, *Scleraxis*-positive tendon progenitors initiate independently of the branchiomeric muscles, although muscle is required for continued maintenance of tendon fate (Grenier et al., 2009). In the limb, tendon progenitor specification is also muscle independent, whereas the maintenance of limb tendon fate is muscle dependent (Brent, Braun, & Tabin, 2005; Edom-Vovard, Schuler, Bonnin, Teillet, & Duprez, 2002; Kardon, 1998; Shellswell & Wolpert, 1977). Divergent from this developmental archetype are the distal tendons of the autopod, which initiate and form in muscleless limbs (Huang et al., 2015; Hurle et al., 1990; Kieny & Chevallier, 1979). Conversely, the myogenic program initiates independently of tendon progenitors in all anatomic locations, but the proper patterning and differentiation of the myofibers are dependent on the tendon and connective tissues (Kardon, 1998; Rinon et al., 2007).

We have shown that a similar program exists in the zebrafish, wherein initiation of *scxa* expression in cranial and pectoral fin regions is muscle independent, but its continued expression is muscle dependent (Chen & Galloway, 2014). Together, the dissection of these interactions suggests a complex interdependence of the musculoskeletal tissues at distinct stages of development. Although FGF signaling has been implicated in some of these interactions (Brent et al., 2003; Brent & Tabin, 2004; Edom-Vovard et al., 2002; Rinon et al., 2007; Smith, Sweetman, Patterson, Keyse, & Munsterberg, 2005), we still have a limited understanding of the molecular cross talk regulating the coordination of musculoskeletal development, assembly, and growth.

The axial musculoskeletal tissues arise from the somites, which differentiate into the dermamyotome (gives rise to the dermis and muscle) and the sclerotome (gives rise to the vertebrae and ribs) (Brand-Saberi & Christ, 2000; Christ & Ordahl, 1995). In amniotes, the somite is comprised predominantly of the medially positioned sclerotome, and vertebral elements develop synchronously with the associated musculature (Christ, Huang, & Wilting, 2000; Grotmol, Kryvi, Nordvik, & Totland, 2003). Axial tendons originate from the syndetome, a somitic compartment that is established by myotomal FGF signals to the adjacent sclerotome to induce formation of tendon progenitors (Brent et al., 2003). Specification of the sclerotome to a tendon-fated syndetome is mutually exclusive of a cartilage fate (Brent et al., 2005). In teleosts, by contrast, the somite is composed predominantly of myotome with a relatively small ventral sclerotome (Stickney, Barresi, & Devoto, 2000). The formation of the axial cartilage and myotome is not synchronized—differentiated myofibers are present in the embryo, whereas the cartilage template forms in the larvae (Bird & Mabee, 2003; Devoto, Melancon, Eisen, & Westerfield, 1996). The myomeres are separated along the dorsal–ventral axis by the horizontal myoseptum and along the anterior–posterior axis by the vertical myosepta. The horizontal and vertical myosepta both function in force transmission during undulatory movement (Nursall, 1956; Westneat, Hoese, Pell, & Wainwright, 1993). Together, they are thought to be homologous to the axial tendons of higher vertebrates based on similarities in architecture and a gene-expression profile enriched in mammalian tendon matrix proteins (Bricard, Ralliere, Lebret, Lefevre, & Rescan, 2014). However, there appear to be some differences in their origins that are distinct from higher vertebrates. Early expression of ECM molecules in zebrafish horizontal myosepta by 24 hpf is thought to derive from the muscle pioneer cells, which originate from the myotome (Devoto et al., 1996; Felsenfeld, Curry, & Kimmel, 1991; Hatta, Bremiller, Westerfield, & Kimmel, 1991; Schweitzer et al., 2005). Although lineage tracing of the myosepta in teleosts is lacking, the sclerotome is postulated to give rise to the *scxa*-expressing cells in the vertical myosepta. Consistent with the mode of amniote axial development, the induction of axial *scxa*-expressing tendon progenitors in zebrafish is muscle dependent (Chen & Galloway, 2014).

The architecture of the vertebrate body axis presents the myosepta as a model for the formation of the myotendinous junction (MTJ), the primary site of force transmission in skeletal muscle. The highly organized collagenous vertical myosepta are a region important in development of the myomeres, functioning specifically as attachment sites for the sarcolemmal basement membrane (Charvet, Malbouyres, Pagnon-Minot, Ruggiero, & Le Guellec, 2011; Henry, McNulty, Durst, Munchel, & Amacher, 2005). Studies in the zebrafish have identified several factors with a role in the stabilization, maintenance, and repair of the MTJ, including *collagen type XXII alpha 1*, *laminin alpha 2*, *thrombospondoin-4b*, *focal adhesion kinase,* and *beta-dystroglycan* (Charvet et al., 2013; Hall et al., 2007; Snow & Henry, 2009; Subramanian & Schilling, 2014). As MTJ-associated defects lead to a subset of muscular dystrophies (Berger, Berger, Hall, Lieschke, & Currie, 2010), a more comprehensive understanding of the ECM-dependent interactions between the muscle sarcolemma and tendon cells should prove fruitful in developing therapeutics for such conditions. Indeed, drug screening approaches for chemical modulators of a muscular phenotype in a zebrafish dystrophy model have identified molecules and pathways that could be candidate therapeutic targets for these diseases (Kawahara et al., 2011; Kawahara & Kunkel, 2013). These findings contribute to our understanding of the cellular and molecular mechanisms involved in development of the MTJ and have the potential to advance therapeutic-based approaches to congenital muscle disorders.

4. METHODS TO STUDY THE EMBRYONIC TENDON PROGRAM IN ZEBRAFISH

Here we describe technologies to assess the development, differentiation, and patterning of tendon and ligament tissues. Our whole-mount protocol combines fluorescent in situ hybridization (FISH) and immunohistochemistry (IHC) to evaluate expression of multiple genes and/or proteins (Fig. 3) in zebrafish embryos. We employ tyramide (Tyr) signal amplification (TSA) technology to robustly label nucleic acid sequences and/or proteins of interest in situ, thereby facilitating spatial resolution at a cellular level using confocal microscopy (Wang, Achim, Hamilton, Wiley, & Soontornniyomkij, 1999). TSA uses enzymatic activity to amplify the signal intensity of a fluorescent-labeled target protein or nucleic acid. We have employed antibodies to stain for muscle and cartilage proteins using the myosin heavy chain (MYH) and collagen type II (Col2) antibodies, respectively (Clement et al., 2008).

Our strategy can be modified to evaluate tendon and ligament development in the context of the musculoskeletal system when assessing gene function using pharmacological agonists and inhibitors, morpholino oligonucleotides to prevent mRNA translation, or CRISPR/Cas9 [clustered regularly interspaced short palindromic

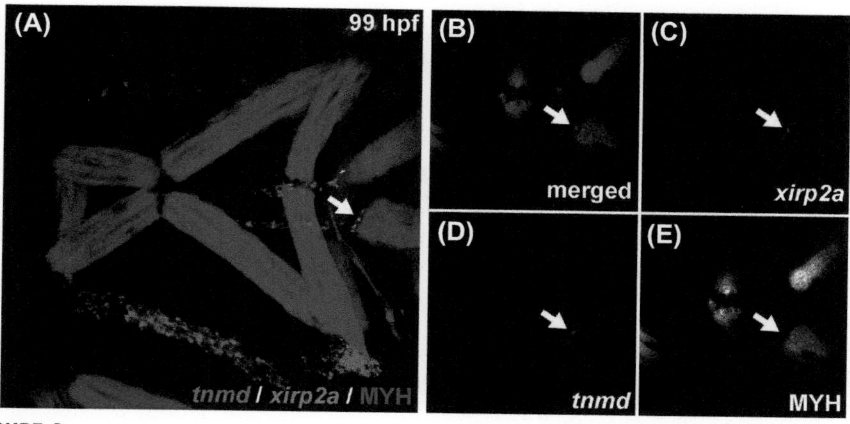

FIGURE 3

Tendon genes [*tenomodulin* (*tnmd*) and *xirp2a*] are coexpressed at the sternohyoideus muscle attachment (*arrow*) in the zebrafish craniofacial region at 99 hpf. (A) Maximum-intensity projection. (B–E) Single channel optical sections from confocal image. Fluorescent in situ hybridization was performed for *tnmd* and *xirp2a* to visualize the tendons. Muscle is indicated by A4.1025 antibody staining for Myosin Heavy Chain (MYH). Ventral view of a flat-mounted embryo.

repeats (CRISPR)/CRISPR-associated protein 9 (Cas9)] knockout/knock-in methods (Kok et al., 2015; Lawson & Wolfe, 2011; Varshney, Sood, & Burgess, 2015). Because the zebrafish provides a robust regenerative model system, our methods could also be applied to cell ablation and regeneration experiments pertinent to musculoskeletal tissues.

4.1 MATERIALS

4.1.1 Preparation of embryos

1. 50X E3 Buffer: 250 mM NaCl, 8.5 mM KCl, 16.5 mM $CaCl_2$, 16.5 mM $MgSO_4$ in distilled water. Prepare as 50X stock and dilute to 1X working solution in distilled water. Store at room temperature.
2. 0.003% *N*-Phenylthiourea (PTU; Sigma–Aldrich #P7629) in E3 Buffer. To dissolve, mix on stir plate overnight at room temperature. Store at room temperature. Use caution—PTU is a toxic chemical and is an irritant to the skin.
3. Pronase (Sigma–Aldrich #10165921001): prepare as 50 mg/mL stock in distilled water. Store at −20°C.
4. 4% paraformaldehyde (PFA): 4% (wt/vol) solution of PFA (Electron Microscopy Sciences #RT19202) in 1X phosphate buffered saline (PBS). To dissolve, heat to 95°C (do not boil) with constant mixing on a stir plate. Store at −20°C. Thaw fresh in 37°C water bath before use. Use caution—PFA is a corrosive, toxic chemical and is an irritant to the skin, eyes, and respiratory tract.

5. 10X PBS: 1.37 M NaCl, 27 mM KCl, 100 mM Na_2HPO_4, 18 mM KH_2PO_4 in distilled water. Adjust the pH to 7.4. Autoclave and store at room temperature. Prepare as 10X stock and dilute to 1X working solution in distilled water.
6. 1X PBS with 0.1% Tween-20. Store at room temperature.
7. Tween-20 (Sigma–Aldrich #P1379).
8. Bleach solution: 0.8% KOH, 0.9% H_2O_2 (Fisher Scientific #H325) in distilled water. Prepare fresh on the day of use. Use caution—hydrogen peroxide (H_2O_2) is a corrosive, toxic chemical and is an irritant to the skin and eyes.
9. Methanol. Use caution—methanol is a flammable, toxic chemical and is an irritant to the skin and eyes; it is harmful to organs if ingested.

4.1.2 Fluorescent in situ hybridization and immunohistochemistry

1. PBS. See previous section.
2. 1X PBS with 0.25% Tween-20 (PBST). Store at room temperature. NB: *A higher concentration of Tween-20 is used for the FISH portion of the protocol.*
3. Methanol. See previous section.
4. Proteinase K (Sigma–Aldrich #3115879001): prepare as 10 mg/mL stock in distilled water. Store at −20°C.
5. 0.2% glutaraldehyde (Sigma–Aldrich #G6257)/4% PFA/1X PBS. Prepare fresh day of use. Use caution—glutaraldehyde is a toxic chemical and is an irritant to the skin, eyes, and respiratory tract.
6. Hybridization solution: 50% formamide (Fisher Scientific #F841), 5X SSC pH 4.5, 2% sodium dodecyl sulfate (SDS), 2% blocking reagent (Sigma–Aldrich #11096176001), 250 μg/mL yeast RNA (Life Technologies #15401029), 100 μg/mL heparin (Fisher Scientific #BP2524100). Store solution at −20°C. Use caution—formamide is a toxic chemical and is an irritant to the skin, eyes, and respiratory tract.
7. 20X SSC Buffer: 3 M NaCl, 300 mM sodium citrate in distilled water. Adjust the pH to 4.5 with citric acid. Prepare as 20X stock and dilute to 5X and 2X working solutions in distilled water. Store at room temperature.
8. SDS: prepare as 10% stock in distilled water. Store at room temperature. Use caution—SDS is a flammable, toxic chemical and is an irritant to the skin, eyes, and respiratory tract; it is harmful to organs if ingested.
9. Dextran sulfate (Sigma–Aldrich #S4030).
10. Solution I: 50% formamide, 2× SSC pH 4.5, 1% SDS in distilled water. Store at 4°C.
11. 2% H_2O_2: dilute 30% stock 1:15 in respective solution. Prepare fresh day of use.
12. TNT: 0.1 M Tris–HCl pH 7.5, 0.15 M NaCl, 0.5% Tween-20 in distilled water. Store at room temperature.
13. TBSTB: TNT with 0.5% blocking reagent. Prepare fresh day of use and store at 4°C for short-term storage.
14. Anti-digoxigenin–peroxidase (Sigma–Aldrich #11207733910).
15. Anti-fluorescein–peroxidase (Sigma–Aldrich #11426346910).

16. TSA plus cyanine 3 (Cy3) and fluorescein system (Perkin Elmer #NEL753001KT).
17. TSA plus Cy5 system (Perkin Elmer #NEL745001KT).
18. Newborn calf serum (NBCS; ThermoFisher Scientific #16010159). Store at −20°C and thaw in 37°C water bath prior to use. Prepare as 10% solution in PBST fresh day of use and store at 4°C for short-term storage.
19. *Optional:* Hoechst 33342 (Invitrogen #H1399).
20. *Optional:* DAPI (Invitrogen #D1306).

4.1.3 Preparation of embryos for imaging

1. Glycerol (Sigma–Aldrich #G7757).
2. Glass microscope slides (Fisher Scientific #12–550-A3).
3. Microscope cover glass, 18 × 18 mm (Fisher Scientific # 12–542A).
4. Modeling clay nonhardening (EZ Shape).
5. Probe (Fine Science Tools #10140-01).
6. Dumont #55 forceps (Fine Science Tools #11255-20).
7. *Optional*: glass bottom culture dishes (MatTek #P50G-1.5–14F).

4.2 METHODS

4.2.1 Breeding of embryos

Zebrafish husbandry is performed as described in *The Zebrafish Book* (Westerfield, 1995). Eggs are collected from mating pairs approximately 10–30 min after laying. The embryos develop in E3 Buffer at 28.5°C. In embryos >24 hpf, pigmentation is inhibited by the prior addition of 0.003% PTU to the E3 Buffer at 15–22 hpf. The chorions are removed by incubation with 1 mg/mL pronase for 20 min at 28.5°C, and embryos are rinsed twice with E3 Buffer. Alternatively, chorions may be removed using forceps. The embryos continue to develop in E3 Buffer at 28.5°C until the desired developmental stage. For observing developing cranial tendons, 48–72 hpf are ideal stages.

4.2.2 Fixation, bleaching, and dehydration of embryos

1. Transfer the embryos to 1.5 mL eppendorf tubes.
2. Embryos are euthanized using a buffered tricaine solution.
3. Fix the euthanized embryos in 4% PFA/1X PBS overnight with gentle rocking (BioExpress GeneMate Rocker #R-3200-1) at 4°C. The fixation step preserves the morphology of the tissue, but overfixation will result in a reduced hybridization signal.
4. Rinse and wash embryos twice in PBS with 0.1% Tween-20 for 5 min each at room temperature. Tween-20 is a mild surfactant that helps to prevent the embryos from sticking to the sides of the tube.
5. Wash embryos in bleach solution with rocking at room temperature until the melanin pigment of the embryos is mostly lost. The bleaching interval

varies between 5 and 20 min depending on the developmental stage of the embryo.

6. Rinse embryos in PBS with 0.1% Tween-20.
7. Fix the embryos in 4% PFA/1X PBS for at least 1 h with rocking at room temperature.
8. Dehydrate the embryos into 100% methanol through a graded methanol/PBS with 0.1% Tween-20 series (25%, 50%, 75% methanol in PBS with 0.1% Tween-20), allowing 5 min with each solution at room temperature. Wash embryos twice in 100% methanol for 5 min each at room temperature.
9. Store in 100% methanol at −20°C for at least 2 h (and up to several months) prior to proceeding with in situ hybridization.

4.2.3 Riboprobe generation

To detect two different RNA transcripts, synthesize the probes with different epitopes: digoxigenin–UTP and fluorescein–UTP. Instructions for synthesis of digoxigenin-labeled (Sigma–Aldrich #11175025910) and fluorescein-labeled (Sigma–Aldrich #11685619910) riboprobes may be obtained from the manufacturer. The fluorescein fluorophore is stronger than the Cy3 fluorophore, so we recommend synthesizing the riboprobe of the transcript with the weakest signal with digoxigenin-labeled NTPs, followed by detection of the riboprobe with anti-digoxigenin–peroxidase and amplification of the signal with fluorescein–Tyr. Conversely, the riboprobe of the transcript with a stronger signal should be synthesized with fluorescein-labeled NTPs, followed by detection of the riboprobe with anti-fluorescein–peroxidase and amplification of the signal with Cy3–Tyr. We perform the protocol with the antibodies in the following order—Tyr–Cy3, Tyr–Fluorescein, and Tyr–Cy5. We have never detected fluorescence from the fluorescein-labeled riboprobe alone, indicating that the fluorescein-labeled riboprobe does not interfere with fluorescein–Tyr signal detection.

4.2.4 Day 1. Hybridization of riboprobe(s)

1. Rehydrate the embryos into PBST through a graded methanol/PBST series (75%, 50%, 25% methanol in PBST) with 5 min each at room temperature. Wash 4× 5 min each in PBST at room temperature.
2. Incubate the embryos with proteinase K in PBST to permeabilize the tissue to allow access to the riboprobe(s). Optimal concentration and duration is dependent on the tissue type, activity of the enzyme, and developmental stage of the embryo (suggestions mentioned in the following are for embryos to be processed for both FISH and IHC). If your procedure does not include antibody staining, we recommend using concentrations and durations known to be effective. Insufficient treatment will reduce hybridization signal; whereas overly aggressive treatment will result in reduced tissue integrity, which may lead to shearing of specimen in subsequent washing steps and/or problems with the antibody detection.

Stage (hpf)	Time	Concentration (μg/mL)
24	10 min	1
30	20 min	1
36	30 min	1
48	45 min	1
55	1 h	1
60	25 min	10
72	30 min	10
>72	~1 h	10

3. Fix the embryos in 0.2% glutaraldehyde/4% PFA/1X PBS for 20 min with rocking at room temperature to stop proteinase K digestion.
4. Wash embryos 4× 5 min each in PBST at room temperature.
5. Incubate embryos with prewarmed hybridization solution containing 5% dextran sulfate for at least 1 h with rocking at 68–70°C. Dextran sulfate increases effective riboprobe concentration, thereby promoting hybridization (Matthiesen & Hansen, 2012; Wahl, Stern, & Stark, 1979).
6. Prepare a mix of 1 μL riboprobe (no more than 1 μg of each probe) per 100 μL prewarmed hybridization solution containing 5% dextran sulfate for each sample. Incubate the riboprobe/hybridization mix for 5–10 min at 68–70°C.
7. Incubate each sample with sufficient riboprobe/hybridization mix to completely submerge the embryos.
8. Incubate samples overnight with rocking at 68–70°C.

4.2.5 Day 2. Posthybridization washes and antibody incubation for the first probe

1. Remove riboprobe/hybridization mix and store at −20°C for future reuse.
2. Rinse and wash samples 4× 30 min each in Solution I with rocking at 68–70°C.
3. Wash samples 2× 10 min each in PBST at room temperature.
4. Incubate samples in 2% H_2O_2/PBST for 1 h with rocking at room temperature.
5. Wash samples 4× 5 min each in TNT at room temperature.
6. Incubate samples in TBSTB blocking buffer for 1–4 h with rocking at room temperature.
7. Replace blocking buffer with anti-fluorescein–POD diluted 1:4000 (if double FISH) or anti-digoxigenin–peroxidase diluted 1:1000 (if single FISH) in TBSTB blocking buffer.
8. Incubate samples overnight with rocking at 4°C.

4.2.6 Day 3. Antibody detection of first probe and antibody incubation for second probe

1. Wash 8X in TNT over the course of 1–2 h at room temperature with rocking.
2. Wash 5 min in 50 μL amplification diluent at room temperature.

3. Prepare Tyr–Cy3 (if double FISH) or Tyr–Fluorescein (if single FISH) diluted 1:50 in amplification diluent, using 50 μL per sample. Centrifuge each solution for 3 min at high speed in a microcentrifuge to pellet any precipitate. Avoid pipetting from bottom of tube in the next step.
4. Incubate each sample in freshly prepared 50 μL Tyr–Cy3 or Tyr–Fluorescein mixture for 1 h (do not exceed time) in the dark with tubes upright on a rocking platform at room temperature. *All subsequent steps in the protocol are performed in the dark* to avoid photobleaching of fluorophores. Incubation time is dependent on the riboprobe and may be shortened to increase signal-to-background ratio.
5. Wash each sample 2× 5 min each in TNT at room temperature.
6. Incubate in 2% H_2O_2/TNT for 1 h with rocking at room temperature.
   (If doing single FISH and IHC, proceed to the IHC section.)
7. Wash samples 4× 5 min each in TNT at room temperature.
8. Incubate in TBSTB blocking buffer for 1–4 h with rocking at room temperature.
9. Replace the blocking buffer with anti-digoxigenin–peroxidase diluted 1:1000 in TBSTB blocking buffer.
10. Incubate samples overnight with rocking at 4°C.

4.2.7 Day 4. Antibody detection of second probe

1. Wash samples 8X with TNT over the course of 1–2 h at room temperature with rocking.
2. Wash samples for 5 min in 50 μL amplification diluent at room temperature.
3. Dilute Tyr–Fluorescein 1:50 with amplification diluent, allowing 50 μL per sample. Centrifuge the solution for 3 min at high speed in a microcentrifuge to pellet any precipitate. Avoid pipetting from bottom of tube in the next step.
4. Incubate each sample in freshly prepared 50 μL Tyr–Fluorescein mixture for 1 h (do not exceed time) in the dark with tubes upright on rocking platform at room temperature. Wash samples 2× 5 min each in TNT at room temperature.
5. Incubate samples in 2% H_2O_2/TNT for 1 h rocking at room temperature.
   (If doing double FISH and IHC, go to IHC section.)
6. Wash samples 2× 5 min each in TNT at room temperature.
7. Wash samples 2× 5 min each in PBST at room temperature.
8. Store the embryos in PBST at 4°C.

4.2.8 Day 5. Immunohistochemistry—primary antibody

Detection of protein begins after completion of the peroxide inactivation step of the FISH protocol. The IHC protocol may need to be optimized for each antibody—troubleshooting may involve adjusting variables such as the blocking buffer, proteinase K digestion, and antibody dilution.

1. Wash samples 2× 5 min each in TNT at room temperature.
2. Wash samples 2× 5 min each in PBST at room temperature.

3. Incubate samples in 10% NBCS/PBST blocking buffer for 1 h with rocking at room temperature.
4. Replace solution with primary antibody diluted in 10% NBCS/PBST blocking buffer. If using peroxidase-conjugated secondary antibodies, which we recommend for proteins that are expressed at low levels, dilute the primary antibodies 1:500. If using Alexa Fluor–conjugated secondary antibodies, dilute the primary antibodies 1:100. These dilutions are optimized for MYH antibodies A4.1025 and MF20 (Developmental Studies Hybridoma Bank, Iowa City, Iowa); other antibodies may require additional optimization.
5. Incubate samples overnight with rocking at 4°C.

4.2.9 Day 6. Immunohistochemistry—secondary antibody

1. Rinse and wash samples 4× 30 min each in PBST at room temperature.
2. Incubate samples in 10% NBCS/PBST blocking buffer for 1 h with rocking at room temperature.
3. Replace solution with secondary antibody diluted in 10% NBCS/PBST blocking buffer. If using peroxidase-conjugated secondary antibody, dilute 1:500. If using Alexa Fluor–conjugated secondary antibodies, dilute 1:400. To confirm the specificity of the secondary antibody for the primary antibody, prepare control samples containing the secondary alone.
4. Incubate overnight with rocking at 4°C.

4.2.10 Day 7. Detection of secondary antibody

1. Rinse and wash samples 4× 30 min each in PBST at room temperature.

   (If using an Alexa Fluor–conjugated secondary antibody, go to step 8. Continue to step 2 if using a peroxidase-conjugated secondary antibody.)
2. Wash samples for 5 min in 50 μL amplification diluent at room temperature.
3. Prepare Tyr–Cy5 diluted 1:50 in amplification diluent, using 50 μL per sample. Spin down for 3 min at high speed in a microcentrifuge to pellet any precipitate. Avoid pipetting from the bottom of the tube in the next step.
4. Incubate each sample in freshly prepared 50 μL Tyr–Cy5 mixture for 30 min in the dark with tubes upright on a rocking platform at room temperature.
5. Wash samples 6× 10 min each in PBST at room temperature.
6. Fix the embryos in 4% PFA/1X PBS for 30 min with rocking at room temperature.
7. Rinse and wash samples 2× 5 min each in PBST at room temperature.
8. *Optional*: To counterstain the embryos, incubate in Hoechst 33342 diluted 1:2000 or DAPI diluted 1:1000 in PBST for 30 min with rocking at room temperature. Wash samples 2× 5 min each in PBST at room temperature.
9. Store the embryos in PBST at 4°C.

4.3 PREPARATION OF EMBRYOS FOR IMAGING

1. Equilibrate the embryos into 100% glycerol for imaging.
2. Dissect the head region of an embryo away from the trunk and yolk sac with forceps.
3. Prepare a coverslip by applying clay around edges to create a slight elevation between the microscope slide and the coverslip when mounting the embryos.
4. Place a small drop of glycerol onto a microscope slide and transfer the head region of the embryo to the center. Orient the embryo to the desired orientation with forceps and apply the clay-covered coverslip.
5. As an alternative, glass bottom culture dishes may be used for mounting on inverted microscopes. This method does not necessitate clay-covered coverslips.
6. High-magnification images are taken with a Zeiss LSM710 NLO or comparable laser-scanning confocal microscope. Digitized images are saved as TIFF files and subsequently processed in Photoshop.

CONCLUSION

This chapter describes the zebrafish as a model for the study of tendon development and details a method to examine the expression of mRNA and protein in the forming zebrafish musculoskeletal system. The recent development of high-throughput platforms has prompted the use of small molecule chemical screening to identify novel regulators of developmental processes in zebrafish—these molecules may have therapeutic benefits (Rennekamp & Peterson, 2015). The genetic tools available in the zebrafish, in combination with transgenic approaches, make the zebrafish system amenable for rapid functional analysis of candidate molecules and target pathways. In addition, the robust regenerative capacity of zebrafish will facilitate future studies aimed at understanding the molecular and cellular mechanisms of tendon healing and regeneration. We believe such studies will lead to new discoveries that can be placed within the larger framework of vertebrate musculoskeletal formation and serve to advance the development of regenerative medicine–based solutions to tendon injuries and disease.

ACKNOWLEDGMENTS

J.WC. is supported by National Institutes of Health grant PO1 DK056246 and by a National Science Foundation Predoctoral Fellowship. J.L.G. is supported by a grant from the Eunice Kennedy Shriver National Institute of Child Health and Human Development (K99/R00HD069533), the Charles H. Hood Foundation, and the National Institute of Dental and Craniofacial Research (R03 DE024771).

REFERENCES

Alberton, P., Popov, C., Pragert, M., Kohler, J., Shukunami, C., Schieker, M., & Docheva, D. (2012). Conversion of human bone marrow-derived mesenchymal stem cells into tendon progenitor cells by ectopic expression of scleraxis. *Stem Cells and Development, 21*(6), 846–858. http://dx.doi.org/10.1089/scd.2011.0150.

Berger, J., Berger, S., Hall, T. E., Lieschke, G. J., & Currie, P. D. (2010). Dystrophin-deficient zebrafish feature aspects of the Duchenne muscular dystrophy pathology. *Neuromuscular Disorders, 20*(12), 826–832. http://dx.doi.org/10.1016/j.nmd.2010.08.004.

Bird, N. C., & Mabee, P. M. (2003). Developmental morphology of the axial skeleton of the zebrafish, *Danio rerio* (Ostariophysi: Cyprinidae). *Developmental Dynamics, 228*(3), 337–357. http://dx.doi.org/10.1002/dvdy.10387.

Birk, D. E., Zycband, E. I., Woodruff, S., Winkelmann, D. A., & Trelstad, R. L. (1997). Collagen fibrillogenesis in situ: fibril segments become long fibrils as the developing tendon matures. *Developmental Dynamics, 208*(3), 291–298. http://dx.doi.org/10.1002/(SICI)1097-0177(199703)208:3<291::AID-AJA1>3.0.CO;2-D.

Blitz, E., Sharir, A., Akiyama, H., & Zelzer, E. (2013). Tendon-bone attachment unit is formed modularly by a distinct pool of Scx- and Sox9-positive progenitors. *Development, 140*(13), 2680–2690. http://dx.doi.org/10.1016/j.devcel.2009.10.010.

Brand-Saberi, B., & Christ, B. (2000). Evolution and development of distinct cell lineages derived from somites. *Current Topics in Developmental Biology, 48*, 1–42.

Brent, A. E., Braun, T., & Tabin, C. J. (2005). Genetic analysis of interactions between the somitic muscle, cartilage and tendon cell lineages during mouse development. *Development, 132*(3), 515–528. http://dx.doi.org/10.1242/dev.01605.

Brent, A. E., Schweitzer, R., & Tabin, C. J. (2003). A somitic compartment of tendon progenitors. *Cell, 113*(2), 235–248.

Brent, A. E., & Tabin, C. J. (2004). FGF acts directly on the somitic tendon progenitors through the Ets transcription factors Pea3 and Erm to regulate scleraxis expression. *Development, 131*(16), 3885–3896. http://dx.doi.org/10.1242/dev.01275.

Bricard, Y., Ralliere, C., Lebret, V., Lefevre, F., & Rescan, P. Y. (2014). Early fish myoseptal cells: insights from the trout and relationships with amniote axial tenocytes. *PLoS One, 9*(3), e91876. http://dx.doi.org/10.1371/journal.pone.0091876.

Butler, D. L., Grood, E. S., Noyes, F. R., & Zernicke, R. F. (1978). Biomechanics of ligaments and tendons. *Exercise and Sport Sciences Reviews, 6*, 125–181.

Charvet, B., Guiraud, A., Malbouyres, M., Zwolanek, D., Guillon, E., Bretaud, S., … Ruggiero, F. (2013). Knockdown of col22a1 gene in zebrafish induces a muscular dystrophy by disruption of the myotendinous junction. *Development, 140*(22), 4602–4613. http://dx.doi.org/10.1242/dev.096024.

Charvet, B., Malbouyres, M., Pagnon-Minot, A., Ruggiero, F., & Le Guellec, D. (2011). Development of the zebrafish myoseptum with emphasis on the myotendinous junction. *Cell and Tissue Research, 346*(3), 439–449. http://dx.doi.org/10.1007/s00441-011-1266-7.

Chen, J. W., & Galloway, J. L. (2014). The development of zebrafish tendon and ligament progenitors. *Development, 141*(10), 2035–2045. http://dx.doi.org/10.1242/dev.104067.

Christ, B., Huang, R., & Wilting, J. (2000). The development of the avian vertebral column. *Anatomy and Embryology, 202*(3), 179–194.

Christ, B., & Ordahl, C. P. (1995). Early stages of chick somite development. *Anatomy and Embryology, 191*(5), 381–396.

Clement, A., Wiweger, M., von der Hardt, S., Rusch, M. A., Selleck, S. B., Chien, C. B., & Roehl, H. H. (2008). Regulation of zebrafish skeletogenesis by ext2/dackel and papst1/pinscher. *PLoS Genetics, 4*(7), e1000136. http://dx.doi.org/10.1371/journal.pgen.1000136.

Connizzo, B. K., Yannascoli, S. M., & Soslowsky, L. J. (2013). Structure-function relationships of postnatal tendon development: a parallel to healing. *Matrix Biology, 32*(2), 106–116. http://dx.doi.org/10.1016/j.matbio.2013.01.007.

Couly, G. F., Coltey, P. M., & Le Douarin, N. M. (1993). The triple origin of skull in higher vertebrates: a study in quail-chick chimeras. *Development, 117*(2), 409–429.

Devoto, S. H., Melancon, E., Eisen, J. S., & Westerfield, M. (1996). Identification of separate slow and fast muscle precursor cells in vivo, prior to somite formation. *Development, 122*(11), 3371–3380.

Docheva, D., Hunziker, E. B., Fassler, R., & Brandau, O. (2005). Tenomodulin is necessary for tenocyte proliferation and tendon maturation. *Molecular and Cellular Biology, 25*(2), 699–705. http://dx.doi.org/10.1128/MCB.25.2.699-705.2005.

Edom-Vovard, F., Schuler, B., Bonnin, M. A., Teillet, M. A., & Duprez, D. (2002). Fgf4 positively regulates scleraxis and tenascin expression in chick limb tendons. *Developmental Biology, 247*(2), 351–366.

Felsenfeld, A. L., Curry, M., & Kimmel, C. B. (1991). The fub-1 mutation blocks initial myofibril formation in zebrafish muscle pioneer cells. *Developmental Biology, 148*(1), 23–30. http://dx.doi.org/10.1016/0012-1606(91)90314-S.

Gaut, L., & Duprez, D. (2016). Tendon development and diseases. *Wiley Interdisciplinary Reviews. Developmental Biology, 5*(1), 5–23. http://dx.doi.org/10.1002/wdev.201.

Gomoll, A. H., Katz, J. N., Warner, J. J., & Millett, P. J. (2004). Rotator cuff disorders: recognition and management among patients with shoulder pain. *Arthritis Rheum, 50*(12), 3751–3761. http://dx.doi.org/10.1002/art.20668.

Grenier, J., Teillet, M. A., Grifone, R., Kelly, R. G., & Duprez, D. (2009). Relationship between neural crest cells and cranial mesoderm during head muscle development. *PLoS One, 4*(2), e4381. http://dx.doi.org/10.1371/journal.pone.0004381.

Grotmol, S., Kryvi, H., Nordvik, K., & Totland, G. K. (2003). Notochord segmentation may lay down the pathway for the development of the vertebral bodies in the Atlantic salmon. *Anatomy and Embryology, 207*(4–5), 263–272. http://dx.doi.org/10.1007/s00429-003-0349-y.

Guerquin, M. J., Charvet, B., Nourissat, G., Havis, E., Ronsin, O., Bonnin, M. A., … Duprez, D. (2013). Transcription factor EGR1 directs tendon differentiation and promotes tendon repair. *Journal of Clinical Investigation, 123*(8), 3564–3576. http://dx.doi.org/10.1172/JCI67521.

Hall, T. E., Bryson-Richardson, R. J., Berger, S., Jacoby, A. S., Cole, N. J., Hollway, G. E., … Currie, P. D. (2007). The zebrafish candyfloss mutant implicates extracellular matrix adhesion failure in laminin alpha2-deficient congenital muscular dystrophy. *Proceedings of the National Academy of Sciences of the United States of America, 104*(17), 7092–7097. http://dx.doi.org/10.1073/pnas.0700942104.

Hatta, K., Bremiller, R., Westerfield, M., & Kimmel, C. B. (1991). Diversity of expression of engrailed-like antigens in zebrafish. *Development, 112*(3), 821–832.

Henry, C. A., McNulty, I. M., Durst, W. A., Munchel, S. E., & Amacher, S. L. (2005). Interactions between muscle fibers and segment boundaries in zebrafish. *Developmental Biology, 287*(2), 346–360. http://dx.doi.org/10.1016/j.ydbio.2005.08.049.

Huang, A. H., Riordan, T. J., Pryce, B., Weibel, J. L., Watson, S. S., Long, F., ... Schweitzer, R. (2015). Musculoskeletal integration at the wrist underlies the modular development of limb tendons. *Development, 142*(14), 2431–2441. http://dx.doi.org/10.1242/dev.122374.

Hurle, J. M., Ros, M. A., Ganan, Y., Macias, D., Critchlow, M., & Hinchliffe, J. R. (1990). Experimental analysis of the role of ECM in the patterning of the distal tendons of the developing limb bud. *Cell Differentiation and Development, 30*(2), 97–108.

Kalson, N. S., Lu, Y., Taylor, S. H., Starborg, T., Holmes, D. F., & Kadler, K. E. (2015). A structure-based extracellular matrix expansion mechanism of fibrous tissue growth. *eLife, 4*. http://dx.doi.org/10.7554/eLife.05958.

Kardon, G. (1998). Muscle and tendon morphogenesis in the avian hind limb. *Development, 125*(20), 4019–4032.

Kaufman, C. K., White, R. M., & Zon, L. (2009). Chemical genetic screening in the zebrafish embryo. *Nature Protocols, 4*(10), 1422–1432. http://dx.doi.org/10.1038/nprot.2009.144.

Kawahara, G., Karpf, J. A., Myers, J. A., Alexander, M. S., Guyon, J. R., & Kunkel, L. M. (2011). Drug screening in a zebrafish model of Duchenne muscular dystrophy. *Proceedings of the National Academy of Sciences of the United States of America, 108*(13), 5331–5336. http://dx.doi.org/10.1073/pnas.1102116108.

Kawahara, G., & Kunkel, L. M. (2013). Zebrafish based small molecule screens for novel DMD drugs. *Drug Discovery Today. Technologies, 10*(1), e91–e96. http://dx.doi.org/10.1016/j.ddtec.2012.03.001.

Kiapour, A. M., Fleming, B. C., & Murray, M. M. (2015). Biomechanical outcomes of bridge-enhanced anterior cruciate ligament repair are influenced by sex in a preclinical model. *Clinical Orthopaedics and Related Research, 473*(8), 2599–2608. http://dx.doi.org/10.1007/s11999-015-4226-9.

Kieny, M., & Chevallier, A. (1979). Autonomy of tendon development in the embryonic chick wing. *Journal of Embryology and Experimental Morphology, 49*, 153–165.

Kok, F. O., Shin, M., Ni, C. W., Gupta, A., Grosse, A. S., van Impel, A., ... Lawson, N. D. (2015). Reverse genetic screening reveals poor correlation between morpholino-induced and mutant phenotypes in zebrafish. *Developmental Cell, 32*(1), 97–108. http://dx.doi.org/10.1016/j.devcel.2014.11.018.

Komi, P. V. (1990). Relevance of in vivo force measurements to human biomechanics. *Journal of Biomechanics, 23*(Suppl. 1), 23–34.

Kontges, G., & Lumsden, A. (1996). Rhombencephalic neural crest segmentation is preserved throughout craniofacial ontogeny. *Development, 122*(10), 3229–3242.

Lawson, N. D., & Wolfe, S. A. (2011). Forward and reverse genetic approaches for the analysis of vertebrate development in the zebrafish. *Developmental Cell, 21*(1), 48–64. http://dx.doi.org/10.1016/j.devcel.2011.06.007.

Le Lievre, C. S. (1978). Participation of neural crest-derived cells in the genesis of the skull in birds. *Journal of Embryology and Experimental Morphology, 47*, 17–37.

Lejard, V., Brideau, G., Blais, F., Salingcarnboriboon, R., Wagner, G., Roehrl, M. H., ... Rossert, J. (2007). Scleraxis and NFATc regulate the expression of the pro-alpha1(I) collagen gene in tendon fibroblasts. *Journal of Biological Chemistry, 282*(24), 17665–17675. http://dx.doi.org/10.1074/jbc.M610113200.

Matthiesen, S. H., & Hansen, C. M. (2012). Fast and non-toxic in situ hybridization without blocking of repetitive sequences. *PLoS One, 7*(7), e40675. http://dx.doi.org/10.1371/journal.pone.0040675.

Murchison, N. D., Price, B. A., Conner, D. A., Keene, D. R., Olson, E. N., Tabin, C. J., & Schweitzer, R. (2007). Regulation of tendon differentiation by scleraxis distinguishes force-transmitting tendons from muscle-anchoring tendons. *Development, 134*(14), 2697–2708. http://dx.doi.org/10.1242/dev.001933.

Noden, D. M. (1978). The control of avian cephalic neural crest cytodifferentiation. I. Skeletal and connective tissues. *Developmental Biology, 67*(2), 296–312.

Nourissat, G., Diop, A., Maurel, N., Salvat, C., Dumont, S., Pigenet, A., … Berenbaum, F. (2010). Mesenchymal stem cell therapy regenerates the native bone-tendon junction after surgical repair in a degenerative rat model. *PLoS One, 5*(8), e12248. http://dx.doi.org/10.1371/journal.pone.0012248.

Nursall, J. R. (1956). The lateral musculature and the swimming of fish. *Proceedings of the Zoological Society of London, 126*, 127–144.

Otten, C., van der Ven, P. F., Lewrenz, I., Paul, S., Steinhagen, A., Busch-Nentwich, E., … Abdelilah-Seyfried, S. (2012). Xirp proteins mark injured skeletal muscle in zebrafish. *PLoS One, 7*(2), e31041. http://dx.doi.org/10.1371/journal.pone.0031041.

Pearse, R. V., 2nd, Scherz, P. J., Campbell, J. K., & Tabin, C. J. (2007). A cellular lineage analysis of the chick limb bud. *Developmental Biology, 310*(2), 388–400. http://dx.doi.org/10.1016/j.ydbio.2007.08.002.

Pelled, G., Snedeker, J. G., Ben-Arav, A., Rigozzi, S., Zilberman, Y., Kimelman-Bleich, N., … Gazit, D. (2012). Smad8/BMP2-engineered mesenchymal stem cells induce accelerated recovery of the biomechanical properties of the Achilles tendon. *Journal of Orthopaedic Research, 30*(12), 1932–1939. http://dx.doi.org/10.1002/jor.22167.

Prockop, D. J., & Kivirikko, K. I. (1995). Collagens: molecular biology, diseases, and potentials for therapy. *Annual Review of Biochemistry, 64*, 403–434. http://dx.doi.org/10.1146/annurev.bi.64.070195.002155.

Rennekamp, A. J., & Peterson, R. T. (2015). 15 years of zebrafish chemical screening. *Current Opinion in Chemical Biology, 24*, 58–70. http://dx.doi.org/10.1016/j.cbpa.2014.10.025.

Rinon, A., Lazar, S., Marshall, H., Buchmann-Moller, S., Neufeld, A., Elhanany-Tamir, H., … Tzahor, E. (2007). Cranial neural crest cells regulate head muscle patterning and differentiation during vertebrate embryogenesis. *Development, 134*(17), 3065–3075. http://dx.doi.org/10.1242/dev.002501.

Schilling, T. F. (1997). Genetic analysis of craniofacial development in the vertebrate embryo. *Bioessays, 19*(6), 459–468. http://dx.doi.org/10.1002/bies.950190605.

Schilling, T. F., & Kimmel, C. B. (1994). Segment and cell type lineage restrictions during pharyngeal arch development in the zebrafish embryo. *Development, 120*(3), 483–494.

Schilling, T. F., & Kimmel, C. B. (1997). Musculoskeletal patterning in the pharyngeal segments of the zebrafish embryo. *Development, 124*(15), 2945–2960.

Schweitzer, J., Becker, T., Lefebvre, J., Granato, M., Schachner, M., & Becker, C. G. (2005). Tenascin-C is involved in motor axon outgrowth in the trunk of developing zebrafish. *Developmental Dynamics, 234*(3), 550–566. http://dx.doi.org/10.1002/dvdy.20525.

Schweitzer, R., Chyung, J. H., Murtaugh, L. C., Brent, A. E., Rosen, V., Olson, E. N., … Tabin, C. J. (2001). Analysis of the tendon cell fate using Scleraxis, a specific marker for tendons and ligaments. *Development, 128*(19), 3855–3866.

Schweitzer, R., Zelzer, E., & Volk, T. (2010). Connecting muscles to tendons: tendons and musculoskeletal development in flies and vertebrates. *Development, 137*(17), 2807–2817. http://dx.doi.org/10.1242/dev.047498.

Sharma, P., & Maffulli, N. (2005). Tendon injury and tendinopathy: healing and repair. *Journal of Bone and Joint Surgery American, 87*(1), 187–202. http://dx.doi.org/10.2106/JBJS.D.01850.

Shellswell, G. B., & Wolpert, L. (1977). The pattern of muscle and tendon development in the chick wing. In D. A. Ede, J. R. Hinchliffe, & M. Balls (Eds.), *Vertebrate limb and somite morphogenesis* (pp. 71–86). Cambridge: Cambridge University Press.

Shukunami, C., Takimoto, A., Oro, M., & Hiraki, Y. (2006). Scleraxis positively regulates the expression of tenomodulin, a differentiation marker of tenocytes. *Developmental Biology, 298*(1), 234–247. http://dx.doi.org/10.1016/j.ydbio.2006.06.036.

Silver, F. H., Freeman, J. W., & Seehra, G. P. (2003). Collagen self-assembly and the development of tendon mechanical properties. *Journal of Biomechanics, 36*(10), 1529–1553.

Smith, T. G., Sweetman, D., Patterson, M., Keyse, S. M., & Munsterberg, A. (2005). Feedback interactions between MKP3 and ERK MAP kinase control scleraxis expression and the specification of rib progenitors in the developing chick somite. *Development, 132*(6), 1305–1314. http://dx.doi.org/10.1242/dev.01699.

Snow, C. J., & Henry, C. A. (2009). Dynamic formation of microenvironments at the myotendinous junction correlates with muscle fiber morphogenesis in zebrafish. *Gene Expression Patterns, 9*(1), 37–42. http://dx.doi.org/10.1016/j.gep.2008.08.003.

Soeda, T., Deng, J. M., de Crombrugghe, B., Behringer, R. R., Nakamura, T., & Akiyama, H. (2010). Sox9-expressing precursors are the cellular origin of the cruciate ligament of the knee joint and the limb tendons. *Genesis, 48*(11), 635–644. http://dx.doi.org/10.1002/dvg.20667.

Stickney, H. L., Barresi, M. J., & Devoto, S. H. (2000). Somite development in zebrafish. *Developmental Dynamics, 219*(3), 287–303. http://dx.doi.org/10.1002/1097-0177(2000)9999:9999<::AID-DVDY1065>3.0.CO;2-A.

Subramanian, A., & Schilling, T. F. (2014). Thrombospondin-4 controls matrix assembly during development and repair of myotendinous junctions. *eLife, 3*. http://dx.doi.org/10.7554/eLife.02372.

Sugimoto, Y., Takimoto, A., Akiyama, H., Kist, R., Scherer, G., Nakamura, T., … Shukunami, C. (2013). Scx+/Sox9+ progenitors contribute to the establishment of the junction between cartilage and tendon/ligament. *Development, 140*(11), 2280–2288. http://dx.doi.org/10.1242/dev.096354.

Varshney, G. K., Sood, R., & Burgess, S. M. (2015). Understanding and editing the zebrafish genome. *Advances in Genetics, 92*, 1–52. http://dx.doi.org/10.1016/bs.adgen.2015.09.002.

Wachtler, F., Christ, B., & Jacob, H. J. (1981). On the determination of mesodermal tissues in the avian embryonic wing bud. *Anatomy and Embryology, 161*(3), 283–289.

Wahl, G. M., Stern, M., & Stark, G. R. (1979). Efficient transfer of large DNA fragments from agarose gels to diazobenzyloxymethyl-paper and rapid hybridization by using dextran sulfate. *Proceedings of the National Academy of Sciences of the United States of America, 76*(8), 3683–3687.

Wang, G., Achim, C. L., Hamilton, R. L., Wiley, C. A., & Soontornniyomkij, V. (1999). Tyramide signal amplification method in multiple-label immunofluorescence confocal microscopy. *Methods, 18*(4), 459–464. http://dx.doi.org/10.1006/meth.1999.0813.

Westerfield, M. (1995). *The zebrafish book. A guide for the laboratory use of zebrafish (*Danio rerio*)* (3rd ed.). Eugene, OR: University of Oregon Press.

Westneat, M. W., Hoese, W., Pell, C. A., & Wainwright, S. A. (1993). The horizontal septum: mechanisms of force transfer in locomotion of Scombrid fishes (Scombridae, Perciformes). *Journal of Morphology, 217*, 183–204.

Wortham, R. A. (1948). The development of the muscles and tendons in the lower leg and foot of chick embryos. *Journal of Morphology, 83*(1), 105–148.

Zhang, G., Young, B. B., Ezura, Y., Favata, M., Soslowsky, L. J., Chakravarti, S., & Birk, D. E. (2005). Development of tendon structure and function: regulation of collagen fibrillogenesis. *Journal of Musculoskeletal & Neuronal Interactions, 5*(1), 5–21.

CHAPTER

Small teleost fish provide new insights into human skeletal diseases

13

P.E. Witten*[,1], M.P. Harris[§], A. Huysseune*, C. Winkler[¶]

Ghent University, Ghent, Belgium

[§]*Harvard Medical School, Boston, MA, United States*

[¶]*National University of Singapore, Singapore, Singapore*

[1]*Corresponding author: E-mail: peckhardwitten@aol.com*

CHAPTER OUTLINE

Abstract

Small teleost fish such as zebrafish and medaka are increasingly studied as models for human skeletal diseases. Efficient new genome editing tools combined with advances in the analysis of skeletal phenotypes provide new insights into fundamental processes of skeletal development. The skeleton among vertebrates is a highly conserved organ system, but teleost fish and mammals have evolved unique traits or have lost particular skeletal elements in each lineage. Several unique features of the skeleton relate to the extremely small size of early fish embryos and the small size of adult fish used as models. A detailed analysis of the plethora of interesting skeletal phenotypes in zebrafish and

Methods in Cell Biology, Volume 138, ISSN 0091-679X, http://dx.doi.org/10.1016/bs.mcb.2016.09.001

medaka pushes available skeletal imaging techniques to their respective limits and promotes the development of new imaging techniques. Impressive numbers of zebrafish and medaka mutants with interesting skeletal phenotypes have been characterized, complemented by transgenic zebrafish and medaka lines. The advent of efficient genome editing tools, such as TALEN and CRISPR/Cas9, allows to introduce targeted deficiencies in genes of model teleosts to generate skeletal phenotypes that resemble human skeletal diseases. This review will also discuss other attractive aspects of the teleost skeleton. This includes the capacity for lifelong tooth replacement and for the regeneration of dermal skeletal elements, such as scales and fin rays, which further increases the value of zebrafish and medaka models for skeletal research.

INTRODUCTION

Small teleost fish such as zebrafish (*Danio rerio*) and medaka (*Oryzias latipes*) are studied as models for human skeletal diseases because basic skeletal components, their development, and essential developmental genes are conserved within osteichthyans; the group that encompasses teleost fish and mammals (Apschner, Schulte-Merker, & Witten, 2011; Harris, Henke, Hawkins, & Witten, 2014). Although variation exists within developmental networks among species, core functions of these networks are often conserved, even across phyla (Elinson & Kezmoh, 2010). This permits the study of genetic functions also between distantly related vertebrates such as teleost fish and mammals (Harris et al., 2014). The availability of mutant lines and transgenes allows unprecedented investigation into the mechanisms of skeletal development. The establishment of gene editing approaches via targeted endonuclease activity such as CRISPR/Cas9 or TALEN, further improves the applicability of organisms amenable to genetics (Hruscha et al., 2013; Hwang et al., 2013). Other advantages of small teleosts as models for skeletal research are external development, the high number of offspring, and short generation times. Yet, these advantages also impose challenges (Bolker, 2014). For instance teleost embryos hatch early and embryonic development continues after hatching outside the egg. This is often wrongly addressed as larval phase (Balon, 1990). Skeletal development in free living embryos is influenced by an array of epigenetic factors (Witten & Hall, 2015), at a time when amniote embryos continue skeletal development inside the egg or the uterus. This example emphasizes the need to understand what aspects of skeletal development may be unique to each lineage due to varied life history strategies (Hall, 2015; Huysseune, 2000; Witten & Huysseune, 2009). Considering the evolutionary relationship between teleost fish and humans is another prerequisite for successful skeletal research. Extinct basal gnathostome bony fish are the common ancestors of tetrapods and ray-finned fish (Fig. 1A). Subsequently parallel lines of evolution gave rise to mammals and teleost fish. The latter is the dominant vertebrate group on the planet with about 30,000 species (Maisey, 2000). Zebrafish and medaka are not in the line of mammalian ancestors. It thus is incorrect to assume that the mammalian skeleton is advanced compared to the teleost skeleton (Metscher & Ahlberg, 1999; Witten & Huysseune, 2009). Characters of the teleost skeleton can be more advanced or elaborate compared to mammals. For example, the teleost skull contains at least twice the number of skeletal elements compared to the mammalian skull (Gregory, 1933; Owen, 1854). The absence of osteocytes in medaka is an advanced character whereas zebrafish and mammals maintain osteocytes, a primitive character on the evolutionary scale (Meunier, 1989; Witten, Hansen, & Hall, 2001). The diversity of skeletal characters among vertebrates permits an analysis of the potential within skeletal developmental programs. Consequently, differences in skeletal development between zebrafish and medaka and humans can reveal alternative pathways worth to be explored to improve our understanding of human skeletal diseases.

1. A FRESH VIEW ON THE TELEOST SKELETON AND ITS SPECIAL CHARACTERS

1.1 THE DERMAL SKELETON

Vertebrates have two skeletal systems, the dermal skeleton (exoskeleton) and the endoskeleton. A mineralized dermal skeleton evolved earlier than a mineralized endoskeleton (Janvier, 2015; Maisey, 2000). The principal unit of the dermal skeleton is the odontode. Odontodes are made from dentin that encloses a pulp cavity and are covered with a hypermineralized layer (enamel or enameloid), and ankylosed to bone (Reif, 1982; Sire & Huysseune, 2003) (Fig. 1B). The teeth of all vertebrates including humans represent odontode derivatives. In addition, teleost fish have scales and dermal fin rays, structures that evolved from odontodes and are lacking in mammals (Sire & Huysseune, 2003). Development of teeth, scales, and fin rays in zebrafish and medaka depends on the expression of ectodysplasin (*eda*) and ectodysplasin receptor (*edar*) genes (Harris et al., 2008; Kondo et al., 2001). All elements of the dermal skeleton can be replaced throughout life. Teleost fish continuously replace their teeth (see Section 5) and are capable of scale and fin ray regeneration (Sire & Akimenko, 2004). It is important to understand that if we study fin ray regeneration in a teleost model we look at dermal skeletal structures that do not occur in mammals (Sire, Donoghue, & Vickaryous, 2009; Smith & Hall, 1990). Despite such caveats, zebrafish is an excellent model to study the basic mechanisms of skeletal regeneration. Recent studies have revealed the similarities between fin ray regeneration in zebrafish and skull vault defect repair (Geurtzen et al., 2014; Knopf et al., 2011; Sousa et al., 2011). These studies provide additional evidence for regeneration via skeletal cell dedifferentiation and redifferentiation and confirm the mechanisms explained and reviewed by Hall (1970).

1.2 ENDOSKELETAL CARTILAGE AND BONE

The endoskeleton of mammals and teleosts is inherited from their common bony fish (osteichthyan) ancestors (Donoghue, 2006; Maisey, 2000) (Fig. 1). The teleost

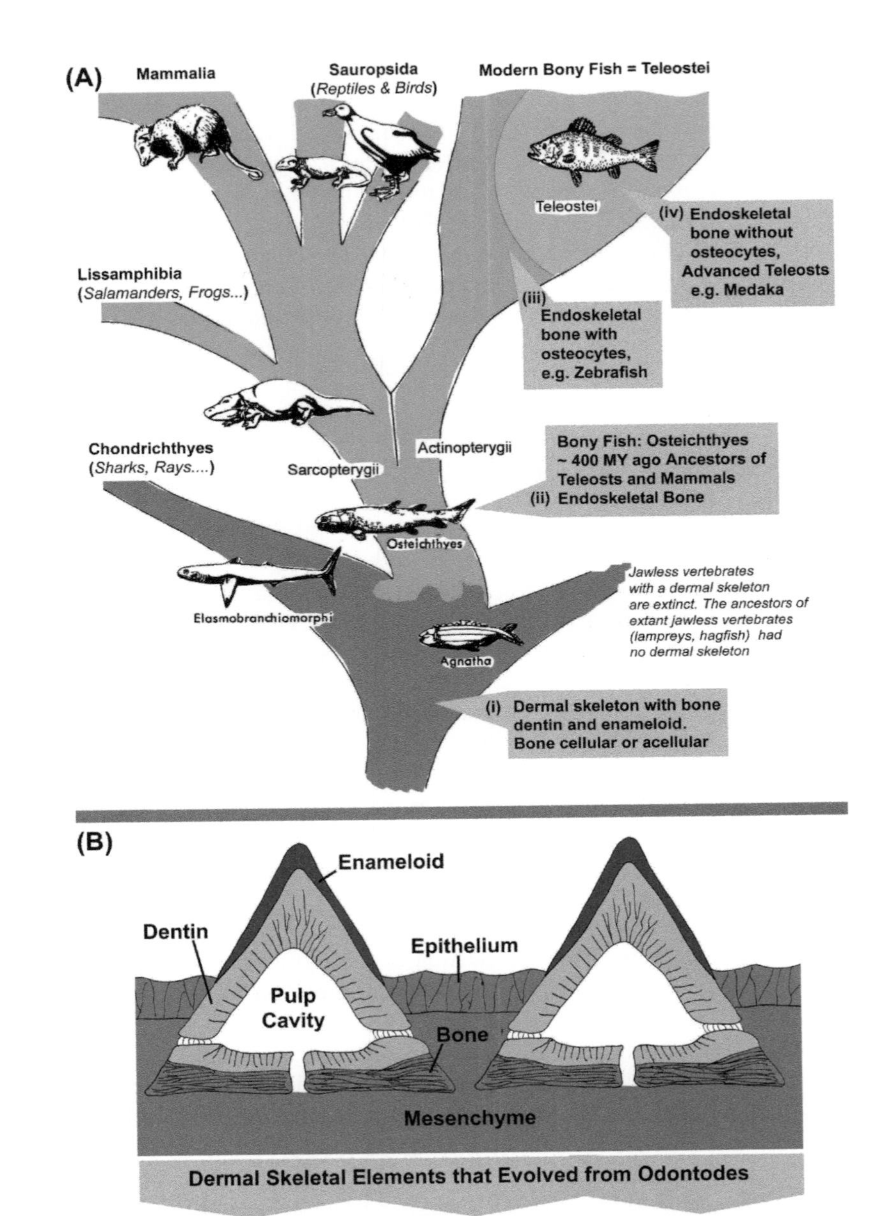

FIGURE 1

(A) Phylogenetic relationships of extant vertebrate groups, modified after Romer (1970) . The figure explains why teleost fish are not in the line of mammalian ancestors. (i) As part of the dermal skeleton, bone dentin and enameloid (see Fig. 1B) were already present in early endoskeleton contains cartilage, bone, and the major categories of skeletal cells: chondroblasts, chondrocytes, osteoblasts, bone lining cells, osteocytes, and osteoclasts (Huysseune, 2000; Witten & Huysseune, 2009). Different from mammals, vertebral body centra have no cartilaginous precursor. Vertebral centra anlagen arise by segmented mineralization of the notochord sheath followed by intramembranous bone formation around the notochord (Arratia, Schultze, & Casciotta, 2001; Huxley, 1859; Kölliker, 1859). In small teleost fish, perichondral bone formation is usually not followed by endochondral bone formation (Fig. 2). Endochondral bone formation occurs, however, in larger species such as in carp (*Cyprinus carpio*), a close relative of zebrafish (Huysseune, 2000; Witten et al., 2001; Witten, Huysseune, & Hall, 2010). A typical element of the zebrafish endoskeleton consists of a perichondral bone tube with cartilage sticking out as a condyle. Membranous apolamellae can extend from the perichondral bone (Witten & Huysseune, 2007). Inside the medullary cavity, adipose tissue has replaced cartilage, along with nerves, blood vessels, and connective tissue cells. Teleosts have no hematopoietic tissue inside the bone marrow. Hematopoiesis takes place in the head kidney (Witten & Huysseune, 2009). Structurally, bone tissue in teleost fish develops first as woven bone, and subsequently, parallel-fibered and lamellar bone develops. Lamellar bone can also form osteons (Amprino & Godina, 1956; Meunier, 2002; 2011; Moss, 1961), yet not in small fish such as zebrafish or medaka. Compared to mammals, a higher number of skeletal tissue subtypes are recognized in teleosts as part of the regular (nonpathological, nonregenerating) skeleton. Several categories of cartilage are intermediate between connective tissue and cartilage, or bone and cartilage (Fig. 2). The best studied teleost "intermediate skeletal" tissue is chondroid bone, for which different variants exist. Chondroid bone contains chondrocyte-like cells (devoid of cell processes) surrounded by bone-like matrix (Huysseune, 1986; Huysseune & Sire, 1990; Huysseune & Verraes, 1990). In addition, secondary cartilage and chondroid

jawless vertebrates. See Janvier (2015) for the relationship of extinct vertebrates. (ii) The common ancestors of teleost fish and mammals, osteichthyans (bony fish), evolved bone in the endoskeleton about 400 million years ago (Maisey, 2000). Extant fish groups closely related to our common ancestors are represented by lungfish, coelacanths, and bichirs. (iii) Endoskeletal bone of extant vertebrates contains osteocytes, also in zebrafish. (iv) In advanced teleosts, such as medaka, osteocytes eventually disappeared after a long marine evolutionary history. (B) Odontodes are the principal units of the dermal skeleton of early jawless vertebrates (ostracoderms). Odontodes are made from dentin, covered with a hypermineralized layer (enamel or enameloid), enclose a pulp cavity and are ankylosed to a piece of bone. Odontodes were structurally maintained (e.g., teeth) or evolved into different types of dermal skeletal elements.

Modified after Reif, W. E. (1982). Evolution of dermal skeleton and dentition in vertebrates. The odontode regulation theory. Evolutionary Biology (New York), 304*(15), 287–368 and Sire, J-Y., & Huysseune, A. (2003). Formation of dermal skeletal and dental tissues in fish: a comparative and evolutionary approach.* Biological Reviews of the Cambridge Philosophical Society, 78, *219–249.*

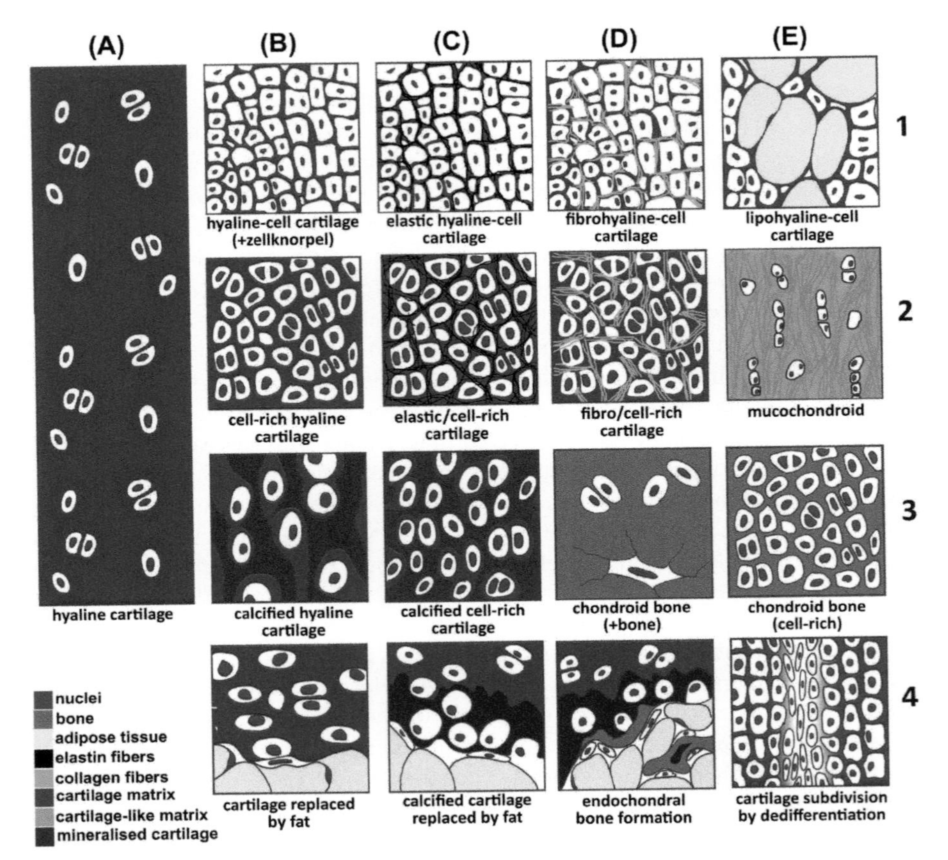

FIGURE 2

Types of differentiated cartilage and cartilaginous tissues in teleost fish modified after Witten et al. (2010). Typical hyaline cartilage (A) is one of many cartilage types. Cell cartilage (B1) and cell-rich cartilage (B2), characterized by a reduced amount of matrix, are common. Hyaline-cell cartilage can intermingle with adipose tissue (E1). Cell-rich cartilages can contain additional fibers made from elastin (C1,2) or collagen type one (D1,2). Hyaline and cell-rich cartilage can permanently mineralize (B3, C3). Different types of chondroid bone (chondrocytes embedded in bone matrix, D3, E3) or chondroid tissues (E2) exist. Endochondral bone formation occurs in large teleost fish (D4). In small teleosts, cartilage is typically replaced by adipose tissue (B4, C4). Cartilage elements can subdivide by dedifferentiation into fibroblasts (E4) (Dewit, Witten, & Huysseune, 2011). In vivo there is a continuum of cartilaginous tissues and cells can have intermediate characters between connective tissue and cartilage, or between cartilage and bone (Hall & Witten, 2007). Beresford (1981) and Hall (2015) also describe intermediate skeletal cells and tissues in mammals.

tissues can develop on cranial dermal bones (Benjamin, 1990; Beresford, 1993; Hall & Witten, 2007; Witten et al., 2010).

1.3 OSTEOCYTES, OSTEOCLASTS, AND ACELLULAR BONE

In mammals, osteocytes represent 95% of all bone cells (Franz-Odendaal, Hall, & Witten, 2006). Osteocytes are mechanoreceptors that regulate bone metabolism and control osteogenic cells on the bone surface (Bonucci, 2009; Boyce, Yao, & Xing, 2009). In teleost fish, osteocytes are not present in all life stages and in all skeletal elements. Early zebrafish bone has no osteocytes (acellular bone) (Fig. 3A), and scales and fin rays remain acellular throughout life (Sire & Akimenko, 2004; Sire, Huysseune, & Meunier, 1990; Witten et al., 2001). Advanced teleosts, such as medaka, have completely lost osteocytes also in the endoskeleton in the course of a long marine evolutionary history (Fig. 1A). As a consequence, alternative regulatory pathways must have evolved that compensate for the absence of osteocytes (Witten & Huysseune, 2009). With or without osteocytes, resorption and remodeling are well documented for teleosts and also for zebrafish and medaka (To et al., 2012; Witten, 1997; Witten et al., 2001). Teleost osteoclasts are often small, mononucleated cells that resorb bone without generating typical resorption lacunae (Witten & Huysseune, 2009), also reported by Parfitt (1988) for human bone. Without special labeling, these osteoclasts are easily overlooked in a search for typical multinucleated giant cells (To et al., 2012; Witten & Huysseune, 2009) (Fig. 3). Teleosts do not depend on their skeleton to maintain plasma calcium homeostasis (Witten et al., 2016). Dominant factors that trigger bone resorption are lifelong growth and continuous tooth replacement. Bone resorption related to these processes should not be mistaken for metabolism-triggered resorption (Witten, 1997; Witten & Huysseune, 2009). Recent studies on transgenic medaka with defective osteoclasts show dramatic consequences for skeletal development. The lumen of neural and hemal arches fails to increase resulting in choking of the neural tube and the interruption of blood vessels (To, Witten, Huysseune & Winkler, 2015) (Fig. 3C,D). Other factors that trigger teleost bone resorption by mono- and multinucleated osteoclasts are inflammation, fracture repair, or extreme mineral deficient conditions. Like in mammals, osteoclasts express tartrate-resistant acid phosphatase (TRAP), a vacuolar proton pump (V-ATPase), and cathepsin K (To et al., 2012; Witten, 1997; Witten, Holliday, Delling, & Hall, 1999; Witten & Huysseune, 2009; Yu et al., 2016). Similar to mammalian osteoclasts, mononucleated osteoclasts in medaka are activated by the receptor activator of NF-κB ligand (RANKL) (To et al., 2012; Yu et al., 2016).

2. ANALYZING SKELETAL PHENOTYPES OF SMALL FISH

The diagnosis of human skeletal diseases relies on the imaging of the skeletal phenotype, at the anatomical and at the cellular level. The interpretation of skeletal

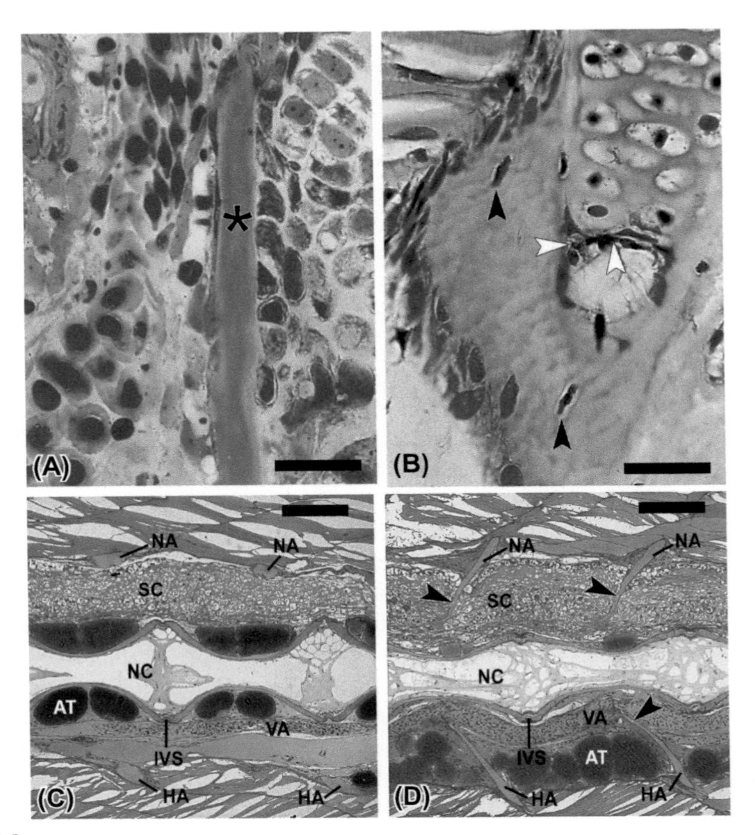

FIGURE 3

(A) Cleithrum (*black asterisk*) of a zebrafish of 15 dpf. Note that the bone is acellular at this stage (no osteocytes) but bone becomes cellular at later stages (see Fig. 3B). Epon embedding, toluidine blue stained 1 μm thick section. Scale bar = 12 μm. (B) Ethmoid of a 40-day-old zebrafish. The bone now contains osteocytes (*black arrowheads*). Flat mononucleated tartrate-resistant acid phosphatase (TRAP)–positive chondroclasts appear at the site of cartilage degradation (*white arrowheads*). Distinction of the small cells from other cells is not possible based on standard staining procedures. Cartilage is removed but not replaced by bone. Scale bar = 15 μm. (C) Parasagittal section through the vertebral column of a juvenile medaka. In growing fish the lumen of neural and hemal arches is continuously enlarged by bone resorption (cells not visible, see text Fig. 3B). (D) In osteoclast defective transgenic medaka, the lumen of neural and hemal arches fails to increase (To et al., 2015) resulting in choking of the neural tube and the interruption of blood vessels (*black arrowheads*). *AT*, adipose tissue; *HA*, hemal arches; *IVS*, intervertebral space; *NA*, neural arches; *NC*, notochord; *SC*, spinal cord; *VA*, ventral aorta. Epon embedding, toluidine blue stained semithin section. Scale bars c, d = 100 μm.

phenotypes should take into account that the evolution of a small body size (miniaturization) is linked to anatomical changes that can also represent new characters (Hanken & Wake, 1993). While advantageous in many aspects, the use of small fish as models also imposes challenges for skeletal imaging and for the interpretation of skeletal phenotypes. Not only are skeletal structures in zebrafish and medaka extremely small, but also, for a different reason, cells are smaller compared to mammalian cells. This is because cell size is tightly correlated to genome size (total nuclear DNA content, expressed in picograms per cell nucleus) (Gregory, 2001). Compared to humans (3.50 pg DNA) zebrafish cells contain about half the amount of DNA per nucleus (1.80 pg DNA) while medaka cells contain only 1.09 pg DNA per nucleus (Harris et al., 2014; Jockusch, 2007). This small cell size requires higher precision in imaging and histological analyses.

2.1 AT THE LIMIT: RADIOGRAPHS AND MICROCOMPUTED TOMOGRAPHY

Radiograph-based images are fundamental diagnostic tools in skeletal research. Conventional 2D radiographs work well for larger fish. The small size of skeletal structures in zebrafish or medaka limits the use of radiographs even for adult specimens (Fisher, Jagadeeswaran, & Halpern, 2003). Ultimately, none of the routine radiological techniques used in human or veterinary medicine are sufficient for detailed analysis of zebrafish or medaka (Bruneel & Witten, 2015). A maximum resolution down to 20 μm can be achieved by using low speed technical X-ray film in combination with low KV emitting X-ray tubes and long exposure times (Witten, Gil-Martens, Huysseune, Takle, & Hjelde, 2009; Witten & Hall, 2003) (Fig. 4A). Instead of 2D radiographs, zebrafish bone and cartilage can also be visualized by microcomputed tomography or micro-CT. Micro-CT uses high-powered X-rays from a single emitter that are detected by gamma cameras (Gregg & Butcher, 2012). Voxel sizes of less than 6 μm (visualizing structures of about 18 μm) can be achieved (Fig. 4B). Using contrasting agents, Metscher (2009) achieved a resolution of 12 μm with a voxel size of 4.6 μm even for soft tissue. Yet, at this resolution some skeletal elements in zebrafish still disappear. The width of bone trabeculae in adult zebrafish can be below 5 μm (Bruneel & Witten, 2015). The use of synchrotron light sources can increase the resolution to around 1.6 μm (Neues, Goerlich, Renn, Beckmann, & Epple, 2007; Zehbe et al., 2010). Electrons are accelerated close to the speed of light in a circular magnetic field. This produces electromagnetic synchrotron radiation from which high-powered X-rays can be filtered. Epple and Neues (2010) explored the resolution limits of synchrotron radiation–based microcomputed tomography for visualizing the medaka skeleton, and Pasco-Viel et al. (2010) applied synchrotron-based micro-CT to the dentition of zebrafish and its relatives. Protocols for micro-CT–based bone mineral density (BMD) measurement have been applied to the zebrafish skeleton (Asharani et al., 2012; Shkil et al., 2014). It is possible to quantify bone mineral density (BMD) and tissue mineral density (TMD) accurately up to a hydroxyapatite density of 800 mg/cm^3. This range can

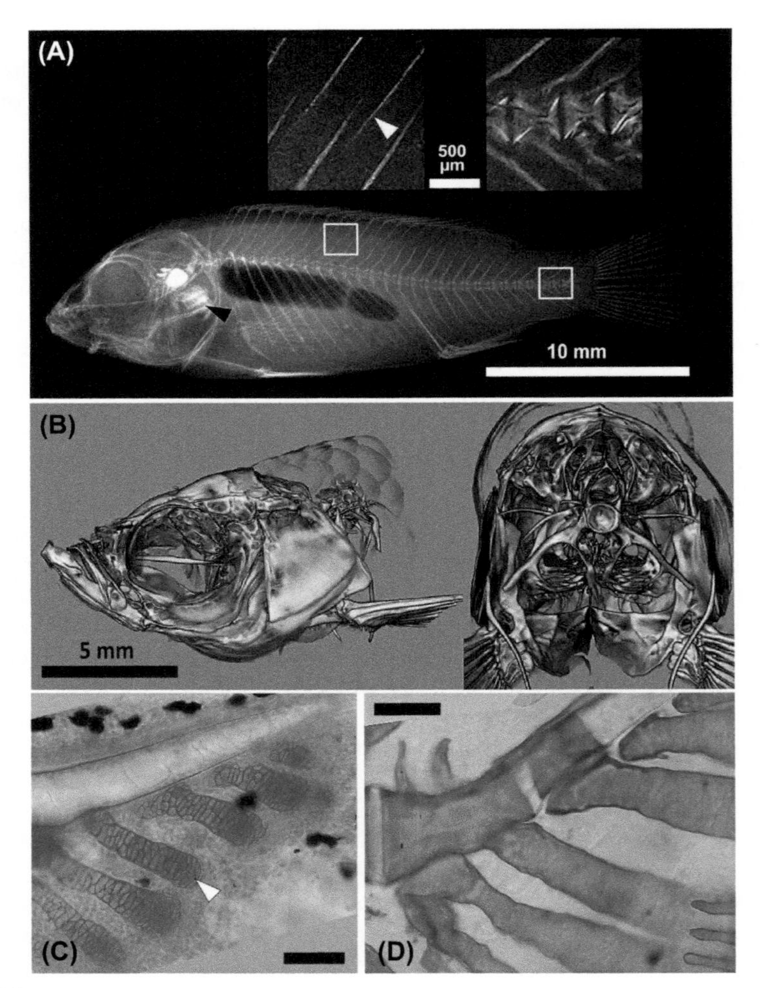

FIGURE 4

(A) X-ray of a 33 mm total length juvenile Nile tilapia (*Oreochromis niloticus*) comparable to the size of a young adult zebrafish. The Nile tilapia has tooth-bearing upper and lower (*black arrowhead*) pharyngeal jaws (the latter similar to the zebrafish pharyngeal jaws). The inserts show higher magnifications and reveal a resolution limit for the size of bone structures that is about 20 μm. Agfa D2 (100ASA) technical X-ray film. (B) 6 μm-voxel size (equals a resolution limit of 18 μm) micro-CT scan from the head of an adult zebrafish. Lateral view (left), and rear view into the head (right). The elements of the shoulder girdle, a vertebral body with elements of the Weberian apparatus and the pharyngeal teeth are well visible. (C and D) Alcian blue and Alizarin red S stained whole mount preparation showing the caudal fin endoskeleton of zebrafish at 10 and 15 dpf, respectively. Chondrocytes are visible. The resolution limit of the microphotographs is about 0.3 μm and thus exceeds the resolution limit of the X-ray (20 μm) or the micro-CT (18 μm) about 60 times. Scale bars = 100 μm.

be extended using high mineral density composite phantoms for calibration (Deuerling, Rudy, Niebur, & Roeder, 2010). Still, size and density of many bone structures in zebrafish and medaka are too small to be accurately quantified. Optical projection tomography (OPT) is an optical equivalent of micro-CT scanning. By rotating the sample embedded in a cylinder of agarose, OPT can transcend the limits imposed by the thickness of the sample in conventional optical microscopy (Sharpe et al., 2002). In this way, fluorescent or nonfluorescent samples with a thickness of up to 15 mm can be visualized to produce 3D images based on a 6 μm voxel size (Walls, Sled, Sharpe, & Henkelman, 2007). OPT has been used to make a whole body atlas of zebrafish development called FishNet: http://www.fishnet.org.au (Bryson-Richardson et al., 2005).

2.2 3D, IN COLOR AND FLUORESCENT: THE NEW WHOLE MOUNT STAINING PROTOCOLS

The most commonly used protocol to visualize the skeleton of small fish in 3D is whole mount staining of cartilage with Alcian blue and mineralized structures with Alizarin red S, followed by making the specimens transparent (Dingerkus & Uhler, 1977; Taylor, 1967; Wassersug, 1976). Combined with a compound microscope (resolution limit 0.2 μm), whole mount staining of small specimens can surpass by far all micro-CT—based methods (Fig. 4C,D). Double staining for bone and cartilage can, however, cause a problem since highly acidic Alcian blue solutions can remove minerals from early mineralized structures. The result is false negative bone staining. The problem was addressed by Vandewalle, Gluckmann, and Wagemans (1998) and was also recognized by others (Bird & Mabee, 2003; Springer & Johnson, 2000). A safe protocol is to stain specimens only with Alizarin red S for mineralized structures (Bensimon-Brito et al., 2016). Walker and Kimmel (2007) proposed an acid-free double staining protocol suitable for early stages of zebrafish. Loizides, Georgiou, Somarakis, Witten, and Koumoundouros (2014) used the Walker and Kimmel (2007) protocol followed by glycol methacrylate embedding for histological analysis. Alizarin red S stained specimens lose staining over time and early developmental stages must be analyzed as soon as possible. Since Alizarin S red is a highly fluorescent dye visualization of weakly stained bones can be greatly enhanced with fluorescent light, a protocol that also permits counterstaining of GFP labeled fish (Bensimon-Brito, Cardeira, Cancela, Huysseune, & Witten, 2012; Bensimon-Brito et al., 2016).

2.3 SMALL SKELETAL CELLS IN ALL DETAILS

Adequate histological analysis of zebrafish skeletal tissues, especially of early developmental stages, requires high-resolution microscopical imaging. Anlagen of skeletal elements may consist only of a limited number of cells and cartilaginous elements can form from a single row of cells (Huysseune & Sire, 1992). Histological standard procedures based on paraffin embedding do not provide sufficient histological details. Embedding of specimens in epoxide or methacrylate resin is recommended. From epoxide blocks semi and ultrathin sections can be obtained for high-resolution light and subsequent transmission electron microscopy. Methacrylate resin permits to use histochemical protocols (Willems et al., 2012; Witten, 1997). High-resolution histology can be complemented with new in vivo imaging techniques. Caution is however advised to strictly consider all aspects of animal welfare before using any in vivo imaging protocol. Confocal microscopy is not suited for studying samples for longer periods of time, as the laser will cause photodamage. Spinning disk confocal microscopy, which uses multiple confocal pinholes and so reduces the time needed to make one image, reduces not only photodamage but also imaging depth. Two-photon excitation fluorescence (TPEF) provides a reduction in photodamage as well as an increase in imaging depth. It is a method that can image in maximum depths ranging from 500 μm to 1.6 mm, depending on the properties of the imaged medium (Balu et al., 2009; Kobat, Horton, & Xu, 2011). The use of longer wavelengths minimizes phototoxicity and light scattering and makes deep-tissue in vivo imaging possible (Christensen & Nedergaard, 2011). Another new technology for in vivo analysis with low phototoxicity is light sheet microscopy. Here, samples are illuminated from the side by a thin sheet of laser light. The illuminated plane coincides with the focal plane of detection, which is placed orthogonal to the light sheet. Photodamage is reduced because only the part of the sample in the light sheet is illuminated (Huisken & Stainier, 2009). The combination of light sheet with two-photon microscopy doubles the imaging depth of one-photon light sheet microscopy (Jemielita, Taormina, DeLaurier, Kimmel, & Parthasarathy, 2013). Second harmonic generation (SHG) microscopy is an alternative to TPEF. SHG signal can be produced by materials that lack generalized mirror symmetry, such as collagens and muscle fibers (Campagnola et al., 2002; Pantazis, Maloney, Wu, & Fraser, 2010). SHG has advantages compared to TPEF. No excited molecules are formed by the production of the SHG signal and no photobleaching or dye-induced phototoxicity occurs. This makes SHG a true noninvasive method (Sun et al., 2004).

3. MUTANT AND TRANSGENIC FISH OPEN NEW DIRECTIONS IN SKELETAL RESEARCH

Compact genomes and the availability of a refined genetic toolbox offer unique approaches to generate transgenic and mutant fish. Genetically modified fish have been used extensively in the past to study skeletal processes by live imaging in the optically clear specimen. Over the past 15 years, a large number of transgenic zebrafish and medaka lines have been generated that express various fluorescent reporters in osteoblast progenitors (Inohaya, Takano, & Kudo, 2007; Knopf et al., 2011; Renn, Büttner, To, Chan, & Winkler, 2013), premature osteoblasts (DeLaurier et al., 2010; Renn & Winkler, 2009), and mature osteoblasts (Geurtzen et al., 2014;

Knopf et al., 2011; Renn & Winkler, 2014), as well as preosteoclasts and osteoclasts (Chatani, Takano, & Kudo, 2011; To et al., 2012). Unfortunately, no reporter lines for osteocytes (Fig. 3B) are yet available, although osteocytes were visualized by RNA in situ hybridization in zebrafish (Jeradi & Hammerschmidt, 2016). Transgenic reporter lines allow to visualize interactions between distinct bone cells at high-resolution during bone modeling and remodeling in intact embryos. Transgenesis has also been used to overexpress bone relevant genes and generate models of human bone diseases. For example, overexpression of RANKL under control of a heat shock–inducible promoter results in the ectopic formation of osteoclasts, which leads to increased bone resorption and an osteoporosis-like phenotype in medaka (To et al., 2012). This can be prevented by incubation of the transgenic larvae in medium containing bisphosphonates (Yu et al., 2016), a clear example of how small teleosts can be used to validate therapies applied for human skeletal disorders.

A multitude of zebrafish and medaka mutants were identified in large scale mutagenesis screens (Furutani-Seiki et al., 2004; Haffter et al., 1996), and a considerable number of these mutants exhibited interesting skeletal phenotypes (e.g., Piotrowski et al., 1996). One of the first zebrafish bone mutants that was characterized in detail was *chihuahua* (Fisher et al., 2003), which carries a deficiency in the *collagen 1* gene and develops skeletal malformations reminiscent of human osteogenesis imperfecta. Other zebrafish mutants helped to identify novel skeletal genes, such as *entpd5*, which is essential for bone mineralization (Huitema et al., 2012), or to provide novel insight into ongoing debates concerning skeletal development. For example, the zebrafish *stocksteif* mutant uncovered the role of retinoic acid metabolism in osteoblast differentiation (Laue, Jänicke, Plaster, Sonntag, & Hammerschmidt, 2008; Spoorendonk et al., 2008; see also Blum & Begemann, 2015).

The recent advent of efficient genome editing tools, such as TALEN and CRISPR/Cas9, allows to introduce targeted deficiencies in selected genes of model teleost fish. The CRISPR/Cas9 approach was recently used to knockout *osterix*/sp7, an important regulator for mammalian osteoblast differentiation, in carp, zebrafish, and medaka. Interestingly, while early bone phenotypes were surprisingly mild in zebrafish and carp (Kague et al., 2016; Zhong et al., 2016), medaka mutants almost completely lacked mineralized bone matrix (C. Winkler, unpublished results). This suggests species-specific differences in the molecular control of osteoblast differentiation in teleosts that require further analysis. Recently, TALEN was used to also generate a medaka *c-fmsa* mutant with deficiencies in preosteoclast differentiation (Mantoku, Chatani, Aono, Inohaya, & Kudo, 2016). Interestingly, these mutants revealed bone hyperplasia and severe defects in the resorption of attachment bone providing further proof for the importance of osteoclasts in teleost bone remodeling. Given the efficiency of targeted genome editing in fish, a large number of studies describing the function of known or novel bone regulatory genes can be expected for the near future. In conclusion, medaka and zebrafish offer valuable approaches as models for functional gene analysis and for live imaging of processes relevant to vertebrate bone modeling and remodeling.

4. GENETIC PHENOCOPIES OF HUMAN SKELETAL DISEASES

One promise of the use of small fishes is the direct modeling of altered gene function that can inform about the causes of human disease. Systematic mutagenesis screens performed in both zebrafish and medaka have defined cases in which loss of gene function is comparable to the effect of disease causing mutations. Analysis between the fish and patient phenotypes can be instructive even if the structures are varied. For example, mutations in *Ectodysplasin* (*EDA*) and its receptor (*EDAR*) have been linked to hypohydrotic ectodermal dysplasia affecting hair, sweat glands, as well as tooth formation (Headon & Overbeek, 1999; Monreal et al., 1999). Loss-of-function mutations identified in the zebrafish and medaka ortholog of *eda* and *edar*, in contrast, lead to a primarily skeletal phenotype of loss of postcranial dermal bone as well as teeth (Harris et al., 2008; Kondo et al., 2001). Analysis of the developmental phenotype of the mutations in both mammals and fishes, however, exposes the shared essential role of these genes in regulating early specification of a developmental signaling center, the epithelial placode. Due to presumed shared ancestry of ectodermal appendage structures, even as diverse as mammalian hair and dermal scales of teleosts, Eda signaling remains an essential component of their formation. Thus, the genetic regulation of placode derived structures then can be modeled in distantly related organisms even if the structural phenotypes are varied. Similarly, there has been quite successful modeling of the function of genes regulating palatogenesis in the zebrafish. Early gene repertoires are maintained in early patterning of the face of both amniotes and fishes (Swartz, Sheehan-Rooney, Dixon, & Eberhart, 2011). Mutations in genes associated with clefting in patients, such as *pdgrf* or *irf6* (McCarthy et al., 2016; Peyrard-Janvid et al., 2014), cause specific alterations in the behavior of neural crest cells of the pharyngeal arches and early periderm signaling leading to patterning defects of the ethmoid during early development. These changes resemble patterning deficiencies seen in facial clefts in patients. However, it is important to note that zebrafish do not have a palate, rather a muscular pad and fishes lack secondary palatogenesis and formation of a palate as seen in amniotes. However, the midline patterning and primary chondrogenic behaviors are retained due to common ancestry. This then permits the use as a model to understand early stages of palate development and the etiology of clefting in patients. Recent work exemplifies this utility (Cvjetkovic et al., 2015; Dougherty et al., 2013; Gfrerer et al., 2014; Liu et al., 2016; Mukherjee et al., 2016; Peyrard-Janvid et al., 2014).

Systematic research into genetic models of human skeletal disease have historically centered on phenotypes observable in early development of posthatching stages due to accessibility of the phenotypes. For example, analysis of the *dackel* and *pincher* mutants identified *ext2* and *papst1* genes as being essential for cartilage morphogenesis and formation of cartilage pathologies similar to multiple hereditary exostoses in patients (Clément et al., 2008; Wiweger et al., 2011). Initial processes of ossification can also be studied during early development and has been useful in studying the local regulation of phosphate precipitation and mechanisms of bone mineralization (Apschner, Huitema, Ponsioen, Peterson-Maduro, & Schulte-Merker, 2014; Huitema et al.,

2012). However, late processes of differentiation and skeletal reworking have not received comparable analysis as screens at adult stages are cumbersome and experimental techniques, including visualization (see above), are more difficult. However, these later phenotypes are important for understanding and modeling clinical phenotypes. For example, osteoporosis, osteoarthritis, scoliosis, and fracture repair are key pathologies that would benefit from new experimental models to investigate their causes and remediation—and small fish models may prove to be useful tools to study these conditions. As support of their utility, zebrafish mutants have been identified affecting genes that are associated with osteogenesis imperfecta, or brittle bone disease, in patients that closely mirror the human phenotype (Asharani et al., 2012; Fisher et al., 2003; Gistelinck, Gioia et al., 2016; Gistelinck, Witten et al., 2016; Martínez-Glez et al., 2012). Additionally, analysis of osteoclastogenesis and function, both in the zebrafish and medaka, have shown that osteoclasts share common developmental specification and genetic control. Recent advances in genetic editing such as TALEN or CRISPR/Cas9 (Bedell et al., 2012; Hwang et al., 2013) provide a direct approach at understanding the effect of patient-specific alleles of genes in skeletogenesis (e.g., microcephaly, Hu et al., 2014). Whereas generation of loss-of-function alleles by creating targetet deletions is very efficient, introducing specific changes by homologous replacement is significantly less so. Homologous replacement, however, remains a powerful approach to generate patient-specific models.

Thus, through conservation of genetic regulation in skeletogenesis, small fishes can model many skeletal disease conditions. Given the experimental utility of these models, the fundamental similarity is useful even in cases where the phenotype is not identical, for instance analysis of ectodermal dysplasia or palatogenesis. Modeling of human disorders can be difficult in the zebrafish and medaka given their small body size, as small size has associated anatomical characteristics that make comparison of mechanisms problematic (Hanken & Wake, 1993). For example, the environment in which the fish skeleton functions imposes different stresses than that of land dwelling amniotes. Teleost muscles that facilitate burst swimming in the dense medium water are stronger than mammalian muscles, and intervertebral joints have to withstand high mechanical loads (Ackerly & Ward, 2016; Johnston, 1981). Conversely, the swim bladder mitigates some effects of gravity and alters the application of force in movement on the maturation of joints in the appendicular and the axial skeleton (Longo, Riccio, & McCune, 2013; Witten & Huysseune, 2009). This difference in force as well as the differences in cartilage structure and composition (Fig. 2) has made good models of osteoarthritis difficult to define despite recent work that suggests that such models are possible (Askary et al., 2016).

5. LIFELONG TOOTH REPLACEMENT

Most nonmammalians replace their teeth lifelong (up to dozens of times) as a natural process, making small teleosts such as zebrafish and medaka attractive models to investigate the cellular and genetic underpinning of this process. The hope is that the mechanism underlying the capacity for continuous replacement, once elucidated, can be translated into strategies that can restore a lost tooth in humans, or can complement the field of regenerative dentistry (Huysseune & Thesleff, 2004).

Because of their evolutionary distance, zebrafish and medaka have dentitions that differ in many ways (Debiais-Thibaud et al., 2007; Stock, 2007). These differences reflect to a certain extent the diversity that is found in teleost dentitions with regard to localization of teeth (e.g., absence of oral teeth in zebrafish; Huysseune, Van der heyden, & Sire, 1998), their number (e.g., a variable but large number of pharyngeal teeth in medaka versus a small and fixed number in zebrafish; Debiais-Thibaud et al., 2007; Van der heyden & Huysseune, 2000), and their size and shape (mild heterodonty both in zebrafish and medaka; Debiais-Thibaud et al., 2007; Wautier, Van der heyden, & Huysseune, 2001). The development of replacement teeth involves budding of the epithelium and reciprocal interactions with the underlying mesenchyme (Huysseune, Delgado, & Witten, 2005). Epithelial budding can be initiated from the oral epithelium (as in medaka, Abduweli et al., 2014), the outer dental epithelium of the predecessor (as in embryonic zebrafish, Van der heyden, Huysseune, & Sire, 2000), or a distinct successional dental lamina (as in adult zebrafish) (Huysseune, 2006).

Irrespective of these differences, the focus in studies attempting to unravel the mechanism of lifelong tooth turnover has been on epithelial stem cells, ever since their involvement in this process was hypothesized (Huysseune & Thesleff, 2004). Evidence for their presence in teleosts has been indirect so far, relying mostly on BrdU label retention and expression of the transcription factor *sox2*. While putative stem cells have been identified in this way in medaka (Abduweli et al., 2014), doubts have been raised for *Polypterus*, a basal actinopterygian, and for Atlantic salmon *Salmo salar* (Vandenplas et al., 2016). Evidence that the Wnt signaling pathway, proposed to be involved in regulating epithelial stem cells in amniotes (Gaete & Tucker, 2013; Jussila, Crespo Yanez, & Thesleff, 2014; Whitlock & Richman, 2013), is implicated also in teleosts has been collected for cichlids (Bloomquist et al., 2015), but has not been convincing so far for zebrafish (Huysseune, Soenens, & Elderweirdt, 2014). Other mechanisms may need to be considered to explain the permanent turnover of teeth in teleosts. These include the role of mesenchymal stem cells, dynamic cell movements involving progenitor cells rather than stem cells, or budding mediated by reaction–diffusion mechanisms. As for the skeleton, the use of reverse genetics, in particular the application of CRISPR/Cas9 mediated genome editing, holds a great promise to understand the genetic underpinning of continuous tooth renewal, and may greatly enhance the use of small teleosts to model human dental disorders (Klein et al., 2013).

CONCLUDING REMARKS

We have discussed the current techniques that allow—and restrict—the visualization and analysis of early, small skeletal structures in zebrafish and medaka. Mutant screening for skeletal phenotypes that mirror human skeletal diseases as well as recent advances in genome editing such as TALEN and CRISPR/Cas9 will further improve our understanding of cellular and molecular mechanisms that underlie skeletal development and remodeling. To translate results obtained on small teleost fish to humans requires an appropriate evolutionary perspective considering the different physiology, anatomy, and life history or the organisms being compared. Nutrition, exercise, the immune system, the nervous system, hormones, and the mineral metabolism are among the many factors that influence the skeleton. Knowledge on how these factors influence the teleost skeleton is also collected outside the zebrafish and medaka community, by disciplines such as palaeontology, evolutionary biology, ichthyology, fish physiology, and aquaculture. Interdisciplinary collaborations will further improve our understanding of the fish skeleton and accelerate the translation of knowledge into tools that can help to cure human skeletal diseases. Eventually the increase of analysis into skeletal development, homeostasis, and regeneration of small teleost fish will increase the value of using these organisms as models for skeletogenesis.

ACKNOWLEDGMENTS

CW receives funding from the Singapore Ministry of Education (grant number 2013-T2-2-126) and the NIH/USA (grant number 1R21AT009452-01A1). AH acknowledges a grant of the Special Research Fund (BOF) of Ghent University (grant number BOF15/24J/058). This work was partially supported by NIH grant U01DE024434 to MPH. PEW acknowledges funding from Skretting ARC (Stavanger, Norway) for studies on the teleost skeletal mineral metabolism.

REFERENCES

Abduweli, D., Baba, O., Tabata, M. J., Higuchi, K., Mitani, H., & Takano, Y. (2014). Tooth replacement and putative odontogenic stem cell niches in pharyngeal dentition of medaka (*Oryzias latipes*). *Microscopy, 63*(2), 141–153.

Ackerly, K. L., & Ward, A. B. (2016). How temperature-induced variation in musculoskeletal anatomy affects escape performance and survival of zebrafish (*Danio rerio*). *Journal of Experimental Zoology, 325A*, 25–40.

Amprino, R., & Godina, G. (1956). Osservazioni sul rinnovamento strutturale dell'osso in pesci teleostei. *Pubblicazioni della Stazione Zoologica di Napoli, 28*, 62–71.

Apschner, A., Huitema, L. F., Ponsioen, B., Peterson-Maduro, J., & Schulte-Merker, S. (2014). Zebrafish enpp1 mutants exhibit pathological mineralization, mimicking features of generalized arterial calcification of infancy (GACI) and pseudoxanthoma elasticum (PXE). *Disease Models & Mechanisms, 7*(7), 811–822.

Apschner, A., Schulte-Merker, S., & Witten, P. E. (2011). Not all bones are created equal – using zebrafish and other teleost species in osteogenesis research. *Methods in Cell Biology, 105*, 239–255.

Arratia, G., Schultze, H. P., & Casciotta, J. (2001). Vertebral column and associated elements in dipnoans and comparison with other fishes: development and homology. *Journal of Morphology, 250*, 101–172.

Asharani, P. V., Keupp, K., Semler, O., Wang, W., Li, Y., Thiele, H., … Carney, T. J. (2012). Attenuated BMP1 function compromises osteogenesis, leading to bone fragility in humans and zebrafish. *American Journal of Human Genetics, 90*(4), 661–674.

Askary, A., Smeeton, J., Paul, S., Schindler, S., Braasch, I., Ellis, N. A., … Crump, J. G. (2016). Ancient origin of lubricated joints in bony vertebrates. *ELife, 5*, e16415. http://dx.doi.org/10.7554/eLife.16415.

Balon, E. K. (1990). Epigenesis of an epigeneticist: the development of some alternative concepts on the early ontogeny and evolution of fishes. *Guelph Ichtyological Reviews, 1*, 1–48.

Balu, M., Baldacchini, T., Carter, J., Krasieva, T. B., Zadoyan, R., & Tromberg, B. J. (2009). Effect of excitation wavelength on penetration depth in nonlinear optical microscopy of turbid media. *Journal of Biomedical Optics, 14*(1). http://dx.doi.org/10.1117/1.3081544.

Bedell, V. M., Wang, Y., Campbell, J. M., Poshusta, T. L., Starker, C. G., Krug, R. G., … Ekker, S. C. (2012). In vivo genome editing using a high-efficiency TALEN system. *Nature, 491*(7422), 114–118.

Benjamin, M. (1990). The cranial cartilages of teleosts and their classification. *Journal of Anatomy, 169*, 153–172.

Bensimon-Brito, A., Cardeira, J., Cancela, M. L., Huysseune, A., & Witten, P. E. (2012). Distinct patterns of notochord mineralization in zebrafish coincide with the localization of osteocalcin isoform 1 during early vertebral centra formation. *BMC Developmental Biology, 12*, 28. http://dx.doi.org/10.1186/1471-213X-12-28.

Bensimon-Brito, A., Cardeira, J., Dionísio, G., Huysseune, A., Cancela, M. L., & Witten, P. E. (2016). Revisiting in vivo staining with Alizarin red S – a valuable approach to analyse zebrafish skeletal mineralization during development and regeneration. *BMC Developmental Biology, 16*, 2. http://dx.doi.org/10.1186/s12861-016-0102-4.

Beresford, W. A. (1981). *Chondroid bone, secondary cartilage and metaplasia*. Baltimore-Munich: Urban & Schwarzenberg.

Beresford, W. A. (1993). Cranial skeletal tissues: diversity and evolutionary trends. In J. Hanken, & B. K. Hall (Eds.), *Patterns of structural and systematic diversity: Vol. 2. The skull* (pp. 69–130). Chicago: University of Chicago Press.

Bird, N. C., & Mabee, P. M. (2003). Developmental morphology of the axial skeleton of the zebrafish *Danio rerio* (Ostariophysi: Cyprinidae). *Developmental Dynamics, 228*, 337–357.

Bloomquist, R. F., Parnell, N. F., Phillips, K. A., Fowler, T. E., Yu, T. Y., Sharpe, P. T., & Streelman, J. T. (2015). Coevolutionary patterning of teeth and taste buds. *Proceedings of the National Academy of Sciences. USA, 112*(44), E5954–E5962.

Blum, N., & Begemann, G. (2015). Osteoblast de- and redifferentiation are controlled by a dynamic response to retinoic acid during zebrafish fin regeneration. *Development, 142*(17), 2894–2903.

Bolker, J. A. (2014). Model species in evo-devo: a philosophical perspective. *Evolution & Development, 16*(1), 49–56.

Bonucci, E. (2009). The osteocyte: the underestimated conductor of the bone orchestra. *Rendiconti Lincei Scienze Fisiche e Naturali, 20*, 237–254.

Boyce, B. F., Yao, Z., & Xing, L. (2009). Osteoclasts have multiple roles in bone in addition to bone resorption. *Critical Reviews in Eukaryotic Gene Expression, 19*(3), 171–180.

Bruneel, B., & Witten, P. E. (2015). Power and challenges of using zebrafish as a model for skeletal tissue imaging. *Connective Tissue Research, 56*(2), 161–173.

Bryson-Richardson, R. J., Cole, N., Hall, T. E., Eckert, S., Sharpe, J., & Currie, P. D. (2005). FishNet, an online 3D database of zebrafish larval anatomy. *Mechanisms of Development, 122*, S147.

Campagnola, P. J., Millard, A. C., Terasaki, M., Hoppe, P. E., Malone, C. J., & Mohler, W. A. (2002). Three-dimensional high-resolution second-harmonic generation imaging of endogenous structural proteins in biological tissues. *Biophysical Journal, 82*(1), 493–508.

Chatani, M., Takano, Y., & Kudo, A. (2011). Osteoclasts in bone modeling, as revealed by in vivo imaging, are essential for organogenesis in fish. *Developmental Biology, 360*(1), 96–109.

Christensen, D. J., & Nedergaard, M. (2011). Two-photon in vivo imaging of cells. *Pediatric Nephrology, 26*, 1483–1489.

Clément, A., Wiweger, M., von der Hardt, S., Rusch, M. A., Selleck, S. B., Chien, C. B., & Roehl, H. H. (2008). Regulation of zebrafish skeletogenesis by ext2/dackel and papst1/pinscher. *PLoS Genetics, 4*(7), e1000136. http://dx.doi.org/10.1371/journal.pgen.1000136.

Cvjetkovic, N., Maili, L., Weymouth, K. S., Hashmi, S. S., Mulliken, J. B., Topczewski, J., … Hecht, J. T. (2015). Regulatory variant in FZD6 gene contributes to nonsyndromic cleft lip and palate in an African-American family. *Molecular Genetics & Genomic Medicine, 3*(5), 440–451.

Debiais-Thibaud, M., Borday-Birraux, V., Germon, I., Bourrat, F., Metcalfe, C. J., Casane, D., & Laurenti, P. (2007). Development of oral and pharyngeal teeth in the medaka (*Oryzias latipes*): comparison of morphology and expression of *eve1* gene. *Journal of Experimental Zoology, 308B*, 693–708.

DeLaurier, A., Eames, B. F., Blanco-Sánchez, B., Peng, G., He, X., Swartz, M. E., … Kimmel, C. B. (2010). Zebrafish sp7:EGFP: Atransgenic for studying otic vesicle formation, skeletogenesis, and bone regeneration. *Genesis, 48*(8), 505–511.

Deuerling, J. M., Rudy, D. J., Niebur, G. L., & Roeder, R. K. (2010). Improved accuracy of cortical bone mineralization measured by polychromatic microcomputed tomography using a novel high mineral density composite calibration phantom. *Medical Physics, 37*(9), 5138–5145.

Dewit, J., Witten, P. E., & Huysseune, A. (2011). The mechanism of cartilage subdivision in the reorganization of the zebrafish pectoral fin endoskeleton. *Journal of Experimental Zoology, 316B*, 584–597.

Dingerkus, G., & Uhler, L. D. (1977). Enzyme clearing of alcian blue stained small vertebrates for demonstration of cartilage. *Stain Technology, 52*, 229–232.

Donoghue, P. C. J. (2006). Early evolution of vertebrate skeletal tissues and cellular interactions, and the canalization of skeletal development. *Journal of Experimental Zoology, 306B*, 278–294.

Dougherty, M., Kamel, G., Grimaldi, M., Gfrerer, L., Shubinets, V., Ethier, R., … Liao, E. C. (2013). Distinct requirements for wnt9a and irf6 in extension and integration mechanisms during zebrafish palate morphogenesis. *Development, 140*(1), 76–81.

Elinson, R. P., & Kezmoh, L. (2010). Molecular Haeckel. *Developmental Dynamics, 239*, 1905–1918.

Epple, M., & Neues, F. (2010). Synchrotron microcomputer tomography for the non-destructive visualization of the fish skeleton. *Journal of Applied Ichthyology, 26*(2), 286–288.

Fisher, S., Jagadeeswaran, P., & Halpern, M. E. (2003). Radiographic analysis of zebrafish skeletal defects. *Developmental Biology, 264*(1), 64–76.
Franz-Odendaal, T., Hall, B. K., & Witten, P. E. (2006). Buried alive: how osteoblasts become osteocytes? *Developmental Dynamics, 235*, 176–190.
Furutani-Seiki, M., Sasado, T., Morinaga, C., Suwa, H., Niwa, K., Yoda, H., ... Kondoh, H. (2004). A systematic genome-wide screen for mutations affecting organogenesis in medaka, *Oryzias latipes*. *Mechanisms of Development, 1*(7–8), 647–658.
Gaete, M., & Tucker, A. S. (2013). Organized emergence of multiple-generations of teeth in snakes is dysregulated by activation of Wnt/Beta-catenin signaling. *PLoS One, 8*, e74484. http://dx.doi.org/10.1371/journal.pone.0074484.
Geurtzen, K., Knopf, F., Wehner, D., Huitema, L. F., Schulte-Merker, S., & Weidinger, G. (2014). Mature osteoblasts dedifferentiate in response to traumatic bone injury in the zebrafish fin and skull. *Development, 141*(11), 2225–2234.
Gfrerer, L., Shubinets, V., Hoyos, T., Kong, Y., Nguyen, C., Pietschmann, P., ... Liao, E. C. (2014). Functional analysis of SPECC1L in craniofacial development and oblique facial cleft pathogenesis. *Plastic and Reconstructive Surgery, 134*(4), 748–759.
Gistelinck, C., Gioia, R., Gagliardi, A., Tonelli, F., Marchese, L., Bianchi, L., ... Forlino, A. (2016). Zebrafish collagen type I: molecular and biochemical characterization of the major structural protein in bone and skin. *Scientific Reports, 6*, 21540. http://dx.doi.org/10.1038/srep21540.
Gistelinck, C., Witten, P. E., Huysseune, A., Symoens, S., Malfait, F., Larionova, D., ... Coucke, P. J. (2016). Loss of type I collagen telopeptide lysyl hydroxylation causes musculoskeletal abnormalities in a zebrafish model of bruck syndrome. *Journal of Bone and Mineral Research*. http://dx.doi.org/10.1002/jbmr.2977 (in press).
Gregg, C. L., & Butcher, J. T. (2012). Quantitative in vivo imaging of embryonic development: opportunities and challenges. *Differentiation; Research in Biological Diversity, 84*(1), 149–162.
Gregory, T. R. (2001). The bigger the C-value, the larger the cell: genome size and red blood cell size in vertebrates. *Blood Cells, Molecules & Diseases, 27*(5), 830–843.
Gregory, W. K. (1933). Fish skulls. A study of the evolution of natural mechanisms. *American Philosophical Society, 20*(2), 75–481.
Haffter, P., Granato, M., Brand, M., Mullins, M. C., Hammerschmidt, M., Kane, D. A., ... Nüsslein-Volhard, C. (1996). The identification of genes with unique and essential functions in the development of the zebrafish, *Danio rerio*. *Development, 123*, 1–36.
Hall, B. K. (1970). Cellular differentiation in skeletal tissues. *Biological Reviews of the Cambridge Philosophical Society, 45*(4), 455–484.
Hall, B. K. (2015). *Bones and cartilage. Developmental and evolutionary skeletal biology* (2nd ed.). San Diego: Elsevier Academic Press.
Hall, B. K., & Witten, P. E. (2007). The origin and plasticity of skeletal tissues in vertebrate evolution and development. In J. S. Anderson, & H.-D. Sues (Eds.), *Major transitions in vertebrate evolution* (pp. 13–56). Bloomington, Indiana: Indiana University Press.
Hanken, J., & Wake, D. B. (1993). Adaptation of bone growth to miniaturisation of body size. In B. K. Hall (Ed.), *Bone* (Vol. 7, pp. 79–104). Boca Raton, Florida: CRC Press.
Harris, M. P., Henke, K., Hawkins, M. B., & Witten, P. E. (2014). Fish is fish: the use of experimental model species to reveal causes of skeletal diversity in evolution and disease. *Journal of Applied Ichthyology, 30*(4), 616–629.
Harris, M. P., Rohner, N., Schwarz, H., Perathoner, S., Konstantinidis, P., & Nüsslein-Volhard, C. (2008). Zebrafish eda and edar mutants reveal conserved and ancestral roles

of ectodysplasin signaling in vertebrates. *PLoS Genetics, 4*(10), e1000206. http://dx.doi.org/10.1371/journal.pgen.1000206.

Headon, D. J., & Overbeek, P. A. (1999). Involvement of a novel Tnf receptor homologue in hair follicle induction. *Nature Genetics, 22*(4), 370–374.

Hruscha, A., Krawitz, P., Rechenberg, A., Heinrich, V., Hecht, J., Haass, C., & Schmid, B. (2013). Efficient CRISPR/Cas9 genome editing with low off-target effects in zebrafish. *Development, 140*, 4982–4987.

Hu, W. F., Pomp, O., Ben-Omran, T., Kodani, A., Henke, K., Mochida, G. H., … Walsh, C. A. (2014). Katanin p80 regulates human cortical development by limiting centriole and cilia number. *Neuron, 84*(6), 1240–1257.

Huisken, J., & Stainier, D. Y. R. (2009). Selective plane illumination microscopy techniques in developmental biology. *Development, 136*(12), 1963–1975.

Huitema, L. F., Apschner, A., Logister, I., Spoorendonk, K. M., Bussmann, J., Hammond, C. L., & Schulte-Merker, S. (2012). Entpd5 is essential for skeletal mineralization and regulates phosphate homeostasis in zebrafish. *Proceedings of the National Academy of Sciences. USA, 109*(52), 21372–21377.

Huxley, T. H. (1859). Observations on the development of some parts of the skeleton of fishes. *Quarterly Journal of Microscopical Sciences (continued as Journal of Cell Sciences), 7*, 33–46.

Huysseune, A. (1986). Late skeletal development at the articulation between upper pharyngeal jaws and neurocranial base in the fish, *Astatotilapia elegans*, with the participation of a chondroid form of bone. *American Journal of Anatomy, 177*, 119–137.

Huysseune, A. (2000). Skeletal system. In G. K. Ostrander (Ed.), *The laboratory fish. Part 4. microscopic functional anatomy* (pp. 307–317). San Diego, CA: Academic Press.

Huysseune, A. (2006). Formation of a successional dental lamina in the zebrafish (*Danio rerio*): support for a local control of replacement tooth initiation. *International Journal of Developmental Biology, 50*, 637–643.

Huysseune, A., Delgado, S., & Witten, P. E. (2005). How to replace a tooth: fish(ing) for answers. *Oral Biosciences & Medicine, 2*, 75–81.

Huysseune, A., & Sire, J.-Y. (1990). Ultrastructural observations on chondroid bone in the teleost fish *Hemichromis bimaculatus*. *Tissue & Cell, 22*, 371–383.

Huysseune, A., & Sire, J.-Y. (1992). Development of cartilage and bone tissues of the anterior part of the mandible in cichlid fish: a light and TEM study. *Anatomical Record, 233*, 357–375.

Huysseune, A., Soenens, M., & Elderweirdt, F. (2014). Wnt signaling during tooth replacement in zebrafish (*Danio rerio*): pitfalls and perspectives. *Frontiers in Physiology, 5*, 386. http://dx.doi.org/10.3389/fphys.2014.00386.

Huysseune, A., & Thesleff, I. (2004). Continuous tooth replacement: the possible involvement of epithelial stem cells. *Bioessays: News and Reviews in Molecular, Cellular and Developmental Biology, 26*, 665–671.

Huysseune, A., Van der heyden, C., & Sire, J.-Y. (1998). Early development of the zebrafish (*Danio rerio*) pharyngeal dentition (Teleostei, Cyprinidae). *Anatomy and Embryology, 198*, 289–305.

Huysseune, A., & Verraes, W. (1990). Carbohydrate histochemistry of mature chondroid bone in *Astatotilapia elegans* (Teleostei: cichlidae) with a comparison to acellular bone and cartilage. *Annales des Sciences Naturelles, Zoologie, Paris, 13*(11), 29–43.

Hwang, W. Y., Fu, Y., Reyon, D., Maeder, M. L., Tsai, S. Q., Sander, J. D., … Joung, J. K. (2013). Efficient genome editing in zebrafish using a CRISPR-Cas system. *Nature Biotechnology, 31*(3), 227–229.

Inohaya, K., Takano, Y., & Kudo, A. (2007). The teleost intervertebral region acts as a growth center of the centrum: in vivo visualization of osteoblasts and their progenitors in transgenic fish. *Developmental Dynamics, 236*(11), 3031–3046.

Janvier, P. (2015). Facts and fancies about early fossil chordates and vertebrates. *Nature, 520*, 483–489.

Jemielita, M., Taormina, M. J., DeLaurier, A., Kimmel, C. B., & Parthasarathy, R. (2013). Comparing phototoxicity during the development of a zebrafish craniofacial bone using confocal and light sheet fluorescence microscopy techniques. *Journal of Biophotonics, 6*(11–12), 920–928.

Jeradi, S., & Hammerschmidt, M. (2016). Retinoic acid-induced premature osteoblast-to-preosteocyte transitioning has multiple effects on calvarial development. *Development, 143*(7), 1205–1216.

Jockusch, E. L. (2007). Genome size. In B. K. Hall, & M. Olson (Eds.), *Keywords and concepts in evolutionary developmental biology* (pp. 152–155). Boston: Harvard University Press.

Johnston, I. A. (1981). Structure and function of fish muscles. *Symposia of the Zoological Society of London, 48*, 71–113.

Jussila, M., Crespo Yanez, X., & Thesleff, I. (2014). Initiation of teeth from the dental lamina in the ferret. *Differentiation; Research in Biological Diversity, 87*, 32–43.

Kague, E., Roy, P., Asselin, G., Hu, G., Simonet, J., Stanley, A., … Fisher, S. (2016). Osterix/Sp7 limits cranial bone initiation sites and is required for formation of sutures. *Developmental Biology, 413*(2), 160–172.

Klein, O. D., Oberoi, S., Huysseune, A., Hovorakova, M., Peterka, M., & Peterkova, R. (2013). Developmental disorders of the dentition: an update. *American Journal of Medical Genetics Part C: Seminars in Medical Genetics, 163C*, 318–332.

Knopf, F., Hammond, C., Chekuru, A., Kurth, T., Hans, S., Weber, C. W., … Weidinger, G. (2011). Bone regenerates via dedifferentiation of osteoblasts in the zebrafish fin. *Developmental Cell, 20*(5), 713–724.

Kobat, D., Horton, G. N., & Xu, C. (2011). In vivo two-photon microscopy to 1.6-mm depth in mouse cortex. *Journal of Biomedical Optics, 16*(10). http://dx.doi.org/10.1117/1.3646209.

Kölliker, A. (1859). On the structure of the chorda dorsalis of the plagiostomes and some other fishes, and on the relation of its proper sheath to the development of the vertebrae. *Proceedings of the Royal Society of London, 10*, 214–222.

Kondo, S., Kuwahara, Y., Kondo, M., Naruse, K., Mitani, H., Wakamatsu, Y., … Shima, A. (2001). The medaka rs-3 locus required for scale development encodes ectodysplasin-A receptor. *Current Biology, 11*(15), 1202–1206.

Laue, K., Jänicke, M., Plaster, N., Sonntag, C., & Hammerschmidt, M. (2008). Restriction of retinoic acid activity by Cyp26b1 is required for proper timing and patterning of osteogenesis during zebrafish development. *Development, 135*(22), 3775–3787.

Liu, H., Leslie, E. J., Jia, Z., Smith, T., Eshete, M., Butali, A., et al. (2016). Irf6 directly regulates Klf17 in zebrafish periderm and Klf4 in murine oral epithelium, and dominant-negative KLF4 variants are present in patients with cleft lip and palate. *Human Molecular Genetics, 25*(4), 766–776.

Loizides, M., Georgiou, A. N., Somarakis, S., Witten, P. E., & Koumoundouros, G. (2014). A new type of lordosis and vertebral body compression in gilthead seabream (*Sparus aurata* Linnaeus, 1758): aetiology, anatomy and consequences for survival. *Journal of Fish Diseases, 37*, 949–957.

Longo, S., Riccio, M., & McCune, A. R. (2013). Homology of lungs and gas bladders: insights from arterial vasculature. *Journal of Morphology, 274*(6), 687–703.

Maisey, J. G. (2000). *Discovering fossil fishes.* Boulder Colorado: Westview Press.

Mantoku, A., Chatani, M., Aono, K., Inohaya, K., & Kudo, A. (2016). Osteoblast and osteoclast behaviors in the turnover of attachment bones during medaka tooth replacement. *Developmental Biology, 409*(2), 370–381.

Martínez-Glez, V., Valencia, M., Caparrós-Martín, J. A., Aglan, M., Temtamy, S., Tenorio, J., … Ruiz-Perez, V. L. (2012). Identification of a mutation causing deficient BMP1/mTLD proteolytic activity in autosomal recessive osteogenesis imperfecta. *Human Mutation, 33*(2), 343–350.

McCarthy, N., Liu, J. S., Richarte, A. M., Eskiocak, B., Lovely, C. B., Tallquist, M. D., & Eberhart, J. K. (2016). Pdgfra and Pdgfrb genetically interact during craniofacial development. *Developmental Dynamics, 245*(6), 641–652.

Metscher, B. D. (2009). MicroCT for developmental biology: a versatile tool for high-contrast 3D imaging at histological resolutions. *Developmental Dynamics, 238*(3), 632–640.

Metscher, B. D., & Ahlberg, P. E. (1999). Zebrafish in context: uses of a laboratory model in comparative studies. *Developmental Biology, 210*, 1–4.

Meunier, F. J. (1989). The acellularisation process in osteichthyan bone. *Progress in Zoology, 35*, 443–445.

Meunier, F. J. (2002). Skeleton. In J. Panfili, H. de Pontual, H. Troadec, & P. J. Wright (Eds.), *Manuel of fish sclerochronology* (pp. 65–88). Brest: Ifremer-IRD coedition.

Meunier, F. J. (2011). The osteichtyes, from the paleozoic to the extant time, through histology and palaeohistology of bony tissues. *Comptes Rendus Palevol, 10*, 347–355.

Monreal, A. W., Ferguson, B. M., Headon, D. J., Street, S. L., Overbeek, P. A., & Zonana, J. (1999). Mutations in the human homologue of mouse dl cause autosomal recessive and dominant hypohidrotic ectodermal dysplasia. *Nature Genetics, 22*(4), 366–369.

Moss, M. L. (1961). Studies of the acellular bone of teleost fish. 1. Morphological and systematic variations. *Acta Anatomica, 46*, 343–462.

Mukherjee, K., Ishii, K., Pillalamarri, V., Kammin, T., Atkin, J. F., Hickey, S. E., … Liao, E. C. (2016). Actin capping protein CAPZB regulates cell morphology, differentiation, and neural crest migration in craniofacial morphogenesis. *Human Molecular Genetics, 25*(7), 1255–1270.

Neues, F., Goerlich, R., Renn, J., Beckmann, F., & Epple, M. (2007). Skeletal Deformations in medaka (*Oryzias latipes*) visualized by synchrotron radiation micro-computer tomography (SR mu CT). *Journal of Structural Biology, 160*(2), 236–240.

Owen, R. (1854). *The principal forms of the skeleton and the teeth as the basis for a system of natural history and comparative anatomy.* New York: William Wood.

Pantazis, P., Maloney, J., Wu, D., & Fraser, S. E. (2010). Second harmonic generating (SHG) nanoprobes for in vivo imaging. *Proceedings of the National Academy of Sciences. USA, 107*(33), 14535–14540.

Parfitt, A. M. (1988). Bone remodelling: relationship to amount and structure of bone, and the pathogenesis and prevention of fractures. In B. L. Riggs, & L. J. Melton (Eds.), *Osteoporosis, etiology, diagnosis, and management* (pp. 45–93). New York: Raven Press.

Pasco-Viel, E., Charles, C., Chevret, P., Semon, M., Tafforeau, P., Viriot, L., & Laudet, V. (2010). Evolutionary trends of the pharyngeal dentition in cypriniformes (Actinopterygii: Ostariophysi). *PLoS One, 5*(6), e11293. http://dx.doi.org/10.1371/journal.pone.0011293.

Peyrard-Janvid, M., Leslie, E. J., Kousa, Y. A., Smith, T. L., Dunnwald, M., Magnusson, M., … Schutte, B. C. (2014). Dominant mutations in GRHL3 cause Van der Woude syndrome and disrupt oral periderm development. *American Journal of Human Genetics, 94*(1), 23–32.

Piotrowski, T., Schilling, T. F., Brand, M., Jiang, Y. J., Heisenberg, C. P., Beuchle, D., ... Nüsslein-Volhard, C. (1996). Jaw and branchial arch mutants in zebrafish II: anterior arches and cartilage differentiation. *Development, 123*, 345–356.

Reif, W. E. (1982). Evolution of dermal skeleton and dentition in vertebrates. The odontode regulation theory. *Evolutionary Biology (New York), 304*(15), 287–368.

Renn, J., Büttner, A., To, T. T., Chan, S. J., & Winkler, C. (2013). A col10a1:nlGFP transgenic line displays putative osteoblast precursors at the medaka notochordal sheath prior to mineralization. *Developmental Biology, 381*(1), 134–143.

Renn, J., & Winkler, C. (2009). Osterix-mCherry transgenic medaka for in vivo imaging of bone formation. *Developmental Dynamics, 238*(1), 241–248.

Renn, J., & Winkler, C. (2014). Osterix/Sp7 regulates biomineralization of otoliths and bone in medaka (*Oryzias latipes*). *Matrix Biology, 34*, 193–204.

Romer, A. S. (1970). *The vertebrate body.* Philadelphia: W. B. Saunders Company.

Sharpe, J., Ahlgren, U., Perry, P., Hill, B., Ross, A., Hecksher-Sorensen, J., ... Davidson, D. (2002). Optical projection tomography as a tool for 3D microscopy and gene expression studies. *Science, 296*(5567), 541–545.

Shkil, F. N., Stolero, B., Sutton, G. A., Belay Abdissa, B., Dmitriev, S. G., & Shahar, R. (2014). Effects of thyroid hormone treatment on the mineral density and mechanical properties of the African barb (*Labeobarbus intermedius*) skeleton. *Journal of Applied Ichthyology, 30*, 814–820.

Sire, J.-Y., & Akimenko, M. A. (2004). Scale development in fish: a review, with description of sonic hedgehog (shh) expression in the zebrafish (*Danio rerio*). *International Journal of Developmental Biology, 48*, 233–247.

Sire, J.-Y., Donoghue, P. C. J., & Vickaryous, M. K. (2009). Origin and evolution of the integumentary skeleton in non-tetrapod vertebrates. *Journal of Anatomy, 214*, 409–440.

Sire, J.-Y., & Huysseune, A. (2003). Formation of dermal skeletal and dental tissues in fish: a comparative and evolutionary approach. *Biological Reviews of the Cambridge Philosophical Society, 78*, 219–249.

Sire, J.-Y., Huysseune, A., & Meunier, J. (1990). Osteoclasts in the teleost fish: light and electron-microscopical observations. *Cell and Tissue Research, 260*, 85–94.

Smith, M. M., & Hall, B. K. (1990). Development and evolutionary origins of vertebrate skeletogenic and odontogenic tissues. *Biological Reviews of the Cambridge Philosophical Society, 65*, 277–373.

Sousa, S., Afonso, N., Bensimon-Brito, A., Fonseca, M., Simões, M., Leon, J., ... Jacinto, A. (2011). Differentiated skeletal cells contribute to blastema formation during zebrafish fin regeneration. *Development, 138*, 3897–3905.

Spoorendonk, K. M., Peterson-Maduro, J., Renn, J., Trowe, T., Kranenbarg, S., Winkler, C., & Schulte-Merker, S. (2008). Retinoic acid and Cyp26b1 are critical regulators of osteogenesis in the axial skeleton. *Development, 135*(22), 3765–3774.

Springer, V. G., & Johnson, G. D. (2000). Use and advantages of ethanol solution of alizarin red S dye for staining bone in fishes. *Copeia, 1*, 300–301.

Stock, D. W. (2007). Zebrafish dentition in comparative context. *Journal of Experimental Zoology, 308B*, 523–549.

Sun, C. K., Chu, S. W., Chen, S. Y., Tsai, T. H., Liu, T. M., Lin, C. Y., & Tsaj, H. J. (2004). Higher harmonic generation microscopy for developmental biology. *Journal of Structural Biology, 147*(1), 19–30.

Swartz, M. E., Sheehan-Rooney, K., Dixon, M. J., & Eberhart, J. K. (2011). Examination of a palatogenic gene program in zebrafish. *Developmental Dynamics, 240*(9), 2204–2220.

Taylor, W. R. (1967). An enzyme method of clearing and staining small vertebrates. *Proceedings of the United States National Museum, 122*(3596), 1–17.

To, T. T., Witten, P. E., Huysseune, A., & Winkler, C. (2015). An adult osteopetrosis model in medaka reveals importance of osteoclast function for bone remodeling in teleost fish. *Comparative Biochemistry and Physiology. Part C: Comparative Pharmacology, 178*, 68–75.

To, T. T., Witten, P. E., Renn, J., Bhattacharya, D., Huysseune, A., & Winkler, C. (2012). Rankl induced osteoclastogenesis leads to loss of mineralization in a medaka osteoporosis model. *Development, 139*, 141–150.

Van der heyden, C., & Huysseune, A. (2000). Dynamics of tooth formation and replacement in the zebrafish (*Danio rerio*) (Teleostei, Cyprinidae). *Developmental Dynamics, 219*, 486–496.

Van der heyden, C., Huysseune, A., & Sire, J.-Y. (2000). Development and fine structure of pharyngeal replacement teeth in juvenile zebrafish (*Danio rerio*) (Teleostei, Cyprinidae). *Cell and Tissue Research, 302*, 205–219.

Vandenplas, S., Willems, M., Witten, P. E., Hansen, T., Fjelldal, P. G., & Huysseune, A. (2016). Epithelial label retaining cells are absent during tooth cycling in *Salmo salar* and *Polypterus senegalus*. *PLoS One*. http://dx.doi.org/10.1371/journal.pone.0152870.

Vandewalle, P., Gluckmann, I., & Wagemans, F. (1998). A critical assessment of the alcian blue/alizarine double staining in fish larvae and fry. *Belgian Journal of Zoology, 128*(1), 93–95.

Walker, M. B., & Kimmel, C. B. (2007). A two-color acid-free cartilage and bone stain for zebrafish larvae. *Biotechnic & Histochemistry, 82*(1), 23–28.

Walls, J. R., Sled, J. G., Sharpe, J., & Henkelman, R. M. (2007). Resolution improvement in emission optical projection tomography. *Physics in Medicine and Biology, 52*(10), 2775–2790.

Wassersug, R. J. (1976). A procedure for differential staining of cartilage and bone in whole formalin-fixed vertebrates. *Stain Technology, 51*(2), 131–134.

Wautier, K., Van der heyden, C., & Huysseune, A. (2001). A quantitative analysis of pharyngeal tooth shape in the zebrafish (*Danio rerio*, Teleostei, Cyprinidae). *Archives of Oral Biology, 46*, 67–75.

Whitlock, J. A., & Richman, J. M. (2013). Biology of tooth replacement in amniotes. *International Journal of Oral Science, 5*, 66–70.

Willems, B., Buettner, A., Huysseune, A., Renn, J., Witten, P. E., & Winkler, C. (2012). Conditional ablation of osteoblasts in medaka. *Developmental Biology, 364*, 128–137.

Witten, P. E. (1997). Enzyme histochemical characteristics of osteoblasts and mononucleated osteoclasts in a teleost fish with acellular bone (*Oreochromis niloticus*, Cichlidae). *Cell and Tissue Research, 287*, 591–599.

Witten, P. E., Gil-Martens, L., Huysseune, A., Takle, H., & Hjelde, K. (2009). Towards a classification and an understanding of developmental relationships of vertebral body malformations in Atlantic salmon (*Salmo salar* L.). *Aquaculture, 295*, 6–14.

Witten, P. E., & Hall, B. K. (2003). Seasonal changes in the lower jaw skeleton in male Atlantic salmon (*Salmo salar* L.): remodelling and regression of the kype after spawning. *Journal of Anatomy, 203*, 435–450.

Witten, P. E., & Hall, B. K. (2015). Teleost skeletal plasticity: modulation, adaptation, and remodelling. *Copeia, 103*(4), 727–739.

Witten, P. E., Hansen, A., & Hall, B. K. (2001). Features of mono- and multinucleated bone resorbing cells of the zebrafish *Danio rerio* and their contribution to skeletal development, remodeling and growth. *Journal of Morphology, 250*, 197–207.

Witten, P. E., Holliday, L. S., Delling, G., & Hall, B. K. (1999). Immunohistochemical identification of a vacuolar proton pump (V-ATPase) in bone-resorbing cells of an advanced teleost species (*Oreochromis niloticus*, Teleostei, Cichlidae). *Journal of Fish Biology, 55*, 1258–1272.

Witten, P. E., & Huysseune, A. (2007). Mechanisms of chondrogenesis and osteogenesis in fins. In B. K. Hall (Ed.), *Fins into limbs: evolution, development, and transformation* (pp. 79–92). Chicago: The University of Chicago Press.

Witten, P. E., & Huysseune, A. (2009). A comparative view on mechanisms and functions of skeletal remodelling in teleost fish, with special emphasis on osteoclasts and their function. *Biological Reviews of the Cambridge Philosophical Society, 84*, 315–346.

Witten, P. E., Huysseune, A., & Hall, B. K. (2010). A practical approach for the identification of the many cartilaginous tissues in teleost fish. *Journal of Applied Ichthyology, 26*(2), 257–262.

Witten, P. E., Owen, M. A. G., Fontanillas, R., Soenens, M., McGurk, C., & Obach, A. (2016). Primary phosphorous-deficiency in juvenile Atlantic salmon: the uncoupling of bone formation and mineralisation. *Journal of Fish Biology, 88*, 690–708.

Wiweger, M. I., Avramut, C. M., de Andrea, C. E., Prins, F. A., Koster, A. J., Ravelli, R. B., & Hogendoorn, P. C. (2011). Cartilage ultrastructure in proteoglycan-deficient zebrafish mutants brings to light new candidate genes for human skeletal disorders. *Journal of Pathology, 3*(4), 531–542.

Yu, T., Witten, P. E., Huysseune, A., Buettner, A., To, T. T., & Winkler, C. (2016). Live imaging of osteoclast inhibition by bisphosphonates in a medaka osteoporosis model. *Disease Models & Mechanisms, 9*, 155–163.

Zehbe, R., Haibel, A., Riesemeier, H., Gross, U., Kirkpatrick, C. J., Schubert, H., & Brochhausen, C. (2010). Going beyond histology. Synchrotron micro-computed tomography as a methodology for biological tissue characterization: from tissue morphology to individual cells. *Journal of The Royal Society Interface, 7*, 49–59.

Zhong, Z., Niu, P., Wang, M., Huang, G., Xu, S., Sun, Y., et al. (2016). Targeted disruption of sp7 and myostatin with CRISPR-Cas9 results in severe bone defects and more muscular cells in common carp. *Scientific Reports, 6*, 22953. http://dx.doi.org/10.1038/srep22953.

CHAPTER

Muscular dystrophy modeling in zebrafish

14

M. Li*, K.J. Hromowyk§, S.L. Amacher§, P.D. Currie*,[1]

**Monash University, Clayton, VIC, Australia*

§The Ohio State University, Columbus, OH, United States

[1]Corresponding author: E-mail: peter.currie@monash.edu

CHAPTER OUTLINE

Methods in Cell Biology, Volume 138, ISSN 0091-679X, http://dx.doi.org/10.1016/bs.mcb.2016.11.004

Abstract

Skeletal muscle performs an essential function in human physiology with defects in genes encoding a variety of cellular components resulting in various types of inherited muscle disorders. Muscular dystrophies (MDs) are a severe and heterogeneous type of human muscle disease, manifested by progressive muscle wasting and degeneration. The disease pathogenesis and therapeutic options for MDs have been investigated for decades using rodent models, and considerable knowledge has been accumulated on the cause and pathogenetic mechanisms of this group of human disorders. However, due to some differences between disease severity and progression, what is learned in mammalian models does not always transfer to humans, prompting the desire for additional and alternative models. More recently, zebrafish have emerged as a novel and robust animal model for the study of human muscle disease. Zebrafish MD models possess a number of distinct advantages for modeling human muscle disorders, including the availability and ease of generating mutations in homologous disease-causing genes, the ability to image living muscle tissue in an intact animal, and the suitability of zebrafish larvae for large-scale chemical screens. In this chapter, we review the current understanding of molecular and cellular mechanisms involved in MDs, the process of myogenesis in zebrafish, and the structural and functional characteristics of zebrafish larval muscles. We further discuss the insights gained from the key zebrafish MD models that have been so far generated, and we summarize the attempts that have been made to screen for small molecules inhibitors of the dystrophic phenotypes using these models. Overall, these studies demonstrate that zebrafish is a useful in vivo system for modeling aspects of human skeletal muscle disorders. Studies using these models have contributed both to the understanding of the pathogenesis of muscle wasting disorders and demonstrated their utility as highly relevant models to implement therapeutic screening regimens.

INTRODUCTION

Skeletal muscle is an important component of animal physiology, generating contractile force on demand and carrying out locomotor function. In humans, skeletal muscle dysfunctions lead to various types of diseases. Inherited muscle diseases are relatively rare, but are often devastating, causing disability and premature death in patients. The mechanistic basis for these muscle disorders is heterogeneous, ranging from defects in cell membrane integrity and structural maintenance to dysfunctional intracellular compartments (Laing, 2012). Inherited muscle diseases can be classified into two distinct groups. Mutations in cytoskeletal proteins that interact with or regulate the contractile apparatus usually result in congenital myopathies (Goebel, 2011), whereas deficiencies in extracellular matrix (ECM) and cytoskeleton links are the causes of most muscular dystrophies (MDs) (Campbell, 1995; Tubridy, Fontaine, & Eymard, 2001). In this review, we will focus on the latter.

MDs are a severe group of muscle disorders that manifest clinically by progressive muscle wasting and weakness. MDs can be subdivided into more than 30 distinct genetic types (Emery, 2002). Unlike other muscle disorders, the hallmarks of MDs are commonly compromised sarcolemma or defective basement membrane integrity of muscle cells. Defective myofibers undergo repetitive cycles of degeneration and regeneration, leading to increased numbers of small muscle fibers with central nuclei, infiltration of inflammatory cells, and fibrosis (reviewed in McNally & Pytel, 2007). One of the most severe and common types of MD is Duchenne MD (DMD), in which disease symptoms begin in early childhood and patients become wheelchair-bound as teenagers and die prematurely in their 20s (Emery, 1993).

Treatment options for MDs are sparse. Currently, there is no cure for the major MDs, despite their devastating clinical consequences. For instance, the use of antiinflammation drugs with strong side effects is the standard clinical management for DMD patients (Emery, 2002; Jay & Vajsar, 2001). Promisingly, alternative therapeutics have been under investigation in the past decades (Beytía, Vry, & Kirschner, 2012). Gene and cell therapy approaches have been applied in MD research, and notably, newly discovered candidate drugs have shown promising effects in clinical trials (Bengtsson, Seto, Hall, Chamberlain, & Odom, 2016; Pichavant et al., 2011). However, as of yet, there is no cure for this disorder.

Zebrafish has emerged as an important vertebrate for disease modeling (Lieschke & Currie, 2007). It offers unique practical advantages, including rapid development, optical transparency of embryos, ease of genetic screening, and transgenesis. Although evolutionarily distant from human, zebrafish share approximately 70% of human protein-coding genes (Howe et al., 2013). Forward and reverse genetic manipulation tools, including N-ethyl N-nitrosourea (ENU)-based mutagenesis, morpholino-induced knockdown, and CRISPR- and transcription activator-like effector nucleases (TALEN)-based genome editing, have further strengthened the use of zebrafish to identify causative genes in human disorders (Berger & Currie, 2012; Lieschke & Currie, 2007; Parant & Yeh, 2016; Santoriello & Zon, 2012).

In addition to the general advantages of using zebrafish for disease modeling, muscle research in zebrafish is also facilitated by and relevant due to additional features: (1) a significant fraction of zebrafish body is composed of skeletal muscle that becomes functional at early embryonic stage (before the end of the first day of development), (2) both slow and fast muscle fiber types are present and organized into myotomes that are derived from the embryonic segments, the somites, and (3) at a cellular level, myofibrillar structure, contractile properties, and metabolic functions are similar between zebrafish and human skeletal muscle (Section 2). Importantly, the optical clarity of zebrafish embryos and larvae allows investigation of muscle function in a living organism. Furthermore, a battery of sophisticated fluorescent transgenic tools has been designed to study distinct muscle compartments. In 2003 the first zebrafish MD model, the *sapje (dmd)* mutant, was molecularly characterized (Bassett et al., 2003). The study demonstrated for the first time that zebrafish mutants could be used to model and investigate the pathogenesis of human MD. It was later reported that *sapje (dmd)* mutants also provide a robust system for compound screening (Kawahara, Karpf, et al., 2011; Peterson, Link, Dowling, & Schreiber, 2000). In addition to small molecule treatment, other therapeutic strategies, such as exon skipping by antisense oligonucleotides (AOs), have been tested in the *sapje (dmd)* model (Berger, Berger, Jacoby, Wilton, & Currie, 2011). Currently, many other zebrafish MD genetic mutants are reported, including *candyfloss (lama2)* (congenital muscular dystrophy type 2), *patchytail (dag1)* (dystroglycanopathy), and others (summarized in Table 1).

Compared to other animal models used in MD research (reviewed in Ng et al., 2012), it is relatively quick and easy to introduce targeted mutations in zebrafish (Parant & Yeh, 2016; Talbot & Amacher, 2014), to examine muscle development and to carry out genetic screens in a large number of animals (Patton & Zon, 2001), and to monitor real-time cell events using live imaging (Vacaru et al., 2014). While disadvantages do exist, including evolutionary distance from mammals and often a lack of sophisticated metabolic and functional assays, the similar phenotypes of zebrafish models and human MDs demonstrate that zebrafish research is likely to inform us about the human pathological condition.

1. MUSCULAR DYSTROPHIES AND THE DYSTROPHIN-ASSOCIATED GLYCOPROTEIN COMPLEX

MDs are a large group of clinically and genetically heterogeneous disorders affecting striated muscles in human patients. Compared to other muscle diseases, most human MDs are caused by deficits in the components of either the ECM or the Dystrophin-associated glycoprotein complex. MDs are manifested by progressive muscle degeneration and weakness (Emery, 2002). DMD results from loss of function of the Dystrophin protein. Partial loss of Dystrophin function leads to Becker muscular dystrophy (BMD), which has milder clinical symptoms and often later onset than DMD (Bieber & Hoffman, 1990). Other major MD forms include congenital, limb-girdle, and Emery–Dreifuss MDs (Emery, 2002).

1.1 DYSTROPHIN-ASSOCIATED GLYCOPROTEIN COMPLEX

In striated muscle cells, the key structure bridging the cytoskeleton and ECM is the Dystrophin-associated glycoprotein complex (DGC) (Campbell, 1995; Guyon et al., 2003). It is composed of multiple proteins, including cytosolic proteins (e.g., Dystrophin, Syntrophin, Dystrobrevin) and transmembrane proteins (Dystroglycan, Sarcoglycan, Sarcospan) (Fig. 1). In the extracellular space, Laminin-2 binds DGC via α-Dystroglycan, connecting the sarcolemma to ECM. As a core component of DGC that maintains sarcolemmal integrity, Dystrophin is found on the cytoplasmic surface and binds F-actin and β-Dystroglycan. The tightly connected α and β-Dystroglycans form an integral membrane subcomplex, providing additional structural support to the DGC.

As a structural complex, the DGC has an essential role in transmitting force and stabilizing the membrane during contraction. Loss of DGC function causes vital sarcolemmal rupture upon mechanical stress (reviewed in Campbell, 1995; Lapidos, Kakkar, & McNally, 2004). In addition to the structural support role, the DGC is also

Table 1 Zebrafish Models of Human Muscular Dystrophies

Gene	Symbol	Localization	Human Disease	Zebrafish Model	Key References
dysferlin	*dysf*	Sarcolemma	Dysferlinopathy (LGMD2B); Miyoshi myopathy (MM)	MO, mutant	Kawahara, Serafini, et al. (2011), Middel et al. (2016), and Roostalu and Strahle (2012)
dystrophin	*dmd*	Cytosol	Duchenne muscular dystrophy (Duchenne MD)	MO, mutant	Bassett et al. (2003), Guyon et al. (2009), and Guyon et al. (2003)
dystroglycan	*dag*	Sarcolemma	Limb-girdle MD (LGMD)	MO, mutant	Gupta et al. (2011) and Parsons et al. (2002)
double homeobox 4	*dux4*	Cytosol	Facioscapulohumeral MD (FSHD)	Overexpression	Mitsuhashi, Mitsuhashi, Lynn-Jones, Kawahara, and Kunkel (2013) and Wallace et al. (2011)
sarcoglycan	*sgc*	Sarcolemma	Sarcoglycanopathy (LGMD2C-F)	MO	Cheng et al. (2006), Guyon et al. (2005), and Vogel et al. (2009)
telethonin	*tcap*	Cytosol	Limb-girdle MD (LGMD2G)	MO	Zhang, Yang, Zhu, and Xu (2009)
lamin A/C	*lmna*	Cytosol	Emery–Dreifuss MD (LGMD1B)	MO	Koshimizu et al. (2011)
laminin alpha2	*lama2*	Sarcolemma	Congenital Merosin-deficient MD, type 1A (MDC1A)	Mutant	Gupta et al. (2012) and Hall et al. (2007)

Continued

Table 1 Zebrafish Models of Human Muscular Dystrophies—cont'd

Gene	Symbol	Localization	Human Disease	Zebrafish Model	Key References
integrin alpha7	*itga7*	Sarcolemma	Congenital MD	MO	Postel et al. (2008)
fukutin	*fktn*	Cytosol	Fukuyama congenital MD (LGMD2M)	MO	Lin et al. (2011) and Wood et al. (2011)
Fukutin-related protein	*fkrp*	Cytosol	Limb-girdle MD, type 2I (LGMD2I)	MO	Kawahara, Guyon, Nakamura, and Kunkel (2010), Lin et al. (2011), Thornhill et al. (2008), and Wood et al. (2011)
caveolin3	*cav3*	Sarcolemma	Limb-girdle MD, type 1C (LGMD1C)	MO	Nixon et al. (2005)
collagen6	*col6*	Extracellular matrix	Ullrich congenital MD (UCMD), Bethlem myopathy (BM)	MO, mutant	Radev et al. (2015) and Telfer et al. (2010)
popeye domain-containing1	*popdc1*	Cytosol	Limb-girdle MD (LGMD)	MO, mutant	Schindler et al. (2016)

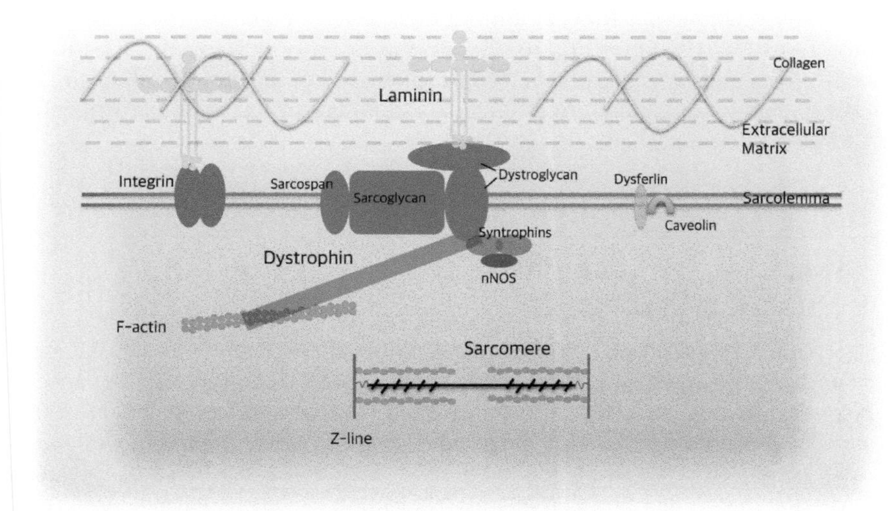

FIGURE 1 Dystrophin-associated glycoprotein complex and other key protein components in striated muscle cell.

a cellular signaling scaffold that anchors signaling proteins to the complex site. A central DGC component, Dystroglycan subcomplex, serves as a membrane receptor. α-Dystroglycan binds several extracellular ligands, including Laminin, Agrin, and Perlecan (Gee, Montanaro, Lindenbaum, & Carbonetto, 1994; Ibraghimov-Beskrovnaya et al., 1992; Peng et al., 1998; Talts, Andac, Gohring, Brancaccio, & Timpl, 1999). Inside the muscle cell, β-Dystroglycan relays the signals by interacting with Dystrophin and Utrophin and components of the extracellular signal-regulated kinase-mitogen-activated protein (ERK-MAP) kinase cascade (reviewed in Barresi & Campbell, 2006; Ervasti & Sonnemann, 2008). Moreover, glycosylation of Dystroglycan is crucial for its function and ligand interactions, indicating that posttranslational modification is involved in congenital muscular disorders (Michele et al., 2002; Muntoni, Brockington, Torelli, & Brown, 2004).

In addition to the DGC core proteins, functional disruption of other molecules is also associated with MDs, including the matrix protein Laminin α2 (congenital MD, Helbling-Leclerc et al., 1995), the transmembrane protein Dysferlin (Limb Girdle MD2B, Miyoshi myopathy, Liu et al., 1998), the calcium-activated protease Calpain (LGMD2A, Richard et al., 1995), Telethonin (Limb Girdle MD2G, Moreira et al., 2000), the intermediate filament protein Lamin A/C (Emery−Dreifuss MD, Limb Girdle MD1B, di Barletta et al., 2000; Muchir et al., 2000), and Collagen VI (Ullrich congenital MD, Bethlem myopathy, Lampe & Bushby, 2005). These genes are highly conserved among species; homologs of the DGC and associated molecules have been identified in mammalian models and in zebrafish.

Zebrafish MD models have been established using different strategies, including forward genetic screens and targeted genetic manipulations (Table 1).

Specifically, technologies for gene editing, such as zinc finger nucleases (Amacher, 2008), TALENs (Huang et al., 2011), and the recently developed CRISPR-Cas system (Hwang et al., 2013), have been applied and optimized in zebrafish. The improvements in genome editing efficiency continue to facilitate the generation of additional disease models.

2. SKELETAL MUSCLE PROPERTIES IN ZEBRAFISH

2.1 MUSCLE DEVELOPMENT

Myogenesis is a vital process in development and maintenance of muscle, and many mechanistic aspects are shared among vertebrate, and even invertebrate, species (Bryson-Richardson & Currie, 2008; Pownall, Gustafsson, & Emerson, 2002). Understanding myogenesis is of great value in MD research, as disease muscle-wasting results from repeated degeneration/regeneration cycles. In MD, impaired regeneration due to muscle satellite cell dysfunction can also contribute to disease progression (Dumont et al., 2015). Thus, knowledge of the molecular and cellular basis of myogenesis and muscle regeneration is crucial for developing potential treatments.

2.1.1 Segmentation and myogenic onset

In zebrafish, like other vertebrates, skeletal muscle derives from the paraxial mesoderm anterior to the extending tail bud called the presomitic mesoderm (PSM) (Holley, 2007). The PSM becomes segmented from head to tail into reiterated blocks of mesodermal tissue called somites, which later differentiate to become the future myotomes. The segmentation process, called somitogenesis, is regulated by pulsatile gene expression that oscillates within PSM cells with the same periodicity as somite formation and by global gradients across the embryo, a mechanism broadly termed the "Clock and Wavefront model" (Cooke & Zeeman, 1976; Yabe & Takada, 2016). Recent advances allowing real-time visualization of oscillating gene expression in live zebrafish embryos have shed new light on clock regulatory dynamics (Delaune, François, Shih, & Amacher, 2012; Schröter et al., 2012; Shih, François, Delaune, & Amacher, 2015; Webb et al., 2016; Webb, Soroldoni, Oswald, Schindelin, & Oates, 2014; Yabe & Takada, 2016). During somitogenesis, the processes of somite polarization and myogenesis are also underway, evident by spatially distinct expression of anterior and posterior somite compartment markers, some of which are myogenic regulatory factors (MRFs) (Kawamura et al., 2005; Oates, Rohde, & Ho, 2005; Sawada et al., 2000; Windner, Bird, Patterson, Doris, & Devoto, 2012; Windner et al., 2015).

It is well established that vertebrate muscle specification is associated with expression of MRF transcription factors (Pownall et al., 2002). The zebrafish MRF genes *myod* and *myf5* are the first MRFs to be expressed in the PSM, with onset occurring in the paraxial mesoderm during gastrulation (Coutelle et al., 2001;

Weinberg et al., 1996). Like in the mouse, function of either *myoD* or *myf5* is required for zebrafish trunk skeletal muscle formation; *myf5;myoD* double mutants, but not singles, lack all trunk muscle (Hinits et al., 2011; Rudnicki et al., 1993). As alluded to above, MRF expression is polarized within the newly formed zebrafish somite. In the anterior somite compartment, *mespb* functions to maintain *pax3* expression, which prevents MRF expression, whereas in the posterior somite compartment, *ripply1* functions to inhibit *tbx6*, which permits MRF expression (Kawamura et al., 2005; Windner et al., 2012, 2015). Establishment of distinct somite compartments during somitogenesis is important for determining the three major myogenic fates found in the developing zebrafish myotome: slow muscle, fast muscle, and the external cell layer (ECL).

2.1.2 Slow/fast muscle specification and morphogenesis

Distinct from many other vertebrates, slow and fast muscle fibers are spatially segregated in the zebrafish myotome (Fig. 2). Elegant lineage tracing experiments showed that slow muscle cell (SMC) and fast muscle cell (FMC) precursors are

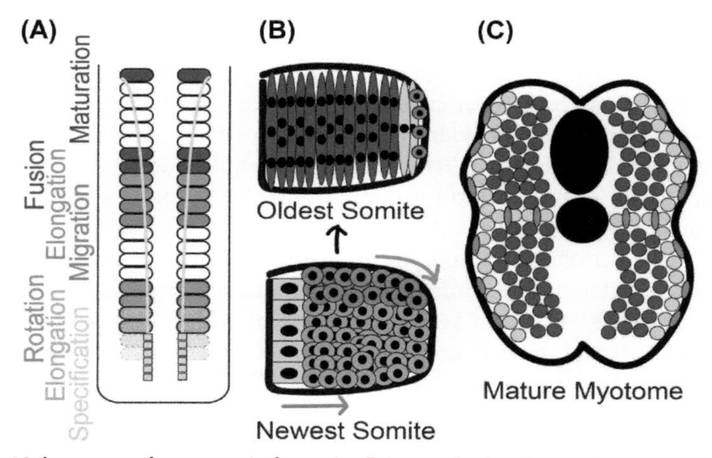

FIGURE 2 Major myogenic events during zebrafish muscle development.

(A) Dorsal view of a 20-somite zebrafish tail depicting the onset of major developmental and morphogenetic events from initial specification to maturation. The *yellow line* represents the position of slow muscle cell (SMC) precursors, also called "adaxial cells," as they migrate from the medial to lateral edge of each somite. (B) Illustration of the distinct cellular compartments found in immature (bottom) and mature (top) somites. Note that the SMCs (in yellow) march laterally (*green arrow*) through the somite to take up their final superficial position, whereas the anterior border cells (ABCs) (in red) adopt their final position in the external cell layer (ECL) through the process of somite rotation (*red arrow*). (C) Cross-sectional representation of muscle cell organization in the larval zebrafish myotome. Fast myofibers are in blue, slow myofibers are in yellow, and Pax7-positive satellite-like cells are in red. Muscle pioneer cells are specialized slow myofibers that do not migrate laterally and form the horizontal myoseptum that separates a somite into dorsal and ventral halves.

morphologically distinguishable even prior to somite formation (Devoto, Melançon, Eisen, & Westerfield, 1996). SMCs are the first muscle cells to mature and elongate into myofibers and are derived from a medial row of *myod*- and *myf5*-expressing PSM "adaxial" cells, located adjacent to the midline (Coutelle et al., 2001; Devoto et al., 1996; Thisse, Thisse, Schilling, & Postlethwait, 1993). Sonic hedgehog (Shh) signaling from the midline is both necessary and sufficient to induce SMC fate (Blagden, Currie, Ingham, & Hughes, 1997; Currie & Ingham, 1996; Du, Devoto, Westerfield, & Moon, 1997; Ingham & Kim, 2005; Lewis et al., 1999). One Shh target is *prdm1a*, which encodes an SMC transcription factor that permits expression of SMC downstream genes in part by inhibiting activity of Sox6, an FMC transcription factor that represses the SMC program (Baxendale et al., 2004; von Hofsten et al., 2008). The FMC gene expression program requires both Sox6 function and the combined action of Myod with Pbx2/4 homeodomain transcription factors at FMC-specific target promoters to drive FMC maturation (von Hofsten et al., 2008; Jackson et al., 2015; Maves et al., 2007).

Once formed, SMCs undergo a dramatic lateral migration through the somite to form the most superficial muscle cell layer (Devoto et al., 1996), although a small subset of Engrailed-expressing SMCs, referred to as muscle pioneers (Hatta, Bremiller, Westerfield, & Kimmel, 1991), remains behind and forms the horizontal myoseptum that separates the myotome into dorsal and ventral halves (Fig. 2). As SMCs migrate laterally, FMCs elongate in their wake (Cortés et al., 2003; Henry & Amacher, 2004) and are dependent upon SMCs for their timely elongation and subsequent maturation (Henry & Amacher, 2004). Unlike embryonic SMCs, which are mononucleate, FMCs fuse to form multinucleated myofibers (Moore, Parkin, Bidet, & Ingham, 2007; Roy, Wolff, & Ingham, 2001; Srinivas, Woo, Leong, & Roy, 2007). The vertebrate-specific *jam2a* and *jam3b* genes, which are expressed in FMCs and encode transmembrane proteins, are required for FMC fusion (Powell & Wright, 2011). Knockdown studies have revealed that muscle fusion factors initially discovered in mouse and flies (Rochlin, Yu, Roy, & Baylies, 2010), such as Tmem8c (Myomaker) (Landemaine, Rescan, & Gabillard, 2014), Crk, Crkl, DOCK1, and DOCK5 (Moore et al., 2007), CKIP-1 (Baas et al., 2012), and *kirrel3l* (Srinivas et al., 2007) are also implicated in zebrafish FMC fusion. The multinucleated fast muscle fibers form the bulk of the zebrafish myotome and are located medially to the superficial slow muscle fibers. Although somitic SMCs and FMCs are largely postmitotic, a third population of myogenic cells, the anterior border cells and their ECL derivatives, continue to proliferate and contribute to continual muscle growth throughout larval stages and adulthood (Hollway et al., 2007; Stellabotte, Dobbs-McAuliffe, Fernández, Feng, & Devoto, 2007).

2.1.3 The dermomyotome-like external cell layer

As SMCs migrate laterally, a third population of somitic cells, the anterior border cells (ABCs; also referred to as Row 1 cells), undergo an Sdf-dependent anterior-to-lateral rotation within each somite to form the somitic external cell layer (ECL)

(Hollway et al., 2007; Stellabotte et al., 2007). ABCs are distinguished from other somitic cells by several criteria; they express markers traditionally associated with the amniote dermomyotome, such as *pax3* and *pax7* homologs and *meox*, they are proliferative, and they contribute to myogenic populations outside the myotome, including appendicular muscle, hypaxial muscle, dermal cells, and dorsal fin cells (Groves, Hammond, & Hughes, 2005; Hammond et al., 2007; Hollway et al., 2007; Stellabotte & Devoto, 2007). In addition, ABCs contribute to both proliferative Pax7-positive cells in the ECL and quiescent Pax7-positive cells in the myotome, suggesting that they give rise to both muscle progenitors and satellite cells (muscle stem cells) (Hammond et al., 2007; Hollway et al., 2007; Pipalia et al., 2016; Stellabotte et al., 2007). Because zebrafish ECL cells share many characteristics with amniote dermomyotome cells, they are also sometimes referred to as dermomyotome cells (Stellabotte & Devoto, 2007; Windner et al., 2012). Thus, although the terms ECL and dermomyotome are used within the literature to describe the same population of cells, we use the former terminology since ECL was the name initially adopted for cells in this position (Waterman, 1969) and because is not yet known whether zebrafish dermomyotome-like cells fully mimic later characteristics of the amniote dermomyotome (Hollway et al., 2007; Stellabotte et al., 2007; Stellabotte & Devoto, 2007).

2.1.4 Hypaxial muscle

Vertebrate hypaxial muscles from ventrally located somitic cells migrate out of the myotomes to form craniofacial, appendicular (limb/fin), and abdominal muscles (Burke & Nowicki, 2003; Hernandez, Patterson, & Devoto, 2005). In zebrafish, hypaxial musculature contributing to the face (Schilling & Kimmel, 1997) and pectoral fin is specifically derived from anterior somites (Haines et al., 2004; Minchin et al., 2013; Neyt et al., 2000; Ochi & Westerfield, 2009). Jaw muscles, the sternohyoid muscle, and oesophageal striated muscle are derived from migratory muscle precursors in somites 1–3, while pectoral fin muscles are derived from the ventral somitic region of somites 2–4 (Hollway et al., 2007; Minchin et al., 2013; Neyt et al., 2000). The posterior hypaxial muscle derived from somites 5 to 6 develops into a skeletal muscle that attaches to the cleithrum, a membranous bone primarily found in fish and primitive tetrapods (Haines et al., 2004). Like mammalian hypaxial muscles, the zebrafish genes *lbx2*, *mox2*, *pax3*, as well as the receptor tyrosine kinase Met and its ligand Hepatocyte Growth Factor have been linked to hypaxial muscle morphogenesis and migration (Haines et al., 2004; Lou, He, Hu, & Yin, 2012; Minchin et al., 2013; Neyt et al., 2000; Ochi et al., 2009). Additionally, the noncanonical Wnt pathway components Wnt5, Disheveled, and RhoA are required downstream of *lbx2* function (Lou et al., 2012).

2.2 MUSCLE STRUCTURAL CHARACTERISTICS

Zebrafish skeletal muscle is segmented into reiterated chevron-shaped myotomes that make up most of the fish body (Fig. 3). Just as mammals, striated myofibers

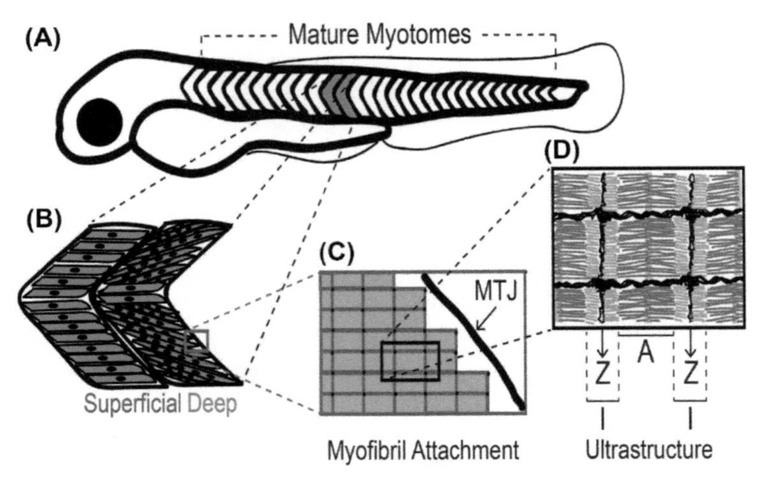

FIGURE 3 Zebrafish muscle organization and ultrastructure.

(A) Diagram illustrating the position and shape of myotomes in a 4 dpf zebrafish. (B) Drawing of two myotomes (4 dpf) at different focal depths depicting the horizontal and angular attachments to the myosepta of the superficial slow myofibers (green) located at the lateral surface and deep fast fibers (red) located more medially. (C) Illustration of an electron micrograph image of the oblique myofibril attachments near the boundary between adjacent myotomes (the myotendinous junction, or MTJ, also called the vertical myoseptum). (D) Depiction of muscle ultrastructure seen by EM. Zebrafish muscle shares typical structural and contractile components with other vertebrates, such as Z-lines (Z), I-bands (I), and A-bands (A).

in zebrafish elongate to span each myotome and connect at each end to a myotendinous junction (MTJ), also known as the vertical myoseptum (Fig. 3A and B). Slow myofibers span the myotome parallel to the long axis of the body, while fast myofibers span the myotome at a more oblique angle (Fig. 3B). Electron microscopy (EM) reveals the oblique attachment of zebrafish sarcomere chains to the MTJ (Bassett et al., 2003) (Fig. 3C). EM also reveals that zebrafish striated muscle is comprised of reiterated sarcomeres and transverse tubules (T-tubules), just as in mammalian muscle (Dowling et al., 2009; Gibbs, Horstick, & Dowling, 2013). A sarcomere is the standard muscle ultra-structural unit and is defined as the segment between two Z-lines, which contains a thick filament (Myosin-containing A-band) and thin filaments (Actin-containing I-bands) (Fig. 3D). T-tubules, which are typically located where the I-band overlaps the A-band, are invaginations of the muscle plasma membrane that connect to the sarcoplasmic reticulum (SR), the major site of calcium storage and release (Block, Imagawa, Campbell, & Franzini-Armstrong, 1988; Hirata et al., 2007; Meissner, 1994).

Mutant studies have shown that sarcomere-associated structural proteins, in addition to the membrane integrin- and dystrophin-associated complexes, are required for proper muscle integrity and function in zebrafish (Berger & Currie, 2012; Gibbs et al., 2013). Zebrafish muscular dystrophy models include mutations in Dystrophin (Bassett et al., 2003; Guyon et al., 2009), Dystroglycan (Gupta et al., 2011), Laminin α2 (Hall et al., 2007), and Titin (Steffen et al., 2007). Additionally, knockdown approaches have implicated additional sarcomere-associated proteins in muscle structure and function (Berger & Currie, 2012; Gibbs et al., 2013; Maves, 2014). T-tubule- and SR-associated components are also required for muscle function. Zebrafish *relatively relaxed (ryr)* mutants, which lack function of the SR-expressed *ryanodine receptor 1 (ryr1)*, are slow swimmers and only weakly contract despite normal neural input, implicating a role for *ryr1* in normal communication between SR and T-tubules and thus for excitation–contraction (E-C) coupling (Hirata et al., 2007). EM has confirmed the intimate association of SR and T-tubules in zebrafish (Dowling et al., 2009), just as in mammals. The similarities between fish and human muscle and the abundance of zebrafish disease models make the zebrafish an ideal system for understanding the mechanisms underlying human muscle disease, particularly the many MDs.

2.3 MECHANICAL FUNCTION

Skeletal muscles are essential for animal physiology and are the organs that generate force and enable physical movements for the whole body. The sarcomeres are the basic units exerting contractile function by sliding and shortening of the thick and thin filaments (Huxley & Niedergerk, 1954). In zebrafish, skeletal muscles comprise a significant percentage of the body mass and are responsible for physical activity throughout entire life span.

Compared to other species, the first muscle contraction in the zebrafish embryo occurs at a very early developmental stage (17 hours post fertilization, hpf), and nerve–muscle coupling becomes progressively more coordinated (Kimmel, Ballard, Kimmel, Ullmann, & Schilling, 1995). Due to the exutero development and transparent chorion, it is easy to monitor the movement and locomotion of the whole zebrafish embryo. The stereotyped progression from spontaneous contractions to swimming initiates with high frequency (0.57 Hz) spontaneous contractions and then advances to touch-evoked coiling and finally to coordinated swimming (Saint-Amant & Drapeau, 1998). Swimming performance can be used to measure mechanical and motor function (Sztal et al., 2015; Telfer, Busta, Bonnemann, Feldman, & Dowling, 2010), especially when performing large-scale screens for muscle phenotypes. Automated imaging and scoring systems, like the Viewpoint Zebrabox (Gibbs et al., 2013), can be used to calculate swimming rate and distance, and thus report on global locomotor activity.

As mentioned previously, the embryonic and larval muscles have morphologically distinct fast and slow territories, with a superficial layer of red (slow) fibers and inner white (fast) fibers. Both fiber types are easily recognized and can be characterized using electrophysiological tools. Experiments carried out on membrane and electrical coupling status demonstrated that both the embryonic red and white fibers were contractile and electrically coupled within each fiber type, allowing synaptic potentials to spread among adjacent myotomes (Buss & Drapeau, 2000).

At the contractile apparatus level, the mechanical properties of early larval muscles were further characterized by Dou, Andersson-Lendahl, and Arner (2008). The maximal isometric force from a single larva reached up to about 1 mN at optimal condition. The larval muscles also displayed a bell-shaped length-active tension behavior, showing a dynamic overlapping of contractile filaments similar to mammalian muscles. Interestingly, like in human and other species (Eston, Mickleborough, & Baltzopoulos, 1995; Warren, Hayes, Lowe, Williams, & Armstrong, 1994), the larval muscles are also vulnerable to eccentric contraction which is a cause for human muscle injuries (Li, Andersson-Lendahl, Sejersen, & Arner, 2013). Thus, it is possible and relevant to use zebrafish tissue to investigate muscle physiology in depth.

Overall, the mechanical properties of fish muscles are similar to other species and perhaps even optimized for rapid swimming behavior (Altringham & Ellerby, 1999). These functional parameters are differentially affected in different muscle diseases and disorders, with the maximal force being a reliable indicator for general muscle weakness. In the zebrafish *sapje (dmd)* mutant, maximal force production is significantly reduced, reflecting the loss of functional myofibers and contractile units (Fig. 4); therapeutic treatment at the optimal dose significantly rescues the force even though structural recovery is less prominent (Li, Andersson-Lendahl, Sejersen, & Arner, 2014).

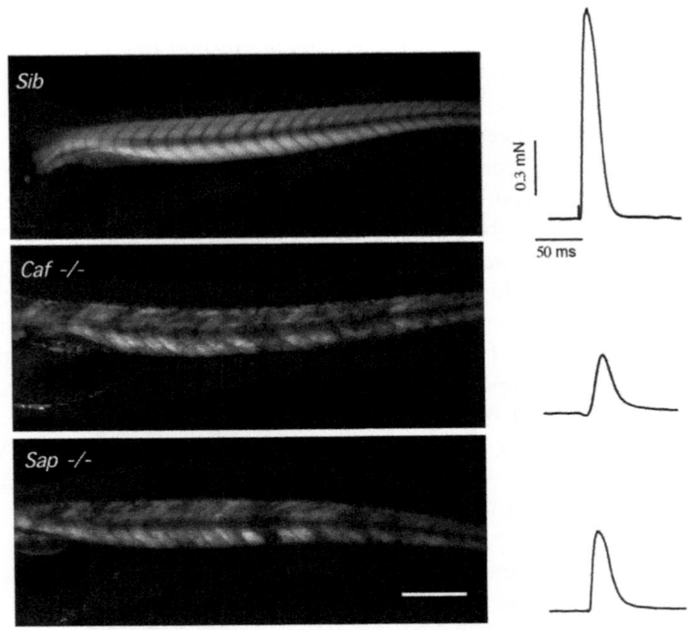

FIGURE 4 Birefringence and mechanical properties of zebrafish larval muscles (5 dpf).

Left: Birefringence images. Right: Maximal isometric contractions from wild-type sibling, *caf (lama2)* and *sapje (dmd)* mutant larvae, respectively.

3. MODELS OF MUSCULAR DYSTROPHY IN ZEBRAFISH

As discussed above, the zebrafish system is an attractive animal for modeling muscle diseases. Optical transparency, rapid development, and availability of direct movement tracking systems allow efficient phenotypic screening and structural and functional analysis. To date, a large number of zebrafish muscular dystrophy models have been generated using different methods including forward and reverse genetics (summarized in Table 1). Many zebrafish muscle mutants, like the *sapje (dmd)* mutant, were isolated from ENU mutagenesis–based forward screening in the 1990s (Granato et al., 1996). Additional target-specific genetic manipulations have also been widely used. For instance, injections of morpholino antisense-oligonucleotides (MOs) offer a quick alternative to assess the phenotype of zebrafish lacking a specific gene (Nasevicius & Ekker, 2000). The transient knockdown of *dystrophin* expression by MOs resulted in a dystrophic phenotype (Guyon et al., 2003). These zebrafish dystrophy models not only facilitate the understanding of disease pathogenesis but also accelerate therapeutic-screening process (see Section 4).

3.1 DUCHENNE MUSCULAR DYSTROPHY

DMD is an X-linked human recessive disorder at the severe end of the clinical spectrum of muscular dystrophy pathologies. The first noticeable onset of disease usually occurs in early childhood with symptoms rapidly progressing. Severe muscle weakness leads to disability during adolescence, and respiratory or cardiac failure often is the cause of death in adults (Emery, 1993). The primary causes of the disease are mutations in the *dystrophin* (*DMD*) gene, resulting in complete loss of functional Dystrophin protein (reviewed in Jay & Vajsar, 2001). In skeletal muscles, the large 427 kDa Dystrophin protein is located on the sarcolemmal cytoplasmic surface. It is a key component of the DGC and is comprised of distinct domains that mediate interactions with other cellular proteins (Fig. 1). The structural support from Dystrophin and the DGC transmits muscle force and protects the sarcolemma from mechanical stress to maintain muscle cell integrity (Ervasti & Sonnemann, 2008). Recently, Dystrophin has been shown to regulate muscle stem cell polarity and division (Dumont et al., 2015). Thus, impaired regeneration due to intrinsic stem cell dysfunction is also an important factor in DMD pathogenesis. In the DMD condition, variable fiber size, inflammatory cell infiltration, progressive myofiber necrosis, and fibrosis are the pathological hallmarks (McNally & Pytel, 2007).

The most common mammalian DMD model is the *mdx* mutant mouse. Like human DMD patients, *mdx* mice have a complete loss of Dystrophin due to X-linked genetic mutation (Bulfield, Siller, Wight, & Moore, 1984). Although the histological and biochemical analysis revealed some features of human DMD, the comparatively mild animal phenotype seen in *mdx* mice differs significantly from the human pathology. A likely explanation for the mild symptoms of *mdx* mice is that Utrophin, a Dystrophin-related protein, partially compensates for Dystrophin loss (Matsumura, Ervasti, Ohlendieck, Kahl, & Campbell, 1992). The *Dystrophin; Utrophin* double knockout mouse more accurately models human DMD (Deconinck et al., 1997; Grady et al., 1997).

The Dystrophin protein is highly conserved between zebrafish and mammals. Like in mammals, Dystrophin is found on the sarcolemmal membrane and is enriched at the myotendinous junction. In the Dystrophin-deficient condition, functional compensation by Utrophin upregulation at sarcolemmal junctions does not occur in zebrafish as it does in the mouse (Bassett et al., 2003). Several DMD phenotypes in zebrafish mutants have been documented using different methods. In a large-scale mutagenesis screen, mutants with locomotor defects were identified (Granato et al., 1996) and later the *sapje (dmd)* mutant was characterized as the first zebrafish mutant model for DMD (Bassett et al., 2003; Bassett & Currie, 2004). The causative mutation in *sapje (dmd)* mutant lies within zebrafish *DMD* gene, causing a Dystrophin null condition. *sapje (dmd)* mutant larvae exhibit muscle degeneration as early as 3 dpf that is more severe than in the *mdx* mouse (Fig. 4). The defects result from compromised cell integrity and subsequent cell detachment and death and are accompanied by inflammation and fibrosis. Muscle progenitor cell regeneration cannot compensate for the massive cell death, and *sapje (dmd)* mutants die at early larval age likely due to feeding failure (Bassett et al., 2003; Berger, Berger, Hall, Lieschke, & Currie, 2010).

Other *DMD* null alleles, including the *sapje-like* mutation, a splice site mutation that causes a premature stop codon, confer a phenotype similar to *sapje (dmd)* mutant larvae (Guyon et al., 2009). Additional alleles carrying different premature stop codons have been isolated in genetic screens (Berger et al., 2011) and they all share the similar structural and functional manifestations. In human patients, allelic mutations might cause a spectrum of clinical manifestations (reviewed in Davies & Nowak, 2006). The availability of various zebrafish DMD models facilitates a deeper investigation of the disease mechanism and therapeutic screens.

Zebrafish DMD mutants model numerous aspects of human DMD and provide distinct advantages for mechanistic research and for therapeutic discovery and validation. In a study by Berger et al. (2011), AOs were designed to cause skipping of the mutation-bearing exon during pre-mRNA splicing, and partially restore Dystrophin function, similar to the condition of BMD, the milder form of DMD. The exon-skipping strategy in this study revealed positive therapeutic rescue (Berger et al., 2011). These results are similar to those reported in human clinical trials showing restoration of dystrophin expression from exon-skipping AOs (Kinali, Arechavala-Gomeza, & Feng, 2009; Kole & Krieg, 2015). Additionally, large-scale chemical screens have been conducted and single candidate molecules have been evaluated in zebrafish DMD models (Section 4). One compound proposed to specifically target nonsense mutations, PTC124 (Ataluren), has been tested on *sapje (dmd)* mutant larvae (Li et al., 2014). In this study, Dystrophin expression and contractile function was significantly restored in the treatment group in a dose-dependent manner, with high doses causing negative side effects in both control and treatment groups, similar to findings from human and mouse myotube culture systems.

3.2 CONGENITAL MUSCULAR DYSTROPHY

As a subgroup of MDs, congenital muscular dystrophy is a heterogeneous disorder with a large disease spectrum and is inherited in autosomal recessive manner. The neuromuscular symptoms in human patients are often severe with early onset. The major genetic subtypes include Laminin α2 chain (Merosin), Integrin α7, and Fukutin-deficient MDs.

3.2.1 Congenital muscular dystrophy type 1A/merosin-deficient muscular dystrophy

Laminins are large, extracellular, trimeric membrane glycoproteins comprised of three nonidentical chains. In skeletal muscle, Laminin-2 is composed of α2, β1, and γ1 chains and binds to α-Dystroglycan and Integrin at the sarcolemma and to the ECM (reviewed in Davies & Nowak, 2006). Laminin α2 chain, also known as merosin, exerting both structural and signaling function, is required for muscle cell integrity and maintenance (Campbell, 1995; Vachon et al., 1997).

LAMA2 gene mutations that induce complete *LAMA2*-deficiency were reported in human merosin-deficient congenital muscular dystrophy type 1A (MDC1A) patients (Helbling-Leclerc et al., 1995). Histological analyses reveal dystrophic features such as fiber necrosis, infiltration of immune cells, and fibrosis. Zebrafish *lama2 (candyfloss, caf)* mutants have a characteristic dystrophic phenotype, readily apparent using muscle birefringence assays (Fig. 4) (Hall et al., 2007). Unlike the zebrafish *sapje (dmd)* mutant, *caf (lama2)* mutant larvae do not show significant signs of membrane rupture, but fiber detachment from the MTJ is instead the most predominant pathological feature. Motor neuron development and early myogenesis are normal in the zebrafish MDC1A model, consistent with the idea that the primary disease mechanism is fiber retraction due to attachment failure (Hall et al., 2007). Further functional dissection of the Laminin complex was carried out in this model, providing evidence for distinct roles of each Laminin subunit (Sztal, Sonntag, Hall, & Currie, 2012). *caf (lama2)* mutants have a significant reduction in force (Li & Arner, 2015). Moreover, mechanical stress plays a major role in disease progression, as immobilization alone can rescue the disease phenotype (Li & Arner, 2015). Another *lama2* allele that affects pre-mRNA splicing confers a severe muscle degeneration phenotype similar to *caf (lama2)* mutants, and thus provides a novel model for study of *LAMA2* splicing defects in CMD patients (Gupta et al., 2012).

3.2.2 Congenital muscular dystrophy with integrin α7 deficiency

Integrin is a major receptor of Laminin at the muscle cell sarcolemma. The Integrin α7 subunit is expressed mostly in striated muscles and has been proposed to influence myogenic cell differentiation and migration (Vachon et al., 1997). Of the different Integrin isoforms, Integrin α7β1 is the predominant form expressed in skeletal muscle that interacts with Laminin (Burkin & Kaufman, 1999). Mutations in the human *ITGA7* gene lead to Integrin α7-related congenital muscular dystrophy (Hayashi et al., 1998), which affects basal lamina integrity and function. The *Itga7*$^{-/-}$ mouse has a similar phenotype that is attributed to impaired function of the MTJ (Mayer et al., 1997). Abnormal expression of Integrin α7 has been reported in MDC1A conditions (Hodges et al., 1997; Vachon et al., 1997), and restoration of the protein partially rescued muscle pathology in MDC1A mouse model (Doe et al., 2011), implicating a role for the Integrin complex in muscle cell integrity. More mechanistic insights have been provided using an *itga7* morphant zebrafish model. *itga7* morphants display muscle fiber retraction and detachment at an early larval stage (3 dpf), showing a dystrophic phenotype with compromised muscle cell stability (Postel, Vakeel, Topczewski, Knoll, & Bakkers, 2008). In addition to Integrin α7 and Laminin, which physically interact, the function of other Integrin binding partners has also been examined. For example, loss of function of Integrin-linked kinase (ILK), the first identified binding partner of Integrin β1, leads to cardiac myopathy and muscular dystrophy (Bendig et al., 2006; Gheyara et al., 2007). Additionally, localization of ILK to the MTJ, as assessed by expression of mRNA encoding an Ilk–GFP fusion protein, is significantly reduced in *itga7* morphants, showing that *itga7* is required for ILK recruitment to the MTJ (Postel et al., 2008). These data demonstrate an important role for Integrin in adhesion complex formation and that loss of Integrin complex function can lead to MDs.

3.3 LIMB-GIRDLE MUSCULAR DYSTROPHY

3.3.1 Dysferlinopathy (LGMD2B, MM)

Human genetic mutations in *Dysferlin* (*DYSF*) gene were first associated with autosomal recessive muscular dystrophy in 1998 (Liu et al., 1998). Dysferlinopathy is often clinically classified as Miyoshi myopathy (MM) and limb-girdle muscular dystrophy type 2B (LGMD2B). Unlike other types of severe, early onset MDs, dysferlinopathy patients usually develop milder symptoms in adulthood (Vainzof et al., 2001). The primary cause for the disease is the loss of functional Dysferlin in muscle tissues, resulting in compromised membrane repair machinery. As a transmembrane protein, Dysferlin has been suggested to be required for calcium-mediated vesicle trafficking and fusion in muscle cells (Bansal & Campbell, 2004). Upon muscle damage, Dysferlin and Dysferlin-associated proteins have been shown to mediate the membrane resealing process (Bansal et al., 2003; Han & Campbell, 2007) and other studies implicate Dysferlin in the inflammation and muscle regeneration response (Chiu et al., 2009; De Luna, Gallardo, & Illa, 2004). In the zebrafish *dysferlin* morphant, larval muscle is severely disorganized, yet other DGC-related proteins like Dystrophin, Laminin, and Dystroglycan are normally localized, suggesting that Dysferlin functions in a distinct, DGC-independent pathway (Kawahara, Serafini, Myers, Alexander, & Kunkel, 2011). Membrane repair machinery was elegantly analyzed in another zebrafish *dysferlin* morphant, which also exhibits severe muscle defects, including misaligned/curved myofibers and gaps that are intensified by evoked physical activity induced by veratridine treatment (Roostalu & Strahle, 2012). Using high-resolution live imaging, Dysferlin was shown to accumulate and define a membrane compartment at sarcolemmal lesions after contraction-induced injury. Subsequently, Annexin family proteins were recruited to the repair patch. Knock-down of *anxa6*, a phospholipid-binding protein, exaggerated the myopathy phenotype of *dysferlin* morphants, demonstrating that both Dysferlin and Annexin A6 are required for muscle cell repair (Roostalu & Strahle, 2012). Recently, investigation of membrane patch removal by macrophages was also carried out in the zebrafish *dysferlin*-deficient model and suggested that the phosphatidylserine sorting process that engages macrophages after muscle injury is dependent on the presence of Dysferlin (Middel et al., 2016).

3.3.2 Sarcoglycanopathy (LGMD 2C-F)

Sarcoglycans are a transmembrane protein family (α-, β-, γ-, δ-, and ε-) that are part of the DGC and support the linkage between cytoskeleton and ECM. In human patients, mutations in the Sarcoglycan α, β, γ, and δ genes (*SGCA*, *SGCB*, *SGCG*, and *SGCD*) have been linked to LGMD Type 2C-F. The loss of function of any gene results in either full or partial impairment of the Sarcoglycan subcomplex and causes similar clinical manifestations to DMD (Hack, Groh, & McNally, 2000; Lim & Campbell, 1998). However, Dystrophin localization and function on plasma membrane is normal after loss of Sarcoglycan function (Hack et al., 1998). Zebrafish *δ-sarcoglycan* morphant larvae show muscle disorganization and loss of muscle activity similar to that observed in human patients, but at an earlier developmental stage. The reduction of δ-Sarcoglycan protein is accompanied by decrease in α- and β-Sarcoglycan, but Dystrophin expression is unaltered (Guyon et al., 2005). In addition to skeletal muscle phenotypes, heart looping and asymmetry are also altered after δ-Sarcoglycan knockdown (Cheng et al., 2006).

3.4 DYSTROGLYCANOPATHY

Dystroglycan is an essential, highly conserved component of DGC in muscle cells and consists of two subunits, the extracellular α-Dystroglycan and the transmembrane β-Dystroglycan glycoprotein, both encoded by a single *dystroglycan* (*DAG1*) gene. It acts as a bridge, connecting Dystrophin to the ECM (Barresi & Campbell, 2006). In addition to this structural role, Dystroglycan also acts as an adhesion and signaling receptor by binding to both intracellular and extracellular ligands, as well as transducing signals in the neuromuscular junction (reviewed in Barresi & Campbell, 2006). Unlike other components of the DGC, posttranslational modifications on dystroglycans, particularly glycosylation, play an important role in pathogenic process (reviewed in Moore & Winder, 2012). Although human patients with recessive *DAG1* mutations are relatively rare, other dystroglycanopathy types caused by defective glycosylation and subsequent loss of Dystroglycan function are more prevalent. For instance, mutations in *glycosyltransferase* genes cause human dystroglycanopathies (secondary dystroglycanopathies) (Mendell, Boue, & Martin, 2006; Thornhill, Bassett, Lochmuller, Bushby, & Straub, 2008). In initial zebrafish studies, *dag1* knockdown did not affect basement membrane formation but led to compromised cell integrity and necrosis (Parsons, Campos, Hirst, & Stemple, 2002). Subsequent work by Gupta et al. (2011) showed that zebrafish *patchytail (dag1)* mutant larvae, in which Dystroglycan is not expressed, exhibit a later onset muscular dystrophy phenotype, with muscle birefringence defects appearing later (7 dpf) than in *dmd* or *lama2* mutants. The structural abnormalities in *patchytail (dag1)* mutant muscle include destabilized fiber attachment due to impaired interaction with Laminin, as well as disorganized T-tubules at early stages. In contrast to the findings from *dag1* knockdown model where abnormalities were highlighted in neuromuscular tissues (Parsons et al., 2002), significant defects were also observed in brain and eye of the mutants, suggesting additional roles for Dystroglycan in brain development (e.g., Walker–Warburg syndrome) (Cormand et al., 2001).

4. SMALL MOLECULE SCREENS

4.1 SCREEN METHODS

Due to the relative ease of whole organism, phenotype-driven chemical screening in zebrafish, this approach has become increasingly popular over the past 15 years (Rennekamp & Peterson, 2015). Zebrafish embryos are small, transparent, and readily absorb chemicals from their environment, greatly facilitating large-scale application and phenotypic assessment. Small molecule screens can be readily and rapidly performed during embryonic and early larval stages, when the fish are small enough to fit in multiwell dishes and are readily imaged, to identify compounds that improve disease phenotypes. Potential therapeutic compounds identified in chemical screens can be completely novel bioactive compounds or previously approved drugs like FDA- and EMA-approved chemicals from the Prestwick Chemical Library. Several zebrafish muscular dystrophy mutants are well established and some show early muscle disease phenotypes by 3–7 days post fertilization (dpf) (Table 1) and thus are amenable to chemical screening. Morphant models can also be used to study disease phenotypes and in chemical screens (see Table 1 for examples); however, there are limitations to morpholino-based knockdown approaches (Kawahara & Kunkel, 2013). One relevant limitation for this discussion is that morpholino knockdown is only effective through 5 dpf at best, which makes chemical screens for compounds that improve late phenotypes or survival unproductive. The Kunkel group tested 2640 bioactive compounds for their ability to ameliorate the phenotype of zebrafish *dystrophin* mutants (Kawahara et al., 2015; Kawahara, Karpf, et al., 2011; Waugh et al., 2014). For the screen, 20 offspring of heterozygous parents were treated between one and four dpf in duplicate with pools of eight bioactive compounds. If a chemical pool caused death of all treated embryos (a "toxic" pool), each chemical in the pool was tested individually. Using a birefringence assay to assess muscle structure, embryos within each treatment pool were scored for the dystrophic mutant phenotype. Because only 25% of the progeny of two heterozygous parents are expected to show the mutant phenotype, pools where >15% of embryos exhibited the expected phenotype were deemed noneffective, while pools where <15% of embryos exhibited the phenotype were deemed effective and subjected to secondary screens. For secondary screening, each chemical within an effective pool was tested individually and evaluated using birefringence and Dystrophin immunostaining, and all treated embryos were genotyped to allow meaningful phenotype/genotype correlations (Kawahara & Kunkel, 2013). Two large-scale chemical screens, along with additional large-scale screens (Waugh et al., 2014) have identified over 20 compounds that can ameliorate the DMD phenotype without restoring Dystrophin expression, highlighting the impact zebrafish small molecule screens can have on identifying new drugs and/or treatment targets for muscle diseases (Kawahara et al., 2015; Kawahara, Karpf, et al., 2011). More targeted chemical approaches have also identified additional small molecules that improve dystrophic phenotypes (Goody et al., 2012; Johnson, Farr III & Maves, 2013; Lipscomb, Piggott, Emmerson, & Winder, 2016).

4.2 PHENOTYPING ASSAYS AND TOOLS

4.2.1 Phenotyping assays and tools to analyze muscle integrity and structure

Since zebrafish MD models display early phenotypes, the initial efficacy of small molecules can be evaluated shortly after drug treatment. One phenotyping method that we have already mentioned is muscle birefringence. Birefringence assays are simple, inexpensive, and take advantage of the highly organized muscle structure and the transparency of zebrafish embryos to assess muscle integrity (Berger & Currie, 2012; Smith, Beggs, & Gupta, 2013). When two polarizing filters are placed perpendicular to each other, very little light is transmitted. However, if a zebrafish embryo with highly ordered, well organized sarcomeres is placed between the two filters, the muscle will exhibit birefringence and appear bright. Not surprisingly, if muscle integrity is compromised due to fiber damage, degeneration, or death, birefringence will be reduced (Berger, Sztal, & Currie, 2012; Smith et al., 2013) (see Fig. 4 for an example). A more detailed analysis of muscle fiber type, organization, and structure can be performed using whole-mount in situ hybridization, immunohistochemistry, histological analysis, EM, and/or muscle transgenic lines (Gibbs et al., 2013; Maves, 2014; Talbot & Maves, 2016). Many muscle transgenic lines are available; one line particularly relevant is the Dystrophin FlipTrap insertion line, $dmd^{ct90aGt}$, which expresses a functional Dystrophin–Citrine fusion protein (Ruf-Zamojski, Trivedi, Fraser, & Trinh, 2015; Trinh et al., 2011). A number of tools and automated technologies are available to facilitate and standardize imaging and fluorescence analysis. Cost-effective plastic molds that generate standardized mounting wells in agarose supports can be used to mount embryos laterally or dorsally for time-lapse imaging for over 24 h (Megason & Fraser, 2003; Wittbrodt, Liebel, & Gehrig, 2014). Similarly, a 96-well plate designed with reflective prisms adjacent to each mounting well is available to allow both lateral and dorsal images of each mounted embryo, increasing data and image consistency between each sample (Rovira et al., 2011). In addition, a fully automated system known as multithread Vertebrate Automated Screening Technology (VAST) can load, detect and position, rotate and image, and unload zebrafish embryos, allowing comparable images to be obtained across multiple samples with over 98% survivability (Chang, Pardo-Martin, Allalou, Wählby, & Yanik, 2012). Because VAST can load, image, and unload multiple embryos simultaneously, processing time is significantly decreased, with the rate-limiting step being image acquisition (Chang et al., 2012). Lastly, high content screening (HCS) imaging software is available allowing automated imaging and processing of 384-well plates of chemically treated embryos. Using customizable parameters, both visual and physical properties can be analyzed by HCS software to assess embryos after chemical treatment (Leet et al., 2014; Yozzo, Isales, Raftery, & Volz, 2013).

4.2.2 Phenotyping assays and tools to analyze muscle function

To assess effectiveness of a potential therapeutic, treated embryos can also be evaluated for survival, muscle function, and contractile strength. Since zebrafish dystrophy models typically die between 7 and 16 dpf, improved survival is a simple and relatively rapid way to gauge treatment success. Motor function can be evaluated by quantifying spontaneous coiling, the escape response, and swimming ability (Farrell et al., 2011; Hirata et al., 2004; Maves, 2014; Smith et al., 2013). Swimming behaviors are typically standardized around 6 dpf and can be monitored and quantified in 96-well plate format (Farrell et al., 2011; Hirata et al., 2004). In addition, contractile strength of whole animals can be assessed by mounting embryos or larvae on a force transducer (Dowling et al., 2012; Martin et al., 2015; Sloboda, Claflin, Dowling, & Brooks, 2013). In all, there are many assays and tools, ranging from simple to elaborate, available to assess zebrafish larval muscle structure and function.

CONCLUSIONS

Since the early 2000s, the zebrafish has emerged as a research model organism for MDs. Zebrafish muscle is similar in structure and function to other vertebrates, including human. Furthermore, the function of many genes that cause human MDs have been shown to be conserved in zebrafish. Well-characterized mutants, the availability of genetic tools, together with the inherent advantages of zebrafish as a research animal model, provide a powerful platform for muscular dystrophy disease modeling and therapeutic screening.

REFERENCES

Altringham, J. D., & Ellerby, D. J. (1999). Fish swimming: patterns in muscle function. *Journal of Experimental Biology, 202*, 3397–3403.

Amacher, S. L. (2008). Emerging gene knockout technology in zebrafish: zinc-finger nucleases. *Briefings of Functional Genomics & Proteomics, 7*(6), 460–464.

Baas, D., Caussanel-Boude, S., Guiraud, A., Calhabeu, F., Delaune, E., Pilot, F., … Rual, J. F. (2012). CKIP-1 regulates mammalian and zebrafish myoblast fusion. *Journal of Cell Science, 125*(16), 3790–3800.

Bansal, D., & Campbell, K. P. (2004). Dysferlin and the plasma membrane repair in muscular dystrophy. *Trends in Cell Biology, 14*, 206–213.

Bansal, D., Miyake, K., Vogel, S. S., Groh, S., Chen, C. C., Williamson, R., … Campbell, K. P. (2003). Defective membrane repair in dysferlin-deficient muscular dystrophy. *Nature, 423*, 168–172.

di Barletta, M. R., Ricci, E., Galluzzi, G., Tonali, P., Mora, M., Morandi, L., … Toniolo, D. (2000). Different mutations in the LMNA gene cause autosomal dominant and autosomal recessive Emery-Dreifuss muscular dystrophy. *American Journal of Human Genetics, 66*, 1407–1412.

Barresi, R., & Campbell, K. P. (2006). Dystroglycan: from biosynthesis to pathogenesis of human disease. *Journal of Cell Science, 119*, 199–207.

Bassett, D. I., Bryson-Richardson, R. J., Daggett, D. F., Gautier, P., Keenan, D. G., & Currie, P. D. (2003). Dystrophin is required for the formation of stable muscle attachments in the zebrafish embryo. *Development, 130*(23), 5851–5860.

Bassett, D., & Currie, P. D. (2004). Identification of a zebrafish model of muscular dystrophy. *Clinical and Experimental Pharmacology & Physiology, 31*(8), 537–540.

Baxendale, S., Davison, C., Muxworthy, C., Wolff, C., Ingham, P. W., & Roy, S. (2004). The B-cell maturation factor Blimp-1 specifies vertebrate slow-twitch muscle fiber identity in response to Hedgehog signaling. *Nature Genetics, 36*(1), 88–93.

Bendig, G., Grimmler, M., Huttner, I. G., Wessels, G., Dahme, T., Just, S., … Rottbauer, W. (2006). Integrin-linked kinase, a novel component of the cardiac mechanical stretch sensor, controls contractility in the zebrafish heart. *Genes & Development, 20*, 2361–2372.

Bengtsson, N. E., Seto, J. T., Hall, J. K., Chamberlain, J. S., & Odom, G. L. (2016). Progress and prospects of gene therapy clinical trials for the muscular dystrophies. *Human Molecular Genetics, 25*(R1), R9–R17.

Berger, J., Berger, S., Hall, T. E., Lieschke, G. J., & Currie, P. D. (2010). Dystrophin-deficient zebrafish feature aspects of the Duchenne muscular dystrophy pathology. *Neuromuscular Disorders, 20*(12), 826–832.

Berger, J., Berger, S., Jacoby, A. S., Wilton, S. D., & Currie, P. D. (2011). Evaluation of exon-skipping strategies for Duchenne muscular dystrophy utilizing dystrophin-deficient zebrafish. *Journal of Cellular and Molecular Medicine, 15*, 2643–2651.

Berger, J., & Currie, P. D. (2012). Zebrafish models flex their muscles to shed light on muscular dystrophies. *Disease Models & Mechanisms, 5*(6), 726–732.

Berger, J., Sztal, T., & Currie, P. D. (July 13, 2012). Quantification of birefringence readily measures the level of muscle damage in zebrafish. *Biochemical and Biophysical Research Communications, 423*(4), 785–788. http://dx.doi.org/10.1016/j.bbrc.2012.06.040.

Beytía, M. L., Vry, J., & Kirschner, J. (2012). Drug treatment of Duchenne muscular dystrophy: available evidence and perspectives. *Acta Myologica, 31*, 4–8.

Bieber, F. R., & Hoffman, E. P. (1990). Duchenne and Becker muscular dystrophies: genetics, prenatal diagnosis, and future prospects. *Clinics in Perinatology, 17*, 845–865.

Blagden, C. S., Currie, P. D., Ingham, P. W., & Hughes, S. M. (1997). Notochord induction of zebrafish slow muscle mediated by Sonic hedgehog. *Genes & Development, 11*(17), 2163–2175.

Block, B. A., Imagawa, T., Campbell, K. P., & Franzini-Armstrong, C. (1988). Structural evidence for direct interaction between the molecular components of the transverse tubule/sarcoplasmic reticulum junction in skeletal muscle. *Journal of Cell Biology, 107*(6), 2587–2600.

Bryson-Richardson, R. J., & Currie, P. D. (2008). The genetics of vertebrate myogenesis. *Nature Reviews. Genetics, 9*(8), 632–646.

Bulfield, G., Siller, W. G., Wight, P. A., & Moore, K. J. (1984). X chromosome-linked muscular dystrophy (mdx) in the mouse. *Proceedings of the National Academy of Sciences of the United States of America, 81*(4), 1189–1192.

Burke, A. C., & Nowicki, J. L. (2003). A new view of patterning domains in the vertebrate mesoderm. *Developmental Cell, 4*(2), 159–165.

Burkin, D. J., & Kaufman, S. J. (1999). The alpha7beta1 integrin in muscle development and disease. *Cell and Tissue Research, 296*, 183–190.

Buss, R. R., & Drapeau, P. (2000). Physiological properties of zebrafish embryonic red and white muscle fibers during early development. *Journal of Neurophysiology, 84*, 1545–1557.

Campbell, K. P. (1995). Three muscular dystrophies: loss of cytoskeleton-extracellular matrix linkage. *Cell, 80*, 675–679.

Chang, T. Y., Pardo-Martin, C., Allalou, A., Wählby, C., & Yanik, M. F. (2012). Fully automated cellular-resolution vertebrate screening platform with parallel animal processing. *Lab on a Chip, 12*(4), 711–716.

Cheng, L., Guo, X. F., Yang, X. Y., Chong, M., Cheng, J., Li, G., … Lu, D. R. (2006). Delta-sarcoglycan is necessary for early heart and muscle development in zebrafish. *Biochemical and Biophysical Research Communications, 344*(4), 1290–1299.

Chiu, Y. H., Hornsey, M. A., Klinge, L., Jorgensen, L. H., Laval, S. H., Charlton, R., … Bushby, K. (2009). Attenuated muscle regeneration is a key factor in dysferlin-deficient muscular dystrophy. *Human Molecular Genetics, 18*, 1976–1989.

Cooke, J., & Zeeman, E. C. (1976). A clock and wavefront model for control of the number of repeated structures during animal morphogenesis. *Journal of Theoretical Biology, 58*(2), 455–476.

Cormand, B., Pihko, H., Bayes, M., Valanne, L., Santavuori, P., Talim, B., … Lehesjoki, A. E. (2001). Clinical and genetic distinction between Walker-Warburg syndrome and muscle-eye-brain disease. *Neurology, 56*, 1059–1069.

Cortés, F., Daggett, D., Bryson-Richardson, R. J., Neyt, C., Maule, J., Gautier, P., … Currie, P. D. (2003). Cadherin-mediated differential cell adhesion controls slow muscle cell migration in the developing zebrafish myotome. *Developmental Cell, 5*(6), 865–876.

Coutelle, O., Blagden, C. S., Hampson, R., Halai, C., Rigby, P. W., & Hughes, S. M. (2001). Hedgehog signalling is required for maintenance of myf5 and myoD expression and timely terminal differentiation in zebrafish adaxial myogenesis. *Developmental Biology, 236*(1), 136–150.

Currie, P. D., & Ingham, P. W. (1996). Induction of a specific muscle cell type by a hedgehog-like protein in zebrafish. *Nature, 382*.

Davies, K. E., & Nowak, K. J. (2006). Molecular mechanisms of muscular dystrophies: old and new players. *Nature Reviews. Molecular Cell Biology, 7*, 762–773.

De Luna, N., Gallardo, E., & Illa, I. (2004). In vivo and in vitro dysferlin expression in human muscle satellite cells. *Journal of Neuropathology and Experimental Neurology, 63*, 1104–1113.

Deconinck, A. E., Rafael, J. A., Skinner, J. A., Brown, S. C., Potter, A. C., Metzinger, L., … Davies, K. E. (1997). Utrophin-dystrophin-deficient mice as a model for Duchenne muscular dystrophy. *Cell, 90*, 717–727.

Delaune, E. A., François, P., Shih, N. P., & Amacher, S. L. (2012). Single-cell-resolution imaging of the impact of Notch signaling and mitosis on segmentation clock dynamics. *Developmental Cell, 23*(5), 995–1005.

Devoto, S. H., Melançon, E., Eisen, J. S., & Westerfield, M. (1996). Identification of separate slow and fast muscle precursor cells in vivo, prior to somite formation. *Development, 122*(11), 3371–3380.

Doe, J. A., Wuebbles, R. D., Allred, E. T., Rooney, J. E., Elorza, M., & Burkin, D. J. (2011). Transgenic overexpression of the alpha7 integrin reduces muscle pathology and improves viability in the dy(W) mouse model of merosin-deficient congenital muscular dystrophy type 1A. *Journal of Cell Science, 124*, 2287–2297.

Dou, Y., Andersson-Lendahl, M., & Arner, A. (2008). Structure and function of skeletal muscle in zebrafish early larvae. *The Journal of General Physiology, 131*, 445–453.

Dowling, J. J., Arbogast, S., Hur, J., Nelson, D. D., McEvoy, A., Waugh, T., … Ferreiro, A. (2012). Oxidative stress and successful antioxidant treatment in models of RYR1-related myopathy. *Brain, 135*(4), 1115–1127.

Dowling, J. J., Vreede, A. P., Low, S. E., Gibbs, E. M., Kuwada, J. Y., Bonnemann, C. G., & Feldman, E. L. (2009). Loss of myotubularin function results in T-tubule disorganization in zebrafish and human myotubular myopathy. *PLoS Genetics, 5*(2), e1000372.

Du, S. J., Devoto, S. H., Westerfield, M., & Moon, R. T. (1997). Positive and negative regulation of muscle cell identity by members of the hedgehog and TGF-β gene families. *Journal of Cell Biology, 139*(1), 145–156.

Dumont, N. A., Wang, Y. X., von Maltzahn, J., Pasut, A., Bentzinger, C. F., Brun, C. E., & Rudnicki, M. A. (2015). Dystrophin expression in muscle stem cells regulates their polarity and asymmetric division. *Nature Medicine, 21*, 1455–1463.

Emery, A. E. H. (1993). *Duchenne muscular dystrophy*. Oxford, New York: Oxford University Press.

Emery, A. E. H. (2002). The muscular dystrophies. *Lancet, 359*, 687–695.

Ervasti, J. M., & Sonnemann, K. J. (2008). Biology of the striated muscle dystrophin-glycoprotein complex. *International Review of Cytology, 265*, 191–225.

Eston, R. G., Mickleborough, J., & Baltzopoulos, V. (1995). Eccentric activation and muscle damage – biomechanical and physiological considerations during downhill running. *British Journal of Sports Medicine, 29*, 89–94.

Farrell, T. C., Cario, C. L., Milanese, C., Vogt, A., Jeong, J. H., & Burton, E. A. (2011). Evaluation of spontaneous propulsive movement as a screening tool to detect rescue of Parkinsonism phenotypes in zebrafish models. *Neurobiology of Disease, 44*(1), 9–18.

Gee, S. H., Montanaro, F., Lindenbaum, M. H., & Carbonetto, S. (1994). Dystroglycan-alpha, a dystrophin-associated glycoprotein, is a functional agrin receptor. *Cell, 77*, 675–686.

Gheyara, A. L., Vallejo-Illarramendi, A., Zang, K., Mei, L., St-Arnaud, R., Dedhar, S., & Reichardt, L. F. (2007). Deletion of integrin-linked kinase from skeletal muscles of mice resembles muscular dystrophy due to alpha 7 beta 1-integrin deficiency. *American Journal of Pathology, 171*, 1966–1977.

Gibbs, E. M., Horstick, E. J., & Dowling, J. J. (2013). Swimming into prominence: the zebrafish as a valuable tool for studying human myopathies and muscular dystrophies. *FEBS Journal, 280*(17), 4187–4197.

Goebel, H. H. (2011). Congenital myopathies. Introduction. *Seminars in Pediatric Neurology, 18*, 213–215.

Goody, M. F., Kelly, M. W., Reynolds, C. J., Khalil, A., Crawford, B. D., & Henry, C. A. (2012). NAD^+ biosynthesis ameliorates a zebrafish model of muscular dystrophy. *PLoS Biology, 10*(10), e1001409.

Grady, R. M., Teng, H., Nichol, M. C., Cunningham, J. C., Wilkinson, R. S., & Sanes, J. R. (1997). Skeletal and cardiac myopathies in mice lacking utrophin and dystrophin: a model for Duchenne muscular dystrophy. *Cell, 90*, 729–738.

Granato, M., van Eeden, F. J., Schach, U., Trowe, T., Brand, M., Furutani-Seiki, M., … Nüsslein-Volhard, C. (1996). Genes controlling and mediating locomotion behavior of the zebrafish embryo and larva. *Development, 123*, 399–413.

Groves, J. A., Hammond, C. L., & Hughes, S. M. (2005). Fgf8 drives myogenic progression of a novel lateral fast muscle fibre population in zebrafish. *Development, 132*(19), 4211–4222.

Gupta, V., Kawahara, G., Gundry, S. R., Chen, A. T., Lencer, W. I., Zhou, Y., … Beggs, A. H. (2011). The zebrafish dag1 mutant: a novel genetic model for dystroglycanopathies. *Human Molecular Genetics, 20*(9), 1712–1725.

Gupta, V. A., Kawahara, G., Myers, J. A., Chen, A. T., Hall, T. E., Manzini, M. C., … Beggs, A. H. (2012). A splice site mutation in laminin-alpha2 results in a severe muscular dystrophy and growth abnormalities in zebrafish. *PLoS One, 7*, e43794.

Guyon, J. R., Goswami, J., Jun, S. J., Thorne, M., Howell, M., Pusack, T., … Kunkel, L. M. (2009). Genetic isolation and characterization of a splicing mutant of zebrafish dystrophin. *Human Molecular Genetics, 18*(1), 202–211.

Guyon, J. R., Mosley, A. N., Jun, S. J., Montanaro, F., Steffen, L. S., Zhou, Y., … Kunkel, L. M. (2005). Delta-sarcoglycan is required for early zebrafish muscle organization. *Experimental Cell Research, 304*(1), 105–115.

Guyon, J. R., Mosley, A. N., Zhou, Y., O'Brien, K. F., Sheng, X., Chiang, K., … Kunkel, L. M. (2003). The dystrophin associated protein complex in zebrafish. *Human Molecular Genetics, 12*(6), 601–615.

Hack, A. A., Groh, M. E., & McNally, E. M. (2000). Sarcoglycans in muscular dystrophy. *Microscopy Research and Technique, 48*, 167–180.

Hack, A. A., Ly, C. T., Jiang, F., Clendenin, C. J., Sigrist, K. S., Wollmann, R. L., & McNally, E. M. (1998). gamma-Sarcoglycan deficiency leads to muscle membrane defects and apoptosis independent of dystrophin. *Journal of Cell Biology, 142*, 1279–1287.

Haines, L., Neyt, C., Gautier, P., Keenan, D. G., Bryson-Richardson, R. J., Hollway, G. E., … Currie, P. D. (2004). Met and Hgf signaling controls hypaxial muscle and lateral line development in the zebrafish. *Development, 131*(19), 4857–4869.

Hall, T. E., Bryson-Richardson, R. J., Berger, S., Jacoby, A. S., Cole, N. J., Hollway, G. E., … Currie, P. D. (2007). The zebrafish candyfloss mutant implicates extracellular matrix adhesion failure in laminin α2-deficient congenital muscular dystrophy. *Proceedings of the National Academy of Sciences, 104*(17), 7092–7097.

Hammond, C. L., Hinits, Y., Osborn, D. P., Minchin, J. E., Tettamanti, G., & Hughes, S. M. (2007). Signals and myogenic regulatory factors restrict pax3 and pax7 expression to dermomyotome-like tissue in zebrafish. *Developmental Biology, 302*(2), 504–521.

Han, R., & Campbell, K. P. (2007). Dysferlin and muscle membrane repair. *Current Opinion in Cell Biology, 19*, 409–416.

Hatta, K., Bremiller, R., Westerfield, M., & Kimmel, C. B. (1991). Diversity of expression of engrailed-like antigens in zebrafish. *Development, 112*(3), 821–832.

Hayashi, Y. K., Chou, F. L., Engvall, E., Ogawa, M., Matsuda, C., Hirabayashi, S., … Arahata, K. (1998). Mutations in the integrin alpha7 gene cause congenital myopathy. *Nature Genetics, 19*, 94–97.

Helbling-Leclerc, A., Zhang, X., Topaloglu, H., Cruaud, C., Tesson, F., Weissenbach, J., … Tryggvason, K. (1995). Mutations in the laminin alpha 2-chain gene (LAMA2) cause merosin-deficient congenital muscular dystrophy. *Nature Genetics, 11*, 216–218.

Henry, C. A., & Amacher, S. L. (2004). Zebrafish slow muscle cell migration induces a wave of fast muscle morphogenesis. *Developmental Cell, 7*(6), 917–923.

Hernandez, L. P., Patterson, S. E., & Devoto, S. H. (2005). The development of muscle fiber type identity in zebrafish cranial muscles. *Anatomy and Embryology, 209*(4), 323–334.

Hinits, Y., Williams, V. C., Sweetman, D., Donn, T. M., Ma, T. P., Moens, C. B., & Hughes, S. M. (2011). Defective cranial skeletal development, larval lethality and haploinsufficiency in Myod mutant zebrafish. *Developmental Biology, 358*(1), 102–112.

Hirata, H., Saint-Amant, L., Waterbury, J., Cui, W., Zhou, W., Li, Q., … Kuwada, J. Y. (2004). accordion, a zebrafish behavioral mutant, has a muscle relaxation defect due to a mutation in the ATPase Ca^{2+} pump SERCA1. *Development, 131*(21), 5457–5468.

Hirata, H., Watanabe, T., Hatakeyama, J., Sprague, S. M., Saint-Amant, L., Nagashima, A., … Kuwada, J. Y. (2007). Zebrafish relatively relaxed mutants have a ryanodine receptor defect, show slow swimming and provide a model of multi-minicore disease. *Development, 134*(15), 2771–2781.

Hodges, B. L., Hayashi, Y. K., Nonaka, I., Wang, W., Arahata, K., & Kaufman, S. J. (1997). Altered expression of the alpha7beta1 integrin in human and murine muscular dystrophies. *Journal of Cell Science, 110*(Pt 22), 2873–2881.

von Hofsten, J., Elworthy, S., Gilchrist, M. J., Smith, J. C., Wardle, F. C., & Ingham, P. W. (2008). Prdm1-and Sox6-mediated transcriptional repression specifies muscle fibre type in the zebrafish embryo. *EMBO Reports, 9*(7), 683–689.

Holley, S. A. (2007). The genetics and embryology of zebrafish metamerism. *Developmental Dynamics, 236*(6), 1422–1449.

Hollway, G. E., Bryson-Richardson, R. J., Berger, S., Cole, N. J., Hall, T. E., & Currie, P. D. (2007). Whole-somite rotation generates muscle progenitor cell compartments in the developing zebrafish embryo. *Developmental Cell, 12*(2), 207–219.

Howe, K., Clark, M. D., Torroja, C. F., Torrance, J., Berthelot, C., Muffato, M., … Stemple, D. L. (2013). The zebrafish reference genome sequence and its relationship to the human genome. *Nature, 496*, 498–503.

Huang, P., Xiao, A., Zhou, M., Zhu, Z., Lin, S., & Zhang, B. (2011). Heritable gene targeting in zebrafish using customized TALENs. *Nature Biotechnology, 29*, 699–700.

Huxley, A. F., & Niedergerk, R. (1954). Structural changes in muscle during contraction; interference microscopy of living muscle fibres. *Nature, 173*, 971–973.

Hwang, W. Y., Fu, Y. F., Reyon, D., Maeder, M. L., Tsai, S. Q., Sander, J. D., … Joung, J. K. (2013). Efficient genome editing in zebrafish using a CRISPR-Cas system. *Nature Biotechnology, 31*(3), 227–229.

Ibraghimov-Beskrovnaya, O., Ervasti, J. M., Leveille, C. J., Slaughter, C. A., Sernett, S. W., & Campbell, K. P. (1992). Primary structure of dystrophin-associated glycoproteins linking dystrophin to the extracellular matrix. *Nature, 355*, 696–702.

Ingham, P. W., & Kim, H. R. (2005). Hedgehog signalling and the specification of muscle cell identity in the zebrafish embryo. *Experimental Cell Research, 306*(2), 336–342.

Jackson, H. E., Ono, Y., Wang, X., Elworthy, S., Cunliffe, V. T., & Ingham, P. W. (2015). The role of Sox6 in zebrafish muscle fiber type specification. *Skeletal Muscle, 5*(1), 1.

Jay, V., & Vajsar, J. (2001). The dystrophy of Duchenne. *Lancet, 357*, 550–552.

Johnson, N. M., Farr, G. H., III, & Maves, L. (2013). The HDAC Inhibitor TSA Ameliorates a zebrafish model of duchenne muscular dystrophy. *PLoS Currents Muscular Dystrophy, 5*.

Kawahara, G., Gasperini, M. J., Myers, J. A., Widrick, J. J., Eran, A., Serafini, P. R., … Kunkel, L. M. (2015). Dystrophic muscle improvement in zebrafish via increased heme oxygenase signaling. *Human Molecular Genetics*. ddv169.

Kawahara, G., Guyon, J. R., Nakamura, Y., & Kunkel, L. M. (February 15, 2010). Zebrafish models for human FKRP muscular dystrophies. *Human Molecular Genetics, 19*(4), 623–633. http://dx.doi.org/10.1093/hmg/ddp528.

Kawahara, G., Karpf, J. A., Myers, J. A., Alexander, M. S., Guyon, J. R., & Kunkel, L. M. (2011). Drug screening in a zebrafish model of Duchenne muscular dystrophy. *Proceedings of the National Academy of Sciences of the United States of America, 108*(13), 5331–5336.

Kawahara, G., & Kunkel, L. M. (2013). Zebrafish based small molecule screens for novel DMD drugs. *Drug Discovery Today. Technologies, 10*(1), e91–96.

Kawahara, G., Serafini, P. R., Myers, J. A., Alexander, M. S., & Kunkel, L. M. (2011). Characterization of zebrafish dysferlin by morpholino knockdown. *Biochemical and Biophysical Research Communications, 413*(2), 358–363.

Kawamura, A., Koshida, S., Hijikata, H., Ohbayashi, A., Kondoh, H., & Takada, S. (2005). Groucho-associated transcriptional repressor ripply1 is required for proper transition from the presomitic mesoderm to somites. *Developmental Cell, 9*(6), 735–744.

Kimmel, C. B., Ballard, W. W., Kimmel, S. R., Ullmann, B., & Schilling, T. F. (1995). Stages of embryonic-development of the zebrafish. *Developmental Dynamics, 203*, 253–310.

Kinali, M., Arechavala-Gomeza, V., & Feng, L. (2009). Local restoration of dystrophin expression with the morpholino oligomer AVI-4658 in Duchenne muscular dystrophy: a single-blind, placebo-controlled, dose-escalation, proof-of-concept study. *Lancet Neurology, 8*, 1083. p. 918.

Kole, R., & Krieg, A. M. (2015). Exon skipping therapy for Duchenne muscular dystrophy. *Advanced Drug Delivery Reviews, 87*, 104–107.

Koshimizu, E., Imamura, S., Qi, J., Toure, J., Valdez, D. M., Jr., Carr, C. E., … Kishi, S. (March 30, 2011). Embryonic senescence and laminopathies in a progeroid zebrafish model. *PLoS One, 6*(3), e17688. http://dx.doi.org/10.1371/journal.pone.0017688.

Laing, N. G. (2012). Genetics of neuromuscular disorders. *Critical Reviews in Clinical Laboratory Sciences, 49*, 33–48.

Lampe, A. K., & Bushby, K. M. (2005). Collagen VI related muscle disorders. *Journal of Medical Genetics, 42*, 673–685.

Landemaine, A., Rescan, P. Y., & Gabillard, J. C. (2014). Myomaker mediates fusion of fast myocytes in zebrafish embryos. *Biochemical and Biophysical Research Communications, 451*(4), 480–484.

Lapidos, K. A., Kakkar, R., & McNally, E. M. (2004). The dystrophin glycoprotein complex: signaling strength and integrity for the sarcolemma. *Circulation Research, 94*, 1023–1031.

Leet, J. K., Lindberg, C. D., Bassett, L. A., Isales, G. M., Yozzo, K. L., Raftery, T. D., & Volz, D. C. (2014). High-content screening in zebrafish embryos identifies butafenacil as a potent inducer of anemia. *PLoS One, 9*(8), e104190.

Lewis, K. E., Currie, P. D., Roy, S., Schauerte, H., Haffter, P., & Ingham, P. W. (1999). Control of muscle cell-type specification in the zebrafish embryo by Hedgehog signalling. *Developmental Biology, 216*(2), 469–480.

Li, M., Andersson-Lendahl, M., Sejersen, T., & Arner, A. (2013). Knockdown of desmin in zebrafish larvae affects interfilament spacing and mechanical properties of skeletal muscle. *The Journal of General Physiology, 141*, 335–345.

Li, M., Andersson-Lendahl, M., Sejersen, T., & Arner, A. (2014). Muscle dysfunction and structural defects of dystrophin-null sapje mutant zebrafish larvae are rescued by ataluren treatment. *FASEB Journal, 28*, 1593–1599.

Li, M., & Arner, A. (2015). Immobilization of dystrophin and laminin alpha2-chain deficient zebrafish larvae in vivo prevents the development of muscular dystrophy. *PLoS One, 10*, e0139483.

Lieschke, G. J., & Currie, P. D. (2007). Animal models of human disease: zebrafish swim into view. *Nature Reviews. Genetics, 8*, 353–367.

Lim, L. E., & Campbell, K. P. (1998). The sarcoglycan complex in limb-girdle muscular dystrophy. *Current Opinion in Neurology, 11*, 443–452.

Lin, Y. Y., White, R. J., Torelli, S., Cirak, S., Muntoni, F., & Stemple, D. L. (2011). Zebrafish Fukutin family proteins link the unfolded protein response with dystroglycanopathies. *Human Molecular Genetics, 20*, 1763–1775.

Lipscomb, L., Piggott, R. W., Emmerson, T., & Winder, S. J. (2016). Dasatinib as a treatment for Duchenne muscular dystrophy. *Human Molecular Genetics, 25*(2), 266–274.

Liu, J., Aoki, M., Illa, I., Wu, C. Y., Fardeau, M., Angelini, C., ... Brown, R. H., Jr. (1998). Dysferlin, a novel skeletal muscle gene, is mutated in Miyoshi myopathy and limb girdle muscular dystrophy. *Nature Genetics, 20*, 31–36.

Lou, Q., He, J., Hu, L., & Yin, Z. (2012). Role of lbx2 in the noncanonical Wnt signaling pathway for convergence and extension movements and hypaxial myogenesis in zebrafish. *Biochimica et Biophysica Acta (BBA)-Molecular Cell Research, 1823*(5), 1024–1032.

Martin, B. L., Gallagher, T. L., Rastogi, N., Davis, J. P., Beattie, C. E., Amacher, S. L., & Janssen, P. M. (2015). In vivo assessment of contractile strength distinguishes differential gene function in skeletal muscle of zebrafish larvae. *Journal of Applied Physiology, 119*(7), 799–806.

Matsumura, K., Ervasti, J. M., Ohlendieck, K., Kahl, S. D., & Campbell, K. P. (1992). Association of dystrophin-related protein with dystrophin-associated proteins in mdx mouse muscle. *Nature, 360*(6404), 588–591.

Maves, L. (2014). Recent advances using zebrafish animal models for muscle disease drug discovery. *Expert Opinion on Drug Discovery, 9*(9), 1033–1045.

Maves, L., Waskiewicz, A. J., Paul, B., Cao, Y., Tyler, A., Moens, C. B., & Tapscott, S. J. (2007). Pbx homeodomain proteins direct Myod activity to promote fast-muscle differentiation. *Development, 134*(18), 3371–3382.

Mayer, U., Saher, G., Fassler, R., Bornemann, A., Echtermeyer, F., von der Mark, H., ... von der Mark, K. (1997). Absence of integrin alpha 7 causes a novel form of muscular dystrophy. *Nature Genetics, 17*(3), 318–323.

McNally, E. M., & Pytel, P. (2007). Muscle diseases: the muscular dystrophies. *Annual Review of Pathology, 2*, 87–109.

Megason, S. G., & Fraser, S. E. (2003). Digitizing life at the level of the cell: high-performance laser-scanning microscopy and image analysis for in toto imaging of development. *Mechanisms of Development, 120*(11), 1407–1420.

Meissner, G. (1994). Ryanodine receptor/Ca^{2+} release channels and their regulation by endogenous effectors. *Annual Review of Physiology, 56*(1), 485–508.

Mendell, J. R., Boue, D. R., & Martin, P. T. (2006). The congenital muscular dystrophies: recent advances and molecular insights. *Pediatric and Developmental Pathology, 9*, 427–443.

Michele, D. E., Barresi, R., Kanagawa, M., Saito, F., Cohn, R. D., Satz, J. S., … Campbell, K. P. (2002). Post-translational disruption of dystroglycan-ligand interactions in congenital muscular dystrophies. *Nature, 418*, 417–422.

Middel, V., Zhou, L., Takamiya, M., Beil, T., Shahid, M., Roostalu, U., … Strähle, U. (2016). Dysferlin-mediated phosphatidylserine sorting engages macrophages in sarcolemma repair. *Nature Communications, 7*, 12875.

Minchin, J. E., Williams, V. C., Hinits, Y., Low, S., Tandon, P., Fan, C. M., … Hughes, S. M. (2013). Oesophageal and sternohyal muscle fibres are novel Pax3-dependent migratory somite derivatives essential for ingestion. *Development, 140*(14), 2972–2984.

Mitsuhashi, H., Mitsuhashi, S., Lynn-Jones, T., Kawahara, G., & Kunkel, L. M. (2013). Expression of DUX4 in zebrafish development recapitulates facioscapulohumeral muscular dystrophy. *Human Molecular Genetics, 22*, 568–577.

Moore, C. A., Parkin, C. A., Bidet, Y., & Ingham, P. W. (2007). A role for the Myoblast city homologues Dock1 and Dock5 and the adaptor proteins Crk and Crk-like in zebrafish myoblast fusion. *Development, 134*(17), 3145–3153.

Moore, C. J., & Winder, S. J. (2012). The inside and out of dystroglycan post-translational modification. *Neuromuscular Disorders, 22*(11), 959–965.

Moreira, E. S., Wiltshire, T. J., Faulkner, G., Nilforoushan, A., Vainzof, M., Suzuki, O. T., … Jenne, D. E. (2000). Limb-girdle muscular dystrophy type 2G is caused by mutations in the gene encoding the sarcomeric protein telethonin. *Nature Genetics, 24*, 163–166.

Muchir, A., Bonne, G., van der Kooi, A. J., van Meegen, M., Baas, F., Bolhuis, P. A., … Schwartz, K. (2000). Identification of mutations in the gene encoding lamins A/C in autosomal dominant limb girdle muscular dystrophy with atrioventricular conduction disturbances (LGMD1B). *Human Molecular Genetics, 9*, 1453–1459.

Muntoni, F., Brockington, M., Torelli, S., & Brown, S. C. (2004). Defective glycosylation in congenital muscular dystrophies. *Current Opinion in Neurology, 17*, 205–209.

Nasevicius, A., & Ekker, S. C. (2000). Effective targeted gene 'knockdown' in zebrafish. *Nature Genetics, 26*, 216–220.

Neyt, C., Jagla, K., Thisse, C., Thisse, B., Haines, L., & Currie, P. D. (2000). Evolutionary origins of vertebrate appendicular muscle. *Nature, 408*(6808), 82–86.

Ng, R., Banks, G. B., Hall, J. K., Muir, L. A., Ramos, J. N., Wicki, J., … Chamberlain, J. S. (2012). Animal models of muscular dystrophy. *Progress in Molecular Biology and Translational Science, 105*, 83–111.

Nixon, S. J., Wegner, J., Ferguson, C., Mery, P. F., Hancock, J. F., Currie, P. D., … Parton, R. G. (2005). Zebrafish as a model for caveolin-associated muscle disease; caveolin-3 is required for myofibril organization and muscle cell patterning. *Human Molecular Genetics, 14*, 1727–1743.

Oates, A. C., Rohde, L. A., & Ho, R. K. (2005). Generation of segment polarity in the paraxial mesoderm of the zebrafish through a T-box-dependent inductive event. *Developmental Biology, 283*(1), 204–214.

Ochi, H., & Westerfield, M. (2009). Lbx2 regulates formation of myofibrils. *BMC Developmental Biology, 9*(1), 13.

Parant, J. M., & Yeh, J. R. (2016). Approaches to inactivate genes in zebrafish. *Advances in Experimental Medicine and Biology, 916*, 61–86.

Parsons, M. J., Campos, I., Hirst, E. M. A., & Stemple, D. L. (2002). Removal of dystroglycan causes severe muscular dystrophy in zebrafish embryos. *Development, 129*, 3505–3512.

Patton, E. E., & Zon, L. I. (2001). The art and design of genetic screens: zebrafish. *Nature Reviews. Genetics, 2*, 956–966.

Peng, H. B., Ali, A. A., Daggett, D. F., Rauvala, H., Hassell, J. R., & Smalheiser, N. R. (1998). The relationship between perlecan and dystroglycan and its implication in the formation of the neuromuscular junction. *Cell Adhesion and Communication, 5*, 475–489.

Peterson, R. T., Link, B. A., Dowling, J. E., & Schreiber, S. L. (2000). Small molecule developmental screens reveal the logic and timing of vertebrate development. *Proceedings of the National Academy of Sciences of the United States of America, 97*, 12965–12969.

Pichavant, C., Aartsma-Rus, A., Clemens, P. R., Davies, K. E., Dickson, G., Takeda, S., … Tremblay, J. P. (2011). Current status of pharmaceutical and genetic therapeutic approaches to treat DMD. *Molecular Therapy, 19*, 830–840.

Pipalia, T. G., Koth, J., Roy, S. D., Hammond, C. L., Kawakami, K., & Hughes, S. M. (2016). Cellular dynamics of regeneration reveals role of two distinct Pax7 stem cell populations in larval zebrafish muscle repair. *Disease Models & Mechanisms, 9*(6), 671–684.

Postel, R., Vakeel, P., Topczewski, J., Knoll, R., & Bakkers, J. (2008). Zebrafish integrin-linked kinase is required in skeletal muscles for strengthening the integrin-ECM adhesion complex. *Developmental Biology, 318*, 92–101.

Powell, G. T., & Wright, G. J. (2011). Jamb and jamc are essential for vertebrate myocyte fusion. *PLoS Biology, 9*(12), e1001216.

Pownall, M. E., Gustafsson, M. K., & Emerson, C. P., Jr. (2002). Myogenic regulatory factors and the specification of muscle progenitors in vertebrate embryos. *Annual Review of Cell and Developmental Biology, 18*, 747–783.

Radev, Z., Hermel, J. M., Elipot, Y., Bretaud, S., Arnould, S., Duchateau, P., … Sohm, F. (2015). A TALEN-exon skipping design for a Bethlem myopathy model in zebrafish. *PLoS One, 10*, e0133986.

Rennekamp, A. J., & Peterson, R. T. (2015). 15 years of zebrafish chemical screening. *Current Opinion in Chemical Biology, 24*, 58–70.

Richard, I., Broux, O., Allamand, V., Fougerousse, F., Chiannilkulchai, N., Bourg, N., … Beckmann, J. S. (1995). Mutations in the proteolytic enzyme calpain 3 cause limb-girdle muscular dystrophy type 2A. *Cell, 81*, 27–40.

Rochlin, K., Yu, S., Roy, S., & Baylies, M. K. (2010). Myoblast fusion: when it takes more to make one. *Developmental Biology, 341*(1), 66–83.

Roostalu, U., & Strahle, U. (2012). In vivo imaging of molecular interactions at damaged sarcolemma. *Developmental Cell, 22*, 515–529.

Rovira, M., Huang, W., Yusuff, S., Shim, J. S., Ferrante, A. A., Liu, J. O., & Parsons, M. J. (2011). Chemical screen identifies FDA-approved drugs and target pathways that induce precocious pancreatic endocrine differentiation. *Proceedings of the National Academy of Sciences, 108*(48), 19264–19269.

Roy, S., Wolff, C., & Ingham, P. W. (2001). The u-boot mutation identifies a Hedgehog-regulated myogenic switch for fiber-type diversification in the zebrafish embryo. *Genes & Development, 15*(12), 1563–1576.

Rudnicki, M. A., Schnegelsberg, P. N., Stead, R. H., Braun, T., Arnold, H. H., & Jaenisch, R. (1993). MyoD or Myf-5 is required for the formation of skeletal muscle. *Cell, 75*(7), 1351–1359.

Ruf-Zamojski, F., Trivedi, V., Fraser, S. E., & Trinh, L. A. (2015). Spatio-temporal differences in dystrophin dynamics at mRNA and protein levels revealed by a novel FlipTrap line. *PLoS One, 10*, e0128944.

Saint-Amant, L., & Drapeau, P. (1998). Time course of the development of motor behaviors in the zebrafish embryo. *Journal of Neurobiology, 37*, 622–632.

Santoriello, C., & Zon, L. I. (2012). Hooked! Modeling human disease in zebrafish. *Journal of Clinical Investigation, 122*, 2337–2343.

Sawada, A., Fritz, A., Jiang, Y., Yamamoto, A., Yamasu, K., Kuroiwa, A., … Takeda, H. (2000). Zebrafish Mesp family genes, mesp-a and mesp-b are segmentally expressed in the presomitic mesoderm, and Mesp-b confers the anterior identity to the developing somites. *Development, 127*(8), 1691–1702.

Schilling, T. F., & Kimmel, C. B. (1997). Musculoskeletal patterning in the pharyngeal segments of the zebrafish embryo. *Development, 124*(15), 2945–2960.

Schindler, R. F., Scotton, C., Zhang, J., Passarelli, C., Ortiz-Bonnin, B., Simrick, S., … Ferlini, A. (January 2016). POPDC1(S201F) causes muscular dystrophy and arrhythmia by affecting protein trafficking. *Journal of Clinical Investigation, 126*(1), 239–253. http://dx.doi.org/10.1172/JCI79562.

Schröter, C., Ares, S., Morelli, L. G., Isakova, A., Hens, K., Soroldoni, D., … Oates, A. C. (2012). Topology and dynamics of the zebrafish segmentation clock core circuit. *PLoS Biology, 10*(7), e1001364.

Shih, N. P., François, P., Delaune, E. A., & Amacher, S. L. (2015). Dynamics of the slowing segmentation clock reveal alternating two-segment periodicity. *Development, 142*(10), 1785–1793.

Sloboda, D. D., Claflin, D. R., Dowling, J. J., & Brooks, S. V. (2013). Force measurement during contraction to assess muscle function in zebrafish larvae. *Journal of Visualized Experiments: JoVE, 77*.

Smith, L. L., Beggs, A. H., & Gupta, V. A. (2013). Analysis of skeletal muscle defects in larval zebrafish by birefringence and touch-evoke escape response assays. *Journal of Visualized Experiments: JoVE, 82*, e50925.

Srinivas, B. P., Woo, J., Leong, W. Y., & Roy, S. (2007). A conserved molecular pathway mediates myoblast fusion in insects and vertebrates. *Nature Genetics, 39*(6), 781–786.

Steffen, L. S., Guyon, J. R., Vogel, E. D., Howell, M. H., Zhou, Y., Weber, G. J., … Kunkel, L. M. (2007). The zebrafish runzel muscular dystrophy is linked to the titin gene. *Developmental Biology, 309*(2), 180–192.

Stellabotte, F., & Devoto, S. H. (2007). The teleost dermomyotome. *Developmental Dynamics, 236*(9), 2432–2443.

Stellabotte, F., Dobbs-McAuliffe, B., Fernández, D. A., Feng, X., & Devoto, S. H. (2007). Dynamic somite cell rearrangements lead to distinct waves of myotome growth. *Development, 134*(7), 1253–1257.

Sztal, T. E., Sonntag, C., Hall, T. E., & Currie, P. D. (2012). Epistatic dissection of laminin-receptor interactions in dystrophic zebrafish muscle. *Human Molecular Genetics, 21*, 4718–4731.

Sztal, T. E., Zhao, M., Williams, C., Oorschot, V., Parslow, A. C., Giousoh, A., … Bryson-Richardson, R. J. (2015). Zebrafish models for nemaline myopathy reveal a spectrum of nemaline bodies contributing to reduced muscle function. *Acta Neuropathologica, 130*, 389–406.

Talbot, J. C., & Amacher, S. L. (2014). A streamlined CRISPR pipeline to reliably generate zebrafish frameshifting alleles. *Zebrafish, 11*(6), 583–585.

Talbot, J., & Maves, L. (2016). Skeletal muscle fiber type: using insights from muscle developmental biology to dissect targets for susceptibility and resistance to muscle disease. *Wiley Interdisciplinary Reviews: Developmental Biology, 5.*

Talts, J. F., Andac, Z., Gohring, W., Brancaccio, A., & Timpl, R. (1999). Binding of the G domains of laminin alpha1 and alpha2 chains and perlecan to heparin, sulfatides, alpha-dystroglycan and several extracellular matrix proteins. *EMBO Journal, 18*, 863–870.

Telfer, W. R., Busta, A. S., Bonnemann, C. G., Feldman, E. L., & Dowling, J. J. (2010). Zebrafish models of collagen VI-related myopathies. *Human Molecular Genetics, 19*, 2433–2444.

Thisse, C., Thisse, B., Schilling, T. F., & Postlethwait, J. H. (1993). Structure of the zebrafish snail1 gene and its expression in wild-type, spadetail and no tail mutant embryos. *Development, 119*(4), 1203–1215.

Thornhill, P., Bassett, D., Lochmuller, H., Bushby, K., & Straub, V. (2008). Developmental defects in a zebrafish model for muscular dystrophies associated with the loss of fukutin-related protein (FKRP). *Brain, 131*, 1551–1561.

Trinh, A., Hochgreb, T., Graham, M., Wu, D., Ruf-Zamojski, F., Jayasena, C. S., … Fraser, S. E. (2011). A versatile gene trap to visualize and interrogate the function of the vertebrate proteome. *Genes & Development, 25*(21), 2306–2320.

Tubridy, N., Fontaine, B., & Eymard, B. (2001). Congenital myopathies and congenital muscular dystrophies. *Current Opinion in Neurology, 14*, 575–582.

Vacaru, A. M., Unlu, G., Spitzner, M., Mione, M., Knapik, E. W., & Sadler, K. C. (2014). In vivo cell biology in zebrafish - providing insights into vertebrate development and disease. *Journal of Cell Science, 127*, 485–495.

Vachon, P. H., Xu, H., Liu, L., Loechel, F., Hayashi, Y., Arahata, K., … Engvall, E. (1997). Integrins (alpha7beta1) in muscle function and survival. Disrupted expression in merosin-deficient congenital muscular dystrophy. *Journal of Clinical Investigation, 100*, 1870–1881.

Vainzof, M., Anderson, L. V. B., McNally, E. M., Davis, D. B., Faulkner, G., Valle, G., … Zatz, M. (2001). Dysferlin protein analysis in limb-girdle muscular dystrophies. *Journal of Molecular Neuroscience, 17*, 71–80.

Vogel, B., Meder, B., Just, S., Laufer, C., Berger, I., Weber, S., … Rottbauer, W. (2009). In-vivo characterization of human dilated cardiomyopathy genes in zebrafish. *Biochemical and Biophysical Research Communications, 390*, 516–522.

Wallace, L. M., Garwick, S. E., Mei, W., Belayew, A., Coppee, F., Ladner, K. J., … Harper, S. Q. (2011). DUX4, a candidate gene for facioscapulohumeral muscular dystrophy, causes p53-dependent myopathy in vivo. *Annals of Neurology, 69*, 540–552.

Warren, G. L., Hayes, D. A., Lowe, D. A., Williams, J. H., & Armstrong, R. B. (1994). Eccentric contraction-induced injury in normal and hindlimb-suspended mouse soleus and EDL muscles. *Journal of Applied Physiology (1985), 77*, 1421–1430.

Waterman, R. E. (1969). Development of the lateral musculature in the Teleost, *Brachydanio rerio*: a fine structural study. *American Journal of Anatomy, 125*, 457–494.

Waugh, T. A., Horstick, E., Hur, J., Jackson, S. W., Davidson, A. E., Li, X., & Dowling, J. J. (2014). Fluoxetine prevents dystrophic changes in a zebrafish model of Duchenne muscular dystrophy. *Human Molecular Genetics*. ddu185.

Webb, A. B., Lengyel, I. M., Jörg, D. J., Valentin, G., Jülicher, F., Morelli, L. G., & Oates, A. C. (2016). Persistence, period and precision of autonomous cellular oscillators from the zebrafish segmentation clock. *eLife, 5*, e08438.

Webb, A. B., Soroldoni, D., Oswald, A., Schindelin, J., & Oates, A. C. (2014). Generation of dispersed presomitic mesoderm cell cultures for imaging of the zebrafish segmentation clock in single cells. *Journal of Visualized Experiments: JoVE, 89*.

Weinberg, E. S., Allende, M. L., Kelly, C. S., Abdelhamid, A., Murakami, T., Andermann, P., ... Riggleman, B. (1996). Developmental regulation of zebrafish MyoD in wild-type, no tail and spadetail embryos. *Development, 122*(1), 271–280.

Windner, S. E., Bird, N. C., Patterson, S. E., Doris, R. A., & Devoto, S. H. (2012). Fss/Tbx6 is required for central dermomyotome cell fate in zebrafish. *Biology Open*. BIO20121958.

Windner, S. E., Doris, R. A., Ferguson, C. M., Nelson, A. C., Valentin, G., Tan, H., ... Devoto, S. H. (2015). Tbx6, Mesp-b and Ripply1 regulate the onset of skeletal myogenesis in zebrafish. *Development, 142*(6), 1159–1168.

Wittbrodt, J. N., Liebel, U., & Gehrig, J. (2014). Generation of orientation tools for automated zebrafish screening assays using desktop 3D printing. *BMC Biotechnology, 14*(1), 1.

Wood, A. J., Muller, J. S., Jepson, C. D., Laval, S. H., Lochmuller, H., Bushby, K., ... Straub, V. (2011). Abnormal vascular development in zebrafish models for fukutin and FKRP deficiency. *Human Molecular Genetics, 20*, 4879–4890.

Yabe, T., & Takada, S. (January 2016). Molecular mechanism for cyclic generation of somites: lessons from mice and zebrafish. *Development, Growth & Differentiation, 58*(1), 31–42. http://dx.doi.org/10.1111/dgd.12249.

Yozzo, K. L., Isales, G. M., Raftery, T. D., & Volz, D. C. (2013). High-content screening assay for identification of chemicals impacting cardiovascular function in zebrafish embryos. *Environmental Science & Technology, 47*(19), 11302–11310.

Zhang, R. L., Yang, J. C., Zhu, J., & Xu, X. L. (2009). Depletion of zebrafish Tcap leads to muscular dystrophy via disrupting sarcomere-membrane interaction, not sarcomere assembly. *Human Molecular Genetics, 18*, 4130–4140.

PART

VI

Central and Sensory Nervous Systems

CHAPTER

Analysis of myelinated axon formation in zebrafish

15

M. D'Rozario*, K.R. Monk*,§, S.C. Petersen¶,1

*Washington University School of Medicine, St. Louis, MO, United States

§Hope Center for Neurological Disorders, Washington University School of Medicine, St. Louis, MO, United States

¶Kenyon College, Gambier, OH, United States

[1]Corresponding author: E-mail: petersens@kenyon.edu

CHAPTER OUTLINE

Abstract

Myelin is a lipid-rich sheath formed by the spiral wrapping of specialized glial cells around axon segments. Myelinating glia allow for rapid transmission of nerve impulses and metabolic support of axons, and the absence of or disruption to myelin results in debilitating motor, cognitive, and emotional deficits in humans. Because myelin is a jawed vertebrate innovation, zebrafish are one of the simplest vertebrate model systems to study the genetics and development of myelinating glia. The morphogenetic cellular movements and genetic program that drive myelination are conserved between zebrafish

Methods in Cell Biology, Volume 138, ISSN 0091-679X, http://dx.doi.org/10.1016/bs.mcb.2016.08.001

and mammals, and myelin develops rapidly in zebrafish larvae, within 3–5 days post-fertilization. Myelin ultrastructure can be visualized in the zebrafish from larval to adult stages via transmission electron microscopy, and the dynamic development of myelinating glial cells may be observed in vivo via transgenic reporter lines in zebrafish larvae. Zebrafish are amenable to genetic and pharmacological screens, and screens for myelinating glial phenotypes have revealed both genes and drugs that promote myelin development, many of which are conserved in mammalian glia. Recently, zebrafish have been employed as a model to understand the complex dynamics of myelinating glia during development and regeneration. In this chapter, we describe these key methodologies and recent insights into mechanisms that regulate myelination using the zebrafish model.

INTRODUCTION

The evolution of myelin, which surrounds and protects axons, was a critically important innovation in the jawed vertebrate lineage (Zalc, Goujet, & Colman, 2008). Myelin is a lipid-rich sheath that facilitates saltatory conduction velocity without a substantial increase in axonal diameter; thus, myelin permits higher brain complexity within a bony skeleton. In humans, the importance of the myelin sheath is most easily realized when it is lost or disrupted, such as in diseases including multiple sclerosis and Charcot-Marie-Tooth. Despite its importance, the cellular and molecular mechanisms that drive the migration, morphogenesis, terminal differentiation, and regeneration of myelinating glia are incompletely understood.

Though the myelin sheath itself is ultrastructurally and biochemically similar in the central and peripheral nervous systems (CNS and PNS, respectively), the cell types that form the myelin sheath are derived from different precursor populations, and the genetic programs and morphogenetic behaviors driving myelin formation are quite different between the CNS and PNS. Myelin is formed in the CNS by oligodendrocytes (OLs) in a multistep developmental process characterized by expression of molecular markers and dramatic changes in cell morphology. OL precursor cells (OPCs; Olig2$^+$ and Sox10$^+$) originate from discrete ventral regions of the neural tube during early embryonic development and in the subventricular region of the brain during adult neurogenesis (Ackerman & Monk, 2015; Emery, 2010). Once specified, OPCs proliferate and migrate remarkable distances to populate the entire CNS. Once migration ceases, a subset of OPCs transition into premyelinating OLs (Nkx2.2$^+$), which extend filapodia-like processes and ensheath axons but do not begin myelination (Kirby et al., 2006; Kucenas et al., 2008; Mitew et al., 2014; Snaidero et al., 2014; Zhu et al., 2014). Finally, premyelinating OLs select axons, terminally differentiate to become myelinating OLs, and iteratively wrap axonal segments (Fig. 1A).

In the PNS, specialized glia called Schwann cells (SCs) form myelin, with both similar and distinct mechanisms as CNS myelination (reviewed in Jessen & Mirsky, 2005). During embryogenesis, SC precursors (SCPs) are proliferative and migrate along with growing axons. Once migration ceases, SCPs differentiate into immature SCs that are associated with many axons. At this point, immature SCs secrete extracellular matrix (ECM) molecules that constitute the basal lamina (Court, Wrabetz, &

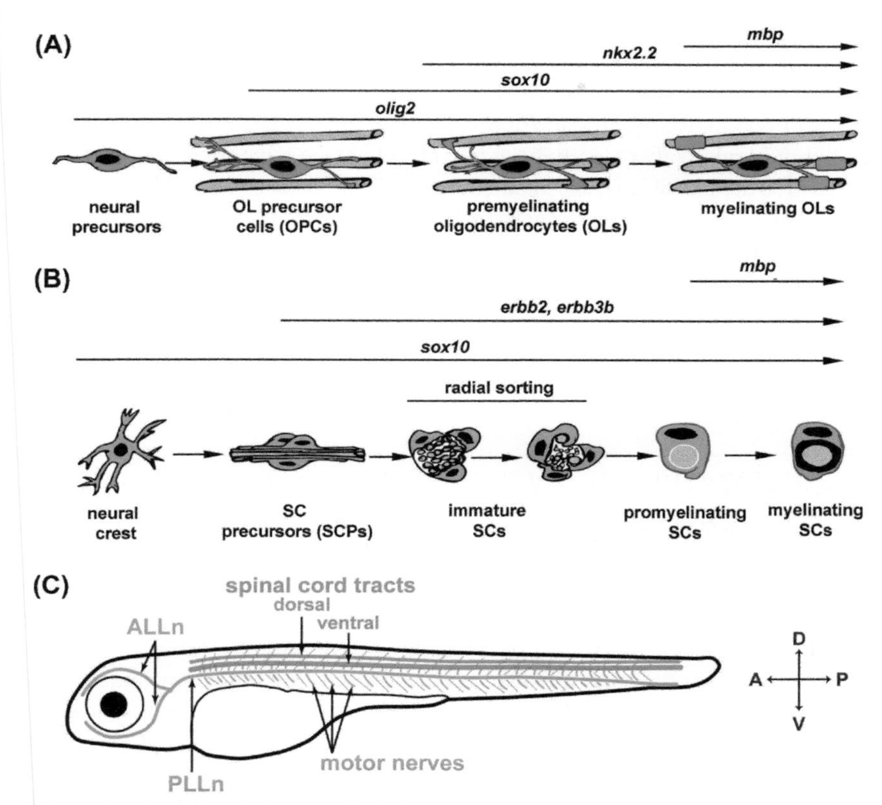

FIGURE 1 Myelinating glia in the zebrafish.

(A) Oligodendrocyte (OL) development. Neural precursor cells, depicted in red, are born within the ventral neural tube during early embryonic development. Once specified, OL precursor cells (OPCs) proliferate and migrate. Once migration ceases, a subpopulation of OPCs begin to extend protrusions, contact axons (premyelinating OLs), and terminally differentiate to become myelinating OLs. (B) Schwann cell (SC) development. Immature SCs associate with bundles of axons, begin the process of radial sorting by segregating individual axons based on diameter, and spiral their cytoplasmic membranes to form myelin. The accompanying arrowheads show the key markers for SC development discussed in this review. (C) Schematized locations of commonly studied myelinated axons in zebrafish larvae at 3–5 days postfertilization. Compass in lower right denotes dorsal (D), ventral (V), anterior (A), and posterior (P). CNS is shown in red and denotes the dorsal and ventral tracts of the spinal cord. PNS is shown in blue and marks the anterior and posterior lateral line nerves (ALLn and PLLn, respectively) and motor nerves that extend ventrally from the spinal cord.

Feltri, 2006). Through a process called radial sorting, immature SCs select and segregate large caliber axons and repeatedly wrap their membranes around individual axonal segments, ultimately generating the myelin sheath (Fig. 1B).

The morphological progression of myelinating glia is well characterized in mammals, but only relatively recently have zebrafish become a premiere tool for in vivo analysis of OLs and SCs (Ackerman & Monk, 2015; Brösamle & Halpern, 2002; Monk & Talbot, 2009). OL development is easily monitored in larval zebrafish CNS, most frequently in the spinal cord, whereas the anterior and posterior lateral line nerves (ALLn and PLLn, respectively) and motor nerves are useful models for SC myelination (Fig. 1C). Here, we describe recent advances in imaging, genetic, and pharmacological tools to investigate conserved mechanisms of myelinating glial cell development and regeneration using the zebrafish.

1. VISUALIZATION OF MYELINATING GLIA IN ZEBRAFISH

A major advantage to using zebrafish as a model system is the ability to perform in vivo imaging to study myelination. Transgenic reporter lines expressing fluorescent proteins under control of glial-specific promoters are ideal tools to study the behavior of myelinating cells in a living vertebrate. Several valuable transgenic reporter lines are available to visualize myelination throughout development (Table 1). Zebrafish transgenic lines are predominantly generated using the Gateway-compatible Tol2Kit (Kwan et al., 2007). Concurrently, the Tol2 sites can also be used to generate enhancer trap lines in which a Gal4 transactivator is placed downstream of a promoter of interest and activates a reporter under Upstream Activating Sequence (UAS) control. Importantly, for the purposes of this review, several transgenic reporter lines are available to study myelination in both the CNS and PNS, at different stages of development, and the expression of these reporters are evolutionarily conserved.

In addition, myelin defects may be scored in fixed tissue with in situ hybridization (ISH) (Thisse, Thisse, Halpern, & Postlethwait, 1994) (Table 2). Whole mount ISH is widely used to describe expression patterns of genes of interest and interrogate myelin gene expression of mutants with defects in myelination. During ISH, an antisense messenger RNA (mRNA) probe tagged with either digoxigenin or fluorescein uridine-5′-triphosphate binds to the endogenous transcript. Following hybridization, the transcript is detected using an antibody specific to the tag and visualized under a light microscope.

The optical transparency and myriad transgenic lines make zebrafish a powerful and tractable system to study myelination. Zebrafish live-cell imaging has opened up new avenues for investigation of the cellular basis of myelination. One of the most exciting advances is the ability to see and study the fine details of individual cells and peer into the cellular and molecular events of migration, process extension, and internode development. During early stages of CNS myelination, there is a great deal of motility as the pro-OL extends actin-rich plasma membrane protrusions to contact multiple axonal segments, and as they mature, OLs myelinate several

Table 1 Available Transgenic Reporters to Visualize Myelinating Glia in Zebrafish

Gene	Labeled Cell Types	Transgenic Lines; References
cldnk	Myelinating OL and SCs	*Tg(cldnk:Gal4)^{ue101Tg}*; Münzel et al. (2012)
foxd3	Immature SCs through myelinating SCs	*Tg(zFoxd3GFP)*; Gilmour, Maischein, and Nüsslein-Volhard (2002)
nkx2.2a	Subset of OPCs and early myelinating oligodendrocytes	*Tg(nkx2.2a:mEGFP)^{vu16Tg}*; Kirby et al. (2006)
mbp	Myelinating OLs and myelinating SCs	*Tg(mbp:EGFP)^{ck1Tg}*; Jung et al. (2010) *Tg(mbp:EGFP)^{ue1Tg}*; Almeida, Czopka, Ffrench-Constant, and Lyons (2011) *Tg(mbp:EGFP-CAAX)^{ue2Tg}*; Almeida et al. (2011) *Tg(mbp:GAL4-VP16)^{co20Tg}*; Hines et al. (2015)
mpz/p0	Myelinating OLs	*Tg(mpz[10kb]:EGFP)^{pt408Tg}*; Bai, Parris, and Burton (2014)
nkx2.2a	Subset of OPCs and early myelinating oligodendrocytes	*Tg(nkx2.2a:mEGFP)^{vu16Tg}*; Kirby et al. (2006)
olig1	OL lineage cells	*Tg(olig1:mEGFP)^{nv150Tg}*; Schebesta and Serluca (2009)
olig2	PMN-derived progenitors, OL lineage, and motor neurons	*Tg(olig2:EGFP)^{vu12Tg}*; Shin, Park, Topczewska, Mawdsley, and Appel (2003) *Tg(olig2:dsRed)^{vu19Tg}*; Kucenas et al. (2008) *Tg(olig2:Kaeda)^{vu85Tg}*; Zannino and Appel (2009)
plp	OL lineage cells	*Tg(Mmu.Plp1:EGFP)^{cc1Tg}*; Yoshida and Macklin (2005)
sox10	OL lineage cells, interneurons, neural crest, and SC	*Tg(sox10:mRFP)^{vu234Tg}*; Kirby et al. (2006) *Tg(sox10:EGFP)^{ba4Tg}*; Dutton et al. (2001) *Tg(sox10:nls-Eos)^{w18Tg}*; Prendergast et al. (2012) *Tg(sox10:Gal4-VP16)^{co19Tg}*; Das and Crump (2012) *Tg(sox10:KalTA4GI)*; Almeida and Lyons (2015)

axon segments. Thus, a number of challenges must be met for OLs to establish and maintain their morphology and function. Meeting these challenges is the work of sophisticated architectural rearrangements within the OL that allow for process extension and wrapping of the plasma membrane and tightly regulated mechanisms of transport and motility. Increasingly, in vivo imaging in zebrafish has begun to add crucial missing pieces to the puzzle of myelin dynamics that cannot be readily observed in living rodent models. For example, using a tagged marker for F-actin, Nawaz et al. (2015) could define actin dynamics during myelin growth in vivo. In

Table 2 Available In Situ Hybridization Probes for Myelinating Glia in Zebrafish

Gene	Cell-Type Expression	References
36K	Myelinating OLs	Morris et al. (2004)
cldnk	Myelinating OL and SCs	Takada, Kucenas, and Appel (2010)
ctnnd2	Myelinating OL and SCs	Takada and Appel (2010)
erbb3	Immature SCs through myelinating SCs	Lyons et al. (2005)
foxd3	Immature SCs through myelinating SCs	Gilmour et al. (2002)
gpr126	SCs	Monk et al. (2009)
krox20	Promyelinating SCs	Lyons et al. (2005)
mbp	Myelinating OLs and myelinating SCs	Brösamle and Halpern (2002) and Lyons et al. (2005)
mid1ip1b	Myelinating OLs and myelinating SCs	Takada and Appel (2010)
nkx2.2a	Subset of OPCs and early myelinating oligodendrocytes	Kirby et al. (2006)
oct6	Promyelinating SCs	Levavasseur et al. (1998)
olig1	OL lineage cells	Schebesta and Serluca (2009)
olig2	PMN-derived progenitors, OL lineage and motor neurons	Park, Mehta, Richardson, and Appel (2002)
mpz/p0	Myelinating OLs	Brösamle and Halpern (2002)
plp1a	Myelinating OLs	Brösamle and Halpern (2002)
plp1b	Myelinating OLs	Brösamle and Halpern (2002)
sox10	OL lineage cells, interneurons, neural crest, and SC	Dutton et al. (2001)
Zwilling-A and *-B*	Myelinating OLs and myelinating SCs	Schaefer and Brösamle (2009)

this study, immediately prior to axonal contact, F-actin was spatially restricted to the furthest edges of the expanding OL processes and following contact localized spirally along the axons as the OL plasma membrane iteratively wrapped the selected axonal segment (Nawaz et al., 2015). Further, these studies were recapitulated in purified mouse OLs in culture (Zuchero et al., 2015), highlighting zebrafish as a complementary approach to studying myelination.

Although the vast majority of OL and SC molecular markers are well conserved from zebrafish to mammals, some differences have been reported. Genomics studies reveal that zebrafish myelin genes such as *myelin basic protein* (*mbp*), *myelin protein zero* (*mpz*), and *proteolipid protein* (*plp*) share significant homology to their mammalian counterparts (Bai, Sun, Stolz, & Burton, 2011; Nawaz, Schweitzer, Jahn, & Werner, 2013; Schweitzer, Becker, Schachner, Nave, & Werner, 2006). However, during early stages of PNS myelination, zebrafish SCs do not encode key myelin compaction genes *mpz* and *plp*, suggesting that myelin compaction may be delayed in zebrafish compared to mammals (Brösamle & Halpern, 2002). Moreover, in zebrafish, *mpz* expression persists in both the CNS and PNS myelinated axons, in contrast to the mammalian homolog P0, which is found exclusively in the PNS (Bai et al., 2011). In addition, several proteins such as Zwilling-A, Zwilling-B, 36k, and Claudin K are distinctively present in zebrafish myelin; their homologs are either absent or not involved in myelin compaction in mammals (reviewed in Ackerman & Monk, 2015).

A deeper understanding of the cell biological functions of myelin proteins can only be achieved using fine structural imaging methods, including electron microscopy. While many aspects of myelin structure are indeed conserved between multiple fish species and mammals, we direct the reader to excellent reviews on important disparities between myelin structure in fish and rodent systems (Avila, Tevlin, Lees, Inouye, & Kirschner, 2007; Möbius, Nave, & Werner, 2016). Differences in myelin protein composition certainly lead to some variations in observed morphology, as electron microscopic analysis in zebrafish is generally performed at time points when myelin is not yet compact. Moreover, conventional chemical fixations are not adequate to preserve the myelin ultrastructure in zebrafish, and so many labs employ microwave-assisted tissue processing techniques (Czopka & Lyons, 2011). Additionally, exciting advances in high-pressure freezing techniques can provide improved resolution in the analysis of myelin ultrastructure in zebrafish as more laboratories acquire expertise in this technology (Möbius et al., 2016).

The myelin ultrastructure of the larval zebrafish can be visualized via transmission electron microscopy (TEM) to study OL and SC development in the spinal cord and the lateral line, respectively (Fig. 2A). A cross section of the larval zebrafish shows the spinal cord situated in the center of the animal and the two PLLns situated on either side (Fig. 2B). A closer magnification of the spinal cord shows several myelinated axons including the Mauthner axon, a critical neural circuit for zebrafish fast escape response (Korn & Faber, 2005; Fig. 2C). Similarly, myelinated axons are also observed in the PLLn (Fig. 2D).

Myelinated axons can also be dissected from adult zebrafish and visualized via TEM. A protocol for TEM preparation of adult lateral line is described in the following section.

1.1 SOLUTIONS REQUIRED

Modified Karnovsky's fixative (4% paraformaldehyde + 2% glutaraldehyde in 0.1-M sodium cacodylate, pH 7.4)
0.1-M sodium cacodylate
2% osmium tetraoxide in 0.1-M sodium cacodylate
25%, 50%, 75%, 95% ethanol in sterile filtered (i.e., MilliQ, MQ) water; 100% ethanol; propylene oxide; EMBED solution (see Section "Recipe" in Czopka & Lyons, 2011).

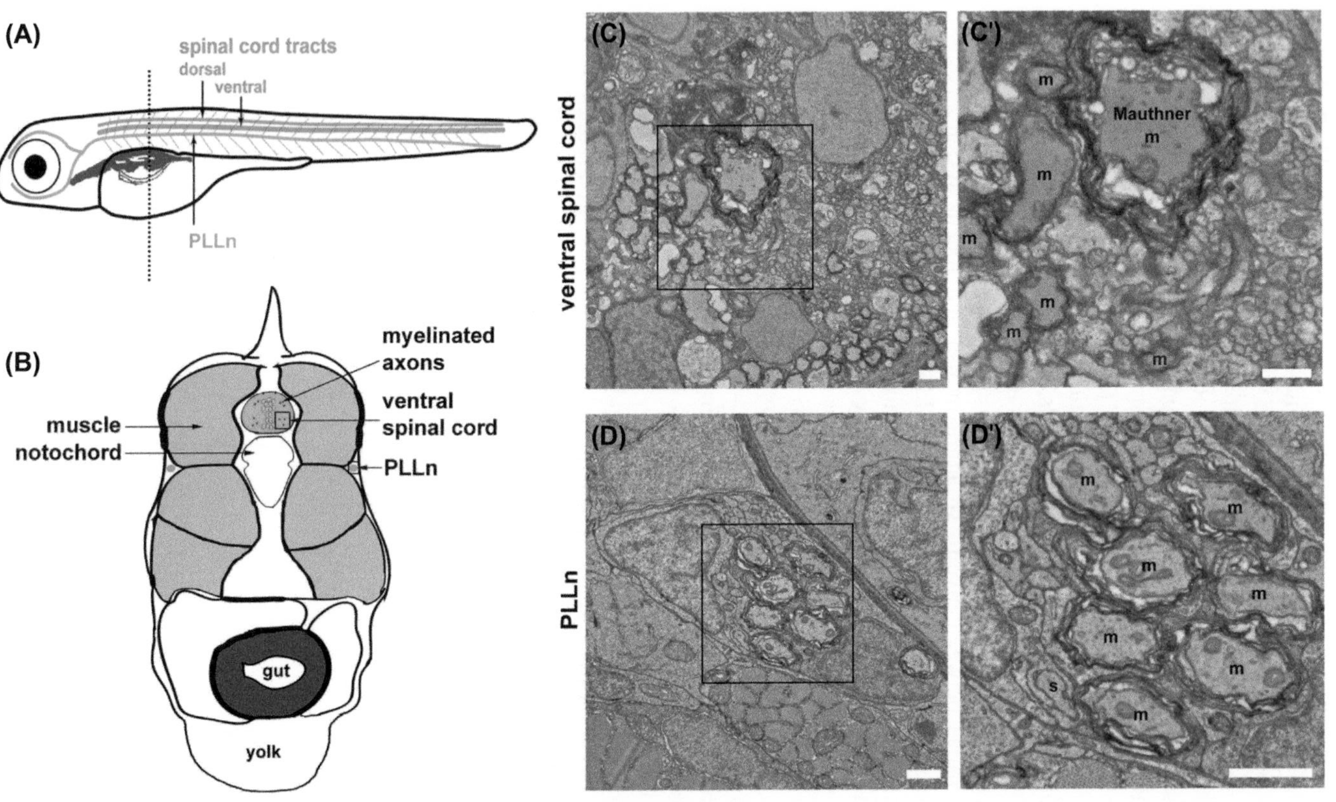

FIGURE 2 Transmission electron microscopy (TEM) for zebrafish myelinated axons.

(A) Schematized cartoon depicting the myelinated axons commonly studied via TEM in larval zebrafish. Dotted line indicates the anterior–posterior position of cross section depicted in panels B–D. (B) Diagram represents the cross section of a larval zebrafish and indicates positions of the spinal cord [red] and the PLLn on either side of the body [blue]. (C–D) Transmission electron micrographs between body segments five to seven at 5 days postfertilization. Boxes in left panels denote regions of interest shown in right panels. All panels, scale bar = 1 μm. (C–C′) Ventral spinal cord with midline to the left. Myelinated axons in right panel are pseudocolored in red and marked with m. Note large caliber Mauthner axon. (D–D′) PLLn with midline to the left. Axons in right panel are pseudocolored in blue to mark myelinated (m) or sorted (s).

1.2 FIXATION AND EMBEDDING OF ADULT NERVE

1. Euthanize individual zebrafish by immediate and prolonged immersion in ice water or by prolonged overdose with tricaine (pH 7.4), until gill movements have ceased. With a clean razor blade, quickly sever the spinal cord.
2. Fix the whole adult fish in Karnovsky's solution in 2-mL Eppendorf tubes. For optimal fixation, the nerves need to be fixed uniformly and as fast as possible. To accelerate tissue processing, the immersed fish is microwaved in a water bath in either a Pelco microwave processor (model 3451; Ted Pella Mountain Lakes, CA) or a countertop microwave with inverter technology. It is crucial to keep the tissue at about 15–18°C to prevent overheating the samples. While the temperature can be regulated via water recirculation in the Pelco system, a water/ice bath may be used as an alternative and the temperature can be monitored with a thermometer and adjusted after each cycle.
   Step 1: 250 W for 1 min, off for 1 min, 250 W for 1 min
   Step 2: 450 W for 1 min, off for 1 min, 450 W for 1 min
3. Upon fixation, the adult PLLn is located and removed using forceps. The ganglia of the PLLn are situated posterior to the ear, and the nerves run along the body on the surface of the fish. The dissected nerve is postfixed in Karnovsky's for 1–2 h at room temperature or 4°C overnight in a 12-well culture plate.
4. Remove Karnovsky's and wash 3 × 15′ with 0.1-M sodium cacodylate at room temperature (RT). Samples can be stored at 4°C, although we have found that prolonged storage (more than a month) at this stage can result in poorer quality images.
5. Postfix in 2% osmium tetraoxide in 0.1-M sodium cacodylate. The optimal incubation time and dilution vary between samples, but we regularly fix for 1 h.
6. At this point, gently move the osmium-treated nerves from the culture dishes to a solvent-proof Eppendorf tube using toothpicks or a paintbrush. Care must be taken to ensure that the samples do not dry.
7. MQ water washes—3 × 15′ at RT.
8. Sequential ethanol dehydration steps:
   a. 25%—20′ at RT
   b. 50%—20′ at RT
   c. 75%—20′ at RT (if necessary, tissue can be stored at 4°C overnight after this step)
   d. 95%—20′ at RT
   e. 100%—20′ at RT × 2
9. Propylene oxide steps:
   a. propylene oxide—15′ at RT
   b. 1:1 propylene oxide: EMBED overnight
   c. 100% EMBED 4 h at RT
10. Embed and polymerize overnight in oven at 65°C or until blocks are solid. Samples are trimmed, sectioned, and stained as previously described (Czopka & Lyons, 2011).

2. GENETIC ANALYSIS OF MYELIN DEVELOPMENT IN ZEBRAFISH

2.1 FORWARD GENETIC SCREENS

A major advantage of the zebrafish model system is the ability to perform large-scale forward genetic screens (Driever et al., 1996; Haffter & Nüsslein-Volhard, 1996). Zebrafish produce large clutches (hundreds of embryos per pair of mating adults), have a reasonable generation time, and can be housed as adults in a relatively small space. To randomly introduce mutations, male zebrafish are exposed to *N*-ethyl-*N*-nitrosourea (ENU), outcrossed to wild-type females, mutations are driven to homozygosity, and larvae are screened for phenotypes of interest in the F3 generation (Fig. 3). In this scheme, transgenes to screen for phenotypes in the F3 progeny may be introduced by outcrossing in the F1 generation. Historically, causative mutations were

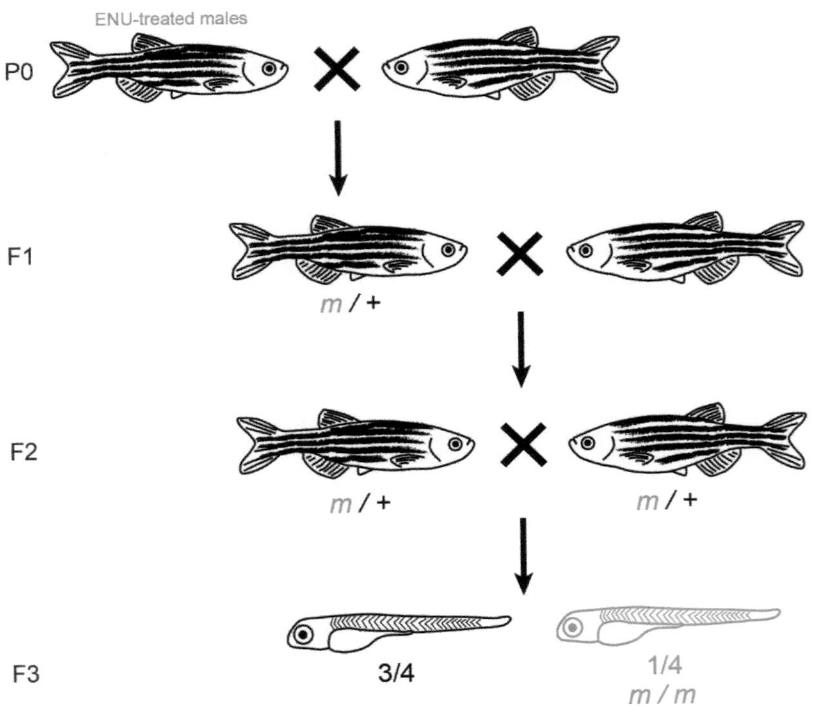

FIGURE 3 Generation of mutants via forward genetic screen.

Schematic for a traditional three-generation screen. Generations (P0, F1, F2, F3) are labeled to the left. P0 males are treated with *N*-ethyl-*N*-nitrosourea (ENU) to induce random mutations, which are driven to homozygosity in the F3 generation using the crossing scheme shown.

identified via a mapcross to an outbred strain, followed by polymerase chain reaction (PCR)-based linkage analysis (Postlethwait & Talbot, 1997). More recently, whole-genome sequencing (WGS) and RNA sequencing have become more affordable and thus more efficient options to rapidly identify a lesion of interest, as they can be used without a map-crossing step or parental information (Henke, Bowen, & Harris, 2013; Hill et al., 2013; Miller, Obholzer, Shah, Megason, & Moens, 2013).

Because of the relatively rapid development of myelinated axons in zebrafish and advent of transgenic tools for in vivo observation, large-scale genetic screens are a tool uniquely suited to the zebrafish to discover novel pathways that drive myelination. In the Appel laboratory, larvae were screened for aberrant *olig2:EGFP* expression in the CNS (Snyder, Kearns, & Appel, 2012), whereas the Talbot group screened for both CNS and PNS mutant phenotypes using ISH for *mbp* (Pogoda et al., 2006), and a complementary screen was performed to analyze sodium channel clustering at the nodes of Ranvier between myelin sheaths along an axon (Monk, Voas, Franzini-Armstrong, Hakkinen, & Talbot, 2013; Voas et al., 2007). The key myelin regulators uncovered in these screens are summarized in Table 3 and in the following sections. Most recently, the Monk laboratory screened for phenotypes using both *mbp:mCherry* and *mbp* ISH in combination with *lhx1:GFP*, which marks axons in the CNS and PNS (Ackerman et al., in preparation; Harty et al., in preparation; Herbert et al., in preparation). Although these screens have been quite expansive (>600 genomes in each of the *mbp*-based screens) and have identified many genes required for myelin development (Table 3), screens for myelination mutants in zebrafish have not yet reached saturation.

As a complementary approach to ENU-based forward genetic screens, Kazakova et al. reasoned that genes required for the establishment of neuronal architecture may also be required in myelin development. Therefore, they assayed *mbp* expression among a subset known early neurodevelopment mutants (Kazakova et al., 2006). One of the mutants, motionless, was later shown to be mediator complex subunit 12 *(med12)* (Wang et al., 2006), which cooperates with *sox10* to promote myelinating glia differentiation in mammals (Vogl et al., 2013).

2.2 REVERSE GENETICS APPROACHES

Classical forward genetic screens for myelin mutants have provided a powerful approach to unraveling new genes that regulate formation of myelinated axons. While forward genetic screens allow multiple entry points to interrogate mechanisms involved in myelination via cloning and sequencing of the mutated genes, reverse genetics is an important complement. The last decade has seen important additions to the gene editing toolbox in zebrafish, particularly with the recent discovery and application of transcription activator-like effector nucleases (TALENs) and the type II prokaryotic clustered regularly interspaced short palindromic repeat (CRISPR/Cas9) system (Gaj, Gersbach, & Barbas, 2013).

Briefly, TALENs consist of bipartite fusion proteins that recognize specific DNA sequences each fused to the FokI restriction endonuclease. We frequently use the

Table 3 Myelination Genes Revealed by Forward Genetic Screening in Zebrafish

Allele	Gene (If Known)	Myelinating Glia Phenotype	Function	Screen/Citation
st14, st48	*erbb3b*	Reduced *mbp* in LLn	Nrg/Erb signaling	Lyons et al. (2005)
st50, st61	*Erbb2*	Reduced *mbp* in LLn	Nrg/Erb signaling	Lyons et al. (2005)
st20	*Foxa2*	Reduced *mbp* in hindbrain	Forkhead transcription factor	Norton et al. (2005)
st47		Reduced *mbp* in CNS		Pogoda et al. (2006)
st64		Reduced *mbp* in ALLn		Pogoda et al. (2006)
i26	*aldh1a2*	PLLn amyelination	Aldehyde dehydrogenase	Kazakova et al. (2006)
M807	*Med12*	PLLn/CNS amyelination	Transcription mediator complex subunit	Kazakova et al. (2006)
tm79a	*dzip1*	Loss of *mbp* in posterior PLLn	DAZ interacting zinc finger protein, regulator of Hedgehog signaling	Kazakova et al. (2006)
tt258		Loss of *mbp* in posterior PLLn		Kazakova et al. (2006)
st25, st53	*Nsfa*	Reduced *mbp* in LLn and CNS	N-ethylmaleimide-sensitive factor, synaptic vesicle fusion	Woods, Lyons, Voas, Pogoda, and Talbot (2006)
st60	*αII-spectrin*	Reduced *mbp* in PNS/CNS	Cortical cytoskeleton component	Voas et al. (2007)
st23	*Kbp*	Reduced *mbp* in LLn, CNS amyelination	Intracellular transport	Lyons, Naylor, Mercurio, Dominguez and Talbot (2008)
st49, st63	*adgrg6/gpr126*	PNS amyelination	GPCR $G\alpha s$ signaling	Monk et al. (2009)
st43	*kif1b*	Reduced *mbp* in LLn, punctate *mbp* in CNS	mRNA transport	Lyons, Naylor, Scholze, and Talbot (2009)
vu56	*fbxw7*	Excess CNS myelination	F-box subunit of E3 ubiquitin ligase	Kearns et al. (2015) and Snyder et al. (2012)
vu76	*dync1h1*	Reduced PNS myelination; reduced OPCs	Dynein heavy chain subunit	Langworthy and Appel (2012) and Yang et al. (2015)
vu166	*Pescadillo*	Reduced OPCs	Ribosome biogenesis, cell proliferation	Simmons and Appel (2012)
st67	*sec63*	Reduced PLLn and CNS myelination	ER translocon machinery	Monk et al. (2013)
st51	*notch3*	Reduced *mbp* in CNS	Notch signaling	Zaucker et al. (2013)
vu57	*hmgcs1*	OPC migration defects, reduced CNS *mbp*	Isoprenoid and cholesterol synthesis	Mathews et al. (2014)
st78	*znf16l*	Reduced *mbp* in CNS	Zinc finger transcription factor	Sidik and Talbot (2015)

TALEN targeter tool (https://tale-nt.cac.cornell.edu/) to find candidate binding sites and assemble TALEN from TALE repeats using the Golden Gate approach. Once assembled, the TALEN is cloned into an appropriate TALEN backbone to generate mRNA expression plasmids. Following transcription, the TALEN mRNA is injected at the one-cell stage. While TALEN assembly is generally more straightforward for genome editing compared to previous technologies, TALENs are labor intensive and maintaining a library of TALE repeats can be expensive. On the other hand, CRISPR technology entails the generation of a guide RNA (gRNA) that is injected along with Cas9 nuclease. There are several online tools to design gRNAs for the target gene including ZiFiT, CRISPR design tool, and CHOPCHOP to name a few (Montague, Cruz, Gagnon, Church, & Valen, 2014; Ran et al., 2013; Sander et al., 2010). The advantages and disadvantages of these techniques are reviewed elsewhere (Gaj et al., 2013); importantly, both approaches allow for targeted, rapid genome editing. Using these facile techniques, it is possible to manipulate the zebrafish genome and recover mutants with small inframe indels that are ideal for structure/function analyses during myelination (Petersen et al., 2015). Excitingly, recent studies in different developmental processes in zebrafish have utilized the CRISPR/Cas9 system to target multiple genomic loci simultaneously via multiplexed pools of gRNAs (Shah, Davey, Whitebirch, Miller, & Moens, 2016). For further reading, we suggest several excellent reviews and seminal research articles (Auer & Del Bene, 2014; Boch et al., 2009; Gagnon et al., 2014; Gaj et al., 2013; Jao, Wente, & Chen, 2013; Talbot & Amacher, 2014).

While TALENs and CRISPR/Cas9 technology allow for precise, targeted mutations in the genome, these techniques are less effective for studying genes that are either important for early embryonic development or have pleiotropic functions at later stages. The constitutive and global loss of these factors may be lethal and preclude further studies of the cellular autonomy of these factors beyond early development. To circumvent this problem, several alternate approaches have been proposed. For instance, cell-specific gene mutations have been carried out by expressing a CRISPR/Cas9 vector under the control of tissue-specific promoters of interest (Ablain, Durand, Yang, Zhou, & Zon, 2015), and knockdowns can be performed in a spatiotemporal manner via photoactivatable caged antisense morpholinos to either block mRNA splicing or protein translation (Ruble, Yeldell, & Dmochowski, 2015). These approaches are technically demanding and are restricted to certain developmental stages. Moreover, recent studies have demonstrated toxic side effects of using morpholinos that have been erroneously attributed to knocking down a specific gene (reviewed in Schulte-Merker & Stainier, 2014). Taken together, morpholinos can be informative in transiently disrupting protein function when used with stringent controls and coupled with corresponding mutant data for the target gene.

In addition to caged morpholinos, expression can be spatially and temporally controlled in zebrafish through site-specific recombination using Cre or Flp recombinases (Hans, Kaslin, Freudenreich, & Brand, 2009), tamoxifen-inducible Cre/Lox technology (Hans et al., 2009), tetracycline-inducible systems (Huang et al., 2005), and the Gal4/UAS system (Halpern et al., 2008). These techniques are conceptually elegant but have minor limitations in zebrafish, as new myelin-specific drivers need to be generated. Nonetheless, the fast generation time and relatively large clutch sizes allow one to screen for mutants within a short period of time. As tools and inducible constructs are developed and become more widely available, in the future, it will be interesting to perform cell-specific gene inactivation to determine autonomy and transient tissue-specific functions during myelination.

2.3 GENETIC REGULATORS OF MYELINATION: LESSONS FROM ZEBRAFISH

The utility of the zebrafish model depends upon faithful recapitulation of the myelination program between teleosts and mammals. To date, microscopic, biochemical, and genetic analyses demonstrate strong similarity of zebrafish myelin to that of rodents, and most of the markers and mutations used to study myelin in mammals have been found to be important in zebrafish myelin development as well (most recently reviewed in Ackerman & Monk, 2015). As an example, Neuregulin-ErbB2/3 signaling is an established, necessary pathway for SC development (Newbern & Birchmeier, 2010), and the conservation of the myelination program in zebrafish is perhaps most profoundly illustrated by the unbiased recovery of mutations in ErbB2/3 signaling in one of the first large-scale forward genetic screens for myelin mutants (Lyons et al., 2005). Subsequent in vivo analyses in zebrafish were able to parse roles for Nrg-ErbB signaling in directed migration, proliferation, and radial sorting of SCs (Lyons et al., 2005; Perlin, Lush, Stephens, Piotrowski, & Talbot, 2011; Raphael, Lyons, & Talbot, 2011), thus illustrating the unique importance of zebrafish in myelination studies.

As mentioned, zebrafish genetic screens have revealed established regulators of myelinating glial cell development. Screens can also uncover genes previously implicated in myelinopathies, which establish new models for disease research. For instance, patients with CADASIL syndrome (cerebral autosomal dominant arteriopathy with subcortical infarcts and leukoencephalopathy, OMIM #125310) present with neurodegeneration accompanied by white matter deficits as a result of an autosomal dominant mutation in *NOTCH3* (Joutel et al., 1996). Recently, a mutant from the Talbot lab screen for *mbp* abnormalities was revealed to a lesion in *notch3*, and the zebrafish adults present with a hemorrhage phenotype similar to the cardiovascular dysfunction in CADASIL patients (Zaucker, Mercurio, Sternheim, Talbot, & Marlow, 2013). In addition to establishing a new zebrafish disease model, this mutant may lend more nuanced insight to how OPC proliferation and specification is regulated by Notch signaling (Park & Appel, 2003).

The true power of zebrafish genetic screens is exemplified by novel, crucial myelination genes discovered therein, which frequently complement the cell biology of myelination known from other systems. For instance, excess OPCs are produced in a mutant for *fbxw7*, which encodes an F-box substrate recognition subunit of E3 ubiquitin ligase. In *fbxw7* mutants, Notch signaling is elevated, and inhibition of Notch signaling is sufficient to suppress the *fbxw7* phenotype (Snyder et al., 2012). Further studies in zebrafish demonstrated that *fbxw7* mutants also have elevated mammalian protein Target Of Rapamycin (mTOR) signaling (Kearns, Ravanelli, Cooper, & Appel, 2015), which results in hypermyelination in mammals (Narayanan, Flores, Wang, & Macklin, 2009). Thus, zebrafish screens have identified a functionally relevant component of the mTOR pathway in myelinating glia. Another example is the discovery of *sec63* as a regulator of both CNS and PNS myelination via its function in the endoplasmic reticulum (ER) translocon machinery. Mutants in *sec63* have fragmented and swollen ER in glia and other tissues, as well as upregulated markers of unfolded protein response and ER stress (Monk et al., 2013). Because myelinating glia must synthesize large amounts of protein and membrane during differentiation and wrapping, deficits in secretory pathways are implicated in myelinopathies, and these effects can be reduced in cultured OLs by treatment with the ER stress blocker guanabenz (reviewed in Clayton & Popko, 2016). Further investigation of *sec63* in zebrafish can examine how cellular stress pathways influence myelination potential.

One of the most critically important myelination genes initially discovered in zebrafish was the adhesion G protein-coupled receptor (aGPCR) *gpr126/adgrg6*, which is essential for SC myelination in the PNS. Though it had been known for decades that elevation of cyclic adenosine monophosphate (cAMP) was essential for SC wrapping and terminal differentiation, the transducer that increased cAMP levels was not known. Starting with a forward genetic screen, the Talbot lab identified two alleles of *gpr126* in which CNS myelination is normal but SCs fail to wrap. This phenotype could be rescued with addition of forskolin, a pharmacological adenylyl cyclase activator (Monk et al., 2009), which suggested that Gpr126 was the sought-after GPCR that elevated cAMP to promote SC wrapping. Subsequent structure-function analysis in zebrafish demonstrated two critically important facets to Gpr126: first, that Gpr126 has bimodal, domain-dependent functions in SC development and second, that Gpr126 contains a cryptic tethered agonist sufficient to activate signaling and promote myelination (Liebscher et al., 2014; Petersen et al., 2015). Furthermore, these studies implicated Laminin-211, a component of the SC basal lamina, as a novel binding partner for Gpr126, which may mechanically activate this a GPCR at a particular phase of SC development. This finding, as well as biochemical experiments demonstrating collagen IV binds zebrafish Gpr126 (Paavola, Sidik, Zuchero, Eckart, & Talbot, 2014), highlights the need to pursue how the ECM cooperates with cell surface receptors in glial cell development in zebrafish.

The discovery of Gpr126 as a critical regulator of SC myelination has spurred reverse genetic analyses to test the role of other adhesion family GPCRs in myelination. Recently, new zebrafish alleles of a related adhesion GPCR, *gpr56,* were created to analyze function in myelinating glia. In complementary papers using both zebrafish and mouse models, Gpr56 was demonstrated to be an evolutionarily conserved regulator of OL development. In *Gpr56* mutants, loss of RhoA signaling results in precocious OPC differentiation at the expense of proliferation such that hypomyelination is observed at later stages (Ackerman, Garcia, Piao, Gutmann, & Monk, 2015; Giera et al., 2015). Together, the critical roles of Gpr126 and Gpr56 in myelination suggest that additional adhesion GPCRs may be required for glial cell development and differentiation, and zebrafish represent as an excellent model to probe the biological functions of adhesion GPCRs in myelinating glia.

Recently, a new zinc finger transcription factor, Znf16l, was identified in a forward genetic screen due to loss of *mbp* expression in the CNS. A combination of imaging studies demonstrated that Znf16l cell autonomously promotes proliferation, migration, and wrapping in the OL lineage. Though Znf16l does not appear conserved in mammals, the *znf16l* phenotype could be rescued by expression of mouse *Zfp488*, which itself appears absent from the zebrafish genome (Sidik & Talbot, 2015). Zfp488 is also a zinc finger transcription factor expressed in OLs and is necessary for CNS myelination and remyelination in mammals (Soundarapandian et al., 2011; Wang et al., 2006). Thus, these experiments highlight the conserved function of Zfp16l and Zfp488 in OL development and place these zinc finger proteins within the transcriptional hierarchy of OL development.

Together, these recent studies and previous work demonstrate the strong advantages of the zebrafish myelination model for gene discovery and for dissecting the conserved function of novel molecular regulators. Given recent advances in reverse genetic techniques in zebrafish, future investigations into known regulators of myelination may prove fruitful for uncovering molecular and cellular mechanisms using this optically and genetically tractable model.

3. PHARMACOLOGICAL MANIPULATION OF MYELINATED AXONS IN ZEBRAFISH

3.1 SMALL MOLECULES AND PEPTIDES ALTER MYELINATION

Pharmacological manipulation of zebrafish development is robust and well established (Bruni, Lakhani, & Kokel, 2014; Rennekamp & Peterson, 2015; Tan & Zon, 2011; Taylor, Grant, Temperley, & Patton, 2010). The ability to treat externally developing, free-swimming zebrafish larvae with small molecules and peptides is a substantial advantage over mammalian models. Although it is necessary to consider that compounds can have off-target effects, drug treatments are rapid compared to generating double or triple mutants and thus can be a powerful approach to identify or validate candidate pathways of interest. For instance, *aldh1a2* was identified in a shelf screen as having CNS *mbp* defects (Kazakova et al., 2006); *aldh1a2* encodes an enzyme involved in synthesis of retinoic acid (RA) (Begemann, Schilling, Rauch, Geisler, & Ingham, 2001), which is required for OPC differentiation and CNS myelination (Huang et al., 2011). Treatment of *aldh1a2* mutant larvae with exogenous RA at 10–16 hpf was sufficient to rescue *mbp* expression in the CNS, while the effect was less potent at a later stage (Kazakova et al., 2006). Together, these data highlight a critical developmental window at which RA signaling is necessary for OL development.

As described previously, the elevation of cAMP via Gpr126 is a critically important step in SC myelination (Jessen & Mirsky, 1991; Mogha et al., 2013; Monk et al., 2009). To demonstrate that cAMP elevation can rescue myelination defects in *gpr126* mutants, larvae were treated with forskolin at 45–52 hpf, when SCs are sorting but typically have not yet myelinated PLLn axons (Monk et al., 2009). Similar to RA signaling, cAMP elevation may also be required at a critical window of SC development as forskolin treatment at earlier or later stages is not sufficient for rescue (Glenn & Talbot, 2013; Monk et al., 2009), and longer treatments (e.g., 24 h) cause lethality, even at lower doses. Importantly, forskolin was used to demonstrate that the genetic differentiation program may be uncoupled from physical wrapping in SCs. In *gpr126* mutants with radial sorting defects, forskolin is unable to rescue sorting or wrapping, but SCs in forskolin-treated *gpr126* mutants nonetheless begin to express *mbp* (Petersen et al., 2015). Interestingly, *gpr126* mutants at 30 days postfertilization (dpf) following a pulsed exposure of forskolin from 50 to 55 hpf had a small number of compact, myelinated axons (Glenn & Talbot, 2013). These data suggest that the lack of myelin seen in the *gpr126* mutants is suppressed by transient elevation of forskolin to initiate myelination, but *gpr126* is not required for myelin maintenance up to 1 month.

Treatment with small peptides is also sufficient to modify myelination via the Gpr126-regulated program. A small tethered agonist termed the *Stachel* (German for "stinger") is encoded within all tested adhesion GPCRs, and in vitro studies suggest that the *Stachel* is necessary and sufficient for activation of signaling (Demberg, Rothemund, Schöneberg, & Liebscher, 2015; Liebscher et al., 2014; Stoveken, Hajduczok, Xu, & Tall, 2015). Given that structure-function analyses demonstrated the *Stachel* is necessary in vivo for myelination in zebrafish, we also wanted to test if the *Stachel* was sufficient for Gpr126 signaling in myelination. Exogenous treatment of hypomorphic *gpr126* larvae with a 16-amino acid *Stachel* peptide fragment suppressed *mbp* expression defects, demonstrating that the *Stachel* is sufficient to activate the Gpr126-regulated myelination program (Liebscher et al., 2014). More recently, Gpr126 was shown to interact with the flexible tail of neuronal prion protein Prp. By treating *gpr126* hypomorphic mutant zebrafish with soluble Prp peptide, we also demonstrated that Prp can partially suppress Mbp defects in these mutants, further exploiting the utility of peptide rescue for myelination and demonstrating the conservation the Prp-Gpr126 interaction in zebrafish (Küffer et al., 2016).

Temporal requirements of Nrg-ErbB signaling in SC development were also revealed via pharmacological manipulation in zebrafish. Mutations in ErbB receptors caused a reduction in myelinated axons, but it was not clear whether this was due to defects in fate specification, migration, proliferation, myelination, or a combination thereof. By pulsing the pan-ErbB inhibitor AG1478 at different stages of larval development, it was revealed that ErbB signaling is continuously required for directed migration of SCPs but that proliferation is not necessary for migration. However, ErbB signaling is also required in postmigratory SCs for proliferation and radial sorting (Lyons et al., 2005). Subsequent experiments with AG1478 demonstrated that both ErbB signaling and proliferation are required for radial sorting but that proliferation itself is not dependent on ErbB (Raphael et al., 2011). In summary, these experiments highlight how the zebrafish myelination model is uniquely suited for temporal analysis of myelin development.

3.2 DRUG SCREENS FOR NOVEL MYELIN REGULATORS

In addition to pathways of interest, zebrafish are an exquisitely powerful system for in vivo drug discovery for therapeutics that can protect or restore the myelin sheath. Myelin is evident in zebrafish by 72 hpf, though cell specification and morphogenetic behaviors to initiate myelination begin earlier. At these time points, zebrafish larvae can be arrayed in 96-well plates using a P1000 pipette with the tip cutoff to create a larger bore; with this technique, several hundred larvae can be precisely arrayed in minutes. We have found that one to three larvae can survive for 24 h in a single well containing 250-μL liquid, allowing multiple larvae to be screened for each compound. Daily water changes are accomplished using a multichannel pipette and permit the larvae to survive in 96-well format up to 5 dpf.

In the first large-scale drug screen for remyelination therapeutics, Buckley et al. screened for enhanced *olig2:GFP* proliferation or migration in larvae treated with 1170 commercially available compounds. Hits from the primary screen, approximately 2% of drugs tested, were subsequently assayed for their effect on *mbp* levels. From this screen, PP2 was identified as a Src kinase inhibitor that decreased *mbp* transcript (Buckley et al., 2010). This effect was likely mediated through Fyn kinase, which has been shown to be an important regulator of OL myelination in mammals and zebrafish (Colognato, Ramachandrappa, Olsen, & Ffrench-Constant, 2004; Czopka, Ffrench-Constant, & Lyons, 2013; Laursen, Chan, & Ffrench-Constant, 2009; Osterhout, Wolven, Wolf, Resh, & Chao, 1999).

New screens for small molecules that promote myelination may instead focus on zebrafish models in which myelination are defective. For instance, hypomorphic *gpr126* mutants have reduced, but not absent, *mbp* expression and myelination (Monk et al., 2009; Petersen et al., 2015), and *gpr56* mutants have a similar hypomorphic phenotype in the CNS (Ackerman et al., 2015). Given that GPCRs are highly druggable and represent significant targets within existing libraries, these proteins represent promising targets for development of future therapies. In this situation, homozygous mutant larvae may be assayed for an increase in *mbp* compared to control via available transgenes. In addition to identifying small molecule targets of aGPCRs, which are known regulators of myelin, this approach may identify compounds that alter the activity of other proteins, lending insight to distinct pathways that drive myelination.

4. PLASTICITY, MAINTENANCE, AND REGENERATION OF MYELINATED AXONS IN ZEBRAFISH

4.1 NEURONAL ACTIVITY AND MYELINATION

Plasticity of the nervous system is evolutionary conserved and necessary for development, learning, and memory across phyla. More recently, it has been shown that not only are synapses highly plastic, but myelin also is altered by changes in neuronal activity during development and in response to the environment (Fields, 2015; Gibson et al., 2014; Makinodan, Rosen, Ito, & Corfas, 2012). Studies in zebrafish have demonstrated how OLs can respond to neuronal activity. Transgenic expression of tetanus toxin (TTX), which prevents synaptic vesicle exocytosis, was sufficient to decrease OPC specification and myelination when expressed in neurons (Hines, Ravanelli, Schwindt, Scott, & Appel, 2015; Mensch et al., 2015). A complementary experiment demonstrated that treatment with pentylenetetrazole, which enhances neuronal signaling, promoted myelination in zebrafish larvae (Mensch et al., 2015). Time-lapse in vivo analyses demonstrated that neuronal activity did not affect OL process extension or "sampling" of axon segments, but rather influenced whether nascent sheaths were stabilized over time. Importantly, stabilized myelin sheaths appeared at sites of synaptic vesicle exocytosis labeled with Syp-EGFP, suggesting that synaptic vesicles are local cues that drive OL myelin maturation (Hines et al., 2015). However, not all neuron subtypes mediate myelination via activity; a recent study demonstrated that TTX expression modifies the myelination pattern of reticulospinal neurons, but not of commissural primary ascending neurons (Koudelka et al., 2016). Together, these results suggest a mechanism for the differences in myelination patterns and activity-dependent myelin changes observed in mammals.

A disadvantage of the zebrafish model system for assaying myelin changes in response to activity compared to mammalian models is the lack of well-established electrophysiological assays for neuronal conduction velocity. Furthermore, there are currently no behavioral tests for zebrafish that model neurological disorders or assess myelin function. Nonetheless, the genetic techniques available in zebrafish, including optogenetics and the recently developed magnetically controlled "Magneto" protein (Wheeler et al., 2016) are uniquely rapid and could complement electrophysiological and behavioral studies in mammals. Future studies in neuronal activity and myelination in the zebrafish might explore these avenues and couple them to in vivo imaging throughout development, rendering the zebrafish model an extremely powerful system to study activity-dependent myelination.

4.2 ANALYSIS OF REMYELINATION IN ZEBRAFISH

Defects in myelin in demyelinating diseases or following injury can lead to devastating symptoms and ultimately paralysis. However, no effective therapeutics exists to prevent demyelination or promote remyelination. Here, we highlight studies investigating the molecular control of remyelination in zebrafish; these studies provide the basis for novel therapies for patients with nerve injury and dysmyelinating diseases. Unlike mammals, zebrafish have the remarkable ability to regenerate their damaged tissues, and this phenomenon has been studied in multiple organs including heart, fin, and the CNS (Gemberling, Bailey, Hyde, & Poss, 2013). As mentioned in Section 3, zebrafish also serve as a suitable chassis for drug screens. Indeed, with respect to CNS remyelination, Buckley et al. (2010) have carried out a pharmacological screen for promyelinating compounds with the goal of opening up an avenue for potential therapeutics to stimulate remyelination in multiple sclerosis (MS).

Spontaneous remyelination occurs in patients with early stages of MS and in animal models of demyelination; however, this regenerative potential declines with disease chronicity and age. Using a focal demyelination paradigm whereby foam soaked in detergent lysophosphatidylcholine was placed next to the zebrafish optic nerve, Munzel et al. observed demyelination followed by remyelination within 4 weeks. Importantly, there was no axonal damage, suggesting an autonomous function in the OLs (Münzel, Becker, Becker, & Williams, 2014). Interestingly, myelin is fully regenerated in young adults (4–7 months) but inefficacious in aged fish (15–18 months), suggesting that the cellular ability to regenerate is limited with increasing age similar to mammals. Moreover, Munzel et al. observed reduced phagocytotic microglia/macrophage recruitment to the site of lesion in old zebrafish suggesting myelin debris clearance could be impaired with age. It remains unknown whether inefficient remyelination in aged fish is due to impaired myelin debris clearance or instead due to reduced intrinsic potential to myelinate.

A major strength of zebrafish in regeneration studies is the ability to perform live-cell imaging to observe cellular behavior following injury. To test the regenerative potential of OLs and track remyelination in vivo, OLs can be genetically ablated using the nitroreductase (NTR)-mediated cell ablation techniques (Chung et al., 2013; Fang et al., 2014; Kim et al., 2015). Briefly, transgenic fish express NTR, a bacterial enzyme, in myelinating glia. When the fish are exposed to a prodrug metronidazole (MTZ), NTR binds and converts MTZ to a cytotoxin leading to targeted cell ablation (Curado, Stainier, & Anderson, 2008). Moreover, the drug can be washed away to allow for regeneration. Using the NTR/MTZ system in OLs, Kim et al. (2015) observed that treating the fish with sulfasalazine, an antiinflammatory compound, promoted remyelination. In addition to uncovering therapeutic targets for coaxing regeneration, the NTR/MTZ system can also be used to elucidate cell lineage relationships during regeneration. During regeneration in the postamputated fin, cell lineage tracing experiments suggest that regenerated tissues arise from preexisting cells and retain their developmental identity (reviewed in Pfefferli & Jaźwińska, 2015). Recent evidence from juvenile and adult mice suggests that the OLs in the CNS are heterogeneous with differential propensity to myelinate (Marques et al., 2016). If the zebrafish CNS also contains distinct subpopulations of OLs, this would open up exciting avenues for cell lineage studies that will examine if the subpopulations of OLs recapitulate their cell fate following ablation and if there are distinct OLs that are more adept at remyelination in the CNS.

In the PNS, following nerve injury, SCs become dedicated "repair SCs," which release neuroprotective factors, elevate cytokines, remove myelin, and axonal debris, and cue macrophage recruitment (reviewed in Jessen & Mirsky, 2016). This repair and regenerative program is orchestrated by bursts of injury-specific genes such as c-Jun, a transcription factor normally downregulated during myelinating SC development (Arthur-Farraj et al., 2012). The role of SCs in peripheral nerve repair has been studied in the PLLn (Graciarena, Dambly-Chaudière, & Ghysen, 2014; Xiao et al., 2015), motor nerves (Rosenberg, Isaacman-Beck, Franzini-Armstrong, & Granato, 2014), and the adult zebrafish maxillary barbels (ZMB) (LeClair & Topczewski, 2010; Moore, Mark, Hogan, Topczewski, & LeClair, 2012). In the larval PLLn, following laser-mediated axonal transection, adjacent SCs extend bridging processes toward the site of injury and begin clearance of axonal debris (Xiao et al., 2015). Similarly, in the transected larval motor nerves, SCs undergo morphological changes and contain fluorescently tagged axonal debris within their membrane, suggesting that SCs are actively involved in debris clearance. However, in mutants lacking SCs, motor growth cone regeneration was unaffected suggesting SCs are dispensible for axonal regrowth (Rosenberg et al., 2014). Moreover, in both the PLLn and motor nerves, in the absence of SCs, regenerated axons are misrouted (Rosenberg et al., 2014; Xiao et al., 2015), indicating that SCs are important for providing directional guidance cues. One of the guidance cues important for directional cue has been suggested to be the collagen modifying enzyme glycosyltransferase lysyl hydroxylase 3 (lh3) (Isaacman-Beck, Schneider, Franzini-Armstrong, & Granato, 2015). While *lh3* mutants have impaired axonal guidance following transfection, SC-specific expression of *lh3* was sufficient to restore regeneration (Isaacman-Beck et al., 2015).

Peripheral nerve remyelination in zebrafish can be studied across development. In order to examine remyelination in adult axons, Moore et al. performed a time-course analysis in ZMBs following amputation (LeClair & Topczewski, 2010; Moore et al., 2012). ZMBs are optically transparent appendages that are innervated by the facial nerve, chemosensory cells, and epithelial nerves (LeClair & Topczewski, 2010). At 10 days post amputation (dpa), few axons myelinate, and a patchy collagen-rich matrix surrounds the axons. By 28 dpa, the matrix thickens, the ultrastructure of the barbel is restored, and axons are remyelinated, albeit myelination is reduced compared to uninjured controls. Regenerating barbels undergo extensive remyelination and several promyelinating transcription factors are upregulated in the distal regenerating barbel tissues. While Moore et al. focused on a candidate gene expression approach to probe the mechanism of remyelination in the ZMB, future WGS approaches would provide the ability to profile relevant, injury-induced targets and compare these large expression data sets to mammalian remyelination paradigms. Taken together, the zebrafish peripheral nerves provide an unprecedented view of the complex cell—cell and cell—matrix interactions involved in SC repair following injury in vivo.

CONCLUSIONS

Zebrafish is a powerful vertebrate system to study myelination using both forward and reverse genetic approaches, in vivo analysis during development and repair, and for performing drug screens to uncover novel drivers of myelination and remyelination. Genetic screens in zebrafish have uncovered several regulators of myelination that are conserved in mammals. Moreover, genome-wide association studies have identified candidate genes associated with schizophrenia (reviewed in Roussos & Haroutunian, 2014), multiple sclerosis (Andlauer et al., 2016), and peripheral neuropathies (Safka Brožková, Nevšímalová, Mazanec, Rautenstrauss, & Seeman, 2012). The advent of genome-editing techniques will allow researchers tools to analyze how single nucleotide polymorphisms contribute to disease progression. The identification of these variants allow for a better molecular handle on the etiology of the several diseases where myelin is impaired. In addition to finding novel genes, zebrafish are amenable to chemical screens demonstrating the feasibility of zebrafish to discover compounds for future therapeutics. Future studies will leverage the many strengths of zebrafish to build a functional network of genetic pathways involved in myelination, a cost-effective and rapid approach to drug discovery, and continue to solidify the importance of zebrafish research in basic science and clinical medicine.

ACKNOWLEDGMENTS

We thank members of the Monk laboratory for helpful discussions and comments, David Lyons for TEM protocol suggestions, and Lisette Mateo and Christina Rozario for assistance with figure generation. We apologize to our colleagues whose primary work we were unable to cite due to space limitations. This work was supported by grants from the National Institutes of Health (NS079445, HD08601 to KRM; NS087786 to SCP), the Muscular Dystrophy Association (MDA293295 to KRM), and KRM is a Harry Weaver Scholar of the National Multiple Sclerosis Society.

REFERENCES

Ablain, J., Durand, E. M., Yang, S., Zhou, Y., & Zon, L. I. (2015). A CRISPR/Cas9 vector system for tissue-specific gene disruption in zebrafish. *Developmental Cell, 32*(6), 756–764. http://doi.org/10.1016/j.devcel.2015.01.032.

Ackerman, S. D., Garcia, C., Piao, X., Gutmann, D. H., & Monk, K. R. (2015). The adhesion GPCR Gpr56 regulates oligodendrocyte development via interactions with Gα12/13 and RhoA. *Nature Communications, 6*, 6122. http://doi.org/10.1038/ncomms7122.

Ackerman, S. D., & Monk, K. R. (2015). The scales and tales of myelination: using zebrafish and mouse to study myelinating glia. *Brain Research*. http://doi.org/10.1016/j.brainres.2015.10.011.

Almeida, R. G., Czopka, T., Ffrench-Constant, C., & Lyons, D. A. (2011). Individual axons regulate the myelinating potential of single oligodendrocytes in vivo. *Development, 138*(20), 4443–4450. http://doi.org/10.1242/dev.071001.

Almeida, R. G., & Lyons, D. A. (2015). Intersectional gene expression in zebrafish using the split KalTA4 system. *Zebrafish, 12*(6), 377–386. http://doi.org/10.1089/zeb.2015.1086.

Andlauer, T. F. M., Buck, D., Antony, G., Bayas, A., Bechmann, L., Berthele, A., et al. (2016). Novel multiple sclerosis susceptibility loci implicated in epigenetic regulation. *Science Advances, 2*(6), e1501678. http://doi.org/10.1126/sciadv.1501678.

Arthur-Farraj, P. J., Latouche, M., Wilton, D. K., Quintes, S., Chabrol, E., Banerjee, A., et al. (2012). c-Jun reprograms Schwann cells of injured nerves to generate a repair cell essential for regeneration. *Neuron, 75*(4), 633–647. http://dx.doi.org/10.1016/j.neuron.2012.06.021.

Auer, T. O., & Del Bene, F. (2014). CRISPR/Cas9 and TALEN-mediated knock-in approaches in zebrafish. *Methods (San Diego, California), 69*(2), 142–150. http://doi.org/10.1016/j.ymeth.2014.03.027.

Avila, R. L., Tevlin, B. R., Lees, J. P. B., Inouye, H., & Kirschner, D. A. (2007). Myelin structure and composition in zebrafish. *Neurochemical Research, 32*(2), 197–209. http://doi.org/10.1007/s11064-006-9136-5.

Bai, Q., Parris, R. S., & Burton, E. A. (2014). Different mechanisms regulate expression of zebrafish myelin protein zero (P0) in myelinating oligodendrocytes and its induction following axonal injury. *Journal of Biological Chemistry, 289*(35), 24114–24128. http://doi.org/10.1074/jbc.M113.545426.

Bai, Q., Sun, M., Stolz, D. B., & Burton, E. A. (2011). Major isoform of zebrafish P0 is a 23.5 kDa myelin glycoprotein expressed in selected white matter tracts of the central nervous system. *The Journal of Comparative Neurology, 519*(8), 1580–1596. http://doi.org/10.1002/cne.22587.

Begemann, G., Schilling, T. F., Rauch, G. J., Geisler, R., & Ingham, P. W. (2001). The zebrafish neckless mutation reveals a requirement for raldh2 in mesodermal signals that pattern the hindbrain. *Development, 128*(16), 3081–3094.

Boch, J., Scholze, H., Schornack, S., Landgraf, A., Hahn, S., Kay, S., et al. (2009). Breaking the code of DNA binding specificity of TAL-type III effectors. *Science, 326*(5959), 1509–1512. http://doi.org/10.1126/science.1178811.

Brösamle, C., & Halpern, M. E. (2002). Characterization of myelination in the developing zebrafish. *Glia, 39*(1), 47–57. http://doi.org/10.1002/glia.10088.

Bruni, G., Lakhani, P., & Kokel, D. (2014). Discovering novel neuroactive drugs through high-throughput behavior-based chemical screening in the zebrafish. *Frontiers in Pharmacology, 5*, 153. http://doi.org/10.3389/fphar.2014.00153.

Buckley, C. E., Marguerie, A., Roach, A. G., Goldsmith, P., Fleming, A., Alderton, W. K., & Franklin, R. J. M. (2010). Drug reprofiling using zebrafish identifies novel compounds with potential pro-myelination effects. *Neuropharmacology, 59*(3), 149–159. http://doi.org/10.1016/j.neuropharm.2010.04.014.

Chung, A.-Y., Kim, P.-S., Kim, S., Kim, E., Kim, D., Jeong, I., et al. (2013). Generation of demyelination models by targeted ablation of oligodendrocytes in the zebrafish CNS. *Molecules and Cells, 36*(1), 82–87. http://doi.org/10.1007/s10059-013-0087-9.

Clayton, B. L. L., & Popko, B. (2016). Endoplasmic reticulum stress and the unfolded protein response in disorders of myelinating glia. *Brain Research.* http://doi.org/10.1016/j.brainres.2016.03.046.

Colognato, H., Ramachandrappa, S., Olsen, I. M., & Ffrench-Constant, C. (2004). Integrins direct Src family kinases to regulate distinct phases of oligodendrocyte development. *The Journal of Cell Biology, 167*(2), 365–375. http://doi.org/10.1083/jcb.200404076.

Court, F. A., Wrabetz, L., & Feltri, M. L. (2006). Basal lamina: Schwann cells wrap to the rhythm of space-time. *Current Opinion in Neurobiology, 16*(5), 501–507. http://doi.org/10.1016/j.conb.2006.08.005.

Curado, S., Stainier, D. Y. R., & Anderson, R. M. (2008). Nitroreductase-mediated cell/tissue ablation in zebrafish: a spatially and temporally controlled ablation method with applications in developmental and regeneration studies. *Nature Protocols, 3*(6), 948–954. http://doi.org/10.1038/nprot.2008.58.

Czopka, T., Ffrench-Constant, C., & Lyons, D. A. (2013). Individual oligodendrocytes have only a few hours in which to generate new myelin sheaths in vivo. *Developmental Cell, 25*(6), 599–609. http://doi.org/10.1016/j.devcel.2013.05.013.

Czopka, T., & Lyons, D. A. (2011). Dissecting mechanisms of myelinated axon formation using zebrafish. *Methods in Cell Biology, 105*, 25–62. http://doi.org/10.1016/B978-0-12-381320-6.00002-3.

Das, A., & Crump, J. G. (2012). Bmps and id2a act upstream of Twist1 to restrict ectomesenchyme potential of the cranial neural crest. *PLoS Genetics, 8*(5), e1002710. http://doi.org/10.1371/journal.pgen.1002710.

Demberg, L. M., Rothemund, S., Schöneberg, T., & Liebscher, I. (2015). Identification of the tethered peptide agonist of the adhesion G protein-coupled receptor GPR64/ADGRG2. *Biochemical and Biophysical Research Communications, 464*(3), 743–747. http://doi.org/10.1016/j.bbrc.2015.07.020.

Driever, W., Solnica-Krezel, L., Schier, A. F., Neuhauss, S. C., Malicki, J., Stemple, D. L., et al. (1996). A genetic screen for mutations affecting embryogenesis in zebrafish. *Development, 123*, 37–46.

Dutton, K. A., Pauliny, A., Lopes, S. S., Elworthy, S., Carney, T. J., Rauch, J., et al. (2001). Zebrafish colourless encodes sox10 and specifies non-ectomesenchymal neural crest fates. *Development, 128*(21), 4113–4125.

Emery, B. (2010). Regulation of oligodendrocyte differentiation and myelination. *Science, 330*(6005), 779–782. http://doi.org/10.1126/science.1190927.

Fang, Y., Lei, X., Li, X., Chen, Y., Xu, F., Feng, X., et al. (2014). A novel model of demyelination and remyelination in a GFP-transgenic zebrafish. *Biology Open, 4*(1), 62–68. http://doi.org/10.1242/bio.201410736.

Fields, R. D. (2015). A new mechanism of nervous system plasticity: activity-dependent myelination. *Nature Reviews Neuroscience, 16*(12), 756–767. http://doi.org/10.1038/nrn4023.

Gagnon, J. A., Valen, E., Thyme, S. B., Huang, P., Akhmetova, L., Ahkmetova, L., et al. (2014). Efficient mutagenesis by Cas9 protein-mediated oligonucleotide insertion and large-scale assessment of single-guide RNAs. *PLoS One, 9*(5), e98186. http://doi.org/10.1371/journal.pone.0098186.

Gaj, T., Gersbach, C. A., & Barbas, C. F. (2013). ZFN, TALEN, and CRISPR/Cas-based methods for genome engineering. *Trends in Biotechnology, 31*(7), 397–405. http://doi.org/10.1016/j.tibtech.2013.04.004.

Gemberling, M., Bailey, T. J., Hyde, D. R., & Poss, K. D. (2013). The zebrafish as a model for complex tissue regeneration. *Trends in Genetics, 29*(11), 611–620. http://doi.org/10.1016/j.tig.2013.07.003.

Gibson, E. M., Purger, D., Mount, C. W., Goldstein, A. K., Lin, G. L., Wood, L. S., et al. (2014). Neuronal activity promotes oligodendrogenesis and adaptive myelination in the mammalian brain. *Science, 344*(6183), 1252304. http://doi.org/10.1126/science.1252304.

Giera, S., Deng, Y., Luo, R., Ackerman, S. D., Mogha, A., Monk, K. R., et al. (2015). The adhesion G protein-coupled receptor GPR56 is a cell-autonomous regulator of oligodendrocyte development. *Nature Communications, 6*, 6121. http://doi.org/10.1038/ncomms7121.

Gilmour, D. T., Maischein, H.-M., & Nüsslein-Volhard, C. (2002). Migration and function of a glial subtype in the vertebrate peripheral nervous system. *Neuron, 34*(4), 577–588.

Glenn, T. D., & Talbot, W. S. (2013). Analysis of Gpr126 function defines distinct mechanisms controlling the initiation and maturation of myelin. *Development, 140*(15), 3167–3175. http://doi.org/10.1242/dev.093401.

Graciarena, M., Dambly-Chaudière, C., & Ghysen, A. (2014). Dynamics of axonal regeneration in adult and aging zebrafish reveal the promoting effect of a first lesion. *Proceedings of the National Academy of Sciences, 111*(4), 1610–1615. http://doi.org/10.1073/pnas.1319405111.

Haffter, P., & Nüsslein-Volhard, C. (1996). Large scale genetics in a small vertebrate, the zebrafish. *The International Journal of Developmental Biology, 40*(1), 221–227.

Halpern, M. E., Rhee, J., Goll, M. G., Akitake, C. M., Parsons, M., & Leach, S. D. (2008). Gal4/UAS transgenic tools and their application to zebrafish. *Zebrafish, 5*(2), 97–110. http://doi.org/10.1089/zeb.2008.0530.

Hans, S., Kaslin, J., Freudenreich, D., & Brand, M. (2009). Temporally-controlled site-specific recombination in zebrafish. *PLoS One, 4*(2), e4640. http://doi.org/10.1371/journal.pone.0004640.

Henke, K., Bowen, M. E., & Harris, M. P. (2013). Identification of mutations in zebrafish using next-generation sequencing. In F. M. Ausubel, et al. (Eds.), *Current protocols in molecular biology* (p. 104) [Unit 7.13] http://doi.org/10.1002/0471142727.mb0713s104.

Hill, J. T., Demarest, B. L., Bisgrove, B. W., Gorsi, B., Su, Y.-C., & Yost, H. J. (2013). MMAPPR: mutation mapping analysis pipeline for pooled RNA-seq. *Genome Research, 23*(4), 687–697. http://doi.org/10.1101/gr.146936.112.

Hines, J. H., Ravanelli, A. M., Schwindt, R., Scott, E. K., & Appel, B. (2015). Neuronal activity biases axon selection for myelination in vivo. *Nature Neuroscience, 18*(5), 683–689. http://doi.org/10.1038/nn.3992.

Huang, J. K., Jarjour, A. A., Nait-Oumesmar, B., Kerninon, C., Williams, A., Krezel, W., et al. (2011). Retinoid X receptor gamma signaling accelerates CNS remyelination. *Nature Neuroscience, 14*(1), 45–53. http://doi.org/10.1038/nn.2702.

Huang, C.-J., Jou, T.-S., Ho, Y.-L., Lee, W.-H., Jeng, Y.-T., Hsieh, F.-J., & Tsai, H.-J. (2005). Conditional expression of a myocardium-specific transgene in zebrafish transgenic lines. *Developmental Dynamics, 233*(4), 1294–1303. http://doi.org/10.1002/dvdy.20485.

Isaacman-Beck, J., Schneider, V., Franzini-Armstrong, C., & Granato, M. (2015). The lh3 glycosyltransferase directs target-selective peripheral nerve regeneration. *Neuron, 88*(4), 691–703. http://doi.org/10.1016/j.neuron.2015.10.004.

Jao, L.-E., Wente, S. R., & Chen, W. (2013). Efficient multiplex biallelic zebrafish genome editing using a CRISPR nuclease system. *Proceedings of the National Academy of Sciences, 110*(34), 13904–13909. http://doi.org/10.1073/pnas.1308335110.

Jessen, K. R., & Mirsky, R. (1991). Schwann cell precursors and their development. *Glia, 4*(2), 185–194. http://doi.org/10.1002/glia.440040210.

Jessen, K. R., & Mirsky, R. (2005). The origin and development of glial cells in peripheral nerves. *Nature Reviews Neuroscience, 6*(9), 671–682. http://doi.org/10.1038/nrn1746.

Jessen, K. R., & Mirsky, R. (2016). The repair Schwann cell and its function in regenerating nerves. *The Journal of Physiology*. http://doi.org/10.1113/JP270874.

Joutel, A., Corpechot, C., Ducros, A., Vahedi, K., Chabriat, H., Mouton, P., et al. (1996). Notch3 mutations in CADASIL, a hereditary adult-onset condition causing stroke and dementia. *Nature, 383*(6602), 707–710. http://doi.org/10.1038/383707a0.

Jung, S.-H., Kim, S., Chung, A.-Y., Kim, H.-T., So, J.-H., Ryu, J., et al. (2010). Visualization of myelination in GFP-transgenic zebrafish. *Developmental Dynamics, 239*(2), 592–597. http://doi.org/10.1002/dvdy.22166.

Kazakova, N., Li, H., Mora, A., Jessen, K. R., Mirsky, R., Richardson, W. D., & Smith, H. K. (2006). A screen for mutations in zebrafish that affect myelin gene expression in Schwann cells and oligodendrocytes. *Developmental Biology, 297*(1), 1–13. http://doi.org/10.1016/j.ydbio.2006.03.020.

Kearns, C. A., Ravanelli, A. M., Cooper, K., & Appel, B. (2015). Fbxw7 limits myelination by inhibiting mTOR signaling. *Journal of Neuroscience, 35*(44), 14861–14871. http://doi.org/10.1523/JNEUROSCI.4968-14.2015.

Kim, S., Lee, Y.-I., Chang, K.-Y., Lee, D.-W., Cho, S. C., Ha, Y. W., et al. (2015). Promotion of remyelination by sulfasalazine in a transgenic zebrafish model of demyelination. *Molecules and Cells, 38*(11), 1013–1021. http://doi.org/10.14348/molcells.2015.0246.

Kirby, B. B., Takada, N., Latimer, A. J., Shin, J., Carney, T. J., Kelsh, R. N., & Appel, B. (2006). In vivo time-lapse imaging shows dynamic oligodendrocyte progenitor behavior during zebrafish development. *Nature Neuroscience, 9*(12), 1506–1511. http://doi.org/10.1038/nn1803.

Korn, H., & Faber, D. S. (2005). The Mauthner cell half a century later: a neurobiological model for decision-making? *Neuron, 47*(1), 13–28. http://doi.org/10.1016/j.neuron.2005.05.019.

Koudelka, S., Voas, M. G., Almeida, R. G., Baraban, M., Soetaert, J., Meyer, M. P., et al. (2016). Individual neuronal subtypes exhibit diversity in CNS myelination mediated by synaptic vesicle release. *Current Biology, 26*(11), 1447–1455. http://doi.org/10.1016/j.cub.2016.03.070.

Kucenas, S., Takada, N., Park, H.-C., Woodruff, E., Broadie, K., & Appel, B. (2008). CNS-derived glia ensheath peripheral nerves and mediate motor root development. *Nature Neuroscience, 11*(2), 143–151. http://doi.org/10.1038/nn2025.

Küffer, A., Lakkaraju, A. K., Mogha, A., Petersen, S. C., Airich, K., Doucerain, C., … Aguzzi, A. (2016). The prion protein is an agonistic ligand of the G protein-coupled receptor Adgrg6. *Nature, 536*(7617), 464–468. http://www.nature.com/nature/journal/v536/n7617/full/nature19312.html.

Kwan, K. M., Fujimoto, E., Grabher, C., Mangum, B. D., Hardy, M. E., Campbell, D. S., et al. (2007). The Tol2kit: a multisite gateway-based construction kit for Tol2 transposon transgenesis constructs. *Developmental Dynamics, 236*(11), 3088–3099. http://doi.org/10.1002/dvdy.21343.

Langworthy, M. M., & Appel, B. (2012). Schwann cell myelination requires Dynein function. *Neural Development, 7*, 37. http://doi.org/10.1186/1749-8104-7-37.

Laursen, L. S., Chan, C. W., & Ffrench-Constant, C. (2009). An integrin-contactin complex regulates CNS myelination by differential Fyn phosphorylation. *Journal of Neuroscience, 29*(29), 9174–9185. http://doi.org/10.1523/JNEUROSCI.5942-08.2009.

LeClair, E. E., & Topczewski, J. (2010). Development and regeneration of the zebrafish maxillary barbel: a novel study system for vertebrate tissue growth and repair. *PLoS One, 5*(1). http://dx.doi.org/10.1371/journal.pone.0008737 C.

Levavasseur, F., Mandemakers, W., Visser, P., Broos, L., Grosveld, F., Zivkovic, D., & Meijer, D. (1998). Comparison of sequence and function of the Oct-6 genes in zebrafish, chicken and mouse. *Mechanisms of Development, 74*(1–2), 89–98.

Liebscher, I., Schön, J., Petersen, S. C., Fischer, L., Auerbach, N., Demberg, L. M., et al. (2014). A tethered agonist within the ectodomain activates the adhesion G protein-coupled receptors GPR126 and GPR133. *Cell Reports, 9*(6), 2018–2026. http://doi.org/10.1016/j.celrep.2014.11.036.

Lyons, D. A., Naylor, S. G., Mercurio, S., Dominguez, C., & Talbot, W. S. (2008). KBP is essential for axonal structure, outgrowth and maintenance in zebrafish, providing insight into the cellular basis of Goldberg-Shprintzen syndrome. *Development, 135*(3), 599–608. http://doi.org/10.1242/dev.012377.

Lyons, D. A., Naylor, S. G., Scholze, A., & Talbot, W. S. (2009). Kif1b is essential for mRNA localization in oligodendrocytes and development of myelinated axons. *Nature Genetics, 41*(7), 854–858. http://doi.org/10.1038/ng.376.

Lyons, D. A., Pogoda, H.-M., Voas, M. G., Woods, I. G., Diamond, B., Nix, R., et al. (2005). erbb3 and erbb2 are essential for schwann cell migration and myelination in zebrafish. *Current Biology, 15*(6), 513–524. http://doi.org/10.1016/j.cub.2005.02.030.

Makinodan, M., Rosen, K. M., Ito, S., & Corfas, G. (2012). A critical period for social experience-dependent oligodendrocyte maturation and myelination. *Science, 337*(6100), 1357–1360. http://doi.org/10.1126/science.1220845.

Marques, S., Zeisel, A., Codeluppi, S., van Bruggen, D., Mendanha Falcão, A., Xiao, L., et al. (2016). Oligodendrocyte heterogeneity in the mouse juvenile and adult central nervous system. *Science, 352*(6291), 1326–1329. http://doi.org/10.1126/science.aaf6463.

Mathews, E. S., Mawdsley, D. J., Walker, M., Hines, J. H., Pozzoli, M., & Appel, B. (2014). Mutation of 3-Hydroxy-3-Methylglutaryl CoA synthase I reveals requirements for isoprenoid and cholesterol synthesis in oligodendrocyte migration arrest, axon wrapping, and myelin gene expression. *Journal of Neuroscience, 34*(9), 3402–3412. http://doi.org/10.1523/JNEUROSCI.4587-13.2014.

Mensch, S., Baraban, M., Almeida, R., Czopka, T., Ausborn, J., El Manira, A., & Lyons, D. A. (2015). Synaptic vesicle release regulates myelin sheath number of individual oligodendrocytes in vivo. *Nature Neuroscience, 18*(5), 628–630. http://doi.org/10.1038/nn.3991.

Miller, A. C., Obholzer, N. D., Shah, A. N., Megason, S. G., & Moens, C. B. (2013). RNA-seq-based mapping and candidate identification of mutations from forward genetic screens. *Genome Research, 23*(4), 679–686. http://doi.org/10.1101/gr.147322.112.

Mitew, S., Hay, C. M., Peckham, H., Xiao, J., Koenning, M., & Emery, B. (2014). Mechanisms regulating the development of oligodendrocytes and central nervous system myelin. *Neuroscience, 276*, 29–47. http://doi.org/10.1016/j.neuroscience.2013.11.029.

Möbius, W., Nave, K.-A., & Werner, H. B. (2016). Electron microscopy of myelin: structure preservation by high-pressure freezing. *Brain Research, 1641*(Pt A), 92–100. http://doi.org/10.1016/j.brainres.2016.02.027.

Mogha, A., Benesh, A. E., Patra, C., Engel, F. B., Schöneberg, T., Liebscher, I., & Monk, K. R. (2013). Gpr126 functions in Schwann cells to control differentiation and myelination via G-protein activation. *Journal of Neuroscience, 33*(46), 17976–17985. http://doi.org/10.1523/JNEUROSCI.1809-13.2013.

Monk, K. R., Naylor, S. G., Glenn, T. D., Mercurio, S., Perlin, J. R., Dominguez, C., et al. (2009). A G protein-coupled receptor is essential for Schwann cells to initiate myelination. *Science, 325*(5946), 1402–1405. http://doi.org/10.1126/science.1173474.

Monk, K. R., & Talbot, W. S. (2009). Genetic dissection of myelinated axons in zebrafish. *Current Opinion in Neurobiology, 19*(5), 486–490. http://doi.org/10.1016/j.conb.2009.08.006.

Monk, K. R., Voas, M. G., Franzini-Armstrong, C., Hakkinen, I. S., & Talbot, W. S. (2013). Mutation of sec63 in zebrafish causes defects in myelinated axons and liver pathology. *Disease Models and Mechanisms, 6*(1), 135–145. http://doi.org/10.1242/dmm.009217.

Montague, T. G., Cruz, J. M., Gagnon, J. A., Church, G. M., & Valen, E. (2014). CHOPCHOP: a CRISPR/Cas9 and TALEN web tool for genome editing. *Nucleic Acids Research, 42*(Web Server Issue), W401–W407. http://doi.org/10.1093/nar/gku410.

Moore, A. C., Mark, T. E., Hogan, A. K., Topczewski, J., & LeClair, E. E. (2012). Peripheral axons of the adult zebrafish maxillary barbel extensively remyelinate during sensory appendage regeneration. *The Journal of Comparative Neurology, 520*(18), 4184–4203. http://doi.org/10.1002/cne.23147.

Morris, J. K., Willard, B. B., Yin, X., Jeserich, G., Kinter, M., & Trapp, B. D. (2004). The 36K protein of zebrafish CNS myelin is a short-chain dehydrogenase. *Glia, 45*(4), 378–391. http://doi.org/10.1002/glia.10338.

Münzel, E. J., Becker, C. G., Becker, T., & Williams, A. (2014). Zebrafish regenerate full thickness optic nerve myelin after demyelination, but this fails with increasing age. *Acta Neuropathologica Communications, 2*, 77. http://doi.org/10.1186/s40478-014-0077-y.

Münzel, E. J., Schaefer, K., Obirei, B., Kremmer, E., Burton, E. A., Kuscha, V., et al. (2012). Claudin k is specifically expressed in cells that form myelin during development of the nervous system and regeneration of the optic nerve in adult zebrafish. *Glia, 60*(2), 253–270. http://doi.org/10.1002/glia.21260.

Narayanan, S. P., Flores, A. I., Wang, F., & Macklin, W. B. (2009). Akt signals through the mammalian target of rapamycin pathway to regulate CNS myelination. *Journal of Neuroscience, 29*(21), 6860–6870. http://doi.org/10.1523/JNEUROSCI.0232-09.2009.

Nawaz, S., Sánchez, P., Schmitt, S., Snaidero, N., Mitkovski, M., Velte, C., et al. (2015). Actin filament turnover drives leading edge growth during myelin sheath formation in the central nervous system. *Developmental Cell, 34*(2), 139–151. http://doi.org/10.1016/j.devcel.2015.05.013.

Nawaz, S., Schweitzer, J., Jahn, O., & Werner, H. B. (2013). Molecular evolution of myelin basic protein, an abundant structural myelin component. *Glia, 61*(8), 1364–1377.

Newbern, J., & Birchmeier, C. (2010). Nrg1/ErbB signaling networks in Schwann cell development and myelination. *Seminars in Cell and Developmental Biology, 21*(9), 922–928. http://doi.org/10.1016/j.semcdb.2010.08.008.

Norton, W. H., Mangoli, M., Lele, Z., Pogoda, H.-M., Diamond, B., Mercurio, S., et al. (2005). Monorail/Foxa2 regulates floorplate differentiation and specification of oligodendrocytes, serotonergic raphé neurones and cranial motoneurones. *Development, 132*(4), 645–658. http://doi.org/10.1242/dev.01611.

Osterhout, D. J., Wolven, A., Wolf, R. M., Resh, M. D., & Chao, M. V. (1999). Morphological differentiation of oligodendrocytes requires activation of Fyn tyrosine kinase. *The Journal of Cell Biology, 145*(6), 1209–1218.

Paavola, K. J., Sidik, H., Zuchero, J. B., Eckart, M., & Talbot, W. S. (2014). Type IV collagen is an activating ligand for the adhesion G protein-coupled receptor GPR126. *Science Signaling, 7*(338), ra76. http://doi.org/10.1126/scisignal.2005347.

Park, H.-C., & Appel, B. (2003). Delta-Notch signaling regulates oligodendrocyte specification. *Development, 130*(16), 3747–3755.

Park, H. C., Mehta, A., Richardson, J. S., & Appel, B. (2002). olig2 is required for zebrafish primary motor neuron and oligodendrocyte development. *Developmental Biology, 248*(2), 356–368. http://www.ncbi.nlm.nih.gov/pubmed/12167410.

Perlin, J. R., Lush, M. E., Stephens, W. Z., Piotrowski, T., & Talbot, W. S. (2011). Neuronal Neuregulin 1 type III directs Schwann cell migration. *Development, 138*(21), 4639–4648. http://doi.org/10.1242/dev.068072.

Petersen, S. C., Luo, R., Liebscher, I., Giera, S., Jeong, S.-J., Mogha, A., et al. (2015). The adhesion GPCR GPR126 has distinct, domain-dependent functions in Schwann cell development mediated by interaction with Laminin-211. *Neuron, 85*(4), 755–769. http://doi.org/10.1016/j.neuron.2014.12.057.

Pfefferli, C., & Jaźwińska, A. (2015). The art of fin regeneration in zebrafish. *Regeneration, 2*(2), 72–83.

Pogoda, H.-M., Sternheim, N., Lyons, D. A., Diamond, B., Hawkins, T. A., Woods, I. G., et al. (2006). A genetic screen identifies genes essential for development of myelinated axons in zebrafish. *Developmental Biology, 298*(1), 118–131. http://doi.org/10.1016/j.ydbio.2006.06.021.

Postlethwait, J. H., & Talbot, W. S. (1997). Zebrafish genomics: from mutants to genes. *Trends in Genetics, 13*(5), 183–190.

Prendergast, A., Linbo, T. H., Swarts, T., Ungos, J. M., McGraw, H. F., Krispin, S., et al. (2012). The metalloproteinase inhibitor Reck is essential for zebrafish DRG development. *Development, 139*(6), 1141–1152. http://doi.org/10.1242/dev.072439.

Ran, F. A., Hsu, P. D., Wright, J., Agarwala, V., Scott, D. A., & Zhang, F. (2013). Genome engineering using the CRISPR-Cas9 system. *Nature Protocols, 8*(11), 2281–2308. http://doi.org/10.1038/nprot.2013.143.

Raphael, A. R., Lyons, D. A., & Talbot, W. S. (2011). ErbB signaling has a role in radial sorting independent of Schwann cell number. *Glia, 59*(7), 1047–1055. http://doi.org/10.1002/glia.21175.

Rennekamp, A. J., & Peterson, R. T. (2015). 15 years of zebrafish chemical screening. *Current Opinion in Chemical Biology, 24*, 58–70. http://doi.org/10.1016/j.cbpa.2014.10.025.

Rosenberg, A. F., Isaacman-Beck, J., Franzini-Armstrong, C., & Granato, M. (2014). Schwann cells and deleted in colorectal carcinoma direct regenerating motor axons towards their original path. *Journal of Neuroscience, 34*(44), 14668–14681. http://doi.org/10.1523/JNEUROSCI.2007-14.2014.

Roussos, P., & Haroutunian, V. (2014). Schizophrenia: susceptibility genes and oligodendroglial and myelin related abnormalities. *Frontiers in Cellular Neuroscience, 8*, 5. http://doi.org/10.3389/fncel.2014.00005.

Ruble, B. K., Yeldell, S. B., & Dmochowski, I. J. (2015). Caged oligonucleotides for studying biological systems. *Journal of Inorganic Biochemistry, 150*, 182–188. http://doi.org/10.1016/j.jinorgbio.2015.03.010.

Safka Brožková, D., Nevšímalová, S., Mazanec, R., Rautenstrauss, B., & Seeman, P. (2012). Charcot-Marie-Tooth neuropathy due to a novel EGR2 gene mutation with mild phenotype—usefulness of human mapping chip linkage analysis in a Czech family. *Neuromuscular Disorders, 22*(8), 742–746. http://doi.org/10.1016/j.nmd.2012.04.002.

Sander, J. D., Maeder, M. L., Reyon, D., Voytas, D. F., Joung, J. K., & Dobbs, D. (2010). ZiFiT (Zinc Finger Targeter): an updated zinc finger engineering tool. *Nucleic Acids Research, 38*(Web Server Issue), W462–W468. http://doi.org/10.1093/nar/gkq319.

Schaefer, K., & Brösamle, C. (2009). Zwilling-A and -B, two related myelin proteins of teleosts, which originate from a single bicistronic transcript. *Molecular Biology and Evolution, 26*(3), 495–499. http://doi.org/10.1093/molbev/msn298.

Schebesta, M., & Serluca, F. C. (2009). olig1 Expression identifies developing oligodendrocytes in zebrafish and requires hedgehog and Notch signaling. *Developmental Dynamics, 238*(4), 887–898. http://doi.org/10.1002/dvdy.21909.

Schulte-Merker, S., & Stainier, D. Y. R. (2014). Out with the old, in with the new: reassessing morpholino knockdowns in light of genome editing technology. *Development, 141*(16), 3103–3104. http://doi.org/10.1242/dev.112003.

Schweitzer, J., Becker, T., Schachner, M., Nave, K.-A., & Werner, H. (2006). Evolution of myelin proteolipid proteins: gene duplication in teleosts and expression pattern divergence. *Molecular and Cellular Neurosciences, 31*(1), 161–177. http://doi.org/10.1016/j.mcn.2005.10.007.

Shah, A. N., Davey, C. F., Whitebirch, A. C., Miller, A. C., & Moens, C. B. (2016). Rapid reverse genetic screening using CRISPR in zebrafish. *Zebrafish, 13*(2), 152–153. http://doi.org/10.1089/zeb.2015.29000.sha.

Shin, J., Park, H.-C., Topczewska, J. M., Mawdsley, D. J., & Appel, B. (2003). Neural cell fate analysis in zebrafish using olig2 BAC transgenics. *Methods in Cell Science: An Official Journal of the Society for In Vitro Biology, 25*(1–2), 7–14. http://doi.org/10.1023/B:MICS.0000006847.09037.3a.

Sidik, H., & Talbot, W. S. (2015). A zinc finger protein that regulates oligodendrocyte specification, migration and myelination in zebrafish. *Development, 142*(23), 4119–4128. http://doi.org/10.1242/dev.128215.

Simmons, T., & Appel, B. (2012). Mutation of pescadillo disrupts oligodendrocyte formation in zebrafish. *PLoS One, 7*(2), e32317. http://doi.org/10.1371/journal.pone.0032317.

Snaidero, N., Möbius, W., Czopka, T., Hekking, L. H. P., Mathisen, C., Verkleij, D., et al. (2014). Myelin membrane wrapping of CNS axons by PI(3,4,5)P3-dependent polarized growth at the inner tongue. *Cell, 156*(1–2), 277–290. http://doi.org/10.1016/j.cell.2013.11.044.

Snyder, J. L., Kearns, C. A., & Appel, B. (2012). Fbxw7 regulates Notch to control specification of neural precursors for oligodendrocyte fate. *Neural Development, 7*, 15. http://doi.org/10.1186/1749-8104-7-15.

Soundarapandian, M. M., Selvaraj, V., Lo, U.-G., Golub, M. S., Feldman, D. H., Pleasure, D. E., & Deng, W. (2011). Zfp488 promotes oligodendrocyte differentiation of neural progenitor cells in adult mice after demyelination. *Scientific Reports, 1*, 2. http://doi.org/10.1038/srep00002.

Stoveken, H. M., Hajduczok, A. G., Xu, L., & Tall, G. G. (2015). Adhesion G protein-coupled receptors are activated by exposure of a cryptic tethered agonist. *Proceedings of the National Academy of Sciences, 112*(19), 6194–6199. http://doi.org/10.1073/pnas.1421785112.

Takada, N., Kucenas, S., & Appel, B. (2010). Sox10 is necessary for oligodendrocyte survival following axon wrapping. *Glia, 58*(8), 996–1006. http://doi.org/10.1002/glia.20981.

Talbot, J. C., & Amacher, S. L. (2014). A streamlined CRISPR pipeline to reliably generate zebrafish frameshifting alleles. *Zebrafish, 11*(6), 583–585. http://doi.org/10.1089/zeb.2014.1047.

Tan, J. L., & Zon, L. I. (2011). Chemical screening in zebrafish for novel biological and therapeutic discovery. *Methods in Cell Biology, 105*, 493–516. http://doi.org/10.1016/B978-0-12-381320-6.00021-7.

Takada, N., & Appel, B. (2010). Identification of genes expressed by zebrafish oligodendrocytes using a differential microarray screen. *Developmental Dynamics, 239*(7), 2041–2047. http://www.ncbi.nlm.nih.gov/pubmed/20549738.

Taylor, K. L., Grant, N. J., Temperley, N. D., & Patton, E. E. (2010). Small molecule screening in zebrafish: an in vivo approach to identifying new chemical tools and drug leads. *Cell Communication and Signaling, 8*, 11. http://doi.org/10.1186/1478-811X-8-11.

Thisse, C., Thisse, B., Halpern, M. E., & Postlethwait, J. H. (1994). Goosecoid expression in neurectoderm and mesendoderm is disrupted in zebrafish cyclops gastrulas. *Developmental Biology, 164*(2), 420–429. http://doi.org/10.1006/dbio.1994.1212.

Voas, M. G., Lyons, D. A., Naylor, S. G., Arana, N., Rasband, M. N., & Talbot, W. S. (2007). alphaII-spectrin is essential for assembly of the nodes of Ranvier in myelinated axons. *Current Biology, 17*(6), 562–568. http://doi.org/10.1016/j.cub.2007.01.071.

Vogl, M. R., Reiprich, S., Küspert, M., Kosian, T., Schrewe, H., Nave, K.-A., & Wegner, M. (2013). Sox10 cooperates with the mediator subunit 12 during terminal differentiation of myelinating glia. *Journal of Neuroscience, 33*(15), 6679–6690. http://doi.org/10.1523/JNEUROSCI.5178-12.2013.

Wang, S.-Z., Dulin, J., Wu, H., Hurlock, E., Lee, S.-E., Jansson, K., & Lu, Q. R. (2006). An oligodendrocyte-specific zinc-finger transcription regulator cooperates with Olig2 to promote oligodendrocyte differentiation. *Development, 133*(17), 3389–3398. http://doi.org/10.1242/dev.02522.

Wheeler, M. A., Smith, C. J., Ottolini, M., Barker, B. S., Purohit, A. M., Grippo, R. M., et al. (2016). Genetically targeted magnetic control of the nervous system. *Nature Neuroscience, 19*(5), 756–761. http://doi.org/10.1038/nn.4265.

Woods, I. G., Lyons, D. A., Voas, M. G., Pogoda, H.-M., & Talbot, W. S. (2006). nsf is essential for organization of myelinated axons in zebrafish. *Current Biology, 16*(7), 636–648. http://doi.org/10.1016/j.cub.2006.02.067.

Xiao, Y., Faucherre, A., Pola-Morell, L., Heddleston, J. M., Liu, T.-L., Chew, T.-L., et al. (2015). High-resolution live imaging reveals axon-glia interactions during peripheral nerve injury and repair in zebrafish. *Disease Models and Mechanisms, 8*(6), 553–564. http://doi.org/10.1242/dmm.018184.

Yang, M. L., Shin, J., Kearns, C. A., Langworthy, M. M., Snell, H., Walker, M. B., & Appel, B. (2015). CNS myelination requires cytoplasmic dynein function. *Developmental Dynamics, 244*(2), 134–145. http://doi.org/10.1002/dvdy.24238.

Yoshida, M., & Macklin, W. B. (2005). Oligodendrocyte development and myelination in GFP-transgenic zebrafish. *Journal of Neuroscience Research, 81*(1), 1–8. http://doi.org/10.1002/jnr.20516.

Zalc, B., Goujet, D., & Colman, D. (2008). The origin of the myelination program in vertebrates. *Current Biology, 18*(12), R511–R512. http://doi.org/10.1016/j.cub.2008.04.010.

Zannino, D. A., & Appel, B. (2009). Olig2+ precursors produce abducens motor neurons and oligodendrocytes in the zebrafish hindbrain. *Journal of Neuroscience, 29*(8), 2322–2333. http://doi.org/10.1523/JNEUROSCI.3755-08.2009.

Zaucker, A., Mercurio, S., Sternheim, N., Talbot, W. S., & Marlow, F. L. (2013). Notch3 is essential for oligodendrocyte development and vascular integrity in zebrafish. *Disease Models and Mechanisms, 6*(5), 1246–1259. http://doi.org/10.1242/dmm.012005.

Zhu, Q., Zhao, X., Zheng, K., Li, H., Huang, H., Zhang, Z., et al. (2014). Genetic evidence that Nkx2.2 and Pdgfra are major determinants of the timing of oligodendrocyte differentiation in the developing CNS. *Development, 141*(3), 548–555. http://doi.org/10.1242/dev.095323.

Zuchero, J. B., Fu, M.-M., Sloan, S. A., Ibrahim, A., Olson, A., Zaremba, A., et al. (2015). CNS myelin wrapping is driven by actin disassembly. *Developmental Cell, 34*(2), 152–167. http://doi.org/10.1016/j.devcel.2015.06.011.

CHAPTER

Zebrafish models of human eye and inner ear diseases

16

B. Blanco-Sánchez, A. Clément, J.B. Phillips, M. Westerfield[1]

University of Oregon, Eugene, OR, United States

[1]*Corresponding author: E-mail: monte@uoneuro.uoregon.edu*

CHAPTER OUTLINE

Abstract

Eye and inner ear diseases are the most common sensory impairments that greatly impact quality of life. Zebrafish have been intensively employed to understand the fundamental mechanisms underlying eye and inner ear development. The zebrafish visual and vestibulo-acoustic systems are very similar to these in humans, and although not yet mature, they are functional by 5 days post-fertilization (dpf). In this chapter, we show how the zebrafish has significantly contributed to the field of biomedical research and how researchers, by establishing disease models and meticulously characterizing their phenotypes, have taken the first steps toward therapies. We review here models for (1) eye

Methods in Cell Biology, Volume 138, ISSN 0091-679X, http://dx.doi.org/10.1016/bs.mcb.2016.10.006

diseases, (2) ear diseases, and (3) syndromes affecting eye and/or ear. The use of new genome editing technologies and high-throughput screening systems should increase considerably the speed at which knowledge from zebrafish disease models is acquired, opening avenues for better diagnostics, treatments, and therapies.

INTRODUCTION

A total of 285 and 360 million people worldwide are affected with vision and hearing impairments, respectively (World Health Organization, 2015), and the prevalence of these diseases continues to rise in our aging population. The medical, social, and economic impacts of vision and hearing loss (HL) make it imperative to understand the cellular and physiological mechanisms underlying these anomalies. Animal models have already contributed significantly to the field. They are essential to understand what goes wrong in each disease and to develop better diagnostics and treatments.

Zebrafish have proven to be excellent models to study vision and hearing impairment for many reasons. First, the mechanisms of eye and inner ear development are well conserved (Baxendale & Whitfield, 2016; Easter & Nicola, 1996; Gestri, Link, & Neuhauss, 2012; Haddon & Lewis, 1996; Malicki, 1999; Schmitt & Dowling, 1994, 1999; Whitfield, Riley, Chiang, & Phillips, 2002). Second, the anatomical and functional compositions of the zebrafish visual and vestibulo-acoustic systems are similar to that of humans. The zebrafish retina contains the classic vertebrate arrangement of photoreceptors, first- and second-order neurons, and glia, as well as the retinal-pigmented epithelium (RPE) and retinal vasculature (Chhetri, Jacobson, & Gueven, 2014; Malicki, Pooranachandran, Nikolaev, Fang, & Avanesov, 2016). Although zebrafish retinas lack a macula and feature, instead, a highly organized mosaic arrangement of photoreceptors, they are, as in humans, cone-rich and optimized for diurnal activity (Link & Collery, 2015; Raymond et al., 2014). Likewise, despite the lack of a cochlea, the structure of the zebrafish inner ear resembles that of other vertebrates (Baxendale & Whitfield, 2016; Whitfield & Hammond, 2007) and the organization and morphology of the supporting cells and hair cells are also comparable to other vertebrates (Baxendale & Whitfield, 2016; Haddon & Lewis, 1996; Nicolson, 2005). Third, visual and hearing systems develop rapidly and are functional by 5 days post-fertilization (dpf) in zebrafish (Baxendale & Whitfield, 2016; Malicki et al., 2016). Finally, behavioral assays have been developed for visual and hearing function that support rapid assessment of retinal and inner ear functions (Table 1).

Zebrafish are becoming increasingly useful in translational research, contributing to the development of new approaches to understand and treat human diseases (Phillips & Westerfield, 2014). This illustrates the value of this organism as an indispensable model of human diseases. In this chapter, we describe current zebrafish models for vision and hearing impairment. We also review the tools currently available and describe their utility in analyzing defective pathways of vision and hearing.

Table 1 Summary of Behavioral Assays

Assay	System	Stage	Measure	Test	References
Optokinetic response	Vision	Larval (from 3 dpf) and adult	Eye movements following perceived motion	Visual acuity, contrast sensitivity, light adaptation, color blindness	Fleisch and Neuhauss (2006), Brockerhoff (2006), and Maurer, Huang, and Neuhauss (2011)
Optomotor response	Vision	Larval (from 6 dpf) and adult	Swimming towards perceived motion	Blindness, contrast sensitivity, color vision	Bilotta (2000), Muto et al. (2005), and Fleisch and Neuhauss (2006)
Startle response/ Visual motor response	Vision	Larval (from 4 dpf)	Body movement following a change in light intensities	Development and maturation of the visual system	Kimmel, Patterson, and Kimmel (1974) and Emran et al. (2008)
Phototactic behavior	Vision	Larval (from 6 dpf)	Swimming towards perceived light	Light intensity, phototaxis	Brockerhoff et al. (1995) and Burgess, Schoch, and Granato (2010)
Electroretinogram	Vision	Larval (from 4 dpf)	Electrical activity	Retinal function	Branchek (1984) and Saszik, Bilotta, and Givin (1999)
Escape response	Vision	Adult	Swimming away from motion	Visual performance and sensitivity	Li and Dowling (1997, 2000)
Auditory-evoked startle response/ Touch response	Hearing/ Balance	Larval (from 4 dpf) and adult	Swimming away from sound, vibration or touch	Ear and lateral line function	Kimmel et al. (1974) and Zeddies and Fay (2005)
Rheotactic behavior	Hearing/ Balance	Larval and adult	Swimming towards a current	Lateral line function	Olszewski, Haehnel, Taguchi, and Liao (2012) and Suli, Watson, Rubel, and Raible (2012)

Continued

Table 1 Summary of Behavioral Assays—cont'd

Assay	System	Stage	Measure	Test	References
Seeker response	Hearing/Balance	Larval (from 5 dpf)	Swimming away from seeker	Lateral line function	Winter et al. (2008)
Vestibulo-ocular reflex	Hearing/Balance	Larval (from 3 dpf) and juvenile	Eye movements following angular/linear acceleration	Vestibular function	Easter and Nicola (1997), Beck, Gilland, Tank, and Baker (2004), Mo et al., 2010, and Bianco et al. (2012)
Microphonic potential recording	Hearing/Balance	Larval (from 2 dpf)	Electrophysiological activity	Sensory hair cells function	Nicolson et al. (1998), Corey et al. (2004), Starr, Kappler, Chan, Kollmar, and Hudspeth (2004), and Kappler, Starr, Chan, Kollmar, and Hudspeth (2004)
Dorsal light reflex	Hearing/Balance	Juvenile and adult	Equilibrium orientation following a change in light position	Vestibular function	Nicolson et al. (1998)
Vestibular righting reflex	Hearing/Balance	Adult	Body repositioning following dropping and a change in orientation	Vestibular function	Whitfield lab, unpublished
Excitatory postsynaptic currents in the Mauthner cells	Hearing	Larval (5–6 dpf)	Acoustic response	Hearing function	Han et al. (2011)
Saccular macula microphonic potential recording	Hearing	Larval (from 3 dpf)	Acoustic response	Hearing function	Yao et al. (2016)

1. ZEBRAFISH MODELS OF EYE DISEASE

Forward and reverse genetic approaches using zebrafish models have supplied valuable insights into hereditary eye diseases for several decades. With new and improved tools for genetic manipulation in zebrafish emerging alongside the increased availability of next generation sequencing in the clinical setting, loss-of-function analyses in zebrafish to validate the pathogenicity of candidate alleles identified in human patients have grown in popularity and have made valuable contributions to the global understanding of a wide range of vision disorders (Table 2).

1.1 DISEASES OF PHOTORECEPTORS AND RETINAL-PIGMENTED EPITHELIUM

Retinal degeneration is a leading cause of progressive blindness worldwide. Due to the extensive molecular interplay between the photoreceptors in the neural retina and the RPE, it is not surprising that mutations in genes that function in either cell type can lead to a common end result of vision loss due to photoreceptor dysfunction and death. Zebrafish models of loci that contribute to dysfunction of RPE or photoreceptors have increased our understanding of the genes and genetic pathways underlying human disease, providing necessary insights for the development of improved diagnostic tools and therapeutic interventions.

Genes that regulate the highly conserved process of phototransduction, whereby photon-activated visual pigment molecules in the photoreceptor outer segments initiate G-protein coupled signaling, are common causes of inherited retinal degeneration (Yau & Hardie, 2009), as are factors involved in the visual cycle that converts 11-cis-retinal to all-trans-retinal and back again (Baehr, Wu, Bird, & Palczewski, 2003).

Zebrafish models of achromatopsia, a cone-rod dystrophy caused by mutations in *PDE6C,* have made significant contributions to our understanding of this disorder. *PDE6C* encodes a cone-specific phosphodiesterase, an enzyme required for regulation of cGMP activity in photoreceptors. Rod degeneration subsequent to the loss of cones indicates a non-cell autonomous action, termed the "bystander effect", which is a variable but common outcome in retinal degenerative diseases with primary effects in either rods or cones.

Stearns, Evangelista, Fadool, and Brockerhoff (2007) identified a zebrafish *pde6c* mutant showing a marked cone degeneration phenotype at 3 dpf, followed by corresponding rod death. Taking advantage of developmental timing in the zebrafish retina, where cone maturation precedes that of rods, Stearns et al. (2007) were able to illustrate that cone:rod density ratios were a critical element of the degenerative pattern. Their data demonstrated that rods in cone-rich regions of the developing retina, and in established central regions of the mature retina, were more susceptible to degeneration than rods occupying regions of higher rod density at the retinal margins in developing and mature, regenerating zebrafish retinas.

Table 2 Zebrafish Models of Eye Disease

Eye Disorder	Human Gene/ OMIM#	Origin of Zebrafish Model (s)	References
Cataracts	*CRYAA* 123580	TALENs	Zou et al., 2015
	SIPA1L3 616655	MO	Greenlees et al. (2015)
Choroideremia	*CHM* 300390	ENU	Krock et al. (2007) and Moosajee et al. (2008, 2009, 2016)
Coloboma	*FZD5* 601723	MO	C. Liu et al. (2016)
	YAP 606608	ENU	Miesfeld et al. (2015)
Cone-rod dystrophies	*GNAT2* 139340	ENU	Brockerhoff et al. (2003) and Kennedy et al. (2007)
	GUCY2D 600179	MO	Stiebel-Kalish et al. (2012)
	PDE6C 600827	ENU	Stearns et al. (2007) and Viringipurampeer et al. (2014)
	RAX 610362	MO	Nelson, Park, and Stenkamp (2009)
	UNC119 604011	MO	Wright et al. (2011) and Rainy et al. (2016)
Congenital stationary night blindness	*CACNA1F* 300110	ENU	Jia et al. (2014)
	GPR179 614515	MO	Peachey et al. (2012)
	GRM6 604096	MO	Huang, Haug, Gesemann, and Neuhauss (2012)
	NYX 300278	MO	Bahadori et al. (2006) and Peachey et al. (2012)
Exudative vitreoretinopathy	*ZNF408* 616454	MO	Collin et al. (2013)
Glaucoma	*FOXC1* 601090	MO	Skarie and Link (2009)
	OPTN 602432	ENU	Paulus and Link (2014)
	SIX6 606326	MO	Carnes et al. (2014)
	WDR36 609669	MO	Skarie and Link (2008)

Table 2 Zebrafish Models of Eye Disease—cont'd

Eye Disorder	Human Gene/ OMIM#	Origin of Zebrafish Model (s)	References
Microphthalmia/ anophthalmia with coloboma	*MAB21L2* 615877	MO, ENU, TALEN	Hartsock, Lee, Arnold, and Gross (2014)[3] and Deml, Kariminejad, et al. (2015)[1,2]
	TMX3 616102	MO	Chao et al. (2010)
Nonsyndromic retinitis pigmentosa	*ARL6* 613575	MO	Pretorius et al. (2010)
	C2ORF71 613425	MO	Nishimura et al. (2010) and Y. P. Liu et al. (2016)
	CERKL 608381	MO	Riera, Burguera, Garcia-Fernandez, and Gonzalez-Duarte (2013) and Li et al. (2014)
	CLUAP1 616787	TALEN, INS	Lee, Wallingford, and Gross (2014)[5]and Soens et al. (2016)[3]
	DHDDS 608172	MO	Zuchner et al. (2011) and Wen et al. (2014)
	NEK2 604043	MO	Nishiguchi et al. (2013)
	PRPF4 607795	MO	Linder et al. (2011, 2014)
	PRPF31 606419	MO	Linder et al. (2011) and Yin, Brocher, Fischer, and Winkler (2011)
	RLBP1 180090	MO	Collery et al. (2008) and Fleisch, Schonthaler, von Lintig, and Neuhauss (2008)
	RP1L1 608581	MO	Y. P. Liu et al. (2016)
	RP2 312600	MO, TALEN	Patil, Hurd, Ghosh, Murga-Zamalloa, and Khanna (2011)[1], Shu et al. (2011)[1] and F. Liu et al. (2015)[3]
	RPGR 312610	MO	Shu et al. (2010)
	SLC7A14 615720	MO	Jin et al. (2014)
	SNRNP200 601664	MO	Y. Liu et al. (2015)

Continued

Table 2 Zebrafish Models of Eye Disease—cont'd

Eye Disorder	Human Gene/ OMIM#	Origin of Zebrafish Model (s)	References
	TOPORS 609507	MO	Chakarova et al. (2011)
	USHRN 608400	MO	Ebermann et al. (2010)
	ZNF513 613598	MO	Li et al. (2010)

Genes represented in the tables were identified by cross-referencing searches in ZFIN (http://zfin.org), the Online Mendelian Inheritance in Man database (www.omim.org), and PubMed (http://www.ncbi.nlm.nih.gov/pubmed). Criteria for inclusion were restricted to zebrafish models of genes associated with human diseases affecting vision and/or hearing for which ocular or otic phenotypes were described. The following numbers indicate the method used to generate the zebrafish model for a given disease. 1, MO, *morpholino; 2,* ENU, N-*ethyl*-N-*nitrosourea; 3,* TALENs, *transcription-activator like effector nucleases; 4,* CRISPRs, *clustered regularly interspaced short palindromic repeats; 5, Viral insertion.*

More recently, Viringipurampeer et al. (2014), reported that cell death in *pde6d* mutants was the result of necroptotic pathway activation, demonstrated by their success in rescuing photoreceptor death by morpholino (MO) knockdown of *rip3*, a key regulator of necroptosis. Taken together, these studies offer key insights into the underlying causes of photoreceptor degeneration in *PDE6D* achromatopsia as well as providing potential avenues for developing treatments.

Choroideremia is a rare, X-linked form of retinal degeneration caused by mutations in *CHM*, which encodes the Rab-escort protein REP1. REP1 is a key component of a catalytic complex in the RPE, Rab GG transferase II, that regulates Rab-mediated melanosome trafficking and phagocytosis. In the absence of functional Rep1, posttranslational modifications of Rab are affected, leading to degeneration of the RPE and the vascular network known as the choroid that overlies the RPE. The zebrafish *chm* mutant has previously provided significant data on the molecular and cellular changes involved in choroideremia (Krock, Bilotta, & Perkins, 2007; Moosajee, Gregory-Evans, Ellis, Seabra, & Gregory-Evans, 2008) and is now contributing to preclinical trials for aminoglycoside drug therapy. Previous studies of *chm* provided Moosajee et al. (2016) with an ideal system in which to demonstrate that nonsense suppression agents could successfully repress retinal cell death and restore biochemical function of *chm* in vivo, bringing this treatment one step closer to clinical trial approval for CHM patients.

1.2 ZEBRAFISH TOOLS FOR DIAGNOSTIC MEDICINE

In addition to testing the pathogenicity of new variants of individual disease genes, zebrafish models have also provided crucial evidence for genetic interactions between previously identified monogenic disorders. More than 50 genes to date have been linked to retinitis pigmentosa (RP), a common form of progressive vision loss due to retinal degeneration. A number of nonsyndromic RP genes (https://sph.uth.edu/retnet/sum-dis.htm) are also involved in syndromic disorders that include RP (Table 2). Two genes, *C2ORF71* and *RP1L1*, both previously known to cause autosomal recessive RP, were recently implicated in a digenic form of syndromic RP, suggesting hitherto unknown haploinsufficiency at both loci (Y.P. Liu et al., 2016). Zebrafish MO and clustered regularly interspaced short palindromic repeats (CRISPRs) models recapitulated the additive loss-of-function effect noted in the retinas and brains of digenic patients, thus validating the rare genetic presentation and providing a new diagnostic option for undiagnosed syndromic RP patients going forward.

Beyond the currently identified disease genes, phenotypic models derived from forward mutagenesis screens as well as reverse genetic approaches targeting components of eye development and function have enhanced the field of eye research tremendously. The avascular zebrafish mutant *cloche* has been a heavily utilized model of vascular development since its discovery (Stainier, Weinstein, Detrich, Zon, & Fishman, 1995), and eye phenotypes resulting from affected blood vessel formation have contributed to better understanding of the role of vascularization in retinal and lens development (Dhakal et al., 2015; Goishi et al., 2006). Recently, *cloche* has been identified molecularly as the bHLH-PAS transcription factor *npas4l* (Reischauer et al., 2016), which has yet to be linked to any type of human disease, but can now be added to screening protocols for vascular disorders of the retina and other systems.

There is a strong precedent for investigations into genetic mechanisms preceding the identification of a gene as causative of human disease. For example, noting that several types of RP are caused by ubiquitously expressed mRNA splicing factors, Linder et al. (2011) and Ruzickova and Stanek (2016) conducted experiments to illuminate the cell-specific effect of these mutations. In this study, they also reported a retina-specific phenotype caused by depletion of another member of the tri-snRNP family, *prfp4*, which had not been previously identified as a human disease gene. Several years later, *PRFP4* mutations were verified as causative of RP (Chen et al., 2014; Linder et al., 2014).

2. ZEBRAFISH MODELS OF EAR DISEASE

A significant portion of the genetic mutations linked to sensorineural deafness are allelic to mutations that cause syndromic disorders. Additionally, numerous genes have been associated with nonsyndromic deafness (Table 3), and their identification has been significantly facilitated by zebrafish research. Zebrafish models of hair cell dysfunction exhibit easily scorable behavioral traits (Table 1), providing an accessible and reproducible method to evaluate loss of function.

Clinically, HL is the most common reported birth defect in developed countries (Hilgert, Smith, & Van Camp, 2009). This condition can have a conductive (outer and/or middle ear), sensorineural (inner ear), mixed (outer, middle, and inner ear), or central auditory origin (Smith, Shearer, Hildebrand, & Van Camp, 1993). Clinical

Table 3 Zebrafish Models of ear Disease

Nonsyndromic Ear Disorder	Human Gene/ OMIM#	Origin of Zebrafish Model	References
DFNA3A, DFNB1A	*GJB2* 605425	MO	Chang-Chien et al. (2014)
DFNA5	*DFNA5* 608798	MO, CRISPR	Busch-Nentwich et al. (2004)[1] and Varshney et al. (2015)[4]
DFNA8/12, DFNB21	*TECTA* 602574	ENU	Whitfield et al. (1996) and Stooke-Vaughan et al. (2015)
DFNA11, DFNB2	*MYO7A* 276903	ENU	Ernest et al. (2000) and Blanco-Sanchez et al. (2014)
DFNA22, DFNB37	*MYO6* 600970	ENU	Seiler et al. (2004)
DFNA25	*SLC17A8* 607557	ENU	Obholzer et al. (2008)
DFNA28	*GRHL2* 608576	MO, INS	Han et al. (2011)
DFNB6	*TMIE* 607237	MO, ENU	Shen et al. (2008)[1] and Gleason et al. (2009)[2]
DFNB9	*OTOF* 603681	MO	Chatterjee et al. (2015)
DFNB12	*CDH23* 605516	ENU	Söllner et al. (2004) and Blanco-Sanchez et al. (2014)
DFNB18A	*USH1C* 605242	MO, ENU	Phillips et al. (2011)[1,2] and Blanco-Sanchez et al. (2014)[2]
DFNB18B	*OTOG* 614945	ENU	Whitfield et al. (1996) and Stooke-Vaughan et al. (2015)
DFNB23	*PCDH15* 605514	ENU	Seiler et al. (2005)
DFNB42	*ILDR1* 609739	MO	Sang et al. (2014)
DFNB48	*CIB2* 605564	MO	Riazuddin et al. (2012)

Table 3 Zebrafish Models of ear Disease—cont'd

Nonsyndromic Ear Disorder	Human Gene/ OMIM#	Origin of Zebrafish Model	References
DFNB66	*DCDC2* 610212	MO	Grati et al. (2015)
DFNB68	*S1PR2* 610419	MO, ENU	Hu et al. (2013)[2] and Santos-Cortez et al. (2016)[1]
DFNB74	MSRB3 613719	MO	Shen et al. (2015)
DFNB84B	*OTOGL* 614925	MO	Yariz et al. (2012)
DFNB99	*TMEM132E* 616178	MO	Li et al. (2015)
DFNB104	*FAM65B* 616515	MO	Diaz-Horta et al. (2014)
DFNB105	*CDC14A* 616958	MO	Delmaghani et al. (2016)

Genes represented in the tables were identified by cross-referencing searches in ZFIN (http://zfin.org), the Online Mendelian Inheritance in Man database (www.omim.org), and PubMed (http://www.ncbi.nlm.nih.gov/pubmed). Criteria for inclusion were restricted to zebrafish models of genes associated with human diseases affecting vision and/or hearing for which ocular or otic phenotypes were described. The following numbers indicate the method used to generate the zebrafish model for a given disease. 1, MO, *morpholino; 2,* ENU, *N-ethyl-N-nitrosourea; 3,* TALENs, *transcription-activator like effector nucleases; 4,* CRISPR, *clustered regularly interspaced short palindromic repeat; 5, Viral insertion.*

manifestation can occur before (prelingual) or after (postlingual) the onset of speech (Smith et al., 1993). Whereas congenital deafness is considered prelingual, not all prelingual deafness is congenital (Smith et al., 1993).

HL prevalence increases as the population ages. It has been estimated that 1 in 500 newborns is affected by moderate to profound bilateral sensorineural hearing loss (SNHL) (Morton & Nance, 2006). However, HL prevalence increases to 3.5 in 1000 by adolescence, showing the relevant significance of postlingual SNHL (Morton & Nance, 2006).

Vestibular disorders can originate in the cerebral cortex (central) or at the level of the eye and/or inner ear (peripheral). Their estimated occurrence is more variable and less accurate than HL, due to difficulty in the diagnosis. As an example, the lack of a neonatal standardized test to measure vestibular function prevents calculation of the congenital incidence. One vestibular epidemiological study projected that, in the United States 35% of adults over 40 have been affected by balance disorders (Agrawal, Carey, Della Santina, Schubert, & Minor, 2009). As with HL prevalence, the frequency of vestibular disorders increases with age, and an estimated 80% of the population over 65 have experienced vestibular dysfunction of some kind (Zalewski, 2015).

Current treatment for SNHL varies according to the severity and age of onset. For severe to profound prelingual disorders, early intervention with cochlear implants yields the best results (Connor, Craig, Raudenbush, Heavner, & Zwolan, 2006; Connor, Hieber, Arts, & Zwolan, 2000; James, Rajput, Brinton, & Goswami, 2008). Treatments for vestibular disorders differ according to the specific diagnosis and include dietary changes, physical therapy, and pharmaceutical and surgical approaches (Driscoll, Kasperbauer, Facer, Harner, & Beatty, 1997; Hain & Yacovino, 2005; Herraiz et al., 2010; McClure, Lycett, & Baskerville, 1982; Ruckenstein, Rutka, & Hawke, 1991; Smith, Sankar, & Pfleiderer, 2005; Strupp et al., 2008, 2011, 2004; Takeda, Morita, Hasegawa, Kubo, & Matsunaga, 1989; Torok, 1977; Whitney & Rossi, 2000). Like the disorders themselves, the success rates of these treatments are highly variable. Moreover, the impact of early childhood vestibular dysfunction has been underestimated, although it can hamper development of motor skills and even literacy due to poor gaze stability (Casselbrant, Villardo, & Mandel, 2008; Christy, Payne, Azuero, & Formby, 2014; Cushing, Papsin, Rutka, James, & Gordon, 2008; Janky & Givens, 2015; Li, Hoffman, Ward, Cohen, & Rine, 2016; Rine, 2009; Rine et al., 2000; Rine, Dannenbaum, Szabo, 2016; Rine, Spielholz, Buchman, 2001; Rine & Wiener-Vacher, 2013; Wiener-Vacher, Ledebt, & Bril, 1996).

Finding and understanding the etiology of human ear diseases are essential for improving and extending therapy success, proposing novel approaches, and finding cures. Due to its accessibility, rapid development, and molecular and genetic manipulations, the zebrafish inner ear is an ideal model to reach these goals as well as to study sensorineural inner ear dysfunction (Table 3). In this section, we discuss the contribution of zebrafish models of genetic inner ear diseases to validation of candidate genes and elucidation of pathogenesis.

2.1 INNER EAR DISEASE

Historically, based on their mode of inheritance, loci associated with nonsyndromic deafness have been grouped into four distinct DFN categories. Loci associated with an autosomal dominant pattern are grouped into the DFNA class, those with an autosomal recessive mode of inheritance are classified as DFNB, and those associated with sex chromosomes are called DFNX and DFNY (Smith et al., 1993). Genetically, such categorization implies that, depending on the mutation, the same gene can be binned into both DFNA and DFNB categories or even be associated with syndromic HL.

Inner ear research has made considerable progress in identifying genes involved in hearing and balance. In 1996, *POU3F4* was the only gene associated with human deafness and 10 loci were linked to each DFNA and DFNB category (Petit, 1996). Similarly, only five genes (*MITF, PAX3, EDNRB, EDN3,* and *MYO7*) were linked to syndromic human deafness (Petit, 1996). During that same year, the Boston and Tübingen forward genetic screens in zebrafish identified 33 genes involved in ear development (Malicki et al., 1996; Whitfield et al., 1996). These genes affect specification of the otic placode, early morphogenesis of the vesicle, patterning, development of otoliths and semicircular canals, or ear size. The Tübingen screen also identified mutants with vestibular dysfunction despite having morphologically normal ear structures, as well as mutants with developmental anomalies in the ear and in other tissues resembling human syndromic conditions (Nicolson et al., 1998; Whitfield et al., 1996). Currently, 141 *DFN* loci are known, and, using both forward and reverse genetic approaches, zebrafish models provide a strong experimental platform for testing and validating candidate genes implicated in otic pathologies.

2.1.1 Sensorineural syndromic deafness

Currently, approximately 400 human syndromes associated with HL have been described (Toriello, Reardon, & Gorlin, 2004). In this section, we present two zebrafish models of SNHL for Branchiootorenal (BOR) and Waardenburg syndromes. Zebrafish models of syndromic eye and inner ear diseases are discussed in section 4 of this chapter.

2.1.1.1 Branchiootorenal/Branchiootic syndrome

EYA1 and *SIX1* encode evolutionary conserved transcriptional cofactors that assemble into a transcriptional unit through direct physical interaction of the SIX and EYA domains (Buller, Xu, Marquis, Schwanke, & Xu, 2001; Grifone et al., 2004; Ikeda, Watanabe, Ohto, & Kawakami, 2002; Ozaki, Watanabe, Ikeda, & Kawakami, 2002; Pignoni et al., 1997; Ruf et al., 2004). In this transcriptional complex, SIX1 provides the DNA binding activity through its homeodomain and EYA1 confers transcriptional activation. In humans, mutations in *EYA1* and *SIX1* result in the autosomal dominant genetic disorders BOR and Branchiootic (BO) syndromes (Abdelhak et al., 1997; Buller et al., 2001; Ozaki et al., 2002; Ruf et al., 2004). BOR/BO patients suffer moderate to severe conductive, sensorineural or mixed HL, branchial fistulae, and variable renal dysfunction. Other symptoms include vestibular aqueduct dilation, arrested development of the cochlea, vestibular canals, and facial nerve, cleft palate and cataracts (Ruf et al., 2004; Sanggaard et al., 2007).

In vertebrates, these genes are expressed in the preplacodal domain (PPD) (Sahly, Andermann, & Petit, 1999; Sato et al., 2010), an anteriorly located horseshoe shaped embryonic field established by the end of gastrulation (Schlosser & Ahrens, 2004). The PPD lies between the neural and non-neural ectodermal borders and partially overlaps with the anterior neural crest progenitor domain. During development, preplacodal cells segregate and sort into subdomains that give rise to the otic, trigeminal, and nasal placodes and the lens (Saint-Jeannet & Moody, 2014). Cranial clinical features of BOR patients correspond to the predicted PPD derivatives and the common embryonic origin shared by preplacodal and anterior neural crest precursors.

In zebrafish, *eya1* and *six1* expression initiates around 9–10 hpf (Bessarab, Chong, & Korzh, 2004; Sahly et al., 1999). Both genes are expressed throughout the otic preplacodal and placodal tissues (Kozlowski, Whitfield, Hukriede, Lam, & Weinberg, 2005). However, their expression is restricted to the ventral half of the otic vesicle by 24 hpf. In 1996, three mutant alleles of the *eya1* (*dog-eared*) gene were identified that truncate the protein at the Eya domain, thus disrupting assembly of the Eya1-Six1 transcriptional unit (Whitfield et al., 1996). In the *eya1* zebrafish mutant at 24 hpf, the otic vesicle is present with no apparent morphological defects. Development of the semicircular canals initiates by 48 hpf in zebrafish, when epithelial projections evaginate from the vesicle walls toward the lumen (Whitfield et al., 2002). At this stage, *eya1* mutants develop a smaller otic vesicle with abnormal morphogenesis of the semicircular canals (Kozlowski et al., 2005; Whitfield et al., 1996). As a result, the inner ear is severely dysmorphic. Additional analysis showed that Eya1 is necessary for promoting the induction and delamination of neuronal precursors that give rise to the statoacoustic ganglion (SAG), as well as hair cell survival (Kozlowski et al., 2005; Whitfield et al., 2002).

In zebrafish, Six1 function has been studied by antisense MO oligonucleotides (Bricaud & Collazo, 2006, 2011). The *six1* morphant shows neuronal lineage defects opposite to that of *eya1*, indicating that Six1 inhibits neuronal specification. Specifically, Bricaud & Collazo (2011) overexpressed a mutant form of Six1 that abrogates its interaction with Eya1 without affecting its DNA binding or ability to form a transcriptional repressor complex with Groucho1. The number of SAG precursor cells was reduced, similar to *six1* morphants. This suggested that Eya1 and Six1 do not cooperate physically to specify the neuronal lineage and that they have both dependent and independent roles during development. These zebrafish models of BOR/BO syndromes have demonstrated that mutations in *eya1* or *six1* are not equivalent and that these two genes differentially affect otic development. Thus, BOR and BO have distinct cellular etiologies, meaning that human patients may need distinct therapeutic approaches depending on their genotype.

2.1.1.2 Waardenburg syndrome types IIE (WS2E) and IV (WS4)

SOX10 is another key gene in otic development. In humans, mutations in this transcription factor lead to Waardenburg syndrome types IIE and IV. This genetic condition causes SNHL, variable Hirschsprung disease, and pigmentation defects, and it can have a recessive or dominant inheritance pattern (Kuhlbrodt et al., 1998; Potterf, Furumura, Dunn, Arnheiter, & Pavan, 2000).

During zebrafish development, *sox10* is expressed in the posterior region of the PPD, the anterior neural crest progenitor domain, and the entire otic placode and vesicle (Dutton et al., 2009). By 24 hpf, the otic vesicle of the *sox10* (*colourless*, *cls*) mutant has no apparent morphological defects (Dutton et al., 2009). However, by 48 hpf, the *cls* otic vesicle is smaller, and development of the semicircular canals is severely affected. The mutant phenotype is variable and ranges from small to swollen ears with or without distinguishable semicircular canal pillars (Dutton et al., 2009). Molecular analysis showed that Sox10 is necessary for correct inner ear patterning, proper macular development, and expression of gap junction coding genes such as *connexin* (*cx*) *33.8* and *connexin 27.5* (Dutton et al., 2009). Prior to the published zebrafish work in 2009, it was proposed that Waardenburg syndrome IIE and IV forms of deafness were due to a deficient contribution of melanocytes to the stria vascularis of the cochlea causing dysregulation of the endolymph ion composition (Matsushima et al., 2002; Tachibana et al., 1992, 2003). However, the zebrafish *sox10* otic expression pattern and mutant phenotype indicated a more complex etiology in which otic patterning, epithelial integrity, and possible compartmentalization play roles in Waardenburg syndromes IIE and IV (Dutton et al., 2009). In 2013, a more detailed evaluation of 14 patients with mutations in *SOX10* showed that all were affected by agenesis or hypoplasia of one or more semicircular canals, enlarged vestibule, reduced cochlear size without compartmentalization defects, and variable defects at the level of cochlear nerve (Elmaleh-Berges et al., 2013), phenotypes predicted from the zebrafish model.

2.1.2 Nonsyndromic deafness

To validate candidate genes involved in nonsyndromic deafness, zebrafish researchers have adopted both forward and reverse genetic approaches based on loss-of-function mutations using *N*-ethyl-*N*-nitrosourea (ENU)-based mutagenesis, MO, viral insertions, or genome editing techniques like transcription-activator like effector nuclease (TALEN) or CRISPR. Some of these models lack obvious morphological defects but others show broad morphological impacts. Based on phenotypes, predicted proteins, and proposed biological functions or requirement, we have established the following categories of zebrafish models.

2.1.2.1 Epithelial integrity

DFNA28 is a form of autosomal dominant deafness. Zebrafish *dfna28* (*grhl2b*) encodes a transcription factor present in the otic vesicle at 24 hpf (Han et al., 2011). It is involved in positive regulation of *claudin b* and *epcam* that both encode components of the apical tight junction complex (Han et al., 2011). Zebrafish homozygous mutants have abnormal swimming behavior characterized by a corkscrew pattern and an increased latency to respond to sound (Han et al., 2011). *grhl2b* mutation affects both otoliths and semicircular canals. By 5 dpf, mutant semicircular canals form but present localized epithelial dentations. Unlike *eya1* or *sox10* mutants, *grhl2b* mutants lack severe patterning defects and develop normal SAG and hair cells (Han et al., 2011). Because *grhl2b* is necessary for epithelial integrity, the authors proposed that vestibular dysfunction is due to a leaky epithelial barrier that leads to dysregulation of endolymph composition.

ILDR1 encodes a transmembrane protein associated with tight junction complex proteins, and mutations in this gene give rise to DFNB42 (Borck et al., 2011). The zebrafish *idlr1b* MO model has semicircular canal and lateral line migration defects (Sang et al., 2014). Interestingly, the authors also showed that this transmembrane protein is necessary for the expression of *atp1b2b*, a gene encoding a Na^{+}/K^{+} ATPase transporter previously shown to promote morphogenesis of the semicircular canal, thus providing potential insight into the etiology of the disease.

Two other genes have been implicated in semicircular canal morphogenesis based on MO studies. The first, *dfna5b*, an ortholog of *DFNA5*, is thought to act as a transcription factor required for expression of enzymes involved in synthesis of extracellular matrix (Busch-Nentwich, Sollner, Roehl, & Nicolson, 2004). *dfna5b* mRNA is detected in the otic vesicle at the columnar epithelial projections that give rise to the semicircular canals (Busch-Nentwich et al., 2004). The second gene, *tmie*, encodes the transmembrane protein Tmie (Dfnb6). *tmie* morphants show no gross patterning defects but exhibit a reproducible failure in the epithelial fusion between the ventral bulge and projection (Shen et al., 2008). These zebrafish mutants have severe (*tmie*) or mild (*dfna5b*) mechanosensory dysfunction, and neither develops gross malformation of the balance organ (Gleason et al., 2009; Varshney et al., 2015).

2.1.2.2 Sensory organ architectural defects

The glycoproteins Otogelin (*OTOG*), Otogelin-like, alpha-Tectorin (*TECTA*), beta-Tectorin, and Otolin are components of the otolith membrane in fish and the otoconial and tectorial membranes in mammals (Deans, Peterson, & Wong, 2010; Goodyear & Richardson, 2002; Yariz et al., 2012). These acellular membranes transmit mechanical stimuli to the hair cell mechanoreceptor (Eatock, Fay, & Popper, 2006; Richardson, de Monvel, & Petit, 2011; Tavazzani et al., 2016). In humans, mutations in *OTOG* have been associated with deafness (DFNB18b) and balance defects, whereas mutations in *TECTA* have been linked to recessive (DFNB21) and dominant (DFNA8/12) forms of deafness (Meyer et al., 2007; Mustapha et al., 1999; Naz et al., 2003; Schraders et al., 2012; Verhoeven et al., 1997).

Genetic analysis of the *otogelin* (*einstein*) and *tecta* (*rolling stone*) zebrafish mutants revealed that these two extracellular proteins act independently and in a sequential mechanism (Stooke-Vaughan, Obholzer, Baxendale, Megason, & Whitfield, 2015). Otogelin plays the initial role in the otolith seeding process, and alpha-Tectorin then functions to maintain tethering. Interestingly, *otogelin* zebrafish mutants have aberrant swimming behavior probably because they develop only one otolith that often tethers at the saccular macula (Stooke-Vaughan et al., 2015; Whitfield et al., 1996). In contrast, *tecta* zebrafish mutants swim normally and have two otoliths, although the saccular otolith tends to be dislodged (Stooke-Vaughan et al., 2015; Whitfield et al., 1996). Double *otogelin;tecta* zebrafish mutants have an additive phenotype; only one otolith forms, and it fails to tether (Stooke-Vaughan et al., 2015).

These zebrafish studies not only provide models for otoconia-associated vestibular dysfunction, like benign paroxysmal positional vertigo, but also contribute to our understanding of the mechanisms of extracellular matrix synthesis and remodeling.

2.1.2.3 Hair cell dysfunction

Mechanoreceptor structural defects and impaired neurotransmission severely affect hearing and balance (Blanco-Sanchez, Clement, Fierro, Washbourne, & Westerfield, 2014; El-Amraoui & Petit, 2005; Ernest et al., 2000; Gopal et al., 2015; Phillips et al., 2011; Reiners, Nagel-Wolfrum, Jurgens, Marker, & Wolfrum, 2006; Seiler et al., 2005; Söllner et al., 2004). In this section, we discuss zebrafish models of mechanoreceptor and synaptic dysfunction.

2.1.2.3.1 Mechanoreceptor dysfunction

In humans, mutations in *MYO7AA* (DFNB2, DFNA11, USH1B), *USH1C* (DFN18A), *CDH23* (DFNB12, USH1D), *PCDH15A* (DFNB23, USH1F) and *CLRN1* (USH3A) genes can give rise to Usher syndrome (USH) or nonsyndromic forms of deafness or RP. These genes encode a variety of proteins such as atypical myosin motor, scaffold, adhesion, and transmembrane molecules that assemble into complexes (Reiners et al., 2006). Mechanoreceptors of zebrafish models for these genes are characterized by the presence of splayed hair bundles, consistent with the role of these proteins in forming the connecting links between stereocilia and kinocilia (Blanco-Sanchez et al., 2014; Ernest et al., 2000; Gopal et al., 2015; Phillips et al., 2011; Seiler et al., 2005; Söllner et al., 2004). Work from our group showed that in the hair cells, harmonin, the product of the *ush1c* gene, Cdh23, and Myo7aa assemble into a complex at the level of the endoplasmic reticulum (ER) (Blanco-Sanchez et al., 2014). Disruption in the assembly affects the homeostasis of the secretory pathway, induces the ER stress response, and promotes cell death through the CDK5 signaling pathway.

The zebrafish *myo6b* mutant is a model for DFNA22 and DFNB37 forms of deafness. Myo6b is another atypical myosin that plays a dual role as a transporter or an anchor depending on ATP concentration (De La Cruz, Ostap, & Sweeney, 2001; Wells et al., 1999). Myo6b has been associated with apical vesicular trafficking, and *myo6b* mutations impair stereocilia growth and result in variable splaying of the hair bundle or occasional stereociliary fusion (Seiler et al., 2004). Detachment of the plasma membrane from the cuticular plate in the apical region and vesicle accumulation have also been reported in the *myo6b* mutant.

Recent MO-based studies in zebrafish have linked the function of *cib2* (DFNB48, USH1J, *fam65b (*DFNB104), *tmem132e* (DFNB99), *dcdc2* (DFN66) and *cdc14a* (DFNB105) with mechanoreceptor development and/or function (Delmaghani et al., 2016; Diaz-Horta et al., 2014; Grati et al., 2015; Li et al., 2015; Riazuddin et al., 2012).

2.1.2.3.2 Synapse dysfunction

In humans, mutations in *SLC17A8* and *OTOFERLIN* (*OTOF*) cause DFNA25 and DFNB9, respectively (Adato, Raskin, Petit, & Bonne-Tamir, 2000; Migliosi et al., 2002; Ruel et al., 2008; Yasunaga et al., 2000). SLC17A8 is a glutamate vesicular transporter and OTOF is a six C-2 domain transmembrane protein postulated to promote vesicular fusion at the presynaptic membrane (Chatterjee et al., 2015; Obholzer et al., 2008; Takamori, Malherbe, Broger, & Jahn, 2002; Yasunaga et al., 1999). Zebrafish has two *otof* copies (*otofa* and *otofb*) that are expressed in a complementary manner at the level of the sensory patches and one copy of *slc17a8* that is expressed in hair cells (Chatterjee et al., 2015; Obholzer et al., 2008). As expected, all three proteins are enriched at the basolateral membrane

(Chatterjee et al., 2015; Obholzer et al., 2008). *slc17a8* zebrafish mutants show the circler phenotype characteristic of other inner ear mutants, i.e., no startle response to dish-tapping, vestibular areflexia, and a corkscrew swimming pattern (Obholzer et al., 2008) (Table 1). In zebrafish models of DFNB9, vestibular defects are present only when both *otof* copies are knocked down with MOs (Chatterjee et al., 2015). In both models, mechanotransduction is unaffected as assayed by uptake of FM1-43 or YO-PRO1 dyes. Action currents in postsynaptic acousticolateralis neurons were absent in *slc17a8* mutants, indicating a failure in neurotransmission (Obholzer et al., 2008).

3. ZEBRAFISH MODELS OF SYNDROMES AFFECTING EYE AND/OR EAR

3.1 USHER SYNDROME

Mutations in a number of genes are known to affect both eye and inner ear function in humans. Zebrafish models of combined eye and inner ear disease can help identify common and single pathological mechanisms involved in the dysfunction of each organ. In this section, we discuss zebrafish models of human diseases in which both eye and ear function are affected.

Zebrafish has recently emerged as an excellent model for studies of USH. Not only can visual and vestibular function be quickly assayed (Table 1), but also the tissues of interest are accessible for live imaging and studies at cellular resolution (Blanco-Sanchez et al., 2014; Ebermann et al., 2010; Glover, Mueller, Sollner, Neuhauss, & Nicolson, 2012; Gopal et al., 2015; Ogun & Zallocchi, 2014; Phillips et al., 2011; Seiler et al., 2005).

USH is the leading cause of hereditary combined deafness and blindness, affecting 1 in 6000 Americans. It is characterized by congenital sensorineural deafness, RP, and, in some cases, balance problems (Mathur & Yang, 2015). Three clinical types (USH1−USH3) are distinguished by symptom onset and severity. USH1 is the most clinically severe with profound congenital deafness, vestibular dysfunction and an onset of retinal degeneration in early childhood. USH2 is the most common subtype and is clinically distinct from USH1 manifestations due to hearing impairment in the moderate to severe range, a later onset of vision loss in the second decade of life, and the absence of vestibular symptoms. USH3 is the most variable subtype with stable progressive or profound HL, variable vestibular dysfunction, and progressive vision loss. In some cases, mutations in genes that cause USH can also lead to RP or nonsyndromic deafness depending on their nature and position in the gene of interest (Tables 2−4). Here, we describe zebrafish models of USH syndrome.

3.1.1 USH1C

The *USH1C* gene encodes a key organizer scaffold protein, harmonin, that binds multiple proteins and organizes them into a complex. The USH proteins are among

Table 4 Zebrafish Models of Syndromes Affecting Eye and/or Ear

Syndrome	Human Gene/ OMIM#	Differential Diagnosis	Zebrafish Model Phenotype		Reference
			Eye	Ear	
Bardet–Biedl syndrome (BBS)	*BBS1* 209901	BBS1	Microphthalmia, retinal organization defects	n/a	Kim et al. (2013)[1]
	BBS5 603650	BBS5	Microphthalmia, retinal organization defects, RP	n/a	Al-Hamed et al. (2014)[1]
	C8ORF37 614477	BBS21, RP, cone-rod dystrophy	Impaired visual behavior	n/a	Heon et al. (2016)[5]
	CEP290 610142	BBS, LCA, Meckel syndrome, NPHP	Disorganized photoreceptors, reduced visual function	n/a	Baye et al. (2011)[1] and Murga-Zamalloa et al. (2011)[1]
	IFT172 607386	BBS, RP71, short-rib thoracic dysplasia 10 with or without polydactyly	Disorganized photoreceptors, thin outer nuclear layer, retinal degeneration	Reduced otic vesicle, kinocilia defects	Lunt, Haynes, and Perkins (2009)[1,2], Sukumaran and Perkins (2009)[2], and Bujakowska et al. (2015)[1]
Branchiootic syndrome (BO/ BOR)	*EYA1* 601653	BO syndrome, BOR syndrome	n/a	Dysmorphogenesis, hair cell death, decrease of SAG precursor number	Whitfield et al. (2002)[2] and Kozlowski et al. (2005)[2]
	SIX1 601205	BO3, DFNA23	n/a	Reduced hair cell number, increase of SAG precursor number	Bricaud and Collazo (2006)[1]
CHARGE syndrome	*CHD7* 608892	CHARGE syndrome	Retinal development defects	Abnormal otoliths	Patten et al. (2012)[1], Balow et al. (2013)[1], and Balasubramanian et al. (2014)[1]

Continued

Table 4 Zebrafish Models of Syndromes Affecting Eye and/or Ear—cont'd

Syndrome	Human Gene/ OMIM#	Differential Diagnosis	Zebrafish Model Phenotype		Reference
			Eye	Ear	
Jalili syndrome	*CNNM4* 607805	Jalili syndrome	Reduction of retinal ganglion cell numbers	n/a	Polok et al. (2009)[1]
Joubert syndrome (JBTS)	*AHI1* 608894	JBTS	Microphthalmia, photoreceptor cilia defects	n/a	Simms et al. (2012)[1] and Elsayed et al. (2015)[1]
	ARL13B 608922	JBTS	Retinal degeneration	n/a	Song, Dudinsky, Fogerty, Gaivin, and Perkins (2016)[2]
	CC2D2A 612013	JBTS, Meckel syndrome, COACH syndrome	Photoreceptor outer segment defects	Hair cell number	Owens et al. (2008)[2] and Bachmann-Gagescu et al. (2011)[2]
	CEP41 610523	JBTS	Microphthalmia	Abnormal otoliths	Lee et al. (2012)[1]
	INPP5E 613037	JBTS,	Microphthalmia, disorganized retina	n/a	Luo, Lu, and Sun (2012)[1]
	PDE6D 602676	JBTS	Disorganized retina, microphthalmia	n/a	Thomas et al. (2014)[1]
	POC1B 614784	JBTS, LCA	Microphthalmia, affected retinal cilia, reduced visual function	n/a	Beck et al. (2014)[1] and Roosing et al. (2014)[2]
Leber congenital amaurosis (LCA)	*CRX* 602225	LCA, cone-rod dystrophy	Retinal dysmorphogenesis, retinal cell death	n/a	Shen and Raymond (2004)[1]
	GDF6 601147	LCA, Klippel–Feil syndrome 1, microphthalmia with coloboma 6	Coloboma, microphthalmia, anophthalmia, photoreceptors morphogenesis defects	n/a	Asai-Coakwell et al. (2007[1], 2013), Valdivia et al. (2016)[2], and Gosse and Baier (2009)[2]

	GUCY2D 600179	LCA, cone-rod dystrophy	Retinal dystrophy, loss-of-preouter segments	n/a	Stiebel-Kalish et al. (2012)[1]
Nephronophthisis (NPHP)	*CEP164* 614848	NPHP	Retinal dysplasia	n/a	Chaki et al. (2012)[1]
	GLIS2 608539	NPHP	Microphthalmia, retinal degeneration	n/a	Kim et al. (2013)[1]
Norrie disease	*NDP* 300658	Norrie disease, exudative vitreoretinopathy	Narrow intraocular vessels	n/a	Wu et al. (2016)[1]
Oculoauricular syndrome	*HMX1* 142992	Oculoauricular syndrome	Microphthalmia	n/a	Schorderet et al. (2008)[1]
Papillorenal syndrome	*PAX2* 167409	Papillorenal syndrome	Coloboma	n/a	Moosajee et al. (2008)[2] and Gregory-Evans, Moosajee, Shan, and Gregory-Evans (2011)[2]
Poretti–Boltshauser syndrome	*LAMA1* 150320	Poretti–Boltshauser syndrome	Coloboma, microphthalmia	n/a	Zinkevich, Bosenko, Link, and Semina (2006)[1], Semina et al. (2006)[1,2], and Bryan, Chien, and Kwan (2016)[2]
Senior–Loken syndrome (SLSN)	*NPHP4* 606966	SLSN	Retinal anomalies	n/a	Burcklé et al. (2011)[1]
	TRAF3IP1 607380	SLSN	Retinal degeneration, prc cilia defects, microphthalmia	n/a	Omori et al. (2008)[1,2] and Bizet et al. (2015)[2]

Continued

Table 4 Zebrafish Models of Syndromes Affecting Eye and/or Ear—cont'd

Syndrome	Human Gene/ OMIM#	Differential Diagnosis	Zebrafish Model Phenotype		Reference
			Eye	Ear	
Syndromic optic atrophy	*RTN4IP1* 610502	Optic atrophy with or without ataxia, mental retardation and seizures	Microphthalmia, absence of retinal ganglion cells and plexiform layers, impaired visual behavior	n/a	Angebault et al. (2015)[1]
	SLC25A46 610826	Syndromic optic atrophy	Optic nerve atrophy	n/a	Abrams et al. (2015)[1]
Syndromic retinitis pigmentosa	*TRNT1* 612907	RP and erythrocytic microcytosis, hearing loss	Microphthalmia, impaired visual behavior	n/a	DeLuca et al. (2016)[1]
Treacher Collins syndrome	*TCOF1* 606847	Treacher Collins syndrome	Microphthalmia	Developmental defects	Weiner, Scampoli, and Calcaterra (2012)[1]
Usher syndrome	*MYO7A* 276903	USH1B, DFNA11 DFNB2	Prc degeneration, reduced visual function	Mechano-transduction defects	Ernest et al. (2000)[2] and Wasfy et al. (2014)[2]
	USH1C 605242	USH1C, DFNB18	Prc degeneration, reduced visual function	Mechano-transduction defects	Phillips et al. (2011)[1,2] and Blanco-Sanchez et al. (2014)[2]
	CDH23 605516	USH1D, DFNB12	n/a	Mechano-transduction defects	Söllner et al. (2004)[2], Glover et al. (2012)[2]; Blanco-Sanchez et al. (2014)[2]
	PCDH15 605514	USH1F, DFNB23	Prc disorganization, reduced visual function	Mechano-transduction defects	Seiler et al. (2005)[1,2]
	CIB2 605564	USH1J, DNFB48	n/a	Mechano-transduction defects	Riazuddin et al. (2012)[1]

	USHRN 608400	USH2A, RP39	Prc degeneration	n.d.	Ebermann et al. (2010)[1]
	ADGRV1 602851	USH2C	Prc degeneration	n.d.	Ebermann et al. (2010)[1]
	CLRN1 606397	USH3A	n/a	Mechano-transduction defects	Ogun and Zallocchi (2014)[1] and Gopal et al. (2015)[2]
	PDZD7 612971	Modifier/digenic mutation in USH2 pathology	Prc degeneration	Hair bundle structural defects	Ebermann et al. (2010)[1]
Syndromes with vision and/or hearing symptoms	*BCL6* 109565	Oculofacio-cardiodental and Lenz microphthalmia syndromes;	Microphthalmia; Coloboma	n/a	Lee, Lee, and Gross (2013)[1]
	BCOR 300485	Oculofacio-cardiodental and Lenz microphthalmia syndromes;	Microphthalmia; Coloboma	n/a	Ng et al. (2004)[1]
	EFTUD2 603892	Mandibulofacial dysostosis, Guion-Almeida type with unusual ocular features	Coloboma, microphthalmia	n/a	Deml, Reis, Muheisen, Bick, and Semina (2015)[3]
	GATA3 131320	Hypopara thyroidism, sensorineural deafness and renal dysplasia	n/a	Reduced number of sensory hair cells	Sheehan-Rooney, Swartz, Zhao, Liu, and Eberhart (2013)[2]

Continued

Table 4 Zebrafish Models of Syndromes Affecting Eye and/or Ear—cont'd

Syndrome	Human Gene/ OMIM#	Differential Diagnosis	Zebrafish Model Phenotype		Reference
			Eye	Ear	
	GJB2 605425	Bart–Pumphrey syndrome, hystrix-like ichthyosis with deafness, keratitis-ichthyosis-deafness syndrome, keratoderma, palmoplantar with deafness, Vohwinkel syndrome, DFNA3, DFNB1	n/a	Sensorineural hearing loss with variable onset and severity	Chang-Chien et al. (2014)[1]
	PLK4 605031	Microcephaly, chorioretinopathy	Microphthalmia, reduced retinal growth	n/a	Martin et al. (2014)[1]
	PNPLA6 603197	Oliver–McFarlane, Laurence–Moon, Boucher–Neuhauser syndromes	Microphthalmia	Small otic vesicle	Song et al. (2013)[1] and Hufnagel et al. (2015)[1]
	RAB18 602207	Warburg micro syndrome	Microphthalmia, coloboma, disrupted retinal development	n/a	Bem et al. (2011)[1]
	SMOC1 608488	Waardenburg anophthalmia syndrome	Microphthalmia, coloboma, disrupted retinal development	n/a	Abouzeid et al. (2011)[1]

	SOX10 602229	Waardenburg syndrome type 2E	n/a.	Semicircular canal dysmorphogenesis, patterning defects	Dutton et al. (2009)[2]
	TUBGCP4 609610	Chorioretinopathy and microcephaly	Microphthalmia, reduced number of pics	n/a	Scheidecker et al. (2015)[1]

Genes represented in the tables were identified by cross-referencing searches in ZFIN (http://zfin.org), the Online Mendelian Inheritance in Man database (www.omim.org), and PubMed (http://www.ncbi.nlm.nih.gov/pubmed). Criteria for inclusion were restricted to zebrafish models of genes associated with human diseases affecting vision and/or hearing for which ocular or otic phenotypes were described. The following numbers indicate the method used to generate the zebrafish model for a given disease. 1, MO, *morpholino; 2,* ENU, N-*ethyl*-N-*nitrosourea; 3,* TALENs, *transcription-activator like effector nucleases; 4,* CRISPRs, *clustered regularly interspaced short palindromic repeats; 5, Viral insertion.* SAG, *Statoacoustic ganglion.*

the proteins with which harmonin interacts (Adato et al., 2005; Bahloul et al., 2010; Pretorius et al., 2010; Reiners, Marker, Jurgens, Reidel, & Wolfrum, 2005; Reiners, van Wijk, et al., 2005; Siemens et al., 2002; Weil et al., 2003; Wu, Pan, Zhang, & Zhang, 2012). In sensory hair cells, the harmonin scaffold protein binds to MYO7A and CDH23. Harmonin is absolutely required for proper trafficking of these proteins from the ER to the tips of the hair bundle stereocilia and is critical for maintenance of the USH protein complex (Blanco-Sanchez et al., 2014; Boeda et al., 2002; Pan, Yan, Wu, & Zhang, 2009). Failure to maintain this complex results in loss of hair bundle integrity and defective mechanotransduction, and ultimately leads to impaired hearing and balance (Grillet et al., 2009; Michalski et al., 2009).

The role of harmonin in the photoreceptors of the eye is more complex. It is not essential for the function of the USH protein complex in vesicle transport at the periciliary region of the photoreceptors. However, it is critical for synaptic integrity and for the connection between the calycal processes and the outer segments of the photoreceptors (Maerker et al., 2008; Phillips et al., 2011; Sahly et al., 2012). In the intestine, similar to its function in the hair cells, harmonin forms a complex with PCDH24, CDHR2, and MYO7b and is necessary for the correct assembly of the intestinal brush border (Crawley et al., 2014; Li, He, Lu, & Zhang, 2016).

In humans, the spectrum of mutations in *USH1C* is large, and most mutations lead to syndromic deafness (USH1C) or nonsyndromic autosomal recessive HL (DFNB18) (Ahmed et al., 2002; Bitner-Glindzicz et al., 2000; Bonnet et al., 2016; Ebermann et al., 2007; Ganapathy et al., 2014; Kimberling et al., 2010; Ouyang et al., 2002; Verpy et al., 2000). In addition, mutations in *USH1C* have been associated with USH2, atypical USH, sector RP with severe HL, and RP with late onset of deafness (Khateb et al., 2012; Le Quesne Stabej et al., 2012; Saihan et al., 2011).

In 5 dpf zebrafish larvae, *ush1c* is present in the sensory hair cells of the inner ear and the neuromasts, the Müller cells of the retina, and the intestine (Blanco-Sanchez et al., 2014; Phillips et al., 2011). It is also found in adult photoreceptors (Phillips et al., 2011). Like other *ush1* zebrafish mutants, vestibular function and hair bundle integrity are disrupted in *ush1c* depleted larvae (Blanco-Sanchez et al., 2014; Phillips et al., 2011) (Fig. 1). Further investigation of the molecular etiology of these phenotypes in *ush1c* mutants showed that harmonin is a critical protein necessary for preassembly of Cdh23 and Myo7aa into a complex at the level of the ER, and as described above, mutations in *ush1c* prevent formation of the complex and its transport to the stereocilia, as well as ER stress, a likely cause of hair cell degeneration (Blanco-Sanchez et al., 2014). Visual function is also affected in *ush1c* mutants, and maturation of the photoreceptor ribbon synapse is abnormal leading to a decrease in synaptic transmission (Phillips et al., 2011) (Fig. 2). Although the expression of *ush1c* in the intestine correlates with a possible conserved role of harmonin in the intestinal brush border, its function has not been studied in this tissue.

In summary, the zebrafish *ush1c* mutant not only recapitulates both the inner ear and retinal phenotypes of USH1 patients, making it an excellent model to study

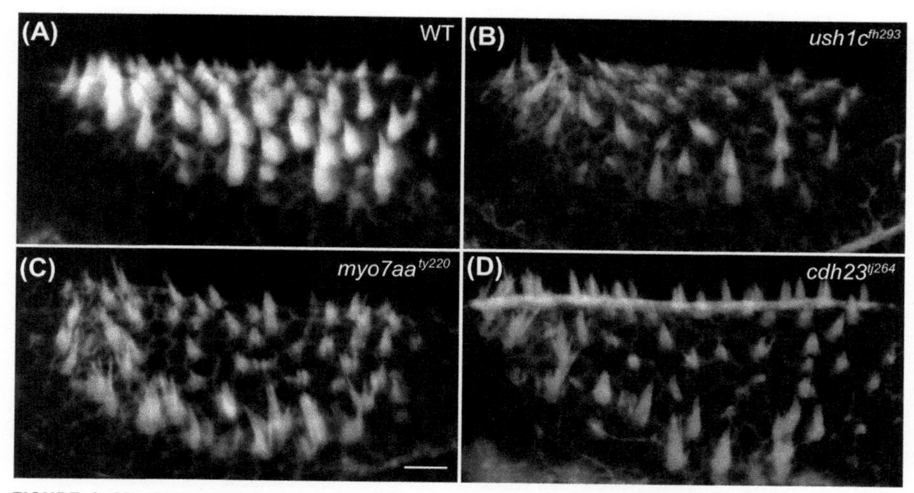

FIGURE 1 Mechanoreceptor structural defects in *ush1* zebrafish mutants.

Phalloidin staining of the anterior macula at 5 dpf in wild-type (A) and *ush1c*^{fh293} (B), *myo7aa*^{ty220} (C), and *cdh23*^{tj264} (D) homozygous mutant larvae.

Modified from Blanco-Sanchez, B., Clement, A., Fierro, J., Jr, Washbourne, P., & Westerfield, M. (2014). Complexes of Usher proteins preassemble at the endoplasmic reticulum and are required for trafficking and ER homeostasis. Disease Models & Mechanisms, *7, 547–559.*

USH1C, but it also provides insights into the pathways affected in this disease and the mechanism of sensory cell degeneration.

*3.1.2 USH1B (*MYO7A*)*

MYO7A is another member of the USH family of genes. Mutations in *MYO7A*, which encodes an unconventional myosin, cause USH1B. In sensory hair cells, *MYO7A* is involved in hair bundle integrity (Ernest et al., 2000; Inoue & Ikebe, 2003; Nicolson et al., 1998; Udovichenko, Gibbs, & Williams, 2002). Together with other USH1 proteins, MYO7A is thought to anchor the tip links (formed by CDH23 and PCDH15) within the stereocilia (El-Amraoui & Petit, 2014; Richardson et al., 2011). Absence or malfunction of this motor protein results in disorganization of the hair bundle stereocilia (Self et al., 1998). In the eye, MYO7A has been implicated in transport of RPE melanosomes, the transport of opsin in the photoreceptor, and phagocytosis of disk membranes (Gibbs et al., 2004; Gibbs, Kitamoto, Williams, 2003; Liu et al., 1998).

Like *USH1C*, mutations in *MYO7A* can give rise to different symptoms based on their nature and position. Mutations typically lead to USHIB, autosomal dominant HL (DFNA11), or autosomal recessive HL (DFNB2) (Bolz et al., 2004; Di Leva et al., 2006; Duman, Sirmaci, Cengiz, Ozdag, & Tekin, 2011; Kimberling et al., 2010; Liu, Newton, Steel, & Brown, 1997; Liu, Walsh, et al., 1997; Mutai et al., 2013; Riazuddin et al., 2008; Roux et al., 2011; Street, Kallman, & Kiemele,

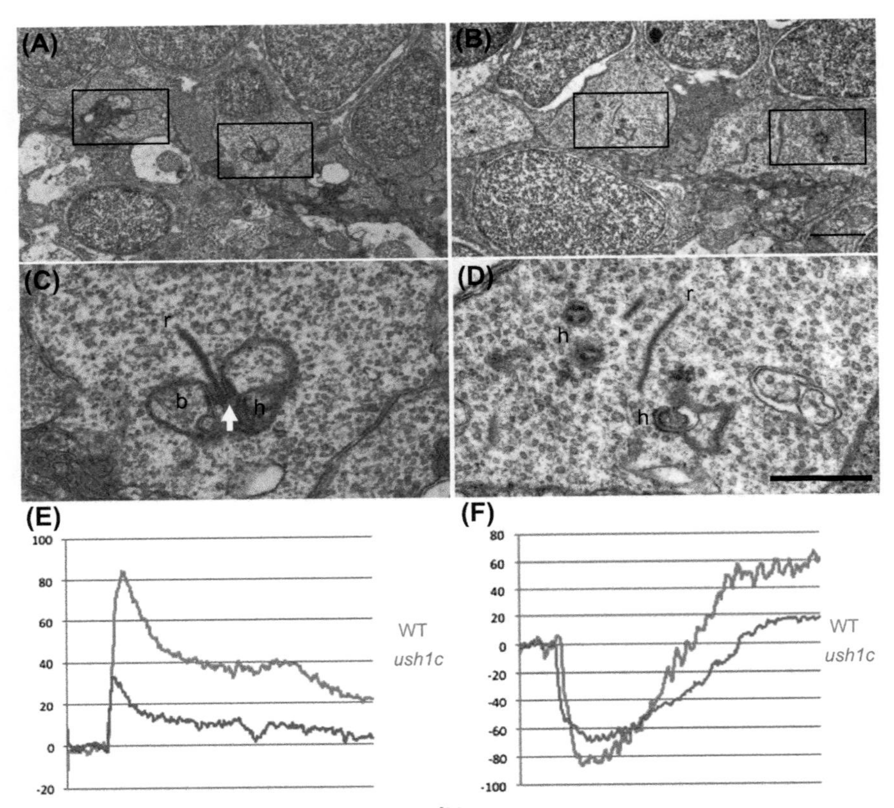

FIGURE 2 Maturation of the photoreceptors ribbon synapses.

Transmission electron micrographs of cone pedicles at 6 dpf in wild-type (A,C) and *ush1c* morphant (B,D) larvae. Multiple triads (boxed areas in A and B and enlargements in C and D) consisting of a synaptic ribbon (r), arciform density (*arrow*) and postsynaptic processes from bipolar (b) and horizontal (h) cells indicate synaptic maturation. Scale bars 1 μm (A,B) and 500 nm (C,D). Electroretinogram recordings at 5 dpf showing b-waves (E) and a-wave (F) amplitudes from wild-type (uninjected control) and *ush1c* morphant larvae.

Modified from Phillips, J. B., Blanco-Sanchez, B., Lentz, J. J., Tallafuss, A., Khanobdee, K., Sampath, S., ..., Westerfield, M. (2011). Harmonin (Ush1c) is required in zebrafish Muller glial cells for photoreceptor synaptic development and function. Disease Models & Mechanisms, 4, *786–800.*

2004; Sun et al., 2011; Weil et al., 1995, 1997). In other cases, however, mutations in *MYO7A* have been associated with USH2 or atypical USH (Bonnet et al., 2011; Le Quesne Stabej et al., 2012; Roberts, George, Greenberg, & Ramesar, 2015; Rong et al., 2014; Roux et al., 2011; Zhai et al., 2015).

Zebrafish have two *MYO7A* genes, *myo7aa* and *myo7ab*. *myo7aa* is better characterized. It is expressed in the sensory hair cells of the inner ear as early as 24 hpf, when the first hair cells start to differentiate, and expression persists throughout their development (Ernest et al., 2000). It is also detected in the neuromasts from 48 hpf (Ernest et al., 2000). Myo7aa is localized in both the cytoplasm and along the stereocilia of the hair bundle (Blanco-Sanchez et al., 2014; Coffin, Dabdoub, Kelley, & Popper, 2007). In the eye, *myo7aa* is expressed in the inner and outer nuclear layers (Wasfy, Matsui, Miller, Dowling, & Perkins, 2014). The protein is also found in the accessory outer segment of cone photoreceptors (Hodel et al., 2014).

Multiple alleles of *myo7aa* (*mariner*) mutants have been analyzed and all display vestibular dysfunction (Nicolson et al., 1998). The splayed stereocilia of the hair bundles and the resulting defective mechanotransduction are responsible for this phenotype (Ernest et al., 2000; Nicolson et al., 1998; Seiler & Nicolson, 1999) (Fig. 1). A recent study described the retinal phenotypes in *myo7aa* mutants that include diminished visual function and degeneration of rod photoreceptors (Wasfy et al., 2014). As observed in *Myo7a* mouse mutants, opsins are not correctly localized; rhodopsin accumulates near the connecting cilium and blue cone opsin partially mislocalizes in the inner segment.

*3.1.3 USH1D (*CDH23*) and USH1F (*PCDH15*)*

CDH23 and *PCDH15* encode transmembrane proteins known for their role in maintaining the integrity of the inner ear hair cell mechanoreceptors (Boeda et al., 2002; Di Palma et al., 2001; Kikkawa, Pawlowski, Wright, & Alagramam, 2008; Seiler et al., 2005; Söllner et al., 2004). Specifically, CDH23 together with PCDH15 forms the tip links that connect stereocilia together (Kazmierczak et al., 2007). Preservation of these links is critical for mechanosensation, and any disruption leads to defective hearing and balance (El-Amraoui & Petit, 2005).

As with other USH genes, mutations in *CDH23* can give rise to syndromic or nonsyndromic deafness. In most cases, splice site, frameshift, and nonsense mutations give rise to USHID, whereas missense mutations are more likely to generate autosomal recessive HL (DFNB12) (Ammar-Khodja et al., 2009; Astuto et al., 2002; Bork et al., 2001). Interestingly, a few patients with atypical USH (milder symptoms in either hearing, balance, or vision), USH2, or USH3 have also been reported with mutations in *CDH23* (Astuto et al., 2002; Besnard et al., 2014; Bonnet et al., 2011; Le Quesne Stabej et al., 2012). Similarly, mutations in *PCDH15* can lead to USH1F or autosomal recessive HL (DFNB23), with missense mutations tending to cause DFNB23 (Ahmed et al., 2003, 2008; Doucette et al., 2009).

In zebrafish, a panoply of mutants for *cdh23* has been generated and studied. Based on several behavioral assays (Table 1), the *cdh23* (*sputnik*) mutants were characterized with auditory and vestibular defects (Nicolson et al., 1998). Further studies of the mutants, focusing on the morphology of the mechanoreceptors in hair cells of the crista, showed that the stereocilia are detached from the kinocilium and splayed (Nicolson et al., 1998; Söllner et al., 2004) (Fig. 1). Microphonic potentials that measure the extracellular voltage response are absent in the mutants, and the FM1-43 dye, a marker of cycling vesicles, does not incorporate into the hair cells by endocytosis (Nicolson et al., 1998; Seiler & Nicolson, 1999), consistent with a disruption in mechanotransduction.

Although all the data point toward zebrafish *cdh23* mutants being excellent models to study mechanotransduction defects, several observations suggest they may not be appropriate for studies of combined deafness and blindness. First, using in situ hybridization techniques, *cdh23* expression was found in a subset of GABAergic amacrine cells, but not in photoreceptors of larval zebrafish eyes (Glover et al., 2012). Second, optokinetic reflex measurements (OKR) showed normal visual function for two alleles of *cdh23* (Glover et al., 2012; Mo, Chen, Nechiporuk, & Nicolson, 2010). One of the alleles, however, was a missense mutation, and as discussed above, missense mutations in humans typically lead to HL only whereas nonsense or frameshift mutations generate USH1D. It is therefore possible that the alleles studied are not adequate models of USH. In addition, the behavioral analysis in this study was conducted at 5 dpf. At the same stage, *myo7aa* mutants do not yet show retinal cell death, and visual function assayed by electroretinogram and the OKR is only diminished (Wasfy et al., 2014). Experiments at later time points could help determine whether *cdh23* mutants can be used as models for USH.

Zebrafish have two orthologs of human *PCDH15*, *pcdh15a* and *pcdh15b*. *pcdh15a* is expressed in the eye, mechanosensory hair cells, and brain, whereas the *pcdh15b* paralog is expressed in photoreceptors, the brain, and weakly in the inner ear and neuromasts (Seiler et al., 2005).

Alleles of *pcdh15a* (*orbiter*) were isolated in the same screen for vestibular dysfunction as the *cdh23* (*sputnik*) mutants (Nicolson et al., 1998). Similar to other *ush1* mutants, *pcdh15a* depleted larvae display splayed stereocilia at 5 dpf and disrupted mechanotransduction (Nicolson et al., 1998; Seiler et al., 2005). At 4 dpf, the OKR is normal. Interestingly, the complementary phenotype is observed in *pcdh15b* morphant larvae. Vestibular function and mechanotransduction are unaffected, but visual function is impaired (Seiler et al., 2005). With the lack of an OKR defect, *pcdh15a* mutants do not seem to recapitulate the vision defects of USH1F patients. However, like *cdh23* mutants, visual function was assayed early in *pcdh15a* mutants, leaving open the question of whether they provide a good model for USH1F.

3.1.4 Potential models of USH

3.1.4.1 Usher syndrome type 2

Zebrafish mutants and morphants of USH2 (*ush2a* and *adgrv1*) and USH3 (*clrn1*) genes have also been generated. Although comprehensive studies are still incomplete (i.e., only visual or vestibulo-acoustic function has been reported), the results are promising and suggest these mutants may potentially be useful models of USH.

Zebrafish injected with MOs against *ush2a* or *adgrv1* (USH2C) (Ebermann et al., 2010) exhibit retinal degeneration in the first week of life. No vestibular defects were described. Depleting either of these genes in conjunction with knockdown of the scaffold protein-encoding gene *pdzd7* exacerbates the retinal symptoms, providing supporting evidence for clinical findings implicating *PDZD7* as a modifier of USH2. USH2 is unique among the three subtypes for its lack of reported vestibular dysfunction, so there is reasonable hope that zebrafish USH2 models may be spared the lethal swimming and balance problems characteristic of the USH1 and USH3 models described here. This could allow for long-term assays of visual function and retinal cell loss as well as analysis of auditory function uncoupled from the vestibular system.

3.1.4.2 Usher syndrome type 3

USH3A is caused by mutations in the tetraspanin protein-encoding gene *CLRN1*. Investigations of zebrafish morphants and mutants (Gopal et al., 2015; Ogun & Zallocchi, 2014) have demonstrated the conserved role of *clrn1* in hearing and balance, with both models recapitulating the progressive onset of symptoms. Retinal assays were not described by either group, but Clrn1 is present in larval and adult eyes (Phillips et al., 2013), suggesting a potential for a retinal phenotype in loss-of-function models. The late onset of hearing and balance defects in USH3 patients presents a novel window for preventative therapies, and the zebrafish *clrn1* model could serve as an important component of drug discovery.

3.2 JOUBERT SYNDROME

Zebrafish were used in a candidate gene study and clinical validation of the centrosomal protein-encoding gene *POC1B* as causative of Joubert syndrome (JBTS) and the nonsyndromic retinal ciliopathy Leber congenital amaurosis. Two studies of JBTS patients with loss-of-function mutations in *POC1B* (Beck et al., 2014; Roosing et al., 2014) used zebrafish to validate genotype–phenotype correlation in ciliopathy patients. Another MO-based study demonstrated retinal defects by *poc1b* depletion in zebrafish (Zhang, Zhang, Wang, & Liu, 2015).

Interestingly, although the clinical symptoms of JBTS and other ciliopathies manifest in a broad range of ciliated cell populations, SNHL is infrequently reported (reviewed in Waters & Beales, 2011). In contrast, there are numerous examples of ciliopathy gene defects affecting zebrafish ear morphology, often including defective otolith development (Beck et al., 2014; Kim et al., 2013; Lee et al., 2012; Omori et al., 2008; Simms et al., 2012; see Table 4). The reason for this discrepancy is unknown, but ciliopathies encompass a range of severe neurological and systemic defects, which may make hearing or balance difficulties more challenging to diagnose. In support of this, a recent study of patients with primary ciliary dyskenesis that was motivated by the otolith defects observed in zebrafish ciliopathy models revealed that PCD patients have measurable vestibular defects (Rimmer et al., 2015).

CONCLUSION

Mutagenesis screens have generated a broad array of loss-of-function models that shed light on numerous mechanisms of eye and ear development and disease. Despite the numerous disease models presented in this chapter, there is still a significant discrepancy between the number of distinct genetic diseases and the availability of tractable mutant or MO models. To date, 250 genes have been associated with vision loss in humans (OMIM, 2016), but only 25 mutant models and 64 MO zebrafish models targeting 79 genes have been studied for vision defects (Tables 2 and 4). Similarly, 115 genes are linked to deafness in humans (OMIM, 2016), but only 22 mutant and 24 MO zebrafish models targeting 40 genes have been studied for hearing defects (Tables 3 and 4). In addition, the rate of discovery of disease-causing genes is accelerating due to advanced sequencing methods in use clinically. Continued efforts of the zebrafish research community are vital not only for gene discovery, but also to increase our understanding of the cellular and molecular processes involved in vision and hearing. The Sanger zebrafish mutation project and the new technologies of genome editing provide essential resources for generating new models (Ata, Clark, & Ekker, 2016; Busch-Nentwich et al., 2013; Kettleborough et al., 2013). Additional resources, such as high-throughput methods for efficient assays of vision and hearing impairment and new tools to study the molecular pathology in mutants, will also be helpful. The utility of behavioral assays for auditory and vestibular dysfunction was demonstrated in forward and reverse screens and mutant analyses (Abbas & Whitfield, 2009; Han et al., 2011; Kindt, Finch, & Nicolson, 2012; Malicki et al., 1996; Nicolson et al., 1998; Varshney et al., 2015; Whitfield et al., 1996; Yao, DeSmidt, Tekin, Liu, & Lu, 2016).

Collectively, increased use and optimization of gene editing techniques in the zebrafish research community, alongside increased genotypic data emerging from human patients and more collaborations between clinicians and model organism researchers, will produce new models of vision and hearing disorders that will lead to better diagnoses and treatments of human disease.

REFERENCES

Abbas, L., & Whitfield, T. T. (2009). Nkcc1 (Slc12a2) is required for the regulation of endolymph volume in the otic vesicle and swim bladder volume in the zebrafish larva. *Development, 136*, 2837–2848.

Abdelhak, S., Kalatzis, V., Heilig, R., Compain, S., Samson, D., Vincent, C., … Petit, C. (1997). A human homologue of the Drosophila eyes absent gene underlies branchio-oto-renal (BOR) syndrome and identifies a novel gene family. *Nature Genetics, 15*, 157–164.

Abouzeid, H., Boisset, G., Favez, T., Youssef, M., Marzouk, I., Shakankiry, N., … Schorderet, D. F. (2011). Mutations in the SPARC-related modular calcium-binding protein 1 gene, SMOC1, cause waardenburg anophthalmia syndrome. *American Journal of Human Genetics, 88*, 92–98.

Abrams, A. J., Hufnagel, R. B., Rebelo, A., Zanna, C., Patel, N., Gonzalez, M. A., … Dallman, J. E. (2015). Mutations in SLC25A46, encoding a UGO1-like protein, cause an optic atrophy spectrum disorder. *Nature Genetics, 47*, 926–932.

Adato, A., Michel, V., Kikkawa, Y., Reiners, J., Alagramam, K. N., Weil, D., ... Petit, C. (2005). Interactions in the network of Usher syndrome type 1 proteins. *Human Molecular Genetics, 14*, 347–356.

Adato, A., Raskin, L., Petit, C., & Bonne-Tamir, B. (2000). Deafness heterogeneity in a Druze isolate from the Middle East: novel OTOF and PDS mutations, low prevalence of GJB2 35delG mutation and indication for a new DFNB locus. *European Journal of Human Genetics, 8*, 437–442.

Agrawal, Y., Carey, J. P., Della Santina, C. C., Schubert, M. C., & Minor, L. B. (2009). Disorders of balance and vestibular function in US adults: data from the National Health and Nutrition Examination Survey, 2001–2004. *Archives of Internal Medicine, 169*, 938–944.

Ahmed, Z. M., Riazuddin, S., Ahmad, J., Bernstein, S. L., Guo, Y., Sabar, M. F., ... Wilcox, E. R. (2003). PCDH15 is expressed in the neurosensory epithelium of the eye and ear and mutant alleles are responsible for both USH1F and DFNB23. *Human Molecular Genetics, 12*, 3215–3223.

Ahmed, Z. M., Riazuddin, S., Aye, S., Ali, R. A., Venselaar, H., Anwar, S., ... Friedman, T. B. (2008). Gene structure and mutant alleles of PCDH15: nonsyndromic deafness DFNB23 and type 1 Usher syndrome. *Human Genetics, 124*, 215–223.

Ahmed, Z. M., Riazuddin, S., Friedman, T. B., Riazuddin, S., Wilcox, E. R., & Griffith, A. J. (2002). Clinical manifestations of DFNB29 deafness. *Advances in Oto-Rhino-Laryngology, 61*, 156–160.

Al-Hamed, M. H., van Lennep, C., Hynes, A. M., Chrystal, P., Eley, L., Al-Fadhly, F., ... Sayer, J. A. (2014). Functional modelling of a novel mutation in BBS5. *Cilia, 3*, 3.

Ammar-Khodja, F., Faugere, V., Baux, D., Giannesini, C., Leonard, S., Makrelouf, M., ... Roux, A. F. (2009). Molecular screening of deafness in Algeria: high genetic heterogeneity involving DFNB1 and the Usher loci, DFNB2/USH1B, DFNB12/USH1D and DFNB23/USH1F. *European Journal of Medical Genetics, 52*, 174–179.

Angebault, C., Guichet, P. O., Talmat-Amar, Y., Charif, M., Gerber, S., Fares-Taie, L., ... Lenaers, G. (2015). Recessive mutations in RTN4IP1 cause isolated and syndromic optic neuropathies. *American Journal of Human Genetics, 97*, 754–760.

Asai-Coakwell, M., March, L., Dai, X. H., Duval, M., Lopez, I., French, C. R., ... Lehmann, O. J. (2013). Contribution of growth differentiation factor 6-dependent cell survival to early-onset retinal dystrophies. *Human Molecular Genetics, 22*, 1432–1442.

Asai-Coakwell, M., French, C. R., Berry, K. M., Ye, M., Koss, R., Somerville, M., ... Lehmann, O. J. (2007). GDF6, a novel locus for a spectrum of ocular developmental anomalies. *American Journal of Human Genetics, 80*, 306–315.

Astuto, L. M., Bork, J. M., Weston, M. D., Askew, J. W., Fields, R. R., Orten, D. J., ... Kimberling, W. J. (2002). CDH23 mutation and phenotype heterogeneity: a profile of 107 diverse families with Usher syndrome and nonsyndromic deafness. *American Journal of Human Genetics, 71*, 262–275.

Ata, H., Clark, K. J., & Ekker, S. C. (2016). The zebrafish genome editing toolkit. *Methods in Cell Biology, 135*, 149–170.

Bachmann-Gagescu, R., Phelps, I. G., Stearns, G., Link, B. A., Brockerhoff, S. E., Moens, C. B., & Doherty, D. (2011). The ciliopathy gene cc2d2a controls zebrafish photoreceptor outer segment development through a role in Rab8-dependent vesicle trafficking. *Human Molecular Genetics, 20*, 4041–4055.

Baehr, W., Wu, S. M., Bird, A. C., & Palczewski, K. (2003). The retinoid cycle and retina disease. *Vision Research, 43*, 2957–2958.

Bahadori, R., Biehlmaier, O., Zeitz, C., Labhart, T., Makhankov, Y. V., Forster, U., ... Neuhauss, S. C. (2006). Nyctalopin is essential for synaptic transmission in the cone dominated zebrafish retina. *European Journal of Neuroscience, 24*, 1664–1674.

Bahloul, A., Michel, V., Hardelin, J. P., Nouaille, S., Hoos, S., Houdusse, A., ... Petit, C. (2010). Cadherin-23, myosin VIIa and harmonin, encoded by Usher syndrome type I genes, form a ternary complex and interact with membrane phospholipids. *Human Molecular Genetics, 19*, 3557–3565.

Balasubramanian, R., Choi, J. H., Francescatto, L., Willer, J., Horton, E. R., Asimacopoulos, E. P., ... Crowley, W. F., Jr. (2014). Functionally compromised CHD7 alleles in patients with isolated GnRH deficiency. *Proceedings of National Academy of Sciences of the United States of America, 111*, 17953–17958.

Balow, S. A., Pierce, L. X., Zentner, G. E., Conrad, P. A., Davis, S., Sabaawy, H. E., ... Scacheri, P. C. (2013). Knockdown of fbxl10/kdm2bb rescues chd7 morphant phenotype in a zebrafish model of CHARGE syndrome. *Developmental Biology, 382*, 57–69.

Baxendale, S., & Whitfield, T. T. (2016). Methods to study the development, anatomy, and function of the zebrafish inner ear across the life course. *Methods in Cell Biology, 134*, 165–209.

Baye, L. M., Patrinostro, X., Swaminathan, S., Beck, J. S., Zhang, Y., Stone, E. M., ... Slusarski, D. C. (2011). The N-terminal region of centrosomal protein 290 (CEP290) restores vision in a zebrafish model of human blindness. *Human Molecular Genetics, 20*, 1467–1477.

Beck, J. C., Gilland, E., Tank, D. W., & Baker, R. (2004). Quantifying the ontogeny of optokinetic and vestibuloocular behaviors in zebrafish, medaka, and goldfish. *Journal of Neurophysiology, 92*, 3546–3561.

Beck, B. B., Phillips, J. B., Bartram, M. P., Wegner, J., Thoenes, M., Pannes, A., ... Bolz, H. J. (2014). Mutation of POC1B in a severe syndromic retinal ciliopathy. *Human Mutation, 35*, 1153–1162.

Bem, D., Yoshimura, S., Nunes-Bastos, R., Bond, F. C., Kurian, M. A., Rahman, F., ... Aligianis, I. A. (2011). Loss-of-function mutations in RAB18 cause Warburg micro syndrome. *American Journal of Human Genetics, 88*, 499–507.

Besnard, T., Garcia-Garcia, G., Baux, D., Vache, C., Faugere, V., Larrieu, L., ... Roux, A. F. (2014). Experience of targeted Usher exome sequencing as a clinical test. *Molecular Genetics and Genomic Medicine, 2*, 30–43.

Bessarab, D. A., Chong, S. W., & Korzh, V. (2004). Expression of zebrafish six1 during sensory organ development and myogenesis. *Developmental Dynamics, 230*, 781–786.

Bianco, I. H., Ma, L. H., Schoppik, D., Robson, D. N., Orger, M. B., Beck, J. C., ... Baker, R. (2012). The tangential nucleus controls a gravito-inertial vestibulo-ocular reflex. *Current Biology, 22*, 1285–1295.

Bilotta, J. (2000). Effects of abnormal lighting on the development of zebrafish visual behavior. *Behavioural Brain Research, 116*, 81–87.

Bitner-Glindzicz, M., Lindley, K. J., Rutland, P., Blaydon, D., Smith, V. V., Milla, P. J., ... Glaser, B. (2000). A recessive contiguous gene deletion causing infantile hyperinsulinism, enteropathy and deafness identifies the Usher type 1C gene. *Nature Genetics, 26*, 56–60.

Bizet, A. A., Becker-Heck, A., Ryan, R., Weber, K., Filhol, E., Krug, P., … Saunier, S. (2015). Mutations in TRAF3IP1/IFT54 reveal a new role for IFT proteins in microtubule stabilization. *Nature Communications, 6*, 8666.

Blanco-Sanchez, B., Clement, A., Fierro, J., Jr., Washbourne, P., & Westerfield, M. (2014). Complexes of Usher proteins preassemble at the endoplasmic reticulum and are required for trafficking and ER homeostasis. *Disease Models & Mechanisms, 7*, 547–559.

Boeda, B., El-Amraoui, A., Bahloul, A., Goodyear, R., Daviet, L., Blanchard, S., … Petit, C. (2002). Myosin VIIa, harmonin and cadherin 23, three Usher I gene products that cooperate to shape the sensory hair cell bundle. *EMBO Journal, 21*, 6689–6699.

Bolz, H., Bolz, S. S., Schade, G., Kothe, C., Mohrmann, G., Hess, M., & Gal, A. (2004). Impaired calmodulin binding of myosin-7A causes autosomal dominant hearing loss (DFNA11). *Human Mutation, 24*, 274–275.

Bonnet, C., Grati, M., Marlin, S., Levilliers, J., Hardelin, J. P., Parodi, M., … Denoyelle, F. (2011). Complete exon sequencing of all known Usher syndrome genes greatly improves molecular diagnosis. *Orphanet Journal of Rare Diseases, 6*, 21.

Bonnet, C., Riahi, Z., Chantot-Bastaraud, S., Smagghe, L., Letexier, M., Marcaillou, C., … Petit, C. (2016). An innovative strategy for the molecular diagnosis of Usher syndrome identifies causal biallelic mutations in 93% of European patients. *European Journal of Human Genetics*.

Borck, G., Ur Rehman, A., Lee, K., Pogoda, H. M., Kakar, N., von Ameln, S., … Kubisch, C. (2011). Loss-of-function mutations of ILDR1 cause autosomal-recessive hearing impairment DFNB42. *American Journal of Human Genetics, 88*, 127–137.

Bork, J. M., Peters, L. M., Riazuddin, S., Bernstein, S. L., Ahmed, Z. M., Ness, S. L., … Morell, R. J. (2001). Usher syndrome 1D and nonsyndromic autosomal recessive deafness DFNB12 are caused by allelic mutations of the novel cadherin-like gene CDH23. *American Journal of Human Genetics, 68*, 26–37.

Branchek, T. (1984). The development of photoreceptors in the zebrafish, *Brachydanio rerio*. II. Function. *Journal of Comparitive Neurology, 224*, 116–122.

Bricaud, O., & Collazo, A. (2006). The transcription factor six1 inhibits neuronal and promotes hair cell fate in the developing zebrafish (*Danio rerio*) inner ear. *Journal of Neuroscience, 26*, 10438–10451.

Bricaud, O., & Collazo, A. (2011). Balancing cell numbers during organogenesis: six1a differentially affects neurons and sensory hair cells in the inner ear. *Developmental Biology, 357*, 191–201.

Brockerhoff, S. E. (2006). Measuring the optokinetic response of zebrafish larvae. *Nature Protocols, 1*, 2448–2451.

Brockerhoff, S. E., Hurley, J. B., Janssen-Bienhold, U., Neuhauss, S. C., Driever, W., & Dowling, J. E. (1995). A behavioral screen for isolating zebrafish mutants with visual system defects. *Proceedings of National Academy of Sciences of the United States of America, 92*, 10545–10549.

Brockerhoff, S. E., Rieke, F., Matthews, H. R., Taylor, M. R., Kennedy, B., Ankoudinova, I., … Hurley, J. B. (2003). Light stimulates a transducin-independent increase of cytoplasmic Ca2+ and suppression of current in cones from the zebrafish mutant nof. *The Journal of Neuroscience, 23*, 470–480.

Bryan, C. D., Chien, C. B., & Kwan, K. M. (2016). Loss of laminin alpha 1 results in multiple structural defects and divergent effects on adhesion during vertebrate optic cup morphogenesis. *Developmental Biology, 416*, 324–337.

Bujakowska, K. M., Zhang, Q., Siemiatkowska, A. M., Liu, Q., Place, E., Falk, M. J., … Pierce, E. A. (2015). Mutations in IFT172 cause isolated retinal degeneration and Bardet-Biedl syndrome. *Human Molecular Genetics, 24*, 230–242.

Buller, C., Xu, X., Marquis, V., Schwanke, R., & Xu, P. X. (2001). Molecular effects of Eya1 domain mutations causing organ defects in BOR syndrome. *Human Molecular Genetics, 10*, 2775–2781.

Burckle, C., Gaude, H. M., Vesque, C., Silbermann, F., Salomon, R., Jeanpierre, C., … Schneider-Maunoury, S. (2011). Control of the Wnt pathways by nephrocystin-4 is required for morphogenesis of the zebrafish pronephros. *Human Molecular Genetics, 20*, 2611–2627.

Burgess, H. A., Schoch, H., & Granato, M. (2010). Distinct retinal pathways drive spatial orientation behaviors in zebrafish navigation. *Current Biology, 20*, 381–386.

Busch-Nentwich, E., Kettleborough, R., Dooley, C. M., Scahill, C., Sealy, I., White, R., … Stemple, D. L. (2013). *Sanger Institute Zebrafish Mutation Project mutant data submission*. ZFIN Direct Data Submission http://zfin.org/.

Busch-Nentwich, E., Sollner, C., Roehl, H., & Nicolson, T. (2004). The deafness gene dfna5 is crucial for ugdh expression and HA production in the developing ear in zebrafish. *Development, 131*, 943–951.

Carnes, M. U., Liu, Y. P., Allingham, R. R., Whigham, B. T., Havens, S., Garrett, M. E., … Hauser, M. A. (2014). Discovery and functional annotation of SIX6 variants in primary open-angle glaucoma. *PLoS Genetics, 10*, e1004372.

Casselbrant, M. L., Villardo, R. J., & Mandel, E. M. (2008). Balance and otitis media with effusion. *International Journal of Audiology, 47*, 584–589.

Chakarova, C. F., Khanna, H., Shah, A. Z., Patil, S. B., Sedmak, T., Murga-Zamalloa, C. A., … Bhattacharya, S. S. (2011). TOPORS, implicated in retinal degeneration, is a cilia-centrosomal protein. *Human Molecular Genetics, 20*, 975–987.

Chaki, M., Airik, R., Ghosh, A. K., Giles, R. H., Chen, R., Slaats, G. G., … Hildebrandt, F. (2012). Exome capture reveals ZNF423 and CEP164 mutations, linking renal ciliopathies to DNA damage response signaling. *Cell, 150*, 533–548.

Chang-Chien, J., Yen, Y. C., Chien, K. H., Li, S. Y., Hsu, T. C., & Yang, J. J. (2014). The connexin 30.3 of zebrafish homologue of human connexin 26 may play similar role in the inner ear. *Hearing Research, 313*, 55–66.

Chao, R., Nevin, L., Agarwal, P., Riemer, J., Bai, X., Delaney, A., … Slavotinek, A. (2010). A male with unilateral microphthalmia reveals a role for TMX3 in eye development. *PLoS One, 5*, e10565.

Chatterjee, P., Padmanarayana, M., Abdullah, N., Holman, C. L., LaDu, J., Tanguay, R. L., & Johnson, C. P. (2015). Otoferlin deficiency in zebrafish results in defects in balance and hearing: rescue of the balance and hearing phenotype with full-length and truncated forms of mouse otoferlin. *Molecular and Cellular Biology, 35*, 1043–1054.

Chen, X., Liu, Y., Sheng, X., Tam, P. O., Zhao, K., Chen, X., … Zhao, C. (2014). PRPF4 mutations cause autosomal dominant retinitis pigmentosa. *Human Molecular Genetics, 23*, 2926–2939.

Chhetri, J., Jacobson, G., & Gueven, N. (2014). Zebrafish—on the move towards ophthalmological research. *Eye, 28*, 367–380.

Christy, J. B., Payne, J., Azuero, A., & Formby, C. (2014). Reliability and diagnostic accuracy of clinical tests of vestibular function for children. *Pediatric Physical Therapy, 26*, 180–189.

Coffin, A. B., Dabdoub, A., Kelley, M. W., & Popper, A. N. (2007). Myosin VI and VIIa distribution among inner ear epithelia in diverse fishes. *Hearing Research, 224*, 15–26.

Collery, R., McLoughlin, S., Vendrell, V., Finnegan, J., Crabb, J. W., Saari, J. C., & Kennedy, B. N. (2008). Duplication and divergence of zebrafish CRALBP genes uncovers novel role for RPE- and Muller-CRALBP in cone vision. *Investigative Ophthalmology & Visual Science, 49*, 3812–3820.

Collin, R. W., Nikopoulos, K., Dona, M., Gilissen, C., Hoischen, A., Boonstra, F. N., … Cremers, F. P. (2013). ZNF408 is mutated in familial exudative vitreoretinopathy and is crucial for the development of zebrafish retinal vasculature. *Proceedings of National Academy of Sciences of the United States of America, 110*(24), 9856–9861.

Connor, C. M., Craig, H. K., Raudenbush, S. W., Heavner, K., & Zwolan, T. A. (2006). The age at which young deaf children receive cochlear implants and their vocabulary and speech-production growth: is there an added value for early implantation? *Ear and Hearing, 27*, 628–644.

Connor, C. M., Hieber, S., Arts, H. A., & Zwolan, T. A. (2000). Speech, vocabulary, and the education of children using cochlear implants: oral or total communication? *Journal of Speech, Language, and Hearing Research, 43*, 1185–1204.

Corey, D. P., Garcia-Anoveros, J., Holt, J. R., Kwan, K. Y., Lin, S. Y., Vollrath, M. A., … Zhang, D. S. (2004). TRPA1 is a candidate for the mechanosensitive transduction channel of vertebrate hair cells. *Nature, 432*, 723–730.

Crawley, S. W., Shifrin, D. A., Jr., Grega-Larson, N. E., McConnell, R. E., Benesh, A. E., Mao, S., … Tyska, M. J. (2014). Intestinal brush border assembly driven by protocadherin-based intermicrovillar adhesion. *Cell, 157*, 433–446.

Cushing, S. L., Papsin, B. C., Rutka, J. A., James, A. L., & Gordon, K. A. (2008). Evidence of vestibular and balance dysfunction in children with profound sensorineural hearing loss using cochlear implants. *The Laryngoscope, 118*, 1814–1823.

De La Cruz, E. M., Ostap, E. M., & Sweeney, H. L. (2001). Kinetic mechanism and regulation of myosin VI. *Journal of Biological Chemistry, 276*, 32373–32381.

Deans, M. R., Peterson, J. M., & Wong, G. W. (2010). Mammalian Otolin: a multimeric glycoprotein specific to the inner ear that interacts with otoconial matrix protein Otoconin-90 and Cerebellin-1. *PLoS One, 5*, e12765.

Delmaghani, S., Aghaie, A., Bouyacoub, Y., El Hachmi, H., Bonnet, C., Riahi, Z., … Petit, C. (2016). Mutations in CDC14A, encoding a protein phosphatase involved in hair cell ciliogenesis, cause autosomal-recessive severe to profound deafness. *American Journal of Human Genetics, 98*, 1266–1270.

DeLuca, A. P., Whitmore, S. S., Barnes, J., Sharma, T. P., Westfall, T. A., Scott, C. A., … Slusarski, D. C. (2016). Hypomorphic mutations in TRNT1 cause retinitis pigmentosa with erythrocytic microcytosis. *Human Molecular Genetics, 25*, 44–56.

Deml, B., Kariminejad, A., Borujerdi, R. H., Muheisen, S., Reis, L. M., & Semina, E. V. (2015). Mutations in MAB21L2 result in ocular Coloboma, microcornea and cataracts. *PLoS Genetics, 11*, e1005002.

Deml, B., Reis, L. M., Muheisen, S., Bick, D., & Semina, E. V. (2015). EFTUD2 deficiency in vertebrates: identification of a novel human mutation and generation of a zebrafish model. *Birth Defects Research. Part A, Clinical and Molecular Teratology, 103*, 630–640.

Dhakal, S., Stevens, C. B., Sebbagh, M., Weiss, O., Frey, R. A., Adamson, S., … Stenkamp, D. L. (2015). Abnormal retinal development in Cloche mutant zebrafish. *Developmental Dynamics, 244*, 1439–1455.

Di Leva, F., D'Adamo, P., Cubellis, M. V., D'Eustacchio, A., Errichiello, M., Saulino, C., … Marciano, E. (2006). Identification of a novel mutation in the myosin

VIIA motor domain in a family with autosomal dominant hearing loss (DFNA11). *Audiology & Neuro-otology, 11*, 157–164.

Di Palma, F., Holme, R. H., Bryda, E. C., Belyantseva, I. A., Pellegrino, R., Kachar, B., ... Noben-Trauth, K. (2001). Mutations in Cdh23, encoding a new type of cadherin, cause stereocilia disorganization in waltzer, the mouse model for Usher syndrome type 1D. *Nature Genetics, 27*, 103–107.

Diaz-Horta, O., Subasioglu-Uzak, A., Grati, M., DeSmidt, A., Foster, J., 2nd, Cao, L., ... Tekin, M. (2014). FAM65B is a membrane-associated protein of hair cell stereocilia required for hearing. *Proceedings of National Academy of Sciences of the United States of America, 111*, 9864–9868.

Doucette, L., Merner, N. D., Cooke, S., Ives, E., Galutira, D., Walsh, V., ... Young, T. L. (2009). Profound, prelingual nonsyndromic deafness maps to chromosome 10q21 and is caused by a novel missense mutation in the Usher syndrome type IF gene PCDH15. *European Journal of Human Genetics, 17*, 554–564.

Driscoll, C. L., Kasperbauer, J. L., Facer, G. W., Harner, S. G., & Beatty, C. W. (1997). Low-dose intratympanic gentamicin and the treatment of Meniere's disease: preliminary results. *The Laryngoscope, 107*, 83–89.

Duman, D., Sirmaci, A., Cengiz, F. B., Ozdag, H., & Tekin, M. (2011). Screening of 38 genes identifies mutations in 62% of families with nonsyndromic deafness in Turkey. *Genetic Testing and Molecular Biomarkers, 15*, 29–33.

Dutton, K., Abbas, L., Spencer, J., Brannon, C., Mowbray, C., Nikaido, M., ... Whitfield, T. T. (2009). A zebrafish model for Waardenburg syndrome type IV reveals diverse roles for Sox10 in the otic vesicle. *Disease Models & Mechanisms, 2*, 68–83.

Easter, S. S., Jr., & Nicola, G. N. (1996). The development of vision in the zebrafish (*Danio rerio*). *Developmental Biology, 180*, 646–663.

Easter, S. S., Jr., & Nicola, G. N. (1997). The development of eye movements in the zebrafish (*Danio rerio*). *Developmental Psychobiology, 31*, 267–276.

Eatock, R., Fay, R. R., & Popper, A. N. (2006). *Vertebrate hair cells*. New York, Berlin: Springer-Verlag.

Ebermann, I., Lopez, I., Bitner-Glindzicz, M., Brown, C., Koenekoop, R. K., & Bolz, H. J. (2007). Deafblindness in French Canadians from Quebec: a predominant founder mutation in the USH1C gene provides the first genetic link with the Acadian population. *Genome Biology, 8*, R47.

Ebermann, I., Phillips, J. B., Liebau, M. C., Koenekoop, R. K., Schermer, B., Lopez, I., ... Bolz, H. J. (2010). PDZD7 is a modifier of retinal disease and a contributor to digenic Usher syndrome. *Journal of Clinical Investigation, 120*, 1812–1823.

El-Amraoui, A., & Petit, C. (2005). Usher I syndrome: unravelling the mechanisms that underlie the cohesion of the growing hair bundle in inner ear sensory cells. *Journal of Cell Science, 118*, 4593–4603.

El-Amraoui, A., & Petit, C. (2014). The retinal phenotype of Usher syndrome: pathophysiological insights from animal models. *Comptes Rendus Biologies, 337*, 167–177.

Elmaleh-Berges, M., Baumann, C., Noel-Petroff, N., Sekkal, A., Couloigner, V., Devriendt, K., ... Pingault, V. (2013). Spectrum of temporal bone abnormalities in patients with Waardenburg syndrome and SOX10 mutations. *American Journal of Neuroradiology, 34*, 1257–1263.

Elsayed, S. M., Phillips, J. B., Heller, R., Thoenes, M., Elsobky, E., Nurnberg, G., ... Bolz, H. J. (2015). Non-manifesting AHI1 truncations indicate localized loss-of-function tolerance in a severe Mendelian disease gene. *Human Molecular Genetics, 24*, 2594–2603.

Emran, F., Rihel, J., & Dowling, J. E. (2008). A behavioral assay to measure responsiveness of zebrafish to changes in light intensities. *Journal of Visualized Experiments*.

Ernest, S., Rauch, G. J., Haffter, P., Geisler, R., Petit, C., & Nicolson, T. (2000). Mariner is defective in myosin VIIA: a zebrafish model for human hereditary deafness. *Human Molecular Genetics, 9*, 2189–2196.

Fleisch, V. C., & Neuhauss, S. C. (2006). Visual behavior in zebrafish. *Zebrafish, 3*, 191–201.

Fleisch, V. C., Schonthaler, H. B., von Lintig, J., & Neuhauss, S. C. (2008). Subfunctionalization of a retinoid-binding protein provides evidence for two parallel visual cycles in the cone-dominant zebrafish retina. *Journal of Neuroscience, 28*, 8208–8216.

Ganapathy, A., Pandey, N., Srisailapathy, C. R., Jalvi, R., Malhotra, V., Venkatappa, M., … Anand, A. (2014). Non-syndromic hearing impairment in India: high allelic heterogeneity among mutations in TMPRSS3, TMC1, USHIC, CDH23 and TMIE. *PLoS One, 9*, e84773.

Gestri, G., Link, B. A., & Neuhauss, S. C. (2012). The visual system of zebrafish and its use to model human ocular diseases. *Developmental Neurobiology, 72*, 302–327.

Gibbs, D., Azarian, S. M., Lillo, C., Kitamoto, J., Klomp, A. E., Steel, K. P., … Williams, D. S. (2004). Role of myosin VIIa and Rab27a in the motility and localization of RPE melanosomes. *Journal of Cell Science, 117*, 6473–6483.

Gibbs, D., Kitamoto, J., & Williams, D. S. (2003). Abnormal phagocytosis by retinal pigmented epithelium that lacks myosin VIIa, the Usher syndrome 1B protein. *Proceedings of National Academy of Sciences of the United States of America, 100*, 6481–6486.

Gleason, M. R., Nagiel, A., Jamet, S., Vologodskaia, M., Lopez-Schier, H., & Hudspeth, A. J. (2009). The transmembrane inner ear (Tmie) protein is essential for normal hearing and balance in the zebrafish. *Proceedings of National Academy of Sciences of the United States of America, 106*, 21347–21352.

Glover, G., Mueller, K. P., Sollner, C., Neuhauss, S. C., & Nicolson, T. (2012). The Usher gene cadherin 23 is expressed in the zebrafish brain and a subset of retinal amacrine cells. *Molecular Vision, 18*, 2309–2322.

Goishi, K., Shimizu, A., Najarro, G., Watanabe, S., Rogers, R., Zon, L. I., & Klagsbrun, M. (2006). AlphaA-crystallin expression prevents gamma-crystallin insolubility and cataract formation in the zebrafish cloche mutant lens. *Development, 133*, 2585–2593.

Goodyear, R. J., & Richardson, G. P. (2002). Extracellular matrices associated with the apical surfaces of sensory epithelia in the inner ear: molecular and structural diversity. *Journal of Neurobiology, 53*, 212–227.

Gopal, S. R., Chen, D. H., Chou, S. W., Zang, J., Neuhauss, S. C., Stepanyan, R., … Alagramam, K. N. (2015). Zebrafish models for the mechanosensory hair cell dysfunction in usher syndrome 3 reveal that Clarin-1 is an essential hair bundle protein. *Journal of Neuroscience, 35*, 10188–10201.

Gosse, N. J., & Baier, H. (2009). An essential role for Radar (Gdf6a) in inducing dorsal fate in the zebrafish retina. *Proceedings of National Academy of Sciences of the United States of America, 106*, 2236–2241.

Grati, M., Chakchouk, I., Ma, Q., Bensaid, M., Desmidt, A., Turki, N., … Masmoudi, S. (2015). A missense mutation in DCDC2 causes human recessive deafness DFNB66, likely by interfering with sensory hair cell and supporting cell cilia length regulation. *Human Molecular Genetics, 24*, 2482–2491.

Greenlees, R., Mihelec, M., Yousoof, S., Speidel, D., Wu, S. K., Rinkwitz, S., ... Jamieson, R. V. (2015). Mutations in SIPA1L3 cause eye defects through disruption of cell polarity and cytoskeleton organization. *Human Molecular Genetics, 24*, 5789–5804.

Gregory-Evans, C. Y., Moosajee, M., Shan, X., & Gregory-Evans, K. (2011). Gene-specific differential response to anti-apoptotic therapies in zebrafish models of ocular coloboma. *Molecular Vision, 17*, 1473–1484.

Grifone, R., Laclef, C., Spitz, F., Lopez, S., Demignon, J., Guidotti, J. E., ... Maire, P. (2004). Six1 and Eya1 expression can reprogram adult muscle from the slow-twitch phenotype into the fast-twitch phenotype. *Molecular and Cellular Biology, 24*, 6253–6267.

Grillet, N., Xiong, W., Reynolds, A., Kazmierczak, P., Sato, T., Lillo, C., ... Muller, U. (2009). Harmonin mutations cause mechanotransduction defects in cochlear hair cells. *Neuron, 62*, 375–387.

Haddon, C., & Lewis, J. (1996). Early ear development in the embryo of the zebrafish, *Danio rerio*. *Journal of Comparitive Neurology, 365*, 113–128.

Hain, T. C., & Yacovino, D. (2005). Pharmacologic treatment of persons with dizziness. *Neurologic Clinics, 23*, 831–853. vii.

Han, Y., Mu, Y., Li, X., Xu, P., Tong, J., Liu, Z., ... Meng, A. (2011). Grhl2 deficiency impairs otic development and hearing ability in a zebrafish model of the progressive dominant hearing loss DFNA28. *Human Molecular Genetics, 20*, 3213–3226.

Hartsock, A., Lee, C., Arnold, V., & Gross, J. M. (2014). In vivo analysis of hyaloid vasculature morphogenesis in zebrafish: a role for the lens in maturation and maintenance of the hyaloid. *Developmental Biology, 394*, 327–339.

Heon, E., Kim, G., Qin, S., Garrison, J. E., Tavares, E., Vincent, A., ... Sheffield, V. C. (2016). Mutations in C8ORF37 cause Bardet Biedl syndrome (BBS21). *Human Molecular Genetics*.

Herraiz, C., Plaza, G., Aparicio, J. M., Gallego, I., Marcos, S., & Ruiz, C. (2010). Transtympanic steroids for Meniere's disease. *Otology & Neurotology, 31*, 162–167.

Hilgert, N., Smith, R. J., & Van Camp, G. (2009). Forty-six genes causing nonsyndromic hearing impairment: which ones should be analyzed in DNA diagnostics? *Mutation Research, 681*, 189–196.

Hodel, C., Niklaus, S., Heidemann, M., Klooster, J., Kamermans, M., Biehlmaier, O., ... Neuhauss, S. C. (2014). Myosin VIIA is a marker for the cone accessory outer segment in zebrafish. *Anatomical Record, 297*, 1777–1784.

Huang, Y. Y., Haug, M. F., Gesemann, M., & Neuhauss, S. C. (2012). Novel expression patterns of metabotropic glutamate receptor 6 in the zebrafish nervous system. *PLoS One, 7*, e35256.

Hufnagel, R. B., Arno, G., Hein, N. D., Hersheson, J., Prasad, M., Anderson, Y., ... Ahmed, Z. M. (2015). Neuropathy target esterase impairments cause Oliver-McFarlane and Laurence-Moon syndromes. *Journal of Medical Genetics, 52*, 85–94.

Hu, Z. Y., Zhang, Q. Y., Qin, W., Tong, J. W., Zhao, Q., Han, Y., ... Zhang, J. P. (2013). Gene miles-apart is required for formation of otic vesicle and hair cells in zebrafish. *Cell Death & Disease, 4*, e900.

Ikeda, K., Watanabe, Y., Ohto, H., & Kawakami, K. (2002). Molecular interaction and synergistic activation of a promoter by Six, Eya, and Dach proteins mediated through CREB binding protein. *Molecular and Cellular Biology, 22*, 6759–6766.

Inoue, A., & Ikebe, M. (2003). Characterization of the motor activity of mammalian myosin VIIA. *Journal of Biological Chemistry, 278*, 5478–5487.

James, D., Rajput, K., Brinton, J., & Goswami, U. (2008). Phonological awareness, vocabulary, and word reading in children who use cochlear implants: does age of implantation explain individual variability in performance outcomes and growth? *Journal of Deaf Studies and Deaf Education, 13*, 117–137.

Janky, K. L., & Givens, D. (2015). Vestibular, visual acuity, and balance outcomes in children with cochlear implants: a preliminary report. *Ear and Hearing, 36*, e364–372.

Jia, S., Muto, A., Orisme, W., Henson, H. E., Parupalli, C., Ju, B., … Taylor, M. R. (2014). Zebrafish Cacna1fa is required for cone photoreceptor function and synaptic ribbon formation. *Human Molecular Genetics, 23*, 2981–2994.

Jin, Z. B., Huang, X. F., Lv, J. N., Xiang, L., Li, D. Q., Chen, J., … Qu, J. (2014). SLC7A14 linked to autosomal recessive retinitis pigmentosa. *Nature Communications, 5*, 3517.

Kappler, J. A., Starr, C. J., Chan, D. K., Kollmar, R., & Hudspeth, A. J. (2004). A nonsense mutation in the gene encoding a zebrafish myosin VI isoform causes defects in hair-cell mechanotransduction. *Proceedings of National Academy of Sciences of the United States of America, 101*, 13056–13061.

Kazmierczak, P., Sakaguchi, H., Tokita, J., Wilson-Kubalek, E. M., Milligan, R. A., Muller, U., & Kachar, B. (2007). Cadherin 23 and protocadherin 15 interact to form tip-link filaments in sensory hair cells. *Nature, 449*, 87–91.

Kennedy, B. N., Alvarez, Y., Brockerhoff, S. E., Stearns, G. W., Sapetto-Rebow, B., Taylor, M. R., & Hurley, J. B. (2007). Identification of a zebrafish cone photoreceptor-specific promoter and genetic rescue of achromatopsia in the nof mutant. *Investigative Ophthalmology & Visual Science, 48*, 522–529.

Kettleborough, R. N., Busch-Nentwich, E. M., Harvey, S. A., Dooley, C. M., de Bruijn, E., van Eeden, F., … Stemple, D. L. (2013). A systematic genome-wide analysis of zebrafish protein-coding gene function. *Nature, 496*, 494–497.

Khateb, S., Zelinger, L., Ben-Yosef, T., Merin, S., Crystal-Shalit, O., Gross, M., … Sharon, D. (2012). Exome sequencing identifies a founder frameshift mutation in an alternative exon of USH1C as the cause of autosomal recessive retinitis pigmentosa with late-onset hearing loss. *PLoS One, 7*, e51566.

Kikkawa, Y. S., Pawlowski, K. S., Wright, C. G., & Alagramam, K. N. (2008). Development of outer hair cells in Ames waltzer mice: mutation in protocadherin 15 affects development of cuticular plate and associated structures. *Anatomical Record, 291*, 224–232.

Kimberling, W. J., Hildebrand, M. S., Shearer, A. E., Jensen, M. L., Halder, J. A., Trzupek, K., … Smith, R. J. (2010). Frequency of Usher syndrome in two pediatric populations: implications for genetic screening of deaf and hard of hearing children. *Genetics in Medicine, 12*, 512–516.

Kim, Y. H., Epting, D., Slanchev, K., Engel, C., Walz, G., & Kramer-Zucker, A. (2013). A complex of BBS1 and NPHP7 is required for cilia motility in zebrafish. *PLoS One, 8*, e72549.

Kimmel, C. B., Patterson, J., & Kimmel, R. O. (1974). The development and behavioral characteristics of the startle response in the zebra fish. *Developmental Psychobiology, 7*, 47–60.

Kindt, K. S., Finch, G., & Nicolson, T. (2012). Kinocilia mediate mechanosensitivity in developing zebrafish hair cells. *Developmental Cell, 23*, 329–341.

Kozlowski, D. J., Whitfield, T. T., Hukriede, N. A., Lam, W. K., & Weinberg, E. S. (2005). The zebrafish dog-eared mutation disrupts eya1, a gene required for cell survival and differentiation in the inner ear and lateral line. *Developmental Biology, 277*, 27–41.

Krock, B. L., Bilotta, J., & Perkins, B. D. (2007). Noncell-autonomous photoreceptor degeneration in a zebrafish model of choroideremia. *Proceedings of National Academy of Sciences of the United States of America, 104*, 4600–4605.

Kuhlbrodt, K., Schmidt, C., Sock, E., Pingault, V., Bondurand, N., Goossens, M., & Wegner, M. (1998). Functional analysis of Sox10 mutations found in human Waardenburg-Hirschsprung patients. *Journal of Biological Chemistry, 273*, 23033–23038.

Le Quesne Stabej, P., Saihan, Z., Rangesh, N., Steele-Stallard, H. B., Ambrose, J., Coffey, A., ... Bitner-Glindzicz, M. (2012). Comprehensive sequence analysis of nine usher syndrome genes in the UK national collaborative usher study. *Journal of Medical Genetics, 49*, 27–36.

Lee, J., Lee, B. K., & Gross, J. M. (2013). Bcl6a function is required during optic cup formation to prevent p53-dependent apoptosis and colobomata. *Human Molecular Genetics, 22*, 3568–3582.

Lee, J. E., Silhavy, J. L., Zaki, M. S., Schroth, J., Bielas, S. L., Marsh, S. E., ... Gleeson, J. G. (2012). CEP41 is mutated in Joubert syndrome and is required for tubulin glutamylation at the cilium. *Nature Genetics, 44*, 193–199.

Lee, C., Wallingford, J. B., & Gross, J. M. (2014). Cluap1 is essential for ciliogenesis and photoreceptor maintenance in the vertebrate eye. *Investigative Ophthalmology & Visual Science, 55*, 4585–4592.

Li, C. M., Hoffman, H. J., Ward, B. K., Cohen, H. S., & Rine, R. M. (2016). Epidemiology of dizziness and balance problems in children in the United States: a population-based study. *Journal of Pediatrics, 171*, 240-247.e1–3.

Li, J., He, Y., Lu, Q., & Zhang, M. (2016). Mechanistic basis of organization of the harmonin/USH1C-mediated brush border microvilli tip-link complex. *Developmental Cell, 36*, 179–189.

Li, L., & Dowling, J. E. (1997). A dominant form of inherited retinal degeneration caused by a non-photoreceptor cell-specific mutation. *Proceedings of National Academy of Sciences of the United States of America, 94*, 11645–11650.

Li, L., & Dowling, J. E. (2000). Disruption of the olfactoretinal centrifugal pathway may relate to the visual system defect in night blindness b mutant zebrafish. *Journal of Neuroscience, 20*, 1883–1892.

Li, L., Nakaya, N., Chavali, V. R., Ma, Z., Jiao, X., Sieving, P. A., ... Hejtmancik, J. F. (2010). A mutation in ZNF513, a putative regulator of photoreceptor development, causes autosomal-recessive retinitis pigmentosa. *American Journal of Human Genetics, 87*, 400–409.

Linder, B., Dill, H., Hirmer, A., Brocher, J., Lee, G. P., Mathavan, S., ... Fischer, U. (2011). Systemic splicing factor deficiency causes tissue-specific defects: a zebrafish model for retinitis pigmentosa. *Human Molecular Genetics, 20*, 368–377.

Linder, B., Hirmer, A., Gal, A., Ruther, K., Bolz, H. J., Winkler, C., ... Fischer, U. (2014). Identification of a PRPF4 loss-of-function variant that abrogates U4/U6.U5 tri-snRNP integration and is associated with retinitis pigmentosa. *PLoS One, 9*, e111754.

Link, B. A., & Collery, R. F. (2015). Zebrafish models of retinal disease. *Annual Review of Vision Science, 1*, 125–153.

Liu, F., Chen, J., Yu, S., Raghupathy, R. K., Liu, X., Qin, Y., ... Liu, M. (2015). Knockout of RP2 decreases GRK1 and rod transducin subunits and leads to photoreceptor degeneration in zebrafish. *Human Molecular Genetics, 24*, 4648–4659.

Liu, Y., Chen, X., Qin, B., Zhao, K., Zhao, Q., Staley, J. P., & Zhao, C. (2015). Knocking down Snrnp200 initiates demorphogenesis of rod photoreceptors in zebrafish. *Journal of Ophthalmology, 2015*, 816329.

Liu, X. Z., Hope, C., Walsh, J., Newton, V., Ke, X. M., Liang, C. Y., … Brown, S. D. (1998). Mutations in the myosin VIIA gene cause a wide phenotypic spectrum, including atypical Usher syndrome. *American Journal of Human Genetics, 63*, 909–912.

Liu, Y. P., Bosch, D. G., Siemiatkowska, A. M., Rendtorff, N. D., Boonstra, F. N., Moller, C., … Cremers, F. P. (2016). Putative digenic inheritance of heterozygous RP1L1 and C2orf71 null mutations in syndromic retinal dystrophy. *Ophthalmic Genetics*, 1–6.

Liu, X. Z., Newton, V. E., Steel, K. P., & Brown, S. D. (1997). Identification of a new mutation of the myosin VII head region in Usher syndrome type 1. *Human Mutation, 10*, 168–170.

Liu, X. Z., Walsh, J., Tamagawa, Y., Kitamura, K., Nishizawa, M., Steel, K. P., & Brown, S. D. (1997). Autosomal dominant non-syndromic deafness caused by a mutation in the myosin VIIA gene. *Nature Genetics, 17*, 268–269.

Li, C., Wang, L., Zhang, J., Huang, M., Wong, F., Liu, X., … Liu, M. (2014). CERKL interacts with mitochondrial TRX2 and protects retinal cells from oxidative stress-induced apoptosis. *Biochimica et Biophysica Acta, 1842*, 1121–1129.

Li, J., Zhao, X., Xin, Q., Shan, S., Jiang, B., Jin, Y., … Liu, Q. (2015). Whole-exome sequencing identifies a variant in TMEM132E causing autosomal-recessive nonsyndromic hearing loss DFNB99. *Human Mutation, 36*, 98–105.

Liu, C., Widen, S. A., Williamson, K. A., Ratnapriya, R., Gerth-Kahlert, C., Rainger, J., … Swaroop, A. (2016). A secreted WNT-ligand-binding domain of FZD5 generated by a frameshift mutation causes autosomal dominant coloboma. *Human Molecular Genetics, 25*, 1382–1391.

Lunt, S. C., Haynes, T., & Perkins, B. D. (2009). Zebrafish ift57, ift88, and ift172 intraflagellar transport mutants disrupt cilia but do not affect hedgehog signaling. *Developmental Dynamics, 238*(7), 1744–1759.

Luo, N., Lu, J., & Sun, Y. (2012). Evidence of a role of inositol polyphosphate 5-phosphatase INPP5E in cilia formation in zebrafish. *Vision Research, 75*, 98–107.

Maerker, T., van Wijk, E., Overlack, N., Kersten, F. F., McGee, J., Goldmann, T., … Wolfrum, U. (2008). A novel Usher protein network at the periciliary reloading point between molecular transport machineries in vertebrate photoreceptor cells. *Human Molecular Genetics, 17*, 71–86.

Malicki, J. (1999). Development of the retina. *Methods in Cell Biology, 59*, 273–299.

Malicki, J., Pooranachandran, N., Nikolaev, A., Fang, X., & Avanesov, A. (2016). Analysis of the retina in the zebrafish model. *Methods in Cell Biology, 134*, 257–334.

Malicki, J., Schier, A. F., Solnica-Krezel, L., Stemple, D. L., Neuhauss, S. C., Stainier, D. Y., … Driever, W. (1996). Mutations affecting development of the zebrafish ear. *Development, 123*, 275–283.

Martin, C. A., Ahmad, I., Klingseisen, A., Hussain, M. S., Bicknell, L. S., Leitch, A., … Jackson, A. P. (2014). Mutations in PLK4, encoding a master regulator of centriole biogenesis, cause microcephaly, growth failure and retinopathy. *Nature Genetics, 46*, 1283–1292.

Mathur, P., & Yang, J. (2015). Usher syndrome: hearing loss, retinal degeneration and associated abnormalities. *Biochimica et Biophysica Acta, 1852*, 406–420.

Matsushima, Y., Shinkai, Y., Kobayashi, Y., Sakamoto, M., Kunieda, T., & Tachibana, M. (2002). A mouse model of Waardenburg syndrome type 4 with a new spontaneous mutation of the endothelin-B receptor gene. *Mammalian Genome, 13*, 30–35.

Maurer, C. M., Huang, Y. Y., & Neuhauss, S. C. (2011). Application of zebrafish oculomotor behavior to model human disorders. *Reviews in the Neurosciences, 22*, 5–16.

McClure, J. A., Lycett, P., & Baskerville, J. C. (1982). Diazepam as an anti-motion sickness drug. *Journal of Otolaryngology, 11*, 253–259.

Meyer, N. C., Alasti, F., Nishimura, C. J., Imanirad, P., Kahrizi, K., Riazalhosseini, Y., … Najmabadi, H. (2007). Identification of three novel TECTA mutations in Iranian families with autosomal recessive nonsyndromic hearing impairment at the DFNB21 locus. *American Journal of Medical Genetics Part A, 143A*, 1623–1629.

Michalski, N., Michel, V., Caberlotto, E., Lefevre, G. M., van Aken, A. F., Tinevez, J. Y., … Petit, C. (2009). Harmonin-b, an actin-binding scaffold protein, is involved in the adaptation of mechanoelectrical transduction by sensory hair cells. *Pflugers Archiv, 459*, 115–130.

Miesfeld, J. B., Gestri, G., Clark, B. S., Flinn, M. A., Poole, R. J., Bader, J. R., … Link, B. A. (2015). Yap and Taz regulate retinal pigment epithelial cell fate. *Development, 142*, 3021–3032.

Migliosi, V., Modamio-Hoybjor, S., Moreno-Pelayo, M. A., Rodriguez-Ballesteros, M., Villamar, M., Telleria, D., … Del Castillo, I. (2002). Q829X, a novel mutation in the gene encoding otoferlin (OTOF), is frequently found in Spanish patients with prelingual non-syndromic hearing loss. *Journal of Medical Genetics, 39*, 502–506.

Mo, W., Chen, F., Nechiporuk, A., & Nicolson, T. (2010). Quantification of vestibular-induced eye movements in zebrafish larvae. *BMC Neuroscience, 11*, 110.

Moosajee, M., Gregory-Evans, K., Ellis, C. D., Seabra, M. C., & Gregory-Evans, C. Y. (2008). Translational bypass of nonsense mutations in zebrafish rep1, pax2.1 and lamb1 highlights a viable therapeutic option for untreatable genetic eye disease. *Human Molecular Genetics, 17*, 3987–4000.

Moosajee, M., Tracey-White, D., Smart, M., Weetall, M., Torriano, S., Kalatzis, V., … Welch, E. (2016). Functional rescue of REP1 following treatment with PTC124 and novel derivative PTC-414 in human choroideremia fibroblasts and the nonsense-mediated zebrafish model. *Human Molecular Genetics*.

Moosajee, M., Tulloch, M., Baron, R. A., Gregory-Evans, C. Y., Pereira-Leal, J. B., & Seabra, M. C. (2009). Single choroideremia gene in nonmammalian vertebrates explains early embryonic lethality of the zebrafish model of choroideremia. *Investigative Ophthalmology & Visual Science, 50*, 3009–3016.

Morton, C. C., & Nance, W. E. (2006). Newborn hearing screening–a silent revolution. *The New England Journal of Medicine, 354*, 2151–2164.

Murga-Zamalloa, C. A., Ghosh, A. K., Patil, S. B., Reed, N. A., Chan, L. S., Davuluri, S., … Khanna, H. (2011). Accumulation of the Raf-1 kinase inhibitory protein (Rkip) is associated with Cep290-mediated photoreceptor degeneration in ciliopathies. *Journal of Biological Chemistry, 286*, 28276–28286.

Mustapha, M., Weil, D., Chardenoux, S., Elias, S., El-Zir, E., Beckmann, J. S., … Petit, C. (1999). An alpha-tectorin gene defect causes a newly identified autosomal recessive form of sensorineural pre-lingual non-syndromic deafness, DFNB21. *Human Molecular Genetics, 8*, 409–412.

Mutai, H., Suzuki, N., Shimizu, A., Torii, C., Namba, K., Morimoto, N., … Matsunaga, T. (2013). Diverse spectrum of rare deafness genes underlies early-childhood hearing loss in Japanese patients: a cross-sectional, multi-center next-generation sequencing study. *Orphanet Journal of Rare Diseases, 8*, 172.

Muto, A., Orger, M. B., Wehman, A. M., Smear, M. C., Kay, J. N., Page-McCaw, P. S., … Baier, H. (2005). Forward genetic analysis of visual behavior in zebrafish. *PLoS Genetics, 1*, e66.

Naz, S., Alasti, F., Mowjoodi, A., Riazuddin, S., Sanati, M. H., Friedman, T. B., ... Riazuddin, S. (2003). Distinctive audiometric profile associated with DFNB21 alleles of TECTA. *Journal of Medical Genetics, 40*, 360–363.

Nelson, S. M., Park, L., & Stenkamp, D. L. (2009). Retinal homeobox 1 is required for retinal neurogenesis and photoreceptor differentiation in embryonic zebrafish. *Developmental Biology, 328*, 24–39.

Ng, D., Thakker, N., Corcoran, C. M., Donnai, D., Perveen, R., Schneider, A., ... Biesecker, L. G. (2004). Oculofaciocardiodental and Lenz microphthalmia syndromes result from distinct classes of mutations in BCOR. *Nature Genetics, 36*, 411–416.

Nicolson, T. (2005). The genetics of hearing and balance in zebrafish. *Annual Review of Genetics, 39*, 9–22.

Nicolson, T., Rusch, A., Friedrich, R. W., Granato, M., Ruppersberg, J. P., & Nusslein-Volhard, C. (1998). Genetic analysis of vertebrate sensory hair cell mechanosensation: the zebrafish circler mutants. *Neuron, 20*, 271–283.

Nishiguchi, K. M., Tearle, R. G., Liu, Y. P., Oh, E. C., Miyake, N., Benaglio, P., ... Rivolta, C. (2013). Whole genome sequencing in patients with retinitis pigmentosa reveals pathogenic DNA structural changes and NEK2 as a new disease gene. *Proceedings of National Academy of Sciences of the United States of America, 110*, 16139–16144.

Nishimura, D. Y., Baye, L. M., Perveen, R., Searby, C. C., Avila-Fernandez, A., Pereiro, I., ... Sheffield, V. C. (2010). Discovery and functional analysis of a retinitis pigmentosa gene, C2ORF71. *American Journal of Human Genetics, 86*, 686–695.

Obholzer, N., Wolfson, S., Trapani, J. G., Mo, W., Nechiporuk, A., Busch-Nentwich, E., ... Nicolson, T. (2008). Vesicular glutamate transporter 3 is required for synaptic transmission in zebrafish hair cells. *Journal of Neuroscience, 28*, 2110–2118.

Ogun, O., & Zallocchi, M. (2014). Clarin-1 acts as a modulator of mechanotransduction activity and presynaptic ribbon assembly. *Journal of Cell Biology, 207*, 375–391.

Olszewski, J., Haehnel, M., Taguchi, M., & Liao, J. C. (2012). Zebrafish larvae exhibit rheotaxis and can escape a continuous suction source using their lateral line. *PLoS One, 7*, e36661.

OMIM, 2016. http://www.omim.org/.

Omori, Y., Zhao, C., Saras, A., Mukhopadhyay, S., Kim, W., Furukawa, T., ... Malicki, J. (2008). Elipsa is an early determinant of ciliogenesis that links the IFT particle to membrane-associated small GTPase Rab8. *Nature Cell Biology, 10*, 437–444.

Ouyang, X. M., Xia, X. J., Verpy, E., Du, L. L., Pandya, A., Petit, C., ... Liu, X. Z. (2002). Mutations in the alternatively spliced exons of USH1C cause non-syndromic recessive deafness. *Human Genetics, 111*, 26–30.

Owens, K. N., Santos, F., Roberts, B., Linbo, T., Coffin, A. B., Knisely, A. J., ... Raible, D. W. (2008). Identification of genetic and chemical modulators of zebrafish mechanosensory hair cell death. *PLoS Genetics, 4*, e1000020.

Ozaki, H., Watanabe, Y., Ikeda, K., & Kawakami, K. (2002). Impaired interactions between mouse Eya1 harboring mutations found in patients with branchio-oto-renal syndrome and Six, Dach, and G proteins. *Journal of Human Genetics, 47*, 107–116.

Pan, L., Yan, J., Wu, L., & Zhang, M. (2009). Assembling stable hair cell tip link complex via multidentate interactions between harmonin and cadherin 23. *Proceedings of National Academy of Sciences of the United States of America, 106*, 5575–5580.

Patil, S. B., Hurd, T. W., Ghosh, A. K., Murga-Zamalloa, C. A., & Khanna, H. (2011). Functional analysis of retinitis pigmentosa 2 (RP2) protein reveals variable pathogenic potential of disease-associated missense variants. *PLoS One, 6*, e21379.

Patten, S. A., Jacobs-McDaniels, N. L., Zaouter, C., Drapeau, P., Albertson, R. C., & Moldovan, F. (2012). Role of Chd7 in zebrafish: a model for CHARGE syndrome. *PLoS One, 7*, e31650.

Paulus, J. D., & Link, B. A. (2014). Loss of optineurin in vivo results in elevated cell death and alters axonal trafficking dynamics. *PLoS One, 9*, e109922.

Peachey, N. S., Ray, T. A., Florijn, R., Rowe, L. B., Sjoerdsma, T., Contreras-Alcantara, S., … Gregg, R. G. (2012). GPR179 is required for depolarizing bipolar cell function and is mutated in autosomal-recessive complete congenital stationary night blindness. *American Journal of Human Genetics, 90*, 331–339.

Petit, C. (1996). Genes responsible for human hereditary deafness: symphony of a thousand. *Nature Genetics, 14*, 385–391.

Phillips, J. B., Blanco-Sanchez, B., Lentz, J. J., Tallafuss, A., Khanobdee, K., Sampath, S., … Westerfield, M. (2011). Harmonin (Ush1c) is required in zebrafish Muller glial cells for photoreceptor synaptic development and function. *Disease Models & Mechanisms, 4*, 786–800.

Phillips, J. B., Vastinsalo, H., Wegner, J., Clement, A., Sankila, E. M., & Westerfield, M. (2013). The cone-dominant retina and the inner ear of zebrafish express the ortholog of CLRN1, the causative gene of human Usher syndrome type 3A. *Gene Expression Patterns, 13*, 473–481.

Phillips, J. B., & Westerfield, M. (2014). Zebrafish models in translational research: tipping the scales toward advancements in human health. *Disease Models & Mechanisms, 7*, 739–743.

Pignoni, F., Hu, B., Zavitz, K. H., Xiao, J., Garrity, P. A., & Zipursky, S. L. (1997). The eye-specification proteins So and Eya form a complex and regulate multiple steps in Drosophila eye development. *Cell, 91*, 881–891.

Polok, B., Escher, P., Ambresin, A., Chouery, E., Bolay, S., Meunier, I., … Schorderet, D. F. (2009). Mutations in CNNM4 cause recessive cone-rod dystrophy with amelogenesis imperfecta. *American Journal of Human Genetics, 84*, 259–265.

Potterf, S. B., Furumura, M., Dunn, K. J., Arnheiter, H., & Pavan, W. J. (2000). Transcription factor hierarchy in Waardenburg syndrome: regulation of MITF expression by SOX10 and PAX3. *Human Genetics, 107*, 1–6.

Pretorius, P. R., Baye, L. M., Nishimura, D. Y., Searby, C. C., Bugge, K., Yang, B., … Slusarski, D. C. (2010). Identification and functional analysis of the vision-specific BBS3 (ARL6) long isoform. *PLoS Genetics, 6*, e1000884.

Rainy, N., Etzion, T., Alon, S., Pomeranz, A., Nisgav, Y., Livnat, T., … Stiebel-Kalish, H. (2016). Knockdown of unc119c results in visual impairment and early-onset retinal dystrophy in zebrafish. *Biochemical and Biophysical Research Communications, 473*, 1211–1217.

Raymond, P. A., Colvin, S. M., Jabeen, Z., Nagashima, M., Barthel, L. K., Hadidjojo, J., … Lubensky, D. K. (2014). Patterning the cone mosaic array in zebrafish retina requires specification of ultraviolet-sensitive cones. *PLoS One, 9*, e85325.

Reiners, J., Marker, T., Jurgens, K., Reidel, B., & Wolfrum, U. (2005). Photoreceptor expression of the Usher syndrome type 1 protein protocadherin 15 (USH1F) and its interaction with the scaffold protein harmonin (USH1C). *Molecular Vision, 11*, 347–355.

Reiners, J., Nagel-Wolfrum, K., Jurgens, K., Marker, T., & Wolfrum, U. (2006). Molecular basis of human Usher syndrome: deciphering the meshes of the Usher protein network provides insights into the pathomechanisms of the Usher disease. *Experimental Eye Research, 83*, 97–119.

Reiners, J., van Wijk, E., Marker, T., Zimmermann, U., Jurgens, K., te Brinke, H., … Wolfrum, U. (2005). Scaffold protein harmonin (USH1C) provides molecular links between Usher syndrome type 1 and type 2. *Human Molecular Genetics, 14*, 3933–3943.

Reischauer, S., Stone, O. A., Villasenor, A., Chi, N., Jin, S. W., Martin, M., … Stainier, D. Y. (2016). Cloche is a bHLH-PAS transcription factor that drives haemato-vascular specification. *Nature, 535*, 294–298.

Riazuddin, S., Belyantseva, I. A., Giese, A. P., Lee, K., Indzhykulian, A. A., Nandamuri, S. P., … Ahmed, Z. M. (2012). Alterations of the CIB2 calcium- and integrin-binding protein cause Usher syndrome type 1J and nonsyndromic deafness DFNB48. *Nature Genetics, 44*, 1265–1271.

Riazuddin, S., Nazli, S., Ahmed, Z. M., Yang, Y., Zulfiqar, F., Shaikh, R. S., … Friedman, T. B. (2008). Mutation spectrum of MYO7A and evaluation of a novel nonsyndromic deafness DFNB2 allele with residual function. *Human Mutation, 29*, 502–511.

Richardson, G. P., de Monvel, J. B., & Petit, C. (2011). How the genetics of deafness illuminates auditory physiology. *Annual Review of Physiology, 73*, 311–334.

Riera, M., Burguera, D., Garcia-Fernandez, J., & Gonzalez-Duarte, R. (2013). CERKL knockdown causes retinal degeneration in zebrafish. *PLoS One, 8*, e64048.

Rimmer, J., Patel, M., Agarwal, K., Hogg, C., Arshad, Q., & Harcourt, J. (2015). Peripheral vestibular dysfunction in patients with primary ciliary dyskinesia: abnormal otoconial development?. *Otology & Neurotology, 36*, 662–669.

Rine, R. M. (2009). Growing evidence for balance and vestibular problems in children. *Audiological Medicine, 7*, 138–142.

Rine, R. M., Cornwall, G., Gan, K., LoCascio, C., O'Hare, T., Robinson, E., & Rice, M. (2000). Evidence of progressive delay of motor development in children with sensorineural hearing loss and concurrent vestibular dysfunction. *Perceptual and Motor Skills, 90*, 1101–1112.

Rine, R. M., Dannenbaum, E., & Szabo, J. (2016). 2015 section on pediatrics knowledge translation lecture: pediatric vestibular-related impairments. *Pediatric Physical Therapy, 28*, 2–6.

Rine, R. M., Spielholz, N. I., & Buchman, C. (2001). *Postural control in children with sensorineural hearing loss and vestibular hypofunction: Deficits in sensory system effectiveness and vestibulospinal function*. Amsterdam: Springer-Verlag.

Rine, R. M., & Wiener-Vacher, S. (2013). Evaluation and treatment of vestibular dysfunction in children. *NeuroRehabilitation, 32*, 507–518.

Roberts, L., George, S., Greenberg, J., & Ramesar, R. S. (2015). A founder mutation in MYO7A underlies a significant proportion of usher syndrome in indigenous South Africans: implications for the African diaspora. *Investigative Ophthalmology & Visual Science, 56*, 6671–6678.

Rong, W., Chen, X., Zhao, K., Liu, Y., Liu, X., Ha, S., … Zhao, C. (2014). Novel and recurrent MYO7A mutations in Usher syndrome type 1 and type 2. *PLoS One, 9*, e97808.

Roosing, S., Lamers, I. J., de Vrieze, E., van den Born, L. I., Lambertus, S., Arts, H. H., … Cremers, F. P. (2014). Disruption of the basal body protein POC1B results in autosomal-recessive cone-rod dystrophy. *American Journal of Human Genetics, 95*, 131–142.

Roux, A. F., Faugere, V., Vache, C., Baux, D., Besnard, T., Leonard, S., … Claustres, M. (2011). Four-year follow-up of diagnostic service in USH1 patients. *Investigative Ophthalmology & Visual Science, 52*, 4063–4071.

Ruckenstein, M. J., Rutka, J. A., & Hawke, M. (1991). The treatment of Meniere's disease: Torok revisited. *The Laryngoscope, 101*, 211–218.

Ruel, J., Emery, S., Nouvian, R., Bersot, T., Amilhon, B., Van Rybroek, J. M., ... Puel, J. L. (2008). Impairment of SLC17A8 encoding vesicular glutamate transporter-3, VGLUT3, underlies nonsyndromic deafness DFNA25 and inner hair cell dysfunction in null mice. *American Journal of Human Genetics, 83*, 278–292.

Ruf, R. G., Xu, P. X., Silvius, D., Otto, E. A., Beekmann, F., Muerb, U. T., ... Hildebrandt, F. (2004). SIX1 mutations cause branchio-oto-renal syndrome by disruption of EYA1-SIX1-DNA complexes. *Proceedings of National Academy of Sciences of the United States of America, 101*, 8090–8095.

Ruzickova, S., & Stanek, D. (2016). Mutations in spliceosomal proteins and retina degeneration. *RNA Biology*, 1–9.

Sahly, I., Andermann, P., & Petit, C. (1999). The zebrafish eya1 gene and its expression pattern during embryogenesis. *Development Genes and Evolution, 209*, 399–410.

Sahly, I., Dufour, E., Schietroma, C., Michel, V., Bahloul, A., Perfettini, I., ... Petit, C. (2012). Localization of Usher 1 proteins to the photoreceptor calyceal processes, which are absent from mice. *Journal of Cell Biology, 199*, 381–399.

Saihan, Z., Stabej Ple, Q., Robson, A. G., Rangesh, N., Holder, G. E., Moore, A. T., ... Webster, A. R. (2011). Mutations in the USH1C gene associated with sector retinitis pigmentosa and hearing loss. *Retina, 31*, 1708–1716.

Saint-Jeannet, J. P., & Moody, S. A. (2014). Establishing the pre-placodal region and breaking it into placodes with distinct identities. *Developmental Biology, 389*, 13–27.

Sanggaard, K. M., Rendtorff, N. D., Kjaer, K. W., Eiberg, H., Johnsen, T., Gimsing, S., ... Tranebjaerg, L. (2007). Branchio-oto-renal syndrome: detection of EYA1 and SIX1 mutations in five out of six Danish families by combining linkage, MLPA and sequencing analyses. *European Journal of Human Genetics, 15*, 1121–1131.

Sang, Q., Zhang, J., Feng, R., Wang, X., Li, Q., Zhao, X., ... Wang, L. (2014). Ildr1b is essential for semicircular canal development, migration of the posterior lateral line primordium and hearing ability in zebrafish: implications for a role in the recessive hearing impairment DFNB42. *Human Molecular Genetics, 23*, 6201–6211.

Santos-Cortez, R. L., Faridi, R., Rehman, A. U., Lee, K., Ansar, M., Wang, X., ... Leal, S. M. (2016). Autosomal-recessive hearing impairment due to rare missense variants within S1PR2. *American Journal of Human Genetics, 98*, 331–338.

Saszik, S., Bilotta, J., & Givin, C. M. (1999). ERG assessment of zebrafish retinal development. *Visual Neuroscience, 16*, 881–888.

Sato, S., Ikeda, K., Shioi, G., Ochi, H., Ogino, H., Yajima, H., & Kawakami, K. (2010). Conserved expression of mouse Six1 in the pre-placodal region (PPR) and identification of an enhancer for the rostral PPR. *Developmental Biology, 344*, 158–171.

Scheidecker, S., Etard, C., Haren, L., Stoetzel, C., Hull, S., Arno, G., ... Dollfus, H. (2015). Mutations in TUBGCP4 alter microtubule organization via the gamma-tubulin ring complex in autosomal-recessive microcephaly with chorioretinopathy. *American Journal of Human Genetics, 96*, 666–674.

Schlosser, G., & Ahrens, K. (2004). Molecular anatomy of placode development in *Xenopus laevis*. *Developmental Biology, 271*, 439–466.

Schmitt, E. A., & Dowling, J. E. (1994). Early eye morphogenesis in the zebrafish, *Brachydanio rerio*. *Journal of Comparitive Neurology, 344*, 532–542.

Schmitt, E. A., & Dowling, J. E. (1999). Early retinal development in the zebrafish, *Danio rerio*: light and electron microscopic analyses. *Journal of Comparitive Neurology, 404*, 515–536.

Schorderet, D. F., Nichini, O., Boisset, G., Polok, B., Tiab, L., Mayeur, H., … Munier, F. L. (2008). Mutation in the human homeobox gene NKX5-3 causes an oculo-auricular syndrome. *American Journal of Human Genetics, 82*, 1178–1184.

Schraders, M., Ruiz-Palmero, L., Kalay, E., Oostrik, J., del Castillo, F. J., Sezgin, O., … Kremer, H. (2012). Mutations of the gene encoding otogelin are a cause of autosomal-recessive nonsyndromic moderate hearing impairment. *American Journal of Human Genetics, 91*, 883–889.

Seiler, C., Ben-David, O., Sidi, S., Hendrich, O., Rusch, A., Burnside, B., … Nicolson, T. (2004). Myosin VI is required for structural integrity of the apical surface of sensory hair cells in zebrafish. *Developmental Biology, 272*, 328–338.

Seiler, C., Finger-Baier, K. C., Rinner, O., Makhankov, Y. V., Schwarz, H., Neuhauss, S. C., & Nicolson, T. (2005). Duplicated genes with split functions: independent roles of protocadherin15 orthologues in zebrafish hearing and vision. *Development, 132*, 615–623.

Seiler, C., & Nicolson, T. (1999). Defective calmodulin-dependent rapid apical endocytosis in zebrafish sensory hair cell mutants. *Journal of Neurobiology, 41*, 424–434.

Self, T., Mahony, M., Fleming, J., Walsh, J., Brown, S. D., & Steel, K. P. (1998). Shaker-1 mutations reveal roles for myosin VIIA in both development and function of cochlear hair cells. *Development, 125*, 557–566.

Semina, E. V., Bosenko, D. V., Zinkevich, N. C., Soules, K. A., Hyde, D. R., Vihtelic, T. S., … Link, B. A. (2006). Mutations in laminin alpha 1 result in complex, lens-independent ocular phenotypes in zebrafish. *Developmental Biology, 299*, 63–77.

Sheehan-Rooney, K., Swartz, M. E., Zhao, F., Liu, D., & Eberhart, J. K. (2013). Ahsa1 and Hsp90 activity confers more severe craniofacial phenotypes in a zebrafish model of hypoparathyroidism, sensorineural deafness and renal dysplasia (HDR). *Disease Models & Mechanisms, 6*, 1285–1291.

Shen, Y. C., Jeyabalan, A. K., Wu, K. L., Hunker, K. L., Kohrman, D. C., Thompson, D. L., … Barald, K. F. (2008). The transmembrane inner ear (tmie) gene contributes to vestibular and lateral line development and function in the zebrafish (*Danio rerio*). *Developmental Dynamics, 237*, 941–952.

Shen, X., Liu, F., Wang, Y., Wang, H., Ma, J., Xia, W., … Ma, D. (2015). Down-regulation of msrb3 and destruction of normal auditory system development through hair cell apoptosis in zebrafish. *International Journal of Developmental Biology, 59*, 195–203.

Shen, Y. C., & Raymond, P. A. (2004). Zebrafish cone-rod (crx) homeobox gene promotes retinogenesis. *Developmental Biology, 269*, 237–251.

Shu, X., Zeng, Z., Gautier, P., Lennon, A., Gakovic, M., Cheetham, M. E., … Wright, A. F. (2011). Knockdown of the zebrafish ortholog of the retinitis pigmentosa 2 (RP2) gene results in retinal degeneration. *Investigative Ophthalmology & Visual Science, 52*, 2960–2966.

Shu, X., Zeng, Z., Gautier, P., Lennon, A., Gakovic, M., Patton, E. E., & Wright, A. F. (2010). Zebrafish Rpgr is required for normal retinal development and plays a role in dynein-based retrograde transport processes. *Human Molecular Genetics, 19*, 657–670.

Siemens, J., Kazmierczak, P., Reynolds, A., Sticker, M., Littlewood-Evans, A., & Muller, U. (2002). The Usher syndrome proteins cadherin 23 and harmonin form a complex by means of PDZ-domain interactions. *Proceedings of National Academy of Sciences of the United States of America, 99*, 14946–14951.

Simms, R. J., Hynes, A. M., Eley, L., Inglis, D., Chaudhry, B., Dawe, H. R., & Sayer, J. A. (2012). Modelling a ciliopathy: Ahi1 knockdown in model systems reveals an essential role in brain, retinal, and renal development. *Cellular and Molecular Life Sciences, 69*, 993–1009.

Skarie, J. M., & Link, B. A. (2008). The primary open-angle glaucoma gene WDR36 functions in ribosomal RNA processing and interacts with the p53 stress-response pathway. *Human Molecular Genetics, 17*, 2474–2485.

Skarie, J. M., & Link, B. A. (2009). FoxC1 is essential for vascular basement membrane integrity and hyaloid vessel morphogenesis. *Investigative Ophthalmology & Visual Science, 50*, 5026–5034.

Smith, W. K., Sankar, V., & Pfleiderer, A. G. (2005). A national survey amongst UK otolaryngologists regarding the treatment of Meniere's disease. *Journal of Laryngology and Otology, 119*, 102–105.

Smith, R. J. H., Shearer, A. E., Hildebrand, M. S., & Van Camp, G. (1993). Deafness and hereditary hearing loss overview. In R. A. Pagon, M. P. Adam, H. H. Ardinger, S. E. Wallace, A. Amemiya, L. J. H. Bean, et al. (Eds.), *GeneReviews(R), Seattle (WA)*.

Soens, Z. T., Li, Y., Zhao, L., Eblimit, A., Dharmat, R., Li, Y., ... Chen, R. (2016). Hypomorphic mutations identified in the candidate Leber congenital amaurosis gene CLUAP1. *Genetics in Medicine*.

Söllner, C., Rauch, G.-J., Siemens, J., Geisler, R., Schuster, S. C., The Tübingen 2000 Screen Consortium, & Nicolson, T. (2004). Mutations in cadherin 23 affect tip links in zebrafish sensory hair cells. *Nature, 428*(6986), 955–959.

Song, P., Dudinsky, L., Fogerty, J., Gaivin, R., & Perkins, B. D. (2016). Arl13b interacts with Vangl2 to regulate cilia and photoreceptor outer segment length in zebrafish. *Investigative Ophthalmology & Visual Science, 57*(10), 4517–4526.

Song, Y., Wang, M., Mao, F., Shao, M., Zhao, B., Song, Z., ... Gong, Y. (2013). Knockdown of Pnpla6 protein results in motor neuron defects in zebrafish. *Disease Models & Mechanisms, 6*, 404–413.

Stainier, D. Y., Weinstein, B. M., Detrich, H. W., 3rd, Zon, L. I., & Fishman, M. C. (1995). Cloche, an early acting zebrafish gene, is required by both the endothelial and hematopoietic lineages. *Development, 121*, 3141–3150.

Starr, C. J., Kappler, J. A., Chan, D. K., Kollmar, R., & Hudspeth, A. J. (2004). Mutation of the zebrafish choroideremia gene encoding Rab escort protein 1 devastates hair cells. *Proceedings of National Academy of Sciences of the United States of America, 101*, 2572–2577.

Stearns, G., Evangelista, M., Fadool, J. M., & Brockerhoff, S. E. (2007). A mutation in the cone-specific pde6 gene causes rapid cone photoreceptor degeneration in zebrafish. *Journal of Neuroscience, 27*, 13866–13874.

Stiebel-Kalish, H., Reich, E., Rainy, N., Vatine, G., Nisgav, Y., Tovar, A., ... Bach, M. (2012). Gucy2f zebrafish knockdown–a model for Gucy2d-related leber congenital amaurosis. *European Journal of Human Genetics, 20*, 884–889.

Stooke-Vaughan, G. A., Obholzer, N. D., Baxendale, S., Megason, S. G., & Whitfield, T. T. (2015). Otolith tethering in the zebrafish otic vesicle requires Otogelin and alpha-Tectorin. *Development, 142*, 1137–1145.

Street, V. A., Kallman, J. C., & Kiemele, K. L. (2004). Modifier controls severity of a novel dominant low-frequency MyosinVIIA (MYO7A) auditory mutation. *Journal of Medical Genetics, 41*, e62.

Strupp, M., Hupert, D., Frenzel, C., Wagner, J., Hahn, A., Jahn, K., ... Brandt, T. (2008). Long-term prophylactic treatment of attacks of vertigo in Meniere's disease–comparison of a high with a low dosage of betahistine in an open trial. *Acta Oto-Laryngologica, 128*, 520–524.

Strupp, M., Thurtell, M. J., Shaikh, A. G., Brandt, T., Zee, D. S., & Leigh, R. J. (2011). Pharmacotherapy of vestibular and ocular motor disorders, including nystagmus. *Journal of Neurology, 258*, 1207–1222.

Strupp, M., Zingler, V. C., Arbusow, V., Niklas, D., Maag, K. P., Dieterich, M., … Brandt, T. (2004). Methylprednisolone, valacyclovir, or the combination for vestibular neuritis. *The New England Journal of Medicine, 351*, 354–361.

Sukumaran, S., & Perkins, B. D. (2009). Early defects in photoreceptor outer segment morphogenesis in zebrafish ift57, ift88 and ift172 Intraflagellar Transport mutants. *Vision Research, 49*, 479–489.

Suli, A., Watson, G. M., Rubel, E. W., & Raible, D. W. (2012). Rheotaxis in larval zebrafish is mediated by lateral line mechanosensory hair cells. *PLoS One, 7*, e29727.

Sun, Y., Chen, J., Sun, H., Cheng, J., Li, J., Lu, Y., … Yuan, H. (2011). Novel missense mutations in MYO7A underlying postlingual high- or low-frequency non-syndromic hearing impairment in two large families from China. *Journal of Human Genetics, 56*, 64–70.

Tachibana, M., Hara, Y., Vyas, D., Hodgkinson, C., Fex, J., Grundfast, K., & Arnheiter, H. (1992). Cochlear disorder associated with melanocyte anomaly in mice with a transgenic insertional mutation. *Molecular and Cellular Neurosciences, 3*, 433–445.

Tachibana, M., Kobayashi, Y., & Matsushima, Y. (2003). Mouse models for four types of Waardenburg syndrome. *Pigment Cell Research, 16*, 448–454.

Takamori, S., Malherbe, P., Broger, C., & Jahn, R. (2002). Molecular cloning and functional characterization of human vesicular glutamate transporter 3. *EMBO Reports, 3*, 798–803.

Takeda, N., Morita, M., Hasegawa, S., Kubo, T., & Matsunaga, T. (1989). Neurochemical mechanisms of motion sickness. *American Journal of Otolaryngology, 10*, 351–359.

Tavazzani, E., Spaiardi, P., Zampini, V., Contini, D., Manca, M., Russo, G., … Masetto, S. (2016). Distinct roles of Eps8 in the maturation of cochlear and vestibular hair cells. *Neuroscience, 328*, 80–91.

Thomas, S., Wright, K. J., Le Corre, S., Micalizzi, A., Romani, M., Abhyankar, A., … Attie-Bitach, T. (2014). A homozygous PDE6D mutation in Joubert syndrome impairs targeting of farnesylated INPP5E protein to the primary cilium. *Human Mutation, 35*, 137–146.

Toriello, H. V., Reardon, W., & Gorlin, R. J. (2004). *Hereditary hearing loss and its syndromes*. New York: Oxford University Press.

Torok, N. (1977). Old and new in Meniere disease. *The Laryngoscope, 87*, 1870–1877.

Udovichenko, I. P., Gibbs, D., & Williams, D. S. (2002). Actin-based motor properties of native myosin VIIa. *Journal of Cell Science, 115*, 445–450.

Valdivia, L. E., Lamb, D. B., Horner, W., Wierzbicki, C., Tafessu, A., Williams, A. M., … Cerveny, K. L. (2016). Antagonism between Gdf6a and retinoic acid pathways controls timing of retinal neurogenesis and growth of the eye in zebrafish. *Development, 143*, 1087–1098.

Varshney, G. K., Pei, W., LaFave, M. C., Idol, J., Xu, L., Gallardo, V., … Burgess, S. M. (2015). High-throughput gene targeting and phenotyping in zebrafish using CRISPR/Cas9. *Genome Research, 25*, 1030–1042.

Verhoeven, K., Van Camp, G., Govaerts, P. J., Balemans, W., Schatteman, I., Verstreken, M., … Willems, P. J. (1997). A gene for autosomal dominant nonsyndromic hearing loss (DFNA12) maps to chromosome 11q22-24. *American Journal of Human Genetics, 60*, 1168–1173.

Verpy, E., Leibovici, M., Zwaenepoel, I., Liu, X. Z., Gal, A., Salem, N., … Petit, C. (2000). A defect in harmonin, a PDZ domain-containing protein expressed in the inner ear sensory hair cells, underlies Usher syndrome type 1C. *Nature Genetics, 26*, 51–55.

Viringipurampeer, I. A., Shan, X., Gregory-Evans, K., Zhang, J. P., Mohammadi, Z., & Gregory-Evans, C. Y. (2014). Rip3 knockdown rescues photoreceptor cell death in blind pde6c zebrafish. *Cell Death and Differentiation, 21*, 665–675.

Wasfy, M. M., Matsui, J. I., Miller, J., Dowling, J. E., & Perkins, B. D. (2014). Myosin 7aa(-/-) mutant zebrafish show mild photoreceptor degeneration and reduced electroretinographic responses. *Experimental Eye Research, 122*, 65–76.

Waters, A. M., & Beales, P. L. (2011). Ciliopathies: an expanding disease spectrum. *Pediatric Nephrology, 26*(7), 1039–1056.

Weil, D., Blanchard, S., Kaplan, J., Guilford, P., Gibson, F., Walsh, J., et al. (1995). Defective myosin VIIA gene responsible for Usher syndrome type 1B. *Nature, 374*, 60–61.

Weil, D., El-Amraoui, A., Masmoudi, S., Mustapha, M., Kikkawa, Y., Laine, S., … Petit, C. (2003). Usher syndrome type I G (USH1G) is caused by mutations in the gene encoding SANS, a protein that associates with the USH1C protein, harmonin. *Human Molecular Genetics, 12*, 463–471.

Weil, D., Kussel, P., Blanchard, S., Levy, G., Levi-Acobas, F., Drira, M., … Petit, C. (1997). The autosomal recessive isolated deafness, DFNB2, and the Usher 1B syndrome are allelic defects of the myosin-VIIA gene. *Nature Genetics, 16*, 191–193.

Weiner, A. M., Scampoli, N. L., & Calcaterra, N. B. (2012). Fishing the molecular bases of Treacher Collins syndrome. *PLoS One, 7*, e29574.

Wells, A. L., Lin, A. W., Chen, L. Q., Safer, D., Cain, S. M., Hasson, T., … Sweeney, H. L. (1999). Myosin VI is an actin-based motor that moves backwards. *Nature, 401*, 505–508.

Wen, R., Dallman, J. E., Li, Y., Zuchner, S. L., Vance, J. M., Pericak-Vance, M. A., & Lam, B. L. (2014). Knock-down DHDDS expression induces photoreceptor degeneration in zebrafish. *Advances in Experimental Medicine and Biology, 801*, 543–550.

Whitfield, T. T., Granato, M., van Eeden, F. J., Schach, U., Brand, M., Furutani-Seiki, M., … Nusslein-Volhard, C. (1996). Mutations affecting development of the zebrafish inner ear and lateral line. *Development, 123*, 241–254.

Whitfield, T. T., & Hammond, K. L. (2007). Axial patterning in the developing vertebrate inner ear. *International Journal of Developmental Biology, 51*, 507–520.

Whitfield, T. T., Riley, B. B., Chiang, M. Y., & Phillips, B. (2002). Development of the zebrafish inner ear. *Developmental Dynamics, 223*, 427–458.

Whitney, S. L., & Rossi, M. M. (2000). Efficacy of vestibular rehabilitation. *Otolaryngologic Clinics of North America, 33*, 659–672.

World Health Organization, 2015. http://www.who.int/mediacentre/factsheets/fs282/en/. http://www.who.int/mediacentre/factsheets/fs300/en/.

Wiener-Vacher, S. R., Ledebt, A., & Bril, B. (1996). Changes in otolith VOR to off vertical axis rotation in infants learning to walk. Preliminary results of a longitudinal study. *Annals of the New York Academy of Sciences, 781*, 709–712.

Winter, M. J., Redfern, W. S., Hayfield, A. J., Owen, S. F., Valentin, J. P., & Hutchinson, T. H. (2008). Validation of a larval zebrafish locomotor assay for assessing the seizure liability of early-stage development drugs. *Journal of Pharmacological and Toxicological Methods, 57*, 176–187.

Wright, K. J., Baye, L. M., Olivier-Mason, A., Mukhopadhyay, S., Sang, L., Kwong, M., … Jackson, P. K. (2011). An ARL3-UNC119-RP2 GTPase cycle targets myristoylated NPHP3 to the primary cilium. *Genes & Development, 25*, 2347–2360.

Wu, J. H., Liu, J. H., Ko, Y. C., Wang, C. T., Chung, Y. C., Chu, K. C., … Chung, M. Y. (2016). Haploinsufficiency of RCBTB1 is associated with Coats disease and familial exudative vitreoretinopathy. *Human Molecular Genetics, 25*, 1637–1647.

Wu, L., Pan, L., Zhang, C., & Zhang, M. (2012). Large protein assemblies formed by multivalent interactions between cadherin23 and harmonin suggest a stable anchorage structure at the tip link of stereocilia. *Journal of Biological Chemistry, 287*, 33460–33471.

Yao, Q., DeSmidt, A. A., Tekin, M., Liu, X., & Lu, Z. (2016). Hearing assessment in zebrafish during the first week postfertilization. *Zebrafish, 13*, 79–86.

Yariz, K. O., Duman, D., Seco, C. Z., Dallman, J., Huang, M., Peters, T. A., … Tekin, M. (2012). Mutations in OTOGL, encoding the inner ear protein otogelin-like, cause moderate sensorineural hearing loss. *American Journal of Human Genetics, 91*, 872–882.

Yasunaga, S., Grati, M., Chardenoux, S., Smith, T. N., Friedman, T. B., Lalwani, A. K., … Petit, C. (2000). OTOF encodes multiple long and short isoforms: genetic evidence that the long ones underlie recessive deafness DFNB9. *American Journal of Human Genetics, 67*, 591–600.

Yasunaga, S., Grati, M., Cohen-Salmon, M., El-Amraoui, A., Mustapha, M., Salem, N., … Petit, C. (1999). A mutation in OTOF, encoding otoferlin, a FER-1-like protein, causes DFNB9, a nonsyndromic form of deafness. *Nature Genetics, 21*, 363–369.

Yau, K. W., & Hardie, R. C. (2009). Phototransduction motifs and variations. *Cell, 139*, 246–264.

Yin, J., Brocher, J., Fischer, U., & Winkler, C. (2011). Mutant Prpf31 causes pre-mRNA splicing defects and rod photoreceptor cell degeneration in a zebrafish model for Retinitis pigmentosa. *Molecular Neurodegeneration, 6*, 56.

Zalewski, C. K. (2015). Aging of the human vestibular system. *Seminars in Hearing, 36*, 175–196.

Zeddies, D. G., & Fay, R. R. (2005). Development of the acoustically evoked behavioral response in zebrafish to pure tones. *Journal of Experimental Biology, 208*, 1363–1372.

Zhai, W., Jin, X., Gong, Y., Qu, L. H., Zhao, C., & Li, Z. H. (2015). Phenotype of Usher syndrome type II associated with compound missense mutations of c.721 C>T and c.1969 C>T in MYO7A in a Chinese Usher syndrome family. *International Journal of Ophthalmology, 8*, 670–674.

Zhang, C., Zhang, Q., Wang, F., & Liu, Q. (2015). Knockdown of poc1b causes abnormal photoreceptor sensory cilium and vision impairment in zebrafish. *Biochemical and Biophysical Research Communications, 465*, 651–657.

Zinkevich, N. S., Bosenko, D. V., Link, B. A., & Semina, E. V. (2006). Laminin alpha 1 gene is essential for normal lens development in zebrafish. *BMC Developmental Biology, 6*, 13.

Zou, P., Wu, S. Y., Koteiche, H. A., Mishra, S., Levic, D. S., Knapik, E., … McHaourab, H. S. (2015). A conserved role of alphaA-crystallin in the development of the zebrafish embryonic lens. *Experimental Eye Research, 138*, 104–113.

Zuchner, S., Dallman, J., Wen, R., Beecham, G., Naj, A., Farooq, A., … Pericak-Vance, M. A. (2011). Whole-exome sequencing links a variant in DHDDS to retinitis pigmentosa. *American Journal of Human Genetics, 88*, 201–206.

PART

VII

Cancer

CHAPTER

A zebrafish xenograft model for studying human cancer stem cells in distant metastasis and therapy response

17

L. Chen[*,a], A. Groenewoud[*,a], C. Tulotta[*], E. Zoni[§,¶], M. Kruithof-de Julio[§,¶], G. van der Horst[¶], G. van der Pluijm[¶], B. Ewa Snaar-Jagalska[*,1]

*Leiden University, Leiden, The Netherlands

§University of Bern, Bern, Switzerland

¶Leiden University Medical Centre, Leiden, The Netherlands

[1]Corresponding author: E-mail: b.e.snaar-jagalska@biology.leidenuniv.nl

CHAPTER OUTLINE

[a]Co-first authors.

Methods in Cell Biology, Volume 138, ISSN 0091-679X, http://dx.doi.org/10.1016/bs.mcb.2016.10.009

Abstract

Lethal and incurable bone metastasis is one of the main causes of death in multiple types of cancer. A small subpopulation of cancer stem/progenitor-like cells (CSCs), also known as tumor-initiating cells from heterogenetic cancer is considered to mediate bone metastasis. Although over the past decades numerous studies have been performed in different types of cancer, it is still difficult to track small numbers of CSCs during the onset of metastasis. With use of noninvasive high-resolution imaging, transparent zebrafish embryos can be employed to dynamically visualize cancer progression and reciprocal interaction with stroma in a living organism. Recently we established a zebrafish CSC-xenograft model to visually and functionally analyze the role of CSCs and their interactions with the microenvironment at the onset of metastasis. Given the highly conserved human and zebrafish genome, transplanted human cancer cells are able to respond to zebrafish cytokines, modulate the zebrafish microenvironment, and take advantage of the zebrafish stroma during cancer progression. This chapter delineates the zebrafish CSC-xenograft model as a useful tool for both CSC biological study and anticancer drug screening.

INTRODUCTION

Bone metastasis is the main cause of death in multiple types of carcinoma. In prostate cancer, for instance, 30% of patients still face a risk of mortal relapse and bone metastasis after initial positive clinical response (Jemal, Center, DeSantis, & Ward, 2010). Over the past years, cancer heterogeneity has been suggested to play an important role in tumor development and metastasis. A small subpopulation of cancer cells is hallmarked by self-renewal capabilities, high motility, tumorigenity, and chemoresistance, and they are defined as tumor-initiating cells (TICs) or cancer stem/progenitor-like cells (CSCs). These TICs/CSCs are believed to mediate relapse and bone metastasis in the final stage of cancer progression (Balic et al., 2006; Hermann et al., 2007; Reya, Morrison, Clarke, & Weissman, 2001). Generally combinations of cell surface markers and elevated aldehyde dehydrogenase (ALDH) activity are used for TICs/CSC identification, e.g., $CD44^{+}/ALDH^{hi}/CD24^{-}$ and $CD133^{+}/\alpha2\beta1^{hi}/CD44^{+}/ALDH^{hi}$ in human breast and prostate cancer respectively (Al-Hajj, Wicha, Benito-Hernandez, Morrison, & Clarke, 2003; Collins, Berry, Hyde, Stower, & Maitland, 2005; van den Hoogen et al., 2010).

Metastasis is a highly dynamic process initiated by cancer cells after they escape from the primary tumor followed by intravasation, survival in circulation, extravasation, and colonization at a secondary site. At each step cancer cells have to deal with changes in the microenvironment such as mechanical stress, oxygenic concentration, immune response, stromal cells, and cytokines. Hence, after escaping the primary tumor most disseminated cells will die and only few survive to subsequently give rise to a metastasis (Hanahan & Weinberg, 2011). Accumulating evidence indicates that metastasis is often triggered by a reversible dynamic process called epithelial to mesenchymal transition (EMT), which inhibits cell—cell junction and endows the cells with enhanced motility and survivability (Singh & Settleman, 2010; Thiery, 2003). Moreover, in some studies EMT is implicated to elicit CSCs in metastasis after extravasation (Tam & Weinberg, 2013). Nevertheless, others suggest a reversed program of EMT called mesenchymal to epithelial transition, which is required for eliciting CSCs that subsequently facilitate further metastatic outgrowth (Celià-Terrassa et al., 2012; Ocaña et al., 2012). Although advanced imaging techniques are utilized to observe dissemination of cancer cells from primary tumor in mice models, it is still difficult to track low amount of CSCs during extravasation, colonization, and interaction with microenvironment at early stage of metastasis in real time.

Since the last decade, the zebrafish model (*Danio rerio*) has been developed as a powerful tool for fundamental research in cancer biology and anticancer drugs screening (Goessling, North, & Zon, 2007; Stoletov & Klemke, 2008). Many molecular and cellular components involved in tumorigenity are highly conserved between zebrafish and human making the platform clinically relevant (Amatruda, Shepard, Stern, & Zon, 2002; Barriuso, Nagaraju, & Hurlstone, 2015; Howe et al., 2013; Zon & Peterson, 2005). Zebrafish embryos xenograft models have been established to study several types of cancers including prostate cancer, breast cancer, colon, melanoma, Ewing sarcoma, and leukemia (Drabsch, He, Zhang, Snaar-Jagalska, & ten Dijke, 2013; van der Ent, Burrello, et al., 2014; van der Ent, Jochemsen, et al., 2014; Ghotra et al., 2015; Konantz et al., 2012; Tulotta et al., 2016). Given that the adaptive immune system in zebrafish does not reach maturity until 4 weeks postfertilization, cell graft—host rejection can be circumvented using early developmental stages (van der Ent, Burrello, et al., 2014; van der Ent, Jochemsen, et al., 2014; Haldi, Ton, Seng, & McGrath, 2006; Lam, Chua, Gong, Lam, & Sin, 2004; Lieschke & Trede, 2009; Trede, Langenau, Traver, Look, & Zon, 2004). One of the main advantages of zebrafish embryos is their optical clarity. Use of transgenic lines with fluorescent vasculature, neutrophils, and macrophages allows noninvasive live imaging of cancer development and interaction with the microenvironment at single-cell resolution within 1 week (Ellett, Pase, Hayman, Andrianopoulos, & Lieschke, 2011; He et al., 2012; Nicoli, Ribatti, Cotelli, & Presta, 2007; Renshaw et al., 2006; Stoletov & Klemke, 2008; Zakrzewska et al., 2010). In our previous studies, we showed that after intravascular injection human cancer cells recapitulate all stages of cancer development including primary growth, tumor-induced angiogenesis, interaction with immune cells, and micrometastasis at caudal hematopoietic tissue (CHT). The CHT is the area where hematopoietic stem cells (HSC) emerge from the dorsal aorta in zebrafish embryos, some of the molecular cues of the CHT also exist in mammalian bone marrow (BM) (Table 1) (Kissa & Herbomel, 2010; Sacco et al. 2016; Tulotta et al., 2016). Due to the fact that the context dependent signaling of hematopoietic niche of the BM in mammals is known to modulate cancer stem plasticity during bone metastasis, the experimental metastasis occurring in the CHT, which recapitulates a BM-like metastatic niche, provides an opportunity to investigate reciprocal interaction between hematopoietic niche and cancer stemness at the early steps of bone metastasis. Moreover, combining the next generation RNA sequencing techniques (RNAseq) with CRISPR-cas9 gene knockout enables understanding of the molecular and cellular components from the hematopoietic niche that facilitate the metastatic growth of CSCs (Amsterdam & Hopkins, 2006; Hwang et al., 2013). Therefore, the zebrafish embryo is an ideal tool for studying metastatic behavior of CSCs within comprehensive microenvironment in vivo. In this chapter we report the zebrafish CSC-xenograft model for investigation of metastatic onset and targeting of tumor-stroma interaction as a novel approach in target discovery.

1. ESTABLISHMENT OF HUMAN CANCER STEM CELLS XENOGRAFT MODEL

1.1 RATIONALE

Over the past 10 years, zebrafish embryo xenograft models have been used to study cancer progression for multiple types of carcinomas (Konantz et al., 2012). The optical transparency of the embryos enable the visualization of metastasis initiation and early tumor progression at single-cell level (He et al., 2012; Stoletov et al., 2010) Selective methods of transplantation are employed in combination with different fluorescent lineage reporter fish lines according to research goals or cancer features. Yolk injection is successfully used for high-throughput anticancer drug screenings in prostate cancer, Ewing sarcoma, and uveal melanoma, and leukemia (Corkery, Dellaire, & Berman, 2011; van der Ent, Burrello, et al., 2014; van der Ent, Jochemsen, et al., 2014; Ghotra et al., 2015; Pruvot et al., 2011). Orthotopic transplantation of human cancer cells into eyes and head cavity of the embryos can be used to study both primary growth and metastatic phenotype of retinoblastoma and pediatric brain tumors (Chen et al., 2015; Eden et al., 2015). Tumor-induced angiogenesis is a crucial pathological process that ensured necessary oxygen and nutrient supplies to expanding tumors. Targeting of neovasculature is a clinical approach for anticancer therapy (Vecchiarelli-Federico et al., 2010). Typically transplanting cancer cells into either perivitelline space or duct of Cuvier (DoC) can elicit strong tumor-induced angiogenesis accompanying localized tumor growth. These

Table 1 Molecular Components Facilitating Zebrafish Hematopoiesis and Mammalian Cancer Metastatic Progression

Molecular Components in Caudal Hematopoietic Tissue	Functions in Zebrafish Hematopoiesis	References	Functions in Cancer Progression	References
PGE2	Regulating HSC division	North et al. (2007)	Modulating EMT and promoting CSCs repopulation	Kurtova et al. (2015)
CXCL12	Regulating HSC homing into CHT	Glass et al. (2011)	Regulating cancer cells homing into bone	Shiozawa et al. (2011)
Cripto-1	Functioning at early embryonic development	Zhang, Talbot, and Schier (1998)	Facilitating tumorigenity	Strizzi, Bianco, Normanno, and Salomon (2005)
TGF-β family	Blocking proliferation of progenitor cells at tail	Casari et al. (2014) and Lengerke et al. (2008)	Driving cancer cell EMT process, invasion, and CSC repopulation	Chaffer et al. (2013)
Wnt family	Regulating HSC division; specifying hematopoietic fate	Goessling et al. (2009)	Regulating cancer stem cells	Reya et al. (2001)
FGF family	Regulating HSC specification	Lee et al. (2014)	Triggering EMT	Thiery et al. (2009)
Notch ligands	Controlling hematopoietic cell fate decisions	Burns, Traver, Mayhall, Shepard, and Zon (2005) and Pouget and Traver (2016)	Regulating self-renewal of CSCs	Takebe et al. (2015)
Hedgehog family	Polarizing HSCs emergence from the dorsal aorta	Wilkinson et al. (2009)	Regulating CSC self-renewal and tumorigenity	Clement, Sanchez, De Tribolet, Radovanovic, and i Altaba (2007)

CHT, *caudal hematopoietic tissue;* CSC, *cancer stem/progenitor-like cells;* EMT, *epithelial to mesenchymal transition;* FGF, *fibroblast growth factor;* HSC, *hematopoietic stem cells;* TGF-β, *transforming growth factor-beta.*

models have been used to study the role of hypoxia and macrophage in tumor-induced angiogenesis (He et al., 2012; Lee et al., 2009).

Metastasis is a long and complex, multistep process requiring the establishment of a primary tumor and subsequent dissemination followed by colonization at a distant site. Cancer cells can be directly transplanted into the blood circulation via the DoC to mimic metastatic colonization and outgrowth in a short time window [normally within 3–6 days postimplantation (dpi)]. The DoC, also known as the embryonic common cardinal vein, is a wide circulation channel on the embryonic yolk sac connecting the heart to the trunk vasculature. When 100–500 cancer cells are transplanted into DoC at 2 days postfertilization (2 dpf) most of the cells will immediately disseminate through the circulation while others will be retained at the injection site where a solid primary-like tumor mass is formed (He et al., 2012; Tulotta et al., 2016). After intravenous injection the majority of the cells die but a subset of the original engrafted population survive and lodge within narrow vessels in eyes, brain, trunk, and CHT. We observed that osteotropic cancer cell lines including prostate cancer cell line PC-3M-Pro4, triple-negative breast cancer cell line MDA-MB-231-B1, and bladder cancer cell line UM-UC-3 predominantly form a micrometastasis around the CHT (Tulotta et al., 2016). Independently of the same colonization sites at CHT, osteotropic cancer cells of different origin induce varying metastatic phenotypes (Fig. 1). For instance, MDA-MB-231-B1 and UM-UC-3 first colonize at the CHT and subsequently extravasate and invade into the neighboring tail fin as single cells. PC-3M-Pro4 develops into perivascular solid tumors, which further display collective invasion. Interestingly, the phenotypes induced by two engrafted human prostate cancer cells, PC-3 and PC-3M-Pro4, are dramatically different (Fig. 1). PC-3M-Pro4 is derived from the androgen-independent prostate cancer cells PC-3 by a fourfold orthotopic serial retransplantation in mice, which endows the cells with a stronger bone-metastatic capacity (Pettaway et al., 1996). In addition, PC-3M-Pro4 cells contain a higher number of CSCs as identified by high ALDH activity underlined by an increased expression of stemness markers (van den Hoogen et al., 2010). Therefore we decided to use this cell line to establish human CSC-xenograft model to mimic CSC-driven metastasis in zebrafish.

1.2 ISOLATION AND TRANSPLANTATION OF CANCER STEM/PROGENITOR-LIKE CELLS

A small subpopulation of cancer cells found in heterogeneous prostate tumors are CSCs. These cells exhibit the capacity of self-renewal, high motility, tumorigenity, and chemoresistance and have been suggested to initiate bone metastasis (Reya et al., 2001). Prostate CSC subpopulation is defined by expression of $CD133^{+}$, $CD44^{+}$, $\alpha 2\beta 1^{hi}$, $Cripto^{hi}$, and $ALDH1A1^{hi}$. The ALDH is a cluster of enzymes catalyzing the oxidation of aldehydes and was recently proven to be a marker of breast and prostate CSCs (Croker et al., 2009; Ginestier et al., 2007; van den Hoogen et al., 2010; Zoni et al., 2016). The cells with high ALDH activity are specifically labeled by ALDEFLUOR reagent, which is a fluorescent substrate, called

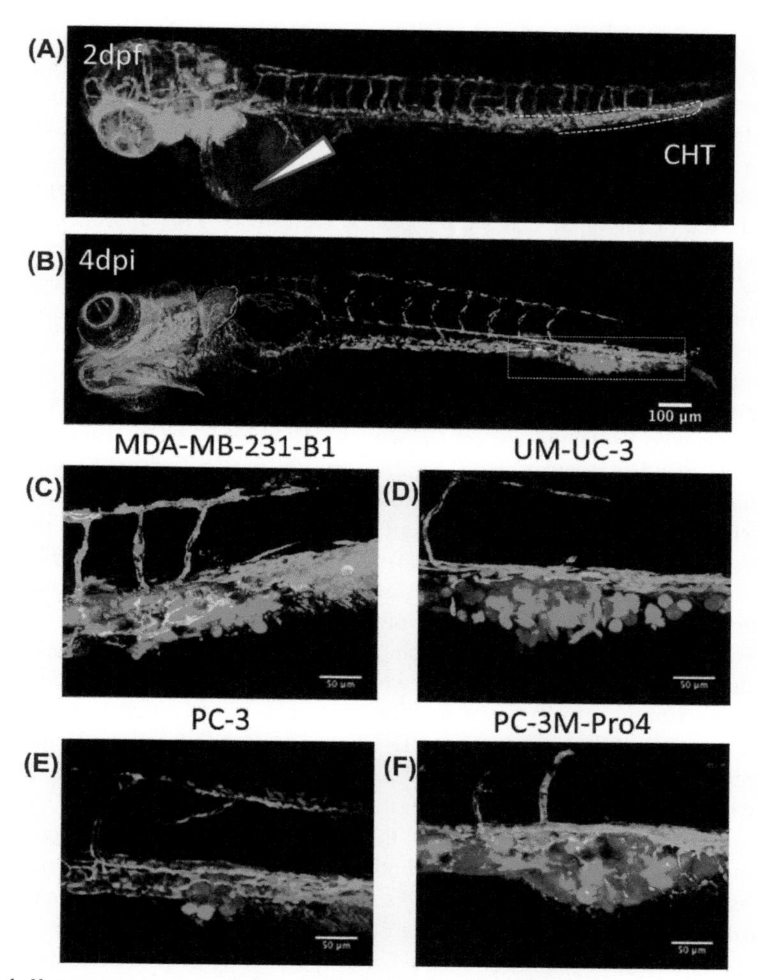

FIGURE 1 Human cancer xenograft model for studying metastatic onset.

(A) Human cancer cells (red) were injected into DoC at 2 days postfertilization (dpf). Cells disseminated into blood circulation (green) and were lodged at caudal hematopoietic tissue (CHT). (B) Perivascular micrometastasis was formed at 4 days postimplantation (dpi) at CHT. (C–D) The osteotropic, human triple-negative breast cancer cell line MDA-MB-231-B1 and the osteotropic bladder cancer cell line UM-UC-3 performed individual extravasation and invasion from CHT into the neighboring tail fin at 4 dpi. (E–F) Highly osteotropic PC-3M-Pro4 generated a perivascular metastasis at the CHT whereas parental PC-3 cells did not form a micrometastasis at 4 dpi.

BODIPY-aminoacetaldehyde that can freely diffuse into living cells and subsequently be converted to BODIPY-aminoacetate (BAA) by ALDH. Once produced, diffusion of BAA from the cells is restricted by incubation on ice, thus staining them allowing the cells to be sorted out with an fluorescence-activated cell sorting (FACS) sorter (Storms et al., 1999).

We isolated CSC cells from PC-3M-Pro4 to establish the CSC-xenograft model to mimic the behavior of CSCs during bone metastasis. This approach enables us functionally analyze how these cells interact with the CHT microenvironment. To visually track the cells, red fluorescent proteins were stably expressed in the cancer cells prior to transplantation into Fli:GFP embryos with green vasculature (Lawson & Weinstein, 2002).

After ALDEFLUOR reagent (stem cell technology) staining FACS sorting was performed following the manufacturer's directions. In brief, after trypsinization, cancer cells were collected and mixed with 1 mL ALDEFLUOR reagent. To define the ALDH-positive population, a negative control was generated by adding 5 μL diethylaminobenzaldehyde (DEAB), an ALDH inhibitor, into 500 μL of the cell suspension with ALDEFLUOR reagent. Afterward the samples and the 500 μL negative control were incubated for 30 min at 37°C respectively. $ALDH^{hi}$ subpopulation was defined as the cells with stronger fluorescein isothiocyanate signal compared to the negative control. Both subpopulations were sorted out by an FACS sorter followed by transplantation of 100−200 cells from both subpopulations into 2 dpf embryos through DoC. The whole injecting procedure should be finished within 2 h. Simultaneously, as a quality control, a clonogenicity assay was performed by seeding the remaining $ALDH^{hi}$ and $ALDH^{low}$ in six wells plates (100 cells/well) to measure the self-renewal capacity (van den Hoogen et al., 2010).

Because the optimal growth temperature of human cancer cells is 37°C but that for embryo is only 28°C, we incubated the transplanted fish at 34°C to compromise between the optimal temperatures of both species. After 3 hours postimplantation, embryos with failed injection were discarded. The behavior of the cancer cells was further followed in the subsequent 6 days (He et al., 2012).

To gain more knowledge about how CSCs specifically contribute to metastasis in a context of heterogenic tumor, $ALDH^{hi}$ cells were visually distinguished from the bulk population in a lineage tracing experiment. The sorted $ALDH^{hi}$ cancer cells (with intrinsic red fluorescence) were additionally labeled with green 5(6)-carboxyfluorescein N-hydroxysuccinimidyl ester dye and mixed with the remaining $ALDH^{low}$ cells with red fluorescence. After coinjection, the behavior of $ALDH^{hi}$ CSCs in the bulk heterogenic cancer cells was followed by confocal analysis (Fig. 2).

1.3 TRACKING MICROMETASTASES INITIATION, PROGRESSION, AND PHENOTYPE QUANTIFICATION

To compare the behavior between CSCs and non-CSCs after transplantation, high-resolution images were taken by a confocal microscope revealing the typical behavior of CSCs in each single process. By transplantation into different fluorescent reporter

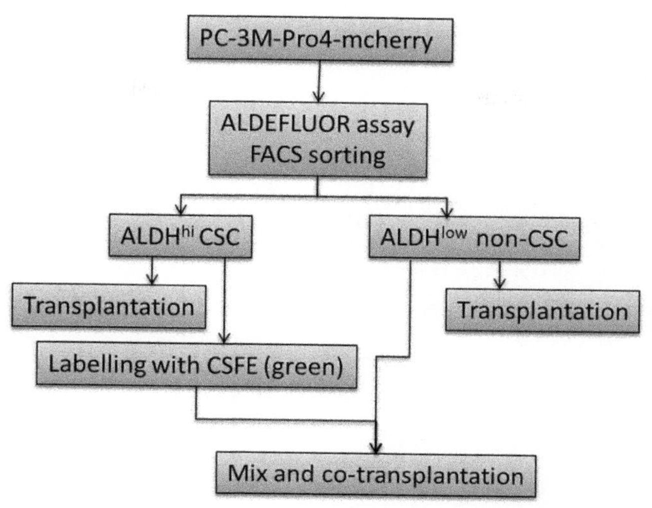

FIGURE 2 Scheme of establishment of human cancer stem/progenitor-like cells (CSC)-xenograft model to visualize behavior of CSCs in vivo.

fish lines, the dynamic interactions between cancer cells and components of microenvironment such as endothelial cells, myeloid cells, and mesenchymal stem cells were observed by time-lapse imaging (He et al., 2012; Tamplin et al., 2015). After transplantation, the CSCs disseminate throughout the whole body. Although some of the disseminated cells lodge in the blood vessels of the eyes, dorsal longitudinal anastomotic vessel (DLAV) and intersegmental blood vessels (ISV), the majority colonized the CHT and form micrometastases (Fig. 3A). Subsequently, the clustered cells either extravasated into the space between the caudal artery and caudal vein or collectively invaded into neighboring tail fin at 1–2 dpi. Subsequently, the extravasated cells started to rapidly proliferate, leading to the formation of a micrometastasis at 4 dpi. In contrast, the ALDH^{low} non-CSCs had significantly lower capacity of extravasation in the first 2 days and they proliferated slower than the ALDH^{hi} (Fig. 3B). To validate the phenotypes induced by ALDH^{hi} and ALDH^{low} cancer cells, a large group of embryos ($n > 25$) was imaged using a stereo fluorescent microscope. Images of whole CHT area were taken with the same settings at day 1, day 2, day 4, and day 6. Afterwards, total cancer cell burden at CHT area was counted with the ZF4 pixel counting program or an ImageJ software (Stoop et al., 2011; Tobin et al., 2010). The amount of total fluorescent pixels in each image was measured and the readouts were normalized as total cancer cell burden (Cui et al., 2011).

A high-speed and precise quantitative bioimaging platform established in our lab was adapted for the quantification. Automated imaging was performed with an automated stage confocal microscope for 6 dpi xenografts in a 96-well glass bottom plate and the images were processed with a predeveloped algorithm for automated analysis of cancer cell burden (Ghotra et al., 2012).

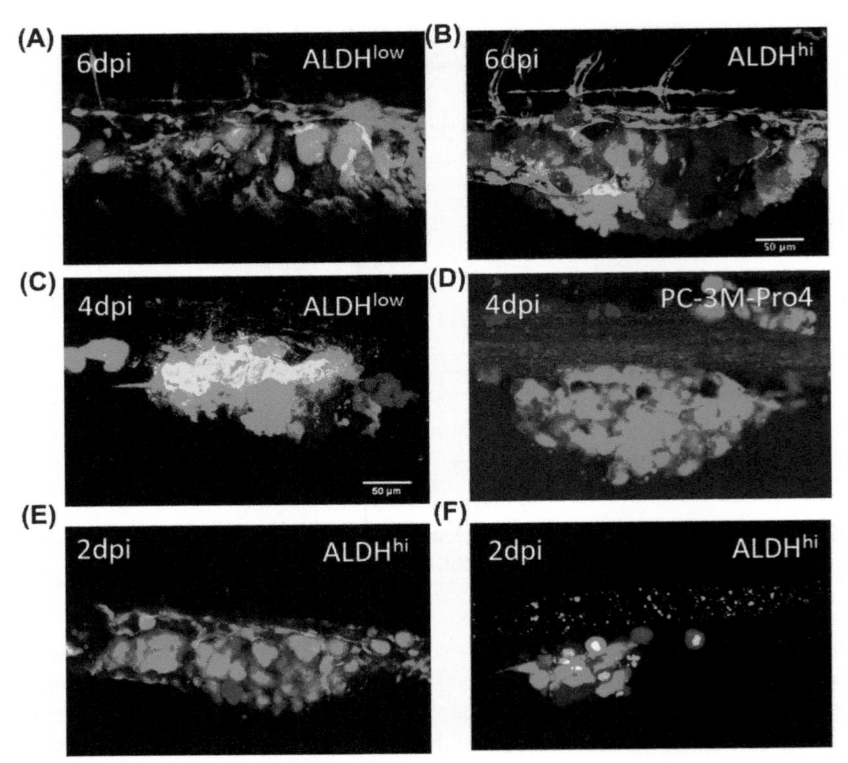

FIGURE 3 Visual analysis of cancer stem/progenitor-like cells (CSC) from PC-3M-Pro4 behavior in xenograft model.

(A–B) Aldehyde dehydrogenase(high) ($ALDH^{hi}$) subpopulation displayed enhanced metastatic capacity compared to the $ALDH^{low}$ cells. (C) CSCs repopulation (green) was observed in vivo by whole mount ALDEFLUOR staining on living embryos after transplantation of $ALDH^{low}$. (D) Recruitment of myeloid cells towards micrometastasis induced by PC-3M-Pro4 was detected by whole mount L-plastin immunostaining (Blue). (E–F) Proliferation of CSCs at caudal hematopoietic tissue was confirmed by phosphorylated Histone3 and ki-67 immunostaining (blue and cyan).

2. ANALYSIS OF INTERACTION BETWEEN CANCER STEM/ PROGENITOR-LIKE CELLS AND MICROENVIRONMENT

2.1 WHOLE MOUNT IMMUNOSTAINING TO DETERMINE CANCER STEM/PROGENITOR-LIKE CELLS PROLIFERATION, APOPTOSIS, AND INFLAMMATORY RESPONSE

The mechanism driving the change in CSCs molecular and cellular characteristics in response to signals from the metastatic niche at the initial stage remains elusive. The established zebrafish xenograft model for CSC study enables transcriptional and translational analysis of the cross-talk between cancer cells and microenvironments at the onset of metastasis. Recruitment of macrophages and neutrophils toward developing metastases, for instance, was observed by a combination of MPX and L-plastin staining (Mathias et al., 2007) (Fig. 3C). The MPX/L-plastin cells were detected to colocalize with neovasculature at primary tumor as well as with the invasive cancer cells at CHT. Genetic and pharmacological depletions of neutrophils and macrophages blocked neovasculature formation and invasion, respectively, indicating that those immune cells functionally participated in both processes (He et al., 2012). Immunohistochemical staining of Ki-67 is widely carried out by pathologists to detect proliferative cancer cells during diagnosis and phosphorylated Histone3 detection is used to observe cancer cells at the mitotic stage (Brenner et al., 2003; Scholzen & Gerdes, 2000). Both stainings were performed on engrafted zebrafish embryos to monitor the proliferative capacity of cancer cells at different stages of metastasis. The majority of p-Histone3 (Fig. 3E) and Ki-67 (Fig. 3F) positive cancer cells were detected in CHT at 4 and 6 dpi but not in other colonization sites like the DLAV or ISV, further confirming CHT is indeed the niche for cancer cell proliferation.

To check dynamical behavior of cancer cells during metastasis, phalloidin staining was performed to visualize the cytoskeleton. Phalloidins are small molecules belonging to a class of toxin which bind to F-actin. After conjugation with fluorescent tags, phalloidins specifically label the actin-based cytoskeleton (Cooper, 1987). The permeability of the embryo enables the phalloidin to freely infiltrate through the surface and stain the whole tumor tissue. As a result, the cytoskeletal remodeling within the cancer cells emphasizes the heterogeneity within the micrometastasis. Actin protrusion–rich cells at the leading edge show an invasive-like morphology stretching deep into tail fin. In contrast, the cells in the center keep a rounded morphology and tightly adhere to each other. This assay suggests that the collective invasion is active behavior led by cancer cells at invasive front of the metastasis.

The TUNEL assay is a well-known method for detecting apoptotic cells by staining nicks in DNA fragments due to activation of caspase signaling pathways (Kraupp et al., 1995). The TUNEL assay can be used as a direct readout for the measurement anticancer therapy efficacy in the zebrafish xenograft model (van der Ent, Burrello, et al., 2014; van der Ent, Jochemsen, et al., 2014).

2.1.1 Whole mount immunostaining in zebrafish embryos

Day 1:

1. Fix embryos for overnight (O/N) in cold 4% paraformaldehyde (PFA) in phosphate-buffered saline (PBS) at 4°C.
2. Wash embryos with phosphate-buffered saline tween (PBST) (0.1% Tween20 in PBS) for 3 × 5 min.
3. Dehydrate embryos using a methanol series (with 25% MeOH for 5 min, 50% MeOH for 10 min, and 75% MeOH for 5 min) subsequently store them in 100% MeOH at −20°C for O/N.

Day 2:

4. Rehydrate embryos in a methanol series with 75% MeOH for 5 min, 50% MeOH for 10 min, 25% MeOH for 5 min, and finally keep them in PBST.
5. Wash embryos for 3 × 10 min with PBS-TX (0.1% Triton-100 in PBS).
6. Permeabilize embryos with 10μg/mL protease K in PBS-TX at 37°C for 10 min and then wash embryos for 3 × 10 min with PBS-TX.
7. Block embryos in blocking buffer (5% goat serum in PBS-TX) at room temperature (RT) for more than 1 h.

For MPX activity detection

8. Wash embryos in 1× Trizmal (from peroxidase (Mpx) Leukocyte Kit, Sigma) containing 0.01% Tween 20 (TT buffer) for 5 min.
9. Incubate embryos in TT buffer containing 1.5 mg/mL substrate (supplied) and 0.015% H_2O_2 for 5–10 min at 37°C.

For regular immunostaining

10. Incubate embryos with first antibody in blocking buffer at RT for 2 h and store at 4°C for O/N.

Day 3:

11. Wash embryos for 4 × 15 min.
12. Incubate embryos with secondary antibody (1:200) and 4',6-diamidino-2-phenylindole (10μg/mL) in blocking buffer at RT for 2 h and store at 4°C for O/N.

Day 4:

13. Wash embryos with PBST for 4 × 15 min.

For phalloidin staining

14. Incubate embryos in 50 μL Alexa Fluor 488 Phalloidin (ThermoFisher) with a concentration of 1:200 dilution in PBX-TX for 1 h at RT in dark.
15. Wash for 4 × 15 min with PBST.
16. Fix embryos with 4% PFA in PBS for 20 min and wash for 3 × 15 min with PBST.
17. Mount embryos on slides with ProLong Antifade Reagent (ThermoFisher).

2.1.2 Whole mount apoptosis TUNEL assay in zebrafish embryos

Day 1:

1. Fix embryos for O/N in 4% PFA in PBS at 4°C.
2. Wash embryos with PBST for 3 × 5 min.
3. Dehydrate embryos in a methanol series with 25% MeOH for 5 min, 50% MeOH for 10 min, and 75% MeOH for 5 min and store them in 100% MeOH at −20°C for O/N.

Day 2:

4. Rehydrate embryos in a methanol series with 75% MeOH for 5 min, 50% MeOH for 10 min, and 25% MeOH for 5 min and keep them in PBST.
5. Permeabilize embryos with 10μg/mL protease K in PBST at 37°C for 10 min and then wash embryos for 3 × 10 min with PBST.

6. Postfix embryos with 4% PFA in PBST and wash them for 3 × 10 min.
7. Incubate embryos in premade ethanol/acetic acid (mixed by 2:1) at −20°C for 15 min.
8. Equilibrate embryos with 75 μL equilibration buffer from the kit for 15 min at RT.
9. Add 17 μL mixture (5 μL of transferase enzyme solution, 10 μL of reaction buffer, and 1.7 μL of 3% Triton-100) into embryos and incubate the mixture for 1 h on ice. Then incubate them at 37°C for another 1 h.
10. Stop the reaction by washing several times with PBST.
11. Block embryos for more than 1 h with blocking buffer (5% sheep serum in PBST).
12. Incubate embryos with antifluorescence conjugated to alkaline phosphatase with 1:5000 dilution in blocking buffer at 4°C for O/N.

Day 3:

13. Wash embryos in PBST for 2 h.
14. Stain embryos with TSA fluorescence buffer (1:50 dilution om PBST).
15. Wash embryos in PBST, store at 4°C.

2.2 IN VIVO ALDEFLUOR STAINING TO VISUALIZE CANCER STEM/PROGENITOR-LIKE CELLS IN XENOGRAFT

Shifting between CSCs and non-CSCs is nowadays a controversial issue. In a theory called one-way directional differentiation (Visvader, 2011), CSCs can be maintained by symmetric and/or asymmetric dividing. Once born from asymmetric division, the differentiated non-CSCs would permanently lose pluripotency and could not dedifferentiate into CSCs any more. In contrast, other evidence indicates that the shift between CSC and non-CSC is bidirectional so that non-CSCs retain the capacity of acquiring pluripotency under certain circumstances (Chaffer & Weinberg, 2015). Chaffer et al. (2013) have found that growth factors from the microenvironment, such as transforming growth factor-beta (TGF-β), stimulate EMT which results in CSC augmentation and increased tumorigenity. They suggest that this cell stem plasticity only functions when cancer cells are held in a certain epigenetic state. During bone metastasis of prostate cancer, a high percentage of CSCs are detected when cancer cells metastasize to BM and compete with HSC for the hematopoietic niche, suggestive that niche factors drive cancer stem plasticity (Shiozawa et al., 2011). To verify if the zebrafish hematopoietic niche enhances cancer stem plasticity when cancer cells colonize CHT, we employ the ALDEFLUOR reagent to stain the xenografts with $ALDH^{low}$ cells. Surprisingly, enhanced $ALDH^{hi}$ positive signals were indeed detected, indicating that molecular components from CHT endow the cancer cells with increased stemlike property (Fig. 3D).

2.2.1 *Whole mount staining with ALDEFLUOR reagent*

1. Anesthetize living embryos with 0.003% tricaine (Sigma) in Daneau buffer, divide them into two groups and load the two groups respectively on the bottom of two confocal petri dishes. Remove excess water.

2. Cover the embryos with a drop of melted low melting temperature agarose and wait for solidification.
3. Cover the embryos with a drop of 0.003% tricaine and remove the solidified agarose around the CHT part with a needle.
4. Stain CHT area of the embryos with 10 μL diluted ALDEFLUOR reagent (1:20 dilution in ALDEFLUOR Assay buffer) at 34°C for 1 h. For the negative control group, stain the CHT at 34°C for 1 h with diluted ALDEFLUOR reagent which has been mixed with DEAB with a ratio of 1:200.
5. Wash embryos three times with ALDEFLUOR reagent and image them using a confocal. A FIJI program will be used to remove background based on the negative control.

2.3 TRANSCRIPTOME ANALYSIS OF TUMOR-STROMA INTERACTIONS

As previously mentioned, interaction between stromal cells and cancer cells within metastatic niche plays a fundamental role in driving stem cells plasticity by facilitating the invasion and supporting metastatic growth. In mammals, osteotropic CSCs have similar molecular characteristics as HSC driving them to preferentially colonize the BM where CSCs compete with HSCs for the niche (Shiozawa et al., 2011). As a positive feedback loop, accumulating cancer cells in the niche modulates the surrounding stroma to produce molecular components, which in turn exacerbates metastatic progression (McAllister & Weinberg, 2014). Therefore, targeting the interaction between cancer cells and stroma in the metastatic niche could be a critical approach to suppress bone metastasis. Employment of next generation RNAseq provides the possibility for transcriptional analysis of cross-species interaction between human cancer cells and murine stromal cells during metastasis, uncovering the molecular components from the bone niche conveying the tumorigenity (Özdemir et al., 2014). Absence of tools for tracing the low amounts of disseminating CSCs in mouse models make transcriptomic analysis of newly formed metastasis difficult. Therefore how a metastatic niche attracts and activates cancer cells at the onset of metastasis has not been elucidated. Using the zebrafish xenograft model RNA of both zebrafish and human cancer cells can be resected and analyzed once the metastatic colonization has occurred.

Based on the results of immunostaining and ALDEFLUOR stain on zebrafish xenografts, we observed that cells in metastatic foci are endowed with enhanced stem-like characteristics. Engrafted cancer cells manipulate the microenvironment in the CHT by recruiting macrophages and remodeling neighboring blood vessels. Thus, the questions arise: how does the interaction between human cancer cells and the zebrafish CHT play a role in both stromal remodeling and CSCs dedifferentiation and whether the cell–stromal interactions are relevant to the interaction in mammalian BM. To this aim, we isolated RNA from PC-3M-Pro4 and MDA-MB-231-B1 cells in culture, from CHTs tissue where metastasis had been developed and from CHTs tissue of nonengrafted embryos. Subsequently RNAseq and qPCR analysis were performed to detect the signaling pathways involved in this process. The final goal of this strategy was to identify new genes in both cancer cells and stroma that regulate stem plasticity and metastatic capacity through stromal interaction. The biological functions of these genes will be further verified through tissue-specific gene knockout by CRISPR on zebrafish stroma and genetic interference on cancer cells (Ablain, Durand, Yang, Zhou, & Zon, 2015; Paddison, Caudy, Bernstein, Hannon, & Conklin, 2002).

2.3.1 RNA isolation from zebrafish caudal hematopoietic tissue

Day 1:

1. Incubate 6 dpi transplanted embryos with metastasis formation in CHT and the negative control embryos (PBS injected at 2 dpf) in sterile Daneau buffer and anesthetize embryos with 0.003% tricaine.
2. Place 10 embryos each time on a petri dish covered with solidified agarose (1.5% in Daneau buffer) and remove excess Daneau buffer.
3. Cut tails including whole CHT with a sterile scissor and remaining tissue.
4. Immediately transfer the cut tissues into ice-cold sterile PBS buffer into a 1.5 mL Eppendorf tube and put the tube on ice.
5. After collecting tissue from approximately 80 embryos, centrifuge the tube with maximum speed for 1 min.
6. Remove the supernatant and resuspend tissues in 500 μL TRIzol1 reagent (Invitrogen) followed with storing in −20°C.

Day 2:

7. Homogenize embryos by using a grinder, such as the MM 301 mixer mill (Retsch, 2 × 30 s at maximum frequency), that places a bead (1 mm diameter, Fabory) in the tube before grinding.
8. Transfer 100 μL chloroform into the tube and mix by vortex. Incubate the tube at RT for 5 min.
9. Centrifuge the tube at 8000 rpm at 4°C for 15 min and transfer the first layer into a new tube.
10. Add 1 volume of 70% ethanol (ETOH) into tubes and mix by vortex.
11. Transfer the upper phase to an RNeasy MinElute Cleanup Kit column (Qiagen). Centrifuge with 8000 rpm for 30 s at RT.
12. Discard the liquid in the collection tube and transfer 500 μL SRP buffer from the kit onto the upper layer of the column. Centrifuge with 8000 rpm for 30 s at RT.
13. Discard the liquid in the collecting tube and transfer 500 μL 80% ETOH onto the upper layer of the column. Centrifuge with 8000 rpm for 2 min at RT.
14. Insert the column into another sterile tube and centrifuge for 5 min at maximum speed at RT with the cap open.
15. Insert the column into another sterile tube and add 30 μL RNase-free water directly onto the membrane. Centrifuge at maximum speed for 1 min.

3. GENETIC AND CHEMICAL TARGETING OF TUMOR-STROMA INTERACTIONS IN CANCER STEM/PROGENITOR-LIKE CELLS—XENOGRAFT

3.1 FUNCTIONAL ANALYSIS OF HUMAN CANCER GENES

Because the multistep process of cancer metastasis is conveniently observed in the embryo, the zebrafish xenograft model delivers a powerful in vivo platform to systematically analyze biological functions of the candidate genes in the whole metastatic cascade.

TGF-β families are crucial EMT and CSC regulators, by inducing high expression of zinc finger E-box binding homeobox 1, an EMT transcriptional factor (Chaffer et al., 2013; Thiery, Acloque, Huang, & Nieto, 2009). Suppressing TGF-β receptors with pharmacological inhibitor SB431542 or genetically depleting Smad4, a key downstream component of the pathways, can significantly reduce breast cancer cell invasion from the CHT (Drabsch et al., 2013). Cripto, also known as TDGF1, is an epidermal growth factor—related protein and has been shown to play a crucial role in vertebrate embryogenesis, homeostasis, and tumorigenity. It is engaged in cancer progression by activating TGF-β subfamily proteins like Nodal, GDF-1, and GDF-3 (Miharada et al., 2011; Parisi et al., 2003; Watanabe et al., 2010). Recently, it has been found that high expression of Cripto associates with EMT and cancer stemness (Francescangeli et al., 2015; Terry et al., 2015).

Also our data showed that knockdown of Cripto and its membrane receptor GRP78, in CSC-enriched PC-3M-Pro4, suppresses extravasation and tumorigenity in zebrafish xenograft model (Zoni et al., 2016). These results underline that metastatic initiation and progression in the Zebrafish CHT is highly reliant on stem-like property of cancer cells (Zoni et al., 2016).

In a study by Zoni et al. (2015), a micro RNA (miRNA), miR-25, was identified as a tumor suppressor. MiRNAs are a class of noncoding small RNA molecules that modulate gene expression by specifically binding to the 3′-untranslated regions (3′-UTR) of mRNAs and subsequently induce mRNA degradation (Bartel, 2009). Despite the detection of high expression of miR-25 in several malignant cancers, miR-25 was found to be downregulated in prostate CSCs in both cell lines and patient tissues. When PC-3M-Pro4 cells with transient overexpression of miR-25 were transplanted into the embryos via DoC, the cells failed to either form clusters or extravasate in the CHT in first 2 days. Subsequent transcriptional analysis indicated that miR-25 facilitates cytoskeletal remodeling by targeting ITGAV and ITGA6. Importantly, these integrins have been uncovered as markers of prostate CSCs and key to skeletal metastasis (Collins et al., 2005; van den Hoogen et al., 2011). By utilizing the zebrafish CSC-xenograft model we further showed that cytoskeleton remodeling in CSCs controlled by integrin pathway signaling is a critical step for metastatic initiation by facilitating cluster formation and extravasation.

In line with this observation another study with a zebrafish xenograft model demonstrated that intravascular locomotion and extravasation of tumor cells are ITGB1 dependent (Stoletov & Klemke, 2008). The important role of integrin-mediated cytoskeleton remodeling and focal adherence in the cancer progression was further emphasized in the study of spleen tyrosine kinase (SYK) (Ghotra et al., 2015). SYK was identified as a novel anticancer target after an RNAi library screening using zebrafish xenograft model by monitoring cancer cell dissemination from yolk sac. We found that SYK inhibition attenuates cancer cell dissemination in zebrafish as well as reduces bone metastasis in a mice model by downregulation of ITGB1, ITGA2, and CD44 expression. Taken together, these zebrafish studies outline that the advanced metastatic capacity of CSCs is, at least partially, governed by integrin pathways.

3.2 FUNCTIONAL ANALYSIS OF STROMAL GENES

Although the fundamental role of stroma in supporting tumorigenity has been recognized for a long time, an in vivo tool to analyze the functions of specific genes of stromal cells in the niche is required. Zebrafish xenograft models are highly suitable for such study. The candidate genes expressed in the CHT and potentially involved in tumor-stroma axis are WNTs, TDGF-1, TGF-β, BMPs, and chemokine (C-X-C motif) ligand 12a/b (CXCL12a/b) among others (Table 1). The crucial role of cross-talk between human cancer cells and zebrafish stroma in regulating metastatic onset is previously underlined by the study of CXCR4/CXCL12a axis (Tulotta et al., 2016). In human, CXCL12 is chemokine protein secreted by mesenchymal stem cells in BM (Aiuti, Webb, Bleul, Springer, & Gutierrez-Ramos, 1997; Ma et al., 1998; Moore et al., 2001). CXCL12 facilitates hematopoiesis by regulating HSC homing into the BM by stimulating the CXCR4 receptor, expressed on the HSC membrane. Similar to HSC, human cancer cells expressing CXCR4 can also target BM by sensing the CXCL12 leading to the metastatic progression (Shiozawa et al., 2011). As in mammals, zebrafish mesenchymal stem cells with CXCL12a expression are present in CHT governing HSC colonization in perivascular niche (Tamplin et al., 2015). When we transplanted MDA-MB-231-B1 cells into CXCL12a$^{-/-}$/b$^{-/-}$ fish, the cells failed to form metastatic lesions at the CHT suggesting that cxcl12a/b is one of the zebrafish stroma signals controlling cancer homing at CHT.

Other genes regulated by tumor-stroma interactions in CSC-xenograft can be identified by Q-PCR and RNAseq after isolation of tissue surrounding the developing tumor (Fig. 4). The tissue-specific CRISPR knockdown approach (Ablain et al., 2015), where either Cas9 or gRNA expression is driven by tissue-specific promoters can be utilized to specifically target selected stromal host genes. Subsequently cancer cells can be transplanted into these mutants for functional analysis of metastatic onset and progression (Fig. 4). CRISPR knockdown approach can uncover how the stromal genes control the metastatic growth.

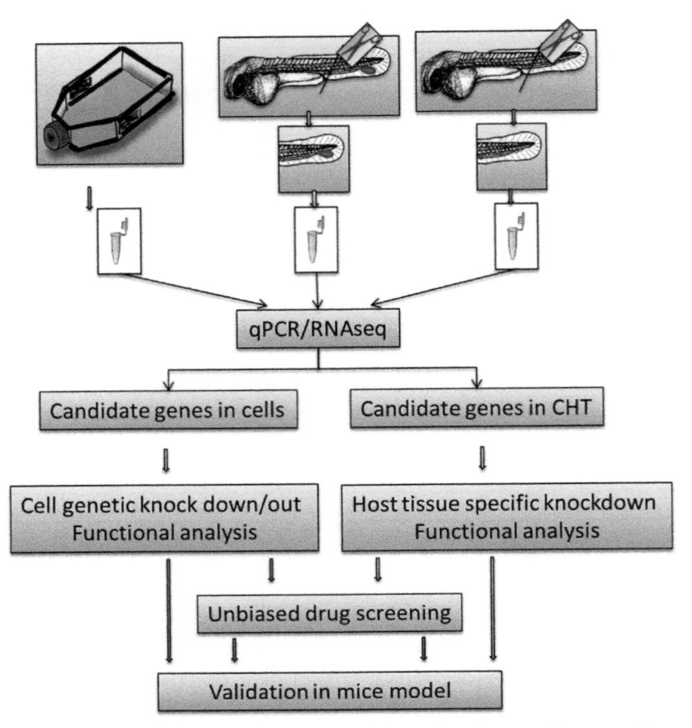

FIGURE 4 Scheme of identifying cancer–stroma interaction by species-specific transcriptomic analysis in situ (i.e., human vs. zebrafish).

RNA samples are collected from cells in culture, caudal hematopoietic tissue (CHT) metastasis and nonengrafted CHT. The biological functions of candidate genes engaging in metastasis can be analyzed by genetic interference and validated in existing preclinical mouse models. The final results can be utilized for drug screening. Assessment of potential gene/protein targets in metastasis can facilitate drug screening aiming at clinical translation.

3.3 DRUG SCREENING FOR NOVEL ANTICANCER STEM/PROGENITOR-LIKE CELL COMPOUNDS

Due to high fecundity, short breeding time, and easy handling, zebrafish provide an excellent in vivo platform for high-throughput drug screening. Hence, performing drug screening using zebrafish CSC-xenograft model would be an ideal strategy to discover novel compounds against the bone metastasis. Potential antitumor compounds can be added to the water as zebrafish embryo can absorb various small molecular weight compounds from water or cancer cells can be treated with selected compounds prior transplantation (Goessling et al., 2007; Tulotta et al., 2016; Zon & Peterson, 2005).

As a validation of this strategy we used iT1t to interfere with the CXCL12a-CXCR4 axis. When we pretreat the human cancer cells with the compound the cells failed to form metastatic lesions at the CHT after injection although cell viability was not affected by the prior treatment. This result shows that targeting CSCs and/or interfering cancer/stromal interactions are high efficient approaches to suppress cancer progression in the zebrafish xenograft model. Importantly, in contrast to screens using cell cultures, the drug-screening system with zebrafish xenograft model is capable of providing a global and comprehensive way to systemically evaluate the real anticancer efficacy of a given drug in a living context and is an ideal complementary platform to the conventional drug-screening strategy.

CONCLUSION

The zebrafish CSC-xenograft model provides a comprehensive platform to study human CSCsand their reciprocal interaction with microenvironment during metastatic progression. Noninvasive high-resolution imaging technologies enable real-time monitoring of CSCs behavior during the entire metastatic process at the organismal level. The zebrafish CSC-xenograft model is relevant for understanding the mechanisms of the early steps of bone metastasis in patients as the molecular and cellular cues from CHT regulate the stemness of human osteotropic cancer cells inducing tumorigenity. Making use of the model, we underlined the crucial role of the CSC cytoskeleton remodeling driven by integrin signaling pathways in CHT colonization and extravasation. We identified miR-25 as a tumor suppressor against metastasis by inhibiting cytoskeleton remodeling through its direct interactions with α_V and α_6-integrin expression. In addition, we showed that human cancer cells expressing CXCR4 receptor respond to zebrafish CXCL12a/b ligands in CHT and therefore performed metastasis. Interruption of CXCR4-CXCL12a/b axis by either suppressing cancer cell CXCR4 or zebrafish CXCL12a/b was sufficient to inhibit the experimental metastasis onset. Importantly, we found that molecular components from zebrafish CHT increased Cripto1 expression in human cancer cells enhancing cancer stem plasticity and invasiveness. Blocking Cripto1 and its coreceptor GRP78 reduced experimental metastasis in zebrafish and mice.

To further elaborate the mechanistic insights about the regulation of cancer stem plasticity by metastatic niche, we have developed a cross-species RNAseq method to identify potential novel candidate genes in tumor-stroma interaction responsible for metastatic onset. This approach provides an opportunity to globally analyze the reciprocal interaction between cancer cells and hematopoietic niche that governs cancer stem plasticity driving metastatic progression. Combining genetic interference and pharmacological inhibition of CSCs or the tumor-stroma axis, the role of a given gene in the metastatic progression can be functionally and dynamically analyzed in vivo. Coupling the advantage of the zebrafish CSC-xenograft model with other conventional systems such as 3D-culture and mouse cancer models would be a highly efficient strategy to uncover the underlying mechanisms of metastasis as well as discover novel therapeutic targets. A future application of this model would be the direct engraftment of patient CSC's for drug response prediction.

ACKNOWLEDGMENTS

The present work was supported by a personalized medicine grant from Alpe D'HuZes (AdH)/KWF PROPER entitled "Near-patient prostate cancer models for the assessment of disease prognosis and therapy" (UL2014-7058), European Union's Horizon 2020 research and innovation programme under grant agreement No 667787, the Netherlands Organization for Scientific Research (TOP GO Grant: 854.10.012), a KWF project grant entitled "Identification and characterization of tumor-initiating cells and supportive stroma involved in human bladder cancer progression and metastasis"(UL-2011-4930) and KWF UL2015-7599.

REFERENCES

Ablain, J., Durand, E. M., Yang, S., Zhou, Y., & Zon, L. I. (2015). A CRISPR/Cas9 vector system for tissue-specific gene disruption in zebrafish. *Developmental Cell, 32*(6), 756–764.

Aiuti, A., Webb, I. J., Bleul, C., Springer, T., & Gutierrez-Ramos, J. C. (1997). The chemokine SDF-1 is a chemoattractant for human CD34+ hematopoietic progenitor cells and provides a new mechanism to explain the mobilization of CD34+ progenitors to peripheral blood. *The Journal of Experimental Medicine, 185*(1), 111–120.

Al-Hajj, M., Wicha, M. S., Benito-Hernandez, A., Morrison, S. J., & Clarke, M. F. (2003). Prospective identification of tumorigenic breast cancer cells. *Proceedings of the National Academy of Sciences of the United States of America, 100*(7), 3983–3988.

Amatruda, J. F., Shepard, J. L., Stern, H. M., & Zon, L. I. (2002). Zebrafish as a cancer model system. *Cancer Cell, 1*(3), 229–231.

Amsterdam, A., & Hopkins, N. (2006). Mutagenesis strategies in zebrafish for identifying genes involved in development and disease. *Trends in Genetics, 22*(9), 473–478.

Balic, M., Lin, H., Young, L., Hawes, D., Giuliano, A., McNamara, G., … Cote, R. J. (2006). Most early disseminated cancer cells detected in bone marrow of breast cancer patients have a putative breast cancer stem cell phenotype. *Clinical Cancer Research, 12*(19), 5615–5621.

Barriuso, J., Nagaraju, R., & Hurlstone, A. (2015). Zebrafish: a new companion for translational research in oncology. *Clinical Cancer Research, 21*(5), 969–975.

Bartel, D. P. (2009). MicroRNAs: target recognition and regulatory functions. *Cell, 136*(2), 215–233.

Brenner, R. M., Slayden, O. D., Rodgers, W. H., Critchley, H. O., Carroll, R., Nie, X. J., & Mah, K. (2003). Immunocytochemical assessment of mitotic activity with an antibody to phosphorylated histone H3 in the macaque and human endometrium. *Human Reproduction, 18*(6), 1185–1193.

Burns, C. E., Traver, D., Mayhall, E., Shepard, J. L., & Zon, L. I. (2005). Hematopoietic stem cell fate is established by the Notch–Runx pathway. *Genes and Development, 19*(19), 2331–2342.

Casari, A., Schiavone, M., Facchinello, N., Vettori, A., Meyer, D., Tiso, N., … Argenton, F. (2014). A Smad3 transgenic reporter reveals TGF-beta control of zebrafish spinal cord development. *Developmental Biology, 396*(1), 81–93.

Celià-Terrassa, T., Meca-Cortés, Ó., Mateo, F., de Paz, A. M., Rubio, N., Arnal-Estapé, A., … Lozano, J. J. (2012). Epithelial-mesenchymal transition can suppress major attributes of human epithelial tumor-initiating cells. *The Journal of Clinical Investigation, 122*(5), 1849–1868.

Chaffer, C. L., Marjanovic, N. D., Lee, T., Bell, G., Kleer, C. G., Reinhardt, F., … Weinberg, R. A. (2013). Poised chromatin at the ZEB1 promoter enables breast cancer cell plasticity and enhances tumorigenicity. *Cell, 154*(1), 61–74.

Chaffer, C. L., & Weinberg, R. A. (2015). How does multistep tumorigenesis really proceed? *Cancer Discovery, 5*(1), 22–24.

Chen, X., Wang, J., Cao, Z., Hosaka, K., Jensen, L., Yang, H., … Cao, Y. (2015). Invasiveness and metastasis of retinoblastoma in an orthotopic zebrafish tumor model. *Scientific Reports, 5*.

Clement, V., Sanchez, P., De Tribolet, N., Radovanovic, I., & i Altaba, A. R. (2007). HEDGEHOG-GLI1 signaling regulates human glioma growth, cancer stem cell self-renewal, and tumorigenicity. *Current Biology, 17*(2), 165–172.

Collins, A. T., Berry, P. A., Hyde, C., Stower, M. J., & Maitland, N. J. (2005). Prospective identification of tumorigenic prostate cancer stem cells. *Cancer Research, 65*(23), 10946–10951.

Cooper, J. A. (1987). Effects of cytochalasin and phalloidin on actin. *The Journal of Cell Biology, 105*(4), 1473–1478.

Corkery, D. P., Dellaire, G., & Berman, J. N. (2011). Leukaemia xenotransplantation in zebrafish—chemotherapy response assay in vivo. *British Journal of Haematology, 153*(6), 786–789.

Croker, A. K., Goodale, D., Chu, J., Postenka, C., Hedley, B. D., Hess, D. A., & Allan, A. L. (2009). High aldehyde dehydrogenase and expression of cancer stem cell markers selects for breast cancer cells with enhanced malignant and metastatic ability. *Journal of Cellular and Molecular Medicine, 13*(8b), 2236–2252.

Cui, C., Benard, E. L., Kanwal, Z., Stockhammer, O. W., van der Vaart, M., Zakrzewska, A., … Meijer, A. H. (2011). Infectious disease modeling and innate immune function in zebrafish embryos. *Methods in Cell Biology, 105*, 273–308.

Drabsch, Y., He, S., Zhang, L., Snaar-Jagalska, B. E., & ten Dijke, P. (2013). Transforming growth factor-β signalling controls human breast cancer metastasis in a zebrafish xenograft model. *Breast Cancer Research, 15*(6), 1.

Eden, C. J., Ju, B., Murugesan, M., Phoenix, T. N., Nimmervoll, B., Tong, Y., … Dapper, J. (2015). Orthotopic models of pediatric brain tumors in zebrafish. *Oncogene, 34*(13), 1736–1742.

Ellett, F., Pase, L., Hayman, J. W., Andrianopoulos, A., & Lieschke, G. J. (2011). mpeg1 promoter transgenes direct macrophage-lineage expression in zebrafish. *Blood, 117*(4), e49–e56.

van der Ent, W., Burrello, C., Teunisse, A. F., Ksander, B. R., van der Velden, P. A., Jager, M. J., … Snaar-Jagalska, B. E. (2014). Modeling of human uveal melanoma in zebrafish xenograft Embryos. *Investigative Ophthalmology and Visual Science, 55*(10), 6612–6622.

van der Ent, W., Jochemsen, A. G., Teunisse, A. F., Krens, S. F., Szuhai, K., Spaink, H. P., … Snaar-Jagalska, B. E. (2014). Ewing sarcoma inhibition by disruption of EWSR1−FLI1 transcriptional activity and reactivation of p53. *The Journal of Pathology, 233*(4), 415−424.

Francescangeli, F., Contavalli, P., De Angelis, M. L., Baiocchi, M., Gambara, G., Pagliuca, A., ... Stassi, G. (2015). Dynamic regulation of the cancer stem cell compartment by Cripto-1 in colorectal cancer. *Cell Death and Differentiation, 22*, 1700−1713.

Ghotra, V. P., He, S., De Bont, H., van Der Ent, W., Spaink, H. P., van De Water, B., ... Danen, E. H. (2012). Automated whole animal bio-imaging assay for human cancer dissemination. *PLoS One, 7*(2), e31281.

Ghotra, V. P., He, S., van der Horst, G., Nijhoff, S., de Bont, H., Lekkerkerker, A., ... Verhoef, E. I. (2015). SYK is a candidate kinase target for the treatment of advanced prostate cancer. *Cancer Research, 75*(1), 230−240.

Ginestier, C., Hur, M. H., Charafe-Jauffret, E., Monville, F., Dutcher, J., Brown, M., ... Schott, A. (2007). ALDH1 is a marker of normal and malignant human mammary stem cells and a predictor of poor clinical outcome. *Cell Stem Cell, 1*(5), 555−567.

Glass, T. J., Lund, T. C., Patrinostro, X., Tolar, J., Bowman, T. V., Zon, L. I., & Blazar, B. R. (2011). Stromal cell−derived factor-1 and hematopoietic cell homing in an adult zebrafish model of hematopoietic cell transplantation. *Blood, 118*(3), 766−774.

Goessling, W., North, T. E., Loewer, S., Lord, A. M., Lee, S., Stoick-Cooper, C. L., ... Zon, L. I. (2009). Genetic interaction of PGE2 and Wnt signaling regulates developmental specification of stem cells and regeneration. *Cell, 136*(6), 1136−1147.

Goessling, W., North, T. E., & Zon, L. I. (2007). New waves of discovery: modeling cancer in zebrafish. *Journal of Clinical Oncology, 25*(17), 2473−2479.

Haldi, M., Ton, C., Seng, W. L., & McGrath, P. (2006). Human melanoma cells transplanted into zebrafish proliferate, migrate, produce melanin, form masses and stimulate angiogenesis in zebrafish. *Angiogenesis, 9*(3), 139−151.

Hanahan, D., & Weinberg, R. A. (2011). Hallmarks of cancer: the next generation. *Cell, 144*(5), 646−674.

He, S., Lamers, G. E., Beenakker, J. W. M., Cui, C., Ghotra, V. P., Danen, E. H., ... Snaar-Jagalska, B. E. (2012). Neutrophil-mediated experimental metastasis is enhanced by VEGFR inhibition in a zebrafish xenograft model. *The Journal of Pathology, 227*(4), 431−445.

Hermann, P. C., Huber, S. L., Herrler, T., Aicher, A., Ellwart, J. W., Guba, M., ... Heeschen, C. (2007). Distinct populations of cancer stem cells determine tumor growth and metastatic activity in human pancreatic cancer. *Cell Stem Cell, 1*(3), 313−323.

van den Hoogen, C., van der Horst, G., Cheung, H., Buijs, J. T., Lippitt, J. M., Guzmán-Ramírez, N., ... Pelger, R. C. (2010). High aldehyde dehydrogenase activity identifies tumor-initiating and metastasis-initiating cells in human prostate cancer. *Cancer Research, 70*(12), 5163−5173.

van den Hoogen, C., van der Horst, G., Cheung, H., Buijs, J. T., Pelger, R. C., & van der Pluijm, G. (2011). Integrin αv expression is required for the acquisition of a metastatic stem/progenitor cell phenotype in human prostate cancer. *The American Journal of Pathology, 179*(5), 2559−2568.

Howe, K., Clark, M. D., Torroja, C. F., Torrance, J., Berthelot, C., Muffato, M., ... McLaren, S. (2013). The zebrafish reference genome sequence and its relationship to the human genome. *Nature, 496*(7446), 498−503.

Hwang, W. Y., Fu, Y., Reyon, D., Maeder, M. L., Tsai, S. Q., Sander, J. D., … Joung, J. K. (2013). Efficient genome editing in zebrafish using a CRISPR-Cas system. *Nature Biotechnology, 31*(3), 227–229.

Jemal, A., Center, M. M., DeSantis, C., & Ward, E. M. (2010). Global patterns of cancer incidence and mortality rates and trends. *Cancer Epidemiology, Biomarkers and Prevention, 19*(8), 1893–1907.

Kissa, K., & Herbomel, P. (2010). Blood stem cells emerge from aortic endothelium by a novel type of cell transition. *Nature, 464*(7285).

Konantz, M., Balci, T. B., Hartwig, U. F., Dellaire, G., André, M. C., Berman, J. N., & Lengerke, C. (2012). Zebrafish xenografts as a tool for in vivo studies on human cancer. *Annals of the New York Academy of Sciences, 1266*(1), 124–137.

Kraupp, B. G., Ruttkay-Nedecky, B., Koudelka, H., Bukowska, K., Bursch, W., & Schulte-Hermann, R. (1995). In situ detection of fragmented DNA (TUNEL assay) fails to discriminate among apoptosis, necrosis, and autolytic cell death: a cautionary note. *Hepatology, 21*(5), 1465–1468.

Kurtova, A. V., Xiao, J., Mo, Q., Pazhanisamy, S., Krasnow, R., Lerner, S. P., … Chan, K. S. (2015). Blocking PGE2-induced tumour repopulation abrogates bladder cancer chemoresistance. *Nature, 517*(7533), 209–213.

Lam, S. H., Chua, H. L., Gong, Z., Lam, T. J., & Sin, Y. M. (2004). Development and maturation of the immune system in zebrafish, *Danio rerio*: a gene expression profiling, in situ hybridization and immunological study. *Developmental and Comparative Immunology, 28*(1), 9–28.

Lawson, N. D., & Weinstein, B. M. (2002). In vivo imaging of embryonic vascular development using transgenic zebrafish. *Developmental Biology, 248*(2), 307–318.

Lee, S. L., Rouhi, P., Jensen, L. D., Zhang, D., Ji, H., Hauptmann, G., … Cao, Y. (2009). Hypoxia-induced pathological angiogenesis mediates tumor cell dissemination, invasion, and metastasis in a zebrafish tumor model. *Proceedings of the National Academy of Sciences of the United States of America, 106*(46), 19485–19490.

Lee, Y., Manegold, J. E., Kim, A. D., Pouget, C., Stachura, D. L., Clements, W. K., & Traver, D. (2014). FGF signalling specifies haematopoietic stem cells through its regulation of somitic Notch signalling. *Nature Communications, 5*.

Lengerke, C., Schmitt, S., Bowman, T. V., Jang, I. H., Maouche-Chretien, L., McKinney-Freeman, S., … Zon, L. I. (2008). BMP and Wnt specify hematopoietic fate by activation of the Cdx-Hox pathway. *Cell Stem Cell, 2*(1), 72–82.

Lieschke, G. J., & Trede, N. S. (2009). Fish immunology. *Current Biology, 19*(16), R678–R682.

Ma, Q., Jones, D., Borghesani, P. R., Segal, R. A., Nagasawa, T., Kishimoto, T., … Springer, T. A. (1998). Impaired B-lymphopoiesis, myelopoiesis, and derailed cerebellar neuron migration in CXCR4-and SDF-1-deficient mice. *Proceedings of the National Academy of Sciences of the United States of America, 95*(16), 9448–9453.

Mathias, J. R., Dodd, M. E., Walters, K. B., Rhodes, J., Kanki, J. P., Look, A. T., & Huttenlocher, A. (2007). Live imaging of chronic inflammation caused by mutation of zebrafish Hai1. *Journal of Cell Science, 120*(19), 3372–3383.

McAllister, S. S., & Weinberg, R. A. (2014). The tumour-induced systemic environment as a critical regulator of cancer progression and metastasis. *Nature Cell Biology, 16*(8), 717–727.

Miharada, K., Karlsson, G., Rehn, M., Rörby, E., Siva, K., Cammenga, J., & Karlsson, S. (2011). Cripto regulates hematopoietic stem cells as a hypoxic-niche-related factor through cell surface receptor GRP78. *Cell Stem Cell, 9*(4), 330–344.

Moore, M. A. S., Hattori, K., Heissig, B., SHIEH, J. H., Dias, S., Crystal, R. G., & Rafii, S. (2001). Mobilization of endothelial and hematopoietic stem and progenitor cells by adenovector-mediated elevation of serum levels of SDF-1, VEGF, and angiopoietin-1. *Annals of the New York Academy of Sciences, 938*(1), 36–47.

Nicoli, S., Ribatti, D., Cotelli, F., & Presta, M. (2007). Mammalian tumor xenografts induce neovascularization in zebrafish embryos. *Cancer Research, 67*(7), 2927–2931.

North, T. E., Goessling, W., Walkley, C. R., Lengerke, C., Kopani, K. R., Lord, A. M., & FitzGerald, G. A. (2007). Prostaglandin E2 regulates vertebrate haematopoietic stem cell homeostasis. *Nature, 447*(7147), 1007–1011.

Ocaña, O. H., Córcoles, R., Fabra, Á., Moreno-Bueno, G., Acloque, H., Vega, S., ... Nieto, M. A. (2012). Metastatic colonization requires the repression of the epithelial-mesenchymal transition inducer Prrx1. *Cancer Cell, 22*(6), 709–724.

Özdemir, B. C., Pentcheva-Hoang, T., Carstens, J. L., Zheng, X., Wu, C. C., Simpson, T. R., ... De Jesus-Acosta, A. (2014). Depletion of carcinoma-associated fibroblasts and fibrosis induces immunosuppression and accelerates pancreas cancer with reduced survival. *Cancer Cell, 25*(6), 719–734.

Paddison, P. J., Caudy, A. A., Bernstein, E., Hannon, G. J., & Conklin, D. S. (2002). Short hairpin RNAs (shRNAs) induce sequence-specific silencing in mammalian cells. *Genes and Development, 16*(8), 948–958.

Parisi, S., D'Andrea, D., Lago, C. T., Adamson, E. D., Persico, M. G., & Minchiotti, G. (2003). Nodal-dependent Cripto signaling promotes cardiomyogenesis and redirects the neural fate of embryonic stem cells. *The Journal of Cell Biology, 163*(2), 303–314.

Pettaway, C. A., Pathak, S., Greene, G., Ramirez, E., Wilson, M. R., Killion, J. J., & Fidler, I. J. (1996). Selection of highly metastatic variants of different human prostatic carcinomas using orthotopic implantation in nude mice. *Clinical Cancer Research, 2*(9), 1627–1636.

Pouget, C., & Traver, D. (2016). Complex regulation of HSC emergence by the Notch signaling pathway. *Developmental Biology, 409*(1), 129–138.

Pruvot, B., Jacquel, A., Droin, N., Auberger, P., Bouscary, D., Tamburini, J., ... Solary, E. (2011). Leukemic cell xenograft in zebrafish embryo for investigating drug efficacy. *Haematologica, 96*(4), 612–616.

Renshaw, S. A., Loynes, C. A., Trushell, D. M., Elworthy, S., Ingham, P. W., & Whyte, M. K. (2006). A transgenic zebrafish model of neutrophilic inflammation. *Blood, 108*(13), 3976–3978.

Reya, T., Morrison, S. J., Clarke, M. F., & Weissman, I. L. (2001). Stem cells, cancer, and cancer stem cells. *Nature, 414*(6859), 105–111.

Sacco, A., Roccaro, A. M., Ma, D., Shi, J., Mishima, Y., Moschetta, M., ... Ghobrial, I. M. (2016). Cancer cell dissemination and homing to the bone marrow in a zebrafish model. *Cancer Research, 76*(2), 463–471.

Scholzen, T., & Gerdes, J. (2000). The Ki-67 protein: from the known and the unknown. *Journal of Cell Physiology, 182*(3), 311–322.

Shiozawa, Y., Pedersen, E. A., Havens, A. M., Jung, Y., Mishra, A., Joseph, J., ... Pienta, M. J. (2011). Human prostate cancer metastases target the hematopoietic stem cell niche to establish footholds in mouse bone marrow. *The Journal of Clinical Investigation, 121*(4), 1298–1312.

Singh, A., & Settleman, J. (2010). EMT, cancer stem cells and drug resistance: an emerging axis of evil in the war on cancer. *Oncogene, 29*(34), 4741–4751.

Stoletov, K., Kato, H., Zardouzian, E., Kelber, J., Yang, J., Shattil, S., & Klemke, R. (2010). Visualizing extravasation dynamics of metastatic tumor cells. *Journal of Cell Science, 123*(13), 2332–2341.

Stoletov, K., & Klemke, R. (2008). Catch of the day: zebrafish as a human cancer model. *Oncogene, 27*(33), 4509–4520.

Stoop, E. J., Schipper, T., Huber, S. K. R., Nezhinsky, A. E., Verbeek, F. J., Gurcha, S. S., … van der Sar, A. M. (2011). Zebrafish embryo screen for mycobacterial genes involved in the initiation of granuloma formation reveals a newly identified ESX-1 component. *Disease Models and Mechanisms, 4*(4), 526–536.

Storms, R. W., Trujillo, A. P., Springer, J. B., Shah, L., Colvin, O. M., Ludeman, S. M., & Smith, C. (1999). Isolation of primitive human hematopoietic progenitors on the basis of aldehyde dehydrogenase activity. *Proceedings of the National Academy of Sciences of the United States of America, 96*(16), 9118–9123.

Strizzi, L., Bianco, C., Normanno, N., & Salomon, D. (2005). Cripto-1: a multifunctional modulator during embryogenesis and oncogenesis. *Oncogene, 24*(37), 5731–5741.

Takebe, N., Miele, L., Harris, P. J., Jeong, W., Bando, H., Kahn, M., … Ivy, S. P. (2015). Targeting Notch, Hedgehog, and Wnt pathways in cancer stem cells: clinical update. *Nature Reviews. Clinical Oncology, 12*(8), 445–464.

Tam, W. L., & Weinberg, R. A. (2013). The epigenetics of epithelial-mesenchymal plasticity in cancer. *Nature Medicine, 19*(11), 1438–1449.

Tamplin, O. J., Durand, E. M., Carr, L. A., Childs, S. J., Hagedorn, E. J., Li, P., … Zon, L. I. (2015). Hematopoietic stem cell arrival triggers dynamic remodeling of the perivascular niche. *Cell, 160*(1), 241–252.

Terry, S., El-Sayed, I. Y., Destouches, D., Maillé, P., Nicolaiew, N., Ploussard, G., … Allory, Y. (2015). CRIPTO overexpression promotes mesenchymal differentiation in prostate carcinoma cells through parallel regulation of AKT and FGFR activities. *Oncotarget, 6*(14), 11994.

Thiery, J. P. (2003). Epithelial–mesenchymal transitions in development and pathologies. *Current Opinion in Cell Biology, 15*(6), 740–746.

Thiery, J. P., Acloque, H., Huang, R. Y., & Nieto, M. A. (2009). Epithelial-mesenchymal transitions in development and disease. *Cell, 139*(5), 871–890.

Tobin, D. M., Vary, J. C., Ray, J. P., Walsh, G. S., Dunstan, S. J., Bang, N. D., … Moens, C. B. (2010). The lta4h locus modulates susceptibility to mycobacterial infection in zebrafish and humans. *Cell, 140*(5), 717–730.

Trede, N. S., Langenau, D. M., Traver, D., Look, A. T., & Zon, L. I. (2004). The use of zebrafish to understand immunity. *Immunity, 20*(4), 367–379.

Tulotta, C., Stefanescu, C., Beletkaia, E., Bussmann, J., Tarbashevich, K., Schmidt, T., & Snaar-Jagalska, B. E. (2016). Inhibition of signaling between human CXCR4 and zebrafish ligands by the small molecule IT1t impairs the formation of triple-negative breast cancer early metastases in a zebrafish xenograft model. *Disease Models and Mechanisms, 9*(2), 141–153.

Vecchiarelli-Federico, L. M., Cervi, D., Haeri, M., Li, Y., Nagy, A., & Ben-David, Y. (2010). Vascular endothelial growth factor—a positive and negative regulator of tumor growth. *Cancer Research, 70*(3), 863–867.

Visvader, J. E. (2011). Cells of origin in cancer. *Nature, 469*(7330), 314–322.

Watanabe, K., Meyer, M. J., Strizzi, L., Lee, J. M., Gonzales, M., Bianco, C., … Vonderhaar, B. K. (2010). Cripto-1 is a cell surface marker for a tumorigenic, undifferentiated subpopulation in human embryonal carcinoma cells. *Stem Cells, 28*(8), 1303–1314.

Wilkinson, R. N., Pouget, C., Gering, M., Russell, A. J., Davies, S. G., Kimelman, D., & Patient, R. (2009). Hedgehog and Bmp polarize hematopoietic stem cell emergence in the zebrafish dorsal aorta. *Developmental Cell, 16*(6), 909–916.

Zakrzewska, A., Cui, C., Stockhammer, O. W., Benard, E. L., Spaink, H. P., & Meijer, A. H. (2010). Macrophage-specific gene functions in Spi1-directed innate immunity. *Blood, 116*(3), e1–e11.

Zhang, J., Talbot, W. S., & Schier, A. F. (1998). Positional cloning identifies zebrafish one-eyed pinhead as a permissive EGF-related ligand required during gastrulation. *Cell, 92*(2), 241–251.

Zon, L. I., & Peterson, R. T. (2005). In vivo drug discovery in the zebrafish. *Nature Reviews. Drug Discovery, 4*(1), 35–44.

Zoni, E., van der Horst, G., van de Merbel, A. F., Chen, L., Rane, J. K., Pelger, R. C., … van der Pluijm, G. (2015). miR-25 modulates invasiveness and dissemination of human prostate cancer cells via regulation of αv-and α6-integrin expression. *Cancer Research, 75*(11), 2326–2336.

CHAPTER

Zebrafish as a model for von Hippel Lindau and hypoxia-inducible factor signaling

18

H.R. Kim*, D. Greenald*, A. Vettori§, E. Markham*, K. Santhakumar¶, F. Argenton§, F. van Eeden*,[1]

**University of Sheffield, Sheffield, United Kingdom*

§Università degli Studi di Padova, Padova, Italy

¶SRM University, Kattankulathur, India

[1]Corresponding author: E-mail: f.j.vaneeden@sheffield.ac.uk

CHAPTER OUTLINE

Methods in Cell Biology, Volume 138, ISSN 0091-679X, http://dx.doi.org/10.1016/bs.mcb.2016.07.001

Abstract

Oxygen is a central molecule in the development of multicellular life, allowing efficient energy generation. Inadequate oxygen supply requires rapid adaptations to prevent cellular damage and the hypoxia-inducible factor (HIF) pathway plays a central role in this adaptation. Numerous diseases and disease processes are influenced by hypoxia and the HIF pathway. One component, von Hippel Lindau (VHL), is a well-known tumor suppressor, which acts at least in part via regulating HIF signaling. The zebrafish has become a central vertebrate model organism in which developmental and disease processes can be studied. In this review, we have tried to bring together knowledge on the HIF/hypoxic signaling pathway in zebrafish, including what is known on VHL functions.

INTRODUCTION

Oxygen is essential for the life of the majority of living multicellular organisms. It allows efficient conversion of carbon-containing molecules to energy in the form of ATP in the mitochondria. It is likely that its vital importance has led to the evolution of the hypoxia-inducible factor (HIF) pathway, which "alerts" and adapts cells to a drop in oxygen levels.

Multicellular organisms first evolved in an aquatic environment, and the solubility of oxygen in water is rather low. This fact, combined with slow mixing and the presence of possibly high concentrations of oxygen using and producing organisms, means that vertebrates living in water will encounter a much more variable environmental oxygen concentration than land vertebrates (see, for example, Erez, Krom, & Neuwirth, 1990; Nikinmaa, 2002). Therefore, even more elaborate adaptive mechanisms may have evolved in aquatic vertebrates to optimize survival. Indeed, crucian carp can tolerate anoxia for extended periods of time relying on glycolysis for ATP production, and surprisingly, converting lactate into ethanol to prevent acidosis (Lutz & Nilsson, 2004; Stenslokken et al., 2014). In zebrafish, it was also reported that the early embryo can survive a limited time of anoxia (Padilla & Roth, 2001) and larvae can survive and grow at 10% air saturation (0.8 mg/L, ~2 kPa O_2) (Marques et al., 2008). Furthermore, similar to higher vertebrates (Verges, Chacaroun, Godin-Ribuot, & Baillieul, 2015), it has been shown that hypoxic preconditioning can protect against further hypoxic exposures in zebrafish larvae and adults (Manchenkov, Pasillas, Haddad, & Imam, 2015; Rees, Sudradjat, & Love, 2001). In this article, we will try to bring together data on zebrafish Hif signaling and Vhl functions. The role of the HIF signaling pathway has also been studied in zebrafish infection, inflammation, and tissue regeneration. However, this was recently reviewed in detail by Elks, Renshaw, Meijer, Walmsley, and van Eeden (2015), and therefore we will not discuss this here.

1. HYPOXIC SIGNALING

The first HIF transcription factor was identified in mammals in a seminal study by the lab of Semenza (Semenza & Wang, 1992), and its regulation has since been elucidated using a variety of model organisms. We have outlined the molecular components of this pathway in Fig. 1 but refer to various excellent recent reviews for details (Semenza, 2009, 2010). In brief, in the absence of sufficient oxygen, the prolyl hydroxylase (PHD) enzymes are unable to hydroxylate HIF-α (Epstein et al., 2001). The PHD enzymes may be the direct oxygen sensors, as they require molecular

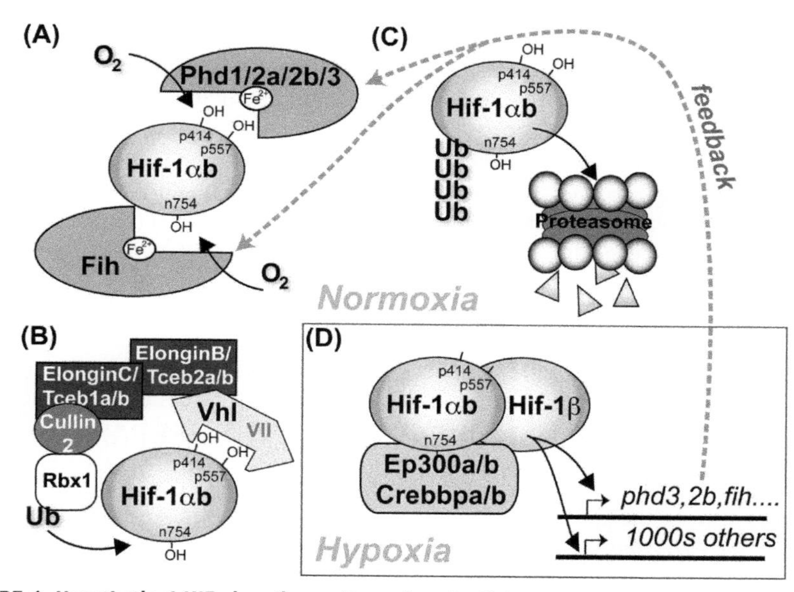

FIGURE 1 Hypothetical HIF signaling pathway in zebrafish.

In the figure, we have substituted human gene names, with names of their likely zebrafish functional orthologues. (A) In zebrafish, we speculate that Hif-1αb is the main sensor of mild hypoxia, with other Hifs being stabilized at lower oxygen concentrations. Hif-1αb can be hydroxylated on proline 414 and 557, the zebrafish equivalents of proline 402 and 564 in human HIF-1α. Current data suggest that Phd2b is most important in hydroxylating Hif under normal conditions. In addition to the Phd genes, Fih may hydroxylate Hif-1αb on Asn754 the zebrafish equivalent of Asn803 in human HIF-1α. (B) Hydroxylation will lead to recognition by Vhl as part of a ubiquitination complex also consisting of ElonginB (Tceb2a/b) ElonginC (Tceb1a/b), Cullin-2 and Rbx1, with Vll playing a minor redundant role. (C) Ubiquitinated Hif is rapidly degraded in the proteasome. (D) Under hypoxic conditions, unhydroxylated Hif-1αb is not recognized by Vhl can enter the nucleus and pair up with Hif-1β and transcriptional coactivators like p300/CBP (Ep300a/b and Crebbpa/b in zebrafish) to bind HREs and activate transcription of possibly thousands of genes. This includes many hydroxylases, like *phd3*, *phd2b,fih*, and others, thereby generating a negative feedback on the hypoxic signal.

oxygen for this reaction (Flashman et al., 2010; Schofield & Ratcliffe, 2005). The hydroxylated form of HIF-α is recognized by the VHL protein (Latif et al., 1993), which is a substrate recognition part of a ubiquitination complex containing ElonginB/C, Cullin-2, and Rbx1 (Duan et al., 1995; Kamura et al., 1999; Kibel, Iliopoulos, DeCaprio, & Kaelin, 1995; Kishida, Stackhouse, Chen, Lerman, & Zbar, 1995; Lonergan et al., 1998; Pause et al., 1997). In normoxia, HIF is therefore rapidly degraded with a reported half-life of minutes (Moroz et al., 2009; Wang, Jiang, Rue, & Semenza, 1995). An additional level of control exists: Factor-inhibiting HIF (FIH) hydroxylates HIF on Asn803 in human HIF-1α, compromising its transactivation capacity. In addition to proteasome-dependent degradation, chaperone-mediated breakdown can also affect overall levels in some cell types (Hubbi, Hu, Kshitiz Ahmed, Levchenko, & Semenza, 2013). Although posttranslational regulation is most likely a major factor, transcriptional regulation of HIF may be important as well; especially nuclear factor κ-light-chain-enhancer of activated B cells (NF-κB) (Rius et al., 2008; van Uden, Kenneth, & Rocha, 2008), which may promote HIF signaling during infection in the absence of physical hypoxia. In hypoxia, hydroxylation is inhibited leading to the accumulation of unhydroxylated HIF-α, which complexes with a constitutively expressed partner, HIF-β (aka. aryl carbon receptor nuclear translocator, ARNT). In mammals, several isoforms of HIF-α and HIF-β have been identified, and HIF-1, -2, and -3α have been studied, which have been shown to have differences in their precise function and regulation.

2. HYPOXIC SIGNALING: OVERVIEW OF THE ZEBRAFISH ORTHOLOGUES

The hypoxic signaling pathway is highly conserved in the zebrafish, its central components: the *PHD* genes 1 and 3 (*PHD1/3*), and *FIH* all have one orthologue, whereas PHD2 has been duplicated. The *VHL* gene has two orthologues (*Vhl* and a more divergent *Vll*). All HIF-α orthologues are duplicated in the zebrafish, resulting in six genes; a geneticist's nightmare. The *HIF-β* gene and the related *ARNT2* gene only have one orthologue each (Fig. 2).

3. THE ZEBRAFISH HIF GENES

mRNA expression of the six HIF paralogues was systematically analyzed by Rytkonen et al. (2013) and Rytkonen, Prokkola, Salonen, & Nikinmaa (2014). Unfortunately naming of these is inconsistent in the literature and often existence of another orthologue is simply ignored by published work. For this review, we will refer to the genes as *hif-1/2/3αa/b*, but we will indicate common alternative names (see Fig. 2A for an overview). Embryonic expression from all except *hif-3αb* can be detected from 8 h postfertilization (hpf) onward by qPCR. For all *hif* genes, expression increases during the first 48 h of development (late pharyngula stage). Whole mount in situ hybridization (WMISH) shows that *hif-1αa* and *b* are expressed in similar patterns with highest expression in the head and lower expression in the neural tube, caudal hematopoietic tissue, and neuromasts, with expression of *hif-1αb* being higher. In late larval stages, expression in the liver and intestine has been observed (Elks et al., 2011; Lin et al., 2014; Rojas et al., 2007). Functional differences were discovered by Elks et al.; dominant active/negative versions of Hif-1αa were inactive in their inflammation assay. In addition, it was found that this gene lacks the hydroxylation site corresponding to P402 in human, thus it might have diverged in its function (Elks et al., 2011). *Hif-2αa (epas1b)* is most strongly expressed in the pharyngeal region, the ear and the retina at late larval stages (Lin et al., 2014), this gene may be expressed more early in the interrenal organ (Rauch et al., 2003). *Hif-1αb* mRNA has the highest expression level in adults, out of all the HIFs (Rytkonen et al., 2014), whereas in 9 dpf larvae, *hif-2αa* was reported to have a higher expression level than *hif-1αb* (Kopp, Koblitz, Egg, & Pelster, 2011). *Hif-2αb (epas1a)* and *hif-3αb* (*hif1al2*) embryonic expression patterns have not been determined by WMISH. *Hif-3αa (hif1al)* is expressed rather generally; higher levels of expression in ear and internal organs were reported in one case, but these might be due to probe trapping (Lin et al., 2014; Thisse & Thisse, 2004; Zhang et al., 2012, 2014).

There is no significant regulation of *hif-1&2αa/b* or *hif-3αb* mRNA expression under environmental hypoxia (5 kPa) up to 48 hpf (Rytkonen et al., 2014). Dimethyloxaloylglycine (DMOG) an activator of the HIF pathway, did upregulate *hif-2αb*, however (Metelo et al., 2015), and *hif-3αb* is upregulated in *vhl* mutants at 4 and 7 days postfertilization (dpf) where HIF signaling is highly activated (Greenald et al., 2015; van Rooijen et al., 2009). The same 4 dpf *vhl* mutant microarray analysis shows that *hif-1αb* mRNA is slightly down. *Hif-3αa* (aka *hif1al*) mRNA expression is weakly induced by hypoxia (but not via *hif-1α*b) at 48 hpf (Rytkonen et al., 2014; Zhang et al., 2012). These findings parallel mammalian studies in many ways, although hypoxic induction of HIF-1 mRNA has been reported in mammals (Heidbreder et al., 2003; Semenza, 2009; Wiesener et al., 2003).

4. HIF PROTEIN EXPRESSION

As regulation of HIFs mainly occurs at the level of the protein, antibody studies on all the HIFs would be very interesting. Unfortunately, few reliable antibodies are available, only recently studies were published. A polyclonal antiserum against Hif-3αa shows strong hypoxic induction at the protein level, and protein degradation was found to be dependent on proline 493 (analogous to P564 in *HsHIF-1α* (Zhang et al., 2012)). *HIF-3α* is often not considered in mammalian studies due to its unclear properties and complex regulation, as it has various splice forms that have different or even opposite functions (Pasanen et al., 2010). In zebrafish, a stabilized full-length form of the gene was capable of activating hypoxia

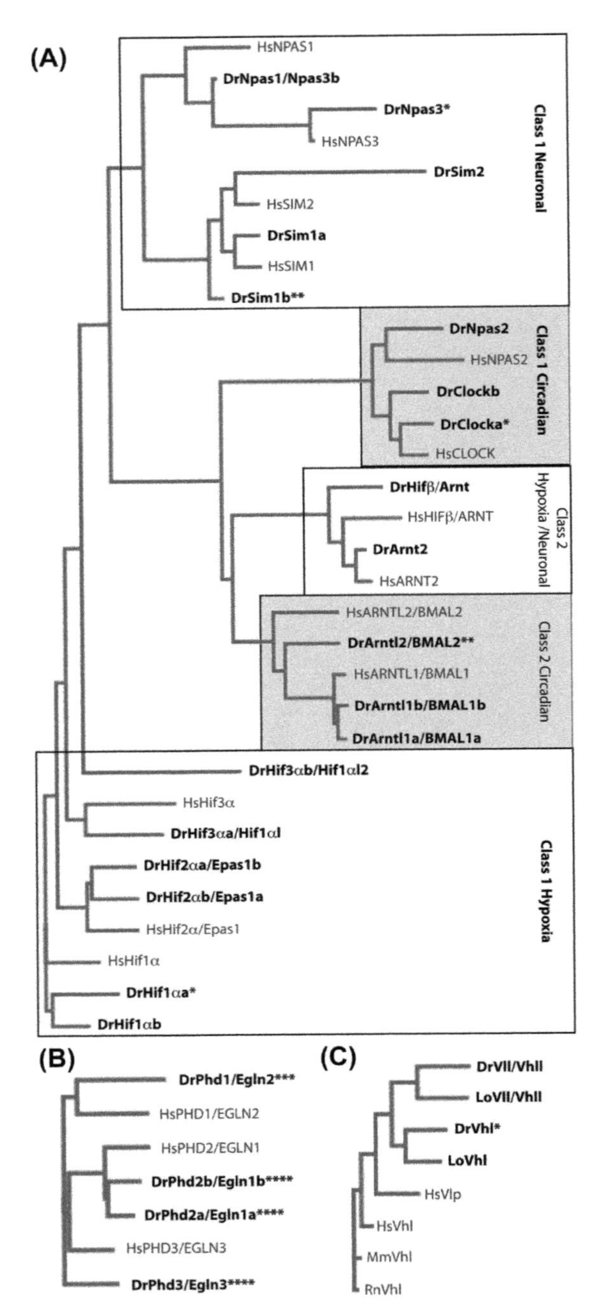

FIGURE 2 Comparison of HIF, PHD, and VHL orthologues.

(A) Cladogram of the basic helix loop helix/Period ARNT Sim (bHLH/PAS) family in zebrafish and human, which includes HIF, CLOCK, and NPAS genes, where class I genes partner with class II genes. The clock genes preferably partner with BMAL genes (both boxed response genes and morpholino (MO) knockdown blunted the activation of a subset of hypoxia response genes, showing its importance in shaping the hypoxic response in zebrafish. Interestingly, in a follow-on study, one of the other isoforms, Hif-3αa2, was identified as an oxygen-insensitive protein-lacking bHLH/PAS domains that nevertheless drives hypoxia-responsive element (HRE)-dependent transcription and, surprisingly, downregulates *wnt* signaling. This was shown using MOs and a mutant that only affected the Hif-3αa2 function (Zhang, Bai, Lu, Li, & Duan, 2016).

A very good polyclonal Hif-1αb antibody was generated and used by Kopp et al., (2011). In 3 and 9 dpf larvae, Hif-1αb is induced by hypoxia (5 kPa O_2) within 24 h but returns to preinduced levels within about 2–4 days under continuous hypoxia (Koblitz, Fiechtner, Baus, Lussnig, & Pelster, 2015; Kopp et al., 2011). More recently, a polyclonal antibody was generated against Hif-2αa (it currently cannot be excluded that this antiserum also recognizes Hif-2αb). In addition, a further polyclonal antiserum against two peptide epitopes in Hif-3αa, that are absent in Hif-3αb, was made (Koblitz et al., 2015). Hif protein levels were compared in embryos incubated in normoxia or hypoxia (5 kPa O_2) at various stages of development. This analysis showed that whereas Hif-1αb levels were stable during normal development, Hif-2αa and Hif-3αa increased from 4–9 dpf. Hypoxia treatment showed that Hif-1αb is the most O_2 responsive in zebrafish: Increases can be observed after 4 h of incubation, whereas Hif-2αa and Hif-3αa were not significantly affected by this level of hypoxia (Koblitz et al., 2015). Such a fast response was also observed using a *phd3:GFP* hypoxia reporter line, where incubation for 4 h (5 kPa O_2) from the 128 cell stage onward was sufficient to get induction of the reporter (Santhakumar et al., 2012). The latter data also suggest that a functional Hif signaling pathway is maternally deposited in the egg. The Hif-2 and Hif-3 data are slightly surprising as this does not match the data of Zhang et al., (2012) and they contrast with mammalian experiments where their stability is regulated by hypoxia (Patel & Simon, 2008; Wiesener et al., 2003). Perhaps, more severe hypoxia is required to stabilize Hif-2 and -3 in zebrafish.

in gray) and the HIF genes with HIF-1β/ARNT2 (boxed in white) (Bersten, Sullivan, Peet, & Whitelaw, 2013). The NPAS1/3 groups are involved in neural development, with both groups thought to act mainly via HIF-1β/ARNT2 (Michaud, DeRossi, May, Holdener, & Fan, 2000; Teh et al., 2006). (B) Cladogram of Phd proteins in zebrafish and human. (C) Cladogram of VHL-like genes in human, zebrafish, spotted gar, mouse, and rat. Trees were generated using clustalW2 at http://www.ebi.ac.uk using with distance correction and otherwise default settings. The sequences were taken from Ensembl (www.ensembl.org) using the major isoform of the respective proteins. If the coding gene contains HREs within the gene or in the surrounding 15 Kb 5′ and 3′ sequence, the protein is marked with the corresponding number of *'s. Common alternative abbreviations for genes are also given. Hs, *Homo sapiens*; Mm, *Mus musculus*; Rn, *Rattus norvegicus*; all in gray. Dr, *Danio rerio*; Lo, *Lepisosteus oculatus*/spotted gar; both in bold/black.

5. THE HIF HYDROXYLASES

The expression of the hydroxylases in the Hif signaling pathway has not been studied in great detail. *Phd1* (*egln2*) and *phd2b* (*egln1b*) are not spatially restricted (Thisse & Thisse, 2004). *Phd2b* is expressed at a higher level than *phd1* and *phd3* (Hyvarinen et al., 2010), suggesting that it may be responsible for keeping the HIF pathway inactive under normoxic conditions, similar to mammals (Appelhoff et al., 2004; Berra et al., 2003). *Phd2b* is strongly induced in *vhl* mutants at 4 and 7 dpf and is associated with a HRE in the promoter and three additional HREs in the first intron *Phd1* is weakly induced in *vhl* mutants at 7 dpf and is also associated with three 5′ HREs. *phd2a/egln1a* has not been analyzed in published studies, but it has four HREs associated in the promoter and introns, thus it is likely to be regulated by Hif. *Phd3* is expressed at a very low level at the MHB region and later in the kidney and internal organs (Santhakumar et al., 2012). Importantly, *phd3* also has four HREs and is highly induced by hypoxia and genetic/chemical induction of HIF signaling. In our hands, it usually has the highest fold induction after activation of HIF signaling by mutation of *vhl* (Greenald et al., 2015; van Rooijen et al., 2009; Santhakumar et al., 2012). This is often more than 100-fold, which may be a reflection of its low baseline expression.

The *fih* (aka. *hifan*) gene is expressed maternally (So et al., 2014). During late somite stages, its expression is focused at the midbrain/hindbrain region but there may be a general expression at basal levels. From 24 hpf, a rather general expression can be observed in the nervous system and ventral mesoderm (Thisse et al., 2001). The *fih* gene is considered a component of negative feedback regulation as it is induced in *vhl* mutants, by activation of HIF signaling of two- and sixfold at 4 and 7 dpf, respectively, consistent with the presence of an HRE at the start of intron 1 (Greenald et al., 2015; van Rooijen et al., 2009). Fih was identified as a target of the mindbomb E3 ubiquitin ligase protein, an essential component of Notch signaling, and its function was characterized by MO experiments (So et al., 2014; Tseng et al., 2014). Morphants show excessive sprouting and proliferation in intersegmental vessels, somewhat similar to *vhl* mutants, and consistent with a negative role in hypoxic signaling. However, this phenotype was not observed in maternal–zygotic null mutants (FvE unpubl.).

6. THE *VHL* GENES

In human, full-length VHL is a 213 amino acid protein (30 kDa). Due to an alternative start, there is another isoform consisting of 160 amino acids (19 kDa). However, both proteins can negatively regulate the transcription factor HIF (Schoenfeld, Davidowitz, & Burk, 1998). The acidic N-terminal extension was reported to be functionally important by some studies (Lolkema et al., 2005), but it is not conserved in zebrafish, and rodents also have a strong reduction in the length of this sequence.

Both human isoforms have tumor suppressor activity (Gossage, Eisen, & Maher, 2015; Schoenfeld et al., 1998; Woodward et al., 2000). In zebrafish, two *VHL*-type genes were identified: *vhl* and a more distantly related *vhl-like* (*vll*, aka *vhll*), which are predicted to encode proteins of 175 (19 kDa) and 158 amino acids, respectively, and are 48% and 33% identical to human VHL when compared by Clustal Omega. A second *vhl-like* gene has also been identified in human, *vlp*, but this is poorly studied and might act as a dominant negative (Qi et al., 2004). According to Ensembl, fish *vll* and *vlp* are the result of independent duplication events. Surprisingly, *vll* is also present in the spotted gar and therefore predates the additional genome duplication that occurred in the teleosts (Braasch et al., 2016). However, it has been only selectively retained in certain fish species. For instance, BLAST detected *vll*-expressed sequence tags (ESTs) in dogfish shark, northern pike, herring, and several others, but it was not found in medaka, stickleback, cod, or fugu sequences. Zebrafish *vhl* is expressed maternally and is broadly expressed at 14 and 24 hpf. Its transcription is weakly induced by loss of *vhl*, according to one study (van Rooijen et al., 2009), and it has a closely associated HRE. *Vll* expression has not been studied, the only data available are "EST sources" in Unigene and these suggest that expression is restricted to the brain. One microarray study found a three-fold induction of *vll* in 4 dpf *vhl* mutants (Greenald et al., 2015).

7. HIF BINDING SITES IN THE GENOME

In addition to a number of directed approaches (Egg et al., 2013; Kajimura, Aida, & Duan, 2005, 2006; Ko et al., 2011; Kulkarni, Tohari, Ho, Brenner, & Venkatesh, 2010; Lin et al., 2014), several microarray/RNAseq studies have identified hypoxic response genes in zebrafish (Greenald et al., 2015; Jopling, Sune, Faucherre, Fabregat, & Izpisua Belmonte, 2012; Kopp et al., 2011; Long et al., 2015; Manchenkov et al., 2015; Marques et al., 2008; Martinovic et al., 2009; van der Meer et al., 2005; van Rooijen et al., 2009; Ton, Stamatiou, & Liew, 2003; Woods & Imam, 2015). In addition, a CHIPseq study that compared wild type *versus vhl* mutants, using the zebrafish Hif-1αb antibody, linked such CHIPseq data directly with a microarray study of these two genotypes (Greenald et al., 2015). This study identified an extraordinarily large number of HREs in the zebrafish genome (5177), larger than any previous study in mammals (Mimura et al., 2012; Mole et al., 2009; Schodel et al., 2011; Tsuchihara et al., 2009; Xia et al., 2009). These CHIPseq peaks were mostly absent in wild-type samples, indicating they were not the result of spurious binding of the Hif-1αb antiserum. A consensus HIF binding site of (A/G)(C/T)GTG was identified, which corresponds well with the mammalian binding consensus RCGTG (Schodel et al., 2011). Surprisingly, the consensus showed that RTGTG also was a possible HRE. This has not been reported in mammals, but it has been functionally confirmed in another fish species (Rees, Figueroa, Wiese, Beckman, & Schulte, 2009).

The reason for the difference in number of HREs is not really clear, it could be evolutionary but also simply experimental; extreme hypoxia and high concentrations of hydroxylase inhibitors have effects other than on the HIF pathway and are therefore problematic in their use. Hence, previous mammalian surveys have generally used short and "mild" treatments. In addition, using embryos harboring many cell types, rather than cell lines, may have contributed to the large number of sites identified. Whatever the reason, it is clear that in the zebrafish, Hif-1 has the potential to be a major architect of the transcriptome. We estimated that 30% of the genes in the zebrafish genome have an HRE within the gene or in the flanking 15 kb (Greenald et al., 2015).

8. HYPOXIC/HIF TARGET GENE COMPARISON

The published microarray/RNAseq studies are difficult to compare because they differ widely in the duration, extent, and way in which hypoxia is applied/simulated, and also because gene names and experimental platforms differ and are evolving. Nevertheless, we have attempted to compare gene sets from seven studies where the experimental sample was supposed to activate signaling (Greenald et al., 2015; Kopp et al., 2011; Long et al., 2015; Marques et al., 2008; van der Meer et al., 2005; van Rooijen et al., 2009; Woods & Imam, 2015) and where fold change was reported. In Suppl. Table 1, we have listed these, striving to update gene names to aid comparison and including data on nearby HREs based on Greenald et al., (2015). From this table, it is clear that the *phd3/egln3* gene is an excellent HIF/hypoxia reporter, upregulated in all but one study, and it is the most consistently upregulated gene in zebrafish. The only study that did not identify *phd3/egln3* analyzed gill arches (van der Meer et al., 2005). Gills will take oxygen directly from the water, therefore it might be a tissue for which it is difficult to achieve a local hypoxic environment.

Other genes that show a hypoxic induction in more than two studies are *adob*, agt, alas2, ankrd37*, bnip3lb*, cp, cpox*, cxcl12a, cyp1b1*, phd2b*, ero1l*, fam117ab, fga gpia*, hbae1, hbbe2, hbz, hdr, hif1al2/hif-3αb, hpx, igfbp1a*, klf4b, mb*, nt5c2l1, p4ha2*, plod1a*, ppp1r15a*, sepn1*, sid4, slc10a4, slc16a3*, slc4a1a*, sptb, sqrdl*, thop1* urod,* and *znf395a**. Stars denote the presence of a clearly associated HRE. Surprisingly, several highly consistent hypoxic response genes were not identified as having an HRE. Several explanations are possible for this. They could be genes that are indirect targets. Indeed, many of these HRE-negative genes are involved with erythrocyte development, thus their upregulation could reflect an increase in cell/precursor number. Alternatively, more distant enhancers could be active, or, interestingly, Hif-2/3-bound enhancers might preferentially regulate some of these genes.

Fkbp5, ncaldb, npas4a*, nr4a1, slc34a2a, sucla2, tceb3/elongin A*, and possibly *cyr61* are downregulated genes. Two of these are associated with a HRE but whether these could be functional is unclear; one for *fkbp5* is located closer to the 5′ start of another gene, and one for *npas4a* is located 3′ to the gene. Overall our data are consistent with mammalian data where there is mainly evidence for HIF acting as a transcriptional activator (Mole et al., 2009; Schodel et al., 2011).

Other useful facts can be gleaned from Suppl. Table 1, for instance, the effect of *vhl* on E2F1 targets at 7 dpf as was reported in Mans et al., (2013), which was not seen in hypoxia or in *vhl* at 4 dpf. Conspicuous are the upregulation of collagen genes and also the downregulation of many proteasome genes in *vhl* mutants at 7 dpf (van Rooijen et al., 2009), and the downregulation of ribosomal proteins in the hypoxic gill study (van der Meer et al., 2005). Interestingly, a large number of lysine demethylase genes (KDM and also *jmjd6*) appear to be upregulated in the absence of *vhl*. These genes require molecular oxygen for activity. It was recently reported that both KDM5c and KDM6a are also often mutated in human VHL tumors (Dalgliesh et al., 2010; van Haaften et al., 2009), thus the upregulation of these genes could be a tumor-suppressing homeostatic function that needs to be overcome to induce malignancy. It suggests that KDM3b, KDM4b/c, KDM5b, and KDM7 could also be involved in such processes.

9. HYPOXIC SIGNALING REPORTERS IN ZEBRAFISH

Currently, two highly useful HIF transgenic reporters have been generated: a reporter made by BAC recombineering, using the *phd3* gene promoter as a driver (*phd3::egfp*) (Santhakumar et al., 2012) and another set of global reporters that employ optimized and concatemerized HREs driving GFP, DsRed, or mCherry (FA unpublished data). We have found that the synthetic HRE-driven reporters most likely pick up lower levels of HIF activity as these fluorescent reporters show clear expression and modulation in normal embryos, larvae as well as organs and tissues of the adult. The *phd3* reporter appears to be more restricted: In normal embryos, expression level is hard to detect by GFP fluorescence but strong induction can be observed by treatments that will activate HIF. The *phd::egfp* line has proven to be a good in vivo model to test drug candidates that act on the HIF pathway. In a search for hydroxylase inhibitors with a more narrow target range, Schofield and colleagues (Chowdhury et al., 2013) used the zebrafish as a convenient organism to get a first in vivo test of their compounds. One compound from this set, IOX2 (compound 6), was a potent activator in zebrafish in the liver and was also found to be active in mice (Chan et al., 2015).

10. LINKS BETWEEN HIF AND THE CIRCADIAN CLOCK

A fascinating set of papers describes linkage between circadian rhythms and HIF signaling. The HIF pathway can be recognized in the most primitive metazoans but not in unicellular organisms (Loenarz et al., 2011). As multicellular organisms evolved in an aquatic environment, and most likely fed on photosynthetic organisms, their microenvironment was probably subject to large circadian variations in oxygen levels; from hyperoxia during the day to hypoxia at night, when photosynthetic organisms become net consumers of oxygen (Erez et al., 1990; Nikinmaa, 2002). Thus oxygen may have been both the chemical to which organisms need to adapt their physiology, as well as a zeitgeber signal that can be exploited to keep the vertebrate circadian clock in tune with the light. The central components of this clock are CLOCK and BMAL heterodimers that activate the transcription of *period* and *cry* genes; their protein products in turn inhibit CLOCK and BMAL function, leading to ~24 h oscillations in expression. Importantly, there is a close evolutionary relationship between *CLOCK* and *HIF* genes (Fig. 2); both belong to the bHLH/PAS group, suggesting common origins. Zebrafish as an aquatic model organism offers an excellent opportunity to study these possibly ancient links between HIF/oxygen and the circadian clock.

A set of papers by the Pelster group has uncovered several links between the two pathways (Egg et al., 2013, 2014; Pelster & Egg, 2015). Long-term hypoxia was shown to increase *period1b* and *clocka* expression but dampens *period1b* oscillations. Hypoxia also leads to a loss of circadian behavior patterns in larvae (Egg et al., 2013). Conversely, Hif accumulation in response to hypoxia showed a biphasic diurnal response (Egg et al., 2013). In zebrafish Z3 cells, misexpression of both dominant active and negative *clockb* (*clock3*) led to a higher induction of Hif.

Several E-boxes were identified in the *period1* (Egg et al., 2013) and *period2* (Pelster & Egg, 2015) promoter region that could be bound by Hif. However, it may be difficult to prove whether Hif bound to these E-boxes/HREs is actually responsible for dampening oscillations. The CHIPseq study by Greenald et al., (2015) independently confirmed two HREs in the *period1b* promoter, but a further one in intron 1 of *period1b*, two upstream of *period1a*, and two further HREs each, near *bmal2* and *cry4*. The latter gene was also picked up by microarray studies (Suppl. Table 1). Finally, one HRE was found near *clocka, period3, cry1aa, cry1ab, cry1bb*, and *cry2*. It is very clear from this data that intimate links exist between HIF and circadian genes. A further study of HIF interaction with these and other circadian regulators like the glucocorticoids is warranted.

11. ZEBRAFISH *VHL* MUTANTS AS MODELS OF HIF HYPERACTIVATION

One very powerful asset of the zebrafish model is the *vhl* mutant. *Vhl* mutants were one of the first knockouts to be created after the development of efficient reverse genetics technology in the zebrafish (van Rooijen et al., 2009). Both induced alleles are predicted null. However, the *vhl* phenotype may not represent a complete loss of function. There is clear evidence for a maternal contribution of the protein, and zebrafish contain the related *vll* gene. For instance, hypoxia (5 kPa O_2) can activate the *phd3::egfp* reporter during late epiboly stages, whereas zygotic *vhl*$^{-/-}$, *phd3::egfp* embryos do not show GFP induction at these stages, and the first signs of HIF activation can only be detected around 22 hpf (Santhakumar et al., 2012). A plausible explanation for this difference is the presence of maternally contributed Vhl protein in *Vhl* mutants. Could *Vll* partially take over the function of *Vhl*? Sequence comparison with human VHL revealed a number of amino acid substitutions that would be predicted to impair its capacity to target HIF (van Rooijen et al., 2009). Yet, in the absence of *vll*, *vhl* mutants have a slightly higher HIF response (FvE unpubl.) suggesting weak redundancy. Nevertheless, *vhl* deficiency is a convenient genetic background to study the consequence of a strong HIF signal; *vhl* mutant larvae live up to late larval stages and show numerous phenotypes that could be expected from such a strong HIF activation. Death in these mutants most likely occurs due to a collapse in circulation or failure to feed. *Vhl* mutation leads to Hif signaling without the compromising effect of loss of oxygen itself and may allow a better "signal-to-noise ratio" with respect to the effects of Hif activation. Chemical methods to activate HIF in the absence of hypoxia, like DMOG, FG-4592, Co^{2+}, or other Phd inhibitors, will in general target other hydroxylases as well, and thus could lead to off-target effects. It should also be stressed, however, that the function of *vhl* is not unique to HIF signaling, and several HIF-independent functions have been noted (Esteban, Harten, Tran, & Maxwell, 2006; Lolkema et al., 2008; Metcalf et al., 2014; Ohh et al., 1998; Thoma et al., 2009, and see below). Thus, caution remains important when using the *vhl* mutant as a hypoxic signaling model.

12. HEMATOPOIETIC AND ANGIOGENIC PHENOTYPES IN *VHL* MUTANTS

Larval *vhl* mutants develop severe polycythemia, which may be comparable to the human VHL-dependent disease, "Chuvash polycythemia" (Ang et al., 2002; van Rooijen et al., 2009). In addition, mutants increase have an increased heart rate and stroke volume, leading to a 15-fold increase at late larval stages (van Rooijen et al., 2009). Analysis of the hematopoietic phenotype showed that HIF signaling most likely promotes early hematopoietic stem cell (HSC) expansion and proliferation, in addition to activating erythropoietin (Epo) signaling, resulting in expansion of both red and white blood cell lineages (van Rooijen et al., 2009). Recently, this idea was confirmed by a study where stimulation of oxidative metabolism by glucose was shown to generate reactive oxygen species that led to a Hif-1αb-dependent increase in HSC formation (Harris et al., 2013).

Although early vessel patterning appears normal up to about 2 dpf, a hyperangiogenic phenotype can clearly be observed by 3 dpf (van Rooijen et al., 2010). Both vessel dilation and ectopic sprouting are observed in many locations, furthermore, vessels were found to be leaky. Raising fish in hypoxia can (in certain studies) elicit similar responses, but the effects are limited by the level of hypoxia that zebrafish can tolerate (Cao, Jensen, Soll, Hauptmann, & Cao, 2008; Schwerte, Uberbacher, & Pelster, 2003) and appear to be generally milder.

An important downstream effector of HIF, with regard to angiogenesis, is vascular endothelial growth factor (VEGF) signaling and indeed several VEGF pathway genes were found to be upregulated (van Rooijen et al., 2010). Importantly, VEGF inhibitors could partially rescue the effects of *vhl* mutants on angiogenesis. Not all effects need to be mediated by VEGF, however, *angiopoietin-like factor 4* may also be induced directly by HIF, and several others of this family are induced (but lack nearby HREs, Suppl. Table 1).

Another important angiogenic regulator that is induced in *vhl*, and by hypoxia, is *chemokine receptor-type 4a (cxcr4a)* and its ligands *cxcl12a* and *b* (Greenald et al., 2015; Marques et al., 2008; van Rooijen et al., 2010). It is interesting to note that *cxcr4a* is normally suppressed by blood flow (Packham et al., 2009), thus HIF signaling must somehow override this negative effect. These genes are most likely indirect targets, as HREs were not seen in their *cis* regions, VEGF being a plausible mediator. When analyzing *vhl* mutants, we surprisingly discovered that when flow is inhibited, the hyperangiogenic phenotype is completely suppressed (Watson et al., 2013). This indicates strong linkages between flow- and HIF-induced angiogenesis. So, HIF signaling can overcome flow-mediated suppression of *cxcr4a*, but it nevertheless needs flow to generate angiogenic responses in the embryo. We suggested Notch signaling initiated by *delta-like 4* (*dll4*) as a possible mediator of this effect. Expression of *dll4* can suppress tip cell formation and angiogenesis (Leslie et al., 2007; Siekmann & Lawson, 2007). Furthermore, *dll4* is a HIF target (Suppl. Table 1) and induced by loss of flow, suggesting a possible synergy (Watson et al., 2013). Unfortunately, a lethal interaction between loss of *dll4*/Notch signaling and *vhl* prevented a direct test of this hypothesis (Watson et al., 2013). It will be interesting to see if hypoxia-driven angiogenesis in tumors has a similarly strong dependence on flow.

13. VHL/HIF EFFECTS ON METABOLISM

It is expected that a strong activation of the HIF pathway will have a profound effect on metabolism, promoting an oxygen-conserving switch to anaerobic glycolysis, and away from oxidative phosphorylation and fatty acid oxidation in the mitochondria (Semenza, 2007). However the precise effect of HIF activation is different depending on the level of activation. For instance, low-level activation of HIF-2a in mouse liver has a beneficial effect on their metabolic profile, as seen in hypomorphic *phd2* mutants (Rahtu-Korpela et al., 2014) whereas high level activation leads to hepatic steatosis (Rankin et al., 2009). Zebrafish *vhl* mutant microarray studies suggest these mutants have strong HIF activation and metabolic change. Many genes involved in glycolysis and lipid processing are modulated (Greenald et al., 2015). *Vhl* mutant embryos use up their yolk more slowly than their wild-type siblings (van Rooijen et al., 2011), and at 8 dpf, they show a 30% reduction in use of oxygen

(Yaqoob & Schwerte, 2010). Their metabolic rate at 7 days is also lower than WT (as measured by acidosis), and they have lower levels of ATP (van der Velden et al., 2011), which all suggest that the *vhl* mutants switch to "energy/oxygen-saving mode." This was elegantly confirmed during the analysis of the *lkb1* mutant, which fails to adapt to low energy conditions. In this mutant the removal of *vhl* could clearly suppress *lkb1* starvation phenotypes (van der Velden et al., 2011). The metabolic shift that occurs under hypoxia, or in the absence of *vhl* function, may have clinical application in certain mitochondrial diseases. In cell culture experiments, loss of VHL was identified as a top suppressor of mitochondrial defects (Jain et al., 2016). Zebrafish *vhl* mutants provided a convenient first in vivo confirmation of this result; they were more resistant to the mitochondrial poison, antimycin, than their wild-type siblings. Similar results were obtained with the chemical HIF activator, FG4592 (Jain et al., 2016). The authors also confirmed their findings in a mouse model of Leigh disease. When found safe, HIF activation could represent a novel treatment option for these and possibly other types of metabolic diseases (Rahtu-Korpela et al., 2014).

14. KIDNEY DEFECTS IN *VHL* MUTANTS AND CANCER

Vhl mutants are known to have hepatic steatosis and accumulation of fatty acids and glycogen in the pronephros (van Rooijen et al., 2011). Recently, pronephros phenotypes were analyzed in-depth by the Iliopoulos group (Noonan et al., 2016). Electron microscopy at 5.5 dpf showed that tubule diameter was increased, but cell numbers per section were decreased, and cilia appeared to be present but were disorganized. Importantly, BODIPY493/503 labeling and PAS staining showed that $vhl^{-/-}$ pronephric tubule cells were packed with both glycogen particles and lipid droplets. The latter cellular phenotypes are also a hallmark of human clear cell renal cell carcinoma (ccRCC). Thus a valid question is whether the *vhl* mutants are showing early signs of tumor formation. Noonan et al. have looked at proliferation in the pronephric ducts and found it to be increased at least from 4.5 dpf to 8.5 dpf. Although this is suggestive, it was also noted that there was no increase in overall cell number and also an increased level of apoptosis was noted. Thus, proliferation could also be a regenerative response to cellular loss, rather than a sign of transformation. Nonetheless, we can state that the pronephric cellular phenotype in *vhl* mutants bears a clear resemblance to human ccRCC.

Could the *vhl* mutant become a model for ccRCC? Human VHL patients are heterozygous and develop cancer as a result of loss of the remaining wild-type copy of the gene (LOH). We tried to address this question using a parallel approach in zebrafish. Heterozygous *vhl* fish are not clearly predisposed to cancer, but their limited lifetime may obscure a tumor suppressor role for *vhl.* To further address this, heterozygous *vhl* fish and wild-type fish that carried the *phd3::GFP* reporter in their background were treated with a wide spectrum carcinogen, DMBA (Spitsbergen et al., 2000), to promote LOH and further second site mutations (Santhakumar et al., 2012). The transgene would allow us to easily recognize tumors that were due to loss of *vhl* by their activation of the *phd3:GFP* reporter. After 14 months, fish were analyzed and *vhl* heterozygotes were clearly more susceptible to tumor formation than their wild-type siblings, and a large number of these tumors were GFP positive. Although second site mutations could not be pinpointed, qPCR analysis of microdissected tumor tissue showed that *vhl* mRNA levels were affected in most cases, pointing toward LOH (Santhakumar et al., 2012). Altogether this indicated that, also in the zebrafish, Vhl is a *bonafide* tumor suppressor. Unfortunately, although abnormal proliferation was observed in the kidney, tumors akin to ccRCC were not detected. Reasons for this can be manifold; possibly the second *vll* gene is important and may need to be inactivated, which is being tested at the moment. Alternative explanations are also likely. Surprisingly, there are no reliable animal models for ccRCC. In human, loss of heterozygosity for VHL is required for ccRCC to develop, and it usually occurs by loss of large pieces of chromosome. Some research suggests this is important, and that the unique human "genomic context" of VHL may be responsible for the preferential formation of kidney tumors (See, for example, Wang et al., 2014). It is clear that more work is required before zebrafish can contribute in this respect.

Despite the lack of tumors in the *vhl* heterozygous and homozygous models, the phenotypes of the homozygote are accessible and easy to quantify; therefore this could provide a surrogate readout for potential therapeutics that might be useful in VHL disease. In ccRCC, it is mainly HIF-2 that drives tumorigenesis. Surprisingly, HIF-1 rather appears to act as a tumor-suppressing gene (Raval et al., 2005; Shen et al., 2011). A recent study by Metelo et al. used Compound 76, which had been identified as a potential inhibitor of Hif-2 signaling (Zimmer et al., 2008) to suppress erythrocytosis and hyperangiogenesis. In addition, significant suppression of *vhl* mutant pronephric phenotypes was shown (Noonan et al., 2016). It will be interesting to see whether this compound will show effectiveness in rodent and xenograft models.

15. LINKS BETWEEN VHL HIF AND P53

HIF is often activated in cancer; therefore its relationship with P53, the most frequently mutated protein in tumors, is of considerable interest. However, the links between VHL, HIF, and P53 are still quite unclear, with various mechanisms being proposed in the literature (Chen, Li, Luo, & Gu, 2003; Galban et al., 2003; Roberts et al., 2009; Roe & Youn, 2006; Sermeus & Michiels, 2011). It is very clear that loss/mutation of the *P53* gene is a surprisingly infrequent event in VHL-deficient ccRCC (Cancer Genome Atlas Research, 2013; Sato et al., 2013), compared to other tumors. This could suggest that P53 function is irrelevant to ccRCC tumors. The high radio- and chemoresistance of ccRCC (Cohen & McGovern, 2005) and the fact that mice mutant for both *p53* and *vhl* do show increased tumor formation, Albers et al., (2013) argue against this. More likely, P53 function is somehow impaired at the protein level or downstream of P53, after loss of *VHL*. Moreover, the stabilization of P53 in hypoxia has been reported since 1994 (Alarcon, Koumenis, Geyer, Maki, & Giaccia, 1999; Graeber et al., 1994), but the precise mechanisms for its stabilization are still not fully understood. This stabilization is not totally dependent on HIF, and MDM2-dependent and -independent pathways have been proposed. One issue in this research has been the frequent use of cultured cell lines, where additional genetic changes may have occurred.

Zebrafish *vhl* mutants will not have such additional changes; their analysis allowed identification of a novel mechanism for the P53 stabilization. P53 is stabilized in the absence of Vhl, or in hypoxia, and becomes hyperstabilized in response to DNA damage (Essers et al., 2015). By mass spectrometry with TAP-tagged VHL, programmed cell death 5 (PDCD5) was identified to interact with pVHL. In normoxic conditions, the direct interaction of pVHL and PDCD5 in the cytoplasm is associated with a high level of MDM2 and a low level of P53 in nuclei. On the contrary, in hypoxia or in the absence of pVHL, PDCD5 translocates into the nucleus, which coincides with the degradation of MDM2 and P53 stabilization. The fact that hypoxia has a similar effect is confusing, and it will need further experiments to establish if this is HIF dependent or possibly a direct HIF-independent effect of hypoxia.

More importantly, despite the significant increase in the P53 protein in *vhl* mutants, embryos do not respond to the irradiation either by cell cycle arrest or by inducing apoptosis. The authors suggest that increased *birc5* (survivin) expression might reduce the apoptosis response. Birc5 has been proposed as a therapeutic target, since suppressing *birc5* restored the expression of caspase-3 in human ccRCC cell lines. Indeed, zebrafish *birc5a/b* have Hif-2α binding sites and *birc5a* was 1.5–2-fold upregulated in *vhl* mutants at 7 and 5 dpf (Essers et al., 2015; van Rooijen et al., 2009) but was not identified by Greenald et al. (2015). Alternatively, loss of *vhl* may somehow reduce acetylation of P53 so it may be less able to activate transcription of downstream genes.

This study reveals a novel HIF-independent role of VHL in the regulation of PDCD5 subcellular localization, and P53 protein stabilization, but stands in contrast to the previous study demonstrating that VHL stabilizes P53 by inhibiting MDM2-mediated degradation and nuclear export (Roe et al., 2006). The latter study also showed that pVHL can interact with Ataxia telangiectasia mutated (ATM) and P300, independently of P53. In the event of DNA damage, VHL strengthens the interaction betweenP53–ATM and P53–P300, thereby stabilizing P53 and enhancing its transactivation (Roe et al., 2006). It is clear that further study into these roles of VHL is warranted. Using *vhl* mutant embryos in contrast to transformed cell lines may help to untangle the complex links between VHL, P53, and genome stability.

APPENDIX A. SUPPLEMENTARY DATA

Supplementary data related to this article can be found at http://dx.doi.org/10.1016/bs.mcb.2016.07.001.

Funding sources, H.R. Kim, E. Markham and F. van Eeden were supported by BBSRC grant BB/M02332X/1. F. van Eeden, F. Argenton and A. Vettori were supported by FP7: HEALTH-F4-2010-242048.

REFERENCES

Alarcon, R., Koumenis, C., Geyer, R. K., Maki, C. G., & Giaccia, A. J. (1999). Hypoxia induces p53 accumulation through MDM2 down-regulation and inhibition of E6-mediated degradation. *Cancer Research, 59*(24), 6046–6051.

Albers, J., Rajski, M., Schonenberger, D., Harlander, S., Schraml, P., von Teichman, A., … Frew, I. J. (2013). Combined mutation of Vhl and Trp53 causes renal cysts and tumours in mice. *EMBO Molecular Medicine, 5*(6), 949–964. http://dx.doi.org/10.1002/emmm.201202231.

Ang, S. O., Chen, H., Hirota, K., Gordeuk, V. R., Jelinek, J., Guan, Y., … Prchal, J. T. (2002). Disruption of oxygen homeostasis underlies congenital Chuvash polycythemia. *Nature Genetics, 32*(4), 614–621. http://dx.doi.org/10.1038/ng1019.

Appelhoff, R. J., Tian, Y. M., Raval, R. R., Turley, H., Harris, A. L., Pugh, C. W., … Gleadle, J. M. (2004). Differential function of the prolyl hydroxylases PHD1, PHD2, and PHD3 in the regulation of hypoxia-inducible factor. *Journal of Biological Chemistry, 279*(37), 38458–38465. http://dx.doi.org/10.1074/jbc.M406026200.

Berra, E., Benizri, E., Ginouves, A., Volmat, V., Roux, D., & Pouyssegur, J. (2003). HIF prolyl-hydroxylase 2 is the key oxygen sensor setting low steady-state levels of HIF-1alpha in normoxia. *EMBO Journal, 22*(16), 4082–4090. http://dx.doi.org/10.1093/emboj/cdg392.

Bersten, D. C., Sullivan, A. E., Peet, D. J., & Whitelaw, M. L. (2013). bHLH-PAS proteins in cancer. *Nature Reviews Cancer, 13*(12), 827–841. http://dx.doi.org/10.1038/nrc3621.

Braasch, I., Gehrke, A. R., Smith, J. J., Kawasaki, K., Manousaki, T., Pasquier, J., … Postlethwait, J. H. (2016). The spotted gar genome illuminates vertebrate evolution and facilitates human-teleost comparisons. *Nature Genetics, 48*(4), 427–437. http://dx.doi.org/10.1038/ng.3526.

Cancer Genome Atlas Research, Network. (2013). Comprehensive molecular characterization of clear cell renal cell carcinoma. *Nature, 499*(7456), 43–49. http://dx.doi.org/10.1038/nature12222.

Cao, R., Jensen, L. D., Soll, I., Hauptmann, G., & Cao, Y. (2008). Hypoxia-induced retinal angiogenesis in zebrafish as a model to study retinopathy. *PLoS One, 3*(7), e2748. http://dx.doi.org/10.1371/journal.pone.0002748.

Chan, M. C., Atasoylu, O., Hodson, E., Tumber, A., Leung, I. K., Chowdhury, R., … Schofield, C. J. (2015). Potent and selective triazole-based inhibitors of the hypoxia-inducible factor prolyl-hydroxylases with activity in the murine brain. *PLoS One, 10*(7), e0132004. http://dx.doi.org/10.1371/journal.pone.0132004.

Chen, D., Li, M., Luo, J., & Gu, W. (2003). Direct interactions between HIF-1 alpha and Mdm2 modulate p53 function. *Journal of Biological Chemistry, 278*(16), 13595–13598. http://dx.doi.org/10.1074/jbc.C200694200.

Chowdhury, R., Candela-Lena, J. I., Chan, M. C., Greenald, D. J., Yeoh, K. K., Tian, Y. M., ... Schofield, C. J. (2013). Selective small molecule probes for the hypoxia inducible factor (HIF) prolyl hydroxylases. *ACS Chemical Biology, 8*(7), 1488–1496. http://dx.doi.org/10.1021/cb400088q.

Cohen, H. T., & McGovern, F. J. (2005). Renal-cell carcinoma. *New England Journal of Medicine, 353*(23), 2477–2490. http://dx.doi.org/10.1056/NEJMra043172.

Dalgliesh, G. L., Furge, K., Greenman, C., Chen, L., Bignell, G., Butler, A., ... Futreal, P. A. (2010). Systematic sequencing of renal carcinoma reveals inactivation of histone modifying genes. *Nature, 463*(7279), 360–363. http://dx.doi.org/10.1038/nature08672.

Duan, D. R., Pause, A., Burgess, W. H., Aso, T., Chen, D. Y., Garrett, K. P., ... Klausner, R. D. (1995). Inhibition of transcription elongation by the VHL tumor suppressor protein. *Science, 269*(5229), 1402–1406.

Egg, M., Koblitz, L., Hirayama, J., Schwerte, T., Folterbauer, C., Kurz, A., ... Pelster, B. (2013). Linking oxygen to time: the bidirectional interaction between the hypoxic signaling pathway and the circadian clock. *Chronobiology International, 30*(4), 510–529. http://dx.doi.org/10.3109/07420528.2012.754447.

Egg, M., Paulitsch, M., Ennemoser, Y., Wustenhagen, A., Schwerte, T., Sandbichler, A. M., ... Pelster, B. (2014). Chronodisruption increases cardiovascular risk in zebrafish via reduced clearance of senescent erythrocytes. *Chronobiology International, 31*(5), 680–689. http://dx.doi.org/10.3109/07420528.2014.889703.

Elks, P. M., van Eeden, F. J., Dixon, G., Wang, X., Reyes-Aldasoro, C. C., Ingham, P. W., ... Renshaw, S. A. (2011). Activation of hypoxia-inducible factor-1alpha (Hif-1alpha) delays inflammation resolution by reducing neutrophil apoptosis and reverse migration in a zebrafish inflammation model. *Blood, 118*(3), 712–722. http://dx.doi.org/10.1182/blood-2010-12-324186. pii:blood-2010-12-324186.

Elks, P. M., Renshaw, S. A., Meijer, A. H., Walmsley, S. R., & van Eeden, F. J. (2015). Exploring the HIFs, buts and maybes of hypoxia signalling in disease: lessons from zebrafish models. *Disease Models & Mechanisms, 8*(11), 1349–1360. http://dx.doi.org/10.1242/dmm.021865.

Epstein, A. C., Gleadle, J. M., McNeill, L. A., Hewitson, K. S., O'Rourke, J., Mole, D. R., ... Ratcliffe, P. J. (2001). *C. elegans* EGL-9 and mammalian homologs define a family of dioxygenases that regulate HIF by prolyl hydroxylation. *Cell, 107*(1), 43–54.

Erez, J., Krom, M. D., & Neuwirth, T. (1990). Daily oxygen variations in marine fish ponds, Elat, Israel. *Aquaculture, 84*(3–4), 289–305. http://dx.doi.org/10.1016/0044-8486(90)90094-4.

Essers, P. B., Klasson, T. D., Pereboom, T. C., Mans, D. A., Nicastro, M., Boldt, K., ... MacInnes, A. W. (2015). The von Hippel-Lindau tumor suppressor regulates programmed cell death 5-mediated degradation of Mdm2. *Oncogene, 34*(6), 771–779. http://dx.doi.org/10.1038/onc.2013.598.

Esteban, M. A., Harten, S. K., Tran, M. G., & Maxwell, P. H. (2006). Formation of primary cilia in the renal epithelium is regulated by the von Hippel-Lindau tumor suppressor protein. *Journal of American Society of Nephrology, 17*(7), 1801–1806. http://dx.doi.org/10.1681/ASN.2006020181.

Flashman, E., Hoffart, L. M., Hamed, R. B., Bollinger, J. M., Jr., Krebs, C., & Schofield, C. J. (2010). Evidence for the slow reaction of hypoxia-inducible factor prolyl hydroxylase 2 with oxygen. *FEBS Journal, 277*(19), 4089–4099. http://dx.doi.org/10.1111/j.1742-4658.2010.07804.x.

Galban, S., Martindale, J. L., Mazan-Mamczarz, K., Lopez de Silanes, I., Fan, J., Wang, W., ... Gorospe, M. (2003). Influence of the RNA-binding protein HuR in pVHL-regulated p53 expression in renal carcinoma cells. *Molecular and Cellular Biology, 23*(20), 7083–7095.

Gossage, L., Eisen, T., & Maher, E. R. (2015). VHL, the story of a tumour suppressor gene. *Nature Reviews Cancer, 15*(1), 55–64. http://dx.doi.org/10.1038/nrc3844.

Graeber, T. G., Peterson, J. F., Tsai, M., Monica, K., Fornace, A. J., Jr., & Giaccia, A. J. (1994). Hypoxia induces accumulation of p53 protein, but activation of a G1-phase checkpoint by low-oxygen conditions is independent of p53 status. *Molecular and Cellular Biology, 14*(9), 6264–6277.

Greenald, D., Jeyakani, J., Pelster, B., Sealy, I., Mathavan, S., & van Eeden, F. J. (2015). Genome-wide mapping of Hif-1alpha binding sites in zebrafish. *BMC Genomics, 16*, 923. http://dx.doi.org/10.1186/s12864-015-2169-x.

van Haaften, G., Dalgliesh, G. L., Davies, H., Chen, L., Bignell, G., Greenman, C., … Futreal, P. A. (2009). Somatic mutations of the histone H3K27 demethylase gene UTX in human cancer. *Nature Genetics, 41*(5), 521–523. http://dx.doi.org/10.1038/ng.349.

Harris, J. M., Esain, V., Frechette, G. M., Harris, L. J., Cox, A. G., Cortes, M., … North, T. E. (2013). Glucose metabolism impacts the spatiotemporal onset and magnitude of HSC induction in vivo. *Blood, 121*(13), 2483–2493. http://dx.doi.org/10.1182/blood-2012-12-471201.

Heidbreder, M., Frohlich, F., Johren, O., Dendorfer, A., Qadri, F., & Dominiak, P. (2003). Hypoxia rapidly activates HIF-3alpha mRNA expression. *FASEB Journal, 17*(11), 1541–1543. http://dx.doi.org/10.1096/fj.02-0963fje.

Hubbi, M. E., Hu, H., Kshitiz, Ahmed, I., Levchenko, A., & Semenza, G. L. (2013). Chaperone-mediated autophagy targets hypoxia-inducible factor-1alpha (HIF-1alpha) for lysosomal degradation. *Journal of Biological Chemistry, 288*(15), 10703–10714. http://dx.doi.org/10.1074/jbc.M112.414771.

Hyvarinen, J., Parikka, M., Sormunen, R., Ramet, M., Tryggvason, K., Kivirikko, K. I., … Koivunen, P. (2010). Deficiency of a transmembrane prolyl 4-hydroxylase in the zebrafish leads to basement membrane defects and compromised kidney function. *Journal of Biological Chemistry, 285*(53), 42023–42032. http://dx.doi.org/10.1074/jbc.M110.145904.

Jain, I. H., Zazzeron, L., Goli, R., Alexa, K., Schatzman-Bone, S., Dhillon, H., … Mootha, V. K. (2016). Hypoxia as a therapy for mitochondrial disease. *Science, 352*(6281), 54–61. http://dx.doi.org/10.1126/science.aad9642.

Jopling, C., Sune, G., Faucherre, A., Fabregat, C., & Izpisua Belmonte, J. C. (2012). Hypoxia induces myocardial regeneration in zebrafish. *Circulation, 126*(25), 3017–3027. http://dx.doi.org/10.1161/CIRCULATIONAHA.112.107888.

Kajimura, S., Aida, K., & Duan, C. (2005). Insulin-like growth factor-binding protein-1 (IGFBP-1) mediates hypoxia-induced embryonic growth and developmental retardation. *Proceedings of National Academy of Sciences of United States of America, 102*(4), 1240–1245. http://dx.doi.org/10.1073/pnas.0407443102.

Kajimura, S., Aida, K., & Duan, C. (2006). Understanding hypoxia-induced gene expression in early development: in vitro and in vivo analysis of hypoxia-inducible factor 1-regulated zebra fish insulin-like growth factor binding protein 1 gene expression. *Molecular and Cellular Biology, 26*(3), 1142–1155. http://dx.doi.org/10.1128/MCB.26.3.1142-1155.2006.

Kamura, T., Koepp, D. M., Conrad, M. N., Skowyra, D., Moreland, R. J., Iliopoulos, O., … Conaway, J. W. (1999). Rbx1, a component of the VHL tumor suppressor complex and SCF ubiquitin ligase. *Science, 284*(5414), 657–661.

Kibel, A., Iliopoulos, O., DeCaprio, J. A., & Kaelin, W. G., Jr. (1995). Binding of the von Hippel-Lindau tumor suppressor protein to Elongin B and C. *Science, 269*(5229), 1444–1446.

Kishida, T., Stackhouse, T. M., Chen, F., Lerman, M. I., & Zbar, B. (1995). Cellular proteins that bind the von Hippel-Lindau disease gene product: mapping of binding domains and the effect of missense mutations. *Cancer Research, 55*(20), 4544–4548.

Koblitz, L., Fiechtner, B., Baus, K., Lussnig, R., & Pelster, B. (2015). Developmental expression and hypoxic induction of hypoxia inducible transcription factors in the zebrafish. *PLoS One, 10*(6), e0128938. http://dx.doi.org/10.1371/journal.pone.0128938.

Kopp, R., Koblitz, L., Egg, M., & Pelster, B. (2011). HIF signaling and overall gene expression changes during hypoxia and prolonged exercise differ considerably. *Physiological Genomics, 43*(9), 506–516. http://dx.doi.org/10.1152/physiolgenomics.00250.2010.

Ko, C. Y., Tsai, M. Y., Tseng, W. F., Cheng, C. H., Huang, C. R., Wu, J. S., … Hu, C. H. (2011). Integration of CNS survival and differentiation by HIF2alpha. *Cell Death and Differentiation, 18*(11), 1757–1770. http://dx.doi.org/10.1038/cdd.2011.44.

Kulkarni, R. P., Tohari, S., Ho, A., Brenner, S., & Venkatesh, B. (2010). Characterization of a hypoxia-response element in the Epo locus of the pufferfish, *Takifugu rubripes*. *Marine Genomics, 3*(2), 63–70. http://dx.doi.org/10.1016/j.margen.2010.05.001.

Latif, F., Tory, K., Gnarra, J., Yao, M., Duh, F. M., Orcutt, M. L., … Lerman, M. I. (1993). Identification of the von Hippel-Lindau disease tumor suppressor gene. *Science, 260*(5112), 1317–1320.

Leslie, J. D., Ariza-McNaughton, L., Bermange, A. L., McAdow, R., Johnson, S. L., & Lewis, J. (2007). Endothelial signalling by the Notch ligand Delta-like 4 restricts angiogenesis. *Development, 134*(5), 839–844. http://dx.doi.org/10.1242/dev.003244.

Lin, T. Y., Chou, C. F., Chung, H. Y., Chiang, C. Y., Li, C. H., Wu, J. L., … Tzou, W. S. (2014). Hypoxia-inducible factor 2 alpha is essential for hepatic outgrowth and functions via the regulation of leg1 transcription in the zebrafish embryo. *PLoS One, 9*(7), e101980. http://dx.doi.org/10.1371/journal.pone.0101980.

Loenarz, C., Coleman, M. L., Boleininger, A., Schierwater, B., Holland, P. W., Ratcliffe, P. J., & Schofield, C. J. (2011). The hypoxia-inducible transcription factor pathway regulates oxygen sensing in the simplest animal, *Trichoplax adhaerens*. *EMBO Reports, 12*(1), 63–70. http://dx.doi.org/10.1038/embor.2010.170.

Lolkema, M. P., Gervais, M. L., Snijckers, C. M., Hill, R. P., Giles, R. H., Voest, E. E., & Ohh, M. (2005). Tumor suppression by the von Hippel-Lindau protein requires phosphorylation of the acidic domain. *Journal of Biological Chemistry, 280*(23), 22205–22211. http://dx.doi.org/10.1074/jbc.M503220200.

Lolkema, M. P., Mans, D. A., Ulfman, L. H., Volpi, S., Voest, E. E., & Giles, R. H. (2008). Allele-specific regulation of primary cilia function by the von Hippel-Lindau tumor suppressor. *European Journal of Human Genetics, 16*(1), 73–78. http://dx.doi.org/10.1038/sj.ejhg.5201930.

Lonergan, K. M., Iliopoulos, O., Ohh, M., Kamura, T., Conaway, R. C., Conaway, J. W., & Kaelin, W. G., Jr. (1998). Regulation of hypoxia-inducible mRNAs by the von Hippel-Lindau tumor suppressor protein requires binding to complexes containing elongins B/C and Cul2. *Molecular and Cellular Biology, 18*(2), 732–741.

Long, Y., Yan, J., Song, G., Li, X., Li, X., Li, Q., & Cui, Z. (2015). Transcriptional events co-regulated by hypoxia and cold stresses in Zebrafish larvae. *BMC Genomics, 16*, 385. http://dx.doi.org/10.1186/s12864-015-1560-y.

Lutz, P. L., & Nilsson, G. E. (2004). Vertebrate brains at the pilot light. *Respiratory Physiology & Neurobiology, 141*(3), 285–296. http://dx.doi.org/10.1016/j.resp.2004.03.013.

Manchenkov, T., Pasillas, M. P., Haddad, G. G., & Imam, F. B. (2015). Novel genes critical for hypoxic preconditioning in zebrafish are regulators of insulin and glucose metabolism. *G3 (Besda), 5*(6), 1107–1116. http://dx.doi.org/10.1534/g3.115.018010.

Mans, D. A., Vermaat, J. S., Weijts, B. G., van Rooijen, E., van Reeuwijk, J., Boldt, K., … Giles, R. H. (2013). Regulation of E2F1 by the von Hippel-Lindau tumour suppressor protein predicts survival in renal cell cancer patients. *Journal of Pathology, 231*(1), 117–129. http://dx.doi.org/10.1002/path.4219.

Marques, I. J., Leito, J. T., Spaink, H. P., Testerink, J., Jaspers, R. T., Witte, F., … Bagowski, C. P. (2008). Transcriptome analysis of the response to chronic constant hypoxia in zebrafish hearts. *Journal of Comparative Physiology B, 178*(1), 77–92. http://dx.doi.org/10.1007/s00360-007-0201-4.

Martinovic, D., Villeneuve, D. L., Kahl, M. D., Blake, L. S., Brodin, J. D., & Ankley, G. T. (2009). Hypoxia alters gene expression in the gonads of zebrafish (*Danio rerio*). *Aquatic Toxicology, 95*(4), 258–272. http://dx.doi.org/10.1016/j.aquatox.2008.08.021.

van der Meer, D. L., van den Thillart, G. E., Witte, F., de Bakker, M. A., Besser, J., Richardson, M. K., … Bagowski, C. P. (2005). Gene expression profiling of the long-term adaptive response to hypoxia in the gills of adult zebrafish. *American Journal of Physiology, Regulatory, Integrative and Comparative Physiology, 289*(5), R1512–R1519. http://dx.doi.org/10.1152/ajpregu.00089.2005.

Metcalf, J. L., Bradshaw, P. S., Komosa, M., Greer, S. N., Stephen Meyn, M., & Ohh, M. (2014). K63-ubiquitylation of VHL by SOCS1 mediates DNA double-strand break repair. *Oncogene, 33*(8), 1055–1065. http://dx.doi.org/10.1038/onc.2013.22.

Metelo, A. M., Noonan, H. R., Li, X., Jin, Y., Baker, R., Kamentsky, L., … Iliopoulos, O. (2015). Pharmacological HIF2alpha inhibition improves VHL disease-associated phenotypes in zebrafish model. *Journal of Clinical Investigation, 125*(5), 1987–1997. http://dx.doi.org/10.1172/JCI73665.

Michaud, J. L., DeRossi, C., May, N. R., Holdener, B. C., & Fan, C. M. (2000). ARNT2 acts as the dimerization partner of SIM1 for the development of the hypothalamus. *Mechanisms of Development, 90*(2), 253–261. http://dx.doi.org/10.1016/S0925-4773(99)00328-7.

Mimura, I., Nangaku, M., Kanki, Y., Tsutsumi, S., Inoue, T., Kohro, T., … Wada, Y. (2012). Dynamic change of chromatin conformation in response to hypoxia enhances the expression of GLUT3 (SLC2A3) by cooperative interaction of hypoxia-inducible factor 1 and KDM3A. *Molecular and Cellular Biology, 32*(15), 3018–3032. http://dx.doi.org/10.1128/MCB.06643-11.

Mole, D. R., Blancher, C., Copley, R. R., Pollard, P. J., Gleadle, J. M., Ragoussis, J., & Ratcliffe, P. J. (2009). Genome-wide association of hypoxia-inducible factor (HIF)-1alpha and HIF-2alpha DNA binding with expression profiling of hypoxia-inducible transcripts. *Journal of Biological Chemistry, 284*(25), 16767–16775. http://dx.doi.org/10.1074/jbc.M901790200.

Moroz, E., Carlin, S., Dyomina, K., Burke, S., Thaler, H. T., Blasberg, R., & Serganova, I. (2009). Real-time imaging of HIF-1alpha stabilization and degradation. *PLoS One, 4*(4), e5077. http://dx.doi.org/10.1371/journal.pone.0005077.

Nikinmaa, M. (2002). Oxygen-dependent cellular functions - why fishes and their aquatic environment are a prime choice of study. *Comparative Biochemistry and Physiology a-Molecular and Integrative Physiology, 133*(1), 1–16. http://dx.doi.org/10.1016/S1095-6433(02) 00132–0. pii:S1095-6433(02)00132-0.

Noonan, H. R., Metelo, A. M., Kamei, C. N., Peterson, R. T., Drummond, I. A., & Iliopoulos, O. (2016 Aug 1). Loss of vhl in the zebrafish pronephros recapitulates early stages of human clear cell renal cell carcinoma. *Dis Model Mech, 9*(8), 873–884. http://dx.doi.org/10.1242/dmm.024380.

Ohh, M., Yauch, R. L., Lonergan, K. M., Whaley, J. M., Stemmer-Rachamimov, A. O., Louis, D. N., … Iliopoulos, O. (1998). The von Hippel-Lindau tumor suppressor protein is required for proper assembly of an extracellular fibronectin matrix. *Molecular Cell, 1*(7), 959–968.

Packham, I. M., Gray, C., Heath, P. R., Hellewell, P. G., Ingham, P. W., Crossman, D. C., … Chico, T. J. (2009). Microarray profiling reveals CXCR4a is downregulated by blood flow in vivo and mediates collateral formation in zebrafish embryos. *Physiological Genomics, 38*(3), 319–327. http://dx.doi.org/10.1152/physiolgenomics.00049.2009.

Padilla, P. A., & Roth, M. B. (2001). Oxygen deprivation causes suspended animation in the zebrafish embryo. *Proceedings of National Academy of Sciences of United States of America, 98*(13), 7331–7335. http://dx.doi.org/10.1073/pnas.131213198.

Pasanen, A., Heikkila, M., Rautavuoma, K., Hirsila, M., Kivirikko, K. I., & Myllyharju, J. (2010). Hypoxia-inducible factor (HIF)-3alpha is subject to extensive alternative splicing in human tissues and cancer cells and is regulated by HIF-1 but not HIF-2. *International Journal of Biochemistry and Cell Biology, 42*(7), 1189–1200. http://dx.doi.org/10.1016/j.biocel.2010.04.008.

Patel, S. A., & Simon, M. C. (2008). Biology of hypoxia-inducible factor-2alpha in development and disease. *Cell Death and Differentiation, 15*(4), 628–634. http://dx.doi.org/10.1038/cdd.2008.17.

Pause, A., Lee, S., Worrell, R. A., Chen, D. Y., Burgess, W. H., Linehan, W. M., & Klausner, R. D. (1997). The von Hippel-Lindau tumor-suppressor gene product forms a stable complex with human CUL-2, a member of the Cdc53 family of proteins. *Proceedings of National Academy of Sciences of United States of America, 94*(6), 2156–2161.

Pelster, B., & Egg, M. (2015). Multiplicity of hypoxia-inducible transcription factors and their connection to the circadian clock in the zebrafish. *Physiological and Biochemical Zoology, 88*(2), 146–157. http://dx.doi.org/10.1086/679751.

Qi, H., Gervais, M. L., Li, W., DeCaprio, J. A., Challis, J. R., & Ohh, M. (2004). Molecular cloning and characterization of the von Hippel-Lindau-like protein. *Molecular Cancer Research, 2*(1), 43–52.

Rahtu-Korpela, L., Karsikas, S., Horkko, S., Blanco Sequeiros, R., Lammentausta, E., Makela, K. A., … Koivunen, P. (2014). HIF prolyl 4-hydroxylase-2 inhibition improves glucose and lipid metabolism and protects against obesity and metabolic dysfunction. *Diabetes, 63*(10), 3324–3333. http://dx.doi.org/10.2337/db14-0472.

Rankin, E. B., Rha, J., Selak, M. A., Unger, T. L., Keith, B., Liu, Q., & Haase, V. H. (2009). Hypoxia-inducible factor 2 regulates hepatic lipid metabolism. *Molecular and Cellular Biology, 29*(16), 4527–4538. http://dx.doi.org/10.1128/MCB.00200-09.

Rauch, G. J., Lyons, D. A., Middendorf, I., Friedlander, B., Arana, N., Reyes, T., & Talbot, W. S. (2003). Submission and curation of gene expression data. *ZFIN Direct Data Submission*. http://zfin.org.

Raval, R. R., Lau, K. W., Tran, M. G., Sowter, H. M., Mandriota, S. J., Li, J. L., … Ratcliffe, P. J. (2005). Contrasting properties of hypoxia-inducible factor 1 (HIF-1) and HIF-2 in von Hippel-Lindau-associated renal cell carcinoma. *Molecular and Cellular Biology, 25*(13), 5675–5686. http://dx.doi.org/10.1128/MCB.25.13.5675-5686.2005.

Rees, B. B., Figueroa, Y. G., Wiese, T. E., Beckman, B. S., & Schulte, P. M. (2009). A novel hypoxia-response element in the lactate dehydrogenase-B gene of the killifish *Fundulus heteroclitus*. *Comparative Biochemistry and Physiology. Part A, Molecular & Integrative Physiology, 154*(1), 70–77. http://dx.doi.org/10.1016/j.cbpa.2009.05.001.

Rees, B. B., Sudradjat, F. A., & Love, J. W. (2001). Acclimation to hypoxia increases survival time of zebrafish, *Danio rerio*, during lethal hypoxia. *Journal of Experimental Zoology, 289*(4), 266–272.

Rius, J., Guma, M., Schachtrup, C., Akassoglou, K., Zinkernagel, A. S., Nizet, V., … Karin, M. (2008). NF-kappaB links innate immunity to the hypoxic response through transcriptional regulation of HIF-1alpha. *Nature, 453*(7196), 807–811. http://dx.doi.org/10.1038/nature06905.

Roberts, A. M., Watson, I. R., Evans, A. J., Foster, D. A., Irwin, M. S., & Ohh, M. (2009). Suppression of hypoxia-inducible factor 2alpha restores p53 activity via Hdm2 and reverses chemoresistance of renal carcinoma cells. *Cancer Research, 69*(23), 9056–9064. http://dx.doi.org/10.1158/0008-5472.CAN-09-1770.

Roe, J. S., Kim, H., Lee, S. M., Kim, S. T., Cho, E. J., & Youn, H. D. (2006). p53 stabilization and transactivation by a von Hippel-Lindau protein. *Molecular Cell, 22*(3), 395–405. http://dx.doi.org/10.1016/j.molcel.2006.04.006.

Roe, J. S., & Youn, H. D. (2006). The positive regulation of p53 by the tumor suppressor VHL. *Cell Cycle, 5*(18), 2054–2056. pii:3247.

Rojas, D. A., Perez-Munizaga, D. A., Centanin, L., Antonelli, M., Wappner, P., Allende, M. L., & Reyes, A. E. (2007). Cloning of hif-1alpha and hif-2alpha and mRNA expression pattern during development in zebrafish. *Gene Expression Patterns, 7*(3), 339–345. http://dx.doi.org/10.1016/j.modgep.2006.08.002.

van Rooijen, E., Santhakumar, K., Logister, I., Voest, E., Schulte-Merker, S., Giles, R., & van Eeden, F. (2011). A zebrafish model for VHL and hypoxia signaling. *Methods in Cell Biology, 105*, 163–190. http://dx.doi.org/10.1016/B978-0-12-381320-6.00007-2.

van Rooijen, E., Voest, E. E., Logister, I., Bussmann, J., Korving, J., van Eeden, F. J., … Schulte-Merker, S. (2010). von Hippel-Lindau tumor suppressor mutants faithfully model pathological hypoxia-driven angiogenesis and vascular retinopathies in zebrafish. *Disease Models & Mechanisms, 3*(5–6), 343–353. http://dx.doi.org/10.1242/dmm.004036. pii:dmm.004036.

van Rooijen, E., Voest, E. E., Logister, I., Korving, J., Schwerte, T., Schulte-Merker, S., … van Eeden, F. J. (2009). Zebrafish mutants in the von Hippel-Lindau tumor suppressor display a hypoxic response and recapitulate key aspects of Chuvash polycythemia. *Blood, 113*(25), 6449–6460. http://dx.doi.org/10.1182/blood-2008-07-167890. pii:blood-2008-07-167890.

Rytkonen, K. T., Akbarzadeh, A., Miandare, H. K., Kamei, H., Duan, C., Leder, E. H., … Nikinmaa, M. (2013). Subfunctionalization of cyprinid hypoxia-inducible factors for roles in development and oxygen sensing. *Evolution, 67*(3), 873–882. http://dx.doi.org/10.1111/j.1558-5646.2012.01820.x.

Rytkonen, K. T., Prokkola, J. M., Salonen, V., & Nikinmaa, M. (2014). Transcriptional divergence of the duplicated hypoxia-inducible factor alpha genes in zebrafish. *Gene, 541*(1), 60–66. http://dx.doi.org/10.1016/j.gene.2014.03.007.

Santhakumar, K., Judson, E. C., Elks, P. M., McKee, S., Elworthy, S., van Rooijen, E., ... van Eeden, F. J. (2012). A zebrafish model to study and therapeutically manipulate hypoxia signaling in tumorigenesis. *Cancer Research, 72*(16), 4017–4027. http://dx.doi.org/10.1158/0008-5472.CAN-11-3148.

Sato, Y., Yoshizato, T., Shiraishi, Y., Maekawa, S., Okuno, Y., Kamura, T., ... Ogawa, S. (2013). Integrated molecular analysis of clear-cell renal cell carcinoma. *Nature Genetics, 45*(8), 860–867. http://dx.doi.org/10.1038/ng.2699.

Schodel, J., Oikonomopoulos, S., Ragoussis, J., Pugh, C. W., Ratcliffe, P. J., & Mole, D. R. (2011). High-resolution genome-wide mapping of HIF-binding sites by ChIP-seq. *Blood, 117*(23), e207–217. http://dx.doi.org/10.1182/blood-2010-10-314427.

Schoenfeld, A., Davidowitz, E. J., & Burk, R. D. (1998). A second major native von Hippel-Lindau gene product, initiated from an internal translation start site, functions as a tumor suppressor. *Proceedings of National Academy of Sciences of United States of America, 95*(15), 8817–8822.

Schofield, C. J., & Ratcliffe, P. J. (2005). Signalling hypoxia by HIF hydroxylases. *Biochemical and Biophysical Research Communications, 338*(1), 617–626. http://dx.doi.org/10.1016/j.bbrc.2005.08.111.

Schwerte, T., Uberbacher, D., & Pelster, B. (2003). Non-invasive imaging of blood cell concentration and blood distribution in zebrafish *Danio rerio* incubated in hypoxic conditions in vivo. *Journal of Experimental Biology, 206*(Pt 8), 1299–1307.

Semenza, G. L. (2007). HIF-1 mediates the Warburg effect in clear cell renal carcinoma. *Journal of Bioenergetics and Biomembranes, 39*(3), 231–234. http://dx.doi.org/10.1007/s10863-007-9081-2.

Semenza, G. L. (2009). Regulation of oxygen homeostasis by hypoxia-inducible factor 1. *Physiology, 24*, 97–106. http://dx.doi.org/10.1152/physiol.00045.2008.

Semenza, G. L. (2010). Oxygen homeostasis. *Wiley Interdisciplinary Reviews. Systems Biology and Medicine, 2*(3), 336–361. http://dx.doi.org/10.1002/wsbm.69.

Semenza, G. L., & Wang, G. L. (1992). A nuclear factor induced by hypoxia via de novo protein synthesis binds to the human erythropoietin gene enhancer at a site required for transcriptional activation. *Molecular and Cellular Biology, 12*(12), 5447–5454.

Sermeus, A., & Michiels, C. (2011). Reciprocal influence of the p53 and the hypoxic pathways. *Cell Death & Disease, 2*, e164. http://dx.doi.org/10.1038/cddis.2011.48.

Shen, C., Beroukhim, R., Schumacher, S. E., Zhou, J., Chang, M., Signoretti, S., & Kaelin, W. G., Jr. (2011). Genetic and functional studies implicate HIF1alpha as a 14q kidney cancer suppressor gene. *Cancer Discovery, 1*(3), 222–235. http://dx.doi.org/10.1158/2159-8290.CD-11-0098.

Siekmann, A. F., & Lawson, N. D. (2007). Notch signalling limits angiogenic cell behaviour in developing zebrafish arteries. *Nature, 445*(7129), 781–784. http://dx.doi.org/10.1038/nature05577.

So, J. H., Kim, J. D., Yoo, K. W., Kim, H. T., Jung, S. H., Choi, J. H., ... Kim, C. H. (2014). FIH-1, a novel interactor of mindbomb, functions as an essential anti-angiogenic factor during zebrafish vascular development. *PLoS One, 9*(10), e109517. http://dx.doi.org/10.1371/journal.pone.0109517.

Spitsbergen, J. M., Tsai, H. W., Reddy, A., Miller, T., Arbogast, D., Hendricks, J. D., & Bailey, G. S. (2000). Neoplasia in zebrafish (*Danio rerio*) treated with 7,12-dimethylbenz[a]anthracene by two exposure routes at different developmental stages. *Toxicologic Pathology, 28*(5), 705–715.

Stenslokken, K. O., Ellefsen, S., Vasieva, O., Fang, Y., Farrell, A. P., Olohan, L., … Cossins, A. R. (2014). Life without oxygen: gene regulatory responses of the crucian carp (*Carassius carassius*) heart subjected to chronic anoxia. *PLoS One, 9*(11), e109978. http://dx.doi.org/10.1371/journal.pone.0109978.

Teh, C. H. L., Lam, K. K. Y., Loh, C. C., Loo, J. M., Yan, T., & Lim, T. M. (2006). Neuronal PAS domain protein 1 is a transcriptional repressor and requires arylhydrocarbon nuclear translocator for its nuclear localization. *Journal of Biological Chemistry, 281*(45), 34617–34629. http://dx.doi.org/10.1074/jbc.M604409200.

Thisse, B., Pflumio, S., Fürthauer, M., Loppin, B., Heyer, V., Degrave, A., … Thisse, C. (2001). Expression of the zebrafish genome during embryogenesis (NIH R01 RR15402). *ZFIN Direct Data Submission*. http://zfin.org.

Thisse, B., & Thisse, C. (2004). Fast release clones: a high throughput expression analysis. *ZFIN Direct Data Submission*. http://zfin.org.

Thoma, C. R., Toso, A., Gutbrodt, K. L., Reggi, S. P., Frew, I. J., Schraml, P., … Krek, W. (2009). VHL loss causes spindle misorientation and chromosome instability. *Nature Cell Biology, 11*(8), 994–1001. http://dx.doi.org/10.1038/ncb1912.

Ton, C., Stamatiou, D., & Liew, C. C. (2003). Gene expression profile of zebrafish exposed to hypoxia during development. *Physiological Genomics, 13*(2), 97–106. http://dx.doi.org/10.1152/physiolgenomics.00128.2002.

Tseng, L. C., Zhang, C., Cheng, C. M., Xu, H., Hsu, C. H., & Jiang, Y. J. (2014). New classes of mind bomb-interacting proteins identified from yeast two-hybrid screens. *PLoS One, 9*(4), e93394. http://dx.doi.org/10.1371/journal.pone.0093394.

Tsuchihara, K., Suzuki, Y., Wakaguri, H., Irie, T., Tanimoto, K., Hashimoto, S., … Sugano, S. (2009). Massive transcriptional start site analysis of human genes in hypoxia cells. *Nucleic Acids Research, 37*(7), 2249–2263. http://dx.doi.org/10.1093/nar/gkp066.

van Uden, P., Kenneth, N. S., & Rocha, S. (2008). Regulation of hypoxia-inducible factor-1alpha by NF-kappaB. *Biochemical Journal, 412*(3), 477–484. http://dx.doi.org/10.1042/BJ20080476.

van der Velden, Y. U., Wang, L., Zevenhoven, J., van Rooijen, E., van Lohuizen, M., Giles, R. H., … Haramis, A. P. (2011). The serine-threonine kinase LKB1 is essential for survival under energetic stress in zebrafish. *Proceedings of National Academy of Sciences of United States of America, 108*(11), 4358–4363. http://dx.doi.org/10.1073/pnas.1010210108.

Verges, S., Chacaroun, S., Godin-Ribuot, D., & Baillieul, S. (2015). Hypoxic conditioning as a new therapeutic modality. *Frontiers in Pediatrics, 3*, 58. http://dx.doi.org/10.3389/fped.2015.00058.

Wang, S. S., Gu, Y. F., Wolff, N., Stefanius, K., Christie, A., Dey, A., … Brugarolas, J. (2014). Bap1 is essential for kidney function and cooperates with Vhl in renal tumorigenesis. *Proceedings of National Academy of Sciences of United States of America, 111*(46), 16538–16543. http://dx.doi.org/10.1073/pnas.1414789111.

Wang, G. L., Jiang, B. H., Rue, E. A., & Semenza, G. L. (1995). Hypoxia-inducible factor 1 is a basic-helix-loop-helix-PAS heterodimer regulated by cellular O2 tension. *Proceedings of National Academy of Sciences of United States of America, 92*(12), 5510–5514.

Watson, O., Novodvorsky, P., Gray, C., Rothman, A. M., Lawrie, A., Crossman, D. C., … Chico, T. J. (2013). Blood flow suppresses vascular Notch signalling via dll4 and is required for angiogenesis in response to hypoxic signalling. *Cardiovascular Research, 100*(2), 252–261. http://dx.doi.org/10.1093/cvr/cvt170.

Wiesener, M. S., Jurgensen, J. S., Rosenberger, C., Scholze, C. K., Horstrup, J. H., Warnecke, C., ... Eckardt, K. U. (2003). Widespread hypoxia-inducible expression of HIF-2alpha in distinct cell populations of different organs. *FASEB Journal, 17*(2), 271–273. http://dx.doi.org/10.1096/fj.02-0445fje.

Woods, I. G., & Imam, F. B. (2015). Transcriptome analysis of severe hypoxic stress during development in zebrafish. *Genomics Data, 6*, 83–88. http://dx.doi.org/10.1016/j.gdata.2015.07.025.

Woodward, E. R., Buchberger, A., Clifford, S. C., Hurst, L. D., Affara, N. A., & Maher, E. R. (2000). Comparative sequence analysis of the VHL tumor suppressor gene. *Genomics, 65*(3), 253–265. http://dx.doi.org/10.1006/geno.2000.6144.

Xia, X., Lemieux, M. E., Li, W., Carroll, J. S., Brown, M., Liu, X. S., & Kung, A. L. (2009). Integrative analysis of HIF binding and transactivation reveals its role in maintaining histone methylation homeostasis. *Proceedings of National Academy of Sciences of United States of America, 106*(11), 4260–4265. http://dx.doi.org/10.1073/pnas.0810067106.

Yaqoob, N., & Schwerte, T. (2010). Cardiovascular and respiratory developmental plasticity under oxygen depleted environment and in genetically hypoxic zebrafish (*Danio rerio*). *Comparative Biochemistry and Physiology. Part A, Molecular & Integrative Physiology, 156*(4), 475–484. http://dx.doi.org/10.1016/j.cbpa.2010.03.033.

Zhang, P., Bai, Y., Lu, L., Li, Y., & Duan, C. (2016). An oxygen-insensitive Hif-3alpha isoform inhibits Wnt signaling by destabilizing the nuclear beta-catenin complex. *eLife, 5*. http://dx.doi.org/10.7554/eLife.08996.

Zhang, P., Lu, L., Yao, Q., Li, Y., Zhou, J., Liu, Y., & Duan, C. (2012). Molecular, functional, and gene expression analysis of zebrafish hypoxia-inducible factor-3alpha. *American Journal of Physiology. Regulatory, Integrative and Comparative Physiology, 303*(11), R1165–R1174. http://dx.doi.org/10.1152/ajpregu.00340.2012.

Zhang, P., Yao, Q., Lu, L., Li, Y., Chen, P. J., & Duan, C. (2014). Hypoxia-inducible factor 3 is an oxygen-dependent transcription activator and regulates a distinct transcriptional response to hypoxia. *Cell Reports, 6*(6), 1110–1121. http://dx.doi.org/10.1016/j.celrep.2014.02.011.

Zimmer, M., Ebert, B. L., Neil, C., Brenner, K., Papaioannou, I., Melas, A., ... Iliopoulos, O. (2008). Small-molecule inhibitors of HIF-2a translation link its 5'UTR iron-responsive element to oxygen sensing. *Molecular Cell, 32*(6), 838–848. http://dx.doi.org/10.1016/j.molcel.2008.12.004.

CHAPTER

19

Discovering novel oncogenic pathways and new therapies using zebrafish models of sarcoma

M.N. Hayes[*,§,¶], **D.M. Langenau**[*,§,¶,1]

[*]*Massachusetts General Hospital, Boston, MA, United States*

[§]*Massachusetts General Hospital, Charlestown, MA, United States*

[¶]*Harvard Stem Cell Institute, Boston, MA, United States*

[1]*Corresponding author: E-mail: dlangenau@partners.org*

CHAPTER OUTLINE

Methods in Cell Biology, Volume 138, ISSN 0091-679X, http://dx.doi.org/10.1016/bs.mcb.2016.11.011

Abstract

Sarcoma is a type of cancer affecting connective, supportive, or soft tissue of mesenchymal origin. Despite rare incidence in adults (<1%), over 15% of pediatric cancers are sarcoma. Sadly, both adults and children with relapsed or metastatic disease have devastatingly high rates of mortality. Current treatment options for sarcoma include surgery, radiation, and/or chemotherapy; however, significant limitations exist with respect to the efficacy of these strategies. Strong impetus has been placed on the development of novel therapies and preclinical models for uncovering mechanisms involved in the development, progression, and therapy resistance of sarcoma. Over the past 15 years, the zebrafish has emerged as a powerful genetic model of human cancer. High genetic conservation when combined with a unique susceptibility to develop sarcoma has made the zebrafish an effective tool for studying these diseases. Transgenic and gene-activation strategies have been employed to develop zebrafish models of rhabdomyosarcoma, malignant peripheral nerve sheath tumors, Ewing's sarcoma, chordoma, hemangiosarcoma, and liposarcoma. These models all display remarkable molecular and histopathological conservation with their human cancer counterparts and have offered excellent platforms for understanding disease progression in vivo. Short tumor latency and the amenability of zebrafish for ex vivo manipulation, live imaging studies, and tumor cell transplantation have allowed for efficient study of sarcoma initiation, growth, self-renewal, and maintenance. When coupled with facile chemical genetic approaches, zebrafish models of sarcoma have provided a strong translational tool to uncover novel drug pathways and new therapeutic strategies.

INTRODUCTION

Sarcomas are mesenchymal in origin and can occur anywhere in the soft tissues of the body including fat, muscle, connective tissue, or nerves. Approximately 30 out of 1 million individuals are affected in their lifetime, with a disproportionate number of cases occurring in children or adolescents, compared to adults (Borden et al., 2003). In the pediatric population, sarcomas represent 15–20% of all cancers with devastatingly high rates of mortality in patients experiencing recurrent disease and/or metastasis (Borden et al., 2003; Burningham, Hashibe, Spector, & Schiffman, 2012). In adults, over 30% of patients develop fatal metastases with a median survival of 15 months (Helman & Meltzer, 2003). To date, more than 30 different types of sarcoma have been identified and are grouped based on histopathology (Borden et al., 2003; Helman & Meltzer, 2003). Despite clear morphologic and histopathological differences observed in some sarcomas, there remains a large discordance among pathologists when classifying specific tumor subtypes or determine grade to predict outcome and assign treatment. Histopathological subtypes also exhibit significant heterogeneity in their clinical behavior and make traditional means of classification a problem in the field, especially since sarcomas are thought to represent a continuum of mesenchymal differentiation with specific oncogene and tumor suppressor pathways driving therapy resistance rather than cell of origin. To date, animal models of sarcoma represent one of the best ways to model these diseases and provide unique avenues to target oncogenic drivers to specific tissue types.

In addition to sarcomas being the result of random, somatic acquired mutations, sarcomas often develop due to underlying genetic syndromes. For example, Li Fraumeni patients have mutations in *TP53* and Costello syndrome results from *NF1* deficiency (Aoki et al., 2005; Fountain et al., 1989; Legius, Marchuk, Collins, & Glover, 1993; Merino & Malkin, 2014). These patients are predisposed to sarcomas including rhabdomyosarcoma (RMS), osteosarcoma, and malignant peripheral nerve sheath tumors (MPNSTs), respectively (Aoki et al., 2005; Fountain et al., 1989; Legius et al., 1993; Merino & Malkin, 2014). These same gene mutations are often found in cases of sporadic cancer, informing biologists of key developmental and molecular pathways that drive sarcomagenesis. Despite a well-studied role for these few genetic-predisposing oncogenic drivers in sarcoma, large-scale genomic analysis has largely failed to identify actionable driver mutations that contribute to disease growth and progression. Tumors often (in)activate a large number of putative oncogenes and/or tumor suppressors that are not highly or recurrently found in a wide array of sarcoma patient samples. Combined with a scarcity of patient-derived tumor samples, it has been challenging to identify novel genetic pathways and drug targets in sarcoma.

Animal models have been valuable for increasing our understanding of molecular mechanisms that drive human cancer and defining genetic abnormalities underlying sarcoma initiation, progression, and therapy responses. For example the use of *Drosophila* and mouse models have uncovered dominant oncogenic roles for PAX3/7-FKHR genomic translocations in alveolar rhabdomyosarcoma (ARMS) and refined potential cells of origin in this devastating pediatric malignancy (Galindo, Allport, & Olson, 2006; Keller et al., 2004). Zebrafish have also emerged as a powerful genetic model of sarcoma. Experimental accessibility combined with a high level of evolutionary conservation in disease pathways has made zebrafish one of the most attractive animal systems to study gene function and model human cancer. Soft tissues, in general, are conserved between fish and humans, making zebrafish an appropriate model for studying genetic pathways involved in sarcomagenesis. Models of RMS, MPNSTs, Ewing's sarcoma, chordoma, hemangiosarcoma, liposarcoma, and fibrosarcoma have all been described in the zebrafish (Table 1) and provide unique opportunities for identifying oncogenic driver mutations and novel drug targets. With limited numbers of human patients available for clinical trials, zebrafish sarcoma models are also ideal platforms for testing novel therapeutic strategies in a tractable animal system.

In this review, we will discuss many zebrafish models of human sarcoma. We will focus on the methodology involved in their establishment and how these models

Table 1 Zebrafish Models of Sarcoma

Genetic Model	Sarcoma Subtype	Method	References
Tg(rag2:KRAS^{G12D})	Embryonal rhabdomyosarcoma (ERMS)	Mosaic transgenesis	Langenau et al. (2007)
Tg(cdh15:KRAS^{G12D})	ERMS	Mosaic transgenesis	Storer et al. (2013)
Tg(mylz2:KRAS^{G12D})	Well-differentiated rhabdomyosarcoma	Mosaic transgenesis	Storer et al. (2013)
Tg(T2KSAG:GFP-HRAS^{G12V})	Rhabdomyosarcoma	Tol2-mediated gene trap	Santoriello et al. (2009)
Tg(actb1:loxP-EGFP-pA-loxP-KRAS^{G12D}) × *Tg(hsp70:Cre)*	ERMS	Stable transgenesis	Le et al. (2007)
	Fibrosarcoma		
tp53$^{M214K/M214K}$	MPNST	*N*-ethyl-*N*-nitrosurea target-selected mutagenesis	Berghmans et al. (2005)
rp$^{+/-}$ (*rps8a*, *rps15a*, *rpl7*, *rpl35*, *rpl36*, *rpl36a*, *rpl13*, *rpl23a*, *rps7*, *rps18*, *rps29*)	MPNST	Retroviral mutagenesis	Amsterdam et al. (2004)
Tg(-8.5nkx2.2a:GFP)^{ia2}	MPNST	Stable transgenesis	Astone et al. (2015)
nf1a$^{+/-}$;nf1b$^{-/-}$;tp53$^{M214K/M214K}$	MPNST	Zinc-finger nuclease targeted mutagenesis	Shin et al. (2012)
nf2$^{+/-}$	MPNST	Retroviral mutagenesis	Amsterdam et al. (2004)
Tg(actb1:EWSR1-FLI1)	Ewing's sarcoma	Tol2-mediated stable transgenesis	Leacock et al. (2012)
Tg(hsp70:EWSR1-FLI1)			
mu4465:Gal4 x *UAS:HRAS^{G12V}*	Chordoma	Gal4/UAS stable transgenesis	Burger et al. (2014)
twhh:Gal4 × *UAS:HRAS^{G12V}*			
ptena$^{+/-}$;ptenb$^{-/-}$	Hemangiosarcoma	*N*-ethyl-*N*-nitrosurea target-selected mutagenesis	Choorapoikayil et al. (2012)
ptena$^{-/-}$;ptenb$^{+/-}$			
Tg(rag2:myr-mAkt2)	Well-differentiated liposarcoma	Mosaic transgenesis	Gutierrez et al. (2011)
Tg(krt4:myr-AKT1)^{cy18}	Lipoma	Tol2-mediated stable transgenesis	Chu et al. (2012)

MPNST, malignant peripheral nerve sheath tumors.

have been used to better understand both genetic and biochemical aspects of human sarcoma with important therapeutic implications for patients.

1. RHABDOMYOSARCOMA

RMS displays phenotypic and biological hallmarks of undifferentiated skeletal muscle and is the most common soft-tissue sarcoma in the pediatric population (Hettmer et al., 2014). RMS affects 4.6/million US children each year and those with unresectable, metastatic, or relapsed RMS have a very poor prognosis, with only 30% of patients surviving their disease (Ardnt & Crist, 1999; Hettmer et al., 2014). Despite being thought of as predominantly a cancer of adolescence, RMS is also found in a small fraction of adult cancers, with ~250 patients being diagnosed annually in the United States (Hettmer et al., 2011). Surgery, radiation, and chemotherapy generally lead to a good outcome for primary tumors, especially in the pediatric setting. Yet, cured patients often have long-term side effects including cognition defects, deformity due to surgery, and elevated rates of secondary treatment–associated cancer (Hettmer et al., 2014).

RMS is divided into two main subtypes. ARMS is characterized by (2; 13) or (1; 13) chromosomal translocations that result in a fusion between the 5′ DNA-binding domain of PAX3 or PAX7 and the 3′ transactivation domain of the fork-head transcription factor (FKHR) (Barr et al., 1993; Davis, D'Cruz, Lovell, Biegel, & Barr, 1994; Galili et al., 1993). Embryonal RMS (ERMS) lacks pathognomonic chromosomal translocations and accounts for 60% of childhood cases. The majority of human ERMS have mutational activation of the RAS pathway (Chen et al., 2006; Langenau et al., 2007; Shern et al., 2014), implicating RAS signaling, and interacting pathway components in tumor onset. ERMS also display a high mutational burden compared to ARMS, suggesting that additional genetic pathways affect the multistep progression to full malignancy (Shern et al., 2014). Significant genetic alterations have been identified in ARMS and ERMS including *TP53* and dystrophin loss and activating mutations in both *NOTCH1* and *MYOD* (Shern et al., 2014; Szuhai, De Jong, Leung, Fletcher, & Hogendoorn, 2014; Wang et al., 2014).

To date, multiple rodent models of RMS have been described in the literature. Keller et al. used a conditional knock-in strategy to express Pax3-Fkhr in skeletal muscle under the control of *Myf6*:Cre. ARMS develop in these mice, but only in an *Ink4a/Arf* or *Tp53* mutant background, indicating that functional impairment of the Tp53 pathway is necessary for the initiation of ARMS in the mouse model (Keller et al., 2004). Tsumura et al. generated conditional knock-in mice where oncogenic K-Ras was expressed in the skeletal muscle of mice heterozygous or homozygous for *Tp53* deletion. K-Ras activation at 10 weeks resulted in 100% RMS tumor incidence in $Tp53^{-/-}$ and $Tp53^{+/-}$ backgrounds (Tsumura, Yoshida, Saito, Imanaka-Yoshida, & Suzuki, 2006). Xenograft models have also uncovered roles for Her2, c-Met, Hedgehog, Myc, and Rb-related signaling pathways in RMS tumorigenesis (Linardic & Counter, 2008); however, additional functional studies are necessary to fully understand the role of these pathways and to uncover novel molecular drivers for use as therapeutic targets in patients with RMS. Zebrafish have become an effective model for studying RMS tumor onset and relapse. They have allowed for the identification of functionally distinct ERMS cell subpopulations, imaging tumor initiation at the single-cell level, assessing relapse through cellular transplantation, and uncovering conserved genetic pathways governing tumorigenesis through genomic resources and bioinformatic approaches.

1.1 TISSUE SPECIFIC TRANSGENE EXPRESSION TO INDUCE RHABDOMYOSARCOMA IN ZEBRAFISH WITHIN A SINGLE GENERATION

Our lab utilizes a robust zebrafish ERMS model that can be generated within a single generation by expressing an activated form of human KRAS (KRAS^{G12D}) under the control of the *rag2* promoter (Langenau et al., 2007). The *rag2* promoter drives transgene expression in T- and B-lymphoid progenitor cells and in mesodermal compartments that include the mononuclear compartment of developing skeletal muscle, consisting of muscle stem cells (referred to as satellite cells) and early myoblasts (Jessen, Jessen, Vogel, & Lin, 2001; Langenau et al., 2007). The transgenic *rag2* promoter drives expression in the developing zebrafish muscle as a result of uncovering a myoD regulatory element found in the truncated 6.5 kB promoter (Langenau et al., 2007). Mosaic *rag2:KRAS^{G12D}*-injected transgenic zebrafish develop ERMS and can be generated through injection of 60–120 ng/μL of linearized DNA suspended in 0.5× *Tris*–EDTA buffer + 100 mM KCl injected into one-cell stage wild-type embryos. Tumor onset can be observed as early as 10 days of life. By 80 days post fertilization, typically 40% of injected zebrafish develop externally visible tumors. Linearized *rag2:GFP* or *rag2:dsRed* DNA can also be coinjected with *rag2:KRAS^{G12D}* to fluorescently label tumor cells. Coinjection of multiple constructs into one-cell stage embryos leads to cosegregation and coexpression of multiple transgenes within developing ERMS (Langenau et al., 2008). This approach has been effectively used in zebrafish ERMS and other sarcoma models to track tumor cells in vivo.

To identify possible cells origin in rhabdomyosarcoma, Storer et al. generated two additional zebrafish RMS models using the mosaic transgene injection strategy described previously. Specifically, human KRAS^{G12D}was expressed from the *cdh15* and *mylz2* promoters and injected at the one-cell stage. Cadherin 15 (Cdh15) is expressed in muscle satellite cells while myosin light chain 2 (mylz2) is expressed in differentiated muscle fibers (Irintchev, Starzinski-Powitz, & Wernig, 1994; Pownall, Gustafsson, & Emerson, 2002; Storer et al., 2013). Tumors derived from injection of *cdh15:KRAS^{G12D}* phenocopied *rag2:KRAS^{G12D}*-derived ERMS and consisted of a large portion of undifferentiated myoblast-like cells, suggesting that an early muscle progenitor can be the cell of origin in ERMS (Storer et al., 2013). By contrast, injection of the *mylz2:KRAS^{G12D}* transgene into one-cell stage fish resulted in the development of tumors that resembled mature skeletal muscle and were histologically classified as well-differentiated RMS (Storer et al., 2013).

In total, zebrafish RMS models highlight that molecular differences within the cell of origin likely impact disease progression and outcome.

Using this transgenic coinjection strategy, tumor cells can also be fluorescently labeled based on differentiation status. The first example of this was when Langenau et al. (2007) coinjected linearized *rag2:KRAS*^{G12D} and *rag2:dsRed* DNA into *α-actin:GFP* (green fluorescent protein) transgenic zebrafish. *rag2:* dsRED was expressed in satellite cells and early myoblasts, while *α-actin*:GFP was expressed in more mature muscle cells (Langenau et al., 2007). To verify that heterogeneous ERMS cell populations could be labeled with fluorescent proteins, tumor cells were dissociated and subjected to fluorescence-activated cell sorting (FACS) based on green and dsRed fluorescence. FACS confirmed that mononuclear tumor cells were composed of four populations: double negative, GFP+/dsRED−, dsRED+/GFP−, and double positive (Langenau et al., 2007). Following sorting of these individual populations, gene expression analysis was performed through RT-PCR (reverse transcription polymerase chain reaction) and microarray-based approaches. The dsRED+/GFP- ERMS cells expressed high levels of satellite cell markers (including *cmet*, *m-cadherin*, and *myf5*), while the GFP + cells expressed more differentiated muscle cell markers [including *myod*, *myogenin*, and *myosin light chain-2 (mylz2)*] (Langenau et al., 2007).

This labeling strategy was further exploited to understand biological characteristics unique to each ERMS cell subpopulation. Stable *myf5:GFP* transgenic zebrafish were coinjected with *rag2:KRAS*^{G12D}, *myogenin:H2B-RFP*, and *myosin: lyn-cyan* to differentially label early progenitors, mid- and late-differentiated ERMS cells, respectively (Fig. 1A–C) (Ignatius et al., 2012). Time-lapse imaging of ERMS tumor cells revealed differential behaviors and dynamic reorganization of tumor cell populations during tumor growth (Fig. 1A–C). First, only *myf5*: GFP + cells divided over the imaging time course, suggesting that *myf5*:GFP+ cells maintain tumor growth in ERMS (Ignatius et al., 2012). Surprisingly, *myogenin+* cells were highly migratory and were the first to seed new areas of tumor growth (Ignatius et al., 2012). These *myogenin+* cells also entered the vasculature of *fli1*: GFP transgenic animals (Ignatius et al., 2012), suggesting that mid-differentiated cells contribute to metastasis during ERMS tumor progression. Human ERMS were also shown to be regionally partitioned within tumors based on ERMS cell differentiation status and provided a plausible cellular understanding of why high MYOGENIN expression predicts poor outcome in human patients (Heerema-McKenney et al., 2008). Therefore, the zebrafish *rag2:KRAS*^{G12D} ERMS model uncovered new roles for molecularly defined tumor cell types in regulating ERMS growth, invasion, and metastasis with clear relevance to human disease.

1.2 CELL TRANSPLANTATION STRATEGIES TO UNDERSTAND TUMOR PROPAGATION AND RELAPSE

The ability to remake tumors in transplant animals is a hallmark of cancer. To establish that sarcoma cells are fully malignant, tumor cells can be transplanted into either irradiated or syngeneic recipients. ERMS cells are isolated by surgical

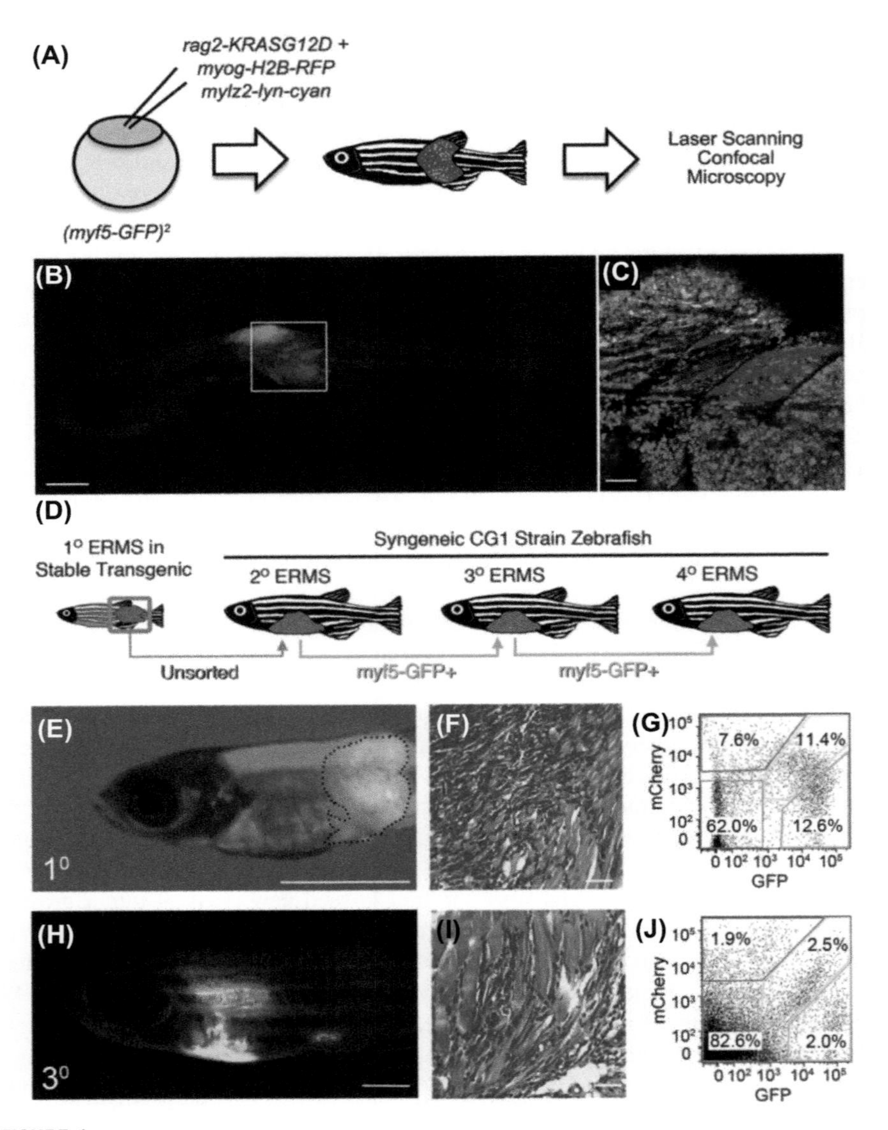

FIGURE 1

Fluorescent-transgenic models of *rag2:KRAS*^{G12D}-induced embryonal rhabdomyosarcoma (ERMS) identified functionally distinct tumor cell populations using live animal imaging and cell transplantation. (A) Schematic of experimental design where linearized *rag2:KRAS*^{G12D} was coinjected with *myogenin:H2B-RFP* and *mylz2:lyn-cyan* into *myf5:GFP* stable transgenic embryos at the one-cell stage. This approach labeled ERMS cells based on their myogenic transcription factor expression and accurately labeled specific stages of muscle development. (B) A triple-fluorescent labeled ERMS at 16 days of life. (C) A high-resolution resection and minced in 5 mL of 0.9× PBS (phosphate-buffered saline)/5% FBS (fetal bovine serum). Samples are filtered through a 40 μm cell strainer, centrifuged at 1000 g for 10 min and resuspended in 500 μL of 0.9× PBS/5% FBS. To define self-renewing and transplantable tumor cell populations, ERMS cells can be isolated by FACS and injected into recipient animals at limiting dilution (10^4−10 cells). Propidium iodine (PI) is used at a concentration of 1 μg/mL to exclude dead cells and debris. Forward and side scatter, combined with GFP and/or dsRed fluorescence allows for sorting of specific ERMS populations to high purity. Sorted cells can be aliquoted into 96-well plates at limiting dilution, spun down and resuspended in 0.9× PBS/5% FBS. Using a Hamilton syringe (#701, Fisher Scientific), 5 μL of cell suspension is then injected into the peritoneal cavity of host animals. The syringe is cleaned with 70% ethanol followed by PBS between injections of different ERMS cell subpopulations of cells. Following transplantation, animals are analyzed for engraftment based on fluorescent reporters typically at 15, 30, and 45 days postinjection. Fluorescent tumor cells can be detected as early as 7 days posttransplantation using a fluorescence-dissecting microscope and by 14 days posttransplantation, large tumor masses can often be visualized at the site of injection (Fig. 1H).

This assay has been used to understand the significance of tumor cell heterogeneity found in zebrafish RAS-induced ERMS. Ignatius et al. injected stable *myf5:GFP;mylz2:mCherry* transgenic zebrafish with *rag2:KRAS*^{G12D} and tumor cell subpopulations were isolated by FACS (Fig. 1D−G). Double negative, *myf5*:GFP+/*mylz2*:mCherry-, *myf5*:GFP+/*mylz2*:mCherry+, and *myf5*:GFP-/*mylz2*:mCherry+ cells were transplanted into syngeneic host animals at limiting

confocal image of the boxed region shown in panel B. ERMS cells were regionally partitioned within the primary tumor based on their muscle differentiation status. (D) Schematic of experimental design used to identify the serially transplantable *myf5*:GFP + ERMS-propagating cells. Linearized *rag2:KRAS*^{G12D} was injected into syngeneic *myf5:GFP; mylz2:mcherry* transgenic embryos at the one-cell stage. Primary (1 degree) ERMS were transplanted into syngeneic animals to expand individual tumor cell populations. ERMS cell fractions were then isolated by fluorescence-activated cell sorting (FACS) and injected into syngeneic recipient fish. Only the *myf5*:GFP+/*mylz2*:mcherry-negative cells could engraft disease into syngeneic zebrafish. (E−G) A primary ERMS arising in syngeneic *myf5:GFP; mylz2:mcherry* transgenic zebrafish at 35 days postfertilization. Merged fluorescent and brightfield images of fish (E), histology of tumor (F), and FACS of fluorescent-labeled ERMS cells (G). *Broken black line* in E denotes tumor area. (H−J) FACS sorted *myf5*:GFP+ /*mylz2*:mcherry-negative ERMS cells engrafted in tertiary syngeneic zebrafish. Merged fluorescent and brightfield images of engrafted fish (H), histology of tumor (I), and FACS of isolated ERMS cells (J). Scale bars, 500 μm (B), 50 μm (C), 2 mm (E,H) and 100 μm (F,I).
Adapted and reprinted with permission from Ignatius, M. S., Chen, E., Elpek, N. M., Fuller, A. Z., Tenente, I. M., Clagg, R., ... Langenau D.M. (2012). In vivo imaging of tumor-propagating cells, regional tumor heterogeneity, and dynamic cell movements in embryonal rhabdomyosarcoma. Cancer Cell, 21*(5), 680−693.*

dilution (1×10^3–10 cells) and assessed over time for engraftment and growth (Fig. 1H–J). Remarkably, only the *myf5*:GFP/*mylz2*:mCherry-cells retained the ability to propagate new tumor, whereas cells that express *mylz2*:mCherry did not (Fig. 1H–J) (Ignatius et al., 2012). Histological analysis revealed that transplanted tumors retained features of the primary malignancy (Fig. 1F and I) (Ignatius et al., 2012), suggesting that ERMS propagating potential was confined to the *myf5*:GFP cell population. Fluorescent-transgenic labeling of tumor cells when combined with cell transplantation provides a unique advantage of zebrafish sarcoma models to identify molecularly defined and functionally distinct tumor cell subpopulations and to directly visualize tumor engraftment in recipient animals. Moreover, larger numbers of transplant recipient animals can be used when compared with traditional rodent models, reflecting lowered cost and reduced space required to house zebrafish.

1.3 COINJECTION STRATEGIES TO IDENTIFY GENETIC MODIFIERS OF RAS-INDUCED ZEBRAFISH EMBRYONAL RHABDOMYOSARCOMA

To determine the genetic contribution of TP53 signaling to ERMS, Langenau et al. incrossed $tp53^{+/-}$ heterozygous mutant zebrafish (Berghmans et al., 2005, described later) and the resulting embryos were injected at the one-cell stage with linearized *rag2:GFP/rag2:KRAS*^{G12D} transgenes. Animals were monitored starting at 10 days of life for tumor onset and genotyped as adults using allele-specific genomic PCR (Berghmans et al., 2005; Langenau et al., 2007). Interestingly, both heterozygous and homozygous *tp53* mutant zebrafish displayed a marked increase in tumor incidence compared to wild-type injected siblings (Langenau et al., 2007). *Tp53* mutant zebrafish also developed more tumors compared to heterozygous and wild-type siblings (Langenau et al., 2007), confirming a significant role for TP53 in ERMS pathogenesis and collaboration with Ras pathway signaling during tumor initiation in vivo.

Using the transgenic coinjection strategy described earlier, researchers have also begun to interrogate genetic modifiers of ERMS initiation and growth. Langenau et al. were the first to demonstrate the efficacy of this approach through coinjection of *rag2:noxa* with *rag2:KRAS*^{G12D}. Noxa is a proapoptotic BH3-only protein and a transcriptional target of Tp53. Noxa expression resulted in significant reduction in both the penetrance and the latency of ERMS (Langenau et al., 2008). To better understand how epigenetic changes affect ERMS, Albacker et al. performed an overexpression screen where plasmids expressing chromatin-modifying genes were coinjected with *rag2:KRAS*^{G12D} and *rag2:GFP* transgenes. Expression of SU39H1, a histone 3-lysine 9-methyl transferase, significantly suppressed ERMS onset and demonstrated the importance of epigenetic modifiers in ERMS (Albacker et al., 2013). Thus, coinjection of linearized transgenes has established an easy and rapid assay to identify enhancers or suppressors of ERMS initiation and growth.

1.4 STABLE TRANSGENIC ZEBRAFISH MODELS OF RHABDOMYOSARCOMA

Costello syndrome is an inherited genetic disorder caused by germ-line mutations in the H-Ras gene (Aoki et al., 2005). Patients experience multiple developmental defects and a tendency to develop various types of neoplasia, including RMS (Aoki et al., 2005). Santoriello et al. attempted to model Costello syndrome in zebrafish using the Tol2 gene trap system (Kawakami, Shima, & Kawakami, 2000). In this strategy, zebrafish were coinjected at the one-cell stage with 100–150 pg of in vitro transcribed Tol2 transposase mRNA and 30–60 pg of plasmid DNA harboring Tol2 recombination sites. Injected fish are raised to adults and mated to identify embryos with transgene stably integrated into the genome (Kawakami et al., 2000). Typically, this identification is made by PCR amplification or by the use of a fluorescent reporter contained within the injected transposon. Through these means, Santoriello et al. generated stable transgenic zebrafish lines carrying constitutively active forms of HRAS ($HRAS^{V12}$) and tagged the protein with GFP. The authors described several phenotypic features related to Costello syndrome, including craniofacial defects, short stature, and cardiovascular defects (Santoriello et al., 2009). A small number of fish developed cancer as juveniles, including a single case of RMS (Santoriello et al., 2009). Although this model is a powerful tool for studying abnormalities related to Costello syndrome, the low penetrance of RMS phenotypes has made this model less attractive for studying RMS-specific pathways associated with tumor initiation.

Combined with a heat shock-inducible Cre-Lox approach, Lee et al. conditionally induced expression of human $KRAS^{G12D}$ in zebrafish. Stable transgenic lines were established that contained an *actb1:loxP-EGFP-STOP-loxP-KRAS^{G12D}* cassette, whereby a floxed EGFP (enhanced green fluorescent protein) was ubiquitously expressed. A strong stop sequence following the polyadenylation site of EGFP prevented $KRAS^{G12D}$ expression until Cre-mediated excision. These fish were bred to stable heat shock-inducible Cre (*hsp70:Cre*) transgenic lines and heat-shock was applied by shifting embryos or larvae to 37°C for approximately 1 h (Le et al., 2007). Heat-shocked zebrafish developed RMS that was characterized by small round blue cells interspersed with atypical terminally differentiated striated muscle fibers (Le et al., 2007). Somewhat surprisingly, nonheat-shocked zebrafish also developed tumors, albeit at a lower frequency, and resulted from aberrant activation of the *hsp70* promoter during development (Le et al., 2007). The observed leakiness of Cre and subsequent $KRAS^{G12D}$ expression has made precise activation of the transgene difficult in this model, making these zebrafish lines less amenable to detailed investigation of RMS initiation. Interestingly, heat-shock of these double transgenic zebrafish lines led to malignancies other than RMS. Of particular note, *hsp70:Cre;actb1:loxP-EGFP-STOP-loxP-KRAS^{G12D}* zebrafish developed one case of MPNST and fibrosarcomas (Le et al., 2007), highlighting the role for activated RAS signaling in the development of multiple sarcomas. MPNST models will be discussed further later on in this review.

1.5 COMPARATIVE GENOMICS TO IDENTIFY MOLECULAR DRIVERS OF HUMAN EMBRYONAL RHABDOMYOSARCOMA

Genomic amplifications or deletions are key prognostic indicators and are oncogenic drivers in a wide range of cancers. Like many other cancers, RMS is characterized by genomic gains and losses (Bridge et al., 2002; Paulson et al., 2011); however, one major challenge has been to identify driver genes among passenger events. Compared to rodent models, zebrafish have smaller blocks of synteny with humans, allowing for the rapid identification of small genomic copy number changes conserved from fish to human cancer. This has made array comparative genomic hybridization (aCGH) a popular tool for identifying conserved genomic changes and has revealed several common drivers shared between zebrafish and human sarcomas. Additionally, zebrafish tumors tend to display very small genomic gains and/or loses making the list of candidate genes quite manageable compared to human and mouse tumors that display genomic alterations that often span several thousand genetic loci.

Using the *rag2:KRAS*^{G12D}-induced ERMS model, tumors were initially assessed for gains and loses using a bacterial artificial chromosome (BAC)-based aCGH platform (Freeman et al., 2009). This first generation platform comprised large genomic fragments that contained zebrafish orthologs of human oncogenes and tumor suppressors and provided the first evidence of amplification and deletions being found in zebrafish cancer (Freeman et al., 2009). However, defining boundaries for specific regional gains and losses was difficult due to the paucity of BAC clones used in these initial arrays and no specific or recurrent drivers were identified when comparing zebrafish ERMS, melanoma, or T-cell leukemia to human cancer (Freeman et al., 2009). To assess copy number alterations (CNAs) on a larger scale, investigators next turned to 400k CGH microarray technology. Specifically, Cy5-labeled genomic DNA isolated from 20 *rag2:KRAS*^{G12D} induced ERMS was hybridized against Cy3-labeled DNA isolated from matched control skeletal muscle on a custom 400k CGH microarray Agilent platform (Chen et al., 2013). A predominance of copy number gains was identified in zebrafish ERMS and uncovered recurrent CNAs across multiple tumors (Chen et al., 2013). These recurrent CNAs mapped to 21 homologous regions in the human genome, with 18 of these regions showing gains in human ERMS (Chen et al., 2013). Four of the genes, *CCND2*, *HOXC6*, *PLXNA1*, and *VEGF* were subsequently shown to play a role in pathogenesis of human ERMS (Chen et al., 2013). VEGF inhibitors were tested in vivo and found to significantly inhibit ERMS tumor growth (Chen et al., 2013). These experiments demonstrated the utility of comparing zebrafish and human tumors to identify conserved CNAs that comprise essential oncogenes required for continued ERMS growth.

1.6 DRUG TESTING

Multiple independent chemical screens have been performed to identify key pathways regulating sarcomagenesis, including ERMS. Le et al. performed a small molecule screen on embryos and assessed effects on Ras target gene expression.

Specifically, stable transgenic *hsp70:HRAS*^{G12V} zebrafish were subjected to heat-shock followed by treatment with a library of small molecules by addition of the compounds to the zebrafish water (Le et al., 2013). Ras activation was assessed through expression of *dusp6* by in situ hybridization (Le et al., 2013). Lead compounds from the screen included TPCK and PD98059 (inhibitors of S6K1 and MEK, respectively) (Le et al., 2013). These drugs were tested alone and in combination in both zebrafish primary ERMS and human RMS cells, uncovering a potent synergy of this drug combination in inhibiting tumor proliferation and growth (Le et al., 2013).

Chen et al. performed similar in vivo analyses following a chemical screen to identify drugs that induce differentiation of human ERMS cells in vitro. In this study, approximately 40,000 compounds were tested on cultured human ERMS cell lines and differentiation status was assessed by myosin heavy chain (*MF20*) expression (Chen et al., 2014). Subsequently, 95 lead compounds were assessed in vivo in a large scale transplantation study, where fluorescently labeled *rag2:*KRAS^{G12D} induced zebrafish ERMS cells were transplanted into 5–6 week old syngeneic host animals (Chen et al., 2014). Compounds were added to the fish water in 6-well plates and incubated for 7 days (Chen et al., 2014). Quantitative fluorescent imaging was performed to quantify total tumor volume and several compounds were identified that affected tumor growth (Chen et al., 2014). One class of lead compounds was GSK3 inhibitors, which inhibited tumor growth through activation of the Wnt/β-catenin signaling pathway (Chen et al., 2014). Further studies indicated conservation of this mechanism in human ERMS cells and revealed a very effective pipeline for drug and/or pathway identification relevant to human disease.

2. MALIGNANT PERIPHERAL NERVE SHEATH TUMORS

MPNSTs are malignant tumors derived from Schwann cells or Schwann cell precursors surrounding peripheral nerves (Dillon, 1997; Gupta, Mammis, & Maniker, 2008). These tumors occur with an estimated incidence of 1:100,000 in the general population and represent 5–10% of all soft-tissue sarcomas (Gupta et al., 2008; Kolberg et al., 2013; Thway & Fisher, 2014). Up to 50% of MPNSTs occur in patients with the neurocutaneous syndrome neurofibromatosis type 1 (NF1) (Cichowski & Jacks, 2001), demonstrating the tendency for this tumor to arise from preexisting neurofibroma. The remaining MPNSTs are found independent of neurofibromatosis and are considered to be sporadic (Thway & Fisher, 2014). Histologically, MPNSTs are pleomorphic tumors displaying epithelial, glandular, and/or cartilaginous components and often display aberrant differentiation. MPNSTs express S100β, GFAP, collagen IV, CD57, PGP 9.5, myelin basic protein, and keratin 8 and 18, suggesting a neural or Schwann cell of origin, while comparative analysis has identified an MPNST-specific gene expression signature, which includes *SOX10* in addition to other genetic factors (Kang, Pekmezci, Folpe, Ersen, & Horvai, 2014; Miller et al., 2006).

Without complete surgical resection, patients with MPNST have extremely poor outcomes. With surgery, 10-year survival is approximately 80% (Casanova et al., 1999). Patients with *NF1* mutations and/or metastases at the time of diagnosis experience very poor disease-associated survival with radiation and chemotherapy having limited efficacy, indicating a strong necessity to identify new molecular targets.

2.1 ZEBRAFISH WITH *Tp53* LOSS OF FUNCTION DEVELOP MALIGNANT PERIPHERAL NERVE SHEATH TUMORS

The first model of zebrafish MPNST was described in homozygous *tp53* mutant animals that had defects in apoptosis and upregulation of p21 following induced DNA damage (Berghmans et al., 2005). Mutations in *TP53* or deregulation of the TP53 signaling pathway are seen in many human MPNSTs and mice homozygous for *Tp53* mutations develop MPNSTs, among other tumor types (Donehower et al., 1992; Jacks et al., 1994). Knowing that more than 90% of human *TP53* inactivating mutations are found in the DNA-binding domain, Berghmans et al. performed a target-selected mutagenesis screen to identify *N*-ethyl-*N*-nitrosurea (ENU)-induced mutations in the DNA-binding domain (exons 4−8) of zebrafish *tp53*. Specifically, 99 adult male zebrafish were treated with 3 mM ENU and crossed to wild-type females to generate a nonmosaic F_1 generation (Wienholds, Schulte-Merker, Walderich, & Plasterk, 2002). Sperm was cryopreserved from 2679 F_1 male fish, while genomic DNA was isolated and arrayed into libraries for future screening. Mutations in target genes were identified using nested PCR amplification of genomic DNA and sequencing (Wienholds et al., 2002). Sequencing of *tp53* exons 4−8 was performed on this library and three missense mutations were identified, two of which were orthologous to *TP53* mutations found in human cancer and therefore prioritized for further analysis (Berghmans et al., 2005). The mutations were recovered through in vitro fertilization of wild-type eggs using cryopreserved sperm, and heterozygous F_2 animals were outcrossed to wild-type fish. F_3 heterozygous fish were incrossed, and their progeny were used for all published experiments (Berghmans et al., 2005).

One mutation, $tp53^{N168K}$, affected the tp53 DNA binding domain; however, this allele was found to behave in a temperature sensitive manner and only inhibited wild-type Tp53 activity at 37°C (Berghmans et al., 2005). Given that zebrafish are normally maintained at 28°C, this allele is generally not preferred for experimental analysis. In contrast, $tp53^{M214K}$ mutant zebrafish displayed no developmental abnormalities and were viable and fertile as homozygous mutant adults (Berghmans et al., 2005). Homozygous mutant embryos display Tp53-associated defects at all temperatures, while in vitro transactivation assays indicated that $tp53^{M214K}$ could inhibit wild-type protein function and behaved as a dominant-negative allele (Berghmans et al., 2005).

Homozygous $tp53^{M214K}$ mutant zebrafish are prone to forming tumors. MPNSTs are the dominant tumor type in this line and arise with an average latency of

8.5 months and a penetrance of approximately 25% (Berghmans et al., 2005). Zebrafish MPNSTs developed in the eye and in the abdominal cavity (Fig. 2A–F). Histological analysis showed that tumors were comprised mainly of spindle cells with tapered nuclei that demonstrated uniform chromatin staining and moderately abundant palely eosinophilic cytoplasm (Fig. 2G) (Berghmans et al., 2005). For zebrafish MPNSTs and other types of zebrafish sarcoma, hematoxylin and eosin (H&E) staining is typically completed on section following fixation in 4% paraformaldehyde (PFA) and embedding in paraffin. These results are consistent with human MPNST and revealed a novel animal model for study of disease progression in vivo. *tp53*^{M214K} MPNSTs can also be serially propagated through cell transplant assays into irradiated host animals, as described above. Fixation and sectioning at 23 days posttransplantation revealed that tumors developed at the site of injection and were highly aggressive, spreading throughout the entire abdominal cavity (Berghmans et al., 2005).

2.2 RIBOSOMAL PROTEIN HAPLOINSUFFICIENCY DRIVES MPNST INITIATION

In the late 1990s–early 2000s, a large-scale retroviral mutagenesis screen was performed to identify zebrafish genes essential for early development. Specifically, pseudotyped mouse retrovirus was injected into blastula stage embryos at the 1000–2000 cell stage and integrated at high efficiency without significant rearrangement of host DNA (Amsterdam et al., 1999). Mutations in over 300 essential zebrafish genes were identified and these lines were maintained in heterozygous carriers over multiple generations (Amsterdam et al., 1999, 2004). In the course of maintaining these lines, it was noted that many lines displayed early mortality and some animals developed gross "lumps", indicative of cancer. Histological analysis revealed diseased fish contained malignant spindle cell tumors with tumor cells aligned to form whirling patterns, highly reminiscent of human MPNSTs (Amsterdam et al., 2004). These tumors were also highly invasive and had a high mitotic index.

Interestingly, these zebrafish lines with predisposition to developing MPNSTs carried mutations in ribosomal proteins (*rp*) (Amsterdam et al., 2004). Tumors developed as early as 8 months in *rp* heterozygous zebrafish for *rps8a*, *rps15a*, *rpl7*, *rpl35*, *rpl36*, *rpl36a*, *rpl13*, *rpl23a*, *rps7*, *rps18*, and *rps29*, with penetrance and severity depending on genotype (Amsterdam et al., 2004). In *rp* mutant zebrafish, other tumor types were often observed and in some cases predominated, however across the 11 tumor susceptible *rp* mutant lines, the majority of tumors were MPNSTs (81% of total tumors), suggesting that *rp* mutation predisposes zebrafish to developing MPNST. Using RT-PCR analysis, it was found that these mutagenic insertions reduced overall *rp* transcript expression, indicating a tumor suppressive role rather than an oncogenic role for RP (Amsterdam et al., 2004). Since MPNSTs developed in heterozygous *rp* lines, loss-of-heterozygosity was also assessed in tumor cells. The corresponding wild-type allele was always detected

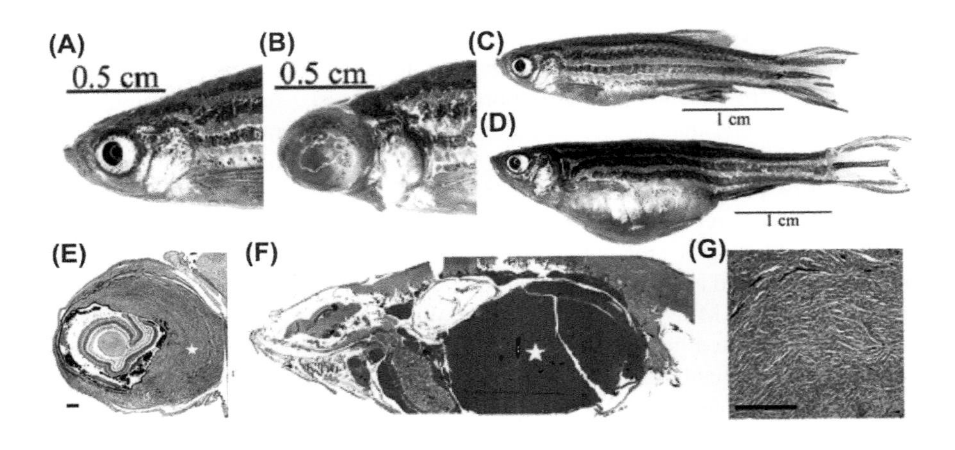

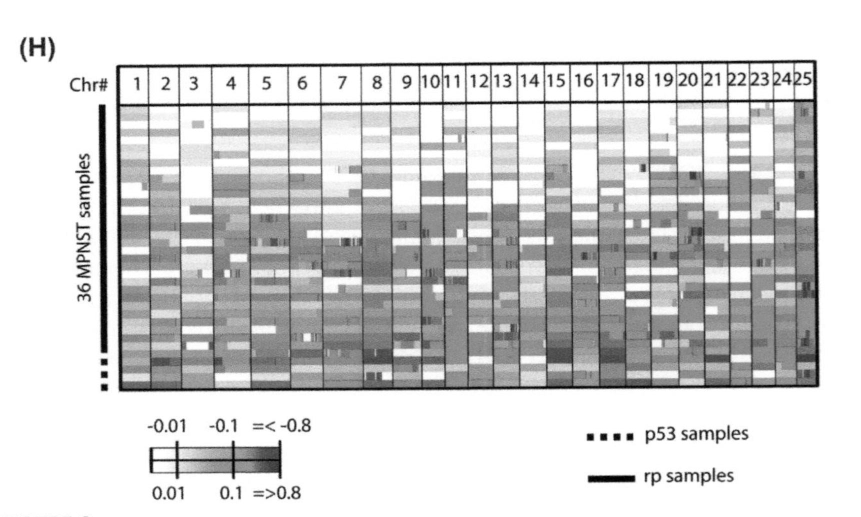

FIGURE 2

Zebrafish models of malignant peripheral nerve sheath tumors (MPNSTs). (A–G) *tp53*^{M214K} homozygous mutant zebrafish develop MPNST. Wild-type controls (A and C). Homozygous *Tp53*^{M214K} fish developed MPNSTs in the eye (B and E) and abdominal cavity (D and F). Histopathology confirmed tumor formation, indicated by the *white stars.* (G) Hematoxylin and eosin stained sections showed that tumors were composed predominantly of spindle cells and to a varying degree of epitheloid cells, consistent with MPNST. Scale bar in G equals 200 μm. (H) Array comparative genomic hybridization (aCGH) analysis revealed conserved copy-number changes in zebrafish *rp*$^{+/-}$ and tp53^{M214K} homozygous mutant MPNSTs. Heatmap of aCGH results with indicated relative chromosomal gains (red) and losses (blue) in tumor/normal pairs. Thirty-one *rp*$^{+/-}$ tumor samples are displayed on the top and five *tp53*^{M214K} samples are shown on the bottom. The heatmap was colored according to the segment mean of the data as processed in zebrafish MPNSTs (Amsterdam et al., 2004), indicating that *rp* haploinsufficiency drove tumor growth in these models.

Association between *rp* mutation and MPNST was surprising, given that *RP* mutations are not commonly associated with human cancer. However, similarities between heterozygous *rp* and *tp53*^{M214K} mutant phenotypes lead MacInnes et al. to interrogate the tp53 signaling pathway in *rp*-deficient zebrafish models. Using monoclonal antibodies raised against zebrafish Tp53, it was discovered that unlike *tp53*^{M214K} mutants where TP53 protein is detected, little or no Tp53 protein was present in *rp* heterozygous MPNSTs (MacInnes, Amsterdam, Whittaker, Hopkins, & Lees, 2008). Tp53 transcript levels were normal and protein loss was not due to global translational downregulation, suggesting an important role for Rp genes in modulating Tp53 protein production and stability in vivo. Given that the majority of work on human TP53 has focused on either mutations or posttranslational mechanisms, this zebrafish model revealed a novel regulatory target that will likely have important diagnostic implications for human patients.

In 2015, Astone et al. described *Tg(-8.5nkx2.2a:GFP)*^{ia2} zebrafish that developed gastrointestinal carcinomas and MPNST. This line was characterized by a 15.2 Mb deletion on chromosome 1 that affected 241 genes (Astone et al., 2015). While homozygous animals displayed severe developmental abnormalities and died by 72 hpf, heterozygous animals were viable and developed MPNSTs by 16 months postfertilization (Astone et al., 2015). Interestingly, *rsp6* was found to be contained within the deleted region and encodes the 40S Ribosomal protein S6 (Astone et al., 2015). Additional characterization of *Tg(-8.5nkx2.2a:GFP)*^{ia2} is necessary to rule out other possible candidate genes at the affected locus; however, this model again suggests important roles for Rp proteins in MPNST biology.

2.3 *Nf1* LOSS-OF-FUNCTION ENHANCES MPNST DEVELOPMENT

The majority of MPNSTs arise in patients with hereditary mutations in *NF1* (Cichowski & Jacks, 2001). NF1 is the most common inherited genetic disease

by circular binary segmentation algorithm, with segment mean values between −0.01 and 0.01 displayed white, and segment mean values between 0.01 (−0.01) and 0.8 (−0.8) displayed in continuous shades of red (blue) on a quasilogarithmic scale (indicated in the color band, bottom left). Segment mean values above (below) 0.8 (−0.8) were rendered in the most saturated color hue.

(A–G) Adapted and reprinted with permission from Berghmans, S., Murphey, R. D., Wienholds, E., Neuberg, D., Kutok, J. L., Fletcher, C. D. M., ... Look A.T. (2005). tp53 mutant zebrafish develop malignant peripheral nerve sheath tumors. Proceedings of the National Academy of Sciences of the United States of America, 102*(2), 407–412. Copyright (2005) National Academy of Sciences, U.S.A. (H) Adapted and reprinted with permission from Zhang, G., Hoersch, S., Amsterdam, A., Whittaker, C.A., Lees, J.A., & Hopkins, N. (2010). Highly aneuploid zebrafish malignant peripheral nerve sheath tumors have genetic alterations similar to human cancers.* Proceedings of the National Academy of Sciences of the United States of America, 107*(39), 16940–16945.*

affecting the nervous system with an estimated incidence of 1 in 3500 live births and heterozygous loss of NF1 increases the risk of developing MPNST from 0.001% to 2%–13% per lifetime (Cichowski & Jacks, 2001; Evans et al., 2002; Legius et al., 1993). NF1 is a GTPase-activating protein (GAP) for the RAS family of proto-oncogenes, with loss of NF1 leading to hyperactivation of RAS pathway signaling (Cichowski & Jacks, 2001). Interestingly, sporadic MPNSTs commonly display somatically acquired mutations in *NF1* and in the context of hereditary *NF1* mutations, neoplastic lesions arise in tissue stem cells that have lost the wild-type allele (Legius et al., 1993). These data indicate a causative role for *NF1* loss in inducing human MPNST. Several mouse models have confirmed this relationship. For example, chimeric $NF1^{-/-}$ cells have been shown to develop into plexiform neurofibromas in mice (Cichowski et al., 1999). When combined with heterozygous loss of *Tp53*, this phenotype progressed into S100-positive MPNST (Cichowski et al., 1999).

Zebrafish contain two copies of the *nf1* gene and early work using morpholino oligonucleotides identified a conserved role for zebrafish *nf1a* and *nf1b* during embryonic development (Lee et al., 2010; Padmanabhan et al., 2009). Morpholino oligonucleotides are antisense nucleotide fragments that bind directly to target RNA and block cell components involved in translation and splicing (Bedell, Westcot, & Ekker, 2011). Morpholinos must be actively delivered by microinjection into one-cell stage zebrafish. Due to their transient nature, morpholinos are not an effective means for studying phenotypes that arise past embryogenesis and therefore, have not be used to model sarcoma in zebrafish. Thus, to study late effects of *nf1* loss in zebrafish, Shin et al. used a zinc-finger nuclease (ZFN) targeting strategy to generate stable *nf1a* and *nf1b* mutant lines. Specifically, ZFNs were engineered with binding specificity to exon 26 of *nf1a* and exon 17 of *nf1b* (Shin et al., 2012). Paired ZFN mRNAs were injected into wild-type embryos at the one-cell stage and mutant alleles were identified by PCR amplification of genomic DNA followed by sequencing. Frameshift mutations that resulted in the generation of a premature stop codon and protein truncation upstream of the GAP regulatory domain were kept for further analysis. Confirming the morpholino results, single $nf1a^{-/-}$ or $nf1b^{-/-}$ mutants are viable, while double mutants were embryonic lethal (Shin et al., 2012). Interestingly, mutant larvae that carried at least one copy *nf1a* or *nf1b* were viable and fertile. In combination with homozygous $tp53^{M214K}$ mutation, 62% of $nf1a^{+/-};nf1b^{-/-};tp53^{M214K/M214K}$ animals went on to develop MPNST starting at 31 weeks post fertilization (Shin et al., 2012). Compared to approximately 28% of homozygous $tp53^{M214K}$ animals that develop MPNST starting at 41 weeks (Berghmans et al., 2005), this result confirmed a potent tumor suppressor role for NF1 during MPNST development.

$nf1a^{-/-};nf1b^{-/-}$ mutant larvae displayed hyperplasia of oligodendrocyte progenitor cells and Schwann cells, as well as melanophore hypoplasia (Shin et al., 2012). These defects mirror those observed in NF1 patients and offer an amenable embryonic model to further interrogate the molecular relationships within both neurofibromatosis and MPNST. Loss of *nf1a* and *nf1b* led to upregulation of Ras signaling in the embryonic spinal cord, confirming the well-described role of Nf1 as a negative regulator of Ras pathway activation (Shin et al., 2012). It should also be noted that in the same retroviral screen used to identify *rp* heterozygous MPNSTs, rare cases of MPNST developed in zebrafish heterozygous for *nf2a* loss-of-function mutation (Amsterdam et al., 2004). Future experiments will likely use these MPNST models to investigate Ras signaling as well as interacting pathways as potential molecular targets for patient therapy.

2.4 COMPARATIVE GENOMIC ANALYSIS TO IDENTIFY MOLECULAR DRIVERS OF MPNST

One major success to come from zebrafish MPNST models was comparative analysis of conserved DNA CNAs and the identification of molecular drivers of tumorigenesis. Zhang et al. isolated tumor cells from *rb* and $p53^{M214K}$ MPNSTs and analyzed viable cells stained with PI using flow cytometry. This analysis showed that zebrafish MPNSTs were aneuploid and on average contained nearly 3N DNA content (Zhang et al., 2010). Chromosomal metaphase spreads confirmed the expected 2N complement with 50 chromosomes found in normal tissue. In contrast, tumors arising from heterozygous *rp* animals had an average of 70 chromosomes, confirming aneuploidy in zebrafish MPNSTs (Zhang et al., 2010). Subsequently, a microarray was designed against the Zv7/danRer5 genome assembly comprised of nearly 15,000 60-mer probes (average separation of $\approx$100 kb). This array was used to analyze DNA from 36 tumors and matched control (nontumor fin) DNA (Zhang et al., 2010). As in human MPNST, genomic changes were found to be nonrandom and permutation-based testing identified regions that were significantly gained or lost in zebrafish MPNST when compared to normal tissue (Fig. 2H) (Zhang et al., 2010). In follow up studies, Zhang et al. assayed an additional 147 zebrafish MPNSTs and performed cross-species comparison with human MPNST. This analysis uncovered new MPNST-associated oncogenes including *CCND2*, *ETV6*, *HGF*, *HSF1*, *KIT*, *MDM2*, *MET*, and *PDGFR* (Zhang et al., 2013). Tumor suppressors were also identified and included *NF1*, *NF2*, *SMARCB1*, and *PTEN* (Zhang et al., 2013). Taken together, these approaches have allowed for the rapid identification of oncogenic drivers in MPNSTs, which can be further explored for functional relevance in driving onset, progression, and therapy response in the human disease.

3. EWING'S SARCOMA

Chromosomal translocations that create fusion oncogenes are well recognized as a significant cause of human sarcoma (Helman & Meltzer, 2003). Ewing's sarcoma is characterized as a malignant bone tumor that most commonly occurs in adolescents and young adults. Ewing's is characterized by a translocation t(11; 22)(q24; q12) that leads to the fusion of the N-terminal portion of Ewing sarcoma (EWS) RNA-binding protein 1 (*EWSR1*) to the C-terminal portion of FLI1, which contains an ETS (E26 transformation-specific) family DNA binding domain (Delattre et al., 1992). Alternative translocations are found in Ewing's, but in all, gene fusions lead to functional changes that promote tumor development.

Histologically, Ewing's sarcomas include small round blue cells tumors (SRBCTs) that are poorly differentiated, contain little cytoplasm, and round nuclei that stain darkly with hematoxylin (Rajwanshi, Srinivas, & Upasana, 2009). In general, Ewing's sarcoma is associated with bone and is the second most common form of pediatric osteosarcoma (Balamuth & Womer, 2010). Patients typically experience a two-year event-free survival of approximately 20% and about 25% of patients present with metastases at the time of diagnosis (Balamuth & Womer, 2010). Like other sarcomas, effective chemotherapies are lacking and a better understanding EWS-ETS fusion related signaling mechanisms could greatly improve patient outcome.

EWSR1-FLI1 is toxic to most cells, however, mouse mesenchymal progenitor cells tolerate expression of the fusion oncoprotein and generate Ewing's-like tumors when transplanted into recipient syngeneic mice (Riggi et al., 2005). Transgenic expression in mesenchymal cells also led to development of undifferentiated sarcoma in *Tp53* deficient mice (Lin et al., 2008). In contrast, expression in mouse blood lineages using *Mx1:Cre* resulted in the onset of aggressive leukemias (Torchia, Boyd, Rehg, Qu, & Baker, 2007). Thus, these mouse models highlight the importance of cellular context in regulating responses to EWSR1-FLI1 expression and demonstrate a need for better models to define Ewing's development in vivo.

Using the Tol2 transposon system (Kawakami et al., 2000), Leacock et al. generated zebrafish stably expressing human EWSR1-FLI1 under the control of a ubiquitous β-actin promoter [Tg*(actb1:EWSR1-FLI1)*] or the inducible heat-shock promoter [Tg*(hsp70:EWSR1-FLI1)*]. EWSR1-FLI1 was FLAG-tagged in both approaches and contained an IRES-GFP expression element to easily allow for the detection of transgene expression (Fig. 3A). $p53^{M214K}$ homozygous mutant embryos were injected at the one-cell stage with plasmid and Tol2 transposase mRNA and within the first generation, robust tumor formation was observed between 6 and 19 months of age (Fig. 3B and C) (Leacock et al., 2012). MPNSTs associated with $tp53^{M214K}$ and leukemias developed; however, 10/48 tumors were characterized as SRBCTs and were not observed in uninjected $tp53^{M214K}$ zebrafish (Fig. 3B and C) (Leacock et al., 2012). This histology was consistent with human Ewing's sarcoma. Transplantation experiments, similar to those described above, revealed that these tumor cells could grow in irradiated recipient animals, confirming their malignant potential.

EWS-ETS fusion proteins are believed to function primarily as aberrant transcription factors; however, microarray studies revealed inconsistencies in the gene targets misregulated among different cellular backgrounds. Leacock et al. used an Affymetrix zebrafish gene expression array and hybridized RNA from three independent zebrafish SRBCTs. Subsequent analysis revealed upregulation of genes

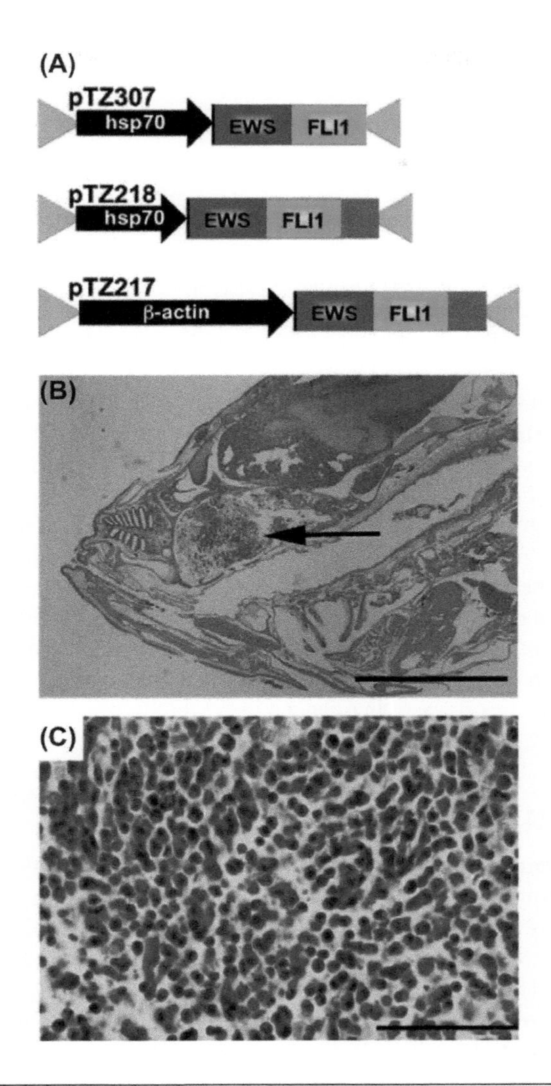

FIGURE 3

Transgenic expression of EWSR1-FLI1 induced Ewing's sarcoma-like tumors in zebrafish. (A) Schematic of Tol2 transposons used by Leacock et al. 2012. *hsp70* or the *actb1a* (*β-actin*) promoter (*black arrows*) were used to drive expression of FLAG-tagged human EWSR1-FLI1 (EWS-FLI1). IRES-GFP (*gray rectangle*) was used to track expression in live animals. To mediate genomic integration, each transposon was flanked by Tol2 recombination sites (*triangle*). (B and C) Histology sections of adult zebrafish injected with EWSR1-FLI1 transgene + Tol2 transposase mRNA at the one-cell stage. Small round blue cell tumors developed in the eye (B, *arrow*). (C) Magnified image of B. Scale bars, 200 μm (B) and 50 μm (C).

Adapted and reprinted with permission from Leacock, S. W., Basse, A. N., Chandler, G. L., Kirk, A. M., Rakheja, D., & Amatruda, J. F. (2012). A zebrafish transgenic model of Ewing's sarcoma reveals conserved mediators of EWS-FLI1 tumorigenesis. Disease Models and Mechanisms, 5*(1), 95–106.*

associated with protein translation and energy production, consistent with the proliferative nature of these sarcoma cells (Leacock et al., 2012). To generate an EWSR1-FLI1-specific gene expression profile, SRBCTs were compared to the MPNSTs that formed over the course of these experiments and attributed to homozygous $p53^{M214K}$ mutation (Leacock et al., 2012). Interestingly, EWSR1-FLI1 expression was associated with upregulation of genes involved in neural development including *nkx2.2*, *olig2*, *sox3*, and *ascla/b*, suggesting a neuronal precursor cell of origin in Ewing's (Leacock et al., 2012). Furthermore, cross-species analysis identified a core EWS-FLI1 expression signature conserved between zebrafish and human Ewing's-like tumor cells and included *NKX2.2*, *MYC*, *MAPT*, *SALL2*, *PADI2*, and *POU3F1* (Leacock et al., 2012).

Consistent with the toxic nature of EWS-FLI1 overexpression, stable Tg*(actb1:EWSR1-FLI1)* zebrafish lines could not be recovered as adults (Leacock et al., 2012). However, stable Tg*(hsp70:EWSR1-FLI1)* mosaic zebrafish were fertile and some transgenic F1 progeny were able to survive until adulthood (Leacock et al., 2012). Thus, this model provides a unique avenue for further interrogation of molecular mechanisms involved in Ewing's sarcoma. The inducible nature of the *hsp70* promoter combined with analysis of developmental phenotypes will allow for a refined dissection of EWS-ETS fusion—related signaling and a better understanding of how these oncogenic translocations lead to tumor development in humans. These zebrafish will also be powerful models for use in chemical genetic screens to identify new drugs to treat this devastating cancer.

4. CHORDOMA

Chordoma is a rare tumor that is thought to arise from remnants of the embryonic notochord, with primary tumors commonly being found in the bones of the axial skeleton (Walcott et al., 2012). In the United States, 300 new cases of chordoma are diagnosed each year with current treatment options limited to surgical resection and/or radiation (McMaster, Goldstein, Bromley, Ishibe, & Parry, 2001; Walcott et al., 2012). Chordomas are highly chemoresistant and prone to metastasis, making chordoma a devastatingly difficult disease to treat (Walcott et al., 2012). Recurrent genetic deletions including those at the *CDKN2A*, *CDKN2B*, and *PTEN* loci have been associated with chordoma and occur in 70—80% of primary tumors (Hallor et al., 2008; Le et al., 2011). *TP53* and *RB* mutations have also been described along with amplifications of the master transcriptional regulator *T* (Brachyury) (Eisenberg, Woloschak, Sen, & Wolfe, 1997; Naka et al., 2005; Yang et al., 2009). To date, no rodent models of chordoma have been described and only a few human cell lines have been established, which limits better genetic understanding of the disease and preclinical drug testing.

The Gal4/*UAS* system was one of the first inducible transgenic methods applied in zebrafish and has provided a unique opportunity to induce the expression of oncogenes in a tissue-specific manner to study molecular drivers of sarcoma in vivo.

Gal4/*UAS* is a binary system, derived from *Saccharomyces cerevisia* and consists of the transcriptional regulator Gal4 that controls gene expression by binding the *UAS* transcriptional motif (Halpern et al., 2008). To obtain tissue-specific gene expression, Gal4 is expressed under the control of a tissue-specific promoter and the gene of interest is placed downstream of *UAS* (Halpern et al., 2008). To study the effect of activated HRAS in the zebrafish intestine, Burger et al. crossed stable *UAS:EGFP-HRAS*^{G12V} zebrafish to the gut-specific enhancer trap line, *Tg(mu4465:Gal;UAS:mcherry)* (Fig. 4A–C). Surprisingly, in addition to intestinal hyperplasia, embryos developed prominent notochord malformations that were characterized by aberrant cellular proliferation with an onset as early as 3 dpf (Fig. 4E and G). One hundred percent of the embryos were affected by 7 dpf and by 14 dpf, all double transgenic animals were dead (Burger et al., 2014).

These tumor-like masses were suggestive of a notochord cellular origin. The *mu4465:Gal4* line utilized the *tiggywinkle hedgehog* (*twhh*) minimal promoter to drive Gal4 expression and was prominently expressed in the developing notochord (Fig. 4D) (Burger et al., 2014). *twhh:Gal4;UAS:EGFP-HRAS*^{G12V} double transgenic zebrafish displayed the same notochord malformations with accelerated onset and increased penetrance (Fig. 4H). Histology in both cases revealed plump cells with vacuolated cytoplasm, focally prominent nuclear pleomorphism and hyperchromasia similar to human chordoma (Fig. 4I–K). Electron microscopy was used to observe multifocal desmosomal cellular junctions, with large "windows" between neighboring cells, similar to human chordoma (Fig. 4L–N). Importantly, the tumor cells stained positive for Brachyury and Cytokeratin, which is used clinically to diagnosis chordoma (Burger et al., 2014).

HRAS mutations have not been reported in human chordoma and unlike the zebrafish model that takes only days to form, human chordoma is a slow-growing sarcoma with an average of 5 years between the onset of symptoms and diagnosis. Despite limited evidence for the involvement of activated RAS signaling in human chordoma, this zebrafish model will likely emerge as an effective tool to assess potential chemical compounds that modify disease-associated phenotypes in vivo. mTORC1/PI3K inhibitors have been used clinically to treat chordoma and, interestingly, both *twhh:Gal4;UAS:HRAS*^{G12V} and *mu4465:Gal4;UAS:HRAS*^{G12V} embryos display a significant delay in tumor onset following treatment with the mTOR (mechanistic target of rapamycin) inhibitor, rapamycin (Burger et al., 2014). Rapamycin increased the life-span of transgenic embryos compared to DMSO controls and correlated with decreased mTOR pathway activation, when assessed by phospho-S6 staining on section (Burger et al., 2014). The activity of rapamycin validated the utility of this zebrafish model and sets the stage for higher-throughput drug screening strategies to identify novel drug candidates for future clinical trials. Early phenotypic onset in this model is especially useful since larval stage zebrafish are highly drug permeable and easily manipulated in small, multiwelled plates. Unfortunately, early lethality limits the degree to which progression could be assessed in primary animals. For example, tumor cells were never observed to enter the circulation and/or establish distant sites of metastasis before the affected animals

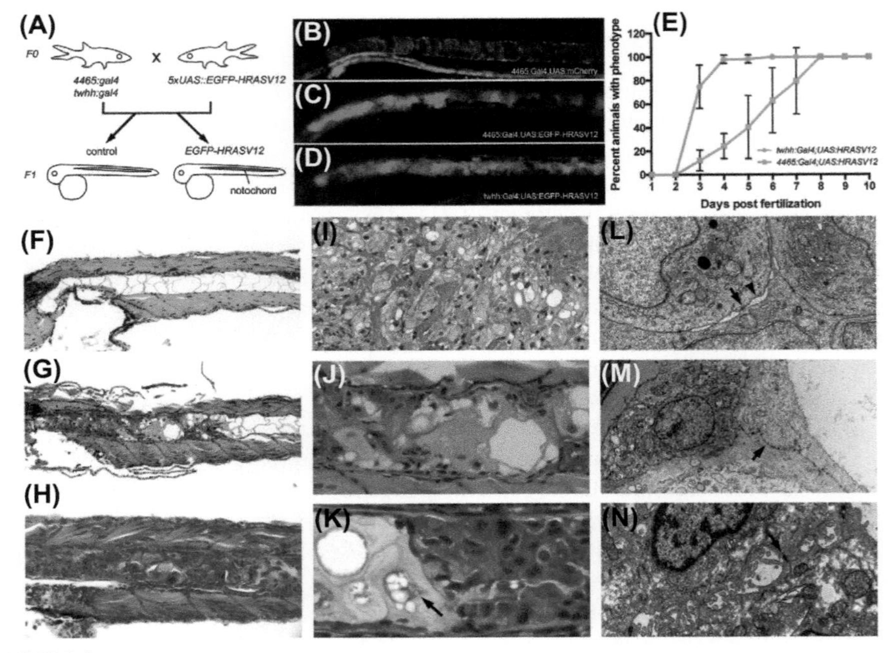

FIGURE 4

Gal4/*UAS*- driven HRAS^{V12} expression in the zebrafish notochord represents the first animal model of human chordoma. (A) Schematic of transgenic breeding design used by Burger et al. (2014). Notochord-specific *Gal4* lines (*4465:Gal4* and *twhh:Gal4*) were independently crossed to *UAS:EGFP-*HRAS^{V12} heterozygous fish resulting in the embryos shown in B–D. (B) Control notochord of *4465:Gal4,UAS:mCherry* embryos; (C) notochord of *4465:Gal4; UAS:EGFP-*HRAS^{V12}; (D) notochord of *twhh:Gal4;UAS:EGFP-*HRAS^{V12}. HRAS^{V12} gene transactivation was monitored via GFP in the notochord (C and D). (B–D) Disorganized growth of notochord tissue was evident in C and D compared with a normal "stack of coins" notochord appearance in *4465:Gal4, UAS:mCherry* embryos (B). The abnormal notochord phenotype was evident as early as 3 dpf and progressively increased with age, with 100% of larvae affected by 8 dpf. (B–D) Pictures are representatives of 10 dpf old animals. (E) The phenotype progressed much faster in *twhh:Gal4;UAS:*HRAS^{V12} compared with *4465:Gal4; UAS:*HRAS^{V12}. (F–N) Histological and ultrastructural examination revealed the presence of a chordoma-like notochord tumor in the transgenic larvae. (F) At 7 dpf, control animals displayed a normal notochord with large vacuolated spaces, thin cytoplasmic septae and bland nuclei. In contrast, *4465:Gal4;UAS:*HRAS^{V12} (G and J) and *twhh:Gal4;UAS:*HRAS^{V12} (H and K) fish showed a replacement of the notochord by a chordoma-like tumor (compare with an example of human chordoma in I). (L) The tumor cells displayed characteristic desmosomal junctions (*arrow*) with the formation of "windows" between neighboring cells (*arrowhead*), which is a common characteristic of human chordomas. In addition, the tumor cells displayed a prominent rough endoplasmic reticulum. (M) The tumor cells often lifted the notochord cells from the basement membrane while still attached to them by numerous desmosomal junctions, which are a part of notochord normal anatomy (*arrow*). (N) An example of human chordoma showing desmosomal junctions (*arrow*).

Reprinted with permission from Burger, A., Vasilyev, A., Tomar, R., Selig, M. K., Nielsen, G. P., Peterson, R. T., ... Haber, D. A. (2014). A zebrafish model of chordoma initiated by notochord-driven expression of HRASV12. Disease Models and Mechanisms, 7(7), 907–913.

died (Burger et al., 2014). However, tumor cell transplantation techniques similar to those previously described could be useful for study of tumor cells at later stages in otherwise wild-type animals.

Thus, notochord-specific expression of $HRAS^{G12V}$ represents the first animal model of human chordoma and offers a novel platform for functionally testing disease-associated oncogenes and for drug screening. For example, the role of Brachyury in chordoma development should be easily accomplished by using any of the transgenic approaches described earlier.

5. HEMANGIOSARCOMA

PTEN is one of the most commonly mutated tumor suppressor genes in human cancer with somatic loss of *PTEN* leading to cancer in a variety of tissue types (Li et al., 1997; Podsypanina et al., 1999; Suzuki et al., 1998). PTEN is a lipid and protein phosphatase that antagonizes the phosphoinositide 3-kinase (PI3K)-Akt pathway with loss of PTEN leading to activation of Akt/PI3K signaling, enhanced cell proliferation and cell survival (Maehama & Dixon, 1998; Myers et al., 1998). In humans, germline mutations of *PTEN* are associated with multiple disorders characterized by the development of noncancerous growths called hamartomas and an increased susceptibility to cancer (Eng, 2003). While loss of *Pten* in mice leads to embryonic lethality, heterozygous animals develop various kinds of hyperplasias and vascular abnormalities including hemangiomas (Alimonti et al., 2010; Podsypanina et al., 1999; Suzuki et al., 1998), indicating an important role for Pten during development and homeostasis of multiple tissue types.

Zebrafish possess two genes encoding Pten, referred to as *ptena* and *ptenb*. Using a morpholino-based knockdown approach, *ptena* and *ptenb* were found to be essential but functionally redundant during embryogenesis (Croushore et al., 2005). To study later stage defects, the same ENU-based target selected gene inactivation screening technique described for $tp53^{M214K}$ was used to identify loss-of-function mutations in both *ptena* and *ptenb* (Faucherre, Taylor, Overvoorde, Dixon, & den Hertog, 2008; Wienholds et al., 2002). Mutants were selected based on premature stop codons well upstream of the essential phosphatase catalytic site. Single mutant embryos were viable and fertile; however, double $ptena^{-/-}$;$ptenb^{-/-}$ mutants died by 6 dpf and displayed endothelial hyperplasia and ectopic vessel sprouting, indicative of enhanced angiogenesis as a result of Pten loss (Choorapoikayil, Weijts, Kers, de Bruin, & den Hertog, 2013; Faucherre et al., 2008).

Functional redundancy between *ptena* and *ptenb* allowed for the study of late-stage nondevelopmental phenotypes, including those related to sarcomagenesis. Zebrafish lacking three of four Pten alleles, $ptena^{+/-}$;$ptenb^{-/-}$ or $ptena^{-/-}$;$ptenb^{+/-}$ spontaneously developed tumors starting at about 3 months of age that typically localized to tissue adjacent to the eye (Fig. 5A–G) (Choorapoikayil, Kuiper, de Bruin, & den Hertog, 2012). Histological analysis revealed that these tumors consisted of variably sized blood-filled spaces lined with CD31 (PECAM-1)-positive endothelial

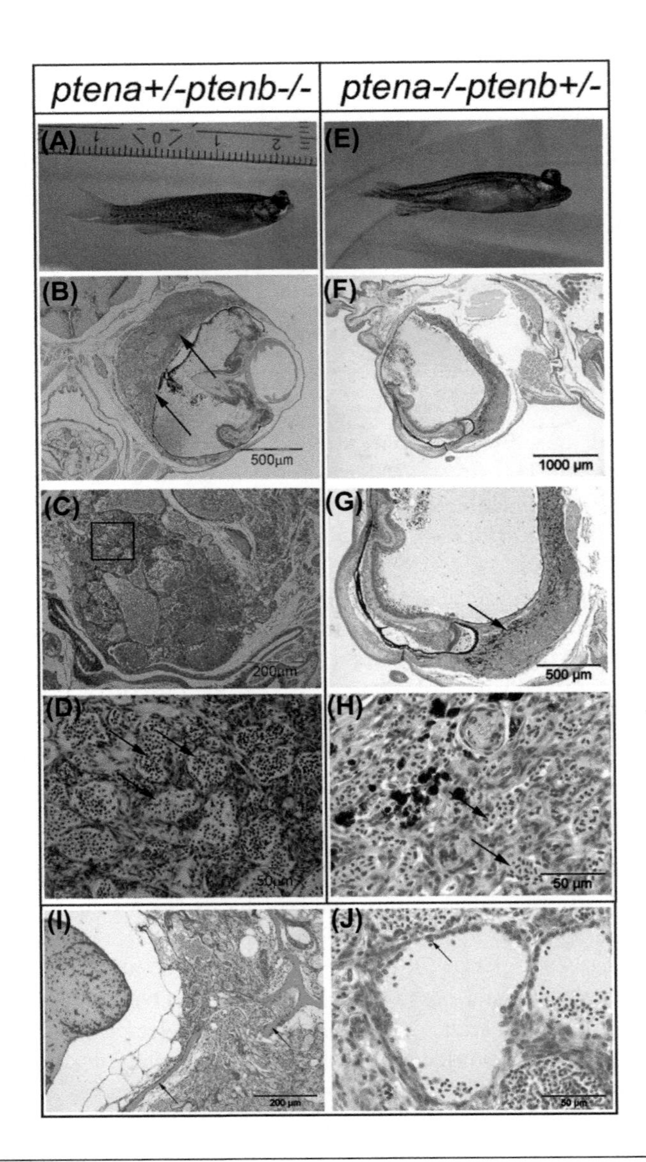

FIGURE 5

Pten$^{+/-}$;*pten*$^{-/-}$ or *pten*$^{-/-}$;*pten*$^{+/-}$ zebrafish develop hemangiosarcoma. (A–D) 3-month-old *pten*$^{+/-}$;*pten*$^{-/-}$ and (E–H) 9-month-old *pten*$^{-/-}$;*pten*$^{+/-}$ mutant zebrafish with ocular tumors. The entire intact fish was fixed and embedded in paraffin. (B–D) Transverse and (F–H) sagittal sections were stained with hematoxylin and eosin (H and E). *Arrows* indicate tumor mass, which is associated with the eye. (B and G) Higher-power magnifications of tumor mass; (D) magnification of the boxed area in C. The tumor consisted of cells that form different sizes of blood-filled spaces (*arrows* in D and H). (I and J) H and E staining of sections from two individuals revealed hemangiosarcoma formation. (I) The tumor was cells, indicative of angiogenic origin (Fig. 5D and H) (Choorapoikayil et al., 2012). As a result, these tumors were classified as hemangiosarcoma. Phospho-Akt staining confirmed activation of AKT/PI3K signaling in endothelial cells of the tumor mass and phospho-GSK-3β was similarly detected (Choorapoikayil et al., 2012). Ptena protein was also detected in *ptena*$^{+/-}$*;ptenb*$^{-/-}$ tumors, suggesting that tumor development is not due to the loss of heterozygosity and supported the observation that hypomorphic PTEN signaling is sufficient for tumorigenesis in both mouse and human disease (Alimonti et al., 2010; Li et al., 1997; Podsypanina et al., 1999; Suzuki et al., 1998).

Using *ptena*$^{-/-}$*;ptenb*$^{-/-}$ mutant zebrafish embryos, Choorapoikayil et al. (2013) found that chemical inhibition of PI3K signaling (LY294002) rescued endothelial hyperplasia and when combined with the angiogenesis inhibitor sunitinib, cooperatively rescued the hypervascularization phenotypes in vivo. These experiments identified potential therapeutic options for neovascularization in cancer and further highlight the utility of zebrafish models for chemical testing in vivo. It is unknown why *ptena*$^{+/-}$*;ptenb*$^{-/-}$ or *ptena*$^{-/-}$*;ptenb*$^{+/-}$ zebrafish are uniquely susceptible to developing hemangiosarcoma, but the ability to manipulate oncogenic signaling in the *pten*-deficient background will likely reveal roles for these important tumor suppressor genes in inducing a wider range of cancers.

6. LIPOSARCOMA

Liposarcoma is a common sarcoma in adults, affecting around 2000 Americans each year (Dalal, Antonescu, & Singer, 2008). Tumors are classified into histopathological subtypes with well-differentiated liposarcoma (WDLPS) representing 46% and dedifferentiated liposarcoma accounting for 18% of all cases, respectively (Dalal et al., 2008). Surgery is the most effective course of treatment; however, high disease-associated morbidity and mortality relates to the fact that tumors often develop in deep anatomical locations surrounding vital structures that often make complete surgical resection impossible. Radiation and chemotherapy have limited efficacy, underscoring the need for improved models for understanding molecular mechanisms involved in liposarcomagenesis.

invasive and penetrated into the brain with enclosing scull elements (*arrows*). (J) Cells with plump morphology (*arrow*) were observed to be detached from surrounding tissue and protruding into the vessel lumen. Sections of representative tumors were featured and reprinted here.

Reprinted with permission from Choorapoikayil, S., Kuiper, R. V., de Bruin, A., & den Hertog, J. (2012). Haploinsufficiency of the genes encoding the tumor suppressor Pten predisposes zebrafish to hemangiosarcoma. Disease Models and Mechanisms, 5, *241–247.*

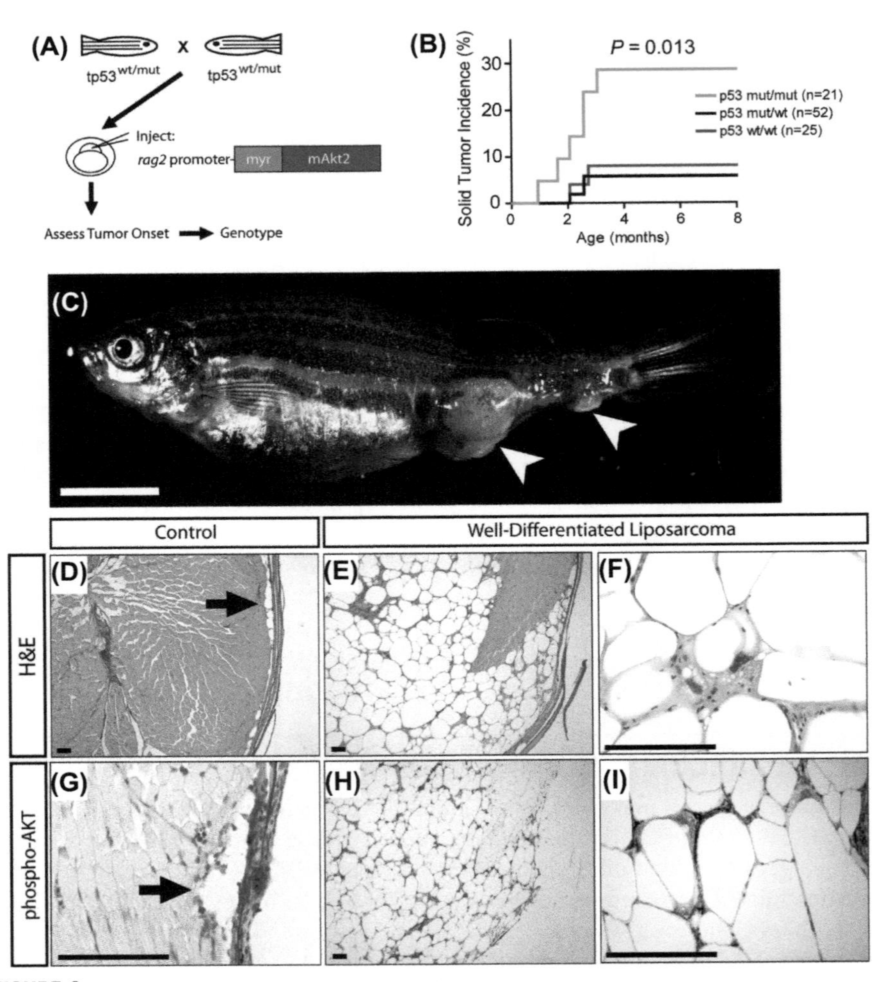

FIGURE 6

Transgenic zebrafish uncover a role for PI3K-AKT signaling in liposarcoma formation. (A) Schematic of breeding and injection scheme used by Gutierrez et al. (2011). *Tp53*^{M214K} heterozygous animals were incrossed and the resulting embryos injected with linearized *rag2:myr-mAkt2* plasmid at the one-cell stage. All injected embryos were raised to adulthood, assessed for tumor formation and genotyped for *tp53*^{M214K} status. (B) Tumor incidence in *tp53* wild-type, heterozygous or *tp53*^{M214K} homozygous mutant zebrafish injected with *rag2:myr-mAkt2*. *p* value was calculated by log-rank test. (C) Representative *rag2:myr-mAkt2*-injected *tp53*^{M214K} homozygous mutant zebrafish that developed two independent tumor masses (*white arrow heads*). (D) Histological analysis of control animal that had normal adipose tissue (*arrow*). Histology of tumor-bearing fish imaged at low (E) and high magnification (F). Multivacuolated cytoplasm and large hyperchromatic nuclei were indicative of well-differentiated liposarcoma. (G–I) Immunohistochemistry of control (G) and

Individuals with germ-line *TP53* mutations are at increased risk for developing liposarcoma at a very young age (Debelenko et al., 2010). The PI3K-AKT signaling pathway is also commonly dysregulated and individuals with germ-line *PTEN* mutations often develop benign lipomas (adipocytic neoplasms), suggesting loss-of-function of PTEN may contribute to liposarcoma in humans (Marsh et al., 1999). AKT activation has been detected in human liposarcoma; however, mouse models of activated PI3K-AKT display no evidence of developing this particular tumor type (Hernando et al., 2007). No other invertebrate or rodent models of liposarcoma have been described, greatly impeding experimental dissection of disease pathology.

To test the role for Akt in liposarcoma, Gutierrez et al. used a mosaic transgenic approach and injected linearized *rag2:myr-mAkt2* expression vector into one-cell stage embryos from an incross between heterozygous $p53^{M214K}$ zebrafish. This expression construct encoded a myristoylated, constitutively active form of mouse *Akt* and was driven by the zebrafish *rag2* promoter (Fig. 6A). Injected animals developed externally visible tumors between 1 and 4 months of age, with an incidence of 29% in *p53*-homozygous mutants and 6–8% in *p53* heterozygous or wild-type siblings, revealing significant cooperation between Akt and *p53* loss during tumor initiation (Fig. 6B and C) (Gutierrez et al., 2011). Histological analysis revealed that the majority of tumors consisted of locally invasive masses of adipocytes of variable size with scattered atypical stromal cells, hyperchromatic nuclei, and lipoblasts, consistent with a diagnosis of WDLPS (Fig. 6D–F). Furthermore, activation of Akt signaling was confirmed in tumor cells through immunohistochemistry of phospho-Akt on section (Fig. 6G–I) (Gutierrez et al., 2011). Finally, analysis of patient samples revealed that AKT signaling is activated in nearly one-third of primary tumors (Gutierrez et al., 2011), uncovering novel roles for AKT in human disease and validating the relevance of the zebrafish model for further study of liposarcoma pathogenesis.

The role for Akt/PI3K signaling was also confirmed by Chu et al. 2012 as a result of a screen of human oncogenes in the zebrafish skin. A Tol2 transposase-based transgenic strategy was used to generate stable lines of zebrafish where potential human oncogenes were placed under the control of the zebrafish skin-specific keratin 4 (*krt4*) promoter (Chu et al., 2012; Kawakami et al., 2000). Interestingly, stable expression of myristoylated human *AKT* led to the formation of lipoma (Chu et al., 2012). $Tg(krt4:myr\text{-}Akt1)^{cy18}$ larvae displayed transformation of skin, bone, and muscle tissue into adipose tissue (Chu et al., 2012), indicating

tumor sections (H and I) revealed strong immunoreactivity for phospho-AKT in tumor cells of *rag2:myr-mAkt2* injected zebrafish. Scale bars, 100 μm.

Reprinted with permission from Gutierrez, A., Snyder, E. L., Marino-Enriquez, A., Zhang, Y.-X., Sioletic, S., Kozakewich, E., ... Look A.T. (2011). Aberrant AKT activation drives well-differentiated liposarcoma. Proceedings of the National Academy of Sciences of United States of America, 108*(39), 16386–16391.*

an important role for AKT signaling in adipose tissue fate specification. Importantly, this transgenic model can now be used to study the contribution of other oncogenic factors that lead to malignant transformation of lipoma to liposarcoma and with significant diagnostic and therapeutic implications.

7. FUTURE PERSPECTIVES AND OPPORTUNITIES

Advanced genomic study of human sarcoma will continue to identify novel genetic mutations; however, challenges exist with this type of analysis in differentiating passenger and driver mutations. Cross-species analyses have already begun to uniquely define potential molecular drivers based on synteny between humans and zebrafish, and will continue to narrow the list of potential oncogenic candidates for further preclinical testing. On a gene-by-gene basis, ease of the transgenic manipulations described in this review will allow for a fast and high-throughput assessment of molecular cooperation between novel oncogenes and tumor suppressors. Continued development of zebrafish models using newly innovated techniques will also likely enhance the accuracy of modeling sarcoma. For example, the recent success of CRISPR/Cas9-induced targeted gene insertion and/or mutagenesis means that new zebrafish models can be developed that harbor patient-specific mutations in oncogenes and tumor suppressors. The accessibility of CRISPR/Cas9 technology also means that large-scale genetic screens can be performed in zebrafish to identify novel molecular pathways and therapeutic targets in sarcoma. Finally, high-throughput drug-screening techniques using zebrafish have led to the discovery of promising new therapeutics for many human diseases, including sarcoma, and will be instrumental to the discovery of relevant drugs for the treatment of these diseases in the future.

In conclusion, the zebrafish models discussed in this review have been instrumental to our understanding of sarcomagenesis and have contributed to the identification of relevant oncogenes, sarcoma cells of origin, and potential new therapeutic options for patients. Zebrafish models have also helped to assess tumor cell heterogeneity and the significance of cellular diversity with respect to relapse, invasion/metastasis, and disease outcome. It is clear that the next decade will continue to see amazing insights gleaned from using zebrafish models of sarcoma, providing novel approaches to uncover new drug targets and therapies for these devastating cancers.

REFERENCES

Albacker, C. E., Storer, N. Y., Langdon, E. M., DiBiase, A., Zhou, Y., Langenau, D. M., & Zon, L. I. (2013). The histone methyltransferase SUV39H1 suppresses embryonal rhabdomyosarcoma formation in zebrafish. *PLoS One, 8*(5), e64969.

Alimonti, A., Carracedo, A., Clohessy, J. G., Trotman, L. C., Nardella, C., Egia, A., ... Pandolfi, P. P. (2010). Subtle variations in Pten dose determine cancer susceptibility. *Nature Genetics, 42*(5), 454–458.

Amsterdam, A., Burgess, S., Golling, G., Chen, W., Sun, Z., Townsend, K., ... Hopkins, N. (1999). A large-scale insertional mutagenesis screen in zebrafish. *Genes and Development, 13*(20), 2713–2724.

Amsterdam, A., Sadler, K. C., Lai, K., Farrington, S., Bronson, R. T., Lees, J. A., & Hopkins, N. (2004). Many ribosomal protein genes are cancer genes in zebrafish. *PLoS Biology, 2*(5), E139.

Aoki, Y., Niihori, T., Kawame, H., Kurosawa, K., Ohashi, H., Tanaka, Y., ... Matsubara, Y. (2005). Germline mutations in HRAS proto-oncogene cause Costello syndrome. *Nature Genetics, 37*(10), 1038–1040.

Ardnt, C. A. S., & Crist, W. M. (1999). Tumors of childhood and adolescence. *New England Journal of Medicine, 341*, 342–352.

Astone, M., Pizzi, M., Peron, M., Domenichini, A., Guzzardo, V., Töchterle, S., ... Vettori, A. (2015). A GFP-tagged gross deletion on chromosome 1 causes malignant peripheral nerve sheath tumors and carcinomas in zebrafish. *PLoS One, 10*(12), e0145178.

Balamuth, N. J., & Womer, R. B. (2010). Ewing's sarcoma: diagnostic, prognostic, and therapeutic implications of molecular abnormalities. *Lancet Oncology, 11*(2), 184–192.

Barr, F. G., Galili, N., Holick, J., Biegel, J. A., Rovera, G., & Emanuel, B. S. (1993). Rearrangement of the PAX3 paired box gene in the paediatric solid tumour alveolar rhabdomyosarcoma. *Nature Genetics, 3*, 113–117.

Bedell, V. M., Westcot, S. E., & Ekker, S. C. (2011). Lessons from morpholino-based screening in zebrafish. *Briefings in Functional Genomics, 10*(4), 181–188.

Berghmans, S., Murphey, R. D., Wienholds, E., Neuberg, D., Kutok, J. L., Fletcher, C. D. M., ... Look, A. T. (2005). tp53 mutant zebrafish develop malignant peripheral nerve sheath tumors. *Proceedings of the National Academy of Sciences of the United States of America, 102*(2), 407–412.

Borden, E. C., Baker, L. H., Bell, R. S., Bramwell, V., Demetri, G. D., Eisenberg, B. L., ... Brennan, M. F. (2003). Soft tissue sarcomas of adults: state of the translational science. *Clinical Cancer Research, 9*(6), 1941–1956.

Bridge, J. A., Liu, J., Qualman, S. J., Suijkerbuijk, R., Wenger, G., Zhang, J., ... Barr, F. G. (2002). Genomic gains and losses are similar in genetic and histologic subsets of rhabdomyosarcoma, whereas amplification predominates in embryonal with anaplasia and alveolar subtypes. *Clinical Cancer Research, 9*(6), 1941–1956.

Burger, A., Vasilyev, A., Tomar, R., Selig, M. K., Nielsen, G. P., Peterson, R. T., ... Haber, D. A. (2014). A zebrafish model of chordoma initiated by notochord-driven expression of HRASV12. *Disease Models and Mechanisms, 7*(7), 907–913.

Burningham, Z., Hashibe, M., Spector, L., & Schiffman, J. D. (2012). The epidemiology of sarcoma. *Clinical Sarcoma Research, 2*(1), 14.

Casanova, M., Ferrari, A., Spreafico, F., Luksch, R., Terenziani, M., Cefalo, G., ... Fossati-Bellani, F. (1999). Malignant peripheral nerve sheath tumors in children: a single-institution twenty-year experience. *Journal of Pediatric Hematology/Oncology, 21*(6), 509–513.

Chen, E. Y., Deran, M. T., Ignatius, M. S., Grandinetti, K. B., Clagg, R., McCarthy, K. M., ... Langenau, D. M. (2014). Glycogen synthase kinase 3 inhibitors induce the canonical WNT/β-catenin pathway to suppress growth and self-renewal in embryonal rhabdomyosarcoma. *Proceedings of the National Academy of Sciences of the United States of America, 111*(14), 5349–5354.

Chen, E. Y., Dobrinski, K. P., Brown, K. H., Clagg, R., Edelman, E., Ignatius, M. S., ... Langenau, D. M. (2013). Cross-species array comparative genomic hybridization identifies novel oncogenic events in zebrafish and human embryonal rhabdomyosarcoma. *PLoS Genetics, 9*(8), e1003727.

Chen, Y., Takita, J., Hiwatari, M., Igarashi, T., Hanada, R., Kikuchi, A., ... Hayashi, Y. (2006). Mutations of the PTPN11 and RAS genes in rhabdomyosarcoma and pediatric hematological malignancies. *Genes, Chromosomes and Cancer, 45*(6), 583–591.

Choorapoikayil, S., Kuiper, R. V., de Bruin, A., & den Hertog, J. (2012). Haploinsufficiency of the genes encoding the tumor suppressor Pten predisposes zebrafish to hemangiosarcoma. *Disease Models and Mechanisms, 5*, 241–247.

Choorapoikayil, S., Weijts, B., Kers, R., de Bruin, A., & den Hertog, J. (2013). Loss of Pten promotes angiogenesis and enhanced vegfaa expression in zebrafish. *Disease Models and Mechanisms, 6*(5), 1159–1166.

Chu, C. Y., Chen, C. F., Rajendran, R. S., Shen, C. N., Chen, T. H., Yen, C. C., ... Hsiao, C. D. (2012). Overexpression of Akt1 enhances adipogenesis and leads to lipoma formation in zebrafish. *PLoS One, 7*(5), e36474.

Cichowski, K., & Jacks, T. (2001). NF1 tumor suppressor gene function. *Cell, 104*, 593–604.

Cichowski, K., Shih, T. S., Schmitt, E., Santiago, S., Reilly, K., McLaughlin, M. E., ... Jacks, T. (1999). Mouse models of tumor development in neurofibromatosis type 1. *Science, 286*(5447), 2172–2176.

Croushore, J. A., Blasiole, B., Riddle, R. C., Thisse, C., Thisse, B., Canfield, V. A., ... Levenson, R. (2005). Ptena and ptenb genes play distinct roles in zebrafish embryogenesis. *Developmental Dynamics, 234*(4), 911–921.

Dalal, K. M., Antonescu, C. R., & Singer, S. (2008). Diagnosis and management of lipomatous tumors. *Journal of Surgical Oncology, 97*(4), 298–313.

Davis, R. J., D'Cruz, C. M., Lovell, M. A., Biegel, J. A., & Barr, F. G. (1994). Fusion of PAX7 to FKHR by the variant t(1;13)(p36;q14) translocation in alveolar rhabdomyosarcoma. *Cancer Research, 54*(11), 2869–2872.

Debelenko, L. V., Perez-Atayde, A. R., Dubois, S. G., Grier, H. E., Pai, S. Y., Shamberger, R. C., & Kozakewich, H. P. (2010). p53+/mdm2– atypical lipomatous tumor/well-differentiated liposarcoma in young children: an early expression of Li–Fraumeni syndrome. *Pediatric and Developmental Pathology, 13*(3), 218–224.

Delattre, O., Zucman, J., Plougastel, B., Desmaze, C., Melot, T., Peter, M., & Rouleau, G. (1992). Gene fusion with an ETS DNA-binding domain caused by chromosome translocation in human tumours. *Nature, 359*(6391), 162–165.

Dillon, P. W. (1997). Nonrhabdomyosarcoma soft tissue sarcomas in children. *Seminars in Pediatric Surgery, 6*(1), 24–28.

Donehower, L. A., Harvey, M., Slagle, B. L., McArthur, M. J., Montgomery, C. A., Butel, J. S., & Bradley, A. (1992). Mice deficient for p53 are developmentally normal but susceptible to spontaneous tumours. *Nature, 356*(6366), 215–221.

Eisenberg, M. B., Woloschak, M., Sen, C., & Wolfe, D. (1997). Loss of heterozygosity in the retinoblastoma tumor suppressor gene in skull base chordomas and chondrosarcomas. *Surgical Neurology, 47*(2), 156–161.

Eng, C. (2003). PTEN: one gene, many syndromes. *Human Mutation, 22*(3), 183–198.

Evans, D. G. R., Baser, M. E., McGaughran, J., Sharif, S., Howard, E., & Moran, A. (2002). Malignant peripheral nerve sheath tumours in neurofibromatosis 1. *Journal of Medical Genetics, 39*(5), 311–314.

Faucherre, A., Taylor, G. S., Overvoorde, J., Dixon, J. E., & den Hertog, J. (2008). Zebrafish pten genes have overlapping and non-redundant functions in tumorigenesis and embryonic development. *Oncogene, 27*(8), 1079–1086.

Fountain, J. W., Wallace, M. R., Bruce, M. A., Seizinger, B. R., Menon, A. G., Gusella, J. F., … Collins, F. S. (1989). Physical mapping of a translocation breakpoint in neurofibromatosis. *Science, 244*(4908), 1085–1087.

Freeman, J. L., Ceol, C., Feng, H., Langenau, D. M., Belair, C., Stern, H. M., … Lee, C. (2009). Construction and application of a zebrafish array comparative genomic hybridization platform. *Genes, Chromosomes and Cancer, 48*(2), 155–170.

Galili, N., Davis, R. J., Fredericks, W. J., Mukhopadhyay, S., Rauscher, F. J., Emanuel, B. S., … Barr, F. G. (1993). Fusion of a fork head domain gene to PAX3 in the solid tumour alveolar rhabdomyosarcoma. *Nature Genetics, 5*(3), 230–235.

Galindo, R. L., Allport, J. A., & Olson, E. N. (2006). A *Drosophila* model of the rhabdomyosarcoma initiator PAX7-FKHR. *Proceedings of the National Academy of Sciences of the United States of America, 103*(36), 13439–13444.

Gupta, G., Mammis, A., & Maniker, A. (2008). Malignant peripheral nerve sheath tumors. *Neurosurgery Clinics of North America, 19*(4), 533–543.

Gutierrez, A., Snyder, E. L., Marino-Enriquez, A., Zhang, Y.-X., Sioletic, S., Kozakewich, E., … Look, A. T. (2011). Aberrant AKT activation drives well-differentiated liposarcoma. *Proceedings of the National Academy of Sciences of the United States of America, 108*(39), 16386–16391.

Hallor, K. H., Staaf, J., Jönsson, G., Heidenblad, M., Vult von Steyern, F., Bauer, H. C. F., … Mertens, F. (2008). Frequent deletion of the CDKN2A locus in chordoma: analysis of chromosomal imbalances using array comparative genomic hybridisation. *British Journal of Cancer, 98*(2), 434–442.

Halpern, M. E., Rhee, J., Goll, M. G., Akitake, C. M., Parsons, M., & Leach, S. D. (2008). Gal4/UAS transgenic tools and their application to zebrafish. *Zebrafish, 5*(2), 97–110.

Heerema-McKenney, A., Wijnaendts, L. C. D., Pulliam, J. F., Lopez-Terrada, D., McKenney, J. K., Zhu, S., … Linn, S. C. (2008). Diffuse myogenin expression by immunohistochemistry is an independent marker of poor survival in pediatric rhabdomyosarcoma: a tissue microarray study of 71 primary tumors including correlation with molecular phenotype. *The American Journal of Surgical Pathology, 32*(10), 1513–1522.

Helman, L. J., & Meltzer, P. (2003). Mechanisms of sarcoma development. *Nature Reviews. Cancer, 3*, 685–694.

Hernando, E., Charytonowicz, E., Dudas, M. E., Menendez, S., Matushansky, I., Mills, J., … Cordon-Cardo, C. (2007). The AKT-mTOR pathway plays a critical role in the development of leiomyosarcomas. *Nature Medicine, 13*(6), 748–753.

Hettmer, S., Li, Z., Billin, A. N., Barr, F. G., Cornelison, D. D. W., Ehrlich, A. R., … Keller, C. (2014). Rhabdomyosarcoma: current challenges and their implications for developing therapies. *Cold Spring Harbor Perspectives in Medicine, 4*(11).

Hettmer, S., Liu, J., Miller, C. M., Lindsay, M. C., Sparks, C.a., Guertin, D. A., & Wagers, A. J. (2011). Sarcomas induced in discrete subsets of prospectively isolated skeletal muscle cells. *Proceedings of the National Academy of Sciences of the United States of America, 108*(50), 20002–20007.

Ignatius, M. S., Chen, E., Elpek, N. M., Fuller, A. Z., Tenente, I. M., Clagg, R., … Langenau, D. M. (2012). In vivo imaging of tumor-propagating cells, regional tumor heterogeneity, and dynamic cell movements in embryonal rhabdomyosarcoma. *Cancer Cell, 21*(5), 680–693.

Irintchev, M., Starzinski-Powitz, A., & Wernig, A. (1994). Expression pattern of M-cadherin in normal, denervated, and regenerating mouse muscles. *Developmental Dynamics, 199*, 326–337.

Jacks, T., Remington, L., Williams, B. O., Schmitt, E. M., Halachmi, S., Bronson, R. T., & Weinberg, R. A. (1994). Tumor spectrum analysis in p53-mutant mice. *Current Biology, 4*(1), 1–7.

Jessen, J. R., Jessen, T. N., Vogel, S. S., & Lin, S. (2001). Concurrent expression of recombination activating genes 1 and 2 in zebrafish olfactory sensory neurons. *Genesis, 29*(4), 156–162.

Kang, Y., Pekmezci, M., Folpe, A. L., Ersen, A., & Horvai, A. E. (2014). Diagnostic utility of SOX10 to distinguish malignant peripheral nerve sheath tumor from synovial sarcoma, including intraneural synovial sarcoma. *Modern Pathology, 27*(1), 55–61.

Kawakami, K., Shima, a, & Kawakami, N. (2000). Identification of a functional transposase of the Tol2 element, an Ac-like element from the Japanese medaka fish, and its transposition in the zebrafish germ lineage. *Proceedings of the National Academy of Sciences of the United States of America, 97*(21), 11403–11408.

Keller, C., Arenkiel, B. R., Coffin, C. M., El-Bardeesy, N., DePinho, R.a, & Capecchi, M. R. (2004). Alveolar rhabdomyosarcomas in conditional Pax3:Fkhr mice: cooperativity of Ink4a/ARF and Trp53 loss of function. *Genes and Development, 18*(21), 2614–2626.

Kolberg, M., Holand, M., Agesen, T. H., Brekke, H. R., Liestol, K., Hall, K. S., … Lothe, R. A. (2013). Survival meta-analyses for >1800 malignant peripheral nerve sheath tumor patients with and without neurofibromatosis type 1. *Neuro-oncology, 15*(2), 135–147.

Langenau, D. M., Keefe, M. D., Storer, N. Y., Guyon, J. R., Kutok, J. L., Le, X., … Zon, L. I. (2007). Effects of RAS on the genesis of embryonal rhabdomyosarcoma. *Genes and Development, 21*(11), 1382–1395.

Langenau, D. M., Keefe, M. D., Storer, N. Y., Jette, C. A., Smith, A. C. H., Ceol, C. J., & Zon, L. I. (2008). Co-injection strategies to modify radiation sensitivity and tumor initiation in transgenic zebrafish. *Oncogene, 27*(30), 4242–4248.

Le, L. P., Nielsen, G. P., Rosenberg, A. E., Thomas, D., Batten, J. M., Deshpande, V., … Iafrate, A. J. (2011). Recurrent chromosomal copy number alterations in sporadic chordomas. *PLoS One, 6*(5), e18846.

Le, X., Langenau, D. M., Keefe, M. D., Kutok, J. L., Neuberg, D. S., & Zon, L. I. (2007). Heat shock-inducible Cre/Lox approaches to induce diverse types of tumors and hyperplasia in transgenic zebrafish. *Proceedings of the National Academy of Sciences of the United States of America, 104*(22), 9410–9415.

Le, X., Pugach, E. K., Hettmer, S., Storer, N. Y., Liu, J., Wills, A.a, … Zon, L. I. (2013). A novel chemical screening strategy in zebrafish identifies common pathways in embryogenesis and rhabdomyosarcoma development. *Development, 140*(11), 2354–2364.

Leacock, S. W., Basse, A. N., Chandler, G. L., Kirk, A. M., Rakheja, D., & Amatruda, J. F. (2012). A zebrafish transgenic model of Ewing's sarcoma reveals conserved mediators of EWS-FLI1 tumorigenesis. *Disease Models and Mechanisms, 5*(1), 95–106.

Lee, J. S., Padmanabhan, A., Shin, J., Zhu, S., Guo, F., Kanki, J. P., & Thomas Look, A. (2010). Oligodendrocyte progenitor cell numbers and migration are regulated by the zebrafish orthologs of the NF1 tumor suppressor gene. *Human Molecular Genetics, 19*(23), 4643–4653.

Legius, E., Marchuk, D. A., Collins, F. S., & Glover, T. W. (1993). Somatic deletion of the neurofibromatosis type 1 gene in a neurofibrosarcoma supports a tumour suppressor gene hypothesis. *Nature Genetics, 3*(2), 122–126.

Li, J., Yen, C., Liaw, D., Podsypanina, K., Bose, S., Wang, S. I., … Parsons, R. (1997). PTEN, a putative protein tyrosine phosphatase gene mutated in human brain, breast, and prostate cancer. *Science, 275*(5308), 1943–1947.

Lin, P. P., Pandey, M. K., Jin, F., Xiong, S., Deavers, M., Parant, J. M., & Lozano, G. (2008). EWS-FLI1 induces developmental abnormalities and accelerates sarcoma formation in a transgenic mouse model. *Cancer Research, 68*(21), 8968–8975.

Linardic, C. M., & Counter, C. M. (2008). Genetic modeling of ras-induced human rhabdomyosarcoma. *Methods in Enzymology, 438*, 419–427.

MacInnes, A. W., Amsterdam, A., Whittaker, C. A., Hopkins, N., & Lees, J. A. (2008). Loss of p53 synthesis in zebrafish tumors with ribosomal protein gene mutations. *Proceedings of the National Academy of Sciences of the United States of America, 105*(30), 10408–10413.

Maehama, T., & Dixon, J. E. (1998). The tumor suppressor, PTEN/MMAC1, dephosphorylates the lipid second messenger, phosphatidylinositol 3,4,5-trisphosphate. *Journal of Biological Chemistry, 273*(22), 13375–13378.

Marsh, D. J., Kum, J. B., Lunetta, K. L., Bennett, M. J., Gorlin, R. J., Ahmed, S. F., … Eng, C. (1999). PTEN mutation spectrum and genotype-phenotype correlations in Bannayan–Riley–Ruvalcaba syndrome suggest a single entity with Cowden syndrome. *Human Molecular Genetics, 8*(8), 1461–1472.

McMaster, M. L., Goldstein, A. M., Bromley, C. M., Ishibe, N., & Parry, D. M. (2001). Chordoma: incidence and survival patterns in the United States, 1973–1995. *Cancer Causes and Control: CCC, 12*(1), 1–11.

Merino, D., & Malkin, D. (2014). p53 and hereditary cancer. *Sub-cellular Biochemistry, 85*, 1–16.

Miller, S. J., Rangwala, F., Williams, J., Ackerman, P., Kong, S., Jegga, A. G., … Ratner, N. (2006). Large-scale molecular comparison of human Schwann cells to malignant peripheral nerve sheath tumor cell lines and tissues. *Cancer Research, 66*(5), 2584–2591.

Myers, M. P., Pass, I., Batty, I. H., Van der Kaay, J., Stolarov, J. P., Hemmings, B. A., … Tonks, N. K. (1998). The lipid phosphatase activity of PTEN is critical for its tumor supressor function. *Proceedings of the National Academy of Sciences of the United States of America, 95*(23), 13513–13518.

Naka, T., Boltze, C., Kuester, D., Schulz, T. O., Schneider-Stock, R., Kellner, A., … Roessner, A. (2005). Alterations of G1-S checkpoint in chordoma: the prognostic impact of p53 overexpression. *Cancer, 104*(6), 1255–1263.

Padmanabhan, A., Lee, J.-S., Ismat, F. A., Lu, M. M., Lawson, N. D., Kanki, J. P., … Epstein, J. A. (2009). Cardiac and vascular functions of the zebrafish orthologues of the type I neurofibromatosis gene NFI. *Proceedings of the National Academy of Sciences of the United States of America, 106*(52), 22305–22310.

Paulson, V., Chandler, G., Rakheja, D., Galindo, R. L., Wilson, K., Amatruda, J. F., & Cameron, S. (2011). High-resolution array CGH identifies common mechanisms that drive embryonal rhabdomyosarcoma pathogenesis. *Genes Chromosomes and Cancer, 50*(6), 397–408.

Podsypanina, K., Ellenson, L. H., Nemes, A., Gu, J., Tamura, M., Yamada, K. M., … Parsons, R. (1999). Mutation of Pten/Mmac1 in mice causes neoplasia in multiple organ systems. *Proceedings of the National Academy of Sciences of the United States of America, 96*(4), 1563–1568.

Pownall, M. E., Gustafsson, M. K., & Emerson, C. P. (2002). Myogenic regulatory factors and the specification of muscle progenitors in vertebrate embryos. *Annual Review of Cell and Developmental Biology, 18*, 747–783.

Rajwanshi, A., Srinivas, R., & Upasana, G. (2009). Malignant small round cell tumors. *Journal of Cytology/Indian Academy of Cytologists, 26*(1), 1–10.

Riggi, N., Cironi, L., Provero, P., Suvà, M.-L., Kaloulis, K., Garcia-Echeverria, C., & Stamenkovic, I. (2005). Development of Ewing's sarcoma from primary bone marrow-derived mesenchymal progenitor cells. *Cancer Research, 65*(24), 11459–11468.

Santoriello, C., Deflorian, G., Pezzimenti, F., Kawakami, K., Lanfrancone, L., d'Adda di Fagagna, F., & Mione, M. (2009). Expression of H-RASV12 in a zebrafish model of Costello syndrome causes cellular senescence in adult proliferating cells. *Disease Models and Mechanisms, 2*(1–2), 56–67.

Shern, J. F., Chen, L., Chmielecki, J., Wei, J. S., Patidar, R., Rosenberg, M., ... Khan, J. (2014). Comprehensive genomic analysis of rhabdomyosarcoma reveals a landscape of alterations affecting a common genetic axis in fusion-positive and fusion-negative tumors. *Cancer Discovery, 4*(2), 216–231.

Shin, J., Padmanabhan, A., de Groh, E. D., Lee, J.-S., Haidar, S., Dahlberg, S., ... Look, A. T. (2012). Zebrafish neurofibromatosis type 1 genes have redundant functions in tumorigenesis and embryonic development. *Disease Models and Mechanisms, 5*(6), 881–894.

Storer, N. Y., White, R. M., Uong, A., Price, E., Nielsen, G. P., Langenau, D. M., & Zon, L. I. (2013). Zebrafish rhabdomyosarcoma reflects the developmental stage of oncogene expression during myogenesis. *Development, 140*(14), 3040–3050.

Suzuki, A., de la Pompa, J. L., Stambolic, V., Elia, A. J., Sasaki, T., del Barco Barrantes, I., ... Mak, T. W. (1998). High cancer susceptibility and embryonic lethality associated with mutation of the PTEN tumor suppressor gene in mice. *Current Biology, 8*(21), 1169–1178.

Szuhai, K., De Jong, D., Leung, W. Y., Fletcher, C. D. M., & Hogendoorn, P. C. W. (2014). Transactivating mutation of the MYOD1 gene is a frequent event in adult spindle cell rhabdomyosarcoma. *Journal of Pathology, 232*(3), 300–307.

Thway, K., & Fisher, C. (2014). Malignant peripheral nerve sheath tumor: pathology and genetics. *Annals of Diagnostic Pathology, 18*(2), 109–116.

Torchia, E. C., Boyd, K., Rehg, J. E., Qu, C., & Baker, S. J. (2007). EWS/FLI-1 induces rapid onset of myeloid/erythroid leukemia in mice. *Molecular and Cellular Biology, 27*(22), 7918–7934.

Tsumura, H., Yoshida, T., Saito, H., Imanaka-Yoshida, K., & Suzuki, N. (2006). Cooperation of oncogenic K-ras and p53 deficiency in pleomorphic rhabdomyosarcoma development in adult mice. *Oncogene, 25*, 7673–7679.

Walcott, B. P., Nahed, B. V., Mohyeldin, A., Coumans, J.-V., Kahle, K. T., & Ferreira, M. J. (2012). Chordoma: current concepts, management, and future directions. *Lancet Oncology, 13*(2), e69–e76.

Wang, Y., Marino-Enriquez, A., Bennett, R. R., Zhu, M., Shen, Y., Eilers, G., ... Fletcher, J. A. (2014). Dystrophin is a tumor suppressor in human cancers with myogenic programs. *Nature Genetics, 46*(6), 601–606.

Wienholds, E., Schulte-Merker, S., Walderich, B., & Plasterk, R. H. (2002). Target-selected inactivation of the zebrafish rag1 gene. *Science, 297*(5578), 99–102.

Yang, X. R., Ng, D., Alcorta, D. A., Liebsch, N. J., Sheridan, E., Li, S., … Kelley, M. J. (2009). T (brachyury) gene duplication confers major susceptibility to familial chordoma. *Nature Genetics, 41*(11), 1176–1178.

Zhang, G., Hoersch, S., Amsterdam, A., Whittaker, C. A., Lees, J. A., & Hopkins, N. (2010). Highly aneuploid zebrafish malignant peripheral nerve sheath tumors have genetic alterations similar to human cancers. *Proceedings of the National Academy of Sciences of the United States of America, 107*(39), 16940–16945.

Zhang, G. J., Hoersch, S., Amsterdam, A., Whittaker, C. A., Beert, E., Catchen, J. M., … Lees, J. A. (2013). Comparative oncogenomic analysis of copy number alterations in human and zebrafish tumors enables cancer driver discovery. *PLoS Genetics, 9*(8), e1003734.

CHAPTER

Zebrafish models of leukemia

20

S. He, C.-B. Jing, A.T. Look[1]
Harvard Medical School, Boston, MA, United States
[1]*Corresponding author: E-mail: thomas_look@dfci.harvard.edu*

CHAPTER OUTLINE

Methods in Cell Biology, Volume 138, ISSN 0091-679X, http://dx.doi.org/10.1016/bs.mcb.2016.11.013

Abstract

The zebrafish, *Danio rerio*, is a well-established, invaluable model system for the study of human cancers. The genetic pathways that drive oncogenesis are highly conserved between zebrafish and humans, and multiple unique attributes of the zebrafish make it a tractable tool for analyzing the underlying cellular processes that give rise to human disease. In particular, the high conservation between human and zebrafish hematopoiesis (Jing & Zon, 2011) has stimulated the development of zebrafish models for human hematopoietic malignancies to elucidate molecular pathogenesis and to expedite the preclinical investigation of novel therapies. While T-cell acute lymphoblastic leukemia was the first transgenic cancer model in zebrafish (Langenau et al., 2003), a wide spectrum of zebrafish models of human hematopoietic malignancies has been established since 2003, largely through transgenesis and genome-editing approaches. This chapter presents key examples that validate the zebrafish as an indispensable model system for the study of hematopoietic malignancies and highlights new models that demonstrate recent advances in the field.

1. T-CELL ACUTE LYMPHOBLASTIC LEUKEMIA

T-cell acute lymphoblastic leukemia (T-ALL) is a high-risk subtype of hematologic malignancy, characterized by infiltration of the bone marrow with immature T-cell progenitors (Durinck et al., 2015; Palomero & Ferrando, 2009). T-ALL accounts for 15% of acute lymphoblastic leukemia (ALL) cases in pediatric patients and 25% of ALL in adults (Durinck et al., 2015). Patients with T-ALL commonly present with large tumor burdens at diagnosis associated with high numbers of circulating leukemic cells in peripheral blood, mediastinal masses, and frequent infiltration of the central nervous system (Palomero & Ferrando, 2009). Although outcomes for these patients have improved, T-ALL remains fatal in 25% of children and over 50% of adults (Roti & Stegmaier, 2014).

T-ALL pathogenesis is a multistep process involving different genetic and epigenetic alterations. At the molecular level, T-ALL can be classified into different subgroups based on both gene expression signatures and aberrant activation of specific oncogenic transcription factors, such as TLX1(HOX11), TLX3(HOX11L2), TAL1, LYL1, LMO1, LMO2, NKX2.1, and HOXA (Ferrando et al., 2002; Van Vlierberghe & Ferrando, 2012). In addition, mutations or rearrangements leading to constitutive activation of NOTCH1 have been detected in more than 60% of T-ALL cases, emphasizing the fundamental role of the NOTCH1 pathway and its downstream target gene, the transcription factor MYC, in this malignancy (Herranz et al., 2014; Van Vlierberghe & Ferrando, 2012; Weng et al., 2004). Additional pathways, including IL7R/JAK/STAT, PI3K/AKT, and RAS/MAPK, have also been implicated in T-ALL pathogenesis (Durinck et al., 2015). Moreover, an early T-cell precursor subtype of T-ALL (ETP-ALL) has been defined based on a unique immunophenotype that represents the transformation of early immigrants into the thymus, corresponding to double-negative thymocytes that often lack T-cell receptor rearrangements. These immature T-cells or early T-cell precursors have distinct genetic alterations and a clinically poor outcome (Durinck et al., 2015; Zhang et al., 2012). Below, we describe the contributions of zebrafish T-ALL models to our understanding of this multifaceted disease.

1.1 TRANSGENIC ZEBRAFISH MODELS OF T-CELL ACUTE LYMPHOBLASTIC LEUKEMIA

1.1.1 MYC-induced zebrafish T-cell acute lymphoblastic leukemia

1.1.1.1 T-cell acute lymphoblastic leukemia by mosaic and stable transgenic expression of Myc

The oncogenic transcription factor MYC, also known as c-MYC, is a key regulator of cell growth, proliferation, metabolism, and survival (Kress, Sabo, & Amati, 2015; McKeown & Bradner, 2014). *MYC* is the most frequently amplified oncogene in human cancers (McKeown & Bradner, 2014) and is overexpressed in most human cases of T-ALL, often downstream of activated NOTCH1 (Gutierrez et al., 2011). The first zebrafish T-ALL model was established in our laboratory by expressing mouse *c-Myc* (*Myc*), or the chimeric enhanced green fluorescent protein (*EGFP*)-*Myc* fusion gene, driven by the lymphocyte-specific *rag2* promoter (Langenau et al., 2003). In this system, the EGFP-labeled lymphocytes were tracked by fluorescence microscopy to detect neoplastic thymic expansion and infiltration into surrounding skeletal muscle and other organs, a well-known characteristic of human leukemia (Fig. 1). Widespread leukemia developed in *Tg*(*rag2:EGFP-Myc*) fish with both mosaic transgene expression (F0) and stable germline transmission (F1), with mean latencies of 52 and 32 days, and leukemia penetrance of 6% and 100%, respectively (Gutierrez, Feng, et al., 2014; Langenau et al., 2003). In this model system, Myc-induced T-ALL can be accelerated by thymic overexpression

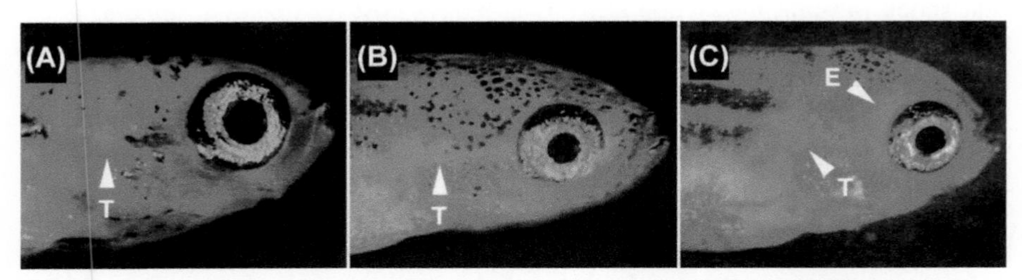

FIGURE 1 Myc overexpression induced T-cell acute lymphoblastic leukemia in zebrafish.

Enhanced green fluorescent protein–positive leukemia progressed from (A) thymic enlargement, to (B) local infiltration of leukemic cells outside the thymus, to (C) extensive infiltration of the head in the *Tg*(*rag2:EGFP-Myc*) zebrafish. *E*, eye; *T*, thymic masses. *Arrowheads* indicate accumulations of leukemic cells.

Modified from Langenau, D. M., Traver, D., Ferrando, A. A., Kutok, J. L., Aster, J. C., Kanki, J. P., ... Look, A. T. (2003). Myc-induced T cell leukemia in transgenic zebrafish. Science, 299*(5608), 887–890. doi:10.1126/science.1080280.*

of either constitutively active Akt2 (Gutierrez et al., 2011) or the intracellular domain of NOTCH1 (Blackburn et al., 2012), but not by overexpression of bcl-2 (Feng et al., 2010) or loss-of-function mutations in *pten* (Gutierrez et al., 2011), *tp53* (Gutierrez, Feng, et al., 2014), or *bim* (Reynolds et al., 2014). Gene expression profiles from transgenic zebrafish revealed that the leukemic cells arise from T-lymphocyte precursors that express zebrafish orthologs of the human T-ALL oncogenes *TAL1/SCL* and *LMO2* (Langenau et al., 2005). MYC-induced zebrafish T-ALL cells are transplantable and can be propagated in irradiated wild-type fish (Langenau et al., 2003), clonal syngeneic strains (Blackburn, Liu, & Langenau, 2011), or immunocompromised *rag2*^{E450fs} mutant fish (Tang et al., 2014). This rapidly developing Myc-induced model of T-ALL therefore established the zebrafish as a promising platform for the preclinical investigation of new therapies for T-ALL.

1.1.1.2 T-cell acute lymphoblastic leukemia by conditional transgenic expression of MYC

1.1.1.2.1 Cre/Lox-mediated system

Due to the rapid development of leukemia in *Tg*(*rag2:EGFP-Myc*) fish, most animals die prior to reaching reproductive maturity thereby necessitating the propagation of this line through in vitro fertilization (Langenau et al., 2003). To circumvent this problem, we developed a Cre/Lox-mediated system to express *EGFP-Myc* conditionally upon excision of an upstream *LoxP-dsRED2-LoxP* cassette by injection of *Cre* RNA (Langenau et al., 2005). *Cre*-injection into the *Tg*(*rag2:loxP-dsRED2-loxP-EGFP-Myc*) [hereafter referred to as "*Tg*(*rag2: LDL-EMyc*)"] fish led to the development of T-ALL with a mean latency of 151 days and 6.5% tumor penetrance (Langenau et al., 2005). Cre-mediated recombination of the *rag2:LDL-EMyc* transgene can also be achieved by heat shock in *Tg*(*hsp70:Cre*; *rag2:LDL-EMyc*) fish (Feng et al., 2007). After optimal heat shock (45-min incubation at 37°C at 3 days post fertilization, dpf), 81% of the compound transgenic fish developed T-lymphoblastic lymphoma (T-LBL) with a mean latency of 120 days, which rapidly progressed to T-ALL (Feng et al., 2007). These modifications therefore helped streamline the potential use of this model system for drug discovery.

1.1.1.2.2 Tamoxifen-inducible system

The *Tg*(*rag2:MYC-ER*) zebrafish line was established by Gutierrez et al. (2011) to provide a system by which to analyze the dependence of MYC-induced T-ALL on different oncogenic pathways. In this model system, the zebrafish *rag2* promoter drives expression of a chimeric fusion between human MYC and the ligand-binding domain of the human estrogen receptor that has been modified such that MYC can be posttranslationally activated by 4-hydroxytamoxifen (4HT) but not by endogenous estrogens. In this system, addition of 4HT into fish water led to the activation of MYC and fully penetrant T-ALL. Tumors were found to be highly dependent on transgenic MYC activity, as 75% of zebrafish with leukemia experienced complete tumor regression upon withdrawal of 4HT (Gutierrez et al., 2011). Using this inducible model, Gutierrez et al. (2011) also revealed a prominent role for PI3K/AKT signaling in MYC-induced T-ALL, as loss of *pten* or gain of constitutively active *Akt2* was sufficient for tumor maintenance independent of sustained MYC activity.

1.1.1.3 Progression of T-lymphoblastic lymphoma to T-cell acute lymphoblastic leukemia

T-LBL and T-ALL represent different clinical presentations of the same early T-lymphoblastic malignancy arising from developing thymocytes. T-LBL cases generally present with a localized mediastinal mass and have either undetectable involvement of the bone marrow or may have involvement that comprises less than 25% of the bone marrow cells. T-ALL is diagnosed when there is greater than 25% involvement of the bone marrow with T lymphoblasts, whether or not a mediastinal mass is evident. Using the *Tg(hsp70:Cre*; *rag2:LDL-EMyc*)-induced T-ALL model described above (see Section 1.1.1.2.1), Feng et al. (2010) analyzed the function of the antiapoptotic protein bcl-2 in the progression of T-LBL to T-ALL and found that while overexpression of bcl-2 significantly accelerated the onset of MYC-induced T-LBL, it profoundly inhibited the progression of T-LBL to T-ALL. Interestingly, overexpression of bcl-2 was found to impair the ability of T-LBL cells to disseminate into the vascular system from the thymus thereby causing the defective T-LBL cells to initiate autophagy. Overexpression of the constitutively activated Akt2 was able to overcome the barrier to intravasation and promote T-LBL dissemination and progression to T-ALL. This finding was reproduced in clinical patient samples and demonstrated the value of zebrafish models to elucidate mechanisms underlying human malignancies and implicate novel pathways for targeted therapy.

1.1.1.4 Survival signaling in MYC-induced zebrafish T-cell acute lymphoblastic leukemia

1.1.1.4.1 pten

Inactivation of the *PTEN* tumor suppressor gene by mutations or deletions leads to constitutive signaling through the PI3K/AKT cascade. These genetic aberrations are found in about 10% of adult T-ALL (Grossmann et al., 2013; Trinquand et al., 2013) and 20–40% of pediatric T-ALL patients (Gutierrez et al., 2009; Jenkinson et al., 2016; Zuurbier et al., 2012) and are often associated with early treatment failure and poor prognosis (Gutierrez et al., 2009; Piovan et al., 2013). To analyze the consequence of *PTEN* loss in the survival and maintenance of T-ALL in vivo, Gutierrez et al. (2011) introduced loss-of-function *pten* mutations (Faucherre, Taylor, Overvoorde, Dixon, & den Hertog, 2007) into the *Tg(rag2:MYC-ER*) zebrafish T-ALL model described above in Section 1.1.1.2.2. In *pten* wild-type fish, the maintenance of MYC-induced T-ALL is dependent upon the constitutive activity of MYC and concomitant inactivation of MYC-induced mitochondrial apoptosis. Gutierrez et al. genetically separated these two events by showing that loss of *pten* mediated the survival and progression of the leukemia cells after withdrawal of 4HT and cessation of MYC activity. They also showed that the survival of these cells was mediated by increased activation of downstream

Akt signaling (Gutierrez et al., 2011). This study established an important role of PTEN and PI3K/AKT signaling in the maintenance of MYC-induced T-ALL and suggested PI3K/AKT inhibition as a new therapeutic strategy to overcome drug resistance in T-ALL.

1.1.1.4.2 *bim*

Inhibition of mitochondrial apoptosis is well known to promote chemoresistance in human cancer (Chonghaile et al., 2011; Deng et al., 2007; Vo et al., 2012). Apoptosis plays a critical role during the development of lymphocytes, and normal T-cell progenitors are primed to undergo apoptosis through both the extrinsic death receptor–mediated and intrinsic mitochondrial pathways (Ryan, Brunelle, & Letai, 2010). The BH3-only protein BIM is a proapoptotic factor of the BCL-2 family that functions in the mitochondrial pathway as a key trigger of apoptosis in thymocytes (Bouillet et al., 1999). An in vivo study using the *Tg(rag2:MYC-ER)* zebrafish line revealed that BIM is likely to be inactivated or downregulated in MYC-induced T-ALL, as loss of *bim* resulted in T-ALL persistence upon MYC inactivation (Reynolds et al., 2014). It was further discovered that repression of *bim* by MYC and AKT mediates survival of tumor cells in this model system. Importantly, dual inhibition of MYC and PI3K/AKT signaling with chemical inhibitors led to an increase in *BIM* and mitochondrial apoptosis in human treatment-resistant T-ALL cells thus representing a promising new therapeutic opportunity for T-ALL patients (Reynolds et al., 2014).

1.1.1.4.3 *tp53*

The TP53 transcription factor is a key tumor suppressor that can initiate apoptosis, cell cycle arrest, and senescence, and its pathway has been found to be inactivated in most human cancers. In human T-ALL, the TP53 pathway is often disabled by deletions in *ARF*, with subsequent acquisition of mutations or deletions in *TP53* in relapse patients (Hof et al., 2011; Zhang et al., 2012). In murine T-ALL models, loss of *Tp53* or *Arf* leads to accelerated tumor onset (Hao et al., 2016; Treanor et al., 2011; Volanakis, Williams, & Sherr, 2009). There are no *ARF* orthologs in the genomes of highly regenerative vertebrates, including zebrafish (Gutierrez, Feng, et al., 2014; Hesse, Kouklis, Ahituv, & Pomerantz, 2015), and loss of *tp53* has no significant impact on the onset of MYC-induced T-ALL. However, wild-type *tp53* has been shown to mediate radiation-induced T-ALL regression, indicating that the *tp53*-dependent DNA damage response is intact in *Myc*-induced zebrafish T-ALL (Gutierrez, Feng, et al., 2014) and that the zebrafish is well suited to study the ARF-independent functions of TP53 in this disease.

1.1.1.4.4 *dlst*

Dihydrolipoamide s-succinyltransferase (DLST) is a component of the α-ketoglutarate (α-KG) dehydrogenase complex, which catalyzes the conversion of α-KG to succinyl-CoA for energy production and macromolecule synthesis in the tricarboxylic acid (TCA) cycle. Using the *Tg(rag2:EGFP-Myc)* zebrafish line, MYC-driven T-ALL cells were found to be metabolically dependent on the TCA cycle since heterozygous loss of *dlst* significantly delayed the onset of MYC-induced T-ALL and impaired the growth and proliferation of leukemia cells (Anderson et al., 2016). This finding was further confirmed in human T-ALL cells and reveals the metabolic dependence of MYC-driven T-ALL and the important role of DLST in leukemia cell growth and survival (Anderson et al., 2016).

1.1.2 NOTCH1-induced zebrafish T-cell acute lymphoblastic leukemia

NOTCH signaling is required during multiple stages of T-cell development and differentiation (Grabher, von Boehmer, & Look, 2006). Aberrant NOTCH signaling has been detected in over 60% of T-ALL through various mechanisms. Gain-of-function mutations in *NOTCH1* include the t(7; 9)(q34; q34.3) chromosomal translocation that fuses the intracellular domain of NOTCH1 (ICN1) to the promoter of the T-Cell Receptor Beta (*TCRB*) gene, as well as mutations in the heterodimerization (HD) and proline, glutamic acid, serine, threonine-rich (PEST) domains. Loss-of-function mutations or deletions are also found in FBXW7, a negative regulator of the NOTCH pathway (Aster, Blacklow, & Pear, 2011; Palomero & Ferrando, 2009; Weng et al., 2004). In mice, a strong gain-of-function mutant form defined by the human *ICN1* can efficiently initiate T-ALL, whereas the weaker but typical cancer-associated *NOTCH1* mutations in the HD and PEST domains are insufficient to induce T-ALL on their own (Chiang et al., 2008). In zebrafish, thymic overexpression of *ICN1* under the zebrafish *rag2* promoter led to T-ALL development in approximately 40% of mosaic *Tg(rag2:ICN1-EGFP)* F0 fish at 3−5 months of age (Blackburn et al., 2012; Chen et al., 2007). These zebrafish T-ALL cells are transplantable and express T-cell lineage−specific markers and NOTCH1 targets such as *her6* and *her9*. Notably, the *myca* and *mycb* genes were not upregulated in ICN1-transformed zebrafish T-ALL likely due to their lack of upstream NOTCH1-driven enhancers (Herranz et al., 2014). In the stable *Tg(rag2-ICN1-EGFP)* F1 fish, the longer latency (starting at 11 months of age) and similar tumor penetrance (40%) were likely due to the low expression level of the ICN1-EGFP fusion in the stable line (Chen et al., 2007). Interestingly, while bcl-2 suppressed MYC-induced zebrafish T-ALL (Feng et al., 2010), its overexpression largely accelerated leukemia onset and increased tumor penetrance to 100% in the *Tg(rag2-ICN1-EGFP)* line (Chen et al., 2007). This model could therefore be used to identify other genes that potentially cooperate with NOTCH1 to induce T-ALL in humans.

1.1.3 Akt2-induced zebrafish T-cell acute lymphoblastic leukemia

The PI3K/AKT signaling pathway, which is negatively regulated by the PTEN tumor suppressor, can be activated by a variety of extracellular signaling molecules, such as cytokines, growth factors, and ligands of G protein−coupled receptors, and is required for cell growth, survival, and oncogenic transformation in a variety of human cancers (Gutierrez et al., 2011; Martini, De Santis, Braccini, Gulluni, & Hirsch, 2014). In human T-ALL, genetic lesions leading to activation of PI3K/AKT or inactivation of PTEN are frequently identified in primary samples and are often associated with MYC overexpression (Gutierrez et al., 2009, 2011). In zebrafish, mosaic thymic expression of a myristoylated constitutively active mouse Akt2 was introduced by zygotic injection of the *Tg(rag2:myr-Akt2)* construct. By this method, expression of MYR-Akt2 induced T-ALL in 17% of injected fish by

20 weeks of age, indicating that constitutive Akt2 activation is sufficient to induce T-ALL (Gutierrez et al., 2011). Moreover, MYR-Akt2 was found to accelerate the onset of MYC-induced T-ALL in the *Tg(rag2:MYC-ER)* line (Blackburn et al., 2014; Gutierrez et al., 2011) and to promote T-ALL growth and survival after downregulation of MYC by 4HT withdrawal. Thus, activation of the AKT pathway was found to circumvent MYC dependence using this model of T-ALL.

1.2 ZEBRAFISH MUTANTS WITH HIGH SUSCEPTIBILITY TO T-LYMPHOBLASTIC LYMPHOMA AND T-CELL ACUTE LYMPHOBLASTIC LEUKEMIA

Zebrafish provide an ideal vertebrate experimental system to perform forward genetic screens, because of their small size, high numbers of offspring per breeding pair, and transparency. Forward genetic screens in zebrafish are one approach to the discovery of cancer-related that may prove relevant in the human disease (Stern & John, 2003). Using the alkylating mutagen N-ethyl-N-nitrosourea (ENU)-mediated mutagenesis, a forward genetic screen was performed in the *lck:GFP* transgenic zebrafish which stably express T-cell-specific GFP fluorescence, in search of T-cell malignancy-prone mutants based on abnormal GFP expression phenotypes (Frazer et al., 2009). Three mutant lines, *hlk, srk*, and *otg*, were identified from this screen, which develop transplantable T-ALL malignancies that phenotypically and histologically resemble oncogene-induced leukemia (Frazer et al., 2009; Rudner et al., 2011).

1.3 GENETIC CONSERVATION BETWEEN ZEBRAFISH AND HUMAN T-CELL ACUTE LYMPHOBLASTIC LEUKEMIA

To detect acquired copy number aberrations (CNAs) in zebrafish T-ALLs, array comparative genomic hybridization (aCGH) was applied to a set of 17 zebrafish T-ALL samples, including Myc-induced T-ALL as well as the spontaneous T-ALL developed in the *hlk, srk*, and *otg* mutant lines (Rudner et al., 2011). Resulted zebrafish T-ALL CNAs were compared with human T-ALL CNAs obtained from 61 patients. The significant overlap between the two CNA datasets suggested the genetic conservation in T-ALL transformation across species.

The genetic conservation between zebrafish and human T-ALL was also examined at transcriptome level. Microarray gene expression profiling and cross-species comparisons were performed among zebrafish, mouse and human T-ALL (Blackburn et al., 2012). Gene Set Enrichment Analysis (GESA) showed that Myc- or NOTCH-induced zebrafish T-ALLs were significantly associated with murine T-ALLs and human T-ALL, but not precursor B-cell acute lymphoblastic leukemia (B-ALL) of human patients. Thus, common molecular pathways involved in T-ALL pathogenesis are highly conserved across species, which validated zebrafish as a valuable model organism to study human T-ALL at the transcriptome level.

1.4 LEUKEMIA-PROPAGATING CELLS

Tumor-cell self-renewal is a defining feature of cancer, and it can be measured by limiting-dilution cell transplantation into immune-matched or immunocompromised animals (Borah, Raveendran, Rochani, Maekawa, & Kumar, 2015). Previous studies showed that zebrafish T-ALLs driven by MYC or NOTCH1 are transplantable (Chen et al., 2007; Langenau et al., 2003), but the existing transplantation protocols required radiation-induced immune suppression of nonimmune-matched recipients. This technical limitation precluded the analysis of the T-ALL cells for self-renewal. The Langenau lab overcame this limitation by establishing MYC- and/or NOTCH1-induced T-ALL using clonal syngeneic zebrafish strains (Smith et al., 2010). Using this system, Smith et al. (2010) showed that while tumor-initiating cells are abundant in zebrafish T-ALL, they have wide differences in tumor-initiating potential. To aid in tumor cell transplantation as well as imaging of tumor cell heterogeneity, the Langenau lab also developed immunocompromised zebrafish strains (Tang et al., 2014, 2016). Combined with multicolor fluorescent reporters, the syngeneic and immunocompromised zebrafish lines streamline large-scale leukemia transplantation and analysis of leukemia-propagating cells (LPCs) in transparent zebrafish. Using these models, the Langenau lab showed that NOTCH1 promotes the expansion of premalignant thymocytes in an MYC-independent manner but does not alter the frequency of LPCs (Blackburn et al., 2012). They also found through a leukemia transplantation screen that Akt signaling enhances T-ALL growth rate and LPC frequency (Blackburn et al., 2014). Furthermore, constitutive activation of Akt established chemoresistance, suggesting that AKT inhibition can be combined with dexamethasone for the treatment of refractory T-ALL (Blackburn et al., 2014). These syngeneic and immunocompromised zebrafish lines represent valuable tools to analyze self-renewal, relapse, and LPCs in T-ALL.

1.5 XENOTRANSPLANTATION OF HUMAN T-CELL ACUTE LYMPHOBLASTIC LEUKEMIA

Xenotransplantation is performed by transplanting human or mouse tumor cells into transparent zebrafish embryos and juveniles allowing for high-resolution, subcellular imaging and in vivo analysis of antitumor drugs (Corkery, Dellaire, & Berman, 2011; Drabsch, He, Zhang, Snaar-Jagalska, & ten Dijke, 2013; He et al., 2012). In a recent study, fluorescently labeled human T-ALL cell lines and patient samples were transplanted into the yolk sac of zebrafish embryos at 2 dpf (Bentley et al., 2015). Four days post transplantation, small molecule drugs were administered to recipient embryos for 2 days followed by enzymatic dissociation of embryos into a single-cell suspension. Proliferation of viable fluorescent cells in the suspension was then quantified in response to different drugs (Corkery et al., 2011). This study showed that primary human T-ALL samples can be successfully engrafted into zebrafish embryos and that response to different drugs in vivo was similar to that observed in vitro. This represents a novel preclinical in vivo platform for the development and evaluation of personalized therapy for T-ALL.

1.6 DRUG DISCOVERY

As a model organism that enables large-scale in vivo phenotype-based analyses, the zebrafish offers unique opportunities for drug discovery (MacRae & Peterson, 2015; Zon & Peterson, 2005). Successful examples include the discovery of prostaglandin E2 for improving hematopoietic stem cell transplantation (North et al., 2007) and leflunomide for suppressing melanoma (White et al., 2011). Indeed, studies over the last two decades have demonstrated that zebrafish leukemia/T-ALL models are highly translational with respect to human disease as they are strikingly similar to human leukemia at the genetic, molecular, and cellular level (Rasighaemi, Basheer, Liongue, & Ward, 2015). Based on the similarities between normally developing thymocytes and transformed thymocytes, Ridges et al. sought to identify potential antileukemic compounds by analyzing the response of normally developing fluorescent thymocytes in the *Tg(lck:EGFP)* line to 26,400 molecules in the ChemBridge DIVERSet Library (Ridges et al., 2012). In this study, five-dpf *Tg(lck:EGFP)* embryos were arrayed in 96-well plates, treated for two days with 10 μM drug and then analyzed for thymic fluorescence. The antileukemic activity of Lenaldekar, a promising hit from the screen, was validated using human T-ALL cell lines, zebrafish with Myc-induced T-ALL, murine T-ALL xenografts, and primary samples from patients with Ph+ B-ALL or Ph+ chronic myelogenous leukemia (CML).

Gutierrez et al. developed a fluorescence-based screen using the *Tg(rag2:MYC-ER)* zebrafish line to identify small molecules with selective activity against MYC-overexpressing preleukemic thymocytes (Gutierrez, Pan, et al., 2014). Three-dpf *Tg(rag2:MYC-ER; rag2:dsRed)* embryos were treated with different drugs at 12.5 μM and analyzed for thymic fluorescence after four days of treatment at 7 dpf (Fig. 2). Perphenazine (PPZ), an FDA-approved phenothiazine antipsychotic, was identified from this screen, and its selective toxicity against MYC-overexpressing thymocytes was confirmed in a complementary screen using human T-ALL cell lines. Surprisingly, the antileukemic activity of PPZ was not due to its well-known ability to inhibit dopamine and calmodulin signaling. To discover the molecular target of PPZ, Gutierrez et al. performed affinity chromatography coupled with mass spectrometry and found that PPZ binds to the Aα subunit of protein phosphatase 2A (PP2A) and causes activation of this tumor suppressor complex. The discovery of this novel antitumor mechanism suggested the therapeutic potential of PP2A activation in T-ALL and other cancers with aberrant phosphorylation of PP2A targets.

2. B-CELL ACUTE LYMPHOBLASTIC LEUKEMIA

B-ALL, a hematologic malignancy derived from immature B-cell precursors, is the most prevalent childhood leukemia and the leading cause of childhood cancer-related deaths (Woo, Alberti, & Tirado, 2014). B-ALL can be divided into several

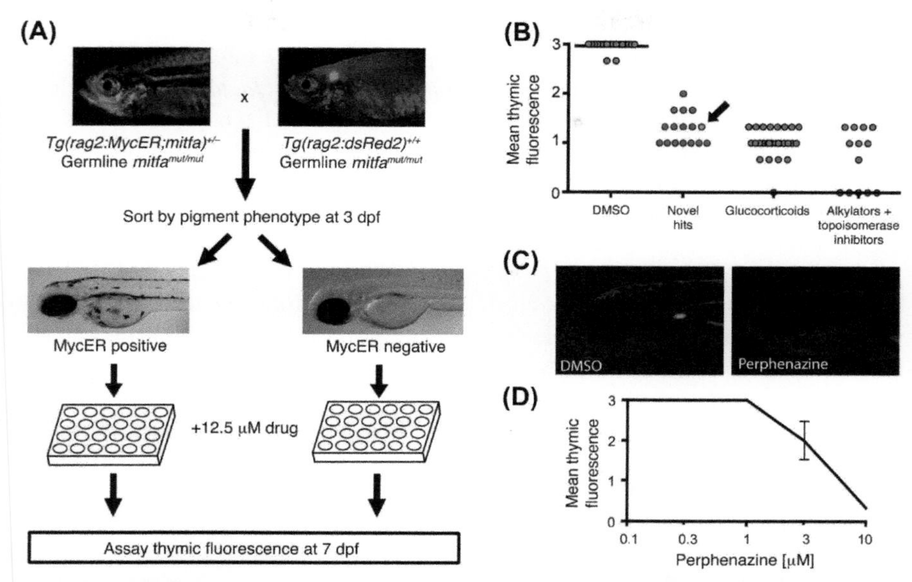

FIGURE 2 Zebrafish screen for small molecules that are toxic to MYC-overexpressing thymocytes.

(A) Primary screen design. (B) Hits from the primary screen. *Arrow* denotes the result obtained with PPZ. (C). Representative images of DMSO (control) or PPZ-treated zebrafish larvae. (D) Dose-response curve from secondary screen of PPZ, with six zebrafish larvae treated per concentration. Drug doses higher than 10 μM induced general toxicity (not shown). Error bars = SD.

Modified from Gutierrez, A., Pan, L., Groen, R. W. J., Baleydier, F., Kentsis, A., Marineau, J., ... Aster, J. C. (2014b). Phenothiazines induce PP2A-mediated apoptosis in T cell acute lymphoblastic leukemia. The Journal of Clinical Investigation, 124*(2), 644–655. doi:10.1172/jci65093.*

subtypes, including pro-B (or early pre-B), pre-B, common and mature B-cell ALL (also known as Burkitt leukemia). In zebrafish, in contrast to the early development of T-cells, B-cells arise at three weeks of age (Page et al., 2013). Different subtypes of zebrafish B-cells, including pro-B, pre-B, and immature/mature-B-cells, can be identified in the adult kidney of the *Tg*(*rag2:DsRed; IgM1:eGFP*) dual fluorescent reporter fish (Page et al., 2013). Unlike T-ALL, however, modeling B-ALL in zebrafish remains challenging to date. For unknown reasons, although the lymphocyte-specific *rag2* gene is expressed in both B- and T-cell lineages, and both lymphoid lineages are labeled by *rag2*-driven fluorescent reporters (Langenau et al., 2004; Moore et al., 2016; Page et al., 2013), only T-ALL results from *rag2*-driven oncogene overexpression (Chen et al., 2007; Feng et al., 2010; Gutierrez et al., 2011; Langenau et al., 2003, 2005; Moore et al., 2016). Indeed, *Tg*(*rag2:EGFP-Myc*) mimics human T-ALL but not pre-B-ALL (Blackburn et al., 2012; Moore et al., 2016), despite the observation that MYC is overexpressed and activated in human Burkitt lymphoma and B-ALL (Da Costa et al., 2013; O'Neil & Look, 2007).

To date, the only successful generation of a zebrafish B-ALL model was achieved by ubiquitous expression of a human *TEL-AML1* (also known as *ETV6-RUNX1*) fusion gene driven by the *Xenopus elongation factor 1α* promoter or the zebrafish *β-actin* promoter (Sabaawy et al., 2006). Each promoter induced lymphoblastic leukemia in approximately 3% of transgenic zebrafish with a latency of 8–12 months. These leukemias were transplantable, negative for TCR-α and IgM, and showed molecular features of CD10+ pre-B-ALL (Sabaawy et al., 2006). Notably, while these leukemias also expressed *rag2*, leukemia was not observed in transgenic zebrafish expressing the same TEL-AML1 fusion driven by the *rag2* promoter (Sabaawy et al., 2006). It therefore remains unclear as to why the *rag2* promoter is insufficient to drive the development of a robust B-cell leukemia in zebrafish using this proven oncogene.

3. MYELOID MALIGNANCIES

Myeloid malignancies are comprised of chronic (including myelodysplastic syndromes, myeloproliferative neoplasms, and chronic myelomonocytic leukemia) and acute (acute myeloid leukemia) subtypes (Murati et al., 2012). They are clonal diseases arising from the hematopoietic stem or progenitor cells (HSPCs) and are characterized by uncontrolled proliferation and/or blockage of differentiation in myeloid progenitor cells. The mutations that drive the development of these diseases occur in genes encoding proteins that primarily belong to five classes: signaling pathways proteins (e.g., CBL, FLT3, JAK2, RAS), transcription factors (e.g., CEBPA, ETV6, RUNX1), epigenetic regulators (e.g., ASXL1, DNMT3A, EZH2, IDH1, IDH2, SUZ12, TET2, UTX), tumor suppressors (e.g.,TP53), and components of the spliceosome (e.g., SF3B1, SRSF2) (Murati et al., 2012). Most of these genes have been shown to be essential for embryonic hematopoiesis in the zebrafish, and the zebrafish has proven to be an excellent system to further understand the roles of gain or loss of function of these genes in human myeloid malignancies (Jing & Zon, 2011).

3.1 MYELOPROLIFERATIVE NEOPLASMS

3.1.1 Myeloproliferative neoplasms models driven by activated JAK/STAT

Myeloproliferative neoplasms (MPN) are a family of diseases with excessive cells of the myeloid lineage, including polycythemia vera (PV), essential thrombocythemia (ET), primary myelofibrosis (PMF), systemic mastocytosis (SM), juvenile myelomonocytic leukemia (JMML), and chronic myelomonocytic leukemia (CMML) (Tefferi, 2016; Tefferi & Gilliland, 2007). Among these diseases, the majority of PV, ET, and PMF have been associated with a gain-of-function mutation in JAK2 (called JAK2-V617F) that leads to a valine-to-phenylalanine missense mutation at position 617 in the JAK2 protein coding sequence (Nangalia & Green, 2014; Viny & Levine, 2014). Hence, hyperactive JAK/STAT signaling is considered to be a common driver of MPN (Skoda, Duek, & Grisouard, 2015; Viny & Levine, 2014).

The JAK/STAT signaling cascade is a key mediator of cytokine-induced signaling and regulates cell survival, growth, proliferation, and differentiation. In zebrafish embryos, *jak2a* represents one of the two JAK2 orthologs and is highly expressed at sites of hematopoiesis (Ma, Ward, Liang, & Leung, 2007; Oates et al., 1999). Suppression of Jak2a by morpholinos or by the JAK inhibitor AG490 impaired hematopoiesis, whereas overexpression of constitutively active Jak2a stimulated erythropoiesis (Ma et al., 2007). Overexpression of Jak2a-V581F (the zebrafish orthologous equivalent to JAK2-V617F) in zebrafish embryos also significantly increased embryonic erythropoiesis and exhibited developmental phenotypes similar to human PV (Ma et al., 2009). Knockdown of Stat5.1, or treatment with the specific JAK2 inhibitor TG101209, ameliorated the effects of Jak2a-V581F overexpression (Ma et al., 2009). In a separate study, enforced expression of gain-of-function mutations in Stat5.1 (H298R, N714F) significantly expanded both early and late myeloid, erythroid, and lymphoid lineages in the embryos, providing a relevant model of the human blood diseases that are normally driven by STAT5 mutations (Lewis, Stephenson, & Ward, 2006).

Besides the JAK2-V617F mutation, constitutive activation of the JAK/STAT pathway can also be induced by TEL-JAK2 (also known as ETV6-JAK2) fusion proteins in different human leukemias such as ALL and CML (Lacronique et al., 1997). To model these separate diseases, the zebrafish Tel-Jak2a fusions, orthologous to the different TEL-JAK2 fusions found in either human T-ALL or CML, were introduced into embryos under the control of either the leukocyte-specific *spi1* promoter or the ubiquitous *CMV* promoter (Onnebo, Rasighaemi, Kumar, Liongue, & Ward, 2012). Interestingly, the T-ALL-derived Tel-Jak2a fusion significantly perturbed lymphopoiesis, with a lesser effect on myelopoiesis, whereas the CML-derived Tel-Jak2a fusion resulted in significant perturbation of the myeloid compartment with elevated numbers of white blood cells and anemia. This result suggests that different TEL-JAK2 fusions may signal through separate downstream pathways in a lineage-specific manner.

3.1.2 Myeloproliferative neoplasms models driven by RAS

Genetic abnormalities leading to hyperactivation of the RAS signaling pathway have been implicated in many types of human cancers (Ward, Braun, & Shannon, 2012; Young et al., 2009). In JMML, a childhood MPN with poor prognosis, 85% of patients have germline and/or somatic mutations in five major components of the RAS signaling pathway: *NF1, NRAS, KRAS, PTPN11,* and *CBL* (Stieglitz et al., 2015; Tefferi & Gilliland, 2007). The oncogenic KRAS G12D mutant is sufficient to induce fully penetrant MPN in mice (Braun et al., 2004; Chan et al., 2004). Similarly, inducible expression of a β-actin-promoter-driven *KRASG12D* transgene by a heat-shock-inducible Cre/Lox system (*Tg(β-actin:LoxP-EGFP-LoxP-kRASG12D; hsp70-Cre)*) resulted in several tumor types in zebrafish, including MPN that was characterized by expansion of myeloid lineage cells and ineffective erythropoiesis (Le et al., 2007). *Ex vivo* heat shock and transplantation of hematopoietic cells derived from the kidney marrow of this double-transgenic line induced MPN in irradiated wild-type recipients. Interestingly, similar to previous observations in mice, KRASG12D-induced MPN could not be serially transplanted in zebrafish suggesting that KRASG12D did not confer self-renewal properties to progenitor cells (Chan et al., 2004; Le et al., 2007).

In another study, Alghisi et al. established a transgenic zebrafish line in which the oncogenic HRASV12 was expressed in the hemogenic endothelium of the dorsal aorta, a tissue that generates hematopoietic cells (Alghisi et al., 2013). Overexpression of HRASV12 in this tissue resulted in hyperproliferation of *cmyb* + HSPCs, expansion of the caudal hematopoietic tissue in zebrafish embryos, and abnormal myeloid cells in the peripheral blood of mosaic juvenile fish. This study revealed that downregulation of the NOTCH pathway following induction of activated HRAS is likely to be an important mechanism driving myelo-erythroid disorders.

CBL (c-CBL) is an E3 ubiquitin ligase that targets a variety of receptor tyrosine kinases for degradation (Swaminathan & Tsygankov, 2006), and loss-of-function mutations in the *CBL* gene can lead to hyperactivation of RAS signaling (Niemeyer et al., 2010; Ward et al., 2012). Moreover, missense mutations in *CBL* have been implicated in hematopoietic malignancies including JMML, CMML, and acute myeloid leukemia (AML). Interestingly, the *c-cbl H382Y* mutation identified from a zebrafish ENU-mutagenesis screen conferred a myeloproliferative phenotype that originated from mutant-*cbl*-induced stabilization of the receptor tyrosine kinase Flt3, an upstream regulator of the RAS pathway (Peng et al., 2015). This study therefore revealed the oncogenic potential for HSPCs harboring a *CBL* mutation.

3.1.3 Myeloproliferative neoplasms model driven by NUP98-HOX9A

Chromosomal rearrangements of the nucleoporin *NUP98* gene have been associated with AML, myelodysplastic syndrome (MDS), and CML (Gough, Slape, & Aplan, 2011). In AML, the t(7;11)(p15;p15) translocation results in a NUP98-HOXA9 fusion protein (Borrow et al., 1996). When expressed under the control of the *spi1* promoter in zebrafish, NUP98-HOXA9 led to significant impairment of myeloid differentiation (Forrester et al., 2011). Twenty-three percent of *Tg(spi1:NUP98-HOXA9)* zebrafish developed malignant myeloid infiltrates into kidney marrow that resemble MPN at 19–23 months of age; however, none progressed to AML (Forrester et al., 2011). Small molecule inhibitors of DNMT or cyclooxygenase (COX) rescued the hematopoietic abnormalities induced by the NUP98-HOXA9 fusion, and synergistic combination of low doses of a histone deacetylase inhibitor with either the DNMT inhibitor or the COX inhibitor produced the same outcome. Thus, Deveau et al. (2015) showed that the oncogenic potential of the NUP98-HOXA9 fusion is dependent upon downstream activation of the transcription factor *meis1*, the PTGS/COX pathway and genome hypermethylation through the DNA methyltransferase *dnmt1*, and that combinatorial therapies at low doses might be beneficial for patients harboring the NUP98-HOXA9 fusion.

3.1.4 Systemic mastocytosis

SM is a rare myeloproliferative disease without curative therapy (Robyn & Metcalfe, 2006). Despite clinical variability, the majority of patients harbor a KIT-D816V missense mutation (Metcalfe, 2008). Adult zebrafish with *β-actin*-promoter-driven expression of human KIT-D816V exhibited a myeloproliferative disease phenotype, including features of aggressive SM in hematopoietic tissues and high expression levels of endopeptidases, consistent with features observed in SM patients (Balci et al., 2014).

3.2 MYELODYSPLASTIC SYNDROME

Tet methylcytosine dioxygenase 2p (*TET2*) is one of the most frequently mutated tumor suppressor genes in MDS, with inactivating mutations found in the bone marrow cells of 20%–30% of MDS patients at the time of diagnosis (Ko et al., 2010; Langemeijer et al., 2009). In HSPCs, mutations in *TET2* help initiate a premalignant state of clonal dominance that drives the acquisition of additional mutations that culminate in MDS. To develop a zebrafish model of MDS, Gjini et al. (2015) established zebrafish lines with loss-of-function mutations in *tet2* using zinc-finger-nuclease-mediated genome-editing technology and showed that *tet2* mutant zebrafish develop progressive clonal myelodysplasia followed by full-blown MDS with anemia at the age of 24 months. The MDS in *tet2* mutant zebrafish is characterized by increased progenitor and myelomonocytic cell numbers and decreased erythrocyte cell numbers within the kidney marrow (Fig. 3). This zebrafish model can be used to identify and exploit pathway dependencies that arise specifically in *tet2*-null HSPCs for the design of new synthetic lethal therapies for MDS patients with mutations in *TET2*.

In humans, the mitochondrial matrix chaperone HSPA9B is located at 5q31, within a commonly deleted region (CDR) associated with MDS (Liu et al., 2007). In a chemical mutagenesis screen for developmental mutants, Craven et al. discovered that a homozygous loss-of-function mutation in *hspa9b* caused developmental phenotypes closely resembling MDS, such as anemia, dysplasia, increased blood cell apoptosis, and multilineage cytopenia (Craven, French, Ye, de Sauvage, & Rosenthal, 2005). However, heterozygous adult fish did not manifest obvious anemia or other characteristic symptoms of MDS, suggesting that loss of heterozygosity or other additional genetic lesions may be required for progression to MDS (Craven et al., 2005).

Loss or deletion of chromosome 7 is frequently observed in MDS patients and is associated with poor prognosis (Haase, 2008). Sundaravel et al. (2015) found that knockdown of dedicator of cytokinesis 4 (*DOCK4*), a gene that maps to a commonly deleted region of 7q in MDS (Sundaravel et al., 2015), resulted in increased numbers of dysplastic erythroid cells in zebrafish embryos through the disruption of actin filaments in red blood cells. Moreover, reexpression of DOCK4 in -7q MDS patient erythroblasts partially reversed their aberrant phenotypes. Loss of DOCK4 may help explain why erythroid dysplasia is a predominant clinical manifestation of MDS.

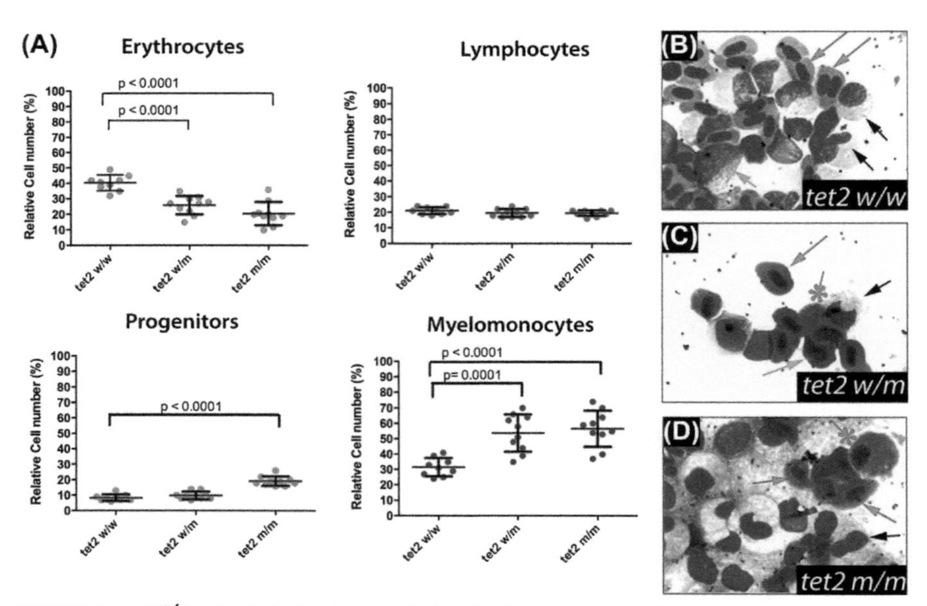

FIGURE 3 ***tet2$^{m/m}$*** **zebrafish develop myelodysplastic syndrome at 24 months of age.**

(A) Forward versus side scatter analysis plots for kidney marrow cell populations in 24-month-old *tet2$^{wt/wt}$*, *tet2$^{wt/m}$*, and *tet2$^{m/m}$* fish. Analysis of the kidney marrow cell populations of 24-month-old fish with loss of *tet2* shows a significant decrease in erythrocytes and a significant increase in the myelomonocyte population in *tet2$^{wt/m}$* and *tet2$^{m/m}$* fish compared with *tet2$^{wt/wt}$* fish. A significant increase in the progenitor cell population is observed only in the *tet2$^{m/m}$* fish. (B–D): May-Grünwald–Giemsa staining of kidney marrow smears at the 24-month-old stage for *tet2$^{wt/wt}$* (B), *tet2$^{wt/m}$* (C), and *tet2$^{m/m}$* (D) fish shows the presence of dysplastic myeloid and progenitor cells and the presence of the basophilic cytoplasm in a subset of *tet2$^{wt/m}$* fish and all *tet2$^{m/m}$* fish. *Red arrows*, mature erythrocytes; *orange arrows*, progenitor cells; *black arrows*, mature myeloid cells; asterisks, dysplastic myeloid cells.

Modified from Gjini, E., Mansour, M. R., Sander, J. D., Moritz, N., Nguyen, A. T., Kesarsing, M., ... Look, A. T. (2015). A zebrafish model of myelodysplastic syndrome produced through tet2 genomic editing. Molecular and Cellular Biology, 35*(5), 789–804. doi:10.1128/mcb.00971-14.*

3.3 ACUTE MYELOID LEUKEMIA

AML is the most common acute leukemia in adults, and the incidence increases with age (Coombs, Tallman, & Levine, 2016). Although many patients with AML achieve complete remission of their disease with induction chemotherapy, most patients treated with chemotherapy will relapse after a median time of about 11 months. AML results from the clonal expansion of undifferentiated myeloid precursors or blast cells, resulting in replacement of normal bone marrow cells and pancytopenia (Papaemmanuil et al., 2016). The genomic aberrations associated with AML include frequent mutations in *NPM1, FLT3, TP53, IDH2*, mutations of genes encoding chromatin and RNA-splicing regulators, and chromosomal aneuploidies (Papaemmanuil et al., 2016).

Mutations in nucleophosmin (*NPM1*), one of the most frequently mutated genes in AML, have been detected in 27% of AML cases (Network, 2013; Papaemmanuil et al., 2016). NPM1 is a ubiquitously expressed phosphoprotein that is primarily localized at nucleoli with continuous shuttling between the nucleus and cytoplasm (Grisendi, Mecucci, Falini, & Pandolfi, 2006). AML-associated NPM1 mutations result in its aberrant cytoplasmic translocation (called NPMc) (Falini et al., 2005). Overexpression of NPMc in zebrafish embryos led to expansion of primitive early myeloid cells, which was significantly enhanced in the absence of functional p53 (Bolli et al., 2010). Importantly, NPMc also led to increased numbers of definitive hematopoietic cells including erythromyeloid progenitors in the posterior blood island and c-myb/cd41(+) cells in the ventral wall of the aorta. As such, this zebrafish model provides a tractable in vivo system to study the pathogenic role of NPMc in myeloid leukemogenesis, mutations in FMS-like tyrosine kinase 3 (*FLT3*) occur in about 30% of AML patients and have been associated with poor prognosis (Zeisig, Kulasekararaj, Mufti, & Eric So, 2012). A tyrosine kinase inhibitor AC220 (also known as quizartinib) is effective at inducing remission of AML harboring mutations in the FLT3 internal tandem duplication domain (FLT3-ITD), but the effect is short-lived as new clones with AC220-resistant FLT3 mutations emerge and expand (Smith et al., 2012). In zebrafish embryos, overexpression of mRNA expressing human *FLT3-ITD* or *FLT3* with a mutation in the tyrosine kinase domain (*FLT3-TKD*) resulted in expansion and clustering of myeloid cells through activation of downstream signaling pathways (He et al., 2014). Treatment with AC220 significantly reduced the myeloid expansion induced by FLT3-ITD, but not FLT3-TKD, expression in zebrafish (He et al., 2014). This study introduced a much-needed animal model for the design of more effective therapies for AMLs harboring common FLT3 mutations.

Approximately 30% of cytogenetically normal AML cells carry mutations of isocitrate dehydrogenase 1 and 2 (*IDH1/2*), suggesting a pathogenic function for IDH1/2 in AML (Gross et al., 2010). Shi et al. (2015) found that Idh1 regulates myelopoiesis and definitive hematopoiesis as loss of *idh1* led to increased myeloid progenitor cells and decreased myeloid differentiation. In addition, overexpression of human IDH1 with an AML-associated mutation (IDH1-R132H), or its zebrafish equivalent, induced pathogenic levels of 2-hydroxyglutarate and expansion of myelopoiesis in zebrafish embryos, suggesting a dominant role for mutant IDH1-R132H in leukemogenesis.

RUNX1, a member of the Runt family of transcription factors, regulates definitive hematopoiesis (Kalev-Zylinska et al., 2002; Okuda et al., 1998). Loss-of-function mutations in *RUNX1* (formerly known as *AML1*) are found in 9%–16% of cases with MDS and about 10% with AML and are associated with poor outcome (Gangat, Patnaik, & Tefferi, 2016; Papaemmanuil et al., 2016). A zebrafish line harboring a truncated allele of *runx1* (*runx1*^{W84X}) was established by ENU mutagenesis (Jin et al., 2009). In *runx1*^{W84X} mutant zebrafish, definitive hematopoiesis is blocked, and the embryos are "bloodless" (Sood et al., 2010). Although about 20% of runx1^{W84X} mutants develop into fertile adults with multilineage hematopoiesis, they exhibit significant reduction in certain myeloid and progenitor cells, suggesting that *runx1* is required for lineage-specific differentiation during hematopoiesis (Sood et al., 2010).

The *RUNX1* gene is commonly disrupted by chromosomal translocations in human myeloid malignancies. In AML, *RUNX1* is often fused to the eight twenty-one gene (*RUNX1-ETO*) (Okuda et al., 1998; Yergeau et al., 1997). To analyze the function of this fusion gene, Yeh et al. (2008) established a transgenic zebrafish line in which expression of *RUNX1-ETO* is controlled by a heat shock promoter. *Tg*(*hsp:RUNX1-ETO*) embryos exhibited an accumulation of immature hematopoietic blast cells, downregulation of *scl/tal1* expression and a shift in myeloerythroid progenitor cell fate, suggesting that *scl/tal1* may contribute to *RUNX1-ETO*-associated leukemia. Besides *scl/tal1* downregulation, additional genetic alterations are probably required for leukemogenesis, as RUNX1-ETO was not sufficient to induce an overt leukemia phenotype in zebrafish or mice (Yeh et al., 2008). Moreover, the effects of RUNX1-ETO were suppressed by the histone deacetylase inhibitor trichostatin A. Subsequent large-scale chemical screens using the *Tg*(*hsp:RUNX1-ETO*) line revealed that inhibitors of the COX-2 and β-catenin pathways, as well as the benzodiazepine Ro5-3335, may also have therapeutic benefit in *RUNX1-ETO*-driven AML (Cunningham et al., 2012; Yeh et al., 2009).

The inv(8)(p11q13) chromosomal rearrangement results in the oncogenic MOZ-TIF2 (MYST3-NCOA2) fusion (Deguchi et al., 2003). To investigate the oncogenic properties of this fusion protein, Zhuravleva et al. (2008) generated a transgenic zebrafish expressing the MOZ-TIF2 fusion under the control of the *pu.1* promoter. Two out of 180 one-cell-stage embryos injected with the *pu.1:MOZ-TIF2-EGFP* transgene developed AML with excessive immature myeloid cells in the kidney marrow, measured at 14 and 26 months post injection (Zhuravleva et al., 2008). The low incidence of leukemia combined with long latency to disease suggested that additional genetic mutations are required during MOZ-TIF2-induced AML.

The *MYCN* transcription factor gene is frequently amplified in multiple hematologic malignancies, including AML, and this genetic aberration is a well-established marker of poor prognosis (Delgado, Albajar, Gomez-Casares, Batlle, & León, 2013). Shen et al. developed a transgenic zebrafish line in which mouse *MycN* expression is regulated by an artificial heat shock promoter (*Tg*(*HSE:MycN, EGFP*)). Embryos from this line that were exposed to heat shock developed excessive hematopoietic cell proliferation at around 60 days and increased blast cell population in adult kidney and spleen (Shen et al., 2013). In addition, massive immature hematopoietic cells emerged in the circulating blood and infiltrated various organs of *Tg*(*HSE:MycN, EGFP*) fish (Shen et al., 2013). This study suggested that *MYCN* amplification contributes to the etiology of AML.

PU.1 is a member of ETS-family of transcription factors and is critical to generate early myeloid progenitors and to determine the fate of the myeloid lineage during normal hematopoietic stem cell development (Dahl et al., 2003; Jin et al., 2012; Scott, Simon, Anastasi, & Singh, 1994). Loss-of-function mutations in *PU.1* or downregulation of PU.1 expression has been associated with the onset and progression of AML in humans (Steidl et al., 2006; Zhu et al., 2012). A zebrafish mutant line carrying a hypomorphic *pu.1* allele (*pu.1*^{G242D}) was identified by the TILLING approach (Sun et al., 2013). At embryonic stages, *pu.1*^{G242D} mutant zebrafish displayed excessive production of immature granulocytes with no effect on erythrocytes. By 18 months, a marked increase in myeloid cells with a concomitant decrease in lymphoid cells was detected in the kidney marrow and peripheral blood of these animals. Moreover, cytological analysis showed that the expanded myeloid population in *pu.1*^{G242D} mutants was largely comprised of immature myeloid cells, thus resembling the phenotypes of human MDS or AML. Importantly, cytarabine, a chemotherapy agent used in AML treatment, showed significant activity against the hyperproliferative myeloid cells in *pu.1*^{G242D} embryos (Sun et al., 2013). The *pu.1*^{G242D} mutant fish provides a valuable animal model system to identify promising new therapies for AML.

CONCLUSIONS

Zebrafish presents a promising preclinical model for understanding the molecular pathogenesis of hematopoietic malignancies and for the development of novel therapeutic strategies to treat these diseases. Over the last two decades, genetically modifies zebrafish have been applied to model human hematopoietic malignancies, including T-ALL, B-ALL, MPN, MDS, and AML, as we have summarized in this chapter. The recent advances in zebrafish genetics and genome editing technologies have further expanded the cancer research toolbox in this model organism. Zebrafish animal models promise to provide robust and genetically tractable systems for in vivo studies to elucidate new molecular pathways in transformation and to provide a means to identify new active small molecules for specific and effective therapies of human hematopoietic malignancies.

REFERENCES

Alghisi, E., Distel, M., Malagola, M., Anelli, V., Santoriello, C., Herwig, L., ... Mione, M. C. (2013). Targeting oncogene expression to endothelial cells induces proliferation of the myelo-erythroid lineage by repressing the notch pathway (Original Article) *Leukemia, 27*(11), 2229–2241. http://dx.doi.org/10.1038/leu.2013.132.

Anderson, N. M., Li, D., Peng, H. L., Laroche, F. J. F., Mansour, M. R., Gjini, E., ... Feng, H. (2016). The TCA cycle transferase DLST is important for MYC-mediated leukemogenesis (Original Article) *Leukemia, 30*(6), 1365–1374. http://dx.doi.org/10.1038/leu.2016.26.

Aster, J. C., Blacklow, S. C., & Pear, W. S. (2011). Notch signalling in T-cell lymphoblastic leukaemia/lymphoma and other haematological malignancies. *The Journal of Pathology, 223*(2), 263–274. http://dx.doi.org/10.1002/path.2789.

Balci, T. B., Prykhozhij, S. V., Teh, E. M., Da'as, S. I., McBride, E., Liwski, R., … Berman, J. N. (2014). A transgenic zebrafish model expressing KIT-D816V recapitulates features of aggressive systemic mastocytosis. *British Journal of Haematology, 167*(1), 48–61. http://dx.doi.org/10.1111/bjh.12999.

Bentley, V. L., Veinotte, C. J., Corkery, D. P., Pinder, J. B., LeBlanc, M. A., Bedard, K., … Dellaire, G. (2015). Focused chemical genomics using zebrafish xenotransplantation as a pre-clinical therapeutic platform for T-cell acute lymphoblastic leukemia. *Haematologica, 100*(1), 70–76. http://dx.doi.org/10.3324/haematol.2014.110742.

Blackburn, J. S., Liu, S., & Langenau, D. M. (2011). Quantifying the frequency of tumor-propagating cells using limiting dilution cell transplantation in syngeneic zebrafish. *Journal of Visualized Experiments*, (53), 2790. http://dx.doi.org/10.3791/2790.

Blackburn, J. S., Liu, S., Raiser, D. M., Martinez, S. A., Feng, H., Meeker, N. D., … Langenau, D. M. (2012). Notch signaling expands a pre-malignant pool of T-cell acute lymphoblastic leukemia clones without affecting leukemia-propagating cell frequency. *Leukemia, 26*(9), 2069–2078. http://www.nature.com/leu/journal/v26/n9/suppinfo/leu2012116s1.html.

Blackburn, J. S., Liu, S., Wilder, J. L., Dobrinski, K. P., Lobbardi, R., Moore, F. E., … Langenau, D. M. (2014). Clonal evolution enhances leukemia-propagating cell frequency in T cell acute lymphoblastic leukemia through Akt/mTORC1 pathway activation. *Cancer Cell, 25*(3), 366–378. http://dx.doi.org/10.1016/j.ccr.2014.01.032.

Bolli, N., Payne, E. M., Grabher, C., Lee, J.-S., Johnston, A. B., Falini, B., … Look, A. T. (2010). Expression of the cytoplasmic NPM1 mutant (NPMc+) causes the expansion of hematopoietic cells in zebrafish. *Blood, 115*(16), 3329–3340. http://dx.doi.org/10.1182/blood-2009-02-207225.

Borah, A., Raveendran, S., Rochani, A., Maekawa, T., & Kumar, D. S. (2015). Targeting self-renewal pathways in cancer stem cells: clinical implications for cancer therapy (Review) *Oncogenesis, 4*, e177. http://dx.doi.org/10.1038/oncsis.2015.35.

Borrow, J., Shearman, A. M., Stanton, V. P., Becher, R., Collins, T., Williams, A. J., … Housman, D. E. (1996). The t(7;11)(p15;p15) translocation in acute myeloid leukaemia fuses the genes for nucleoporin NUP96 and class I homeoprotein HOXA9. *Nature Genetics, 12*(2), 159–167. http://dx.doi.org/10.1038/ng0296-159.

Bouillet, P., Metcalf, D., Huang, D. C. S., Tarlinton, D. M., Kay, T. W. H., Köntgen, F., … Strasser, A. (1999). Proapoptotic Bcl-2 relative bim required for certain apoptotic responses, leukocyte homeostasis, and to preclude autoimmunity. *Science, 286*(5445), 1735–1738. http://dx.doi.org/10.1126/science.286.5445.1735.

Braun, B. S., Tuveson, D. A., Kong, N., Le, D. T., Kogan, S. C., Rozmus, J., … Shannon, K. M. (2004). Somatic activation of oncogenic Kras in hematopoietic cells initiates a rapidly fatal myeloproliferative disorder. *Proceedings of the National Academy of Sciences of the United States of America, 101*(2), 597–602. http://dx.doi.org/10.1073/pnas.0307203101.

Chan, I. T., Kutok, J. L., Williams, I. R., Cohen, S., Kelly, L., Shigematsu, H., … Gilliland, D. G. (2004). Conditional expression of oncogenic K-ras from its endogenous promoter induces a myeloproliferative disease. *The Journal of Clinical Investigation, 113*(4), 528–538. http://dx.doi.org/10.1172/jci20476.

Chen, J., Jette, C., Kanki, J. P., Aster, J. C., Look, A. T., & Griffin, J. D. (2007). NOTCH1-induced T-cell leukemia in transgenic zebrafish. *Leukemia, 21*(3), 462–471. http://www.nature.com/leu/journal/v21/n3/suppinfo/2404546s1.html.

Chiang, M. Y., Xu, L., Shestova, O., Histen, G., L'Heureux, S., Romany, C., … Pear, W. S. (2008). Leukemia-associated NOTCH1 alleles are weak tumor initiators but accelerate K-ras–initiated leukemia. *The Journal of Clinical Investigation, 118*(9), 3181–3194. http://dx.doi.org/10.1172/jci35090.

Chonghaile, T. N., Sarosiek, K. A., Vo, T.-T., Ryan, J. A., Tammareddi, A., Moore, V. D. G., … Letai, A. (2011). Pretreatment mitochondrial priming correlates with clinical response to cytotoxic chemotherapy. *Science, 334*(6059), 1129–1133. http://dx.doi.org/10.1126/science.1206727.

Coombs, C. C., Tallman, M. S., & Levine, R. L. (2016). Molecular therapy for acute myeloid leukaemia (Review) *Nature Reviews Clinical Oncology, 13*(5), 305–318. http://dx.doi.org/10.1038/nrclinonc.2015.210.

Corkery, D. P., Dellaire, G., & Berman, J. N. (2011). Leukaemia xenotransplantation in zebrafish – chemotherapy response assay in vivo. *British Journal of Haematology, 153*(6), 786–789. http://dx.doi.org/10.1111/j.1365-2141.2011.08661.x.

Craven, S. E., French, D., Ye, W., de Sauvage, F., & Rosenthal, A. (2005). Loss of Hspa9b in zebrafish recapitulates the ineffective hematopoiesis of the myelodysplastic syndrome. *Blood, 105*(9), 3528–3534. http://dx.doi.org/10.1182/blood-2004-03-1089.

Cunningham, L., Finckbeiner, S., Hyde, R. K., Southall, N., Marugan, J., Yedavalli, V. R. K., … Liu, P. (2012). Identification of benzodiazepine Ro5-3335 as an inhibitor of CBF leukemia through quantitative high throughput screen against RUNX1–CBFβ interaction. *Proceedings of the National Academy of Sciences of the United States of America, 109*(36), 14592–14597. http://dx.doi.org/10.1073/pnas.1200037109.

Da Costa, D., Agathanggelou, A., Perry, T., Weston, V., Petermann, E., Zlatanou, A., … Stankovic, T. (2013). BET inhibition as a single or combined therapeutic approach in primary paediatric B-precursor acute lymphoblastic leukaemia (Original Article) *Blood Cancer Journal, 3*, e126. http://dx.doi.org/10.1038/bcj.2013.24.

Dahl, R., Walsh, J. C., Lancki, D., Laslo, P., Iyer, S. R., Singh, H., … Simon, M. C. (2003). Regulation of macrophage and neutrophil cell fates by the PU.1:C/EBP[alpha] ratio and granulocyte colony-stimulating factor. *Nature Immunology, 4*(10), 1029–1036. http://www.nature.com/ni/journal/v4/n10/suppinfo/ni973_S1.html.

Deguchi, K., Ayton, P. M., Carapeti, M., Kutok, J. L., Snyder, C. S., Williams, I. R., … Gilliland, D. G. (2003). MOZ-TIF2-induced acute myeloid leukemia requires the MOZ nucleosome binding motif and TIF2-mediated recruitment of CBP. *Cancer Cell, 3*(3), 259–271. http://dx.doi.org/10.1016/S1535-6108(03)00051-5.

Delgado, M. D., Albajar, M., Gomez-Casares, M. T., Batlle, A., & León, J. (2013). MYC oncogene in myeloid neoplasias (Journal Article) *Clinical and Translational Oncology, 15*(2), 87–94. http://dx.doi.org/10.1007/s12094-012-0926-8.

Deng, J., Carlson, N., Takeyama, K., Dal Cin, P., Shipp, M., & Letai, A. (2007). BH3 profiling identifies three distinct classes of apoptotic blocks to predict response to ABT-737 and conventional chemotherapeutic agents. *Cancer Cell, 12*(2), 171–185. http://dx.doi.org/10.1016/j.ccr.2007.07.001.

Deveau, A. P., Forrester, A. M., Coombs, A. J., Wagner, G. S., Grabher, C., Chute, I. C., … Berman, J. N. (2015). Epigenetic therapy restores normal hematopoiesis in a zebrafish model of NUP98-HOXA9-induced myeloid disease (Original Article) *Leukemia, 29*(10), 2086–2097. http://dx.doi.org/10.1038/leu.2015.126.

Drabsch, Y., He, S., Zhang, L., Snaar-Jagalska, B. E., & ten Dijke, P. (2013). Transforming growth factor-β signalling controls human breast cancer metastasis in a zebrafish xenograft model (Journal Article) *Breast Cancer Research, 15*(6), 1–13. http://dx.doi.org/10.1186/bcr3573.

Durinck, K., Goossens, S., Peirs, S., Wallaert, A., Van Loocke, W., Matthijssens, F., … Van Vlierberghe, P. (2015). Novel biological insights in T-cell acute lymphoblastic leukemia. *Experimental Hematology, 43*(8), 625–639. http://dx.doi.org/10.1016/j.exphem.2015.05.017.

Falini, B., Mecucci, C., Tiacci, E., Alcalay, M., Rosati, R., Pasqualucci, L., … Martelli, M. F. (2005). Cytoplasmic nucleophosmin in acute myelogenous leukemia with a normal karyotype. *New England Journal of Medicine, 352*(3), 254–266. http://dx.doi.org/10.1056/NEJMoa041974.

Faucherre, A., Taylor, G. S., Overvoorde, J., Dixon, J. E., & den Hertog, J. (2007). Zebrafish pten genes have overlapping and non-redundant functions in tumorigenesis and embryonic development. *Oncogene, 27*(8), 1079–1086. http://www.nature.com/onc/journal/v27/n8/suppinfo/1210730s1.html.

Feng, H., Langenau, D. M., Madge, J. A., Quinkertz, A., Gutierrez, A., Neuberg, D. S., … Thomas Look, A. (2007). Heat-shock induction of T-cell lymphoma/leukaemia in conditional Cre/lox-regulated transgenic zebrafish. *British Journal of Haematology, 138*(2), 169–175. http://dx.doi.org/10.1111/j.1365-2141.2007.06625.x.

Feng, H., Stachura, D. L., White, R. M., Gutierrez, A., Zhang, L., Sanda, T., … Look, A. T. (2010). T-lymphoblastic lymphoma cells express high levels of BCL2, S1P1, and ICAM1, leading to a blockade of tumor cell intravasation. *Cancer Cell, 18*(4), 353–366. http://dx.doi.org/10.1016/j.ccr.2010.09.009.

Ferrando, A. A., Neuberg, D. S., Staunton, J., Loh, M. L., Huard, C., Raimondi, S. C., … Look, A. T. (2002). Gene expression signatures define novel oncogenic pathways in T cell acute lymphoblastic leukemia. *Cancer Cell, 1*(1), 75–87. http://dx.doi.org/10.1016/S1535-6108(02)00018-1.

Forrester, A. M., Grabher, C., McBride, E. R., Boyd, E. R., Vigerstad, M. H., Edgar, A., … Berman, J. N. (2011). NUP98-HOXA9-transgenic zebrafish develop a myeloproliferative neoplasm and provide new insight into mechanisms of myeloid leukaemogenesis. *British Journal of Haematology, 155*(2), 167–181. http://dx.doi.org/10.1111/j.1365-2141.2011.08810.x.

Frazer, J. K., Meeker, N. D., Rudner, L., Bradley, D. F., Smith, A. C. H., Demarest, B., … Trede, N. S. (2009). Heritable T-cell malignancy models established in a zebrafish phenotypic screen. *Leukemia, 23*(10), 1825–1835. http://www.nature.com/leu/journal/v23/n10/suppinfo/leu2009116s1.html.

Gangat, N., Patnaik, M. M., & Tefferi, A. (2016). Myelodysplastic syndromes: Contemporary review and how we treat. *American Journal of Hematology, 91*(1), 76–89. http://dx.doi.org/10.1002/ajh.24253.

Gjini, E., Mansour, M. R., Sander, J. D., Moritz, N., Nguyen, A. T., Kesarsing, M., … Look, A. T. (2015). A zebrafish model of myelodysplastic syndrome produced through tet2 genomic editing. *Molecular and Cellular Biology, 35*(5), 789–804. http://dx.doi.org/10.1128/mcb.00971-14.

Gough, S. M., Slape, C. I., & Aplan, P. D. (2011). NUP98 gene fusions and hematopoietic malignancies: common themes and new biologic insights. *Blood, 118*(24), 6247–6257. http://dx.doi.org/10.1182/blood-2011-07-328880.

Grabher, C., von Boehmer, H., & Look, A. T. (2006). Notch 1 activation in the molecular pathogenesis of T-cell acute lymphoblastic leukaemia. *Nature Reviews Cancer, 6*(5), 347–359. http://dx.doi.org/10.1038/nrc1880.

Grisendi, S., Mecucci, C., Falini, B., & Pandolfi, P. P. (2006). Nucleophosmin and cancer. *Nature Reviews Cancer, 6*(7), 493–505. http://www.nature.com/nrc/journal/v6/n7/suppinfo/nrc1885_S1.html.

Gross, S., Cairns, R. A., Minden, M. D., Driggers, E. M., Bittinger, M. A., Jang, H. G., … Mak, T. W. (2010). Cancer-associated metabolite 2-hydroxyglutarate accumulates in acute myelogenous leukemia with isocitrate dehydrogenase 1 and 2 mutations. *The Journal of Experimental Medicine, 207*(2), 339–344. http://dx.doi.org/10.1084/jem.20092506.

Grossmann, V., Haferlach, C., Weissmann, S., Roller, A., Schindela, S., Poetzinger, F., … Kohlmann, A. (2013). The molecular profile of adult T-cell acute lymphoblastic leukemia: mutations in RUNX1 and DNMT3A are associated with poor prognosis in T-ALL. *Genes, Chromosomes and Cancer, 52*(4), 410–422. http://dx.doi.org/10.1002/gcc.22039.

Gutierrez, A., Feng, H., Stevenson, K., Neuberg, D. S., Calzada, O., Zhou, Y., … Look, A. T. (2014). Loss of function tp53 mutations do not accelerate the onset of myc-induced T-cell acute lymphoblastic leukaemia in the zebrafish. *British Journal of Haematology, 166*(1), 84–90. http://dx.doi.org/10.1111/bjh.12851.

Gutierrez, A., Grebliunaite, R., Feng, H., Kozakewich, E., Zhu, S., Guo, F., … Look, A. T. (2011). Pten mediates Myc oncogene dependence in a conditional zebrafish model of T cell acute lymphoblastic leukemia. *The Journal of Experimental Medicine, 208*(8), 1595–1603. http://dx.doi.org/10.1084/jem.20101691.

Gutierrez, A., Pan, L., Groen, R. W. J., Baleydier, F., Kentsis, A., Marineau, J., … Aster, J. C. (2014). Phenothiazines induce PP2A-mediated apoptosis in T cell acute lymphoblastic leukemia. *The Journal of Clinical Investigation, 124*(2), 644–655. http://dx.doi.org/10.1172/jci65093.

Gutierrez, A., Sanda, T., Grebliunaite, R., Carracedo, A., Salmena, L., Ahn, Y., … Look, A. T. (2009). High frequency of PTEN, PI3K, and AKT abnormalities in T-cell acute lymphoblastic leukemia. *Blood, 114*(3), 647–650. http://dx.doi.org/10.1182/blood-2009-02-206722.

Haase, D. (2008). Cytogenetic features in myelodysplastic syndromes. *Annals of Hematology, 87*(7), 515–526. http://dx.doi.org/10.1007/s00277-008-0483-y.

Hao, Z., Cairns, R. A., Inoue, S., Li, W. Y., Sheng, Y., Lemonnier, F., … Mak, T. W. (2016). Idh1 mutations contribute to the development of T-cell malignancies in genetically engineered mice. *Proceedings of the National Academy of Sciences of the United States of America, 113*(5), 1387–1392. http://dx.doi.org/10.1073/pnas.1525354113.

He, B.-L., Shi, X., Man, C. H., Ma, A. C. H., Ekker, S. C., Chow, H. C. H., … Leung, A. Y. H. (2014). Functions of flt3 in zebrafish hematopoiesis and its relevance to human acute myeloid leukemia. *Blood, 123*(16), 2518–2529. http://dx.doi.org/10.1182/blood-2013-02-486688.

He, S., Lamers, G. E. M., Beenakker, J.-W. M., Cui, C., Ghotra, V. P. S., Danen, E. H. J., … Snaar-Jagalska, B. E. (2012). Neutrophil-mediated experimental metastasis is enhanced by VEGFR inhibition in a zebrafish xenograft model. *The Journal of Pathology, 227*(4), 431–445. http://dx.doi.org/10.1002/path.4013.

Herranz, D., Ambesi-Impiombato, A., Palomero, T., Schnell, S. A., Belver, L., Wendorff, A. A., ... Ferrando, A. A. (2014). A NOTCH1-driven MYC enhancer promotes T cell development, transformation and acute lymphoblastic leukemia (Article) *Nature Medicine, 20*(10), 1130–1137 http://www.nature.com/nm/journal/v20/n10/abs/nm.3665.html#supplementary-information.

Hesse, R. G., Kouklis, G. K., Ahituv, N., & Pomerantz, J. H. (2015). The human ARF tumor suppressor senses blastema activity and suppresses epimorphic tissue regeneration. *eLife, 4*, e07702. http://dx.doi.org/10.7554/eLife.07702.

Hof, J., Krentz, S., van Schewick, C., Körner, G., Shalapour, S., Rhein, P., ... Kirschner-Schwabe, R. (2011). Mutations and deletions of the TP53 gene predict nonresponse to treatment and poor outcome in first relapse of childhood acute lymphoblastic leukemia. *Journal of Clinical Oncology, 29*(23), 3185–3193. http://dx.doi.org/10.1200/jco.2011.34.8144.

Jenkinson, S., Kirkwood, A. A., Goulden, N., Vora, A., Linch, D. C., & Gale, R. E. (2016). Impact of PTEN abnormalities on outcome in pediatric patients with T-cell acute lymphoblastic leukemia treated on the MRC UKALL2003 trial. *Leukemia, 30*(1), 39–47. http://dx.doi.org/10.1038/leu.2015.206.

Jin, H., Li, L., Xu, J., Zhen, F., Zhu, L., Liu, P. P., ... Wen, Z. (2012). Runx1 regulates embryonic myeloid fate choice in zebrafish through a negative feedback loop inhibiting Pu.1 expression. *Blood, 119*(22), 5239–5249. http://dx.doi.org/10.1182/blood-2011-12-398362.

Jin, H., Sood, R., Xu, J., Zhen, F., English, M. A., Liu, P. P., ... Wen, Z. (2009). Definitive hematopoietic stem/progenitor cells manifest distinct differentiation output in the zebrafish VDA and PBI. *Development, 136*(4), 647–654. http://dx.doi.org/10.1242/dev.029637.

Jing, L., & Zon, L. I. (2011). Zebrafish as a model for normal and malignant hematopoiesis. *Disease Models and Mechanisms, 4*(4), 433–438. http://dx.doi.org/10.1242/dmm.006791.

Kalev-Zylinska, M. L., Horsfield, J. A., Flores, M. V. C., Postlethwait, J. H., Vitas, M. R., Baas, A. M., ... Crosier, K. E. (2002). Runx1 is required for zebrafish blood and vessel development and expression of a human RUNX1-CBF2T1 transgene advances a model for studies of leukemogenesis. *Development, 129*(8), 2015–2030.

Ko, M., Huang, Y., Jankowska, A. M., Pape, U. J., Tahiliani, M., Bandukwala, H. S., ... Rao, A. (2010). Impaired hydroxylation of 5-methylcytosine in myeloid cancers with mutant TET2. *Nature, 468*(7325), 839–843. http://www.nature.com/nature/journal/v468/n7325/abs/nature09586.html#supplementary-information.

Kress, T. R., Sabo, A., & Amati, B. (2015). MYC: connecting selective transcriptional control to global RNA production (Review) *Nature Reviews Cancer, 15*(10), 593–607. http://dx.doi.org/10.1038/nrc3984.

Lacronique, V., Boureux, A., Della Valle, V., Poirel, H., Quang, C. T., Mauchauffé, M., ... Bernard, O. A. (1997). A TEL-JAK2 fusion protein with constitutive kinase activity in human leukemia. *Science, 278*(5341), 1309–1312. http://dx.doi.org/10.1126/science.278.5341.1309.

Langemeijer, S. M. C., Kuiper, R. P., Berends, M., Knops, R., Aslanyan, M. G., Massop, M., ... Jansen, J. H. (2009). Acquired mutations in TET2 are common in myelodysplastic syndromes. *Nature Genetics, 41*(7), 838–842. http://www.nature.com/ng/journal/v41/n7/suppinfo/ng.391_S1.html.

Langenau, D. M., Feng, H., Berghmans, S., Kanki, J. P., Kutok, J. L., & Look, A. T. (2005). Cre/lox-regulated transgenic zebrafish model with conditional myc-induced T cell acute lymphoblastic leukemia. *Proceedings of the National Academy of Sciences of the United States of America, 102*(17), 6068–6073. http://dx.doi.org/10.1073/pnas.0408708102.

Langenau, D. M., Ferrando, A. A., Traver, D., Kutok, J. L., Hezel, J.-P. D., Kanki, J. P., … Trede, N. S. (2004). In vivo tracking of T cell development, ablation, and engraftment in transgenic zebrafish. *Proceedings of the National Academy of Sciences of the United States of America, 101*(19), 7369–7374. http://dx.doi.org/10.1073/pnas.0402248101.

Langenau, D. M., Traver, D., Ferrando, A. A., Kutok, J. L., Aster, J. C., Kanki, J. P., … Look, A. T. (2003). Myc-induced T cell leukemia in transgenic zebrafish. *Science, 299*(5608), 887–890. http://dx.doi.org/10.1126/science.1080280.

Le, X., Langenau, D. M., Keefe, M. D., Kutok, J. L., Neuberg, D. S., & Zon, L. I. (2007). Heat shock-inducible Cre/Lox approaches to induce diverse types of tumors and hyperplasia in transgenic zebrafish. *Proceedings of the National Academy of Sciences of the United States of America, 104*(22), 9410–9415. http://dx.doi.org/10.1073/pnas.0611302104.

Lewis, R. S., Stephenson, S. E. M., & Ward, A. C. (2006). Constitutive activation of zebrafish Stat5 expands hematopoietic cell populations in vivo. *Experimental Hematology, 34*(2), 179–187. http://dx.doi.org/10.1016/j.exphem.2005.11.003.

Liu, T. X., Becker, M. W., Jelinek, J., Wu, W.-S., Deng, M., Mikhalkevich, N., … Look, A. T. (2007). Chromosome 5q deletion and epigenetic suppression of the gene encoding [alpha]-catenin (CTNNA1) in myeloid cell transformation. *Nature Medicine, 13*(1), 78–83. http://www.nature.com/nm/journal/v13/n1/suppinfo/nm1512_S1.html.

Ma, A. C. H., Fan, A., Ward, A. C., Liongue, C., Lewis, R. S., Cheng, S. H., … Leung, A. Y. H. (2009). A novel zebrafish jak2aV581F model shared features of human JAK2V617F polycythemia vera. *Experimental Hematology, 37*(12), 1379–1386.e1374. http://dx.doi.org/10.1016/j.exphem.2009.08.008.

Ma, A. C. H., Ward, A. C., Liang, R., & Leung, A. Y. H. (2007). The role of jak2a in zebrafish hematopoiesis. *Blood, 110*(6), 1824–1830. http://dx.doi.org/10.1182/blood-2007-03-078287.

MacRae, C. A., & Peterson, R. T. (2015). Zebrafish as tools for drug discovery (Review) *Nature Reviews Drug Discovery, 14*(10), 721–731. http://dx.doi.org/10.1038/nrd4627.

Martini, M., De Santis, M. C., Braccini, L., Gulluni, F., & Hirsch, E. (2014). PI3K/AKT signaling pathway and cancer: an updated review. *Annals of Medicine, 46*(6), 372–383. http://dx.doi.org/10.3109/07853890.2014.912836.

McKeown, M. R., & Bradner, J. E. (2014). Therapeutic strategies to inhibit MYC. *Cold Spring Harbor Perspectives in Medicine, 4*(10). http://dx.doi.org/10.1101/cshperspect.a014266.

Metcalfe, D. D. (2008). Mast cells and mastocytosis. *Blood, 112*(4), 946–956. http://dx.doi.org/10.1182/blood-2007-11-078097.

Moore, F. E., Garcia, E. G., Lobbardi, R., Jain, E., Tang, Q., Moore, J. C., … Langenau, D. M. (2016). Single-cell transcriptional analysis of normal, aberrant, and malignant hematopoiesis in zebrafish. *The Journal of Experimental Medicine, 213*(6), 979–992. http://dx.doi.org/10.1084/jem.20152013.

Murati, A., Brecqueville, M., Devillier, R., Mozziconacci, M.-J., Gelsi-Boyer, V., & Birnbaum, D. (2012). Myeloid malignancies: mutations, models and management (Journal Article) *BMC Cancer, 12*(1), 1–15. http://dx.doi.org/10.1186/1471-2407-12-304.

Nangalia, J., & Green, T. R. (2014). The evolving genomic landscape of myeloproliferative neoplasms. *ASH Education Program Book, 2014*(1), 287–296. http://dx.doi.org/10.1182/asheducation-2014.1.287.

Network, T. C. G. A. R. (2013). Genomic and epigenomic landscapes of adult de novo acute myeloid leukemia. *New England Journal of Medicine, 368*(22), 2059–2074. http://dx.doi.org/10.1056/NEJMoa1301689.

Niemeyer, C. M., Kang, M. W., Shin, D. H., Furlan, I., Erlacher, M., Bunin, N. J., … Loh, M. L. (2010). Germline CBL mutations cause developmental abnormalities and predispose to juvenile myelomonocytic leukemia. *Nature Genetics, 42*(9), 794–800. http://www.nature.com/ng/journal/v42/n9/abs/ng.641.html#supplementary-information.

North, T. E., Goessling, W., Walkley, C. R., Lengerke, C., Kopani, K. R., Lord, A. M., … Zon, L. I. (2007). Prostaglandin E2 regulates vertebrate haematopoietic stem cell homeostasis. *Nature, 447*(7147), 1007–1011. http://www.nature.com/nature/journal/v447/n7147/suppinfo/nature05883_S1.html.

O'Neil, J., & Look, A. T. (2007). Mechanisms of transcription factor deregulation in lymphoid cell transformation. *Oncogene, 26*(47), 6838–6849.

Oates, A. C., Brownlie, A., Pratt, S. J., Irvine, D. V., Liao, E. C., Paw, B. H., … Wilks, A. F. (1999). Gene duplication of zebrafish JAK2 homologs is accompanied by divergent embryonic expression patterns: only jak2a is expressed during erythropoiesis. *Blood, 94*(8), 2622–2636.

Okuda, T., Cai, Z., Yang, S., Lenny, N., Lyu, C-j., van Deursen, J. M. A., … Downing, J. R. (1998). Expression of a knocked-in AML1-ETO leukemia gene inhibits the establishment of normal definitive hematopoiesis and directly generates dysplastic hematopoietic progenitors. *Blood, 91*(9), 3134–3143.

Onnebo, S. M. N., Rasighaemi, P., Kumar, J., Liongue, C., & Ward, A. C. (2012). Alternative TEL-JAK2 fusions associated with T-cell acute lymphoblastic leukemia and atypical chronic myelogenous leukemia dissected in zebrafish. *Haematologica, 97*(12), 1895–1903. http://dx.doi.org/10.3324/haematol.2012.064659.

Page, D. M., Wittamer, V., Bertrand, J. Y., Lewis, K. L., Pratt, D. N., Delgado, N., … Traver, D. (2013). An evolutionarily conserved program of B-cell development and activation in zebrafish. *Blood, 122*(8), e1–e11. http://dx.doi.org/10.1182/blood-2012-12-471029.

Palomero, T., & Ferrando, A. (2009). Therapeutic targeting of NOTCH1 signaling in T-ALL. *Clinical Lymphoma and Myeloma, 9*(Suppl. 3), S205. http://dx.doi.org/10.3816/CLM.2009.s.013.

Papaemmanuil, E., Gerstung, M., Bullinger, L., Gaidzik, V. I., Paschka, P., Roberts, N. D., … Campbell, P. J. (2016). Genomic classification and prognosis in acute myeloid leukemia. *New England Journal of Medicine, 374*(23), 2209–2221. http://dx.doi.org/10.1056/NEJMoa1516192.

Peng, X., Dong, M., Ma, L., Jia, X. E., Mao, J., Jin, C., … Chen, S. (2015). A point mutation of zebrafish c-cbl gene in the ring finger domain produces a phenotype mimicking human myeloproliferative disease (Original Article) *Leukemia, 29*(12), 2355–2365. http://dx.doi.org/10.1038/leu.2015.154.

Piovan, E., Yu, J., Tosello, V., Herranz, D., Ambesi-Impiombato, A., Da Silva, A. C., … Ferrando, A. A. (2013). Direct reversal of glucocorticoid resistance by AKT inhibition in acute lymphoblastic leukemia. *Cancer Cell, 24*(6), 766–776. http://dx.doi.org/10.1016/j.ccr.2013.10.022.

Rasighaemi, P., Basheer, F., Liongue, C., & Ward, A. C. (2015). Zebrafish as a model for leukemia and other hematopoietic disorders (Journal Article) *Journal of Hematology and Oncology, 8*(1), 1–10. http://dx.doi.org/10.1186/s13045-015-0126-4.

Reynolds, C., Roderick, J. E., LaBelle, J. L., Bird, G., Mathieu, R., Bodaar, K., … Gutierrez, A. (2014). Repression of BIM mediates survival signaling by MYC and AKT in high-risk T-cell acute lymphoblastic leukemia (Original Article) *Leukemia, 28*(9), 1819–1827. http://dx.doi.org/10.1038/leu.2014.78.

Ridges, S., Heaton, W. L., Joshi, D., Choi, H., Eiring, A., Batchelor, L., … Trede, N. S. (2012). Zebrafish screen identifies novel compound with selective toxicity against leukemia. *Blood, 119*(24), 5621–5631. http://dx.doi.org/10.1182/blood-2011-12-398818.

Robyn, J., & Metcalfe, D. D. (2006). Systemic mastocytosis. In F. W. Alt, K. F. Austen, T. Honjo, F. Melchers, J. W. Uhr, & E. R. Unanue (Eds.), *Advances in immunology* (Vol 89, pp. 169–243). Academic Press.

Roti, G., & Stegmaier, K. (2014). New approaches to target T-ALL (Review) *Frontiers in Oncology, 4*. http://dx.doi.org/10.3389/fonc.2014.00170.

Rudner, L. A., Brown, K. H., Dobrinski, K. P., Bradley, D. F., Garcia, M. I., Smith, A. C. H., … Frazer, J. K. (2011). Shared acquired genomic changes in zebrafish and human T-ALL. *Oncogene, 30*(41), 4289–4296. http://www.nature.com/onc/journal/v30/n41/suppinfo/onc2011138s1.html.

Ryan, J. A., Brunelle, J. K., & Letai, A. (2010). Heightened mitochondrial priming is the basis for apoptotic hypersensitivity of CD4+ CD8+ thymocytes. *Proceedings of the National Academy of Sciences of the United States of America, 107*(29), 12895–12900. http://dx.doi.org/10.1073/pnas.0914878107.

Sabaawy, H. E., Azuma, M., Embree, L. J., Tsai, H.-J., Starost, M. F., & Hickstein, D. D. (2006). TEL-AML1 transgenic zebrafish model of precursor B cell acute lymphoblastic leukemia. *Proceedings of the National Academy of Sciences of the United States of America, 103*(41), 15166–15171. http://dx.doi.org/10.1073/pnas.0603349103.

Scott, E., Simon, M., Anastasi, J., & Singh, H. (1994). Requirement of transcription factor PU.1 in the development of multiple hematopoietic lineages. *Science, 265*(5178), 1573–1577. http://dx.doi.org/10.1126/science.8079170.

Shen, L.-J., Chen, F.-Y., Zhang, Y., Cao, L.-F., Kuang, Y., Zhong, M., … Zhong, H. (2013). *MYCN* transgenic zebrafish model with the characterization of acute myeloid leukemia and altered hematopoiesis. *PLoS One, 8*(3), e59070. http://dx.doi.org/10.1371/journal.pone.0059070.

Shi, X., He, B.-L., Ma, A. C. H., Guo, Y., Chi, Y., Man, C. H., … Leung, A. Y. H. (2015). Functions of idh1 and its mutation in the regulation of developmental hematopoiesis in zebrafish. *Blood, 125*(19), 2974–2984. http://dx.doi.org/10.1182/blood-2014-09-601187.

Skoda, R. C., Duek, A., & Grisouard, J. (2015). Pathogenesis of myeloproliferative neoplasms. *Experimental Hematology, 43*(8), 599–608. http://dx.doi.org/10.1016/j.exphem.2015.06.007.

Smith, A. C. H., Raimondi, A. R., Salthouse, C. D., Ignatius, M. S., Blackburn, J. S., Mizgirev, I. V., … Langenau, D. M. (2010). High-throughput cell transplantation establishes that tumor-initiating cells are abundant in zebrafish T-cell acute lymphoblastic leukemia. *Blood, 115*(16), 3296–3303. http://dx.doi.org/10.1182/blood-2009-10-246488.

Smith, C. C., Wang, Q., Chin, C.-S., Salerno, S., Damon, L. E., Levis, M. J., … Shah, N. P. (2012). Validation of ITD mutations in FLT3 as a therapeutic target in human acute myeloid leukaemia. *Nature, 485*(7397), 260–263. http://www.nature.com/nature/journal/v485/n7397/abs/nature11016.html#supplementary-information.

Sood, R., English, M. A., Belele, C. L., Jin, H., Bishop, K., Haskins, R., … Liu, P. P. (2010). Development of multilineage adult hematopoiesis in the zebrafish with a runx1 truncation mutation. *Blood, 115*(14), 2806–2809. http://dx.doi.org/10.1182/blood-2009-08-236729.

Steidl, U., Rosenbauer, F., Verhaak, R. G. W., Gu, X., Ebralidze, A., Otu, H. H., … Tenen, D. G. (2006). Essential role of Jun family transcription factors in PU.1 knockdown-induced leukemic stem cells. *Nature Genetics, 38*(11), 1269–1277. http://www.nature.com/ng/journal/v38/n11/suppinfo/ng1898_S1.html.

Stern, H. M., & Zon, L. I. (July 2003). Cancer genetics and drug discovery in the zebrafish. *Nature Reviews Cancer, 3*, 533–539. http://dx.doi.org/10.1038/nrc1126.

Stieglitz, E., Taylor-Weiner, A. N., Chang, T. Y., Gelston, L. C., Wang, Y.-D., Mazor, T., … Loh, M. L. (2015). The genomic landscape of juvenile myelomonocytic leukemia (Article) *Nature Genetics, 47*(11), 1326–1333 http://www.nature.com/ng/journal/v47/n11/abs/ng.3400.html#supplementary-information.

Sun, J., Liu, W., Li, L., Chen, J., Wu, M., Zhang, Y., … Liao, W. (2013). Suppression of Pu.1 function results in expanded myelopoiesis in zebrafish (Letter to the Editor) *Leukemia, 27*(9), 1913–1917. http://dx.doi.org/10.1038/leu.2013.67.

Sundaravel, S., Duggan, R., Bhagat, T., Ebenezer, D. L., Liu, H., Yu, Y., … Wickrema, A. (2015). Reduced DOCK4 expression leads to erythroid dysplasia in myelodysplastic syndromes. *Proceedings of the National Academy of Sciences of the United States of America, 112*(46), E6359–E6368. http://dx.doi.org/10.1073/pnas.1516394112.

Swaminathan, G., & Tsygankov, A. Y. (2006). The Cbl family proteins: ring leaders in regulation of cell signaling. *Journal of Cellular Physiology, 209*(1), 21–43. http://dx.doi.org/10.1002/jcp.20694.

Tang, Q., Abdelfattah, N. S., Blackburn, J. S., Moore, J. C., Martinez, S. A., Moore, F. E., … Langenau, D. M. (2014). Optimized cell transplantation using adult rag2 mutant zebrafish (Brief Communication) *Nature Methods, 11*(8), 821–824 http://www.nature.com/nmeth/journal/v11/n8/abs/nmeth.3031.html#supplementary-information.

Tang, Q., Moore, J. C., Ignatius, M. S., Tenente, I. M., Hayes, M. N., Garcia, E. G., … Langenau, D. M. (2016). Imaging tumour cell heterogeneity following cell transplantation into optically clear immune-deficient zebrafish (Article) *Nature Communications, 7*. http://dx.doi.org/10.1038/ncomms10358.

Tefferi, A. (2016). Myeloproliferative neoplasms: a decade of discoveries and treatment advances. *American Journal of Hematology, 91*(1), 50–58. http://dx.doi.org/10.1002/ajh.24221.

Tefferi, A., & Gilliland, D. G. (2007). Oncogenes in myeloproliferative disorders. *Cell Cycle, 6*(5), 550–566. http://dx.doi.org/10.4161/cc.6.5.3919.

Treanor, L. M., Volanakis, E. J., Zhou, S., Lu, T., Sherr, C. J., & Sorrentino, B. P. (2011). Functional interactions between Lmo2, the Arf tumor suppressor, and Notch1 in murine T-cell malignancies. *Blood, 117*(20), 5453–5462. http://dx.doi.org/10.1182/blood-2010-09-309831.

Trinquand, A., Tanguy-Schmidt, A., Ben Abdelali, R., Lambert, J., Beldjord, K., Lengliné, E., … Asnafi, V. (2013). Toward a NOTCH1/FBXW7/RAS/PTEN–based oncogenetic risk classification of adult T-cell acute lymphoblastic leukemia: a Group for Research in Adult Acute Lymphoblastic Leukemia Study. *Journal of Clinical Oncology, 31*(34), 4333–4342. http://dx.doi.org/10.1200/jco.2012.48.5292.

Van Vlierberghe, P., & Ferrando, A. (2012). The molecular basis of T cell acute lymphoblastic leukemia. *The Journal of Clinical Investigation, 122*(10), 3398–3406. http://dx.doi.org/10.1172/jci61269.

Viny, A. D., & Levine, R. L. (2014). Genetics of myeloproliferative neoplasms. *The Cancer Journal, 20*(1), 61–65. http://dx.doi.org/10.1097/ppo.0000000000000013.

Vo, T.-T., Ryan, J., Carrasco, R., Neuberg, D., Rossi, D. J., Stone, R. M., … Letai, A. (2012). Relative mitochondrial priming of myeloblasts and normal HSCs determines chemotherapeutic success in AML. *Cell, 151*(2), 344–355. http://dx.doi.org/10.1016/j.cell.2012.08.038.

Volanakis, E. J., Williams, R. T., & Sherr, C. J. (2009). Stage-specific Arf tumor suppression in Notch1-induced T-cell acute lymphoblastic leukemia. *Blood, 114*(20), 4451–4459. http://dx.doi.org/10.1182/blood-2009-07-233346.

Ward, A. F., Braun, B. S., & Shannon, K. M. (2012). Targeting oncogenic Ras signaling in hematologic malignancies. *Blood, 120*(17), 3397–3406. http://dx.doi.org/10.1182/blood-2012-05-378596.

Weng, A. P., Ferrando, A. A., Lee, W., Morris, J. P., Silverman, L. B., Sanchez-Irizarry, C., … Aster, J. C. (2004). Activating mutations of NOTCH1 in human T cell acute lymphoblastic leukemia. *Science, 306*(5694), 269–271. http://dx.doi.org/10.1126/science.1102160.

White, R. M., Cech, J., Ratanasirintrawoot, S., Lin, C. Y., Rahl, P. B., Burke, C. J., … Zon, L. I. (2011). DHODH modulates transcriptional elongation in the neural crest and melanoma. *Nature, 471*(7339), 518–522. http://www.nature.com/nature/journal/v471/n7339/abs/10.1038-nature09882-unlocked.html#supplementary-information.

Woo, J. S., Alberti, M. O., & Tirado, C. A. (2014). Childhood B-acute lymphoblastic leukemia: a genetic update (Journal Article) *Experimental Hematology and Oncology, 3*(1), 1–14. http://dx.doi.org/10.1186/2162-3619-3-16.

Yeh, J.-R. J., Munson, K. M., Chao, Y. L., Peterson, Q. P., MacRae, C. A., & Peterson, R. T. (2008). AML1-ETO reprograms hematopoietic cell fate by downregulating scl expression. *Development, 135*(2), 401–410. http://dx.doi.org/10.1242/dev.008904.

Yeh, J.-R. J., Munson, K. M., Elagib, K. E., Goldfarb, A. N., Sweetser, D. A., & Peterson, R. T. (2009). Discovering chemical modifiers of oncogene-regulated hematopoietic differentiation. *Nature Chemical Biology, 5*(4), 236–243. http://www.nature.com/nchembio/journal/v5/n4/suppinfo/nchembio.147_S1.html.

Yergeau, D. A., Hetherington, C. J., Wang, Q., Zhang, P., Sharpe, A. H., Binder, M., … Zhang, D.-E. (1997). Embryonic lethality and impairment of haematopoiesis in mice heterozygous for an AML1-ETO fusion gene. *Nature Genetics, 15*(3), 303–306. http://dx.doi.org/10.1038/ng0397-303.

Young, A., Lyons, J., Miller, A. L., Phan, V. T., Alarcón, I. R., & McCormick, F. (2009). Chapter 1 Ras signaling and therapies. In *Advances in cancer research* (Vol. 102, pp. 1–17). Academic Press.

Zeisig, B. B., Kulasekararaj, A. G., Mufti, G. J., & So, E. C. W. (2012). SnapShot: acute myeloid leukemia. *Cancer Cell, 22*(5), 698–698.e691. http://dx.doi.org/10.1016/j.ccr.2012.10.017.

Zhang, J., Ding, L., Holmfeldt, L., Wu, G., Heatley, S. L., Payne-Turner, D., … Mullighan, C. G. (2012). The genetic basis of early T-cell precursor acute lymphoblastic leukaemia. *Nature, 481*(7380), 157–163. http://www.nature.com/nature/journal/v481/n7380/abs/nature10725.html#supplementary-information.

Zhuravleva, J., Paggetti, J., Martin, L., Hammann, A., Solary, E., Bastie, J.-N., ... Delva, L. (2008). MOZ/TIF2-induced acute myeloid leukaemia in transgenic fish. *British Journal of Haematology, 143*(3), 378–382. http://dx.doi.org/10.1111/j.1365-2141.2008.07362.x.

Zhu, X., Zhang, H., Qian, M., Zhao, X., Yang, W., Wang, P., ... Wang, K. (2012). The significance of low PU.1 expression in patients with acute promyelocytic leukemia (Journal Article) *Journal of Hematology and Oncology, 5*(1), 1–6. http://dx.doi.org/10.1186/1756-8722-5-22.

Zon, L. I., & Peterson, R. T. (2005). In vivo drug discovery in the zebrafish. *Nature Reviews Drug Discovery, 4*(1), 35–44. http://dx.doi.org/10.1038/nrd1606.

Zuurbier, L., Petricoin, E. F., Vuerhard, M. J., Calvert, V., Kooi, C., Buijs-Gladdines, J. G., ... Meijerink, J. P. P. (2012). The significance of PTEN and AKT aberrations in pediatric T-cell acute lymphoblastic leukemia. *Haematologica, 97*(9), 1405–1413. http://dx.doi.org/10.3324/haematol.2011.059030.

CHAPTER

Investigating microglia-brain tumor cell interactions in vivo in the larval zebrafish brain

21

K.R. Astell, D. Sieger[1]

University of Edinburgh, Edinburgh, United Kingdom

[1]*Corresponding author: E-mail: Dirk.Sieger@ed.ac.uk*

CHAPTER OUTLINE

Methods in Cell Biology, Volume 138, ISSN 0091-679X, http://dx.doi.org/10.1016/bs.mcb.2016.10.001

Abstract

Glioblastoma is the most frequent and aggressive primary malignant brain tumor. Gliomas exhibit high genetic diversity in addition to complex and variable clinical features. Glioblastoma tumors are highly resistant to multimodal therapies and there is significant patient mortality within the first two years after prognosis. At present clinical treatments are palliative, not curative.

Glioblastomas contain a high number of microglia and infiltrating macrophages, which are positively correlated with glioma grade and invasiveness. Microglia are the resident macrophages of the central nervous system. These cells constantly scan the brain and react promptly to any abnormality, removing detrimental factors and safeguarding the central nervous system against further damage. Microglia and macrophages that have colonized the glioblastoma display protumoral functions and promote tumor growth.

The optically transparent zebrafish larva facilitates imaging of fluorescently labeled cells at high spatial and temporal resolution in vivo. It is therefore an excellent model to investigate microglia-glioma cell interactions at the early stages of tumor development. Here we provide several methods that can be used to study the early stages of microglia-glioma cell interactions in the zebrafish. We present a technique for the xenotransplantation of mammalian oncogenic cells into the zebrafish brain and provide advice for image capture and analysis.

INTRODUCTION

GLIOBLASTOMA

Glioblastoma is the most common and aggressive tumor of the central nervous system (Wen & Kesari, 2008). Glioblastoma is classified as a World Health Organization grade IV neoplasm (Louis et al., 2007). These tumors are malignant and mitotically active with widespread infiltration of the surrounding tissue. They are associated with rapid disease progression and a fatal outcome (Louis et al., 2007).

Glioblastomas are histologically and genetically heterogeneous, which may be partly due to the cell type of origin. Originally glioblastomas were thought to originate from glial cells, as a high proportion of glioblastomas express the glial-specific marker GFAP (Jones, Bigner, Schold, Eng, & Bigner, 1981). However, current evidence suggests that gliomas can originate from a variety of cells of the central nervous system including neural stem cells, astrocytes, and oligodendrocyte precursor cells (Zong, Parada, & Baker, 2015). In addition, there is growing evidence that gliomas can be maintained by a population of cancerous cells that have similar properties to neuronal stem cells (Hemmati et al., 2003; Singh et al., 2003). These cells display stem cell characteristics, including long-term self-renewal and a capacity to differentiate. This is thought to contribute to the complex heterogeneity and invasiveness of glioblastomal tumors (Pollard et al., 2009) as well as posttreatment recurrence (Jun, Bronson, & Charest, 2014).

Genes commonly mutated in glioblastoma have been found to affect several signaling pathways including activation of the RTK/RAS/PI-3K pathway (translation, cell survival, and proliferation), inactivation of the p53 tumor suppressor pathway (apoptosis and senescence), and inactivation of the RB tumor suppressor pathway (G1/S progression; Cancer Genome Atlas Research Network, 2008).

At present, the standard treatment for glioblastoma includes maximal tumor resection followed by radiotherapy and temozolomide chemotherapy (Stupp et al., 2005; Weller et al., 2014). However, despite advancements in neurooncology, the prognosis for patients with glioblastoma is still extremely poor. It was recently reported that the overall survival of glioblastoma patients was 46.4% at 1 year, 22.5% at 2 years, and 14.4% at 3 years (Gramatzki et al., 2016), while a previous study reported a 5% survival rate at 5 years post diagnosis (Ostrom et al., 2014).

MICROGLIA

Microglia are a member of the innate immune system and the resident macrophages of the central nervous system (Kettenmann, Hanisch, Noda, & Verkhratsky, 2011). Microglia have a very dynamic morphology. Under normal physiological conditions mature microglia display a ramified morphology (Fig. 1A), with a small stationary

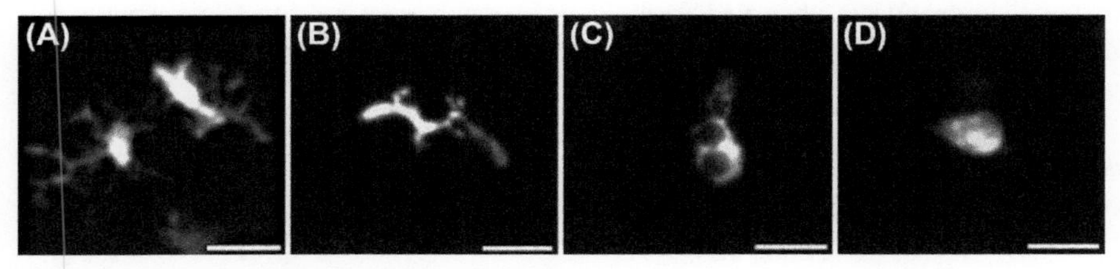

FIGURE 1 Microglia exhibit a dynamic morphology.

(A) Ramified microglia have a small stationary cell body with multiple thin processes that are continuously extended and retracted to survey the surrounding tissue. (B) Microglia with an intermediate phenotype exhibit fewer, thicker processes and may have a larger cell body when compared to ramified microglia. (C) Phagocytic microglia exhibit phagosomes within the cytosol that contain engulfed material. (D) Amoeboid microglia have a rounded morphology, with a large cell body and no processes. Maximum intensity projections of 3D z-stacks. Scale bars are 20 μm.

cell body and fine motile processes that are continuously extended and retracted to survey the surrounding tissue (Nimmerjahn, Kirchhoff, & Helmchen, 2005). In response to an injury or pathogen, the microglia assume an amoeboid morphology (Fig. 1D) and migrate to the site of infection or the lesion. Here, they phagocytose and destroy the pathogen or remove damaged cells and detrimental factors, safeguarding the central nervous system against further damage (Davalos et al., 2005; Haynes et al., 2006; Neumann, Kotter, & Franklin, 2009; Sieger, Moritz, Ziegenhals, Prykhozhij, & Peri, 2012). Microglia can also exhibit intermediate morphologies (Fig. 1B), with shorter and thicker processes and a cell body that is larger than that seen for a ramified microglia (Karperien, Ahammer, & Jelinek, 2013), as well as phagocytic phenotypes where the phagosomes are visible (Fig. 1C).

During zebrafish development macrophage precursor cells originate in the rostralmost lateral mesoderm. These cells then migrate to the yolk sac and in the process start expressing leucocyte-specific l-plastin. By 24 hours post fertilization (hpf) the premacrophages have accumulated in the yolk sac and begin to exhibit macrophage-like morphologies (Herbomel, Thisse, & Thisse, 1999). The young macrophages migrate, through the zebrafish tissues, into the head mesenchyme between 23 and 30 hpf and have begun to colonize the brain itself by 35 hpf (Herbomel, Thisse, & Thisse, 2001). By 48 hpf young macrophages can be detected in the forebrain, midbrain, and hindbrain. At 60 hpf the number of macrophages specifically in the optic tectum increases, which coincides with differentiation into early microglia. Expression of apolipoprotein E, in addition to morphological features, confirms that these cells have gained a microglia phenotype at 72 hpf (Herbomel et al., 2001). Between 3 and 5 dpf the microglia exhibit high motility and a phagocytic amoeboid phenotype. At 5 dpf the microglia motility slows and the cells begin to adopt a ramified phenotype. However, they continue to morph between the amoeboid and ramified phenotypes until 10 dpf (Svahn et al., 2013). During this time zebrafish microglia form direct contacts with neurons and the developing vasculature as well as each other (Svahn et al., 2013). A second increase in tectal microglia is observed from 15 dpf, which may be due to microglia cell division and the infiltration of a new cell population derived from the ventral wall of the dorsal aorta (Svahn et al., 2013; Xu et al., 2015).

As well as injury responses and immune surveillance, microglia have several additional roles in brain development and homeostasis. These include clearance of apoptotic cells in the central nervous system (Peri & Nüsslein-Volhard, 2008; Sierra et al., 2010), supporting neurogenesis and brain wiring (Shigemoto-Mogami, Hoshikawa, Goldman, Sekino, & Sato, 2014; Squarzoni et al., 2014), synaptic refinement (Paolicelli et al., 2011; Schafer et al., 2012), as well as growth and patterning of the vasculature (Fantin et al., 2010; Rymo et al., 2011). The functional plasticity of microglia to perform different tasks can be classified as proinflammatory (M1) and antiinflammatory (M2). This M1/M2 classification, applicable to all macrophages, is a basic biochemical definition based on the molecules that macrophages (and hence, microglia) produce while performing their different functions (reviewed by Mills, 2012). It should be noted that the M1/M2 classification defines opposing responses; however, in reality macrophages and microglia may exhibit a spectrum of M1/M2 molecules in a single cell as well as a mixture of M1/M2 phenotypes within the cell population (Hsieh et al., 2013; Hu et al., 2015).

MICROGLIA PROMOTE GLIOBLASTOMA GROWTH

Microglia and circulating macrophages are attracted to glioblastoma tumors (Badie & Schartner, 2001; Graeber, Scheithauer, & Kreutzberg, 2002). Glioblastomas may attract microglia by expressing chemoattractants such as monocyte chemotactic protein-3 (MCP-3; Okada et al., 2009), colony-stimulating factor 1 (CSF-1; Alterman & Stanley, 1994), and hepatocyte growth factor/scatter factor (Badie, Schartner, Klaver, & Vorpahl, 1999). These tumor-associated microglia/macrophages (TAM/Ms) can account for up to a third of the tumor mass (Hambardzumyan, Gutmann, & Kettenmann, 2016; Morantz, Wood, Foster, Clark, & Gollahon, 1979) and the number of TAM/Ms within a glioblastoma tumor is positively correlated with tumor grade and invasiveness (Komohara, Ohnishi, Kuratsu, & Takeya, 2008).

TAM usually have an amoeboid morphology associated with the proinflammatory (M1) phenotype (Graeber et al., 2002); however, they exhibit an M2 antiinflammatory polarization and protumoral activity. It is thought that glioblastomas actively suppress the proinflammatory phenotype in microglia by producing antiinflammatory cytokines, such as IL-10, IL-4, IL-6, TGF-b, and prostaglandin E2 (Charles, Holland, Gilbertson, Glass, & Kettenmann, 2012). At the same time, glioblastoma-derived factors such as M-CSF actively support the microglial M2, antiinflammatory phenotype (Komohara et al., 2008). In addition TAM seem to be less efficient at initiating an antitumor T-cell response due to impaired MHC class II expression (Flügel, Labeur, Grasbon-Frodl, Kreutzberg, & Graeber, 1999; Schartner et al., 2005).

Microglia support glioblastoma tumor growth in several different ways. Microglia promote the invasiveness and growth of glioblastoma. Glioblastoma cells produce metalloprotease 2, which activates microglia-derived type 1 metalloprotease and degrades the extracellular matrix, promoting invasion of glioblastoma cells into the surrounding brain (Markovic, Glass, Synowitz, Rooijen, & Kettenmann, 2005; Markovic et al., 2009). As a result, glioblastoma cell invasion was significantly reduced in microglia-depleted mouse brain slices (Markovic et al., 2005) and tumor size was significantly reduced when microglia were depleted in an in vivo model (Markovic et al., 2009). In addition, microglia promote the migration of mouse GL261 glioma cells in vitro (Bettinger, Thanos, & Paulus, 2002). Microglia are also a source of proangiogenic growth factors and cytokines, such as CXCL2, and depletion of microglia caused reduced vasculature in tumors (Brandenburg et al., 2016). Finally, the sum of all these activities defines the protumoral activity of microglia. Recent evidence suggests that microglia can be "re-educated" within the glioma environment, which might open the door for new therapies. In a study based on a mouse glioma model, inhibition of the CSF-1R led to a decrease in M2 marker expression on microglia, an increase in their phagocytic activity, and finally to a decrease in tumor volume (Pyonteck et al., 2013).

Given the ineffectiveness of current therapies against gliomas and the promoting role of macrophages/microglia during tumor progression, there is a genuine clinical need to understand the signaling mechanisms that regulate these cells within the tumor environment. Understanding these signaling mechanisms, in vivo, is the first step towards developing therapeutic strategies that either inhibit protumoral activity or promote an antitumoral microglia response. Here we present protocols for the investigation of glioblastoma–microglia interactions in vivo using the zebrafish larval brain as a model. We give information about suitable transgenic lines for visualizing microglia as well as mutant lines that have either reduced numbers of microglia, or that lack microglia completely during early development. We present a technique for the xenotransplantation of mammalian glioblastoma cells, into the zebrafish brain, to study glioblastoma–microglia interactions in vivo with high special and temporal resolution, and provide advice for image capture and analysis.

1. METHODS

1.1 THE ZEBRAFISH IS AN EXCELLENT MODEL TO STUDY IMMUNOLOGY AND CANCER BIOLOGY

Several attributes make the zebrafish larva an excellent model for studying the interactions between microglia and brain tumors. (1) The larvae are optically transparent, facilitating noninvasive live imaging at high spatial and temporal resolution. (2) There is good conservation of the genome and molecular pathways between the zebrafish and mammals (Gates et al., 1999; Oosterhof, Boddeke, & van Ham, 2015). (3) The larval zebrafish can be used as a xenograft model without undesirable immune responses or the need for immunosuppression because the zebrafish adaptive immune system does not mature until 28 dpf (Lam, Chua, Gong, Lam, & Sin, 2004). (4) Larvae can absorb small molecular weight compounds directly from the surrounding water, enabling high throughput screening of potential chemotherapeutics (Terriente & Pujades, 2013).

1.2 VISUALIZING MICROGLIA IN THE ZEBRAFISH LARVA

1.2.1 Neutral red staining

Microglia and macrophages can be identified by phagocytosis of the eurhodin dye neutral red (Herbomel et al., 2001; Shiau, Kaufman, Meireles, & Talbot, 2015):

1. Incubate larvae in 2.5 μg/mL neutral red in embryo medium (6.4 mM KCl, 0.22 mM NaCl, 0.33 mM $CaCl_2 \cdot 2H_2O$, 0.33 mM $MgSO4 \cdot 7H_2O$) for 2–5 h at 25–30°C in the dark;

2. Remove the neutral red with two changes of the embryo medium;
3. Analyze anaesthetized larvae immediately or at desired time points using a stereomicroscope. To anaesthetize larvae, incubate in 2.5 mM Ethyl 3-amino-benzoate methanesulfonate (tricaine) in embryo medium until nonresponsive to touch.

1.2.2 Transgenic reporter lines and microglia mutant zebrafish

There are several transgenic zebrafish lines (Table 1) in which microglia and/or macrophages express a fluorescent protein. This allows them to be visualized by fluorescent microscopy. There are also several transgenic lines that allow visualization of microglia activation states (Table 2) as well as microglia mutant lines in which microglia are reduced or absent at particular stages of development (Table 3).

1.2.3 Whole mount immunohistochemistry

It is not always appropriate to use transgenic zebrafish lines in which microglia are labeled. In these circumstances, the microglia can be detected by performing immunohistochemistry (IHC) on fixed specimens. An antibody against the zebrafish lymphocyte cytosolic protein 1 (L-plastin; Feng, Santoriello, Mione, Hurlstone, & Martin, 2010) can be used to detect cells from the monocyte lineage, including microglia at early stages of development. L-plastin is a leukocyte-specific actin-binding protein that is involved in leukocyte adhesion and activation (Jones, Wang, Turck, & Brown, 1998). The 4C4 antibody can be used to specifically detect microglia in the zebrafish. The epitope that is recognized by the 4C4 antibody has not been identified; however, based on morphological criteria and location, the cells labeled by this antibody are microglia (Becker & Becker, 2001). Here we describe the reagents and the whole mount IHC protocol that we use to detect microglia within zebrafish larvae aged between 0 and 12 dpf.

Reagents

- Phosphate buffered saline (PBS): 1× PBS (0.01 M phosphate buffer, 0.0027 M potassium chloride, and 0.137 M sodium chloride, pH 7.4)
- PFA fixative: 4% paraformaldehyde + 1% DMSO in PBS
- PBST: 0.2% Triton-X100 in PBS
- Collagenase-PBS: 2 mg/mL crude collagenase from *Clostridium histolyticum* (Sigma-Aldrich) in PBS (we use a 1:5 dilution of a 10 mg/mL stock solution)
- Glycine-PBST: 50 mM glycine
- Whole mount IHC blocking buffer: (1% DMSO +1% normal goat serum +1% bovine serum albumin +0.7% Triton-X100 in PBS)
- Glycerol-PBS: 70% glycerol in PBS

Primary antibodies

- Anti-mouse, 4C4 (Becker & Becker, 2001).
  1:50 dilution; 8 μL antibody +392 μL whole mount IHC blocking buffer
- Anti-rabbit, L-Plastin (Feng et al., 2010)
  1:500 dilution; 0.8 μL antibody +399.2 μL whole mount IHC blocking buffer

Table 1 Transgenic Zebrafish Lines for Visualization of Microglia and/or Macrophages

Tg Line	Cell Type	Reporter	Primary References
Tg(spi1b: GAL4,UAS:GFP)	Myeloid cells	GFP expressed under the control of the spi1/pU1 promoter	Peri and Nüsslein-Volhard (2008)
Tg(spi1b: GAL4,UAS: TagRFP)	Myeloid cells	RFP expressed under the control of the spi1/pU1 promoter	Sieger et al. (2012)
Tg(mpeg1:EGFP)	Macrophages & microglia	EGFP expressed under control of the mpeg1 promoter (fragment)	Ellett, Pase, Hayman, Andrianopoulos, and Lieschke (2011)
Tg(mpeg1: mCherry)	Macrophages & microglia	mCherry expressed under the control of the mpeg1 promoter (fragment)	Ellett et al. (2011)
Tg(mpeg1:GAL4/ UAS:Kaede)	Macrophages & microglia	The photoconvertible Kaede protein under the control of the mpeg1 promoter (fragment)	Ellett et al. (2011)
Tg(apoeb:LY-EGFP)	Microglia	Membrane-bound GFP expressed from the apolipoprotein-E locus (generated by BAC HR in bacteria)	Peri and Nüsslein-Volhard (2008)
Tg(p2ry12:p2ry12-GFP)	Microglia	GFP fused to P2Y12 receptor (generated by BAC HR in bacteria)	Sieger et al. (2012)

BAC, bacterial artificial chromosome; *EGFP,* enhanced green fluorescent protein; *GFP,* green fluorescent protein; *HR,* homologous recombination; *RFP,* red fluorescent protein; *Tg,* transgenic.

Secondary antibodies

- Goat anti-mouse IgG (H + L) secondary antibody, Alexa Fluor 405 conjugate
- Goat anti-mouse IgG (H + L) secondary antibody, Alexa Fluor 488 conjugate
- Goat anti-mouse IgG (H + L) secondary antibody, Alexa Fluor 647 conjugate

Table 2 Transgenic Zebrafish Lines to Visualize the Activation State of Microglia

Tg Line	Application	Primary References
Tg(mpeg1:mCherryF/tnfa:eGFP-F)	eGFP-F expression in cells expressing the inflammatory cytokine, tumor necrosis factor alpha Marker of microglia/macrophages with an M1 polarization	Nguyen-Chi et al. (2015)
Tg(il1b:GFP-F)	eGFP-F expression in cells expressing the interleukin 1 beta (Il1β) proinflammatory cytokine Il1b is expressed in microglia, macrophages, and neutrophils in response to "danger signals"	Nguyen-Chi et al. (2014)

eGFP, enhanced green fluorescent protein; *F,* the farnesylation sequence of p21(Ras); *GFP,* green fluorescent protein; *Tg,* transgenic.

- Goat anti-rabbit IgG (H + L) secondary antibody, Alexa Fluor 405 conjugate
- Goat anti-rabbit IgG (H + L) secondary antibody, Alexa Fluor 488 conjugate
- Goat anti-rabbit IgG (H + L) secondary antibody, Alexa Fluor 647 conjugate

All secondary antibodies are diluted 1:200 prior to incubation (2 μL secondary antibody +388 μL whole mount IHC blocking buffer).

Day 1, fixation, digestion, blocking reactive sites, and primary antibody incubation

1. Anaesthetize the larvae using 2.5 mM tricaine in embryo medium;
2. Transfer the larvae to 1.5 mL microtubes and remove the excess tricaine-embryo medium;
   Note: We remove the supernatant from the microtubes using a 10 mL syringe fitted with a 23G × 1″ -Nr. 16 hypodermic needle (BD Microlance 3). We have found this as the most effective way to remove the supernatant without disturbing the larvae.
3. Fix larvae in 500 μl PFA fixative for 2 h (h) at room temperature (RT);
4. Wash larvae 4 × 5 min in 1 mL PBS, with gentle agitation, at RT;
5. Wash larvae 2 × 5 min in 1 mL PBST, with gentle agitation, at RT;
6. Incubate the larvae in 400 μl collagenase-PBS, without agitation for 25–45 min at RT;
   Note: We incubate larvae aged between 0 and 4 dpf for 25 min and we incubate larvae aged between 5 and 12 dpf for 45 min.
7. Wash the larvae 3 × 5 min in 1 mL PBST, with gentle agitation, at RT;
8. Incubate the larvae in 1 mL 50 mM glycine for 10 min at RT;

Table 3 Zebrafish Mutants With Reduced or Absent Microglia

Tg/Mutant Line	Mutation	Application	Primary References
irf8$^{st95/st95}$;*gl22*	A frameshift mutation 3′ of the *irf8* translational initiation codon that generates a premature stop codon	*irf8* is a transcription factor and a member of the interferon regulatory factor (IRF) family *irf8*$^{-/-}$ zebrafish lack macrophages until around 7 dpf and microglia until 31 dpf	Shiau et al. (2015)
nlrc3l$^{st73/st73}$	A point mutation in *nlrc3-like* that causes a premature stop codon	*nlrc3-like* is a negative regulator of macrophage activation and inflammation Aberrant activation of microglia precursor cells prevents migration and the cells undergo premature cell death Microglia are absent in mutants until 9 dpf	Shiau, Monk, Joo, and Talbot (2013)
NO067^{t30713}	A to T change in the splice acceptor of intron 4 of *slc7a7* causing deletion of exon 5, a frameshift and premature stop codon	Scl7a7, a transporter for arginine and leucine, is expressed in a macrophage sublineage that gives rise to the microglia in the brain Homozygous mutants survive until 7 dpf and lack microglia in the eyes and brain throughout this time	Rossi, Casano, Henke, Richter, and Peri (2015)
Panther^{j4blue}	A loss-of-function mutation of *fms*	*fms* codes for the colony-stimulating factor 1 receptor (Csf1r), a tyrosine kinase transmembrane receptor	Herbomel et al. (2001), Parichy, Ransom, Paw, Zon, and Johnson (2000)

Table 3 Zebrafish Mutants With Reduced or Absent Microglia—cont'd

Tg/Mutant Line	Mutation	Application	Primary References
		Macrophage and microglia precursor cells in *panther* mutants exhibit failed/delayed migration from the yolk sac during development	
		Microglia are either absent or greatly reduced in the brain until 7.5 dpf	

dpf, days post fertilization; *st,* stop; *Tg,* transgenic.

Note: Aldehyde fixatives, such as formalin or PFA, can be problematic for immunofluorescence as they produce free aldehyde groups that bind to the primary or secondary antibodies and produce high background signals in the tissue. Glycine binds to these free aldehyde groups and prevents them binding to the antibodies.

9. Wash the larvae 1 × 5 min in 1 mL PBST, with gentle agitation, at RT;
10. Incubate the larvae in 1 mL whole mount IHC blocking buffer for 2 h at RT;
11. Incubate the larvae in primary antibody, diluted in whole mount IHC blocking buffer (400 μL), overnight, with gentle agitation, at 4°C.
    Note: We place the tubes, in a rack, on an orbital shaker in the cold room

Day 2, washes and incubation with secondary antibody

12. At the beginning of the afternoon of the second day, wash the larvae 3 × 15 min in 1 mL PBS with Tween 20 (PBST), with gentle agitation, at RT;
13. Incubate the larvae in 1 mL PBST, at RT, until the end of the working day;
14. At the end of the working day, incubate the larvae in the secondary antibody, diluted in whole mount IHC blocking buffer (400 μL), overnight at 4°C with gentle agitation.
    Note: As fluorophores are light sensitive, this step and all subsequent steps must be performed with the larvae protected from light.

Day 3, washes and preparation for imaging.

15. The following morning, wash the larvae 2 × 15 min in 1 mL PBST, with gentle agitation, at RT;
16. Wash the larvae 3 × 15 min in 1 mL PBS, with gentle agitation, at RT;
17. Incubate the larvae in 70% glycerol-PBS for 30 min at RT;

Note: The glycerol functions as an aqueous mounting medium. It protects the larvae from damage and adds contrast during microscopy. Larvae can be stored in 70% glycerol-PBS at 4°C, protected from light, for up to one month.

18. To prepare the larvae for imaging, add 500 µl PBS to each microtube to dilute the glycerol and then immediately remove the supernatant;
19. Add 1 mL PBS and then immediately remove the supernatant and mount the larvae in low melting point agarose for imaging (Section 1.4).

    Note: Do not keep the larvae in PBS for prolonged periods of time as this is detrimental to the fluorescent signal.

1.3 THE XENOGRAFT TECHNIQUE TO STUDY MICROGLIA-GLIOBLASTOMA CELL INTERACTIONS IN VIVO

A xenotransplantation or xenograft is the transplantation of cells from one species into a recipient of a different species. The zebrafish tumor xenotransplantation model has emerged as an excellent model system to investigate tumor growth, invasiveness, angiogenesis, and inflammation (reviewed by Konantz et al., 2012; Veinotte, Dellaire, & Berman, 2014). We have developed a xenotransplantation protocol to specifically study in vivo interactions between glioblastoma cells and microglia (Fig. 2) that could easily be used for additional types of brain cancer cells.

1.3.1 Obtaining larvae for xenografting

There are several important factors to consider when raising zebrafish to embryonic (Kimmel, Ballard, Kimmel, Ullmann, & Schilling, 1995) and postembryonic stages (Parichy, Elizondo, Mills, Gordon, & Engeszer, 2009). To obtain larvae for xenografting:

1. Mate the desired zebrafish lines to obtain eggs by natural spawning. Maintain eggs in petri dishes, with a maximum of 50 eggs per 50 mL embryo medium at 28.5°C;
2. To maintain optical transparency treat the embryos with 0.2 mM 1-phenyl 2-thiourea (PTU) from 8 hpf until the conclusion of the experiment.

   Note: PTU prevents pigmentation by inhibiting tyrosinase activity; however, it also causes developmental defects at concentrations exceeding 75 µM (Elsalini & Rohr, 2003; Karlsson, von Hofsten, & Olsson, 2001). As an alternative to chemical inhibition, there are several mutant zebrafish lines in which pigmentation has been disrupted including Golden (a slc24A5 mutant; Lamason et al., 2005), nacre (a mitfa mutant; Lister, Robertson, Lepage, Johnson, & Raible, 1999), and casper (a nacre and roy orbison (roy) double mutant; White et al., 2008). However, pigmentation mutants have been linked to immunodeficiency and mitfa is involved in immune signaling (Gutknecht et al., 2015; Stinchcombe, Bossi, & Griffiths, 2004). In addition, the roy mutation has not been characterized and may have unknown effects on the immune system and cancer biology.
3. Xenografts are performed when the larvae are 3 or 4 dpf. Dechorionate the larvae and anaesthetize with 2.5 mM tricaine.

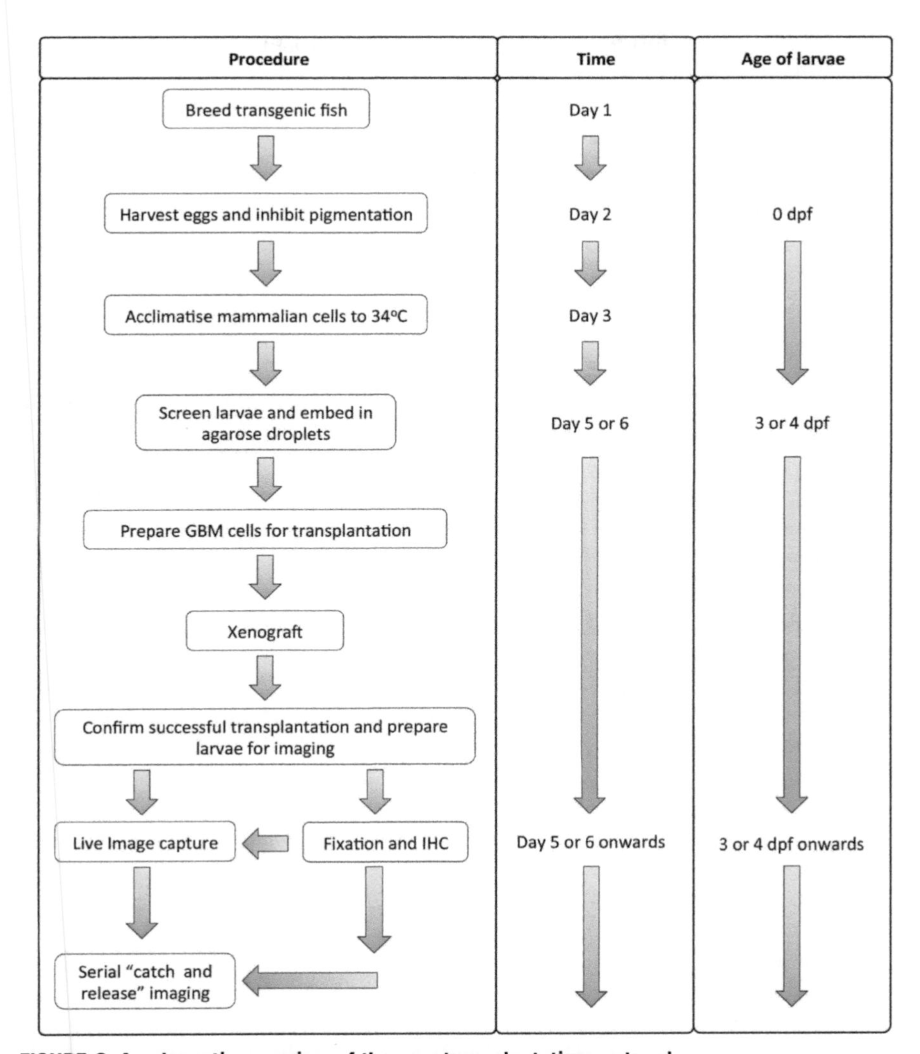

FIGURE 2 A schematic overview of the xenotransplantation protocol.

Xenotransplantation of mammalian glioblastoma cells into zebrafish larvae, with labeled microglia, can be used to investigate microglia-glioma cell interactions in vivo. *dpf*, days post fertilisation; *GBM*, glioblastoma; *IHC*, immunohistochemistry.

4. Once the larvae are nonresponsive to touch, use a fluorescent stereomicroscope to sort the larvae according to transgene expression. Select larvae with robust transgene expression in the microglia (and any additional cells/structures of interest).

Note: If PTU has been used to inhibit pigmentation, only the healthy looking larvae should be selected for xenografting. In addition, 4 dpf larvae are screened to confirm inflation of the swim bladder and "float-up." Failure to initiate swim bladder inflation at 4 dpf is suggestive of a developmental issue and these weaker larvae may not survive the injection/imaging procedure.

1.3.2 Immobilizing zebrafish larvae prior to the xenotransplantation procedure

To minimize injury and thus prevent a microglia injury response, it is important to immobilize the larvae throughout the transplantation procedure.

1. Dissolve low melting point (LMP) agarose in embryo medium to a final concentration of 1.5%.
   Note: LMP agarose can be made in advance and aliquots can be stored at 4°C.
2. Prepare mounting solution: Add tricaine (2.5 mM) and PTU (0.2 mM, if required) to molten 1.5% LMP agarose, mix, and incubate at 40°C.
3. Using a Pasteur pipette or 1000 μL pipette tip with the point removed, transfer several anaesthetized larvae into the mounting solution.
   Note: Take care to transfer minimal volumes of embryo medium as this will dilute the agarose. With practice, it is possible to prepare 5/6 larvae per droplet of agarose.
4. Transfer the larvae with approximately 300 μL of mounting solution (for 5 or 6 larvae) to a 10 cm petri dish.
   Note: The agar droplets should be positioned in one half of the petri dish.
5. Return the mounting solution to 40°C so that it remains molten and can be reused.
6. Use a dissecting needle, or a straight metal pin attached to a suitable handle, orientate the larvae within the molten agarose droplet. The ventral side of the body should be against the bottom of the petri dish and the heads should all point towards the midline of the petri dish and, therefore, the direction of the capillary (Fig. 3A and B).
   Note: The molten LMP agarose may take several minutes to solidify. Therefore, it is important to check that the larvae remain in the desired orientation, especially if the swim bladder is present.
7. Repeat steps 2 to 7 for the desired number of embryos per droplet per dish.
8. Once the agarose has solidified add embryo medium, containing 2.5 mM tricaine, so that the larvae are submerged.
9. Use a hypodermic needle to carefully remove the agarose from around the head area (Fig. 3C).
   Note: This prevents blockage of the transplantation needle.
10. Return the immobilized larvae to 28.5°C while preparing the mammalian cells for transplantation.

1.3.3 Glioblastoma cell lines

There are several established and well-characterized brain tumor cell lines available for transplantation studies. For example, there are the human glioblastoma U87MG

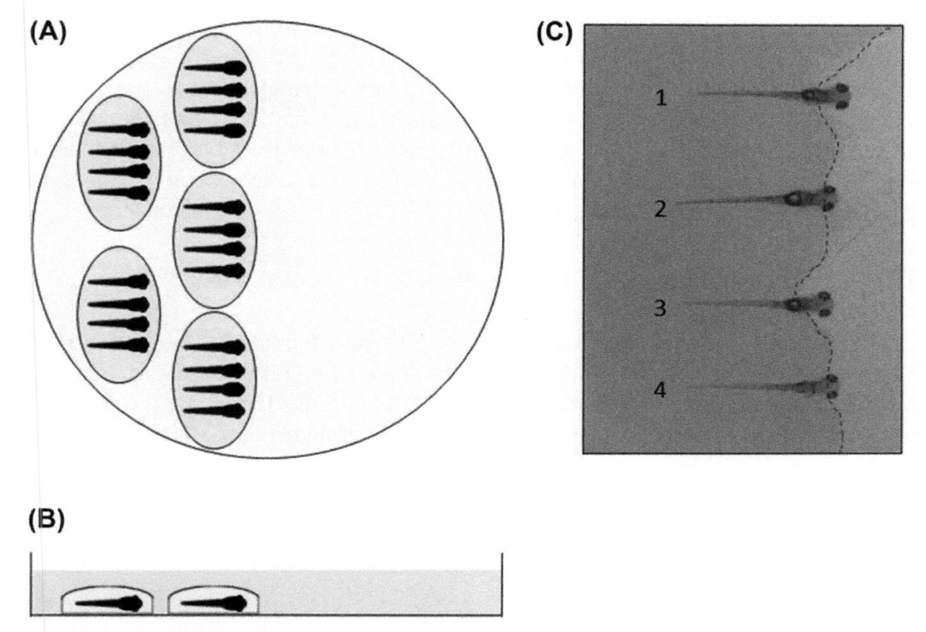

FIGURE 3 Preparation of larvae for xenografting.

(A) To minimize the injury caused by the xenotransplantation procedure the 3 or 4 dpf larvae are immobilized in agarose droplets positioned on one side of a 10 cm petri dish. (B) Once the agarose has solidified, the larvae are submerged in embryo medium containing 2.5 mM tricaine and 0.2 mM PTU if required. (C) To prevent blockage of the glass capillary, the agarose is removed from around the head region using a hypodermic needle. The agarose should be removed so that it is in level with the eyes (second and fourth larvae). If too much agarose is removed the head will move during injection and this will inhibit the penetration of the capillary through the skin and into the brain (first and third larvae).

and U251MG cell lines (Ponten, 1975) as well as the mouse glioblastoma model cell line GL261 (Szatmári et al., 2006) and mouse neural stem cells harboring glioblastoma promoting mutations (Bachoo et al., 2002). However, all of these cell lines have been in continuous culture for many years and may no longer retain a true glioblastoma phenotype. For this reason, many research groups are now using genome editing techniques to introduce cancer-promoting mutations into wild type cell lines or turning to primary patient brain tumor cells.

1.3.3.1 Labeling of glioblastoma cells prior to transplantation

To visualize glioblastoma cells within the zebrafish larvae, synthetic dyes such as CM-Dil can be used to label the cell membranes in vitro prior to transplantation. However, after transplantation of CM-Dil glioblastoma cells we observed many foci, believed to be dye aggregates, which spread throughout the brain tissue in all dimensions and were rapidly engulfed by microglia. The CM-Dil fluorescent signal remained stable within phagosomes and was detectable for several hours as the microglia migrated through the brain tissue. Hence, we would discourage the use of membrane-binding synthetic dyes for investigating the invasiveness of oncogenic cells as they may cause misleading results and an overestimation of invasiveness. Instead, we would recommend generating cell lines that stably express a suitable fluorescent marker. We have successfully used a lentivirus to transform glioblastoma cells and generate stable cell lines that express GFP or mCherry.

1.3.3.2 Acclimatizing mammalian cells to 34°C prior to xenotransplantation

Zebrafish larvae are maintained at 28.5°C, while mammalian cells are cultured at 37°C. Therefore, it is common practice to incubate xenografted larvae, containing mammalian cells, at 33–34°C. At this temperature, both the larvae and the glioblastoma cells survive. We acclimatize the mammalian glioblastoma cells to 34°C for a minimum of 48 h prior to transplantation as we have found that this improves cell survival post transplantation. However, it is important to (1) optimize the time required for acclimatization and (2) establish how a reduced growth temperature affects the in vitro growth of each cell line prior to transplantation.

1.3.3.3 Ensuring that PTU is not detrimental to mammalian cell growth

If xenografted larvae are to be incubated in PTU, it is necessary to ensure that PTU is not detrimental to the growth of the transplanted cells in vitro prior to transplantation.

1.3.4 Preparation of glioblastoma cells for xenotransplantation

We maintain mycoplasma-free mammalian glioblastoma cells as a monolayer culture and harvest them immediately before transplantation using the following protocol:

1.3.4.1 Hoechst staining prior to cell harvest (optional)

Hoechst is a DNA stain that is used to label the nuclei of living cells. Nuclear stains can be beneficial for improving the accuracy of cell counts during image analysis.

1. Incubate cells in complete medium supplemented with 2.5–5 μg/mL Hoechst 33342 for 30 min at 34°C.

   Note: It may not be appropriate to use Hoechst with neural stem cells as they can effectively efflux the dye. The dye can then stain the nuclei of surrounding zebrafish cells, in vivo, after transplantation.
2. Wash 2 × PBS (or appropriate wash medium) and proceed to harvesting the cells.

1.3.4.2 Harvesting glioblastoma cells for transplantation

1. Glioblastoma cells are harvested during the exponential growth phase, when less than 80% confluent.

Note: This ensures that the cells are proliferating and promotes cell survival in the zebrafish.

2. Wash cells 2 × 10 mL PBS and 1 × 5 mL EDTA-PBS (see step 3).
3. Incubate the cells in a suitable volume of 0.48–2 mM EDTA-PBS until they have detached from the surface of the culture flask/dish.

   Note: We do not use an enzymatic treatment (such as Accutase or Trypsin) to detach the cells as these may cleave cell surface proteins that are involved in microglia-glioma cell interactions. To achieve a single cell suspension, the EDTA concentration as well as the incubation time must be optimized for each cell type.
4. Transfer the cells to a 15 mL tube and pellet the cells at 200 g for 3.5 min.
5. Wash the cells with 3 × 10 mL PBS or wash medium appropriate for the cell line.

   Note: Ensure that all the EDTA has been removed as it is a Ca^{2+} chelating agent and calcium has been shown to be involved in cell signaling to microglia (Sieger et al., 2012).
6. Quantify the cell number using either a haemocytometer or coulter counter.
7. Pellet the required volume of cell suspension at 200 g for 3.5 min.
8. Resuspend the cells in 25–100 μL of complete culture medium supplemented with 5 mM phenol red.

   Note: The concentration of the cell suspension must be optimized for each cell type to ensure consistent flow through the glass capillary. We have found that 5×10^6 cells resuspended in 50 μL works well for human glioblastoma cell lines. At this final stage it is important to ensure that a single cell suspension has been achieved, as cell clumps will block the glass capillary. The cells should be kept at 34°C and used immediately.

1.3.5 Xenotransplantation equipment and method

1.3.5.1 Equipment for xenotransplantation (Fig. 4)

- Femtojet 4i micro-injector (Eppendorf, Hamburg, Germany)
- Stereomicroscope (for example a Nikon SMZ1500)
- A standard manual mechanical micromanipulator (for example an MM 33 right; Marzhauser, Germany)
- A 34°C water bath
- Borosilicate glass capillaries without a central filament
- Tip Microloader 0.5–20 μL (Eppendorf)
- P10/P20 pipette
- Watch makers forceps

1.3.5.2 Borosilicate glass capillaries

To transplant cells into the zebrafish larvae we use borosilicate glass capillaries that have been prepared using a P-97 Flaming/Brown micropipette horizontal needle puller according to the manufacturer's instructions (Sutter Instrument Co.). See Table 4 for needle pulling parameters.

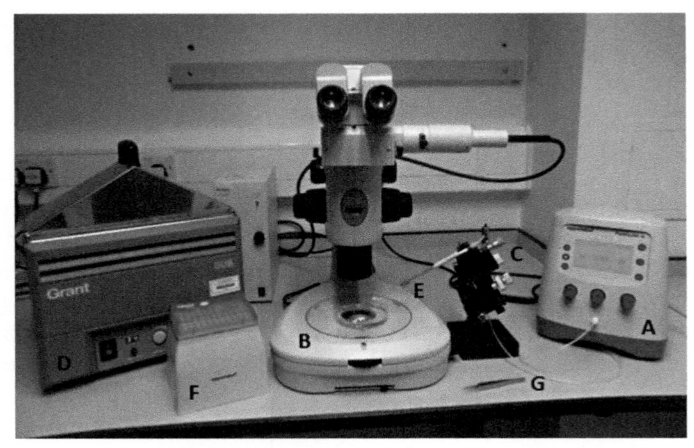

FIGURE 4 Equipment used for xenotransplantation.

The transplantation of mammalian oncogenic cells into zebrafish larvae is performed using a Femtojet 4i (A) with a stereomicroscope (B) to view the injection process. The mammalian cells are maintained in a 34°C water bath (D) and loaded into a glass capillary using a tip microloader (F). The capillary is inserted into the universal capillary holder (E), which is mounted on a standard mechanical micromanipulator (C). The petri dish containing the larvae is positioned on the stereoscope stage and the capillary is lowered into position. Watch makers' forceps (G) are then used to break the end of the capillary at a suitable diameter depending on cell size.

1.3.5.3 Transplanting mammalian cells into zebrafish larvae

1. Keep the mammalian single cell suspension(s) at 34°C in a water bath, or heat block, throughout the xenograft procedure.
2. Ensure that the injection tube is not attached to the Femtojet. Switch on the Femtojet and allow it to perform the self-test and then build up operating pressure.
3. Following the manufacturer's instructions attach the injecting tube to the Femtojet using the bayonet coupling.
4. Use a tip microloader to load the capillary with 3–5 μL of the cell suspension. Attach the capillary to the universal capillary holder according to the manufacturer's instructions and place in the tool holder on the micromanipulator.

Table 4 Parameters for Pulling Glass Capillaries

Parameter	Value
Heat	552
Pull	250
Velocity	55
Time	100

Note: As cell clumps can block the capillary it is not wise to load the entire cell suspension volume into a single capillary. It is also useful to have a bit of blue tac on the side of the stereomicroscope stock, where filled capillaries can be temporarily stored.

5. The petri dish containing the immobilized larvae is positioned on the stereomicroscope stage to obtain a suitable field of view. Confirm that the larvae are unresponsive to touch and look healthy with strong heartbeats. The capillary can then be lowered into position ready for injecting.
6. Use watch makers forceps, or similar, to break the capillary at an angle to create a sharp point.
7. The capillary does not have an internal filament. Therefore, move the cell suspension to the tip of the capillary using puffs of air from the Femtojet.
8. Maintaining a sharp point, slowly trim the capillary, using the forceps, until a suitable flow of cells is achieved.

   Note: At this stage it is important not to remove too much of the capillary as a larger diameter will cause damage to the larvae and a microglia injury response.
9. While trimming back the capillary, use the rotary knobs on the Femtojet to obtain the correct pressure parameters for injecting:
   1. Compensation pressure: Prevents the embryo medium in the petri dish being drawn into the capillary by capillary action.
   2. Injection pressure: Defines the pressure used to inject the cell suspension into the larval brain. The injection pressure will be applied for the duration of the injection time.
   3. Injection time: Defines the time period of the injection pressure.

      Note: Each capillary will function slightly differently depending on the evenness of the pull when it was prepared and the diameter of the hole required to allow the cells to pass through it. Therefore, at this stage some time must be taken to optimize the injection pressure, injection time, and compensation pressure to ensure a consistent release of cells at a suitable flow rate.
10. Once the capillary is suitable for use, insert the capillary into the left optic tectum of a few larvae, but do not inject cells.

    Note: These larvae serve as a control, to determine the microglia injury response to this particular capillary diameter.
11. Inject the mammalian oncogenic cells into the optic tectum within the left hemisphere of the brain (or the region of interest; Fig. 5A and B).

    Notes: The noninjected right hemisphere serves as an internal control. It is possible to see the cells move along the capillary and enter the brain tissue, to form a cell mass. Due to the phenol red, it is also possible to visualize a rupture of the brain membranes (Fig. 5C). This normally occurs if the injection pressure is set too high, or if medium alone passes along the capillary. If the membranes of the brain are ruptured, remove this larva from the experiment as the loss of compartmentalization may produce a

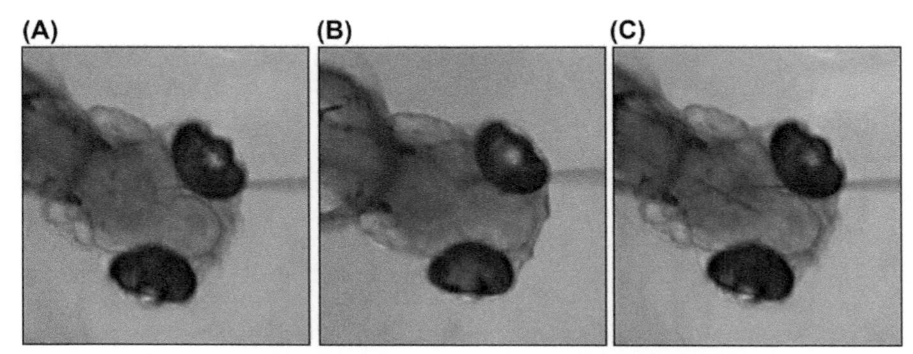

FIGURE 5 Transplantation of mammalian oncogenic cells into the optic tectum of zebrafish larvae.

Zebrafish larvae at 3 or 4 dpf are immobilized in agarose droplets prior to injection. The head of the larvae is liberated from the agarose and the mammalian oncogenic cells are transplanted into the left optic tectum using a borosilicate glass capillary (A). The right optic tectum serves as an internal control. The cell suspension is supplemented with phenol red so that the transplanted cells can be observed within the optic tectum (B). If the optic tectum is ruptured during the injection process the phenol red will spread to other regions of the brain, such as the ventricle (C), and these larvae should be removed from the experiment.

misrepresentation of invasiveness. If the needle becomes blocked, it can sometimes be cleared using the "clean" function on the Femtojet, otherwise it will have to be replaced.

12. It is necessary to determine how the microglia respond to the growth medium that is injected along with the cells. Therefore, for each individual mammalian cell line to be injected, it is important to inject several larvae with the corresponding growth medium alone so the microglia response can be assessed.
13. Once xenotransplantation is complete, visualize the larvae/oncogenic cells using a fluorescent stereomicroscope. Confirm the presence of oncogenic cells and ensure that the cells look healthy and have formed a tight cell mass (Fig. 6A).
    Note: Discard larvae that contain few transplanted cells as well as larvae in which the transplanted cells have ruptured the brain compartment of interest (Fig. 6B and C).
14. Select larvae for image capture and analysis. Using a hypodermic needle, liberate the larvae from the agarose droplets. Carefully cut along one side of the larvae, ensuring that the fins are not damaged and peel the agarose away to release the larvae.
15. Immediately prepare the selected larvae for imaging.

1.3.6 Staining xenotransplanted cells for additional traits

The zebrafish IHC protocol (Section 1.2.3.) can be used to stain the transplanted mammalian cells, within the zebrafish, to investigate additional traits. These may

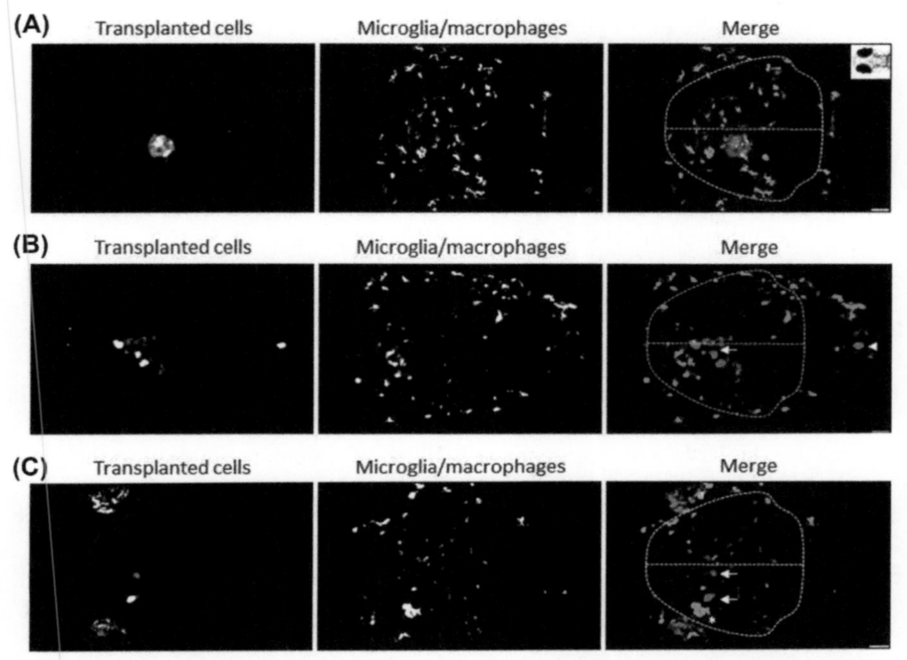

FIGURE 6 Xenografted larvae may not be suitable for further analysis and should be removed from the experiment.

(A) Mouse neural progenitor cells, harboring glioblastoma-promoting mutations, were transplanted into the left optic tectum of Tg(mpeg1:mCherry) zebrafish larvae at 4 dpf. The larvae were screened using a fluorescent stereomicroscope. The transplanted cells have formed a cell mass contained within the optic tectum. This larva would be retained for further analysis. The inset indicates the field of view and the orientation of the larval head. (B and C) Human glioblastoma cells were transplanted into the left optic tectum of Tg(mpeg1:EGFP) zebrafish larvae at 4 dpf. After injection the larvae were screened using a fluorescent stereomicroscope. (B) This larva was removed from the experiment because the compartments of the brain were ruptured due to excessive injection pressure. Glioblastoma cells have ruptured through the boundary of the optic tectum into the ventricle (*white arrow*). The injection pressure has propelled several cells from the optic tectum, through the ventricle and into the hind brain (*white arrow head*). This could produce an overrepresentation of invasiveness. (C) This larva was removed from the experiment as a low number of glioblastoma cells had been transplanted into the optic tectum (*white arrows*) and they have not formed a tight cell mass. These cells would fail to engraft. The Langerhans cells within the skin have responded to the injury caused by the capillary and are clustered around the puncture hole (*white asterisk*). Maximum intensity projections of 3D z-stacks. Mouse/human oncogenic cells are shown in *magenta*; microglia/macrophages are shown in *green*. The outline of the optic tectum and the midline of the ventricle are indicated by the *yellow dashed line*. Scale bars are 50 μm.

include staining for proliferation markers (Ki-67 or PCNA), apoptosis markers (Caspases, Bcl-2 family members or fragmented DNA), as well as cell surface proteins and receptors.

1.4 PREPARING LARVAE FOR LIVE IMAGING

1. Prepare mounting solution: Add tricaine (2.5 mM) and PTU (200 μM, if necessary) to molten 1.5% LMP agarose, mix and incubate at 40°C.
2. Anaesthetize the larvae in embryo medium containing 2.5 mM tricaine.
   Note: The larvae should be nonresponsive to touch.
3. Using a Pasteur pipette or 200 μL pipette tip with the point removed, transfer individual larva to the mounting solution.
   Note: Take care to transfer minimal volumes of embryo medium as this will dilute the agarose.
4. Transfer the larva with approximately 50 μL of mounting solution to the glass microwell of a MatTek petri dish.
5. Return the mounting solution to 40°C so that it remains molten and can be reused.
6. Use a dissecting needle to orientate the larvae, within the molten agarose droplet, so that the dorsal side of the head is positioned against the glass at the bottom of the microwell. This will maximize the objective working distance. If the swim bladder has started to inflate, care must be taken to ensure that the larvae do not "float-up" in the molten agarose.
   Note: This mounting method is relevant for an inverted microscope. The molten LMP agarose may take several minutes to solidify. Therefore, it is important to check that the larvae remain in the desired orientation.
7. Repeat steps 2 to 7 for the desired number of embryos per dish (Table 5).
8. Once the agarose has solidified, half fill the dish with embryo medium, containing tricaine (and PTU if necessary), so that the larvae are submerged. Transport the larvae to the confocal microscope stage.
9. Carefully add embryo medium containing tricaine (and PTU if required) to the maximum capacity of the dish (Table 5). Replace the lid on the dish to minimize evaporation.

Table 5 Parameters for Mounting Live Zebrafish Larvae for Imaging

Dish Diameter (mm)	Glass Microwell Diameter (mm)	Glass Thickness	Embryo Medium Volume (Maximum) (mL)	Maximum Larvae (≤5 dpf)	Maximum Larvae (≥6 dpf)
30	14	No. 1.5	6	6	4
50	30	No. 1.5	10	12	8

1.4.1 Additional considerations for imaging live zebrafish larvae

To image cellular processes in vivo, it is essential that the larvae are kept in an environment that will not induce stress responses, as this may alter the behavior of the cells of interest. Good water quality, with suitable oxygen levels must be maintained. See Table 5 for the recommended maximum number of larvae per dish. The volume of the agarose droplet should be kept to a minimum to prevent excessive pressure upon the larvae and the tails of the larvae should be liberated from the agarose so that growth is not inhibited. If a larva dies during imaging it will rapidly decompose and toxins will spread through the agarose and may cause the death of neighboring larvae. It is possible to avoid this outcome by embedding single larvae in isolated agarose droplets.

1.5 VISUALIZING MICROGLIA AND ONCOGENIC CELLS WITHIN THE BRAIN

To visualize microglia and oncogenic cells within the tissues of the larval brain, confocal or multiphoton excitation microscopy is used to capture optical sections through the tissue. Line scanning confocal microscopy and spinning disk confocal microscopy both use a pinhole to prevent detection of the out-of-focus background fluorescence. However, the excitation light may cause photobleaching and phototoxicity in living specimens. In addition, absorption of the excitation light combined with scattering of the emitted light can limit penetration depth. Multiphoton microscopy allows optical sectioning without absorption and therefore, has increased penetrance with reduced photobleaching and phototoxicity in live specimens. We use several confocal microscopes to image larvae. A LSM710 line scanning microscope (Zeiss) is used to capture images of fixed samples and single time points using live specimens. A spinning disk confocal microscope (Andor) is used for fast image capture of live specimens at multiple time points and the LSM880 confocal microscope (Zeiss) is used for high-resolution image capture.

1.5.1 A basic in vivo confocal imaging protocol

1. Open the imaging software (e.g., Zen, Zeiss), select a suitable objective, and use the transmitted light source to locate the larvae and find the region of interest.
2. Select the desired fluorophores from the "Dye List" (BFP, GFP, RFP/mCherry) and adjust the emission ranges to obtain clean separation of the reporter signals with maximum signal to noise ratios.

   Note: These settings will need to be defined for each experiment to minimize crosstalk while maximizing signal.
3. Select desired pinhole aperture and gradually raise the laser power to achieve a suitable image intensity while avoiding pixel saturation.

   Note: A smaller pinhole aperture will require higher laser power and a larger aperture will require less laser power. However, if the aperture is set above 1 airy unit image resolution will be lost. For in vivo imaging and time-lapse imaging it is important to use low laser powers and fast scan speeds to reduce photobleaching and phototoxicity. However, these settings will reduce image resolution.

4. Ensure that there is no crosstalk between channels by monitoring each channel while manually turning each laser on and off.
   Note: Each laser should only produce detectable signal in the corresponding channel. In addition a sequential imaging mode can be used to reduce crosstalk; however, this will cause delayed image acquisition.
5. Scan averaging can be used to improve image resolution. However, scan averaging can increase photobleaching and may not be suitable for time-lapse imaging as it will introduce a temporal delay through the z-stack.
6. Set the z-stack dimensions and step size depending on experimental requirements.
   Note: To image the entire optic tectum in a larva between 4 and 10 dpf we set a z-stack depth of 150–200 μm from immediately below the dorsal surface of the head. If Nyquist sampling is not required we set a step size between 1.5 and 2.5 μm. Signal intensity will decrease with z-depth; therefore, it is best practice to image the deeper tissues first to acquire the weaker signals when minimal photobleaching has occurred.
7. If live specimens are being imaged with a line scanning confocal microscope, it is important to capture each channel sequentially for each optical plane and not each channel sequentially for each z-stack. Otherwise interaction artifacts may be generated.
   Note: This is not necessary for fixed specimens.
8. Acquire and save images for each larvae.
9. Once image capture of live larvae is complete, we use a sharp hypodermic needle to carefully liberate the larvae from the agarose. For repeated imaging, on consecutive days, we house the embryos individually in 12-well plates containing 2–4 mL embryo medium and 0.2 mM PTU at 34°C. We refresh the embryo medium and PTU daily and feed the larvae from 5 dpf onwards.

1.5.2 Time-lapse in vivo confocal imaging protocol

For time-lapse imaging it is essential that photobleaching and phototoxicity are minimal; therefore, laser power and duration must be kept to a minimum while still maintaining an adequate signal to noise ratio. Interactions between microglia and oncogenic cells are highly dynamic and if possible, image capture of multiple channels should be performed simultaneously, rather than sequentially.

1. For prolonged periods of imaging it is essential that the temperature is maintained at 28.5°C (34°C for xenografted larvae) using a heated stage.
2. Minimize evaporation by ensuring that lids are replaced on dishes (for inverted microscopes).
3. For prolonged automated acquisition the z-stack size must allow for larval growth and/or dimensional drift.
   Note: Dimensional drift can be corrected during image analysis by using software, such as the "registration" plug-in, available for ImageJ.

4. The time intervals between acquisitions will be determined by several factors including, (1) the number of larvae, (2) the rate of photobleaching, (3) phototoxic sensitivity, (4) the z-depth, (5) the number of steps within the z-stack, (6) the number of larvae, and (7) the biological event of interest.
   Note: Some time may need to be spent optimizing the acquisition times for in vivo time lapse imaging.
5. Acquire images for the desired period of time and save.

1.6 IMAGE ANALYSIS OF MICROGLIA AND ONCOGENIC CELLS WITHIN THE ZEBRAFISH LARVAE

The captured images are analyzed using the Imaris software package (Bitplane) as it has robust 3D and 4D rendering capabilities that are superior to many of the open source and commercial analysis software packages currently available. If images have been captured with a suitable pinhole aperture, laser power, and optical slice depth it is possible to obtain data for analysis without deconvolution.

1.6.1 Oncogenic cell survival and proliferation

To determine transplanted cell survival and growth we manually count the fluorescently labeled cells. Immediately after transplantation, it can be difficult to determine individual cells within the cell mass. The addition of a nuclear marker can increase the ease and accuracy of these cell counts. Alternatively a cell number can be obtained using an average pixel number per cell or average cell volume. However, we have found that these methods can lack accuracy as the size of each transplanted cell within the population can vary dramatically depending on factors such as cell cycle, process extension, and migration through the brain tissues.

1.6.2 Oncogenic cell migration and invasiveness

The migration, or invasiveness, of a cell line can be determined in two ways. Either the cell can be tracked over a time course (4D) and the distance travelled and speed of migration can be calculated by the Imaris software. Alternatively the spread of the cells from the initial cell mass can be determined using 3D stacks captured at single daily time points. For this approach, a bounding sphere is generated around the entire cell population, including the cell processes, and the diameter is determined (Fig. 7). By calculating the fold-increase between spheres on sequential days a valuation of invasiveness can be obtained. An important prerequisite to measure invasiveness precisely is the additional labeling of microglia. Depending on the cell type, transplanted cells may be engulfed by microglia and signals will be detected within microglial phagosomes. These signals do not represent viable cells and thus do not reflect true infiltration. The additional microglial labeling allows the engulfed cell debris to be identified and removed from the analysis.

1.6.3 Microglia interactions and activation state

It is essential that a 3D image, constructed from the z-slice data, is used to determine true interactions between microglia and the transplanted cells. Both multiple

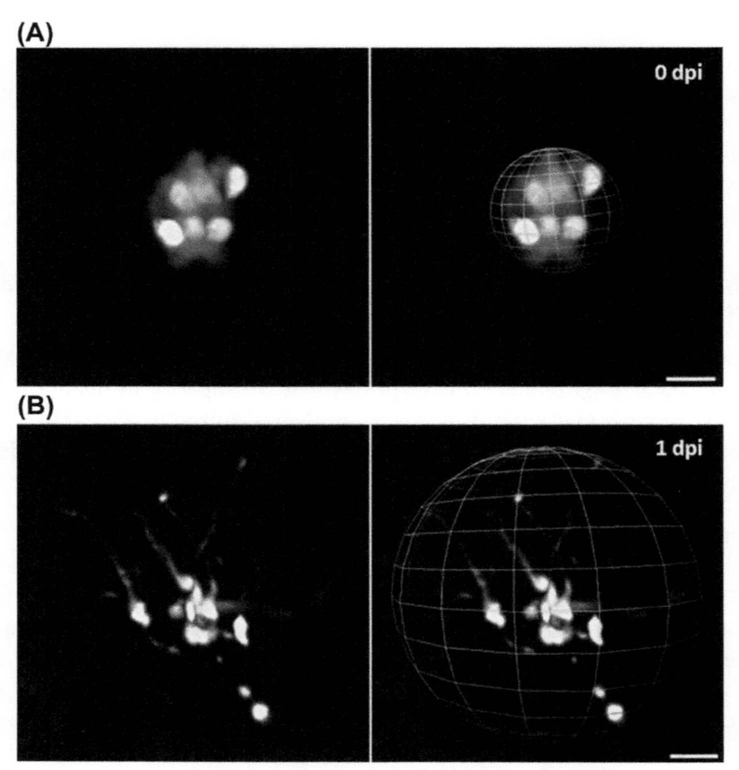

FIGURE 7 Measuring oncogenic cell migration and invasiveness in vivo.

Mouse neural progenitor cells, harboring glioblastoma-promoting mutations, were transplanted into the left optic tectum of Tg(mpeg1:mCherry) zebrafish larvae at 4 dpf. The larvae were imaged by spinning disk confocal microscopy within (A) 2 hours post injection (0 dpi) and at (B) 1 day post injection (1 dpi). The invasiveness/migration of the oncogenic cells from the initial cell mass is determined using the Imaris image analysis software. First the 3D images are rebuilt from the z-stacks that were captured for each larva on consecutive days. A bounding 3D sphere is then generated around the entire cell population, including any cell processes, and the diameter of the sphere is determined. For this example the diameter of the sphere at 0 dpi is 62 μm and the diameter of the sphere at 1 dpi is 133 μm. By calculating the fold-increase between spheres on sequential days a valuation of invasiveness can be obtained (for this example the fold increase is 2.12) and compared for different cell lines. Maximum intensity projections of 3D z-stacks. Scale bars are 20 μm.

intensity projections of flattened z-stacks, as well as 2D images captured by wide field microscopy generate false positive interactions as cells located in different z-planes can appear to interact with one another. It is often possible to determine the activation status of microglia based on morphology, especially if the cells are in either the ramified or amoeboid states (Fig. 8). However, to quantify intermediate

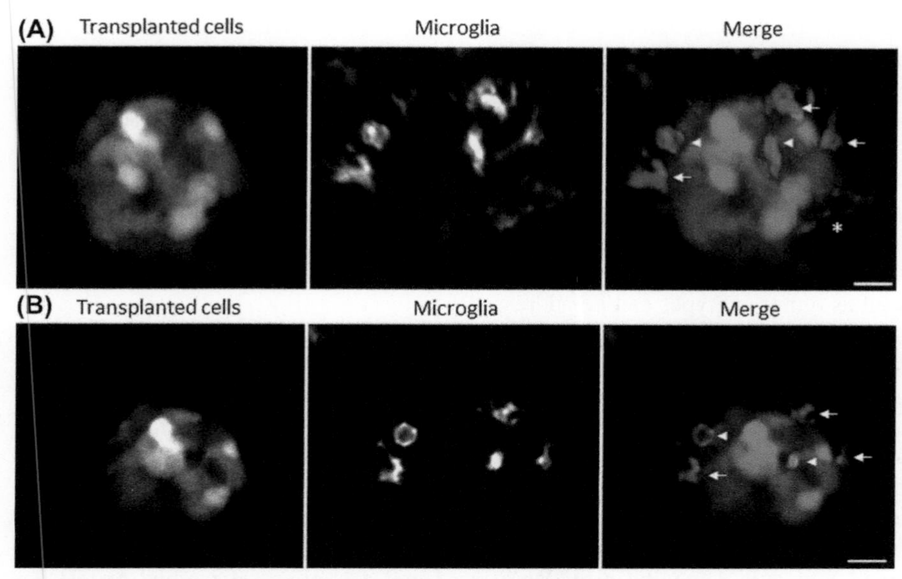

FIGURE 8 Microglia interact with transplanted oncogenic cells.

Mouse neural progenitor cells, harboring glioblastoma-promoting mutations were transplanted into the left optic tectum of Tg(mpeg1:mCherry) zebrafish larvae at 4 dpf. The larvae were imaged at 2 hours post injection. Microglia were attracted to the oncogenic cell mass and formed direct surface interactions with the oncogenic cells. (A) Maximum intensity projection of the 3D z-stack shows that microglia with different morphologies, including ramified (*white asterisk*), intermediate (*white arrows*), and amoeboid (*white arrow heads*), interact with the oncogenic cells. The duration of the interactions as well as microglia functions, such as phagocytosis, can be monitored by time-lapse live imaging of zebrafish larvae. Microglia-glioblastoma cell interactions must be investigated in 3D. The maximum intensity projection shown in (A) is a flattened image of 199 optical slices. To rule out false positive interactions and confirm that microglia are interacting with oncogenic cells in the same optical plane, individual z-slices must be examined. (B) The 126th optical slice through the oncogenic cell mass confirms that intermediate (*white arrows*) and amoeboid (*white arrow heads*) microglia are interacting with oncogenic cells in the same optical plane. The ramified microglia (*asterisk* in A) is also interacting with the oncogenic cells and can be observed in an optical plane at a different depth. Oncogenic mouse neural progenitor cells are shown in *green*; microglia are shown in *magenta*. Scale bars are 20 μm.

activation states it may be necessary to calculate the cell surface area to volume ratios and set thresholds for different categories of interest.

1.7 APPLICATIONS

Our xenotransplantation and imaging protocol is suitable for a variety of applications. The analysis of microglia-tumor cell interactions in vivo provides direct insights into the function of microglia within the tumor environment. This allows a direct read out of the microglia activation status based on morphology. The ramified microglia that are surveying the surrounding environment are in the nonactivated mode, while the amoeboid microglia can be considered activated. Based on this technique, the differential activation of microglia within tumors can be studied without the need to fix the tissues or stain for putative activation markers. This will provide further insight into the heterogeneity of microglial activation, in vivo, in relation to their location within the tumor. Furthermore, putative differences in microglial activation and function in response to different types of brain tumor as well as tumor subtypes can be addressed.

The opportunity to perform time lapse in vivo microcopy allows direct assessment of the protumoral as well as the antitumoral activity of microglia. Microglia are the professional phagocytes of the central nervous system. Therefore, antitumoral activity is directly correlated with an increase in microglial phagocytic activity towards the oncogenic cells. Visualization and quantification of these phagocytic events allows both the circumstances and the mechanisms that lead to antitumoral activity within microglia to be determined.

Finally, the combination of this xenotransplantation and imaging technique with small molecule screening provides a rapid and cost-effective method to test the effects of different compounds specifically on microglial activity within brain tumors. This will help to accelerate the development of immunotherapeutics for brain tumors in the future.

2. SUMMARY

The zebrafish has proved to be a valuable vertebrate model organism for investigating cancer biology and immunology. We have described techniques that utilize xenotransplantation of oncogenic cells, in combination with the optical transparency of the zebrafish larvae, to investigate the interactions between microglia and glioblastoma cells in vivo at the initial stages of tumor growth. Understanding the underlying cell signaling mechanisms within the tumor environment is the first step towards developing future therapies to interfere with the microglia-glioma interactions that promote tumor growth. In future, these methodologies could be combined with high throughput screens of small molecules to produce a powerful technique for identifying potential chemotherapeutics that induce antitumoral activity within microglia.

ACKNOWLEDGMENTS

We would like to thank the BRR zebrafish facility (QMRI, University of Edinburgh) for maintenance and care of the zebrafish. We are grateful to Jennifer Richardson and Lloyd Hamilton for help with the establishment of the xenograft technique. D.S. was supported by a Cancer Research UK Career Establishment Award.

REFERENCES

Alterman, R. L., & Stanley, E. R. (1994). Colony stimulating factor-1 expression in human glioma. *Molecular and Chemical Neuropathology/Sponsored by the International Society for Neurochemistry and the World Federation of Neurology and Research Groups on Neurochemistry and Cerebrospinal Fluid, 21*(2–3), 177–188. http://dx.doi.org/10.1007/BF02815350.

Bachoo, R. M., Maher, E. A., Ligon, K. L., Sharpless, N. E., Chan, S. S., You, M. J., … DePinho, R. A. (2002). Epidermal growth factor receptor and Ink4a/Arf: convergent mechanisms governing terminal differentiation and transformation along the neural stem cell to astrocyte axis. *Cancer Cell, 1*(3), 269–277.

Badie, B., Schartner, J., Klaver, J., & Vorpahl, J. (1999). In vitro modulation of microglia motility by glioma cells is mediated by hepatocyte growth factor/scatter factor. *Neurosurgery, 44*(5), 1077–1082. discussion 1082–1083.

Badie, B., & Schartner, J. (July 15 2001). Role of microglia in glioma biology. *Microsc Res Tech, 54*(2), 106–113.

Becker, T., & Becker, C. G. (2001). Regenerating descending axons preferentially reroute to the gray matter in the presence of a general macrophage/microglial reaction caudal to a spinal transection in adult zebrafish. *The Journal of Comparative Neurology, 433*(1), 131–147.

Bettinger, I., Thanos, S., & Paulus, W. (2002). Microglia promote glioma migration. *Acta Neuropathologica, 103*(4), 351–355. http://dx.doi.org/10.1007/s00401-001-0472-x.

Brandenburg, S., Müller, A., Turkowski, K., Radev, Y. T., Rot, S., Schmidt, C., … Vajkoczy, P. (2016). Resident microglia rather than peripheral macrophages promote vascularization in brain tumors and are source of alternative pro-angiogenic factors. *Acta Neuropathologica, 131*(3), 365–378. http://dx.doi.org/10.1007/s00401-015-1529-6.

Cancer Genome Atlas Research Network. (2008). Comprehensive genomic characterization defines human glioblastoma genes and core pathways. *Nature, 455*(7216), 1061–1068. http://dx.doi.org/10.1038/nature07385.

Charles, N. A., Holland, E. C., Gilbertson, R., Glass, R., & Kettenmann, H. (2012). The brain tumor microenvironment. *Glia, 60*(3), 502–514.

Davalos, D., Grutzendler, J., Yang, G., Kim, J. V., Zuo, Y., Jung, S., … Gan, W.-B. (2005). ATP mediates rapid microglial response to local brain injury in vivo. *Nature Neuroscience, 8*(6), 752–758. http://dx.doi.org/10.1038/nn1472.

Ellett, F., Pase, L., Hayman, J. W., Andrianopoulos, A., & Lieschke, G. J. (2011). mpeg1 promoter transgenes direct macrophage-lineage expression in zebrafish. *Blood, 117*(4), e49–56. http://dx.doi.org/10.1182/blood-2010-10-314120.

Elsalini, O. A., & Rohr, K. B. (2003). Phenylthiourea disrupts thyroid function in developing zebrafish. *Development Genes and Evolution, 212*(12), 593–598. http://dx.doi.org/10.1007/s00427-002-0279-3.

Fantin, A., Vieira, J. M., Gestri, G., Denti, L., Schwarz, Q., Prykhozhij, S., … Ruhrberg, C. (2010). Tissue macrophages act as cellular chaperones for vascular anastomosis downstream of VEGF-mediated endothelial tip cell induction. *Blood, 116*(5), 829–840. http://dx.doi.org/10.1182/blood-2009-12-257832.

Feng, Y., Santoriello, C., Mione, M., Hurlstone, A., & Martin, P. (2010). Live imaging of innate immune cell sensing of transformed cells in zebrafish larvae: parallels between tumor initiation and wound inflammation. *PLoS Biology, 8*(12), e1000562. http://dx.doi.org/10.1371/journal.pbio.1000562.

Flügel, A., Labeur, M. S., Grasbon-Frodl, E. M., Kreutzberg, G. W., & Graeber, M. B. (1999). Microglia only weakly present glioma antigen to cytotoxic T cells. *International Journal of Developmental Neuroscience, 17*(5–6), 547–556.

Gates, M. A., Kim, L., Egan, E. S., Cardozo, T., Sirotkin, H. I., Dougan, S. T., ... Talbot, W. S. (1999). A genetic linkage map for zebrafish: comparative analysis and localization of genes and expressed sequences. *Genome Research, 9*(4), 334–347.

Graeber, M. B., Scheithauer, B. W., & Kreutzberg, G. W. (2002). Microglia in brain tumors. *Glia, 40*(2), 252–259. http://dx.doi.org/10.1002/glia.10147.

Gramatzki, D., Dehler, S., Rushing, E. J., Zaugg, K., Hofer, S., Yonekawa, Y., ... Weller, M. (2016). Glioblastoma in the Canton of Zurich, Switzerland revisited: 2005 to 2009. *Cancer.* http://dx.doi.org/10.1002/cncr.30023.

Gutknecht, M., Geiger, J., Joas, S., Dörfel, D., Salih, H. R., Müller, M. R., ... Rittig, S. M. (2015). The transcription factor MITF is a critical regulator of GPNMB expression in dendritic cells. *Cell Communication and Signaling: CCS, 13*, 19. http://dx.doi.org/10.1186/s12964-015-0099-5.

Hambardzumyan, D., Gutmann, D. H., & Kettenmann, H. (2016). The role of microglia and macrophages in glioma maintenance and progression. *Nature Neuroscience, 19*(1), 20–27. http://dx.doi.org/10.1038/nn.4185.

Haynes, S. E., Hollopeter, G., Yang, G., Kurpius, D., Dailey, M. E., Gan, W.-B., & Julius, D. (2006). The P2Y12 receptor regulates microglial activation by extracellular nucleotides. *Nature Neuroscience, 9*(12), 1512–1519. http://dx.doi.org/10.1038/nn1805.

Hemmati, H. D., Nakano, I., Lazareff, J. A., Masterman-Smith, M., Geschwind, D. H., Bronner-Fraser, M., & Kornblum, H. I. (2003). Cancerous stem cells can arise from pediatric brain tumors. *Proceedings of the National Academy of Sciences of the United States of America, 100*(25), 15178–15183. http://dx.doi.org/10.1073/pnas.2036535100.

Herbomel, P., Thisse, B., & Thisse, C. (1999). Ontogeny and behaviour of early macrophages in the zebrafish embryo. *Development, 126*(17), 3735–3745.

Herbomel, P., Thisse, B., & Thisse, C. (2001). Zebrafish early macrophages colonize cephalic mesenchyme and developing brain, retina, and epidermis through a M-CSF receptor-dependent invasive process. *Developmental Biology, 238*(2), 274–288. http://dx.doi.org/10.1006/dbio.2001.0393.

Hsieh, C. L., Kim, C. C., Ryba, B. E., Niemi, E. C., Bando, J. K., Locksley, R. M., ... Seaman, W. E. (2013). Traumatic brain injury induces macrophage subsets in the brain. *European Journal of Immunology, 43*(8), 2010–2022. http://dx.doi.org/10.1002/eji.201243084.

Hu, X., Leak, R. K., Shi, Y., Suenaga, J., Gao, Y., Zheng, P., & Chen, J. (2015). Microglial and macrophage polarization—new prospects for brain repair. *Nature Reviews. Neurology, 11*(1), 56–64. http://dx.doi.org/10.1038/nrneurol.2014.207.

Jones, S. L., Wang, J., Turck, C. W., & Brown, E. J. (1998). A role for the actin-bundling protein L-plastin in the regulation of leukocyte integrin function. *Proceedings of the National Academy of Sciences of the United States of America, 95*(16), 9331–9336.

Jones, T. R., Bigner, S. H., Schold, S. C., Eng, L. F., & Bigner, D. D. (1981). Anaplastic human gliomas grown in athymic mice. Morphology and glial fibrillary acidic protein expression. *The American Journal of Pathology, 105*(3), 316–327.

Jun, H. J., Bronson, R. T., & Charest, A. (2014). Inhibition of EGFR induces a c-MET-driven stem cell population in glioblastoma. *Stem Cells, 32*(2), 338–348. http://dx.doi.org/10.1002/stem.1554.

Karlsson, J., Von Hofsten, J., & Olsson, P. E. (2001). Generating transparent zebrafish: a refined method to improve detection of gene expression during embryonic development. *Marine Biotechnology, 3*(6), 522–527.

Karperien, A., Ahammer, H., & Jelinek, H. F. (2013). Quantitating the subtleties of microglial morphology with fractal analysis. *Frontiers in Cellular Neuroscience, 7*, 3. http://dx.doi.org/10.3389/fncel.2013.00003.

Kettenmann, H., Hanisch, U.-K., Noda, M., & Verkhratsky, A. (2011). Physiology of microglia. *Physiological Reviews, 91*(2), 461–553. http://dx.doi.org/10.1152/physrev.00011.2010.

Kimmel, C. B., Ballard, W. W., Kimmel, S. R., Ullmann, B., & Schilling, T. F. (1995). Stages of embryonic development of the zebrafish. *Developmental Dynamics, 203*(3), 253–310. http://dx.doi.org/10.1002/aja.1002030302.

Komohara, Y., Ohnishi, K., Kuratsu, J., & Takeya, M. (2008). Possible involvement of the M2 anti-inflammatory macrophage phenotype in growth of human gliomas. *The Journal of Pathology, 216*(1), 15–24. http://dx.doi.org/10.1002/path.2370.

Konantz, M., Balci, T. B., Hartwig, U. F., Dellaire, G., André, M. C., Berman, J. N., & Lengerke, C. (2012). Zebrafish xenografts as a tool for in vivo studies on human cancer. *Annals of the New York Academy of Sciences, 1266*, 124–137. http://dx.doi.org/10.1111/j.1749-6632.2012.06575.x.

Lam, S. H., Chua, H. L., Gong, Z., Lam, T. J., & Sin, Y. M. (2004). Development and maturation of the immune system in zebrafish, *Danio rerio*: a gene expression profiling, in situ hybridization and immunological study. *Developmental and Comparative Immunology, 28*(1), 9–28.

Lamason, R. L., Mohideen, M.-A., Mest, J. R., Wong, A. C., Norton, H. L., Aros, M. C., … Cheng, K. C. (2005). SLC24A5, a putative cation exchanger, affects pigmentation in zebrafish and humans. *Science, 310*(5755), 1782–1786. http://dx.doi.org/10.1126/science.1116238.

Lister, J. A., Robertson, C. P., Lepage, T., Johnson, S. L., & Raible, D. W. (1999). Nacre encodes a zebrafish microphthalmia-related protein that regulates neural-crest-derived pigment cell fate. *Development, 126*(17), 3757–3767.

Louis, D. N., Ohgaki, H., Wiestler, O. D., Cavenee, W. K., Burger, P. C., Jouvet, A., … Kleihues, P. (2007). The 2007 WHO classification of tumours of the central nervous system. *Acta Neuropathologica, 114*(2), 97–109. http://dx.doi.org/10.1007/s00401-007-0243-4.

Markovic, D. S., Glass, R., Synowitz, M., van Rooijen, N., & Kettenmann, H. (2005). Microglia stimulate the invasiveness of glioma cells by increasing the activity of metalloprotease-2. *Journal of Neuropathology and Experimental Neurology, 64*(9), 754–762.

Markovic, D. S., Vinnakota, K., Chirasani, S., Synowitz, M., Raguet, H., Stock, K., … Kettenmann, H. (2009). Gliomas induce and exploit microglial MT1-MMP expression for tumor expansion. *Proceedings of the National Academy of Sciences of the United States of America, 106*(30), 12530–12535. http://dx.doi.org/10.1073/pnas.0804273106.

Mills, C. D. (2012). M1 and M2 macrophages: oracles of health and disease. *Critical Reviews in Immunology, 32*(6), 463–488.

Morantz, R. A., Wood, G. W., Foster, M., Clark, M., & Gollahon, K. (1979). Macrophages in experimental and human brain tumors. Part 2: studies of the macrophage content of human brain tumors. *Journal of Neurosurgery, 50*(3), 305–311. http://dx.doi.org/10.3171/jns.1979.50.3.0305.

Neumann, H., Kotter, M. R., & Franklin, R. J. M. (2009). Debris clearance by microglia: an essential link between degeneration and regeneration. *Brain: a Journal of Neurology, 132*(Pt 2), 288–295. http://dx.doi.org/10.1093/brain/awn109.

Nguyen-Chi, M., Laplace-Builhe, B., Travnickova, J., Luz-Crawford, P., Tejedor, G., Phan, Q. T., ... Djouad, F. (2015). Identification of polarized macrophage subsets in zebrafish. *eLife, 4*, e07288. http://dx.doi.org/10.7554/eLife.07288.

Nguyen-Chi, M., Phan, Q. T., Gonzalez, C., Dubremetz, J.-F., Levraud, J.-P., & Lutfalla, G. (2014). Transient infection of the zebrafish notochord with *E. coli* induces chronic inflammation. *Disease Models and Mechanisms, 7*(7), 871–882. http://dx.doi.org/10.1242/dmm.014498.

Nimmerjahn, A., Kirchhoff, F., & Helmchen, F. (2005). Resting microglial cells are highly dynamic surveillants of brain parenchyma in vivo. *Science, 308*(5726), 1314–1318. http://dx.doi.org/10.1126/science.1110647.

Okada, M., Saio, M., Kito, Y., Ohe, N., Yano, H., Yoshimura, S., ... Takami, T. (2009). Tumor-associated macrophage/microglia infiltration in human gliomas is correlated with MCP-3, but not MCP-1. *International Journal of Oncology, 34*(6), 1621–1627.

Oosterhof, N., Boddeke, E., & Van Ham, T. J. (2015). Immune cell dynamics in the CNS: learning from the zebrafish. *Glia, 63*(5), 719–735. http://dx.doi.org/10.1002/glia.22780.

Ostrom, Q. T., Gittleman, H., Liao, P., Rouse, C., Chen, Y., Dowling, J., ... Barnholtz-Sloan, J. (2014). CBTRUS statistical report: primary brain and central nervous system tumors diagnosed in the United States in 2007–2011. *Neuro-oncology, 16*(Suppl. 4), iv1–iv63. http://dx.doi.org/10.1093/neuonc/nou223.

Paolicelli, R. C., Bolasco, G., Pagani, F., Maggi, L., Scianni, M., Panzanelli, P., ... Gross, C. T. (2011). Synaptic pruning by microglia is necessary for normal brain development. *Science, 333*(6048), 1456–1458. http://dx.doi.org/10.1126/science.1202529.

Parichy, D. M., Elizondo, M. R., Mills, M. G., Gordon, T. N., & Engeszer, R. E. (2009). Normal table of postembryonic zebrafish development: staging by externally visible anatomy of the living fish. *Developmental Dynamics, 238*(12), 2975–3015. http://dx.doi.org/10.1002/dvdy.22113.

Parichy, D. M., Ransom, D. G., Paw, B., Zon, L. I., & Johnson, S. L. (2000). An orthologue of the kit-related gene fms is required for development of neural crest-derived xanthophores and a subpopulation of adult melanocytes in the zebrafish, *Danio rerio*. *Development, 127*(14), 3031–3044.

Peri, F., & Nüsslein-Volhard, C. (2008). Live imaging of neuronal degradation by microglia reveals a role for v0-ATPase a1 in phagosomal fusion in vivo. *Cell, 133*(5), 916–927. http://dx.doi.org/10.1016/j.cell.2008.04.037.

Pollard, S. M., Yoshikawa, K., Clarke, I. D., Danovi, D., Stricker, S., Russell, R., ... Dirks, P. (2009). Glioma stem cell lines expanded in adherent culture have tumor-specific phenotypes and are suitable for chemical and genetic screens. *Cell Stem Cell, 4*(6), 568–580. http://dx.doi.org/10.1016/j.stem.2009.03.014.

Ponten, J. (1975). Neoplastic human glia cells in culture. In J. Fogh (Ed.), *Human tumour cells in vitro*. New York: Plenum Press.

Pyonteck, S. M., Akkari, L., Schuhmacher, A. J., Bowman, R. L., Sevenich, L., Quail, D. F., ... Joyce, J. A. (2013). CSF-1R inhibition alters macrophage polarization and blocks glioma progression. *Nature Medicine, 19*(10), 1264–1272. http://dx.doi.org/10.1038/nm.3337.

Rossi, F., Casano, A. M., Henke, K., Richter, K., & Peri, F. (2015). The SLC7A7 transporter identifies microglial precursors prior to entry into the brain. *Cell Reports, 11*(7), 1008–1017. http://dx.doi.org/10.1016/j.celrep.2015.04.028.

Rymo, S. F., Gerhardt, H., Wolfhagen Sand, F., Lang, R., Uv, A., & Betsholtz, C. (2011). A two-way communication between microglial cells and angiogenic sprouts regulates angiogenesis in aortic ring cultures. *PLoS One, 6*(1), e15846. http://dx.doi.org/10.1371/journal.pone.0015846.

Schafer, D. P., Lehrman, E. K., Kautzman, A. G., Koyama, R., Mardinly, A. R., Yamasaki, R., … Stevens, B. (2012). Microglia sculpt postnatal neural circuits in an activity and complement-dependent manner. *Neuron, 74*(4), 691–705. http://dx.doi.org/10.1016/j.neuron.2012.03.026.

Schartner, J. M., Hagar, A. R., Van Handel, M., Zhang, L., Nadkarni, N., & Badie, B. (2005). Impaired capacity for upregulation of MHC class II in tumor-associated microglia. *Glia, 51*(4), 279–285. http://dx.doi.org/10.1002/glia.20201.

Shiau, C. E., Kaufman, Z., Meireles, A. M., & Talbot, W. S. (2015). Differential requirement for irf8 in formation of embryonic and adult macrophages in zebrafish. *PLoS One, 10*(1), e0117513. http://dx.doi.org/10.1371/journal.pone.0117513.

Shiau, C. E., Monk, K. R., Joo, W., & Talbot, W. S. (2013). An anti-inflammatory NOD-like receptor is required for microglia development. *Cell Reports, 5*(5), 1342–1352. http://dx.doi.org/10.1016/j.celrep.2013.11.004.

Shigemoto-Mogami, Y., Hoshikawa, K., Goldman, J. E., Sekino, Y., & Sato, K. (2014). Microglia enhance neurogenesis and oligodendrogenesis in the early postnatal subventricular zone. *The Journal of Neuroscience, 34*(6), 2231–2243. http://dx.doi.org/10.1523/JNEUROSCI.1619-13.2014.

Sieger, D., Moritz, C., Ziegenhals, T., Prykhozhij, S., & Peri, F. (2012). Long-range Ca^{2+} waves transmit brain-damage signals to microglia. *Developmental Cell, 22*(6), 1138–1148. http://dx.doi.org/10.1016/j.devcel.2012.04.012.

Sierra, A., Encinas, J. M., Deudero, J. J. P., Chancey, J. H., Enikolopov, G., Overstreet-Wadiche, L. S., … Maletic-Savatic, M. (2010). Microglia shape adult hippocampal neurogenesis through apoptosis-coupled phagocytosis. *Cell Stem Cell, 7*(4), 483–495. http://dx.doi.org/10.1016/j.stem.2010.08.014.

Singh, S. K., Clarke, I. D., Terasaki, M., Bonn, V. E., Hawkins, C., Squire, J., & Dirks, P. B. (2003). Identification of a cancer stem cell in human brain tumors. *Cancer Research, 63*(18), 5821–5828.

Squarzoni, P., Oller, G., Hoeffel, G., Pont-Lezica, L., Rostaing, P., Low, D., … Garel, S. (2014). Microglia modulate wiring of the embryonic forebrain. *Cell Reports, 8*(5), 1271–1279. http://dx.doi.org/10.1016/j.celrep.2014.07.042.

Stinchcombe, J., Bossi, G., & Griffiths, G. M. (2004). Linking albinism and immunity: the secrets of secretory lysosomes. *Science, 305*(5680), 55–59. http://dx.doi.org/10.1126/science.1095291.

Stupp, R., Mason, W. P., Van den Bent, M. J., Weller, M., Fisher, B., Taphoorn, M. J. B., … National Cancer Institute of Canada Clinical Trials Group. (2005). Radiotherapy plus concomitant and adjuvant temozolomide for glioblastoma. *The New England Journal of Medicine, 352*(10), 987–996. http://dx.doi.org/10.1056/NEJMoa043330.

Svahn, A. J., Graeber, M. B., Ellett, F., Lieschke, G. J., Rinkwitz, S., Bennett, M. R., & Becker, T. S. (2013). Development of ramified microglia from early macrophages in the zebrafish optic tectum. *Developmental Neurobiology, 73*(1), 60–71. http://dx.doi.org/10.1002/dneu.22039.

Szatmári, T., Lumniczky, K., Désaknai, S., Trajcevski, S., Hídvégi, E. J., Hamada, H., & Sáfrány, G. (2006). Detailed characterization of the mouse glioma 261 tumor model for experimental glioblastoma therapy. *Cancer Science, 97*(6), 546–553. http://dx.doi.org/10.1111/j.1349-7006.2006.00208.x.

Terriente, J., & Pujades, C. (2013). Use of zebrafish embryos for small molecule screening related to cancer. *Developmental Dynamics, 242*(2), 97–107. http://dx.doi.org/10.1002/dvdy.23912.

Veinotte, C. J., Dellaire, G., & Berman, J. N. (2014). Hooking the big one: the potential of zebrafish xenotransplantation to reform cancer drug screening in the genomic era. *Disease Models and Mechanisms, 7*(7), 745–754. http://dx.doi.org/10.1242/dmm.015784.

Weller, M., Van den Bent, M., Hopkins, K., Tonn, J. C., Stupp, R., Falini, A., ... European Association for Neuro-Oncology (EANO) Task Force on Malignant Glioma. (2014). EANO guideline for the diagnosis and treatment of anaplastic gliomas and glioblastoma. *The Lancet. Oncology, 15*(9), e395–403. http://dx.doi.org/10.1016/S1470-2045(14)70011-7.

Wen, P. Y., & Kesari, S. (2008). Malignant gliomas in adults. *The New England Journal of Medicine, 359*(5), 492–507. http://dx.doi.org/10.1056/NEJMra0708126.

White, R. M., Sessa, A., Burke, C., Bowman, T., LeBlanc, J., Ceol, C., ... Zon, L. I. (2008). Transparent adult zebrafish as a tool for in vivo transplantation analysis. *Cell Stem Cell, 2*(2), 183–189. http://dx.doi.org/10.1016/j.stem.2007.11.002.

Xu, J., Zhu, L., He, S., Wu, Y., Jin, W., Yu, T., ... Wen, Z. (2015). Temporal-spatial resolution fate mapping reveals distinct origins for embryonic and adult microglia in zebrafish. *Developmental Cell, 34*(6), 632–641. http://dx.doi.org/10.1016/j.devcel.2015.08.018.

Zong, H., Parada, L. F., & Baker, S. J. (2015). Cell of origin for malignant gliomas and its implication in therapeutic development. *Cold Spring Harbor Perspectives in Biology, 7*(5). http://dx.doi.org/10.1101/cshperspect.a020610.

PART

Transplantation VIII

CHAPTER

Transplantation in zebrafish

22

J.M. Gansner, M. Dang, M. Ammerman, L.I. Zon[1]

Harvard Medical School, Boston, MA, United States

[1]*Corresponding author: E-mail: zon@enders.tch.harvard.edu*

CHAPTER OUTLINE

Abstract

Tissue or cell transplantation is an invaluable technique with a multitude of applications including studying the developmental potential of certain cell populations,

Methods in Cell Biology, Volume 138, ISSN 0091-679X, http://dx.doi.org/10.1016/bs.mcb.2016.08.006

dissecting cell-environment interactions, and identifying stem cells. One key technical requirement for performing transplantation assays is the capability of distinguishing the transplanted donor cells from the endogenous host cells and tracing the donor cells over time. The zebrafish has emerged as an excellent model organism for performing transplantation assays, thanks in part to the transparency of embryos and even adults when pigment mutants are employed. Using transgenic techniques and fast-evolving imaging technology, fluorescence-labeled donor cells can be readily identified and studied during development in vivo. In this chapter, we will discuss the rationale of different types of zebrafish transplantation in both embryos and adults and then focus on four detailed methods of transplantation: blastula/gastrula transplantation for mosaic analysis, hematopoietic stem cell transplantation, chemical screening using a transplantation model, and tumor transplantation.

INTRODUCTION

One common theme in developmental, stem cell, and cancer biology studies is to isolate a small population of cells of interest and explore how these cells maintain their cell fate and/or become other types of cells. Transplantation techniques are well suited for these types of studies. Transferring a group of cells from a donor embryo into a recipient of the same or different species and age was used more than half a century ago to map the developmental fate of individual cells and tissues. Following the introduction of molecular approaches in developmental biology, this method became the gold standard for distinguishing cell-autonomous versus non–cell-autonomous effects of a given gene mutation (Carmany-Rampey & Moens, 2006). Transplantation is also the gold standard for identifying adult somatic and cancer stem cells based on their ability to self-renew and differentiate even after serial transplantation into more than one generation of recipients.

Several features of the zebrafish make it an extraordinary vertebrate model for transplantation experiments. First, the transparency of embryos and certain adults (e.g., *casper* mutants) grants imaging power unparalleled by any other vertebrate model. Donor cells can be directly tracked in vivo with a simple fluorescence dissection microscope, or with a confocal or more specialized microscope if higher resolution is desired (Tamplin et al., 2015; White et al., 2008). Second, efficient transgenic techniques for zebrafish enable rapid construction of stable transgenic donor animals with fluorescently labeled cells of interest. Third, zebrafish eggs are externally fertilized, which greatly facilitates the transplantation procedure and posttransplantation observation compared to avian and mammalian embryos. Last but not least, relatively low husbandry costs and high fecundity make zebrafish the most compelling candidate for large-scale transplantation-based screening.

In this chapter, we will review the current status of using transplantation for mosaic analysis in zebrafish embryos, as well as stem cell and tumor transplantation, both in embryos and adults. We will further discuss the challenges, solutions, and potential future applications of transplantation approaches in zebrafish.

1. RATIONALE

1.1 MOSAIC ANALYSIS AND TRANSPLANTATION IN DEVELOPMENTAL BIOLOGY

A mosaic organism consists of cells of different genotypes (Rossant & Spence, 1998); there are multiple ways to construct a mosaic embryo. Injecting DNA into one-cell stage zebrafish embryos results in genetic mosaics as injections lead to stochastic distribution and integration of the transgene in the descendent cells (Westerfield, Wegner, Jegalian, DeRobertis, & Püschel, 1992). This technique, however, does not assure quality and can only be applied to DNA-based transgene overexpression. A refinement using a single genomic attP integration site is available but is still limited to transgene overexpression (Mosimann et al., 2013). If genetic deletion or editing is desired, a number of systems can be employed. For instance, inducible and/or tissue-specific promoters can drive FLP recombinase to recombine FRT sites, thereby creating a somatic mosaic in a spatiotemporally controllable manner (Theodosiou & Xu, 1998; Wong, Draper, & Van Eenennaam, 2011). More recently, revolutionary genome-editing technology such as CRISPR/Cas9 has permitted the creation of mosaic embryos (Hwang et al., 2013). Despite these significant advances, using transplantation to transfer cells from a donor embryo to a recipient embryo is still an important method for creating mosaic zebrafish embryos, especially in the following scenarios:

- To determine if the function of a gene is cell-autonomous or non—cell-autonomous. To understand whether a genetic mutation causes an aberrant mutant phenotype through a defect in the cell of interest or a problem in the surrounding cells that then affects the cell of interest, it is necessary to create a chimera with mutant cells in a wild-type environment and a chimera with wild-type cells in a mutant environment. If the mutant cells exhibit the mutant phenotype in a wild-type environment and wild-type cells behave normally in a mutant environment (thus rescuing the aberrant phenotype), the gene is considered to function in a cell-autonomous fashion. In comparison, if the wild-type environment makes the mutant cells behave normally, or the mutant environment can cause the wild-type cells to adopt the mutant phenotype, then the gene functions in a non—cell-autonomous fashion (Carmany-Rampey & Moens, 2006).
- To test the potential of a cell to commit to different fates by challenging it in a different environment with different developmental cues. For example, malignant teratocarcinoma cells, after being introduced into normal blastocysts, can give rise to tissues from all three germ layers, even tissues not seen in mouse teratomas such as liver, thymus, and kidney; this revealed that mouse teratocarcinoma cells are totipotent (Mintz & Illmensee, 1975).
- To analyze maternal and less common paternal effects of zygotic lethal mutations. Primordial germ cells (PGCs) from homozygous mutant embryos can be transplanted into wild-type embryos in which specific antisense morpholinos

block the development of host PGCs. This can result in the development of viable homozygous mutant adults. These chimeric adults can help in elucidating the maternal or paternal functions of known homozygous zygotic lethal genes. They can also be used to generate large batches of entirely homozygous mutants if two chimeras are crossed; this is particularly useful for experiments where analysis of mutant embryos early in development is desired, but the onset of the mutant phenotype has not yet occurred (Ciruna et al., 2002).

- To study the effects of a gene in later developmental stages. If a genetic mutation causes early lethality or severe phenotypes, mosaic embryos may circumvent these problems and permit studies of mutant cells in vivo at later developmental stages.

1.2 TRANSPLANTATION IN STEM CELL BIOLOGY

A long-standing use of the transplantation technique is to identify adult stem cells based on their key abilities to differentiate into various descendent cell types with specialized functions and at the same time proliferate/self-renew. In the past two decades, the zebrafish has developed into a versatile platform for stem cell research due to its conserved stem cell regulatory pathways, superior imaging characteristics, and extraordinary regenerative capability compared to mammals. Multiple types of stem and progenitor cells have been studied in zebrafish embryos and adults, including hematopoietic stem cells (HSCs) (Bertrand, Kim, Teng, & Traver, 2008; Ma, Zhang, Lin, Italiano, & Handin, 2011; Tamplin et al., 2015; Traver et al., 2003, 2004), neural crest stem cells (Dorsky, Moon, & Raible, 1998), neural stem cells (Barbosa et al., 2015; Chapouton et al., 2006), and muscle stem cells (Xu et al., 2013). The homeostasis of most adult tissues is maintained by adult stem cells; therefore, zebrafish can quickly regenerate damaged tissues upon injury, as seen in the caudal tail fin, which includes multiple different tissue and stem cell types, including vascular progenitors, mesenchymal stem cells, and melanocyte stem cells (Johnson & Bennett, 1999). Transplantation of a subpopulation of cells from these tissues can test the "stemness" of each subpopulation, leading to the identification and purification of tissue-specific stem cells.

1.3 TRANSPLANTATION IN CANCER BIOLOGY

The earliest transplantation of primary human tumors into other mammalian recipients was described by Harry Greene (1938, 1941). Transplantation has now become an indispensable approach to answer a panoply of questions in cancer biology, including: (1) Evaluation of tumor malignancy: positive correlations have been drawn between human tumor malignancy and transplantability in animal models (Greene, 1941); (2) Evaluation of self-renewal capability, usually via serial transplantation: serial transplantation is defined as when a donor tissue is used for another transplant after engraftment in a recipient. This method is especially useful in distinguishing leukemia from hematopoietic overproliferative disorders, since the latter cannot self-renew, and are thus not serially transplantable; (3) Identification of the cell population with self-renewal capability within a tumor: although, currently, it is controversial whether the self-renewal capability of tumors is confined to a subpopulation of cells (the cancer stem cell model) or all the tumor cells have the potential to self-renew (the stochastic model), limiting dilution is the gold standard for quantitatively measuring tumor self-renewal potential (Ignatius & Langenau, 2009); (4) Creation of tumor-bearing animal models, including metastatic models, when genetic models are not available or applicable; and (5) Studies of the in vivo behavior of human tumor cells versus in vitro cell culture.

In the past two decades, numerous publications have reported that zebrafish can develop tumors with pathological and molecular similarity to human tumors. The power of transplantation experiments in understanding the nature of tumors made the development and refinement of zebrafish transplantation protocols a major technical focus. As with adult stem cell transplantation, the inherent transparency of zebrafish embryos provides an unparalleled model to study the dynamic aspects of tumor growth and metastatic dissemination. In addition, transplanting human tumor cells into surrogate immunocompromised animal models (known as xenotransplantation) has been an important approach to study human tumors in vivo, as well as propagating tumor cells (Greene, 1941). The late maturation of a functional immune system in zebrafish facilitates the transplantation of human tumor cells into zebrafish larvae to study tumor angiogenesis and metastasis (Veinotte, Dellaire, & Berman, 2014).

1.4 ASSESSMENT OF ENGRAFTED POPULATIONS IN THE *CASPER* STRAIN

One of the major limitations in either stem cell or tumor transplantation is the sensitivity by which engrafted cells may be detected. For example, it is well recognized that single transplanted cells can fully recapitulate the hematopoietic hierarchy or even a heterogeneous tumor mass (Quintana et al., 2008). Nonetheless, it is difficult to monitor the initial steps of engraftment, self-renewal, and differentiation with in vivo mouse models, even by using highly sensitive bioluminescent techniques. To address this limitation and to extend the utility of zebrafish as both a genetic and an imaging tool, the *casper* zebrafish strain was created by mating two pigment mutants, *nacre* and *roy*. The *nacre* strain encodes a homozygous mutation in the *mitfa* gene (Lister, Robertson, Lepage, Johnson, & Raible, 1999) that results in the loss of melanocytes. The *roy* strain is a homozygous mutant in the *mpv17* gene (Richard M. White, unpublished observation) that results in the loss of reflective iridophores (Ren, McCarthy, Zhang, Adolph, & Li, 2002). The loss of both melanocytes and iridophores renders adults relatively transparent: the melanocytes normally absorb incident light, and the iridophores reflect it away.

The transparency of this line as an adult allows for detection of cell engraftment with much greater sensitivity than usual, particularly if the cells are fluorescently labeled. This can be visualized at the single-cell level in a living animal, something generally not achievable in mouse or adult wild-type zebrafish. When crossed with transgenic reporter lines, *casper* allows for real-time in vivo analysis of how cells interact with the host. For example, transplantation of mCherry-labeled lymphoma cells into a *fli1:EGFP casper* background allowed for direct visualization of the way in which tumor cells intravasate into the vasculature (Feng et al., 2010). This study elucidated the observations that *bcl2* overexpression leads to a blockade of tumor intravasation in lymphoma, a striking example of how this strain may be used to gain insight into tumor and stem cell biology.

2. METHODS

2.1 BLASTULA/GASTRULA TRANSPLANTATION IN ZEBRAFISH EMBRYOS FOR MOSAIC ANALYSIS

In the scenarios mentioned earlier, where mosaic analysis is desired, transplantation is used to create mosaic zebrafish embryos. Usually the donor and/or recipient embryos are genetically manipulated and labeled with fluorescent markers to distinguish donor and recipient cells posttransplant. Labeling of the donor can be done by injecting a fluorescent dye into the embryos before the eight-cell stage, which will uniformly label all cells of the donor embryo (Ho & Kimmel, 1993). Alternatively, transgenic embryos expressing a tissue-specific fluorescent reporter can be used to track cell populations. Once the donor and/or recipient embryos are labeled, transplantation can be performed at either the blastula or the gastrula stages, before the cells are fully committed to a specific cell type. For example, to study the role of nitric oxide synthase (*nos1*) in HSC specification, *Tg(cmyb:GFP)* donors, in which HSCs are GFP positive, were injected with control or *nos1* morpholino (North et al., 2009). Cells from the donors were then transplanted into *Tg(lmo2:DsRed)* recipients. At 36 h post fertilization (hpf), embryos were assessed for GFP-positive cells in the aorta–gonad-mesonephros (AGM) region; these are putative embryonic HSCs derived from donor cells. This experiment showed that *nos1* is required for AGM HSC in a cell-autonomous way. The reciprocal experiment, transplanting wild-type donor cells into morpholino-injected recipients, was also performed and is necessary for drawing the conclusion about cell autonomy.

2.1.1 Blastula transplantation

The term blastula refers to the period between the 128-cell stage (2.25 hpf, 8th zygotic cell cycle) and the beginning of gastrulation (5.25 hpf, 14th zygotic cell cycle, around 50% epiboly) (Kimmel, Ballard, Kimmel, Ullmann, & Schilling, 1995). The optimal time window for transplantation lies between the 1000-cell stage (3 hpf) and the dome stage (4.3 hpf), before the blastoderm spreads across the yolk. Clonal lineage tracing studies employing single-cell tracer dye injections revealed that cells descending from a single blastomere remain closely associated throughout the blastula stage, but scatter and become dispersed among unlabeled cells from the onset of epiboly (Kimmel & Law, 1985). Therefore, in embryo transplantation experiments, the low resolution of the fate map at the blastula stage and the extensive cell movement later on make it extremely difficult to predict what tissue or cell type the donor cells will become. Thus, a large number of transplanted embryos need to be screened to select the desired chimera.

One unique case of blastula transplantation is germ-line replacement. Donor embryos are homozygous mutants injected with fluorescent dextran at the one-cell stage, which effectively labels all donor cells. The prospective hosts are injected with antisense morpholino against the *dead end* gene, which completely inhibits host PGC development without affecting host viability (Ciruna et al., 2002). At mid-blastula stages, 50–100 cells can be removed through aspiration from the margin of donor embryos and transplanted into the animal pole of similarly staged hosts. After transplantation, donor embryos are screened for the expected homozygous mutant phenotype, and host embryos are screened at 30 hpf for fluorescent PGCs as an indicator of successful transplantation. Only chimeric embryos that have successfully incorporated homozygous mutant PGCs are raised to adulthood. In crossing such chimeric adults can generate large batches of 100% homozygous mutant embryos. This approach circumvents the inconsistency and nonspecific phenotypes associated with morpholino antisense injection.

2.1.2 Gastrula transplantation

During the gastrula period (5.25–10 hpf), cells undergo extensive movements of involution, convergence, and extension, producing the primary germ layers and the embryonic axis (Kimmel et al., 1995). While blastula transplants can be performed on a dissecting microscope, gastrula transplants require a fixed-stage or adjustable-stage compound microscope, which is more expensive and time consuming. However, the trade-off is an ability to target donor cells to a more precise anatomical position than in blastula-stage transplants. With gastrula transplants, donor embryos are preferably slightly younger than recipients, at the 30–50% epiboly stage. Shield stage (6 hpf) is the first time when both the anterior–posterior (AP) and dorsal–ventral (DV) axes can be reliably determined. The embryonic shield marks the eventual dorsal side of the embryo, and cells at the animal pole will develop into head structures. Together with blastula transplantation, these techniques provide a powerful method to investigate questions of cell autonomy.

2.1.2.1 Resources and protocols

- Detailed protocols for both blastula and gastrula transplants (Carmany-Rampey & Moens, 2006; Deschene & Barresi, 2009; Kemp, Carmany-Rampey, & Moens, 2009).
- A comprehensive description of germ-line replacement technique (Ciruna et al., 2002).
- A representative live-imaging protocol of retinal and brain development (Zou & Wei, 2010).

2.2 HEMATOPOIETIC STEM CELL TRANSPLANTATION IN ZEBRAFISH

Transplantation techniques to identify stem cells have attracted significant efforts from multiple labs, especially in the HSC field. Among a variety of different tissue-specific stem cells, HSCs are the best-understood stem cell type and have been harnessed in transplantation therapies for treating patients with leukemia and other bone marrow or immune disorders. During embryonic development, these stem cells gradually obtain the full capacity of self-renewal and differentiation that lasts throughout adult life. In zebrafish, the adult hematopoietic tissue is the kidney marrow, the equivalent of the mammalian hematopoietic bone marrow. Whole kidney marrow (WKM) contains HSCs, progenitors, and mature blood cells that can be distinguished to some degree by forward- and side-scatter characteristics using flow cytometry (Traver et al., 2003). Transplantation of zebrafish HSCs has been performed in three different manners: embryo to embryo (Bertrand et al., 2008; Tamplin et al., 2015), adult to embryo (Traver et al., 2003), and adult to adult (Ma et al., 2011; Tamplin et al., 2015; Traver et al., 2004). Importantly, donor cells can be harvested from fluorescent reporter zebrafish, which can help define the HSC population and also be used to confirm multilineage engraftment (Bertrand et al., 2008; de Jong et al., 2011; Ma et al., 2011; Tamplin et al., 2015; Traver et al., 2003). To date, antibodies to zebrafish cell surface antigens that permit the enrichment of HSCs have not been described.

Of note, HSC engraftment has been measured in different ways and at different time points. In one publication, engraftment was defined as the presence of one or more donor-derived cells circulating in the peripheral blood, as measured by detection of *cd41:GFP* expression using an inverted fluorescence microscope at 30 days posttransplantation (Ma et al., 2011). In another publication, recipient WKM was subjected to analysis by flow cytometry at 3 months posttransplantation, and engraftment was considered to have occurred if donor-derived *ubi:GFP* expression was three standard deviations above the mean background GFP fluorescence (de Jong et al., 2011). In yet another publication, recipient WKM was subjected to analysis by flow cytometry at 3 months posttransplantation, but engraftment was now considered to have occurred if any donor-derived *ubi:GFP* or *ubi:mCherry* expression was detectable above background (Tamplin et al., 2015). The leading explanation for why low chimerism is obtained in many zebrafish transplantation experiments relates to the fact that donors and recipients are not matched at appropriate major histocompatibility complex (MHC) loci and that rejection of the graft therefore occurs.

2.2.1 Transplanting hematopoietic stem cells into zebrafish embryos

Large-scale mutagenesis screens have generated diverse zebrafish mutants with embryonic hematopoietic defects. Transplanting wild-type HSCs into these blood mutants can help understand the cellular origins of these mutants. For example, *bloodless* is an unmapped mutant with a decreased number of primitive hematopoietic cells and severe embryonic anemia (Liao et al., 2002). Transplantation of adult

Tg(bactin2:EGFP;gata1:DsRed) WKM cells into embryonic *bloodless* recipients results in robust expansion of circulating donor-derived blood cells that can be detected in adulthood (Traver et al., 2003). Transplantation into embryonic donors has also been used to identify populations enriched for HSCs. For example, putative embryonic HSCs from 3-day-old embryos, which were *cd41:GFP* positive and *gata1:DsRed* negative, robustly colonized the caudal hematopoietic tissues 1 day after transplantation into age-matched wild-type embryos (Bertrand et al., 2008). Since the *cd41:GFP*-positive cells carried a *gata1:DsRed* transgene, the erythroid cells derived from the donor HSCs could be visualized, thereby providing a functional readout of the donor HSCs.

To obtain single-cell suspensions of donor cells, younger embryos can be processed using mechanical dissociation followed by filtering. For older embryos, the combination of mechanical and protease-assisted dissociation followed by filtering has been effective (Tamplin et al., 2015). Cells are injected into the sinus venosus of recipient embryos using borosilicate glass capillary needles. HSCs then migrate to an appropriate microenvironment and differentiate into mature blood lineages after transplantation.

2.2.2 Transplanting hematopoietic stem cells into adult zebrafish

Transplanting transgene-labeled adult zebrafish WKM cells into gamma-irradiated wild-type recipients can rescue the recipients by repopulating all blood lineages (Langenau et al., 2004; Traver et al., 2004). The engraftment can be quantified by flow cytometric analysis of transgene-positive cells in recipient WKM or by microscopic imaging of transgene-derived fluorescence. The latter method has been facilitated by the development of the transparent *casper* mutant zebrafish (White et al., 2008). A number of refinements have been made to the transplantation technique over time, including a switch from intracardiac to retro-orbital injection, optimization of the conditioning regiment, and improved posttransplantation care (de Jong & Zon, 2012; Pugach, Li, White, & Zon, 2009). Transplantation of HSCs into adult zebrafish has been used to define the frequency of HSCs in adult WKM, which has been estimated at between 1 in 38,000 and 1 in 65,000 cells (Hess, Iwanami, Schorpp, & Boehm, 2013; de Jong et al., 2011). It has also been used as a functional readout in experiments seeking to isolate subpopulations of WKM that are enriched in adult HSCs (Ma et al., 2011; Tamplin et al., 2015). Recently, there has been interest in using the zebrafish in xenograft HSC transplants (Hess & Boehm, 2016; Staal, Spaink, & Fibbe, 2016).

2.2.2.1 Resources and protocols

- Protocols for HSC transplantation involving zebrafish embryos (Bertrand et al., 2008; Traver et al., 2003).
- Detailed protocols of dissecting kidney marrow cells, transplantation, and posttransplantation analysis (Gerlach, Schrader, & Wingert, 2011; LeBlanc, Bowman, & Zon, 2007; Pugach et al., 2009).

- Examples of limiting dilution transplantation experiments using enriched stem cell populations (Ma et al., 2011; Tamplin et al., 2015).

2.3 CHEMICAL SCREENING USING TRANSPLANTATION

A truly unique advantage of zebrafish as a model organism is its large-scale screening potential. Through unbiased screening techniques, it is possible to understand the genetic basis of diseases and to develop new therapeutics. For example, anticancer drug discovery and development is recognized as historically being highly inefficient. The inefficiency is rooted in the high rate of compound attrition, the relative small number of patients available for Phase I/II testing, and the finite research and development budgets of the biopharmaceutical industry. To ameliorate these problems, the high rate of ineffective compounds entering clinical trials must be decreased, requiring more stringent preclinical screening. Mouse xenograft testing has been an invaluable step in almost all the successful cancer therapies developed in the modern era. However, mouse xenograft models are poor predictors of the outcome of clinical trials, partially because of a failure to represent the enormous genetic diversity of tumors in patients (Sharpless & Depinho, 2006). In this regard, a zebrafish xenotransplantation model can be less labor intensive and can accommodate a more diverse range of tumors, which is useful for large-scale assessment of tumor response and modeling resistance. Such large-scale approaches can potentially increase the success of cancer drug development.

In zebrafish, transplantation-based screening has proven to be advantageous. Using WKM competitive transplantation with quantitative imaging as the readout of engraftment, novel compounds with the capability of enhancing marrow cell engraftment were recently discovered (Li et al., 2015). In addition, a high-throughput fluorescence microscopy–based screening system based on T-cell acute lymphoblastic leukemia cell transplantation has been described (Smith et al., 2010). This system was used to quantify the frequency of tumor-propagating cells (Blackburn, Liu, & Langenau, 2011).

2.3.1 Resources and protocols

- A protocol for conducting competitive HSC transplantation is available (Li et al., 2015).

2.4 TUMOR TRANSPLANTATION IN ZEBRAFISH

The zebrafish has emerged as an important animal model in cancer biology studies over the past two decades. Teleost fish can acquire spontaneous cancers in almost all organs after exposure to waterborne carcinogens (Spitsbergen et al., 2000a, 2000b). In addition, transgenic tumor models of tissue-specific, promoter-driven oncogene expression have been developed, such as BRAF(V600E) melanoma (Ceol et al., 2011; Patton et al., 2005), T/B-cell acute lymphoid leukemia (Langenau et al., 2003; Sabaawy et al., 2006), and embryonal rhabdomyosarcoma

(RMS) (Langenau et al., 2007). At the same time, a variety of transplantation techniques have been developed to study the nature of these tumors and compare them with human tumors. The currently available transplantation techniques include the following:

- Serial transplantation to identify the self-renewal potential of overproliferative cells. This approach is especially useful in distinguishing leukemia from hematopoietic overproliferative disorders, since the latter cannot self-renew and thus are not serially transplantable. For example, homozygous *shrek* (*srk*) mutant zebrafish develop T-ALL in adulthood. About 100,000 kidney marrow cells from *srk* mutants can be serially transplanted into sublethally irradiated wild-type adult zebrafish, even up to tertiary recipients. With each passage, an increasingly malignant phenotype is acquired, as measured by the mean survival of the recipients (Frazer et al., 2009). In contrast, kidney marrow cells from *hsp70:kRASG12D* transgenics induce a myeloproliferative disorder (MPD) and can only be effectively transplanted into primary recipients; the cells fail to cause an MPD in secondary recipients (Le et al., 2007). This result distinguishes MPDs from myeloid leukemia, in which the leukemic progenitor cells can self-renew.
- Limiting dilution transplantation to estimate the frequency of self-renewing cells in a tumor. Cells from a given tumor are serially diluted and different numbers of tumor cells (typically 10^1-10^5 cells) are transplanted into recipients (Frazer et al., 2009; Langenau et al., 2007). The percentage of recipients that are engrafted with tumor cells at each dosage at a predetermined time point post-transplantation is then determined. Based on the results, the frequency of self-renewing cells can be calculated through established mathematical models, such as Bonnefoix limiting dilution method (Bonnefoix, Bonnefoix, Verdiel, & Sotto, 1996), or the L-Calc statistical software (STEMCELL Technologies). Several groups have used this method to estimate the self-renewal frequency in zebrafish T-ALL and embryonal RMS (Frazer et al., 2009; Langenau et al., 2007; Smith et al., 2010). However, one caveat requires great attention, namely the transplantation conditions, as they can greatly affect the numbers obtained in such experiments, as apparent by single-cell transplantation of human melanoma into NOD-SCID mice (Quintana et al., 2008). Optimizing the cell handling protocol and donor/host immune system compatibility are critical parameters that determine efficient engraftment.
- Purification of subpopulations versus single-cell transplantation to identify specific cancer stem cells. A tumor is a bulk of heterogeneous cells. Transplantation in combination with advanced cell-labeling/sorting methods allows the isolation of a subpopulation or even single cells capable of transferring the malignancy. In a *rag2:kRASG12D*-induced embryonal RMS model, the transgenics also carry *b-actin:GFP* and *rag2:dsRed* transgenes. Using fluorescence-activated cell sorting (FACS), different subpopulations of the tumor can be purified based on GFP and dsRed positivity. Limiting dilution transplantation

with these different subpopulations shows that the dsRed-positive population transplants more efficiently than the dsRed/GFP double-positive population, which correlates with a more immature gene expression signature found in the dsRed-positive population (Langenau et al., 2007). In another study, individual tumor cells from fish T-ALL were sorted and successfully transplanted (Smith et al., 2010).

- Using the immunocompromised *rag2(E450fs) casper* mutant zebrafish. This is the first immunocompromised zebrafish recipient allowing long-term engraftment of several tissues and cancers. Mutant *rag2(E450fs)* zebrafish harbor reduced functional T and B cells and do not require radiation or chemical ablation of the immune system prior to transplantation (Tang et al., 2014, 2016).
- Using transparent recipients to directly visualize tumor cell proliferation and dissemination. Similar to GFP-positive WKM transplantation, melanoma cells have been transplanted into *casper* zebrafish via intraperitoneal injection, and tumor progression was monitored in vivo over a 2-week period (White et al., 2008). Recent studies using high-resolution imaging of transplanted recipients illustrate the various stages of metastasis at a single-cell resolution and report a quantitative algorithm to calculate the metastasis initiating cell frequency (Heilmann et al., 2015).

In addition to zebrafish-derived tumors, xenotransplantation of human tumor cells into larval or immunosuppressed adult zebrafish has greatly broadened the spectrum of cancer biology questions that can be studied in zebrafish. Xenograft testing in immunocompromised mice has been a necessary step in almost all the successful cancer therapies developed in the modern era. Usually, 0.5–1.0 million cultured human tumor cells are injected subcutaneously into SCID and nude mice, subsequently forming palpable tumor nodules within 2–6 weeks of transplantation (Sharpless & Depinho, 2006).

The Hendrix lab pioneered xenotransplantation in zebrafish by injecting 50–100 cultured metastatic human melanoma cell lines into blastula-stage embryos. These cells did not form tumors; instead, transplanted melanoma cells responded to developmental cues, spread to different locations, and divided (Lee, Seftor, Bonde, Cornell, & Hendrix, 2005). In some cases, the transplanted tumor cells can induce ectopic formation of the embryonic axis in the host; this observation enabled the discovery of a critical role for Nodal signaling in melanoma progression (Topczewska et al., 2006). Primary human tumors of the pancreas, stomach, and colon have also been injected into the yolk sac of 2-day-old embryos and have shown invasiveness and micrometastasis formation within 24 h, which is much faster than the equivalent mouse models (Marques et al., 2009). Current theories in the mouse xenotransplantation field propose that orthotopic transplantation might generate more physiological tumor–stroma interactions than subcutaneous injections (Sharpless & Depinho, 2006). In orthotopic transplantation, cell lines are transplanted into the relevant tissue for the tumor of interest. Marques and coworkers also demonstrated the feasibility of implanting primary human pancreas tumor cells into the liver of 5-day-old zebrafish (Marques et al., 2009). Transplanted tumor cells invaded the embryo and formed distant metastases (Marques et al., 2009). In the zebrafish tumor xenograft model, tumor-induced angiogenesis processes have also been observed: human or murine tumor cells that express the angiogenic fibroblast growth factor 2 (FGF2) and/or vascular endothelial growth factor (VEGF) rapidly induce neovascularization in the host (Nicoli, Ribatti, Cotelli, & Presta, 2007). Xenotransplantation of patient-derived multiple myeloma cells into zebrafish larvae has provided a preclinical platform to test drug sensitivity. While patient-derived multiple myeloma cells are difficult to propagate in vitro or even in murine in vivo models, the cells robustly engraft and proliferate in 48 hpf zebrafish embryos (Lin et al., 2016). While xenotransplantation of human tumor cells into zebrafish larvae has been previously described, recent work shows that larval immunotolerance permits adult zebrafish to accept human tumor cells without immunosuppression. In this model, irradiated human tumor cells are transplanted into uncompromised zebrafish larvae. Three months later, nonirradiated human tumor cells are transplanted, and successfully engraft, proliferate, and metastasize in the same preconditioned adult zebrafish recipient. The development of this technique provides a model to study xenograft transplantations with a tumor microenvironment supporting normal immunity (Zhang et al., 2016).

2.4.1 Resources and protocols

- A comprehensive summary of tumor transplant experiments performed in zebrafish (Taylor & Zon, 2009).
- A zebrafish tumor transplantation protocol, including an immunosuppressive regimen, tumor tissue preparation, and injection procedures (Dovey & Zon, 2009).
- A protocol of human tumor xenotransplantation into zebrafish and an angiogenesis assay (Nicoli et al., 2007).

3. DISCUSSION

3.1 ZEBRAFISH IMMUNOLOGY AND CLONAL ZEBRAFISH

The immune system protects organisms from foreign, "nonself" bodies such as bacteria, parasites, and viruses; therefore, it also creates a potential barrier for allogeneic transplantation. The molecular cues responsible for distinguishing "self" from "nonself" are mediated by proteins of the MHC. In human HSC transplantation, certain donor and recipient MHC loci presumably need to be matched to reduce the risks of graft rejection and graft-versus-host diseases. In addition, "space" needs to be created for the new graft, generally using radiation or chemotherapy to ablate existing marrow. The zebrafish immune system is incompletely characterized, but MHC loci that impact HSC engraftment have been identified (de Jong et al., 2011; de Jong & Zon, 2012; Lieschke & Trede, 2009). Several approaches have been taken to improve engraftment rates of both HSCs and tumor cells in zebrafish, including the following:

- Sublethal irradiation to suppress the recipient's immune system before transplantation (de Jong et al., 2011; Ma et al., 2011; Traver et al., 2004).
- Using embryonic recipients since they are immunologically immature and do not appear to reject donor cells as readily (Bertrand et al., 2008; Traver et al., 2003).
- Using immunocompromised recipients (Hess & Boehm, 2016; Hess et al., 2013; Tang et al., 2014).
- Using MHC-matched donors and recipients (de Jong et al., 2011).
- Using chemical treatment with compounds such as cyclosporine A pre-transplantation (Shayegi et al., 2014).
- Using clonally derived zebrafish strains (Smith et al., 2010), which have been difficult to generate due to a loss of fecundity and skewed sex ratios in offspring (LaFave, Varshney, Vemulapalli, Mullikin, & Burgess, 2014). Another consideration for this approach is that any transgenes required for marking cells would ideally need to be generated in the clonal strain.

3.2 IDENTIFYING *BONA FIDE* HEMATOPOIETIC STEM CELLS

In mouse, subpopulations of transplantable cells, especially a variety of stem cells, can be purified and characterized by flow cytometry. For example, long-term HSCs can be enriched with combinations of multiple antibodies with or without the use of Hoescht dye (Mayle, Luo, Jeong, & Goodell, 2013). In zebrafish, the existence of long-term HSCs is widely recognized but there are very few antibodies to zebrafish cell surface proteins and none that have been used to enrich for HSCs. Generating additional antibodies against zebrafish hematopoietic markers for use in HSC enrichment is therefore critically important for realizing the full potential of this model organism in stem cell research.

In the absence of cell surface markers, enrichment of zebrafish HSCs has been attained using transgenic techniques to express fluorescent proteins under the control of stem cell–specific promoters or enhancers (Bertrand et al., 2008; Ma et al., 2011; Tamplin et al., 2015). In zebrafish, the murine *Runx1* +23 HSC enhancer (Bee et al., 2009; Nottingham et al., 2007) has been used to create a transgenic line—*Tg(+23.5Mmu.Runx1:GFP)*—where approximately 1 in 35 GFP-positive cells in adults and 1 in 3 GFP-positive cells in embryos is a HSC based on limiting dilution transplantation experiments (Tamplin et al., 2015). If the same calculation is applied to data from *cd41:GFP*^{low}-positive cells, a lower frequency of these cells are HSCs (Ma et al., 2011). The generation of multicolor transgenic zebrafish employing a combination of multiple stem cell markers is a feasible approach to increase the purity of HSCs. Multicolor transgenic zebrafish also facilitate the live tracking and imaging of cell populations in vivo through fluorescence microscopy (Tamplin et al., 2015). This benefit has not yet been fully realized in other vertebrate models.

ACKNOWLEDGMENTS

The authors gratefully acknowledge the contributions of previous authors, Pulin Li and Richard M. White, to the development and writing of this chapter. We also thank Elliott J. Hagedorn for helpful comments.

REFERENCES

Barbosa, J. S., Sanchez-Gonzalez, R., Di Giaimo, R., Baumgart, E. V., Theis, F. J., Gotz, M., & Ninkovic, J. (2015). Live imaging of adult neural stem cell behavior in the intact and injured zebrafish brain. *Science, 348*, 789–793.

Bee, T., Ashley, E. L., Bickley, S. R., Jarratt, A., Li, P.-S., Sloane-Stanley, J., … de Bruijn, M. F. (2009). The mouse Runx1 +23 hematopoietic stem cell enhancer confers hematopoietic specificity to both Runx1 promoters. *Blood, 113*, 5121–5124.

Bertrand, J. Y., Kim, A. D., Teng, S., & Traver, D. (2008). CD41+ cmyb+ precursors colonize the zebrafish pronephros by a novel migration route to initiate adult hematopoiesis. *Development, 135*, 1853–1862.

Blackburn, J. S., Liu, S., & Langenau, D. M. (2011 Jul 14). Quantifying the frequency of tumor-propagating cells using limiting dilution cell transplantation in syngeneic zebrafish. *Journal of Visualized Experiments*, (53), e2790. http://dx.doi.org/10.3791/2790.

Bonnefoix, T., Bonnefoix, P., Verdiel, P., & Sotto, J. J. (1996). Fitting limiting dilution experiments with generalized linear models results in a test of the single-hit Poisson assumption. *Journal of Immunological Methods, 194*, 113–119.

Carmany-Rampey, A., & Moens, C. B. (2006). Modern mosaic analysis in the zebrafish. *Methods, 39*, 228–238.

Ceol, C. J., Houvras, Y., Jane-Valbuena, J., Bilodeau, S., Orlando, D. A., Battisti, V., … Ferré, F. (2011). The histone methyltransferase SETDB1 is recurrently amplified in melanoma and accelerates its onset. *Nature, 471*, 513–517.

Chapouton, P., Adolf, B., Leucht, C., Tannhäuser, B., Ryu, S., Driever, W., & Bally-Cuif, L. (2006). her5 expression reveals a pool of neural stem cells in the adult zebrafish midbrain. *Development, 133*, 4293–4303.

Ciruna, B., Weidinger, G., Knaut, H., Thisse, B., Thisse, C., Raz, E., & Schier, A. F. (2002). Production of maternal-zygotic mutant zebrafish by germ-line replacement. *Proceedings of the National Academy of Sciences of the United States of America, 99*, 14919–14924.

Deschene, E. R., & Barresi, M. J. (2009 Sep 11). Tissue targeted embryonic chimeras: zebrafish gastrula cell transplantation. *Journal of Visualized Experiments*, (31), pii: 1422. http://dx.doi.org/10.3791/1422.

Dorsky, R. I., Moon, R. T., & Raible, D. W. (1998). Control of neural crest cell fate by the Wnt signalling pathway. *Nature, 396*, 370–373.

Dovey, M. C., & Zon, L. I. (2009). Defining cancer stem cells by xenotransplantation in zebrafish. *Methods in Molecular Biology, 568*, 1–5.

Feng, H., Stachura, D. L., White, R. M., Gutierrez, A., Zhang, L., Sanda, T., … Langenau, D. M. (2010). T-lymphoblastic lymphoma cells express high levels of BCL2, S1P1, and ICAM1, leading to a blockade of tumor cell intravasation. *Cancer Cell, 18*, 353–366.

Frazer, J. K., Meeker, N. D., Rudner, L., Bradley, D. F., Smith, A. C. H., Demarest, B., ... Tripp, S. (2009). Heritable T-cell malignancy models established in a zebrafish phenotypic screen. *Leukemia, 23*, 1825–1835.

Gerlach, G. F., Schrader, L. N., & Wingert, R. A. (2011 Aug 29). Dissection of the adult zebrafish kidney. *Journal of Visualized Experiments*, (54), pii: 2839. http://dx.doi.org/10.3791/2839.

Greene, H. S. N. (1938). Heterologous transplantation of human and other mammalian tumors. *Science, 88*, 357–358.

Greene, H. S. N. (1941). Heterologous transplantation of mammalian TUMORS: II. The transfer of human tumors to ALIEN species. *The Journal of Experimental Medicine, 73*, 475–486.

Heilmann, S., Ratnakumar, K., Langdon, E. M., Kansler, E. R., Kim, I. S., Campbell, N. R., ... van Rooijen, E. (2015). A quantitative system for studying metastasis using transparent zebrafish. *Cancer Research, 75*, 4272–4282.

Hess, I., & Boehm, T. (2016). Stable multilineage xenogeneic replacement of definitive hematopoiesis in adult zebrafish. *Scientific Reports, 6*, 19634.

Hess, I., Iwanami, N., Schorpp, M., & Boehm, T. (2013). Zebrafish model for allogeneic hematopoietic cell transplantation not requiring preconditioning. *Proceedings of the National Academy of Sciences of the United States of America, 110*, 4327–4332.

Ho, R. K., & Kimmel, C. B. (1993). Commitment of cell fate in the early zebrafish embryo. *Science, 261*, 109–111.

Hwang, W. Y., Fu, Y., Reyon, D., Maeder, M. L., Tsai, S. Q., Sander, J. D., ... Joung, J. K. (2013). Efficient genome editing in zebrafish using a CRISPR-Cas system. *Nature Biotechnology, 31*, 227–229.

Ignatius, M. S., & Langenau, D. M. (2009). Zebrafish as a model for Cancer self-renewal. *Zebrafish, 6*, 377–387.

Johnson, S. L., & Bennett, P. (1999). Growth control in the ontogenetic and regenerating zebrafish fin. *Methods in Cell Biology, 59*, 301–311.

de Jong, J. L. O., Burns, C. E., Chen, A. T., Pugach, E., Mayhall, E. A., Smith, A. C. H., ... Zon, L. I. (2011). Characterization of immune-matched hematopoietic transplantation in zebrafish. *Blood, 117*, 4234–4242.

de Jong, J.L.O., & Zon, L. I. (2012). Histocompatibility and Hematopoietic Transplantation in the Zebrafish. *Advances in Hematology, 2012*, 8. http://dx.doi.org/10.1155/2012/282318. Article ID 282318.

Kemp, H. A., Carmany-Rampey, A., & Moens, C. (2009 Jul 17). Generating chimeric zebrafish embryos by transplantation. *Journal of Visualized Experiments*, (29), pii: 1394. http://dx.doi.org/10.3791/1394.

Kimmel, C. B., Ballard, W. W., Kimmel, S. R., Ullmann, B., & Schilling, T. F. (1995). Stages of embryonic development of the zebrafish. *Developmental Dynamics: An Official Publication of the American Association of Anatomists, 203*, 253–310.

Kimmel, C. B., & Law, R. D. (1985). Cell lineage of zebrafish blastomeres. III. Clonal analyses of the blastula and gastrula stages. *Developmental Biology, 108*, 94–101.

LaFave, M. C., Varshney, G. K., Vemulapalli, M., Mullikin, J. C., & Burgess, S. M. (2014). A defined zebrafish line for high-throughput genetics and genomics: NHGRI-1. *Genetics, 198*, 167–170.

Langenau, D. M., Ferrando, A. A., Traver, D., Kutok, J. L., Hezel, J.-P. D., Kanki, J. P., ... Trede, N. S. (2004). In vivo tracking of T cell development, ablation, and engraftment in transgenic zebrafish. *Proceedings of the National Academy of Sciences of the United States of America, 101*, 7369–7374.

Langenau, D. M., Keefe, M. D., Storer, N. Y., Guyon, J. R., Kutok, J. L., Le, X., … Zon, L. I. (2007). Effects of RAS on the genesis of embryonal rhabdomyosarcoma. *Genes & Development, 21*, 1382–1395.

Langenau, D. M., Traver, D., Ferrando, A. A., Kutok, J. L., Aster, J. C., Kanki, J. P., … Zon, L. I. (2003). Myc-induced T cell leukemia in transgenic zebrafish. *Science, 299*, 887–890.

LeBlanc, J., Bowman, T. V., & Zon, L. (2007). Transplantation of whole kidney marrow in adult zebrafish. *Journal of Visualized Experiments*, 159.

Le, X., Langenau, D. M., Keefe, M. D., Kutok, J. L., Neuberg, D. S., & Zon, L. I. (2007). Heat shock-inducible Cre/Lox approaches to induce diverse types of tumors and hyperplasia in transgenic zebrafish. *Proceedings of the National Academy of Sciences of the United States of America, 104*, 9410–9415.

Lee, L. M. J., Seftor, E. A., Bonde, G., Cornell, R. A., & Hendrix, M. J. C. (2005). The fate of human malignant melanoma cells transplanted into zebrafish embryos: assessment of migration and cell division in the absence of tumor formation. *Developmental Dynamics: An Official Publication of the American Association of Anatomists, 233*, 1560–1570.

Li, P., Lahvic, J. L., Binder, V., Pugach, E. K., Riley, E. B., Tamplin, O. J., … Heffner, G. C. (2015). Epoxyeicosatrienoic acids enhance embryonic haematopoiesis and adult marrow engraftment. *Nature, 523*, 468–471.

Liao, E. C., Trede, N. S., Ransom, D., Zapata, A., Kieran, M., & Zon, L. I. (2002). Non-cell autonomous requirement for the bloodless gene in primitive hematopoiesis of zebrafish. *Development, 129*, 649–659.

Lieschke, G. J., & Trede, N. S. (2009). Fish immunology. *Current Biology, 19*, R678–R682.

Lin, J., Zhang, W., Zhao, J.-J., Kwart, A. H., Yang, C., Ma, D., … Handin, R. I. (2016 Jul 14). A clinically relevant in vivo zebrafish model of human multiple myeloma (MM) to study preclinical therapeutic efficacy. *Blood, 128*(2), 249–252. http://dx.doi.org/10.1182/blood-2016-03-704460. Epub 2016 May 18.

Lister, J. A., Robertson, C. P., Lepage, T., Johnson, S. L., & Raible, D. W. (1999). Nacre encodes a zebrafish microphthalmia-related protein that regulates neural-crest-derived pigment cell fate. *Development, 126*, 3757–3767.

Ma, D., Zhang, J., Lin, H.-F., Italiano, J., & Handin, R. I. (2011). The identification and characterization of zebrafish hematopoietic stem cells. *Blood, 118*, 289–297.

Marques, I. J., Weiss, F. U., Vlecken, D. H., Nitsche, C., Bakkers, J., Lagendijk, A. K., … Bagowski, C. P. (2009). Metastatic behaviour of primary human tumours in a zebrafish xenotransplantation model. *BMC Cancer, 9*, 128.

Mayle, A., Luo, M., Jeong, M., & Goodell, M. A. (2013). Flow cytometry analysis of murine hematopoietic stem cells. *Cytometry. Part A: The Journal of International Society for Analytical Cytology, 83*, 27–37.

Mintz, B., & Illmensee, K. (1975). Normal genetically mosaic mice produced from malignant teratocarcinoma cells. *Proceedings of the National Academy of Sciences of the United States of America, 72*, 3585–3589.

Mosimann, C., Puller, A.-C., Lawson, K. L., Tschopp, P., Amsterdam, A., & Zon, L. I. (2013). Site-directed zebrafish transgenesis into single landing sites with the phiC31 integrase system. *Developmental Dynamics: An Official Publication of the American Association of Anatomists, 242*, 949–963.

Nicoli, S., Ribatti, D., Cotelli, F., & Presta, M. (2007). Mammalian tumor xenografts induce neovascularization in zebrafish embryos. *Cancer Research, 67*, 2927–2931.

North, T. E., Goessling, W., Peeters, M., Li, P., Ceol, C., Lord, A. M., ... Huang, P. (2009). Hematopoietic stem cell development is dependent on blood flow. *Cell, 137*, 736–748.

Nottingham, W. T., Jarratt, A., Burgess, M., Speck, C. L., Cheng, J.-F., Prabhakar, S., ... Kong-A-San, J. (2007). Runx1-mediated hematopoietic stem-cell emergence is controlled by a Gata/Ets/SCL-regulated enhancer. *Blood, 110*, 4188–4197.

Patton, E. E., Widlund, H. R., Kutok, J. L., Kopani, K. R., Amatruda, J. F., Murphey, R. D., ... Fletcher, C. D. M. (2005). Braf mutations are sufficient to promote nevi formation and cooperate with p53 in the genesis of melanoma. *Current Biology, 15*, 249–254.

Pugach, E. K., Li, P., White, R., & Zon, L. (2009 Dec 7). Retro-orbital injection in adult zebrafish. *Journal of Visualized Experiments*, (34), pii: 1645. http://dx.doi.org/10.3791/1645.

Quintana, E., Shackleton, M., Sabel, M. S., Fullen, D. R., Johnson, T. M., & Morrison, S. J. (2008). Efficient tumour formation by single human melanoma cells. *Nature, 456*, 593–598.

Ren, J. Q., McCarthy, W. R., Zhang, H., Adolph, A. R., & Li, L. (2002). Behavioral visual responses of wild-type and hypopigmented zebrafish. *Vision Research, 42*, 293–299.

Rossant, J., & Spence, A. (1998). Chimeras and mosaics in mouse mutant analysis. *Trends in Genetics, 14*, 358–363.

Sabaawy, H. E., Azuma, M., Embree, L. J., Tsai, H.-J., Starost, M. F., & Hickstein, D. D. (2006). TEL-AML1 transgenic zebrafish model of precursor B cell acute lymphoblastic leukemia. *Proceedings of the National Academy of Sciences of the United States of America, 103*, 15166–15171.

Sharpless, N. E., & Depinho, R. A. (2006). The mighty mouse: genetically engineered mouse models in cancer drug development. *Nature Reviews. Drug Discovery, 5*, 741–754.

Shayegi, N., Meyer, C., Lambert, K., Ehninger, G., Brand, M., & Bornhäuser, M. (2014). CXCR4 blockade and Sphingosine-1-phosphate activation facilitate engraftment of haematopoietic stem and progenitor cells in a non-myeloablative transplant model. *British Journal of Haematology, 164*, 409–413.

Smith, A. C. H., Raimondi, A. R., Salthouse, C. D., Ignatius, M. S., Blackburn, J. S., Mizgirev, I. V., ... Zhou, Y. (2010). High-throughput cell transplantation establishes that tumor-initiating cells are abundant in zebrafish T-cell acute lymphoblastic leukemia. *Blood, 115*, 3296–3303.

Spitsbergen, J. M., Tsai, H. W., Reddy, A., Miller, T., Arbogast, D., Hendricks, J. D., & Bailey, G. S. (2000a). Neoplasia in zebrafish (*Danio rerio*) treated with *N*-methyl-*N'*-nitro-*N*-nitrosoguanidine by three exposure routes at different developmental stages. *Toxicologic Pathology, 28*, 716–725.

Spitsbergen, J. M., Tsai, H. W., Reddy, A., Miller, T., Arbogast, D., Hendricks, J. D., & Bailey, G. S. (2000b). Neoplasia in zebrafish (*Danio rerio*) treated with 7,12-dimethylbenz[a]anthracene by two exposure routes at different developmental stages. *Toxicologic Pathology, 28*, 705–715.

Staal, F. J. T., Spaink, H. P., & Fibbe, W. E. (2016). Visualizing human hematopoietic stem cell trafficking in vivo using a zebrafish xenograft model. *Stem Cells and Development, 25*, 360–365.

Tamplin, O. J., Durand, E. M., Carr, L. A., Childs, S. J., Hagedorn, E. J., Li, P., ... Zon, L. I. (2015). Hematopoietic stem cell arrival triggers dynamic remodeling of the perivascular niche. *Cell, 160*, 241–252.

Tang, Q., Abdelfattah, N. S., Blackburn, J. S., Moore, J. C., Martinez, S. A., Moore, F. E., … Berman, J. N. (2014). Optimized cell transplantation using adult rag2 mutant zebrafish. *Nature Methods, 11*, 821–824.

Tang, Q., Moore, J. C., Ignatius, M. S., Tenente, I. M., Hayes, M. N., Garcia, E. G., … Blackburn, J. S. (2016). Imaging tumour cell heterogeneity following cell transplantation into optically clear immune-deficient zebrafish. *Nature Communications, 7*, 10358.

Taylor, A. M., & Zon, L. I. (2009). Zebrafish tumor assays: the state of transplantation. *Zebrafish, 6*, 339–346.

Theodosiou, N. A., & Xu, T. (1998). Use of FLP/FRT system to study *Drosophila* development. *Methods, 14*, 355–365.

Topczewska, J. M., Postovit, L.-M., Margaryan, N. V., Sam, A., Hess, A. R., Wheaton, W. W., … Hendrix, M. J. C. (2006). Embryonic and tumorigenic pathways converge via Nodal signaling: role in melanoma aggressiveness. *Nature Medicine, 12*, 925–932.

Traver, D., Paw, B. H., Poss, K. D., Penberthy, W. T., Lin, S., & Zon, L. I. (2003). Transplantation and in vivo imaging of multilineage engraftment in zebrafish bloodless mutants. *Nature Immunology, 4*, 1238–1246.

Traver, D., Winzeler, A., Stern, H. M., Mayhall, E. A., Langenau, D. M., Kutok, J. L., … Zon, L. I. (2004). Effects of lethal irradiation in zebrafish and rescue by hematopoietic cell transplantation. *Blood, 104*, 1298–1305.

Veinotte, C. J., Dellaire, G., & Berman, J. N. (2014). Hooking the big one: the potential of zebrafish xenotransplantation to reform cancer drug screening in the genomic era. *Disease Models & Mechanisms, 7*, 745–754.

Westerfield, M., Wegner, J., Jegalian, B. G., DeRobertis, E. M., & Püschel, A. W. (1992). Specific activation of mammalian Hox promoters in mosaic transgenic zebrafish. *Genes & Development, 6*, 591–598.

White, R. M., Sessa, A., Burke, C., Bowman, T., LeBlanc, J., Ceol, C., … Burns, C. E. (2008). Transparent adult zebrafish as a tool for in vivo transplantation analysis. *Cell Stem Cell, 2*, 183–189.

Wong, A. C., Draper, B. W., & Van Eenennaam, A. L. (2011). FLPe functions in zebrafish embryos. *Transgenic Research, 20*, 409–415.

Xu, C., Tabebordbar, M., Iovino, S., Ciarlo, C., Liu, J., Castiglioni, A., … Kahn, C. R. (2013). A zebrafish embryo culture system defines factors that promote vertebrate myogenesis across species. *Cell, 155*, 909–921.

Zhang, B., Shimada, Y., Hirota, T., Ariyoshi, M., Kuroyanagi, J., Nishimura, Y., & Tanaka, T. (2016). Novel immunologic tolerance of human cancer cell xenotransplants in zebrafish. *Translational Research: The Journal of Laboratory and Clinical Medicine, 170*, 89–98. e1–3.

Zou, J., & Wei, X. (2010 Jul 19). Transplantation of GFP-expressing blastomeres for live imaging of retinal and brain development in chimeric zebrafish embryos. *Journal of Visualized Experiments*, (41), pii: 1924. http://dx.doi.org/10.3791/1924.

PART

Chemical Screening

CHAPTER

Chemical screening in zebrafish for novel biological and therapeutic discovery

23

D.S. Wiley, S.E. Redfield, L.I. Zon[1]

Stem Cell Program and Division of Hematology and Oncology, Childrens' Hospital Boston, Dana-Farber Cancer Institute, Howard Hughes Medical Institute and Harvard Medical School, Boston, MA, United States

[1]*Corresponding author: E-mail: zon@enders.tch.harvard.edu*

CHAPTER OUTLINE

Methods in Cell Biology, Volume 138, ISSN 0091-679X, http://dx.doi.org/10.1016/bs.mcb.2016.10.004

Abstract

Zebrafish chemical screening allows for an in vivo assessment of small molecule modulation of biological processes. Compound toxicities, chemical alterations by metabolism, pharmacokinetic and pharmacodynamic properties, and modulation of cell niches can be studied with this method. Furthermore, zebrafish screening is straightforward and cost effective. Zebrafish provide an invaluable platform for novel therapeutic discovery through chemical screening.

INTRODUCTION

In the past 15 years, many successful therapeutics have been efficiently discovered by cell-based and biochemical drug screening. However, these screening methods do not consider in vivo small molecule activity. Potential therapeutics from such screens often do not pass in vivo testing in live organisms such as mice, since they have inherent toxicity and poor pharmacoproperties undetectable by the screening process. Also, small molecules may act differently in whole organisms due to their complex biology, as compared to more straightforward biology in cell cultures and purified proteins. Such screens are encountering problems with proteins that are difficult to target, such as transcription factors. These proteins are termed "undruggable", since they are inept in binding small molecules and often carry out their functions through protein–protein or protein–DNA/RNA interactions.

Zebrafish chemical screening can address the problems inherent in cell-based and biochemical screens. Screening in a whole organism context means drug toxicity and in vivo drug effects are addressed concurrently. Whole organism screening has the advantage of being less targeted then cell-based and biochemical screens, allowing the drug to interact with any biological pathway. The readout is an alteration of a whole organism phenotype that relates well to disease. In contrast, protein–compound binding or cell-based reporters give little indication of disease phenotype modulation. Furthermore, technological advances have made zebrafish screens straightforward and cost effective. It has been 15 years since the first zebrafish screen was attempted, and already, a number of potential therapeutics have been discovered that target processes ranging from hematopoiesis to cancer (Table 1). Zebrafish screening might also provide the ability to discover therapeutic modulators of "undruggable" processes, as it explores biology to a complexity unseen in cell-based or biochemical screens. Overall, zebrafish screening is a convenient and ideal technology for novel therapeutic discovery.

Table 1 Phenotypic Readouts

Screening Type	References	Phenotype	Fish
Morphology	Cao et al. (2009)	Polycystic kidney disease (PKD)	PKD mutants
	Colanesi et al. (2012)	Pigment cell patterning and number	Mifta mutant
	Das et al. (2010)	Body axis/cardiac defects	Wild type
	Hao et al. (2013)	Embryo dorsalization	Wild type
	Ishizaki et al. (2015)	Pigmentation and notochord defects	Wild type
	Jin et al. (2013)	Survival in presence of organophosphates	Wild type
	Jung et al. (2005)	Pigmentation	Wild type
	Khersonsky et al. (2003)	Brain and eye	Wild type
	Mathew et al. (2007)	Fin regeneration	Wild type
	Milan et al. (2003)	Heart rate	Wild type
	Moon et al. (2002)	Microtubule disruption	Wild type
	Nishiya et al. (2014)	Presence of eyes in 6-bromoindirubin-3′-oxime–treated zebrafish	Wild type
	Oppedal and Goldsmith (2010)	Fin regeneration	Wild type
	Padilla et al. (2012)	Toxicity in developing embryos	Wild type
	Peal et al. (2011)	Atrioventricular heart rhythm	tb218 mutant
	Peterson et al. (2000)	Multiple organs	Wild type
	Peterson et al. (2004)	Coarctation	Gridlock mutant
	Sachidanandan et al. (2008)	Multiple organs	Wild type
	Sandoval et al. (2013)	Morphological defects in the embryo	Wild type
	Spring et al. (2002)	Multiple organs	Wild type
	Torregroza et al. (2009)	Body axis/cardiac defects	Wild type
	Truong et al. (2014)	Toxicity	Wild type
	Williams et al. (2015)	Embryonic morphology	Wild type
	Yu et al. (2008)	Dorsal–ventral axis	Wild type
	Wong et al. (2004)	Cardiac defects	Wild type

Continued

Table 1 Phenotypic Readouts—cont'd

Screening Type	References	Phenotype	Fish
Cell state	Alvarez et al. (2009)	Angiogenesis	fli1:EGFP
	Asimaki et al. (2014)	Normalization of natriuretic peptide levels	Plakoglobin mutant; nppb: luciferase
	Becker et al. (2012)	Modifiers of hypertrophic cardiomyopathy signaling	nppb:luciferase
	Clifton et al. (2010)	Lipid absorption	Wild type
	Evason et al. (2015)	Liver size	fabp10a:pt-β-cat; fabp10a: EGFP
	Gallardo et al. (2015)	Migration of labeled lateral line primordium	cldnb:EGFP
	Gebruers et al. (2013)	Inducers of ectopic tail formation	Wild type; cmlc2: eGFP
	Gut et al. (2013)	Activated fasting-like energy state	pck1:Luc2; cryaa:mCherry
	Gutierrez et al. (2014)	Death of Avian myelocytomatosis virus oncogene cellular homolog (MYC)-expressing thymocytes	rag2:Myc-ER; rag2:dsRed2; mitf mutant
	Hong, Peterson, Hong, and Peterson (2006)	Coarctation	Gridlock mutant
	Kitambi, McCulloch, Peterson, and Malicki (2009)	Angiogenesis	fli1:EGFP
	Kong et al. (2014)	Craniofacial morphology	Wild type
	Le et al. (2013)	Modified RAS activity/ Dusp6 expression	hsp70: HRASG12V
	Li et al. (2015)	Imaging of fluorescent stem cell grafts	Casper
	Liu et al. (2013)	Migration of leukocytes to wound	zlyz:EGFP
	Liu et al. (2014)	Rescue of cardiac function	Myl7:EGFP
	Molina et al. (2009)	Dusp6 expression	dusp6:EGFP
	Murphey et al. (2006)	Cell cycle	crb mutant
	Namdaran, Reinhart, Owens, Raible, and Rubel (2012)	GFP expression in hair cells after ablation	pou4f3:gap43-GFP

Table 1 Phenotypic Readouts—cont'd

Screening Type	References	Phenotype	Fish
	Nath et al. (2013)	Survival of cyanide exposure	Wild type
	Nath et al. (2015)	Biochemical measurement of glucose	Wild type
	North et al. (2007)	Hematopoiesis	Wild type
	Owens et al. (2008)	Hair cells	Wild type
	Paik et al. (2010)	Hematopoiesis	cdx4 mutant
	Ridges et al. (2012)	Selective leukemia toxicity	p56lck:EGFP
	Rovira et al. (2011)	Number of fluorescent β-cells in pancreas	Tp1:hmgb1-mCherry; pax6b: GFP
	Saydmohammed, Vollmer, Onuoha, Vogt, and Tsang (2011)	Fibroblast growth factor signaling, dusp6 expression	dusp6:EGFP
	Shafizadeh, Peterson, and Lin (2004)	Hematopoiesis	gata1:EGFP
	Stern et al. (2005)	Cell cycle	crb mutant
	Tran et al. (2007)	Angiogenesis	VEGFR:GRCFP
	Tsuji et al. (2014)	In vivo cell cycle indicator technology	ins:mAG-zGeminin
	Wang et al. (2010)	Angiogenesis	flk1:EGFP
	Wang et al. (2015)	Number of fluorescent β-cells in pancreas	ins:PhiYFP-2A-nsfB; sst2: TagRFP
	Weger, Weger, Nusser, Brenner-Weiss, and Dickmeis (2012)	Glucocorticoid signaling reporter	AB.9 GRE:Luc
	White et al. (2011)	Neural crest	Wild type
	Xu et al. (2010)	Angiogenesis	fli1:EGFP
	Yeh et al. (2009)	Leukemia (AML1-ETO)	hsp:AML1-ETO
Behavior	Baraban et al. (2013)	Inhibition of convulsive behaviors	Nav1.1 mutant
	Kokel et al. (2010)	Photomotor response	Wild type
	Kokel et al. (2013)	Photoactivation of motor behaviors	Wild type
	Rihel et al. (2010)	Rest/wake	Wild type
	Wolman et al. (2011)	Habituation to acoustic startle	Wild type

1. RATIONALE

Zebrafish screening allows for high-throughput chemical genetics in vivo. This is its greatest advantage over cell-based and biochemical screening. Screening chemicals in the context of the whole organism allows for unique phenotypes to be screened for, other than the traditional alteration of cell state in cell-based assays or target identification (target ID) in protein-binding biochemical assays. Furthermore, small molecules are screened in the context of the complex biology of the whole organism. This allows for assessment of (1) compound toxicity, (2) chemical alteration by metabolism, (3) drug pharmacokinetics and pharmacodynamics, and (4) drug modulation of cell niches (MacRae & Peterson, 2015; Rennekamp & Peterson, 2015; Wheeler & Brändli, 2009; Zon & Peterson, 2005).

In addition, zebrafish embryonic screening is reasonably cost effective, straightforward and biologically relevant (MacRae & Peterson, 2015; Rennekamp & Peterson, 2015; Wheeler & Brändli, 2009; Zon & Peterson, 2005). Firstly, fish husbandry requirements are straightforward and embryos are easily obtained in large numbers of 200–300 per mating pair. Secondly, embryos develop *ex utero* so their development can be monitored easily. Thirdly, embryos are more easily manipulated under a microscope. Fourthly, embryos can be screened at stages with no pigment, so phenotypes are easily observed. Fifthly, drug targets between humans and zebrafish are conserved. Sixthly, in vivo toxicity is observed, eliminating hits that are poor drug candidates.

With the feasibility of high-throughput screening and the advantages associated with in vivo drug assessment, the zebrafish are an ideal organism for whole organism-based therapeutic drug discovery.

2. MATERIALS AND METHODS

2.1 ZEBRAFISH SCREEN SCORING PHENOTYPES

2.1.1 Specific versus nonspecific phenotypes

Most often screens are conducted to generate hypotheses on a specific biological question. These screens are scored based on a chosen morphology change of interest and the aim is to discover specific chemical modifiers of disease or biological pathways. Less frequently conducted are nonspecific screens that score any morphological change observed. These have been carried out to determine compound bioactivity in broad terms; the readouts being any observable perturbation of development(Das, McCartin, Liu, Peterson, & Evans, 2010; Jung et al., 2005; Khersonsky et al., 2003; Moon et al., 2002; Peterson, Link, Dowling, & Schreiber, 2000; Sachidanandan, Yeh, Peterson, & Peterson, 2008; Spring, Krishnan, Blackwell, & Schreiber, 2002; Sternson, Louca, Wong, & Schreiber, 2001; Torregroza, Evans, & Das, 2009; Wong, Sternson, Louca, Hong, & Schreiber, 2004). Often, in-house synthetic libraries containing one specific pharmacophore are screened on wild-type zebrafish embryos. Phenotypes are scored and characterized by eye. For example, Das et al. conducted a screen on wild-type embryos using a synthetic retinoid analogue library. Their aim was to discover novel retinoids that showed bioactivity in vivo. Hence, they nonspecifically scored any developmental defect they observed. This led to the discovery of BT10, which caused cardiovascular defects in fish and which bound specifically to retinoic acid receptors. This example highlights that undirected-phenotype screens are typically conducted with pharmacophore analogues, to discover better tools or drugs in pathways already known to be modulated by small molecules.

2.2 TYPES OF SCORING PHENOTYPES

Zebrafish are easily manipulated to produce a number of different readouts in chemical screens (Table 1). The types of readout have been subdivided here into three categories: (1) phenotypes that are scored by bright-field (gross morphology scoring), (2) phenotypes that are observable by cellular, genetic, or biochemical manipulation of the zebrafish (cell, genetic, and biochemical scoring), and (3) behavioral phenotypes (behavioral scoring).

2.2.1 Gross morphology scoring

A variety of observable developmental phenotypes have been characterized in zebrafish due to genetic or chemical perturbation. Here we define gross morphology scoring as phenotypic readout scored under a bright-field microscope without the assistance of any fluorescent markers or biochemical assays. Chemicals that effect early developmental patterning processes often produce gross morphological defects that are amenable to this screening process. Dorsomorphin, a bone morphogenic protein (BMP) antagonist, was identified in a screen for factors that disrupt dorsoventral patterning during early embryogenesis (Yu et al., 2008). Often, morphological changes are scored manually but certain morphologies can be scored by automation. For example, in a chemical suppressor screen for inhibitors of polycystic kidney disease (PKD) zebrafish models, Cao et al. designed a computer algorithm that could identify modulation of laterality and curvature in embryos. The error rate was low at 2.2%, suggesting that automating morphology scoring is highly possible (Cao et al., 2009). The screen identified histone deacetylase inhibitors as suppressors of the PKD phenotype, eliciting viable drug candidates for treating PKD. As illustrated by this example, investigating specific morphology changes focuses in on one or a few disease pathways and allows for a more directed and automated screening approach.

2.2.2 Cell state scoring

Cell state is defined here as a molecular phenotype not evident to the naked eye. Examples include mRNA expression levels, protein phosphorylation and cell mitotic state. When scoring a change in cell state, one requires a consequent secondary assay after chemical screening. The three most common secondary assays that have been applied to zebrafish cell state screens are (1) in situ hybridization (North et al., 2007; Paik, de Jong, Pugach, Opara, & Zon, 2010; Yeh et al., 2009), (2) immunohistochemistry (Murphey, Stern, Straub, & Zon, 2006; Stern et al., 2005), and (3) fluorescent protein reporter expression (Molina et al., 2009).

In situ hybridization involves hybridizing an mRNA-specific probe to expressed mRNA transcripts in fixed embryos. A color reaction with the probe localizes expressed transcripts to specific tissues. In addition, color intensity provides a semi-quantitative assessment of transcript levels in the tissue(s) of interest. North et al. utilized in situ hybridization to assess the expression levels of *cmyb* and *runx1*, two genes required for hematopoietic stem cell (HSC) development. They sought to discover modulators of HSC formation in their chemical screen. Thirty-five and forty-seven compounds increased or decreased *cmyb* or *runx1* expression respectively in the screen. This resulted in the discovery that compounds which modulate prostaglandin E2 levels modulate overall HSC homeostasis (North et al., 2007).

Immunohistochemistry can be used to identify levels of modified proteins via specific antibodies. Two screens have been carried out with immunohistochemical readouts to serine-10–phosphorylated histone H3 protein (Murphey et al., 2006; Stern et al., 2005). Histone H3 serine-10 phosphorylation occurs in late G2 to early M phase and is dephosphorylated in anaphase (Hendzel et al., 1997). Both screens were conducted on a *bmyb* zebrafish mutant to identify chemical suppressors of the *bmyb* phenotype. The *bmyb* mutant phenotype entails decreased cyclin B1, mitotic arrest, and genomic instability (Shepard et al., 2005). Mitotic arrest in *bmyb* mutants results in an accumulation of antibody-detectable histone H3 phosphorylation (Murphey et al., 2006; Stern et al., 2005). The screen by Stern et al. identified a small molecule, persynthamide, which reduced histone H3 phosphorylation to wild-type levels, suppressing the *bmyb* phenotype.

It is also possible to screen transgenic zebrafish with a fluorescent reporter for compounds that modulate a pathway of interest. Molina et al. (2009) designed a screen based on Dual specificity phosphatase 6 (Dusp6) expression using Tg(*dusp6*:*EGFP*)^{pt6} embryos. This transgenic line reports on the fibroblast growth factor (FGF) signaling pathway, since Dusp6 is involved in feedback attenuation of this pathway (Thisse & Thisse, 2005; Tsang & Dawid, 2004). Molina et al. discovered a small molecule, 2-benzylidene-3-(cyclohexylamino)-2,3-dihydro-1H-inden-1-one (BCI), which increased Enhanced Green Fluorescent Protein (EGFP) fluorescence in the embryos and further characterized BCI as a Dusp6 inhibitor.

Zebrafish cell state screening is very versatile and not limited to scoring assays mentioned above. For example, fluorescent lipid analogues were used to in a chemical screen for modulators of dietary lipid absorption (Clifton et al., 2010), and a luciferase reporter of *pck1* was used to identify chemical regulators of whole body energy control (Gut et al., 2013). As new cellular, genetic, and biochemical tools are applied to zebrafish, the assays for screening chemical screen hits will expand.

2.2.3 Behavioral scoring

Zebrafish movement, in response to stimuli, can be characterized. Changes in such movements can be scored for alteration by chemical perturbation. Zebrafish have been screened for psychotropic and neuroactive drugs by characterizing changes in their photomotor response (PMR) (Kokel et al., 2010, 2013), rest/wake behavior (Rihel et al., 2010), habituation to acoustic startle (Wolman, Jain, Liss, & Granato, 2011), and convulsive behaviors and electrographic seizures (Baraban, Dinday, & Hortopan, 2013). This type of screen showcases the robustness of the zebrafish in identifying drugs that target complex pathways in vivo. Such drug discovery is impossible in in vitro screens that cannot recapitulate the biology of an entire organism. Phenotypes are scored by camera recordings and computer analyses. A behavioral screen conducted by Kokel et al. (2010) overcame the inability of cell-based and biochemical chemical screening to identify compounds that modulate the central nervous system (CNS). CNS biology manifests itself in organism behavior, so a convincing study involves the intact whole organism. Kokel et al. discovered that light-stimulating zebrafish embryos resulted in a PMR that could be easily bar-coded. The PMR was recorded by a camera and bar coding was performed by custom computer scripts. A diverse collection of libraries, including neurotransmitters and ion channel binders were screened and scored for perturbation of the PMR. This study discovered novel neuroactive compounds at a highly efficient rate, illustrating the usefulness of the zebrafish in neuroactive and psychotropic drug discovery.

2.3 ADVANCES SCORING PHENOTYPES

2.3.1 Adult chemical screens

Nearly all chemical screens have been performed on embryos at various stages of development. Screening in adults requires more time and resources while the fish develop to adulthood. The workflow for adult screens often involves anesthetizing individual fish for experimental procedures. Finally, adults are also mobile and pigmented making it difficult to automate these screens and limiting the size of these chemical screens.

Despite these limitations successful adult zebrafish screens have been implemented. Chemical screens for fin regeneration have been performed in adult zebrafish (Mathew et al., 2007; Oppedal & Goldsmith, 2010). In another screen, competitive hematopoietic stem and progenitor cell transplants were performed in adult zebrafish and identified epoxyeicosatrienoic acids as a potent inducer of bone marrow engraftment (Li et al., 2015). Recent advances in administering drugs to adult fish have become more feasible with the use of oral gavage and an anesthetic combination of MS-222 and isoflurane (Dang, Henderson, Garraway, & Zon, 2016). In addition, the transparent fish line *casper*, where pigmentation is ablated, makes the adult zebrafish more amenable to screening (White et al., 2011). As the tool box for manipulating adult zebrafish expands the prevalence of screening adult phenotypes will likely follow.

2.3.2 *Suppressor chemical screens*

The emergence of gene editing tools, such as Clustered regularly interspaced short palindromic repeats and Transcription activator-like effector nucleases, has expanded the possibilities for developmental biology and disease modeling in zebrafish. Chemical suppressor screens have been designed to suppress the phenotypes associated with PKD (Cao et al., 2009), Dravet syndrome (Baraban et al., 2013), and long QT (LQT) syndrome (Peal et al., 2011) in zebrafish disease models. In addition to suppressing genetic phenotypes, chemical screens can also be designed to suppress chemical induced phenotypes. Nishiya et al. (2014) preformed a chemical screen on embryos treated with a small molecule that activates Wingless-Type MMTV Integration Site Family (WNT) signaling, 6-bromoindirubin-3′-oxime, and found a novel inhibitor of the WNT signaling pathway. Suppressor screens designed to repress chemical or genetic phenotypes are a useful tool for researches that are limited by the phenotypes available in wild-type fish.

2.4 CHOICE OF SMALL MOLECULE LIBRARY

A number of chemical libraries have been tested in zebrafish (Table 2). Three broad categories of small molecule libraries are available: commercial vendor libraries, natural product libraries, and synthetic libraries. The majority of zebrafish screens have utilized commercial libraries, specifically the subcategory of bioactive, annotated small molecules. A small number of screens have used personalized synthetic libraries to address specific issues. A description of each library is as follows.

2.4.1 *Commercial libraries*

Libraries consist of small molecules adhering to Lipinski's rules. These rules describe chemical aspects of small molecules that give them good pharmacokinetics and dynamics. Compounds have low molecular weight, partition coefficient values that afford efficient membrane absorption, and a total number of hydrogen bond donors and acceptors within appropriate limits. These properties predict good bioavailability in organisms (Lipinski, Lombardo, Dominy, & Feeney, 2001).

2.4.2 *Bioactive libraries*

These libraries are a subset of commercial libraries that are annotated with known protein targets, drug-like molecules, and bioactivity. These libraries are extremely useful in identifying small molecule targets after screening, since the target or pathway is already known. Also, screening such libraries can yield valuable information on multiple pathways in a disease phenotype. Some of the bioactive libraries used in zebrafish screening include the DIVERSet E (ChemBridge), the ICCB Known Bioactives (Biomol), the LOPAC1280 (Sigma–Aldrich), the NINDS Custom Collection (NIH/National Institute of Neurological Disease and Stroke), and the Spectrum Collection (MicroSource) (Table 2). The number of libraries is constantly expanding and companies, such as Selleckchem, ChemDiv, and ChemBridge, all have diverse libraries.

Table 2 List of Chemical Libraries

Library	# Chemicals	Times Used
LOPAC library (Sigma–Aldrich)	1,280	14
Prestwick chemical library (Prestwick)	1,280	9
DIVERSet (ChemBridge)	10,000	14
Spectrum collection (Microsource)	2,560	12
Biomol ICCB known bioactives (Enzo Life Sciences)	472	10
NINDS custom collection II (Microsource)	1,040	4
FDA-approved drug library (Enzo Life Sciences)	640	2
Johns Hopkins drug library		2
NatProd library (MicroSource Discovery Systems Inc.)	800	2
InhibitorSelect 384-well Protein kinase inhibitor library I (EMD Millipore/ Calbiochem)	160	1
Actiprobe library (TimTec)	10,000	1
NIH clinical collection (Evotec)	719	1
Chemistry and marine natural product libraries (University of Utah)		1
Screen-Well kinase library (Enzo Life Sciences)	80	1
Screen-Well phosphatase library (Enzo Life Sciences)	33	1
GSK published kinase inhibitor set (PKIS)	376	1
International drug collection (MicroSource Discovery System)	400	1
Maybridge screening collection (Fisher Scientific International)	53,000	1
US drug collection (MicroSource Discovery Systems)	1,360	1
Molecular screening Centers Network	100,000	1
Nuclear receptor ligand library (Enzo Life Sciences)	74	1
Phosphatase targeted (ChemDiv)	15,000	1
Small molecule library from Vanderbilt HTF	160,000	1
Natural products library (University of Strathclyde)	5,000	1
TocrisScreen mini library	1,120	1
ToxCast EPA phase I chemicals	293	1
Diversity set (NCI)	1,593	1
Neurotransmitter library (Enzo Life Sciences)	661	1
Ion channel ligand library (Enzo Life Sciences)	70	1
Orphan ligand library (Enzo Life Sciences)	84	1
Screening Committee of Anticancer Drugs library		1
Serotonergic ligand library (Enzo Life Sciences)	79	1
InhibitorSelect 96-well protein kinase inhibitor library II (EMD Millipore)	80	1

2.4.3 In-house synthetic libraries

Some labs synthesize their own compounds depending on their specific goals. In-house libraries are advantageous in (1) discovering novel bioactive small molecules and (2) straightforward target ID. Such libraries are based on known pharmacophores, so one can shortlist candidates of possible targets. The outcome of such a library screen is usually the discovery of novel bioactivity of an analogue of a known pharmacophore. An example of such a study is a zebrafish screen carried out with novel retinoid analogues that discovered a novel retinoid with retinoid receptor specificity. This lead compound is useful for probing the biology of the retinoic acid signaling pathways (Das et al., 2010). A second aspect of a synthetic library is that compounds can be designed with tags, allowing for target ID through protein pull-down. Tagged compounds are screened in the zebrafish to confirm that the tag does not interfere with the bioactivity of the molecule. One such screen with a tagged-triazine library identified a novel inhibitor of mitochondrial ATPase, which induces pigmentation in early zebrafish embryos (Jung et al., 2005). The pull-down protocol was straightforward since the tagged compounds were chemically ideal for binding resin. Also, the tag was already confirmed as noninterfering with target binding by the screen.

2.4.4 Diversity-oriented synthesis libraries

Diversity-oriented synthesis (DOS) libraries expand the boundaries of chemical space by the synthesis of novel pharmacophores (Schreiber, 2000). Such libraries encompass chemical space that is not covered by commercial libraries, hence providing greater potential for novel modulation of "undruggable" pathways and targets if screened.

2.4.5 Natural product libraries

These libraries consist of compounds extracted from nature (Clardy & Walsh, 2004). Famous examples of therapeutics derived from natural products are cancer drugs such as Taxol from the Pacific yew tree or antibiotics such as penicillin from *Penicillium* fungi. Like DOS compounds, natural product libraries increase the potential of novel discovery in screening.

2.5 CHEMICAL SCREENING PLATFORM

The actual screen is performed in a specific order: first, a small-scale optimization screen is conducted to determine the appropriate screening parameters. Second, sufficient zebrafish embryos are generated. Third, the screen is carried out either by hand, for a small screen, or by automation, for high-throughput. Fourth, hits are rescreened for validation.

The optimization screen is performed with a small number of embryos and compounds to determine optimal parameters such as desired plate format, compound concentrations, number of embryos per well, and embryonic stage. It is helpful if a compound that causes a positive phenotype in the assay is available. This would provide an ideal positive control and allow for compound concentrations to be fine tuned, generating an obvious scoring phenotype without causing embryonic lethality. Determining the developmental stage at which embryos are screened is also important since this can affect the phenotypic readout.

Large numbers of zebrafish embryos are needed for high-throughput screens. Traditionally, these are generated by setting up large numbers of mating pairs in multiple tanks. This method takes up a lot of space, makes embryo collection tedious, and may not yield synchronized embryos. Recently, this bottleneck in zebrafish screening has been solved by the introduction of the zebrafish spawning vessel technology (Fig. 1). The zebrafish spawning vessel allows for over 200 fish of any given strain to be spawned simultaneously. This allows for collection of a maximum of 10,500 highly synchronized embryos with a typical spawning time of 10 min. In addition, the apparatus has a small footprint, saving lab space (Adatto, Lawrence, Thompson, & Zon, 2011). Obtaining large numbers of synchronized embryos is now efficient and no longer limits the scale of chemical screening.

In addition to improving embryo collection, advances in technology have also made the handling of large compound libraries easy. Liquid handling robots, such as the TECAN robot (Tecan, Durham, NC), are used to distribute media and chemicals into plate wells rapidly and accurately. These robots are easily calibrated to operate for a range of compound volumes and plate formats. Distributing embryos into plate wells is usually preformed by hand and is a tedious and rate limiting step in

FIGURE 1

iSpawn breeding cage.

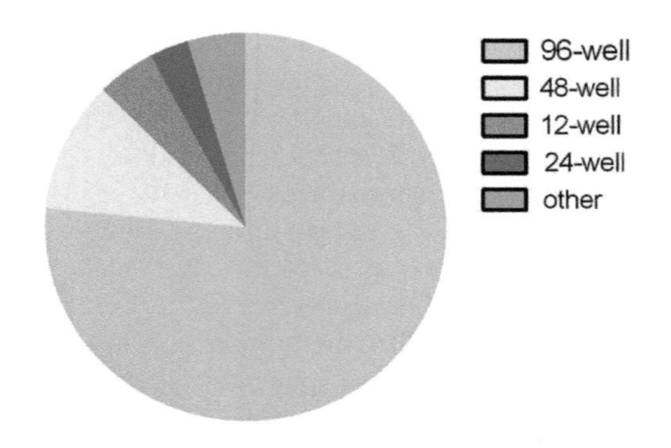

FIGURE 2

Frequency of screening plate format.

the chemical screening process, however some recent chemical screens have automated this process (Truong et al., 2014; Wang et al., 2015).

Depending on the aim of the study, the mechanics of the screen can be conducted in a number of ways. There exist variations in the plate format, the number of embryos per well, and the compound concentration in each well. Embryos are distributed into a variety of plate sizes, ranging from 6-well up to 384-well transparent plates, with the common practice being ~3 embryos per 96-well plate (Fig. 2). Such plates are amenable to liquid handling robotics for compound library addition as well as automated image recording for high-throughput phenotypic readouts. Compound libraries are added to plates with well concentrations ranging from 1 to 100 μM with the average being ~20 μM the most frequent concentration being 10 μM (Fig. 3). Present library sizes range from 10 to 160,000 (Table 2). The size and compound composition frequently depend on the respective study goals.

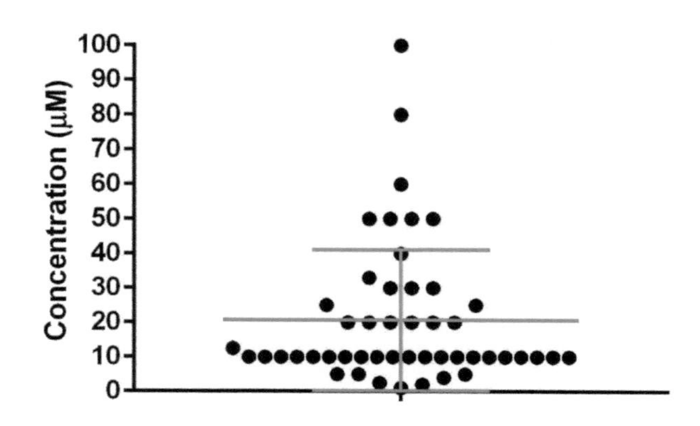

FIGURE 3

Concentration (μM).

2.6 TARGET IDENTIFICATION

To develop effective therapeutics, a near complete assessment of the drug candidate is required. This involves identifying the compound's toxicity, pharmacokinetic dynamics, and very importantly, its molecular target(s). A range of target ID methods have been applied to zebrafish chemical screens. These include the traditional techniques of protein pull-down, cell-based assays, in vitro biochemical assays and computer docking simulations. However, the most commonly used target ID method in zebrafish screens is candidate-based ID, which utilizes (1) annotated bioactive libraries, (2) chemoinformatics, and/or (3) genetics to elicit the target.

2.6.1 Candidate-based identification

Most screens address a specific biological question, such as perturbations to the hematopoietic system, pigmentation, or cardiovascular system. This allows one to narrow down the list of possible pathways modulated by the small molecules screened. However, pinpointing the exact protein being targeted is difficult in the context of a whole organism, since traditional biochemical and cell-based target ID methods are not feasible due to the complexity of the organism. Candidate-based ID is achieved by using drug libraries with known targets, chemoinformatic analysis to infer targets and/or genetic experimentation to infer modulated pathways. One or a combination of these methods is used to achieve a more complete understanding of the detailed biological effect(s) exerted by the drug candidate in question.

2.6.1.1 Annotated bioactive libraries

Annotated bioactive libraries are often used in chemical screens. These libraries contain therapeutic compounds with known targets and/or drug-like compounds. With these libraries, compound effects and targets are already known or can be easily predicted. This allows for straightforward target ID upon hit confirmation.

2.6.1.2 Chemoinformatics

When a hit with unknown targets is identified, it can be subjected to chemoinformatic analysis to predict its possible target(s). Often, from a chemical screen with diverse compound libraries, one finds interesting hit compounds that are either not well annotated or unannotated. The simplest way to hypothesize the target pathway or protein of such compounds is to find well-annotated compounds with similar structural features. Classes of compounds with similar chemical structures typically target similar pathways. Chemoinformatics can compile structural similarity information for the hit compound of interest, allowing the researcher to use preexisting structure—activity information from other similar molecules to hypothesize the possible activity of the compound of interest.

Chemoinformatic analysis can be applied by utilizing the many chemical databases with integrated search options. These databases compile useful information such as structural information, bioactivity, 3D molecular models, literature and patent links, material safety data sheets, and commercial availability. Liao, Sitzmann, Pugliese, and Nicklaus (2011) reviewed the software and databases available for drug design and discovery. Some examples of the useful database include PubChem (https://pubchem.ncbi.nlm.nih.gov/), ChemBank (http://chembank.broadinstitute.org), chemical identifier resolver (https://cactus.nci.nih.gov/chemical/structure), Fiehn Lab (http://cts.fiehnlab.ucdavis.edu/conversion/batch), Mcule (https://mcule.com/), CrossFire Beilstein, and DiscoveryGate (http://www.discoverygate.com). DiscoveryGate uses a proprietary algorithm to search its compound databases, which encompass many sources such as journal articles, patent information, and commercial and proprietary chemical databases. One can perform structural or text-based searches on DiscoveryGate to obtain vast amounts of pharmacological and biological information on chemicals related to the search compound. This tool requires a flat fee for usage (Trompouki & Zon, 2010). Searches can be performed with the chemical structure of interest to find analogous chemicals with known bioactivity. This might elicit related structures such as known pharmacophores or reactivity groups with biochemical activity. Also, hits obtained from screens performed by others can be compared for similar compound activity. In this way, the bioactivity of the hit compound can be predicted, so that the appropriate validation experiments can be conducted (Brown, 2005; Parker & Schreyer, 2004; Trompouki & Zon, 2010).

Chemical structures of search compounds can be entered into chemoinformatics tools in a variety of ways. The most common molecular format that gives detailed structural information in a highly simplified manner is the Simplified Molecular Input Line Entry Specification (SMILES) format. The SMILES format involves a simple textual representation of chemical features such as bonds, aromaticity, stereochemistry, branching, and isotopes without the use of complicated chemical drawing software. This format allows for rapid data interpretation by computers with little ambiguity in the chemical structure. Other chemical formats used are the Chemical Markup Language, GROMACS, CHARMM, the chemical file format, and the SYBYL Line Notation. To convert between the formats, one can use open source tools such as Open Babel and JOELib. Integrated applets for drawing traditional two-dimensional chemical structures into the search engine are also available (Trompouki & Zon, 2010).

There are different search algorithms one can use to search chemical databases. Examples include the commonly used Tanimoto similarity scoring, the Similarity Ensemble Approach algorithm, or the Tversky similarity algorithm. Some databases also use proprietary algorithms unique to their services. Different algorithms often lead to varying results so one might need to test more than one algorithm should the first one prove unsuccessful. Although not always consistent with each other, these algorithms are all equally important in hypothesis generation through the compilation of structurally similar compounds to the small molecule of interest. This information allows the researcher to test if the shortlisted compounds can phenocopy the compound of interest, thus narrowing down the activity of the compound of interest (Brown, 2005; Parker & Schreyer, 2004; Trompouki & Zon, 2010).

Chemoinformatics was used for target ID by Hong et al. In their chemical screen, Hong et al. obtained a hit compound GS4898 that rescued tail and trunk circulation in zebrafish *gridlock* mutant embryos. GS4898 had not been previously characterized so Hong et al. used chemoinformatics to predict that GS4898 might be a protein kinase inhibitor, since the molecule was structurally related to flavone kinase inhibitors. They then tested structurally related flavones kinase inhibitors and found that a specific phosphatidylinositol-3 kinase (PI3K) inhibitor phenocopies the rescue of the *gridlock* mutation by GS4898. This allowed Hong et al. to further validate that GS4898 did indeed inhibit PI3K.

2.6.1.3 Genetics

If the small molecule of interest is poorly annotated and chemoinformatics does not provide a viable hypothesis, one can perform genetic studies to predict the mode of action of the small molecule. Target ID via genetics first involves identifying a candidate pathway or pathways most likely modulated by the chemical. Secondly, the expression levels of genes in each pathway are analyzed to see if they are perturbed by the chemical. Thirdly, if chemical inhibitors of various steps of the pathway are available, these can be used to see if known inhibitor treatment phenocopies the effect of the chemical of interest. If inhibitors are not available, gene knockdowns and knockouts can be performed to try and phenocopy the effect of the chemical of interest. Genetic studies give evidence for what pathway the chemical of interest is modulating and serves as the basis for other follow-up biochemical and cell-based experiments directed at a specific pathway.

Gene expression changes can be measured in a number of ways. Microarrays are used to generate a large data set for gene expression changes encompassing a myriad of pathways. This is useful when either (1) a significant number of pathways are responsible for the phenotypic change or (2) the pathways responsible are unclear. In cases where a small number of candidate pathways can be shortlisted, real-time PCR is used to evaluate the expression changes of individual genes in the pathways of interest. In situ hybridization, which allows for visual observation of gene expression changes in an intact zebrafish, allows for assessment of spatial changes in gene expression if present.

Yu et al. (2008) took a multistep genetics approach to identify the target of dorsomorphin, a chemical they obtained from their screen that looked for small molecule effectors of zebrafish embryo dorsalization. The dorsoventral axis is established by BMP signaling gradients, and excess BMP signaling causes ventralization while reduced BMP signaling causes dorsalization (Fürthauer, Thisse, & Thisse, 1999; Mintzer et al., 2001; Mullins et al., 1996; Nguyen et al., 1998). The screen by Yu et al. was based on the hypothesis that BMP signaling antagonists would cause dorsalization in zebrafish. To identify the target of their chosen molecule, dorsomorphin, they first performed in situ hybridization to investigate the effects of dorsomorphin on dorsal and ventral gene markers. They noticed that the level of ventral marker *eve1* was reduced while dorsal markers such as *egr2b* and *pax2a* underwent lateral expansion during dorsomorphin treatment. Since dorsomorphin phenocopies BMP antagonism in fish, the second experiment they performed was to use dorsomorphin treatment to rescue zebrafish that were deficient in the endogenous BMP antagonist, chordin. Dorsomorphin was able to rescue the phenotype of chordin morphants, thus validating that dorsomorphin is indeed a BMP antagonist. From this genetic evidence, Yu et al. (2008) proceeded with biochemical and cell-based assays to show that dorsomorphin inhibits Suppressor of Mothers Against Decapentaplegic-dependent BMP signals and BMP type I receptor function.

2.6.2 General target identification methods applied to zebrafish screens

2.6.2.1 Protein pull-down

This method is relatively straightforward for target ID. It involves immobilizing a compound of interest to resin, by means of a chemical linker group, and incubating the resin with cell lysates. This allows intracellular binders to associate tightly with the immobilized compound, pulling it out from suspension. The resin is then washed to remove nonspecific binders, and the bound proteins are analyzed by mass spectrometry to identify them. Protein pull-down is often used with synthetic libraries, which are screened with a linker group already present on the compounds. This ensures that in vivo activity is unaffected by the linker. Compounds from other libraries are more difficult to modify for resin linkage.

2.6.2.2 Cell-based assays

Cell-based assays are used to confirm that drug candidates can bind their target in vivo. Such an assay would involve addressing a specific target of interest in a cellular environment. For example, if the hit compound is hypothesized to inhibit an enzyme that phosphorylates a certain substrate, then one can design an experiment that indicates that compound addition inhibits substrate phosphorylation. Cell-based assays are used as a confirmation for target ID, rather than for broad spectra target discovery.

2.6.2.3 In vitro biochemical assays

Similar to cell-based assays, in vitro biochemical assays are more suited to confirm targets than for discovering them. Biochemical assays are nearly identical to cell-based assays but do not involve complex cell biology. These assays require purification of the target of interest and if necessary, the target's substrate. The ability of the target to carry out its function on its substrate is then assessed under in vitro conditions with and without the small molecule.

2.6.2.4 Computer docking simulations

If crystal structures of targets are available, computer modeling can be performed to study the possibility of small molecules binding to their targets. Molina et al. (2009) use this method in their analysis of Dusp6 inhibition by BCI. Since the crystal structures of Dusp catalytic sites exist (Almo et al., 2007; Jeong et al., 2006; Stewart, Dowd, Keyse, & McDonald, 1999), Molina et al. could ascertain the probable binding site of BCI to Dusp6 with a program called ORCHESTRAR (Tripos).

3. DISCUSSION/CAVEATS

3.1 BIOLOGICAL RELEVANCE OF ZEBRAFISH SCREENING

Multiple phenotypes can be observed in zebrafish chemical screening. One can observe behavioral changes such as sleep/wake patterns or a movement response to light. In addition, one can also observe changes in gene expression either overall or in specific tissues due to chemical action. Such observations are not possible in cell-based or biochemical screening platforms. Zebrafish screening allows for exploration into the behavioral effects of small molecules, something only whole organism screens can achieve. In addition, gene expression changes are observed in vivo so this reflects accurately the biologically relevant action of the molecules. In zebrafish screening, there is no doubt that small molecules exert an effect in the context of a multicellular organism with active metabolism. This cannot be said for cell-based or biochemical platforms where further in vivo testing is required to confirm biological relevance. Compound toxicity or side effects are also not readily apparent in cell-based screens, unlike in zebrafish screens (Zon & Peterson, 2005).

Zebrafish chemical screening accounts for the biological response of cell niches. This allows one to conclude that phenotypic changes resulting from the chemical are relevant in a multicellular environment. Conversely, traditional biochemical and cell-based screens only indicate chemical activity on a specific target or cell type, ignoring the interactions of the cellular niche and the metabolic activity of the whole organism. Some chemicals undoubtedly exert phenotype change due to modulation of the surrounding cells, rather than the cell type in question. Also, phenotypic change can occur by modulation of a wide range of cell types. In vivo screening can elicit hits that in vitro screens cannot pick up. One might therefore observe different sets of hits and/or unexpected outcomes when comparing in vivo and in vitro screens. Hits from the zebrafish screen are more biologically relevant since phenotypes are due to chemical action on the whole organism instead of on one protein or one cell type.

Drug effects in humans are largely conserved in zebrafish; so zebrafish are ideal for human therapeutic discovery. Data from previous zebrafish chemical screens have shown a high degree of conservation between mammals in zebrafish, in terms of drug effects and toxicities. Cardiotoxicity screening has shown a high degree of correlation between humans and zebrafish (Milan, Peterson, Ruskin, Peterson, & MacRae, 2003). In addition, screens to discover novel neuroactive and psychotropic drugs detected known human drugs with similar effects on zebrafish (Baraban et al., 2013; Kokel et al., 2010, 2013; Rihel et al., 2010; Wolman et al., 2011). Furthermore, 50–70% of chemicals in a zebrafish cell cycle screen show similar effects when tested in a mammalian cell culture assay (Zon & Peterson, 2005). As such, zebrafish screening is very relevant when applied to novel human therapeutic discovery.

3.2 SCREENING TECHNOLOGY CAVEATS

Compound concentrations are fixed per screen; so some compounds that should be classified as hits may be missed because they are not effective at the screening concentration. Others might cause toxicity at the screening concentration and are also ignored, even if they might be nontoxic and effective at lower doses. In addition, certain phenotypes are not amenable to high-throughput screens due to specific scoring requirements.

3.3 HIT DETECTION CAVEATS

Hit rates from all zebrafish studies range from <1% to 70% (Fig. 4). This can be influenced by scoring system and zebrafish biology. As mentioned previously, there are three different scoring types: morphological, behavioral, or cell state alteration. Using a different scoring type can lead to vastly different results. Also, human judgment plays an important role in what is scored as a hit. Since it is difficult to find two or more individuals with identical interpretation of scoring phenotype, human error introduces more variability into the screen.

In addition, variability in screening results is observed when similar chemicals are applied at different stages of development. There are a number of possible entry sites for small molecules into the fish. In embryos, chemicals permeate through the chorion and uptake can occur through the epidermal layer or through the digestive system. In the larval and adult stage, entry points are similar with the inclusion of uptake through gills. The epidermal layer of the larvae is much less permeable than that in developing embryos, so one can surmise that the distribution of compound uptake by each of the aforementioned means is different depending on the stage of development. This distribution affects the action of small molecules since differing chemical modifications occur when compounds are subjected to varying cellular environments. Compounds that enter the digestive tract are subject to first-pass metabolism that often alters their chemical properties. Compounds that

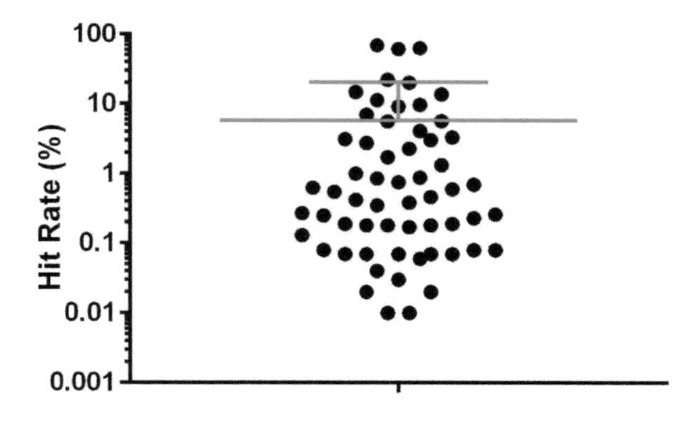

FIGURE 4

Percent hit rate (log10 scale).

permeate through the epidermis or that are administered intravenously typically undergo little chemical change. The biological activity of certain chemicals requires activation of these compounds by chemical modification in vivo. On the other hand, the biological activity of certain other chemicals is abolished by in vivo chemical modification. The influence of the whole organism biological system can alter screening results and lead to inconsistencies that require further study.

Biological organism screening variability is also affected by penetrance. The genetic background of the fish exerts a considerable influence on screening results and can affect confirmation assays. Even in wild-type strains, genetic variability leads to different levels of drug penetrance in different clutches. Often, one can look for consistent penetrance percentages to determine true hits.

SUMMARY

Zebrafish chemical screening is very useful for therapeutic and bioprobe discovery. It provides a medium- to high-throughput manner of assessing the phenotypic effects of small molecule libraries on an in vivo system. This allows for toxicity, pharmacoproperties and effects of compounds to be studied in a complex biological system, taking into account metabolism and cell–cell interactions. Also, zebrafish provide a wide variety of scoring phenotypes which can be adapted to specific study aims. In addition, compound library choices are abundant and although the largest library used so far was ~160,000 compounds, technological advances can potentially increase this number. Scaling up embryo generation to a larger scale should also not be problematic.

However, there are some caveats to note when screening zebrafish. Using juvenile/adult fish restricts the throughput since fish at this stage take longer to accumulate in large numbers. Also, juvenile/adult manipulation is more challenging. Variable hit rates are observed in screens due to compound libraries used, human error, scoring type, genetic penetrance, and fish developmental stage. Target ID is another challenge in zebrafish screening. The complexity of the whole organism means that traditional target ID methods are not ideal, and the main target ID method is candidate-based inference.

Overall, zebrafish chemical screening is an indispensible tool in therapeutic discovery. The "low hanging fruit" of drug discovery has already been taken and the focus is now on "undruggable" targets such as transcription factors and protein–protein interactions. Traditional cell-based and biochemical drug discovery screens are no longer efficient in finding therapeutics to "undruggable" targets. Also, these methods do not consider in vivo drug interactions, which could result in unwanted side-effects. Zebrafish screening has the added advantages of assessing drug toxicity at an early stage of drug development (Zon & Peterson, 2005). Also, any drug processing by metabolism is taken into consideration. Furthermore, pharmacokinetics and pharmacodynamics can be studied in the fish. Screening in zebrafish can also discover drugs that modulate the cell niche, rather than the target cell-type directly. This type of drug target would not be detected in cell-based and biochemical screens, which focus on a specific cell-type or protein target. With the aforementioned advantages, whole organism screens are undoubtedly the next step forward in chemical screening for therapeutic discovery.

ACKNOWLEDGMENTS

We would like to thank Justin L. Tan for writing the last edition of this chapter, Thorsten Schlaeger and Richard White for input on chemoinformatics, and Isaac Adatto and Christian Lawrence for their input on the zebrafish spawning vessel technology. L.I. Zon is an investigator of the Howard Hughes Medical Institute. L.I. Zon is a founder and stock holder of Fate, Inc. and a scientific advisor for Stemgent.

REFERENCES

Adatto, I., Lawrence, C., Thompson, M., & Zon, L. I. (2011). A new system for the rapid collection of large numbers of developmentally staged zebrafish embryos. *PLoS One, 6*(6), e21715. http://dx.doi.org/10.1371/journal.pone.0021715.

Almo, S. C., Bonanno, J. B., Sauder, J. M., Emtage, S., Dilorenzo, T. P., Malashkevich, V., … Burley, S. K. (2007). Structural genomics of protein phosphatases. *Journal of Structural and Functional Genomics, 8*(2–3), 121–140. http://dx.doi.org/10.1007/s10969-007-9036-1.

Alvarez, Y., Astudillo, O., Jensen, L., Reynolds, A. L., Waghorne, N., Brazil, D. P., … Kennedy, B. N. (2009). Selective inhibition of retinal angiogenesis by targeting PI3 kinase. *PLoS One, 4*(11). http://dx.doi.org/10.1371/journal.pone.0007867.

Asimaki, A., Kapoor, S., Plovie, E., Karin Arndt, A., Adams, E., Liu, Z., … Saffitz, J. E. (2014). Identification of a new modulator of the intercalated disc in a zebrafish model of arrhythmogenic cardiomyopathy. *Science Translational Medicine, 6*(240), 240ra74. http://dx.doi.org/10.1126/scitranslmed.3008008.

Baraban, S. C., Dinday, M. T., & Hortopan, G. A. (2013). Drug screening in Scn1a zebrafish mutant identifies clemizole as a potential Dravet syndrome treatment. *Nature Communications, 4*, 2410. http://dx.doi.org/10.1038/ncomms3410.

Becker, J. R., Robinson, T. Y., Sachidanandan, C., Kelly, A. E., Coy, S., Peterson, R. T., & MacRae, C. A. (2012). In vivo natriuretic peptide reporter assay identifies chemical modifiers of hypertrophic cardiomyopathy signalling. *Cardiovascular Research, 93*(3), 463–470. http://dx.doi.org/10.1093/cvr/cvr350.

Brown, F. (2005). Editorial opinion: chemoinformatics – a ten year update. *Current Opinion in Drug Discovery and Development, 8*(3), 298–302. http://www.ncbi.nlm.nih.gov/pubmed/15892243.

Cao, Y., Semanchik, N., Lee, S. H., Somlo, S., Barbano, P. E., Coifman, R., & Sun, Z. (2009). Chemical modifier screen identifies HDAC inhibitors as suppressors of PKD models. *Proceedings of the National Academy of Sciences of the United States of America, 106*(51), 21819–21824. http://dx.doi.org/10.1073/pnas.0911987106.

Clardy, J., & Walsh, C. (2004). Lessons from natural molecules. *Nature, 432*(7019), 829–837. http://dx.doi.org/10.1038/nature03194.

Clifton, J. D., Lucumi, E., Myers, M. C., Napper, A., Hama, K., Farber, S. A., … Pack, M. (2010). Identification of novel inhibitors of dietary lipid absorption using zebrafish. *PLoS One, 5*(8). http://dx.doi.org/10.1371/journal.pone.0012386.

Colanesi, S., Taylor, K. L., Temperley, N. D., Lundegaard, P. R., Liu, D., North, T. E., … Patton, E. E. (2012). Small molecule screening identifies targetable zebrafish pigmentation pathways. *Pigment Cell & Melanoma Research, 25*(2), 131–143. http://dx.doi.org/10.1111/j.1755-148X.2012.00977.x.

Dang, M., Henderson, R. E., Garraway, L. A., & Zon, L. I. (2016). Long-term drug administration in the adult zebrafish using oral gavage for cancer preclinical studies. *Disease Models & Mechanisms, 9*(7), 811–820. http://dmm.biologists.org/content/9/7/811.abstract.

Das, B. C., McCartin, K., Liu, T. C., Peterson, R. T., & Evans, T. (2010). A forward chemical screen in zebrafish identifies a retinoic acid derivative with receptor specificity. *PLoS One, 5*(4). http://dx.doi.org/10.1371/journal.pone.0010004.

Evason, K. J., Francisco, M. T., Juric, V., Balakrishnan, S., del Pilar Lopez Pazmino, M., Gordan, J. D., … Stainier, D. Y. R. (2015). Identification of chemical inhibitors of β-catenin-driven liver tumorigenesis in zebrafish. *PLoS Genetics, 11*(7). http://dx.doi.org/10.1371/journal.pgen.1005305.

Fürthauer, M., Thisse, B., & Thisse, C. (1999). Three different noggin genes antagonize the activity of bone morphogenetic proteins in the zebrafish embryo. *Developmental Biology, 214*(1), 181–196. http://dx.doi.org/10.1006/dbio.1999.9401.

Gallardo, V. E., Varshney, G. K., Lee, M., Bupp, S., Xu, L., Shinn, P., … Burgess, S. M. (2015). Phenotype-driven chemical screening in zebrafish for compounds that inhibit collective cell migration identifies multiple pathways potentially involved in metastatic invasion. *Disease Models & Mechanisms, 8*(6), 565–576. http://dx.doi.org/10.1242/dmm.018689.

Gebruers, E., Cordero-Maldonado, M. L., Gray, A. I., Clements, C., Harvey, A. L., Edrada-Ebel, R., … Esguerra, C. V. (2013). A phenotypic screen in zebrafish identifies a novel small-molecule inducer of ectopic tail formation suggestive of alterations in non-canonical Wnt/PCP signaling. *PLoS One, 8*(12), e83293.

Gut, P., Baeza-Raja, B., Andersson, O., Hasenkamp, L., Hsiao, J., Hesselson, D., … Stainier, D. Y. (2013). Whole-organism screening for gluconeogenesis identifies activators of fasting metabolism. *Nature Chemical Biology, 9*(2), 97–104. http://dx.doi.org/10.1038/nchembio.1136.

Gutierrez, A., Pan, L., Groen, R. W. J., Baleydier, F., Kentsis, A., Marineau, J., … Aster, J. C. (2014). Phenothiazines induce PP2A-mediated apoptosis in T cell acute lymphoblastic leukemia. *Journal of Clinical Investigation, 124*(2), 644–655. http://dx.doi.org/10.1172/JCI65093.

Hao, J., Ao, A., Zhou, L., Murphy, C. K., Frist, A. Y., Keel, J. J., … Hong, C. C. (2013). Selective small molecule targeting β-catenin function discovered by in vivo chemical genetic screen. *Cell Reports, 4*(5), 898–904. http://dx.doi.org/10.1016/j.celrep.2013.07.047.

Hendzel, M. J., Wei, Y., Mancini, M. A., Van Hooser, A., Ranalli, T., Brinkley, B. R., … Allis, C. D. (1997). Mitosis-specific phosphorylation of histone H3 initiates primarily within pericentromeric heterochromatin during G2 and spreads in an ordered fashion coincident with mitotic chromosome condensation. *Chromosoma, 106*(6), 348–360. http://dx.doi.org/10.1007/s004120050256.

Hong, C. C., Peterson, Q. P., Hong, J. Y., & Peterson, R. T. (2006). Artery/vein specification is governed by opposing phosphatidylinositol-3 kinase and MAP kinase/ERK signaling. *Current Biology, 16*(13), 1366–1372. http://dx.doi.org/10.1016/j.cub.2006.05.046.

Ishizaki, H., Spitzer, M., Wildenhain, J., Anastasaki, C., Zeng, Z., Dolma, S., ... Patton, E. E. (2015). Combined zebrafish-yeast chemical-genetic screens reveal gene-copper-nutrition interactions that modulate melanocyte pigmentation. *Disease Models & Mechanisms, 3*(9–10), 639–651. http://dx.doi.org/10.1242/dmm.005769.

Jeong, D. G., Yoon, T.-S., Kim, J. H., Shim, M. Y., Jung, S. K., Son, J. H., ... Kim, S. J. (2006). Crystal structure of the catalytic domain of human MAP kinase phosphatase 5: structural insight into constitutively active phosphatase. *Journal of Molecular Biology, 360*(5), 946–955. http://dx.doi.org/10.1016/j.jmb.2006.05.059.

Jin, S., Sarkar, K. S., Jin, Y. N., Liu, Y., Kokel, D., Van Ham, T. J., ... Peterson, R. T. (2013). An in vivo zebrafish screen identifies organophosphate antidotes with diverse mechanisms of action. *Journal of Biomolecular Screening, 18*(1), 108–115. http://dx.doi.org/10.1177/1087057112458153.

Jung, D.-W., Williams, D., Khersonsky, S. M., Kang, T. W., Heidary, N., Chang, Y. T., & Orlow, S. J. (2005). Identification of the F1F0 mitochondrial ATPase as a target for modulating skin pigmentation by screening a tagged triazine library in zebrafish. *Molecular Biosystems, 1*, 85–92. http://dx.doi.org/10.1039/b417765g.

Khersonsky, S. M., Jung, D. W., Kang, T. W., Walsh, D. P., Moon, H. S., Jo, H., ... Chang, Y. T. (2003). Facilitated forward chemical genetics using a tagged triazine library and zebrafish embryo screening. *Journal of the American Chemical Society, 125*(39), 11804–11805. http://dx.doi.org/10.1021/ja035334d.

Kitambi, S. S., McCulloch, K. J., Peterson, R. T., & Malicki, J. J. (2009). Small molecule screen for compounds that affect vascular development in the zebrafish retina. *Mechanisms of Development, 126*(5–6), 464–477.

Kokel, D., Bryan, J., Laggner, C., White, R., Cheung, C. Y., Mateus, R., ... Peterson, R. T. (2010). Rapid behavior-based identification of neuroactive small molecules in the zebrafish. *Nature Chemical Biology, 6*(3), 231–237. http://dx.doi.org/10.1038/nchembio.307.

Kokel, D., Cheung, C. Y. J., Mills, R., Coutinho-Budd, J., Huang, L., Setola, V., ... Peterson, R. T. (2013). Photochemical activation of TRPA1 channels in neurons and animals. *Nature Chemical Biology, 9*(4), 257–263. http://dx.doi.org/10.1038/nchembio.1183.

Kong, Y., Grimaldi, M., Curtin, E., Dougherty, M., Kaufman, C., White, R. M., ... Liao, E. C. (2014). Neural crest development and craniofacial morphogenesis is coordinated by nitric oxide and histone acetylation. *Chemistry & Biology, 21*(4), 488–501. http://dx.doi.org/10.1016/j.chembiol.2014.02.013.

Le, X., Pugach, E. K., Hettmer, S., Storer, N. Y., Liu, J., Wills, A. A., ... Zon, L. I. (2013). A novel chemical screening strategy in zebrafish identifies common pathways in embryogenesis and rhabdomyosarcoma development. *Development, 140*(11), 2354–2364. http://dx.doi.org/10.1242/dev.088427.

Li, P., Lahvic, J. L., Binder, V., Pugach, E. K., Riley, E. B., Tamplin, O. J., ... Zon, L. I. (2015). Epoxyeicosatrienoic acids enhance embryonic haematopoiesis and adult marrow engraftment. *Nature, 523*(7561), 468–471. http://dx.doi.org/10.1038/nature14569.

Liao, C., Sitzmann, M., Pugliese, A., & Nicklaus, M. C. (2011). Software and resources for computational medicinal chemistry. *Future Medicinal Chemistry, 3*(8), 1057–1085. http://dx.doi.org/10.4155/fmc.11.63.

Lipinski, C. A., Lombardo, F., Dominy, B. W., & Feeney, P. J. (2001). Experimental and computational approaches to estimate solubility and permeability in drug discovery and development settings1. *Advanced Drug Delivery Reviews, 46*(1–3), 3–26. http://dx.doi.org/10.1016/S0169-409X(00)00129-0.

Liu, Y., Asnani, A., Zou, L., Bentley, V. L., Yu, M., Wang, Y., … Peterson, R. T. (2014). Visnagin protects against doxorubicin-induced cardiomyopathy through modulation of mitochondrial malate dehydrogenase. *Science Translational Medicine, 6*(266), 266ra170. http://dx.doi.org/10.1126/scitranslmed.3010189.

Liu, Y. J., Fan, H. B., Jin, Y., Ren, C. G., Jia, X. E., Wang, L., … Ren, R. (2013). Cannabinoid receptor 2 suppresses leukocyte inflammatory migration by modulating the JNK/c-Jun/Alox5 pathway. *Journal of Biological Chemistry, 288*(19), 13551–13562. http://dx.doi.org/10.1074/jbc.M113.453811.

MacRae, C. A., & Peterson, R. T. (2015). Zebrafish as tools for drug discovery. *Nature Reviews. Drug Discovery, 14*(10), 721–731. http://dx.doi.org/10.1038/nrd4627.

Mathew, L. K., Sengupta, S., Kawakami, A., Andreasen, E. A., Löhr, C. V., Loynes, C. A., … Tanguay, R. L. (2007). Unraveling tissue regeneration pathways using chemical genetics. *Journal of Biological Chemistry, 282*(48), 35202–35210. http://dx.doi.org/10.1074/jbc.M706640200.

Milan, D. J., Peterson, T. A., Ruskin, J. N., Peterson, R. T., & MacRae, C. A. (2003). Drugs that induce repolarization abnormalities cause bradycardia in zebrafish. *Circulation, 107*(10), 1355–1358. http://dx.doi.org/10.1161/01.CIR.0000061912.88753.87.

Mintzer, K. A., Lee, M. A., Runke, G., Trout, J., Whitman, M., & Mullins, M. C. (2001). Lost-a-fin encodes a type I BMP receptor, Alk8, acting maternally and zygotically in dorsoventral pattern formation. *Development, 128*(6), 859–869.

Molina, G., Vogt, A., Bakan, A., Dai, W., Queiroz de Oliveira, P., Znosko, W., … Tsang, M. (2009). Zebrafish chemical screening reveals an inhibitor of Dusp6 that expands cardiac cell lineages. *Nature Chemical Biology, 5*(9), 680–687. http://dx.doi.org/10.1038/nchembio.190.

Moon, H. S., Jacobson, E. M., Khersonsky, S. M., Luzung, M. R., Walsh, D. P., Xiong, W., … Chang, Y. T. (2002). A novel microtubule destabilizing entity from orthogonal synthesis of triazine library and zebrafish embryo screening. *Journal of the American Chemical Society, 124*(39), 11608–11609. http://dx.doi.org/10.1021/ja026720i.

Mullins, M. C., Hammerschmidt, M., Kane, D. A., Odenthal, J., Brand, M., van Eeden, F. J., … Nüsslein-Volhard, C. (1996). Genes establishing dorsoventral pattern formation in the zebrafish embryo: the ventral specifying genes. *Development, 123*(1995), 81–93.

Murphey, R. D., Stern, H. M., Straub, C. T., & Zon, L. I. (2006). A chemical genetic screen for cell cycle inhibitors in zebrafish embryos. *Chemical Biology & Drug Design, 68*(4), 213–219. http://dx.doi.org/10.1111/j.1747-0285.2006.00439.x.

Namdaran, P., Reinhart, K. E., Owens, K. N., Raible, D. W., & Rubel, E. W. (2012). Identification of modulators of hair cell regeneration in the zebrafish lateral line. *The Journal of Neuroscience, 32*(10), 3516–3528. http://dx.doi.org/10.1523/JNEUROSCI.3905-11.2012.

Nath, A. K., Roberts, L. D., Liu, Y., Mahon, S. B., Kim, S., Ryu, J. H., … Peterson, R. T. (2013). Chemical and metabolomic screens identify novel biomarkers and antidotes for cyanide exposure. *FASEB Journal, 27*(5), 1928–1938. http://dx.doi.org/10.1096/fj.12-225037.

Nath, A. K., Ryu, J. H., Jin, Y. N., Roberts, L. D., Dejam, A., Gerszten, R. E., & Peterson, R. T. (2015). PTPMT1 inhibition lowers glucose through succinate dehydrogenase phosphorylation. *Cell Reports, 10*(5), 694–701. http://dx.doi.org/10.1016/j.celrep.2015.01.010.

Nguyen, V. H., Schmid, B., Trout, J., Connors, S. A., Ekker, M., & Mullins, M. C. (1998). Ventral and lateral regions of the zebrafish gastrula, including the neural crest progenitors, are established by a bmp2b/swirl pathway of genes. *Developmental Biology, 199*(1), 93–110. http://dx.doi.org/10.1006/dbio.1998.8927.

Nishiya, N., Oku, Y., Kumagai, Y., Sato, Y., Yamaguchi, E., Sasaki, A., ... Uehara, Y. (2014). A zebrafish chemical suppressor screening identifies small molecule inhibitors of the Wnt/beta-catenin pathway. *Chemistry & Biology, 21*(4), 530–540. http://dx.doi.org/10.1016/j.chembiol.2014.02.015.

North, T. E., Goessling, W., Walkley, C. R., Lengerke, C., Kopani, K. R., Lord, A. M., ... Zon, L. I. (2007). Prostaglandin E2 regulates vertebrate haematopoietic stem cell homeostasis. *Nature, 447*(7147), 1007–1011. http://dx.doi.org/10.1038/nature05883.

Oppedal, D., & Goldsmith, M. I. (2010). A chemical screen to identify novel inhibitors of fin regeneration in zebrafish. *Zebrafish, 7*(1), 53–60. http://dx.doi.org/10.1089/zeb.2009.0633.

Owens, K. N., Santos, F., Roberts, B., Linbo, T., Coffin, A. B., Knisely, A. J., ... Raible, D. W. (2008). Identification of genetic and chemical modulators of zebrafish mechanosensory hair cell death. *PLoS Genetics, 4*(2), e1000020.

Padilla, S., Corum, D., Padnos, B., Hunter, D. L., Beam, A., Houck, K. A., ... Reif, D. M. (2012). Zebrafish developmental screening of the ToxCast™ Phase I chemical library. *Reproductive Toxicology, 33*(2), 174–187. http://dx.doi.org/10.1016/j.reprotox.2011.10.018.

Paik, E. J., de Jong, J. L. O., Pugach, E., Opara, P., & Zon, L. I. (2010). A chemical genetic screen in zebrafish for pathways interacting with cdx4 in primitive hematopoiesis. *Zebrafish, 7*(1), 61–68. http://dx.doi.org/10.1089/zeb.2009.0643.

Parker, C. N., & Schreyer, S. K. (2004). Application of chemoinformatics to high-throughput screening: practical considerations. *Methods in Molecular Biology, 275*, 85–110. http://dx.doi.org/10.1385/1-59259-802-1:085.

Peal, D. S., Mills, R. W., Lynch, S. N., Mosley, J. M., Lim, E., Elllinor, P. T., ... Milan, D. J. (2011). Novel chemical suppressors of long QT syndrome identified by an in vivo functional screen. *Circulation, 123*(1), 23–30. http://dx.doi.org/10.1161/CIRCULATIONAHA.110.003731.

Peterson, R. T., Link, B. A., Dowling, J. E., & Schreiber, S. L. (2000). Small molecule developmental screens reveal the logic and timing of vertebrate development. *Proceedings of the National Academy of Sciences of the United States of America, 97*(24), 12965–12969. http://dx.doi.org/10.1073/pnas.97.24.12965.

Peterson, R. T., Shaw, S. Y., Peterson, T. A., Milan, D. J., Zhong, T. P., Schreiber, S. L., ... Fishman, M. C. (2004). Chemical suppression of a genetic mutation in a zebrafish model of aortic coarctation. *Nature Biotechnology, 22*(5), 595–599.

Rennekamp, A. J., & Peterson, R. T. (2015). 15 years of zebrafish chemical screening. *Current Opinion in Chemical Biology, 24*, 58–70. http://dx.doi.org/10.1016/j.cbpa.2014.10.025.

Ridges, S., Heaton, W. L., Joshi, D., Choi, H., Eiring, A., Batchelor, L., ... Trede, N. S. (2012). Zebrafish screen identifies novel compound with selective toxicity against leukemia. *Blood, 119*(24), 5621–5631. http://dx.doi.org/10.1182/blood-2011-12-398818.

Rihel, J., Prober, D. A., Arvanites, A., Lam, K., Zimmerman, S., Jang, S., … Schier, A. F. (2010). Zebrafish behavioral profiling links drugs to biological targets and rest/wake regulation. *Science, 327*(5963), 348–351.

Rovira, M., Huang, W., Yusuff, S., Shim, J. S., Ferrante, A. A., Liu, J. O., & Parsons, M. J. (2011). Chemical screen identifies FDA-approved drugs and target pathways that induce precocious pancreatic endocrine differentiation. *Proceedings of the National Academy of Sciences of the United States of America, 108*(48), 19264–19269. http://dx.doi.org/10.1073/pnas.1113081108.

Sachidanandan, C., Yeh, J. R. J., Peterson, Q. P., & Peterson, R. T. (2008). Identification of a novel retinoid by small molecule screening with zebrafish embryos. *PLoS One, 3*(4). http://dx.doi.org/10.1371/journal.pone.0001947.

Sandoval, I. T., Manos, E. J., Van Wagoner, R. M., Delacruz, R. G., Edes, K., Winge, D. R., … Jones, D. A. (2013). Juxtaposition of chemical and mutation-induced developmental defects in zebrafish reveal a copper-chelating activity for kalihinol F. *Chemistry & Biology, 20*(6), 753–763. http://dx.doi.org/10.1016/j.chembiol.2013.05.008.

Saydmohammed, M., Vollmer, L. L., Onuoha, E. O., Vogt, A., & Tsang, M. (2011). A high-content screening assay in transgenic zebrafish identifies two novel activators of fgf signaling. *Birth Defects Research Part C – Embryo Today: Reviews, 93*(3), 281–287. http://dx.doi.org/10.1002/bdrc.20216.

Schreiber, S. L. (2000). Target-oriented and diversity-oriented organic synthesis in drug discovery. *Science, 287*(5460), 1964–1969. http://dx.doi.org/10.1126/science.287.5460.1964.

Shafizadeh, E., Peterson, R. T., & Lin, S. (2004). Induction of reversible hemolytic anemia in living zebrafish using a novel small molecule. *Comparative Biochemistry and Physiology – Part C Toxicology and Pharmacology, 138*, 245–249. http://dx.doi.org/10.1016/j.cca.2004.05.003.

Shepard, J. L., Amatruda, J. F., Stern, H. M., Subramanian, A., Finkelstein, D., Ziai, J., … Zon, L. I. (2005). A zebrafish bmyb mutation causes genome instability and increased cancer susceptibility. *Proceedings of the National Academy of Sciences of the United States of America, 102*(37), 13194–13199. http://dx.doi.org/10.1073/pnas.0506583102.

Spring, D. R., Krishnan, S., Blackwell, H. E., & Schreiber, S. L. (2002). Diversity-oriented synthesis of biaryl-containing medium rings using a one bead/one stock solution platform. *Journal of the American Chemical Society, 124*(7), 1354–1363.

Stern, H. M., Murphey, R. D., Shepard, J. L., Amatruda, J. F., Straub, C. T., Pfaff, K. L., … Zon, L. I. (2005). Small molecules that delay S phase suppress a zebrafish bmyb mutant. *Nature Chemical Biology, 1*(7), 366–370. http://dx.doi.org/10.1038/nchembio749.

Sternson, S. M., Louca, J. B., Wong, J. C., & Schreiber, S. L. (2001). Split - pool synthesis of 1,3-dioxanes leading to arrayed stock solutions of single compounds sufficient for multiple phenotypic and protein-binding assays. *Journal of the American Chemical Society, 123*(8), 1740–1747. http://dx.doi.org/10.1021/ja0036108.

Stewart, A. E., Dowd, S., Keyse, S. M., & McDonald, N. Q. (1999). Crystal structure of the MAPK phosphatase Pyst1 catalytic domain and implications for regulated activation. *Nature Structural Biology, 6*(2), 174–181. http://dx.doi.org/10.1038/5861.

Thisse, B., & Thisse, C. (2005). Functions and regulations of fibroblast growth factor signaling during embryonic development. *Developmental Biology, 287*(2), 390–402. http://dx.doi.org/10.1016/j.ydbio.2005.09.011.

Torregroza, I., Evans, T., & Das, B. C. (2009). A forward chemical screen using zebrafish embryos with novel 2-substituted 2H-chromene derivatives. *Chemical Biology & Drug Design, 73*(3), 339–345. http://dx.doi.org/10.1111/j.1747-0285.2009.00782.x.

Tran, T. C., Sneed, B., Haider, J., Blavo, D., White, A., Aiyejorun, T., … Sandberg, E. M. (2007). Automated, quantitative screening assay for antiangiogenic compounds using transgenic zebrafish. *Cancer Research, 67*(23), 11386–11392. http://dx.doi.org/10.1158/0008-5472.CAN-07-3126.

Trompouki, E., & Zon, L. I. (2010). Small molecule screen in zebrafish and HSC expansion. *Methods in Molecular Biology, 636*, 301–316. http://dx.doi.org/10.1007/978-1-60761-691-7_19.

Truong, L., Reif, D. M., Mary, L. S., Geier, M. C., Truong, H. D., & Tanguay, R. L. (2014). Multidimensional in vivo hazard assessment using zebrafish. *Toxicological Sciences, 137*(1), 212–233. http://dx.doi.org/10.1093/toxsci/kft235.

Tsang, M., & Dawid, I. B. (2004). Promotion and attenuation of FGF signaling through the Ras-MAPK pathway. *Sciences's STKE, 2004*(228), pe17. http://dx.doi.org/10.1126/stke.2282004pe17.

Tsuji, N., Ninov, N., Delawary, M., Osman, S., Roh, A. S., Gut, P., & Stainier, D. Y. (2014). Whole organism high content screening identifies stimulators of pancreatic beta-cell proliferation. *PLoS One, 9*(8). http://dx.doi.org/10.1371/journal.pone.0104112.

Wang, G., Rajpurohit, S. K., Delaspre, F., Walker, S. L., White, D. T., Ceasrine, A., … Mumm, J. S. (July 2015). First quantitative high-throughput screen in zebrafish identifies novel pathways for increasing pancreatic β-cell mass. *eLife*, 4. http://dx.doi.org/10.7554/eLife.08261.001.

Wang, C., Tao, W., Wang, Y., Bikow, J., Lu, B., Keating, A., … Wen, X. Y. (2010). Rosuvastatin, identified from a zebrafish chemical genetic screen for antiangiogenic compounds, suppresses the growth of prostate cancer. *European Urology, 58*(3), 418–426. http://dx.doi.org/10.1016/j.eururo.2010.05.024.

Weger, B. D., Weger, M., Nusser, M., Brenner-Weiss, G., & Dickmeis, T. (2012). A chemical screening system for glucocorticoid stress hormone signaling in an intact vertebrate. *ACS Chemical Biology, 7*(7), 1178–1183. http://dx.doi.org/10.1021/cb3000474.

Wheeler, G. N., & Brändli, A. W. (2009). Simple vertebrate models for chemical genetics and drug discovery screens: lessons from zebrafish and Xenopus. *Developmental Dynamics, 238*(6), 1287–1308. http://dx.doi.org/10.1002/dvdy.21967.

White, R. M., Cech, J., Ratanasirintrawoot, S., Lin, C. Y., Rahl, P. B., Burke, C. J., … Zon, L. I. (2011). DHODH modulates transcriptional elongation in the neural crest and melanoma. *Nature, 471*, 518–522. http://dx.doi.org/10.1038/nature09882.

Williams, C. H., Hempel, J. E., Hao, J., Frist, A. Y., Williams, M. M., Fleming, J. T., … Hong, C. C. (2015). An in vivo chemical genetic screen identifies phosphodiesterase 4 as a pharmacological target for hedgehog signaling inhibition. *Cell Reports, 11*(1), 43–50. http://dx.doi.org/10.1016/j.celrep.2015.03.001.

Wolman, M. A., Jain, R. A., Liss, L., & Granato, M. (2011). Chemical modulation of memory formation in larval zebrafish. *Proceedings of the National Academy of Sciences of the United States of America, 108*(37), 15468–15473. http://dx.doi.org/10.1073/pnas.1107156108.

Wong, J. C., Sternson, S. M., Louca, J. B., Hong, R., & Schreiber, S. L. (2004). Modular synthesis and preliminary biological evaluation of stereochemically diverse 1,3-

dioxanes. *Chemistry & Biology, 11*(9), 1279–1291. http://dx.doi.org/10.1016/j.chembiol.2004.07.012.

Xu, Z., Li, Y., Xiang, Q., Pei, Z., Liu, X., Lu, B., … Lin, Y. (2010). Design and synthesis of novel xyloketal derivatives and their vasorelaxing activities in rat thoracic aorta and angiogenic activities in zebrafish angiogenesis screen. *Journal of Medicinal Chemistry, 53*(12), 4642–4653. http://dx.doi.org/10.1021/jm1001502.

Yeh, J.-R. J., Munson, K. M., Elagib, K. E., Goldfarb, A. N., Sweetser, D. A., & Peterson, R. T. (2009). Discovering chemical modifiers of oncogene-regulated hematopoietic differentiation. *Nature Chemical Biology, 5*(4), 236–243. http://dx.doi.org/10.1038/nchembio.147.

Yu, P. B., Hong, C. C., Sachidanandan, C., Babitt, J. L., Deng, D. Y., Hoyng, S. A., … Peterson, R. T. (2008). Dorsomorphin inhibits BMP signals required for embryogenesis and iron metabolism. *Nature Chemical Biology, 4*(1), 33–41. http://dx.doi.org/10.1038/nchembio.2007.54.

Zon, L. I., & Peterson, R. T. (2005). In vivo drug discovery in the zebrafish. *Nature Reviews. Drug Discovery, 4*(1), 35–44. http://dx.doi.org/10.1038/nrd1606.

Index

'*Note:* Page numbers followed by "f" indicate figures and "t" indicate tables.'

B

C

D

F

J

K

L

M

N

O

P

Q

R

S

Volumes in Series

Founding Series Editor
DAVID M. PRESCOTT

Volume 1 (1964)
Methods in Cell Physiology
Edited by David M. Prescott

Volume 2 (1966)
Methods in Cell Physiology
Edited by David M. Prescott

Volume 3 (1968)
Methods in Cell Physiology
Edited by David M. Prescott

Volume 4 (1970)
Methods in Cell Physiology
Edited by David M. Prescott

Volume 5 (1972)
Methods in Cell Physiology
Edited by David M. Prescott

Volume 6 (1973)
Methods in Cell Physiology
Edited by David M. Prescott

Volume 7 (1973)
Methods in Cell Biology
Edited by David M. Prescott

Volume 8 (1974)
Methods in Cell Biology
Edited by David M. Prescott

Volume 9 (1975)
Methods in Cell Biology
Edited by David M. Prescott

Volume 10 (1975)
Methods in Cell Biology
Edited by David M. Prescott

Volume 11 (1975)
Yeast Cells
Edited by David M. Prescott

Volume 12 (1975)
Yeast Cells
Edited by David M. Prescott

Volume 13 (1976)
Methods in Cell Biology
Edited by David M. Prescott

Volume 14 (1976)
Methods in Cell Biology
Edited by David M. Prescott

Volume 15 (1977)
Methods in Cell Biology
Edited by David M. Prescott

Volume 16 (1977)
Chromatin and Chromosomal Protein Research I
Edited by Gary Stein, Janet Stein, and Lewis J. Kleinsmith

Volume 17 (1978)
Chromatin and Chromosomal Protein Research II
Edited by Gary Stein, Janet Stein, and Lewis J. Kleinsmith

Volume 18 (1978)
Chromatin and Chromosomal Protein Research III
Edited by Gary Stein, Janet Stein, and Lewis J. Kleinsmith

Volume 19 (1978)
Chromatin and Chromosomal Protein Research IV
Edited by Gary Stein, Janet Stein, and Lewis J. Kleinsmith

Volume 20 (1978)
Methods in Cell Biology
Edited by David M. Prescott

Advisory Board Chairman
KEITH R. PORTER

Volume 21A (1980)
Normal Human Tissue and Cell Culture, Part A: Respiratory, Cardiovascular, and Integumentary Systems
Edited by Curtis C. Harris, Benjamin F. Trump, and Gary D. Stoner

Volume 21B (1980)
Normal Human Tissue and Cell Culture, Part B: Endocrine, Urogenital, and Gastrointestinal Systems
Edited by Curtis C. Harris, Benjamin F. Trump, and Gray D. Stoner

Volume 22 (1981)
Three-Dimensional Ultrastructure in Biology
Edited by James N. Turner

Volume 23 (1981)
Basic Mechanisms of Cellular Secretion
Edited by Arthur R. Hand and Constance Oliver

Volume 24 (1982)
The Cytoskeleton, Part A: Cytoskeletal Proteins, Isolation and Characterization
Edited by Leslie Wilson

Volume 25 (1982)
The Cytoskeleton, Part B: Biological Systems and In Vitro Models
Edited by Leslie Wilson

Volume 26 (1982)
Prenatal Diagnosis: Cell Biological Approaches
Edited by Samuel A. Latt and Gretchen J. Darlington

Series Editor
LESLIE WILSON

Volume 27 (1986)
Echinoderm Gametes and Embryos
Edited by Thomas E. Schroeder

Volume 28 (1987)
***Dictyostelium discoideum*: Molecular Approaches to Cell Biology**
Edited by James A. Spudich

Volume 29 (1989)
Fluorescence Microscopy of Living Cells in Culture, Part A: Fluorescent Analogs, Labeling Cells, and Basic Microscopy
Edited by Yu-Li Wang and D. Lansing Taylor

Volume 30 (1989)
Fluorescence Microscopy of Living Cells in Culture, Part B: Quantitative Fluorescence Microscopy—Imaging and Spectroscopy
Edited by D. Lansing Taylor and Yu-Li Wang

Volume 31 (1989)
Vesicular Transport, Part A
Edited by Alan M. Tartakoff

Volume 32 (1989)
Vesicular Transport, Part B
Edited by Alan M. Tartakoff

Volume 33 (1990)
Flow Cytometry
Edited by Zbigniew Darzynkiewicz and Harry A. Crissman

Volume 34 (1991)
Vectorial Transport of Proteins into and across Membranes
Edited by Alan M. Tartakoff

Selected from Volumes 31, 32, and 34 (1991)
Laboratory Methods for Vesicular and Vectorial Transport
Edited by Alan M. Tartakoff

Volume 35 (1991)
Functional Organization of the Nucleus: A Laboratory Guide
Edited by Barbara A. Hamkalo and Sarah C. R. Elgin

Volume 36 (1991)
***Xenopus laevis*: Practical Uses in Cell and Molecular Biology**
Edited by Brian K. Kay and H. Benjamin Peng

Series Editors
LESLIE WILSON AND PAUL MATSUDAIRA

Volume 37 (1993)
Antibodies in Cell Biology
Edited by David J. Asai

Volume 38 (1993)
Cell Biological Applications of Confocal Microscopy
Edited by Brian Matsumoto

Volume 50 (1995)
Methods in Plant Cell Biology, Part B
Edited by David W. Galbraith, Don P. Bourque, and Hans J. Bohnert

Volume 51 (1996)
Methods in Avian Embryology
Edited by Marianne Bronner-Fraser

Volume 52 (1997)
Methods in Muscle Biology
Edited by Charles P. Emerson, Jr. and H. Lee Sweeney

Volume 53 (1997)
Nuclear Structure and Function
Edited by Miguel Berrios

Volume 54 (1997)
Cumulative Index

Volume 55 (1997)
Laser Tweezers in Cell Biology
Edited by Michael P. Sheetz

Volume 56 (1998)
Video Microscopy
Edited by Greenfield Sluder and David E. Wolf

Volume 57 (1998)
Animal Cell Culture Methods
Edited by Jennie P. Mather and David Barnes

Volume 58 (1998)
Green Fluorescent Protein
Edited by Kevin F. Sullivan and Steve A. Kay

Volume 59 (1998)
The Zebrafish: Biology
Edited by H. William Detrich III, Monte Westerfield, and Leonard I. Zon

Volume 60 (1998)
The Zebrafish: Genetics and Genomics
Edited by H. William Detrich III, Monte Westerfield, and Leonard I. Zon

Volume 61 (1998)
Mitosis and Meiosis
Edited by Conly L. Rieder

Volume 84 (2007)
Biophysical Tools for Biologists, Volume One: In Vitro Techniques
Edited by John J. Correia and H. William Detrich, III

Volume 85 (2008)
Fluorescent Proteins
Edited by Kevin F. Sullivan

Volume 86 (2008)
Stem Cell Culture
Edited by Dr. Jennie P. Mather

Volume 87 (2008)
Avian Embryology, 2nd Edition
Edited by Dr. Marianne Bronner-Fraser

Volume 88 (2008)
Introduction to Electron Microscopy for Biologists
Edited by Prof. Terence D. Allen

Volume 89 (2008)
Biophysical Tools for Biologists, Volume Two: In Vivo Techniques
Edited by Dr. John J. Correia and Dr. H. William Detrich, III

Volume 90 (2008)
Methods in Nano Cell Biology
Edited by Bhanu P. Jena

Volume 91 (2009)
Cilia: Structure and Motility
Edited by Stephen M. King and Gregory J. Pazour

Volume 92 (2009)
Cilia: Motors and Regulation
Edited by Stephen M. King and Gregory J. Pazour

Volume 93 (2009)
Cilia: Model Organisms and Intraflagellar Transport
Edited by Stephen M. King and Gregory J. Pazour

Volume 94 (2009)
Primary Cilia
Edited by Roger D. Sloboda

Volume 95 (2010)
Microtubules, in vitro
Edited by Leslie Wilson and John J. Correia

Volume 96 (2010)
Electron Microscopy of Model Systems
Edited by Thomas Müeller-Reichert

Volume 97 (2010)
Microtubules: In Vivo
Edited by Lynne Cassimeris and Phong Tran

Volume 98 (2010)
Nuclear Mechanics & Genome Regulation
Edited by G.V. Shivashankar

Volume 99 (2010)
Calcium in Living Cells
Edited by Michael Whitaker

Volume 100 (2010)
The Zebrafish: Cellular and Developmental Biology, Part A
Edited by: H. William Detrich III, Monte Westerfield and Leonard I. Zon

Volume 101 (2011)
The Zebrafish: Cellular and Developmental Biology, Part B
Edited by: H. William Detrich III, Monte Westerfield and Leonard I. Zon

Volume 102 (2011)
Recent Advances in Cytometry, Part A: Instrumentation, Methods
Edited by Zbigniew Darzynkiewicz, Elena Holden, Alberto Orfao, William Telford and Donald Wlodkowic

Volume 103 (2011)
Recent Advances in Cytometry, Part B: Advances in Applications
Edited by Zbigniew Darzynkiewicz, Elena Holden, Alberto Orfao, Alberto Orfao and Donald Wlodkowic

Volume 104 (2011)
The Zebrafish: Genetics, Genomics and Informatics 3rd Edition
Edited by H. William Detrich III, Monte Westerfield, and Leonard I. Zon

Volume 105 (2011)
The Zebrafish: Disease Models and Chemical Screens 3rd Edition
Edited by H. William Detrich III, Monte Westerfield, and Leonard I. Zon

Volume 106 (2011)
Caenorhabditis elegans: Molecular Genetics and Development 2nd Edition
Edited by Joel H. Rothman and Andrew Singson

Volume 107 (2011)
Caenorhabditis elegans: Cell Biology and Physiology 2nd Edition
Edited by Joel H. Rothman and Andrew Singson

Volume 108 (2012)
Lipids
Edited by Gilbert Di Paolo and Markus R Wenk

Volume 109 (2012)
Tetrahymena thermophila
Edited by Kathleen Collins

Volume 110 (2012)
Methods in Cell Biology
Edited by Anand R. Asthagiri and Adam P. Arkin

Volume 111 (2012)
Methods in Cell Biology
Edited by Thomas Müler Reichart and Paul Verkade

Volume 112 (2012)
Laboratory Methods in Cell Biology
Edited by P. Michael Conn

Volume 113 (2013)
Laboratory Methods in Cell Biology
Edited by P. Michael Conn

Volume 114 (2013)
Digital Microscopy, 4th Edition
Edited by Greenfield Sluder and David E. Wolf

Volume 115 (2013)
Microtubules, in Vitro, 2nd Edition
Edited by John J. Correia and Leslie Wilson

Volume 116 (2013)
Lipid Droplets
Edited by H. Robert Yang and Peng Li

Volume 117 (2013)
Receptor-Receptor Interactions
Edited by P. Michael Conn

Volume 118 (2013)
Methods for Analysis of Golgi Complex Function
Edited by Franck Perez and David J. Stephens

Volume 119 (2014)
Micropatterning in Cell Biology Part A
Edited by Matthieu Piel and Manuel Théry

Volume 120 (2014)
Micropatterning in Cell Biology Part B
Edited by Matthieu Piel and Manuel Théry

Volume 121 (2014)
Micropatterning in Cell Biology Part C
Edited by Matthieu Piel and Manuel Théry

Volume 122 (2014)
Nuclear Pore Complexes and Nucleocytoplasmic Transport - Methods
Edited by Valérie Doye

Volume 123 (2014)
Quantitative Imaging in Cell Biology
Edited by Jennifer C. Waters and Torsten Wittmann

Volume 124 (2014)
Correlative Light and Electron Microscopy II
Edited by Thomas Müller-Reichert and Paul Verkade

Volume 125 (2015)
Biophysical Methods in Cell Biology
Edited by Ewa K. Paluch

Volume 126 (2015)
Lysosomes and Lysosomal Diseases
Edited by Frances Platt and Nick Platt

Volume 127 (2015)
Methods in Cilia & Flagella
Edited by Renata Basto and Wallace F. Marshall

CPI Antony Rowe
Chippenham, UK
2017-03-07 19:07